Sports Rehabilitation and Injury Prevention

Sports Rehabilitation and Injury Prevention

Edited by

Paul Comfort
School of Health, Sport & Rehabilitation Sciences, University of Salford, Salford, UK

Earle Abrahamson
London Sport Institute at Middlesex University, UK

WILEY-BLACKWELL

A John Wiley & Sons, Ltd., Publication

This edition first published 2010, © 2010 John Wiley & Sons, Ltd

Wiley-Blackwell is an imprint of John Wiley & Sons, formed by the merger of Wiley's global Scientific, Technical and Medical business with Blackwell Publishing.

Registered Office
John Wiley & Sons, Ltd, The Atrium, Southern Gate, Chichester, West Sussex, PO19 8SQ, UK

Other Editorial Offices
9600 Garsington Road, Oxford, OX4 2DQ, UK
111 River Street, Hoboken, NJ 07030-5774, USA

For details of our global editorial offices, for customer services and for information about how to apply for permission to reuse the copyright material in this book please see our website at www.wiley.com/wiley-blackwell

The right of the author to be identified as the author of this work has been asserted in accordance with the Copyright, Designs and Patents Act 1988.

All rights reserved. No part of this publication may be reproduced, stored in a retrieval system, or transmitted, in any form or by any means, electronic, mechanical, photocopying, recording or otherwise, except as permitted by the UK Copyright, Designs and Patents Act 1988, without the prior permission of the publisher.

Wiley also publishes its books in a variety of electronic formats. Some content that appears in print may not be available in electronic books.

Designations used by companies to distinguish their products are often claimed as trademarks. All brand names and product names used in this book are trade names, service marks, trademarks or registered trademarks of their respective owners. The publisher is not associated with any product or vendor mentioned in this book. This publication is designed to provide accurate and authoritative information in regard to the subject matter covered. It is sold on the understanding that the publisher is not engaged in rendering professional services. If professional advice or other expert assistance is required, the services of a competent professional should be sought.

Library of Congress Cataloging-in-Publication Data
Sports rehabilitation and injury prevention / edited by Paul Comfort, Earle Abrahamson.
　　p. ; cm.
　Includes bibliographical references and index.
　　ISBN 978-0-470-98562-5 (cloth)
1. Sports injuries.　I. Comfort, Paul.　II. Abrahamson, Earle.
　[DNLM: 1. Athletic Injuries – prevention & control.　2. Athletic Injuries – rehabilitation. QT 261 S7676 2010]
　RD97.S785　2010
　617.1$'$027 – dc22
　　2010005619

ISBN: 9780470985625 (HB)
　　　 9780470985632 (PB)

A catalogue record for this book is available from the British Library.

Set in 10/11.5pt Times by Aptara Inc., New Delhi, India.

Contents

Preface ... ix
Acknowledgements ... xi
About the editors ... xiii
List of contributors ... xv
How to use this book ... xvii

PART 1   INTRODUCTION TO SPORTS REHABILITATION ... 1

1   **Introduction to sport injury management** ... 3
    Jeffrey A. Russell

PART 2   INJURY SCREENING AND ASSESSMENT OF PERFORMANCE ... 13

2   **Injury prevention and screening** ... 15
    Phil Barter

3   **Assessment and needs analysis** ... 39
    Paul Comfort and Martyn Matthews

PART 3   PATHOPHYSIOLOGY OF MUSCULOSKELETAL INJURIES ... 65

4   **Pathophysiology of skeletal muscle injuries** ... 67
    Dr Lee Herrington and Paul Comfort

5   **Tendons** ... 79
    Dr Stephen Pearson

6   **Pathophysiology of ligament injuries** ... 95
    Dror Steiner

7   **Pathophysiology of skeletal injuries** ... 105
    Sarah Catlow

8 Peripheral nerve injuries
Elizabeth Fowler
119

PART 4 EFFECTIVE CLINICAL DECISION MAKING
143

9 An introduction to periodisation
Paul Comfort and Martyn Matthews
145

10 Management of acute sport injury
Jeffrey A. Russell
163

11 Musculoskeletal assessment
Julian Hatcher
185

12 Progressive systematic functional rehabilitation
Earle Abrahamson, Victoria Hyland, Sebastian Hicks, and Christo Koukoullis
199

13 Strength and conditioning
Paul Comfort and Martyn Matthews
223

14 Nutritional considerations for performance and rehabilitation
Helen Matthews and Martyn Matthews
245

15 Psychology and sports rehabilitation
Rhonda Cohen, Dr Sanna M. Nordin and Earle Abrahamson
275

16 Clinical reasoning
Earle Abrahamson and Dr Lee Herrington
297

PART 5 JOINT SPECIFIC INJURIES AND PATHOLOGIES
307

17 Shoulder injuries in sport
Ian Horsley
309

18 The elbow
Angela Clough
337

19 Wrist and hand injuries in sport
Luke Heath
365

20 The groin in sport
John Allen and Stuart Butler
385

21 The knee
Nicholas Clark and Dr Lee Herrington
407

22	**Ankle complex injuries in sport** *David Joyce*	**465**
23	**The foot in sport** *John Allen*	**497**

Index 517

Preface

The concept for this book is based on the expanding field of sports rehabilitation and injury prevention. Evidence of this expansion includes an increasing amount of research and publications related to sports rehabilitation and allied fields of practice such as sports therapy, athletic training and sports physiotherapy.

Despite the number and volume of publications in sports rehabilitation, there appears to be limited resources that accurately and effectively account for evidence-based practices. Whilst some resources expand evidence-based practice knowledge, there is a need to develop a complete resource that fully explains and articulates these important principles. This current text has used an evidence-based practice approach to fully acknowledge the many diverse areas, applications and management strategies that are often unique to sports rehabilitation, but distinctly different from similar fields of practice and study.

Few sports rehabilitation programmes currently provide students with the breadth of information and practical application required for professional practice. This text has attempted to bridge the knowledge and practice gap, by considering the functional development of the sports rehabilitator's knowledge and practice requirements for professional competency. The text provides an up-to-date look at different evidence-based practice protocols and initial assessment strategies for the screening of injury and pathological conditions.

The first few chapters introduce the scope of practice for sports rehabilitation, and then describe, explain and evaluate the initial assessment and screening procedures necessary for decision making and clinical practice. These chapters further provide analysis on musculoskeletal function and dysfunction in relation to systemic organisation. The next set of chapters combine a useful integration of applied areas and practices of study relevant to sports rehabilitation practice. These include, amongst others, nutritional analysis, psychological considerations in injury management and prevention, clinical reasoning development, and strength and conditioning principles. The book concludes with a range of chapters devoted to different injury conditions and body regions. These chapters detail the more common injuries and pathologies and argue for best management strategies based on research and applied evidence.

Each chapter also contains several practical application boxes that provide additional information summarising unique chapter-specific information. The majority of chapters contain applied examples and case studies to illustrate the processes and decisions necessary for clinical action and management. Each case study has been carefully developed to facilitate group discussion in the classroom, or for the clinician to consider as part of continued professional development.

In addition to serving as an upper level undergraduate or graduate textbook for students or clinicians in practice, the book is an excellent resource guide, filled with useful information and evidence-based practice considerations and applications. You will want to have this textbook on your desk or bookshelf. The features of consistent organisation, case studies, discussion questions, up-to-date references, research evidence and practical application boxes are designed to provide information required for effective study as well as directing clinical practice.

The design of this text can be compared to building a house, in that each component of both the text and house building can be modelled on individual building blocks. In the case of the house building these units are represented by the bricks, whereas in the text, the individual chapters are synonymous with these units. Before one commences the building process, there is a carefully constructed visual or diagrammatic plan to navigate the process; so too does this planning apply to the design and shaping of this text. In the building process, consideration is given to the foundation, in terms of its shape, depth, form, and length. This text has a number of foundation chapters that secure the content for future development of the other chapters. The main foundation knowledge is the understanding of anatomical application, and using this knowledge to guide assessment. This anatomical foundation knowledge informs the decisions necessary for clinical action in terms of injury management. Whilst bricks are important in terms of informing the structure of a building, it is the cement that ensures that each brick is secured and articulates with other bricks and structures. In this text, the cement is represented by underpinning themes, such as clinical reasoning skills and abilities, that traverse the chapters and ensures that each chapter although perceptively different, is able to articulate with other chapters and develop this consortium of knowledge.

After completion, houses take on a new shape and design, one which may have transformed the original landscape; however there is always room for change, improvement or refinement. This text, in its final form, has orchestrated the journey of clinical practice from consideration of the scope of practice, through to the essential skills necessary for decision making, and concluding with a consideration of how to manage a range of injuries and pathologies. The text is coated with an evidence-based approach to using and applying knowledge. The true advantage of developing the text within an evidence-based context is that it allows the reader to consider the existing knowledge and evidence; challenge the research; and move towards asking different types of questions to consider new ways of dealing with client management issues. As new research becomes available, clinical practice will be questioned. The contents of this text will evolve and change to accommodate and explore new ideas and advances in clinical research. This book provides the architecture necessary to consider the real issues current to clinical practices. It is important to use it as a map for navigating the concepts, principles, challenges and decisions of clinical practice.

We hope that this book is a valuable resource both for teaching and as a reference for sports rehabilitators and clinicians.

Paul Comfort
Earle Abrahamson

Acknowledgements

Thank you to all of the authors involved with the development of this text, including those who provided advice and feedback on each of the many drafts. Without the expertise, dedication and effort of each of these individuals, this text would not have been possible.

Thank you to my family, especially my children, for putting up with my 'absences' and long hours staring at the laptop, during the development of this book. Your support and understanding has been more than I should have asked for.

Paul Comfort

A special thanks to the many contributors who worked so diligently, often under difficult and pressurised circumstances, to write this text and to those who provided expert reviews. Also to my many students who taught me so much about how to articulate concepts, theories and applications in a learner friendly manner, which helped shape the landscape of this book.

To my wonderful wife, Emma, and my adorable son, Benjamin, thanks for putting up with me and providing much love, support and understanding.

To my father, Charles, and my brother, Michael, thanks for always believing in me and encouraging me to succeed and achieve in life.

Last but not least, I would like to dedicate my contribution to this book, to the memory of my late mother, Josephine, whose support, inspiration, kindness and generosity, will forever be cherished and respected. Thank you for believing in me and supporting my academic and professional development.

Earle Abrahamson

About the editors

Paul Comfort (BSc (Hons), MSc, PGCAP, CSCS*D, ASCC) is a senior lecturer, programme leader for the MSc Strength and Conditioning programme at the University of Salford. Paul is also currently Head of Sports Science Support for Salford City Reds Rugby League Football Club and co-ordinates the Strength and Conditioning for England Lacrosse (men's squad). He is a Certified Strength and Conditioning Specialist (Recertified with Distinction) (CSCS*D) with the National Strength and Conditioning Association and a founder member and Accredited Strength and Conditioning Coach with the United Kingdom Strength and Conditioning Association. He is also currently completing a part-time PhD.

Earle Abrahamson (B Phys Ed, BA Hons, MA, BPS, BASRaT, FRSM, BRCP, AHPCSA, HPCSA, PsySSA) is a principal lecturer, teaching fellow and programme leader for the Sports Rehabilitation and Injury Prevention programme at Middlesex University. Through his programme leadership and teaching fellowship duties, Earle has developed an interest in student learning and thinking. Earle spent the majority of his life in South Africa, studying and working, and moved to the UK in 2002. He is a South African-registered therapist and psychologist and has membership and professional registration with a number of UK authorities. Earle has worked extensively as a sports rehabilitator with national and international teams, including the world strongest man event. Earle sits on the executive committee of the British Association of Sports Rehabilitators and Trainers (BASRaT), as their student liaison officer. In this role he deals with and promotes the BASRaT student experience. Earle is the Middlesex University representative for the higher education academy's hospitality, leisure, sport and tourism sector. He is currently working on a professional doctorate investigating different learning approaches in the development of clinical reasoning skills on undergraduate sports rehabilitation programmes.

Earle is married to Emma and has a son, Benjamin. In his spare time he enjoys sport and is an active cricketer and tennis player. He further enjoys reading and music.

List of contributors

John Allen
Lead Physiotherapist
England Athletics
UK

Phil Barter
Senior Lecturer and Programme Leader
 for Sport Science
London Sport Institute at Middlesex University,
 London
UK

Stuart Butler
Physiotherapist
Allen Physiotherapy Rehabilitation and Sports
 Medicine
England Athletics
UK

Sarah Catlow
University College Plymouth St Mark & St John,
 Plymouth
UK

Nicholas Clark
Clinical Director and Lower Limb Rehabilitation
 Consultant
Integrated Physiotherapy & Conditioning Ltd,
 London
UK

Angela Clough
Senior Lecturer, Programme Leader Sport
 Rehabilitation,
Fellow Society of Orthopaedic Medicine
University of Hull
UK

Rhonda Cohen
Head
London Sport Institute at Middlesex University,
 London
UK

Elezabeth Fowler
Lecturer
University of Salford, Greater Manchester
UK

Julian Hatcher
Senior Lecturer and Programme Leader (Bsc
 (Hons) Sports Rehabilitation)
University of Salford, Greater Manchester
UK

Luke Heath
Graduate Sports Rehabilitator

Dr Lee Herrington
Senior Lecturer and Programme Leader
 (MSc Sports Injury Rehabilitation)
University of Salford, Manchester
UK
Lead Physiotherapist Great Britain Womens
 Basketball

Sebastian Hicks
Graduate Sports Rehabilitator

Ian Horsley
Lead Physiotherapist
English Institute of Sport
UK

Victoria Hyland
Lecturer
London Sport Institute, Middlesex University
UK

David Joyce
Chartered Sports Physiotherapist
Blackburn Rovers FC
The University of Bath
UK

Christo Koukoullis
Graduate Sport Rehabilitator

Helen Matthews
Senior Lecturer and Associate Dean (Teaching and Learning)
University of Salford, Greater Manchester
UK

Martyn Matthews
Senior Lecturer
University of Salford, Greater Manchester
UK

Dr Sanna M. Nordin
Research Fellow, Dance Science, Trinity Laban

Dr Stephen Pearson
Senior Lecturer
University of Salford, Greater Manchester
UK

Jeffrey A. Russell
Assistant Professor of Dance Science
University of California, Irvine
USA

Dror Steiner
Chartered Osteopath

How to use this book

The text has been designed to allow the reader to consider and understand important themes, principles and applications that inform clinical practice. Each chapter begins with an introductory paragraph (see below) that identifies and outlines the aims and outcomes for that chapter.

> The chapter aims and objectives will be emphasised at the beginning. Use these to confirm your understanding of the chapter content.

uses a schema diagram to illustrate how the sports rehabilitator works with other sport medicine practitioners to manage injury. When reading this initial chapter, consider how your scope of practice and professional identity is formed. Use the chapter to help you reinforce your code of practice and reflect

> This chapter provides an overview, analysis, and application of clinical reasoning and problem solving skills in the development of professional competencies within the health care profession generally and more specifically sports rehabilitation. The chapter is important as it will help you develop your thinking skills as you progress your reading throughout the book. By the end of this chapter the reader will be able to locate and explain the role and efficacy of clinical reasoning skills within a professional practice domain. This will inform an appreciation for the complex nature of knowledge construction in relation to clinical explanation and judgement. By considering clinical reasoning as a functional skill set, the reader will further be in a position to explain different models of reasoning and ask structured questions in an attempt to better formulate and construct answers to clinical questions, issues, and decisions. The chapter will further encourage the reader to use problem solving and clinical reasoning skills to justify substantially, through research evidence, professional practice actions and outcomes.

The first chapter provides an overview of the scope of practice for the sports rehabilitator and/or allied health care professional. Within this chapter careful consideration has been given to the position of the sports rehabilitator within a sport and exercise medicine team. The chapter further deals with issues around medical, ethical and legal concerns, and on the medical ethical and legal requirements for your profession.

The following chapters deal with issues around injury screening and performance assessment. These chapters introduce and debate issues concerning assessment and screening, and present research evidence to validate claims. It is useful when reading these chapters to consider how screening and assessment work to accommodate a range of athletes from different sports. Clinicians who simply follow a set programme or protocol for assessment may find it difficult to defend clinical actions and decisions should the athlete not improve following the intervention delivered. It is important to be able to relate the content of the chapter and decide on how best to screen or assess an athlete based on evidence from research studies.

Chapters 4–8 introduce and evaluate the pathophysiology of musculoskeletal components. These chapters are crucial when considering injury management as well as prevention strategies. Each of these chapters makes use of diagrammatic representations of the key musculoskeletal components (see below) and highlights the healing and repair stages of musculoskeletal injuries.

preciate the sport sciences and how an understanding of principles of strength and conditioning, psychology, nutrition, performance assessment and clinical reasoning could be used to highlight areas of concern and move the practitioner to a more complete evaluation and treatment of the athlete. The design of these chapters, have been carefully considered to ensure that you, as reader and clinician, can use important conceptual applications in the management of the client. The themes explored within these chapters are not unique to the chapter per se, but rather form an important thread throughout the text. Exploring the themes within these chapters will hopefully allow the reader to conceptualise sports rehabilitation and injury prevention as a functional ongoing and working operation that requires thought and research evidence to fully appreciate the merit of treatment and rehabilitation.

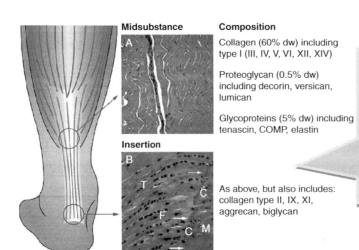

The pathophysiological chapters make use of diagrams and illustrations to highlight key anatomical landmarks and pathological concerns that could impact healing and prolong recovery.

Reference to later chapters and consideration of specific treatment strategies supported by research is evident. When reading these pathophysiological chapters it is useful to consider the primary anatomy of the structure and its normal functional state. Consider how this functional state changes or compensates movement as a result of trauma or pathology. Use this knowledge as a precursor to injury management and a way to shape clinical decisions and actions.

The next seven chapters encompass important themes necessary for effective clinical decisions and management options. Use these chapters to help ap-

The final section of the text is dedicated to joint-specific injuries and pathologies. These chapters introduce the injuries and specific assessment techniques by considering evidence-based practice protocols. These chapters tie together the important consideration for injury prevention and management. The chapters culminate in applied case studies (see below) that are used to illustrate the thought process and clinical decision mapping necessary for effective injury management. It is important to consider how decisions are reached and what processes need to be examined as opposed to simply reaching a decision.

Case Study 20.2

A 24 year old male sprinter with left sided groin discomfort since a plyometric session three months before this initial consultation had resulted in discomfort after every training session.

- Lower abdominal and medial anterior groin pain following activity that is becoming progressively longer to improve with rest.

- Becomes very low grade and almost unnoticeable with rest.

- There is irritable pain when coughing and sneezing.

- Feels 'sore' in the groin when sitting upright for a while.

- Pain in the deep inner groin when squeezing the legs together, particularly in bed.

Pain was described as exercise related and variable between 1 and 7 on the 10 point scale.

> Each injury-specific chapter makes use of an applied case study to frame the clinical issues and consider appropriate and evidence-based treatment and rehabilitation programmes. Use these studies to check your own understanding and decide on whether you agree with the clinical management and/or decisions discussed within the study.

There were minimal impingement signs with hip flexion-adduction.

On inverting the scrotum and placing the little finger in both superficial inguinal rings, the left side appeared more tender and dilated than the right, with a cough impulse.

The left adductor was relatively weaker than the right and painful in resisted adduction lying with straight legs, but not with legs bent in flexion.

There was no discomfort on stretch.

Stork views of the pelvis, standing on one leg and then the other excluded pelvic instability, pubic symphysis and hip pathology.

The patient was referred to a surgeon for opinion.

During surgery the following groin disruption was identified in the operative report:

- torn external oblique aponeurosis

- the conjoined tendon was torn from pubic tubercle

- dehiscence between conjoined tendon and inguinal ligament

Each element of this groin disruption was repaired surgically.

Treatment and rehabilitation

Normal protocol for the first day post operation included stand and walking with gentle stretching and stability exercises.

Five days post operative ultrasound ascertained core stability to be poor and Transversus Abdominis activation (Cowan 2004) was achieved with practice, using patient visualisation of the ultrasound real-time image for re-education.

Adductor exercises (Figures 20.4–5) were encouraged one week post op, several times per day.

Closed chain exercises for stability (e.g., Figures 20.6–9) combined with slow controlled squats progressing to single leg squats, were developed two weeks post op with hydrotherapy for flexibility and stability.

Swimming, cycling and cross-trainer elliptical exercise developed in the third week.

After four weeks he started straight line running build ups alternate days.

Conclusion

This athlete returned to relatively full training after two months and competed internationally six months after the surgery.

Discussion

- At what time should an athlete with groin discomfort be referred to a surgeon to consider operative intervention.

- Should a longer period of conservative treatment and rehabilitation take place before referral for surgery.

- Should the patient have been referred for other investigations, e.g. ultrasound scan or MRI.

- What other areas of the body may contribute towards this athletes injury.

In summary, the contents of this book, are designed to evoke clinical decisions based on research evidence. The chapters are sequenced to allow the reader to develop an appreciation for understanding and analysing clinical practice and actions. Individually the chapters provide a framework for conceptualising different scientific applications and practices, but collectively they form a compendium of clinical knowledge, cemented by clinical practice and framed within an evidence-based context.

Part 1
Introduction to sports rehabilitation

1
Introduction to sport injury management

Jeffrey A. Russell
University of California–Irvine, USA

Introduction and aims

The popularity of physical activity in all of its forms continues to steadily increase. More than just the domain of elite or professional athletes, the populace enjoys a variety of recreational pursuits from hiking and running to skiing and surfing, from badminton and tennis to cricket and hockey. In such endeavours many participants find that injury is inevitable. Unfortunate circumstances are not confined to those engaging in rugby or "X games", daredevil sports like Parkour, kitesurfing or acrobatic bicycle jumping, although clearly these carry a high cost in physical trauma (Young 2002; Spanjersberg and Schipper 2007; Miller and Demoiny 2008). Young footballers and senior golfers alike are prone to injury, as are Olympic performers and "weekend warriors" because injury does not discriminate (Delaney et al. 2009; Falvey et al. 2009). Likewise, non-traditional athletes such as dancers (Fitt 1996; Stretanski 2002; Koutedakis and Jamurtas 2004) will not escape injury (Bowling 1989; Garrick and Lewis 2001; Bronner, Ojofeitimi and Spriggs 2003; Laws 2005).

Whether they are pursuing gold medals or leisure, those who participate in physical activity require both proper preventive training and proper healthcare; they will benefit greatly from experts who can deliver these. Sport rehabilitators and other allied health professionals have much to offer physically active people. This chapter aims to:

- define the role of the sport rehabilitator as a member of the sport injury care team;

- promote individual and organisational professionalism within the field of sport rehabilitation;

- provide a framework for ethical conduct of sport rehabilitators and related professionals;

- describe legal parameters that must be considered by those in sport rehabilitation and related fields.

The role of the sport rehabilitator

Preparing an individual to successfully participate in sport requires, by its very nature, expertise from multiple specialities. Managing the injuries that occur to sport participants also requires input from many specialists. Thus, at any given point the athlete may be surrounded by a team of professionals, including the coach, club manager, conditioning specialist, biomechanist, physiotherapist, nutritionist, exercise physiologist, chiropodist, chiropractor,

4 INTRODUCTION TO SPORT INJURY MANAGEMENT

Table 1.1 The variety of sport medicine team members who work with athletes (see also Figure 1.1)

Medicals and surgeons	Para-medicals	Sport scientists	Sport educators
GP	Sport rehabilitator	Biomechanist	Coach
Chiropodist	Physiotherapist	Exercise	Conditioning specialist
Sport dentist	Osteopath	physiologist	Physical educator
Consultants:	Chiropractor	Sport psychologist	Club manager
Orthopaedic surgeon	Massage therapist	Nutritionist	
General surgeon	Sport optometrist	Kinesiologist	
Neurosurgeon	Acupuncturist		
Cardiologist	First responder		
Radiologist	Alternative therapy practitioner		
Physiatrist			
Neurologist			

osteopath, sport optometrist, sport psychologist, sport dentist, GP, consultant and, indeed, sport rehabilitator (Table 1.1 and Figure 1.1). Depending on the sport, an athlete's level in the sport and the venue, all of the listed professionals may not be involved in care. Further, some professionals may be qualified to administer more than one care speciality. However, regardless of the situation the management of sport injury is a team activity, and the sport rehabilitator plays a key role.

The British Association of Sport Rehabilitators and Trainers (BASRaT) administer the credential "Graduate Sport Rehabilitator," which is abbreviated to "GSR." According to this professional society, "a Graduate Sport Rehabilitator is a graduate level autonomous healthcare practitioner specialising in musculoskeletal management, exercise based rehabilitation and fitness" (British Association of Sport Rehabilitators and Trainers 2009b). Further, BASRaT outline the skill domains of a Graduate Sport Rehabilitator as being:

- professional responsibility and development

- prevention

- recognition and evaluation of the individual

- management of the individual–therapeutic intervention, rehabilitation and performance enhancement

- immediate care

Whilst prevention of injury is certainly desirable, the reality that athletes will be injured is part of sport participation. Thus, the sport rehabilitator must always be prepared to administer the care for which they are trained. The ideal place to begin providing this care is pitchside or courtside where the circumstances surrounding the injury have been observed and evaluation of the injury can be performed prior to the onset of complicating factors such as muscle spasm. Any sport rehabilitator who expects to offer this type of care must possess the proper qualification and additional credentials to support it. Minimum

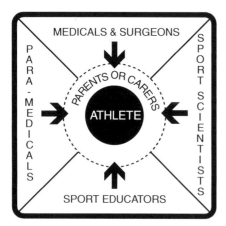

Figure 1.1 Diagram showing the breadth of sport injury management. Note that in the situation of an athlete who is a minor child, the parents or carers become part of the management scenario.

Table 1.2 Components of the British Association of Sport Rehabilitators and Trainers (2009b) skill domains

Skill Domain	Components
Professional responsibility and development	Record keeping
	Professional practice – conduct and ethical issues
	Professional practice – performance issues
Prevention	Risk assessment and management
	Pre-participation screening
	Prophylactic interventions
	Health and safety
	Risks associated with environmental factors
Recognition and evaluation of the individual	Subjective evaluation
	Neuromusculoskeletal evaluation
	Physiological and biomechanical evaluation
	Nutritional, pharmacological, and psychosocial factors
	Health and lifestyle evaluation
	Clinical decision making
	Dissemination of assessment findings
Management of the individual – therapeutic intervention, rehabilitation and performance enhancement	Therapeutic intervention
	Exercise based rehabilitation
	Performance enhancement
	Factors affecting recovery and performance
	Monitoring
	Health promotion and lifestyle management
Immediate care	Emergency first aid
	Evaluation
	Initiation of care

abilities include cardiopulmonary resuscitation, first aid, blood-borne pathogen safeguards, strapping and bracing, and practical experience (in a proper clinical education programme) with the variety of traumatic injuries that accompany sport participation. Furthermore, working with certain sports – such as cricket, ice hockey and North American football – requires specialised understanding of protective equipment that includes how to administer care in emergency situations when the injured athlete is encumbered by such equipment.

BASRaT's (2009b) *Role Delineation of the Sport Rehabilitator* document details the implementation of the skill domains listed above into a scope of practice. Table 1.2 outlines the components of each domain; these are further subdivided into knowledge components and skill components to create a framework both for the education of sport rehabilitators and the extent of their capabilities to serve as healthcare professionals.

A brief introduction to a similar type of sport healthcare provider in the United States of America is useful here as a comparison. Certified Athletic Trainers (denoted by the qualification "ATC") are "health care professionals who collaborate with physicians to optimize activity and participation of patients and clients. Athletic training encompasses the prevention, diagnosis, and intervention of emergency, acute, and chronic medical conditions involving impairment, functional limitations, and disabilities" (National Athletic Trainers' Association 2009b). The National Athletic Trainers' Association, the professional body of Certified Athletic Trainers, has existed since 1950. Standards of practice are set and a certification examination is administered by the Board of Certification (2009) to ensure that the profession is properly regulated. Most individual states in the USA also require possession of a licence in order to practice as an athletic trainer. Comparable to the role delineation skill domains for sport rehabilitators listed above, the requisite skills of Certified Athletic Trainers are categorised into 13 content areas (National Athletic Trainers' Association 2009a):

1. foundational behaviours of professional practice
2. risk management and injury prevention
3. pathology of injuries and illnesses
4. orthopaedic clinical examination and diagnosis
5. medical conditions and disabilities
6. acute care of injuries and illnesses
7. therapeutic modalities
8. conditioning and rehabilitative exercise
9. pharmacology
10. psychosocial intervention and referral
11. nutritional aspects of injuries and illnesses
12. health care administration
13. professional development and responsibility

These content areas define how Certified Athletic Trainers are educated and how they retain the ATC credential via continuing professional development hours (called continuing education in the USA, with the participation increments called CEUs, or continuing education units). As with Graduate Sport Rehabilitators, accountability to such standards is imperative for sustaining the integrity of the profession.

Continuing professional development

There is no place pitchside for healthcare practitioners who cannot perform the required duties that arise under the pressure of managing injury during sporting competition. Therefore, a fundamental responsibility of the sport rehabilitator – or any other healthcare practitioner – is to secure a high standard in their education. Certainly this encompasses the undergraduate and postgraduate courses and the motivation to embrace diligence and excellence in all required modules, work placements, internships and the like. The knowledge required and tasks allowed for specific professional qualifications are usually dictated by professional organisations. As mentioned above, BASRaT hold sport rehabilitators to a high standard of education. Once a qualification is attained, however, another educational process ensues: professionals must engage in continuing professional development (CPD). The importance of this cannot be overstated. CPD helps the sport rehabilitator not only maintain their skills, but acquire new ones that broaden one's ability to offer high quality healthcare to athletes, clients and patients. Moreover, knowledge in sport science and sport medicine is constantly evolving as further basic and applied research is undertaken. Adequate CPD helps the sport rehabilitator stay abreast of these developments.

CPD courses afford exciting opportunities for personal enrichment. Many topics are germane to the field and a veritable subculture exists to provide adequate chances for professionals to enlist in training courses that match every ability, need and desire. Most professional societies, including BASRaT, advise their members about suitable courses and the required quantity of CPD hours. Advanced life support, manual therapy, pitchside emergency care, strength training, exercise testing, specialised joint examinations, rehabilitative exercise and management of non-orthopaedic injuries and conditions are only a few topics representative of the wide gamut of offerings.

A qualification in basic cardiopulmonary resuscitation for healthcare providers (i.e. BLS/AED – Basic Life Support/Automated External Defibrillation) is considered a minimal credential that should be kept up to date by periodic skills retraining. The Resuscitation Council (UK) and the European Resuscitation Council publish the appropriate standards for BLS and AED training (European Resuscitation Council 2009; Resuscitation Council (UK) 2009); the latter also maintains a calendar of many life support courses offered around Europe, including the United Kingdom.

Knowledge, ability and wisdom

It is important for professional healthcare providers to distinguish amongst knowledge, ability and wisdom. These are distinct, yet interrelated, characteristics that all sport rehabilitators must strive for as they provide care to the public. Knowledge is the learning and understanding of facts that form the basis for practice. It provides the information on which a successful career is built. Ability is the application of knowledge. Thus, knowledge really is not useful until a person accomplishes a task by applying it.

Wisdom, though, is like the glue that holds a professional career together. It is the most difficult – but also the most significant – of the three to garner because it is gained over time as one matures and is exposed to an ever-widening variety of experiences. Wisdom considers both the available knowledge and ability, mixing them in the right proportion to elicit the best result within a given set of present circumstances. Whilst this may seem somewhat esoteric, the three characteristics are fundamental to success and all healthcare professionals draw on each of them everyday.

Ethical considerations

Ethics refers to a set of concepts, principles and laws that inform people's moral obligation to behave with decency. Part of this is the necessity to protect people who are in a relatively vulnerable position, such as a patient or client in a healthcare setting. Similar to other professionals, each sport rehabilitator must consider themselves a healthcare practitioner and, therefore, under an ethical obligation for inscrutable professional conduct. Sport medicine presents challenging parameters within which to apply an ethical framework (Dunn et al. 2007; Salkeld 2008), due largely to the high public visibility of sport itself. This is perhaps an even more significant reason for the sport rehabilitator to ardently ensure that their practice falls under appropriate accountability.

Unfortunately ethical dilemmas do not always lend themselves to clear, objective dispensation; thus, governing bodies codify guiding principles for conduct. The Code of Ethics of the British Association of Sport Rehabilitators and Trainers, shown in Table 1.3, is an example of guidelines that promote proper behaviour.

In healthcare the field of ethics sets appropriate and acceptable standards to protect the public from damages incurred at the hands of unscrupulous or incompetent practitioners and the deleterious effects of unwarranted or dangerous diagnostic or therapeutic interventions. Respect for the dignity of humans is placed foremost and healthcare practice must accommodate to this high standard. There are a number of circumstances that occur in sport that can strain the typical application of ethics; areas where difficulties arise include:

- decisions about return to sport activity with a persisting injury

- pharmaceutical therapies to assist participation

- participation of children, especially in high-risk sport

- sharing of confidential athlete medical information amongst practitioners, or between practitioners and public representatives, such as the press

- ergogenic aids, such as anabolic steroids and blood "doping."

Of these, treating an athlete's medical information with confidentiality is likely to be the most difficult and frequently compromised, particularly in the pitchside environment (Salkeld 2008). Salkeld suggests that several competing challenges and pressures collide pitchside to create ethical dilemmas: the close proximity of an injured player to other players and coaches when being examined, the public visibility of an injury, the interests of the sporting club and the desire of the coaching staff to receive information about the injury coupled with the concomitant desire of the player to shield this information from the coaches. Additional areas of contemporary ethical challenges for practitioners caring for athletes include informed consent for care, drug prescription and use of innovative or emerging technologies (Dunn et al. 2007).

The most appropriate way for the sport rehabilitator to manage potentially difficult ethical predicaments is to practise diligently under an approved ethical code, such as that of the British Association for Sport Rehabilitators and Trainers, and to decide how individual ethical quandaries will be handled *prior* to being confronted by them. The consequences of infractions are severe and have resulted in revoked professional licences, registrations and certifications, and have ended careers in particularly egregious cases.

Table 1.3 The Code of Ethics of the British Association of Sport Rehabilitators and Trainers (2009a)

PRINCIPLE 1: Members shall accept responsibility for their scope of practice
1.1 Members shall not misrepresent in any manner, either directly or indirectly, their skills, training, professional credentials, identity or services
1.2 Members shall provide only those services of assessment, analysis and management for which they are qualified and by pertinent legal regulatory process
1.3 Members have a professional responsibility to maintain and manage accurate medical records
1.4 Members should communicate effectively with other healthcare professionals and relevant outside agencies in order to provide an effective and efficient service to the client
Supporting Legislation: Data Protection Act 1998; Human Rights Act 1998

PRINCIPLE 2: Members shall comply with the laws and regulations governing the practice of musculoskeletal management in sport and related occupational settings
2.1 Members shall comply with all relevant legislation
2.2 Members shall be familiar with and adhere to all British Association of Sport Rehabilitators and Trainers' Guidelines and Code of Ethics
2.3 Members are required to report illegal or unethical practice detrimental to musculoskeletal management in sport and related occupational settings

PRINCIPLE 3: Members shall respect the rights, welfare and dignity of all individuals
3.1 Members shall neither practice nor condone discrimination on the basis of race, creed, national origin, sex, age, handicap, disease entity, social status, financial status or religious affiliation. Members shall comply at all times with relevant anti-discriminatory legislation
3.2 Members shall be committed to providing competent care consistent with both the requirements and limitations of their profession
3.3 Members shall preserve the confidentiality of privileged information and shall not release such information to a third party not involved in the client's care unless the person consents to such release or release is permitted or required by law

PRINCIPLE 4: Members shall maintain and promote high standards in the provision of services
4.1 Members shall recognise the need for continuing education and participation in various types of educational activities that enhance their skills and knowledge
4.2 Members shall educate those whom they supervise in the practice of musculoskeletal management in sport and related occupational settings with regard to the code of ethics and encourage their adherence to it
4.3 Whenever possible, members are encouraged to participate and support others in the conduct and communication of research and educational activities, that may contribute to improved client care, client or student education and the growth of evidence-based practice in musculoskeletal management in sport and related occupational settings
4.4 When members are researchers or educators, they are responsible for maintaining and promoting ethical conduct in research and education

PRINCIPLE 5: Members shall not engage in any form of conduct that constitutes a conflict of interest or that adversely reflects on the profession
5.1 The private conduct of the member is a personal matter to the same degree as is any other person's, except when such conduct compromises the fulfillment of professional responsibilities
5.2 Members shall not place financial gain above the welfare of the client being treated and shall not participate in any arrangement that exploits the client
5.3 Members may seek remuneration for their services that is commensurate with their services and in compliance with applicable law

Legal considerations

An additional concern when providing care to athletes is the increasingly litigious aura that pervades much of Western society. Sport rehabilitators and other practitioners of sport injury care are subject to lawsuits brought by athletes and their representatives (e.g. parents, carers). As previously mentioned, consistently following an appropriate code of ethics and continually educating yourself via CPD are two ways to ameliorate the risk. It is also crucial that sport injury professionals maintain malpractice and liability insurance cover, a caveat for which BASRaT ensures compliance of its member Graduate Sport Rehabilitators.

The discussion of legal liability first needs a directive citing the proper way of acting that is acknowledged by courts when deriving judgments. "The man on the Clapham omnibus" is a common phrase in English law that denotes a person who acts truly and fairly (Glynn and Murphy 1996) with all faculties that would be expected under the circumstances. (An American equivalent is "a reasonable and prudent person.") A structure of accountability is fundamental to application of this concept. Within a given context it may be modified appropriately; healthcare is only one realm to which it pertains (Glynn and Murphy 1996). Whilst being afraid of the potential for litigation in a sport healthcare environment would unnecessarily constrain a well-qualified professional, undeniably sport rehabilitators and other healthcare practitioners must be cognisant of the inherent risk of being sued for wrong actions (acts of commission) or for inaction when action is warranted (acts of omission). Instead of being intimidated, one should take all necessary steps to reduce the likelihood of a lawsuit as much as possible.

The tenet of a "public right to expertise" was proposed for the sport and physical education fields more than 25 years ago (Baker 1980, 1981). The general concept states that members of the public have the right to expect that those who offer themselves as professionals in a given field of endeavour are qualified as experts in that field. In the context of sport rehabilitation, affording the public this right is paramount because of the potential for severe consequences when healthcare providers are inadequately skilled or make errors in practice or judgement (Goodman 2001).

Countless legal cases transcend recent decades (Appenzeller 2005) as plaintiffs (people filing a lawsuit) persist in claiming negligence by defendants (people being sued) such as healthcare providers, coaches and institutions. Generally a negligence claim must show the following (Champion 2005):

- there is a verifiable standard of care to which the defendant should be held

- the defendant had a duty to care for the plaintiff

- the defendant breached their duty

- the plaintiff sustained damages or injury

- the damages or injury were caused by the defendant's breach of the duty.

Risk of exposure to legal liability related to healthcare in sport usually occurs in four main areas, the first three of which are related to one another (Kane and White 2009):

1. Pre-participation physical examination – A screening process to evaluate the athlete's physical and mental status prior to engaging in sport should be a fundamental requirement before such engagement occurs.

2. Determination of an athlete's ability to participate – Whether confronted with signs and symptoms pitchside, courtside, in a first aid facility, in a polyclinic, or elsewhere, proper decision making about an athlete's fitness to participate must be made in accordance with current healthcare practice.

3. Evaluation and care of significant injuries on the pitch or court – Healthcare professionals not only must be well-qualified, they must deliver care that is appropriate for a given situation. Concussions, spinal cord injuries and hyperthermia are three examples of injuries requiring urgent, specialised diagnostic and treatment procedures. A sponsoring club, university, school or organisation must ensure that a plan is in place to adequately respond to emergency situations that may arise in sport.

Table 1.4 Some examples of negligence that can lead to injury litigation in sport

Area of potential negligence	Examples
Facility safety	Poor condition of the surface of the pitch, court, track, etc. (e.g. holes, uneven surfaces)
	Unsafe equipment (e.g. exposed sharp edges, broken or rusted parts)
	Unsafe practices (e.g. reduced visibility if lights are not used when training held at night)
	Impeding objects that are not part of the sport activity
Warning of (or unnecessary) risk or danger	Failure to teach safe techniques for the sport
	Failure to disclose potential injury consequences of playing and of not playing using safe techniques
	Failure to intervene when players do not use safe techniques
	Mismatched players (e.g. adult players participating together with young players)
Protective equipment	Failure to provide proper protective equipment
	Failure to require use of protective equipment
	Improper fit of protective equipment
Documentation of injury	Failure to maintain injury records
	Failure to maintain treatment and rehabilitation records
	Failure to maintain confidentiality of records
	Falsifying or altering medical records
Appropriate care	Failure to follow proper care protocols
	Failure to refer injured player to healthcare professional of greater experience or higher qualification
	Failure to remove injured player from participation

4. Disclosure of personal medical record information – Confidentiality is a fundamental right and expectation of all patients and clients, including athletes. The sport rehabilitator must take care to not convey – even unwittingly – information about an athlete's case to others without the athlete's permission.

Additional concerns for the sport rehabilitator that relate to potential injury circumstances in these general categories are accumulated in Table 1.4 (Anderson 2002; Champion 2005; Kane and White 2009).

Following a review of pertinent legal cases, Goodman (2001) corroborated that those who supervise teams could be liable if they or their sport healthcare providers failed to perform properly in any of these specific areas:

- Provide appropriate training instruction.

- Maintain or purchase safe equipment.

- Hire or supervise competent and responsible personnel.

- Give adequate warning to participants concerning dangers inherent in a sport.

- Provide prompt and proper medical care.

- Prevent the injured athlete from further competition that could aggravate an injury (Goodman 2001, p.449).

Finally, Konin and Frederick (2005, p.38) identified six common mistakes sport healthcare providers make in caring for athletes; these are shown below and provide key areas for attention by sport rehabilitators:

1. Not establishing baseline (i.e. "normal" uninjured) data with respect to a patient/athlete

2. Accidentally verbally breaching a patient's privacy

3. Not knowing rules and regulations related to confidentiality of patient information and medical records

4. Making decisions based on experience and instincts rather than seeking appropriate authoritative advice

5. Not educating a patient/athlete about a therapeutic modality intervention

6. Underestimating the amount of documentation required with catastrophic injury events

In short, sadly there are virtually no limits to what one can be sued for with respect to managing sport injury. This should be so sobering that the prudent sport rehabilitator will prepare accordingly to reduce as much as possible the likelihood of this occurring.

Conclusion

The sport rehabilitator is a key member of the sport injury management team. As such, you must adhere to several important professional, practical, ethical and legal principles. Properly equipping yourself to administer acute injury management in the venues where practice will be undertaken – whether pitchside, courtside, trackside, in a clinic or elsewhere – is vitally important. However, simply being prepared to deliver care required by sport participants does not sufficiently qualify a sport rehabilitator, or any other sport health professional for that matter. Proper ethical and legal frameworks are integral to success, as well. Without these underpinnings the most skillful healthcare worker will not be able to sustain their practice under the guidelines deemed appropriate by civilised societies.

In summary, this entire textbook is devoted to ensuring the reader's success in sport rehabilitation or a related field. It is a welcome instructional resource to the student, but it is a valuable informational reference to the clinician, too. There is a wealth of material presented where the authors offer insights from their knowledge, abilities and wisdom in order to equip the reader for excellence in their career post.

References

Anderson, M.K. (2002) *Fundamentals of Sports Injury Management*. Philadelphia, PA: Lippincott Williams and Wilkins.

Appenzeller, H. (2005) Risk management in sport. In Appenzeller, H. (Ed.) *Risk Management in Sport: Issues and Strategies*, 2nd edn. Durham, NC: Carolina Academic Press, pp. 5–10.

Baker, B.B. (1980) The public right to expertise (part 1). *Interscholastic Athletic Administration*, 7 (2), 21–23.

Baker, B.B. (1981) The public right to expertise (part 2). *Interscholastic Athletic Administration*, 7 (3), 22–25.

Board of Certification (2009) *What is the BOC?* Omaha, NE: Board of Certification (accessed 14th August 2009), <http://bocatc.org/index.php?option=com_content&task=view&id=27&Itemid=29>

Bowling, A. (1989) Injuries to dancers: prevalence, treatment and perception of causes. *British Medical Journal*, 298, 731–734.

British Association of Sport Rehabilitators and Trainers (2009a) *Role Delineation and Definition of Graduate Sport Rehabilitator (GSR)*. Salford: British Association for Sport Rehabilitators and Trainers (accessed 27th July 2009) <http://www.basrat.org/role.asp>

British Association of Sport Rehabilitators and Trainers (2009b) *Role Delineation of the Sport Rehabilitator*. Salford: British Association for Sport Rehabilitators and Trainers (accessed 14th August 2009) <http://basrat.org/docs/basrat_role_delineation.pdf>

Bronner, S., Ojofeitimi, S. and Spriggs, J. (2003) Occupational musculoskeletal disorders in dancers. *Physical Therapy Reviews*, 8, 57–68.

Champion, W.T., Jr. (2005) *Sports Law in a Nutshell*. St. Paul, MN: Thomson/West.

Delaney, R.A., Falvey, E., Kalimuthu, S., Molloy, M.G. and Fleming, P. (2009) Orthopaedic admissions due to sports and recreation injuries. *Irish Medical Journal*, 102 (2), 40–42.

Dunn, W.R., George, M.S., Churchill, L. and Spindler, K.P. (2007) Ethics in sports medicine. *American Journal of Sports Medicine*, 35 (5), 840–844.

European Resuscitation Council (2009) *European Resuscitation Council*. Edegem, Belgium: European Resuscitation Council (accessed 27th July 2009) <https://www.erc.edu/new/>

Falvey, E.C., Eustace, J., Whelan, B., Molloy, M.S., Cusack, S.P., Shanahan, F. and Molloy, M.G. (2009) Sport and recreation-related injuries and fracture occurrence among emergency department attendees: implications for exercise prescription and injury prevention. *Emergency Medicine Journal*, 26 (8), 590–595.

Fitt, S.S. (1996) *Dance Kinesiology*. New York: Schirmer Books.

Garrick, J.G. and Lewis, S.L. (2001) Career hazards for the dancer. *Occupational Medicine*, 16 (4), 609–618.

Glynn, J.J. and Murphy, M.P. (1996) Failing accountabilities and failing performance review. *International Journal of Public Sector Management*, 9 (5/6), 125–137.

Goodman, R.S. (2001) Sports medicine. In Sanbar, S.S., Gibofsky, A., Firestone, M.H., LeBlang, T.R., Liang, B.A. and Snyder, J.W. (Eds) *Legal Medicine*, 5th edn. St. Louis: Mosby, pp. 448–450.

Kane, S.M. and White, R.A. (2009) Medical malpractice and the sports medicine clinician. *Clinical Orthopaedics and Related Research*, *467* (2), 412–419.

Konin, J.G. and Frederick, M.A. (2005) *Documentation for Athletic Training*. Thorofare, NJ: Slack.

Koutedakis, Y. and Jamurtas, A. (2004) The dancer as a performing athlete. *Sports Medicine*, *34* (10), 651–661.

Laws, H. (2005) *Fit to Dance 2*. London: Dance UK.

Miller, J.R. & Demoiny, S.G. (2008) Parkour: a new extreme sport and a case study. *Journal of Foot and Ankle Surgery*, *47* (1), 63–65.

National Athletic Trainers' Association. (2009a) *Competencies*. [online]. Dallas, TX, USA: National Athletic Trainers' Association. [accessed 14th August 2009]. <http://www.nata.org/education/competencies.htm>.

National Athletic Trainers' Association. (2009b) *What is an Athletic Trainer?* Dallas, TX: National Athletic Trainers' Assocation (accessed 14th August 2009) <http://www.nata.org/about_AT/whatisat.htm>

Resuscitation Council (UK) (2009) *Guidelines, medical information and reports*. London: Resuscitation Council (UK) (accessed 27th July 2009) <http://www.resus.org.uk/pages/mediMain.htm>

Salkeld, L.R. (2008) Ethics and the pitchside physician. *Journal of Medical Ethics*, *34* (6), 456–457.

Spanjersberg, W.R. and Schipper, I.B. (2007) Kitesurfing: when fun turns to trauma – the dangers of a new extreme sport. *Journal of Trauma*, *63* (3), E76–E80.

Stretanski, M.F. (2002) Classical ballet: the full contact sport. *American Journal of Physical Medicine and Rehabilitation*, *81* (5), 392–393.

Young, C.C. (2002) Extreme sports: injuries and medical coverage. *Current Sports Medicine Reports*, *1* (5), 306–311.

Part 2

Injury screening and assessment of performance

2
Injury prevention and screening

Phil Barter
London Sport Institute at Middlesex University, London, UK

Introduction

The main aims of this chapter are to introduce musculoskeletal screening and outline the available methods and the related reliability and validity issues. This chapter will allow the reader to gain an understanding of musculoskeletal screening and its role in injury prevention, identify the musculoskeletal screening methods available including a discussion of the validity and reliability of screening methods. The chapter will finally recommend a screening procedure for injury risk identification.

With the need for athletes to play an ever-increasing number of fixtures, the enforced breaks due to injury need to be decreased. Several approaches can be taken to ensure that the athlete is trained and prepared so that any possible problems are either dealt with before they arise or measures are in place so that treatment can be administered rapidly upon injury. Injury prevention is a process whereby the athlete is screened through a variety of tests to identify any potential problems with their musculoskeletal composition. These problems can then be identified and training practices put in place to either eradicate these problems or reduce their possible impact. Several procedures are used by sports practitioners with varying degrees of success as the need for one common procedure for musculoskeletal screening becomes apparent. Several researchers have attempted to identify which methods offer the highest degree of accuracy and validity (Gabbe et al. 2004; Miller and Callister 2009; McClean et al. 2005).

Pre-habilitation can often be overlooked in the makeup of a sports support team, which can often lead to problems being overlooked and the team or individual not performing to their potential throughout their season due to injury. In contrast some professional clubs spend too much time on remedial level pre-habilitation and not enough time on high intensity training that meets the demands of the sport. Procedures need to be implemented to ensure the amount of training days and competitive sessions missed are minimised. Practitioners need to be proactive with their treatments plans and not rely on the traditional reactive plans. In order for this to be the case practitioners need to be fully aware of the latest research and methods in the area of need through continued professional development. These plans often commence during the offseason or the early part of pre-season. During this period the athlete can be assessed without the demands of competition, which will enable the practitioner to gain the knowledge needed to plan for the upcoming season.

Screening can be completed through a variety of tests including physical activity tests, functional assessment and questionnaires. These all have varying strengths and weaknesses and are also dependent on the practitioner who is carrying out the screening.

Sports Rehabilitation and Injury Prevention   Edited by Paul Comfort and Earle Abrahamson
© 2010 John Wiley & Sons, Ltd

Regardless of the method chosen the aim is to identify a series of risk factors that will enable any potential problems to be identified and diagnosed. The findings of the athlete screening can be assessed for risk of injury so that plans can be made to reduce the level of risk for the athlete.

Screening methods

The approach taken by the leisure industry as a whole towards injury prevention is one that involved a health screening questionnaire. The questionnaire can often be modified to include a few general musculoskeletal questions which, if answered negatively, can then result in the athlete being referred to their GP. The method of screening which includes this GP referral approach is very general and mainly focuses on reduced liability of the administering facility or practitioner. Although this approach would not be recommended when working with athletes in sport rehabilitation, the use of a questionnaire is often overlooked, even though research has shown it to be a useful screening tool. Research by Dawson et al. (2009) suggests that through the use of the Extended Nordic Musculoskeletal Questionnaire (NMQ-E) (see Figure 2.1) potential pain areas and consequential problems can be identified. The questionnaire needs to be administered by a suitable practitioner and not the athlete involved, ensuring the results are valid. Research suggested that in conjunction with relevant functional assessments this questionnaire was a useful starting point in the screening process. The results found that prevalence of musculoskeletal problems could be correctly diagnosed and treated effectively. The reliability of the questionnaire used was tested over a series of trials with the same results shown, indicating the validity and repeatability (Dawson et al. 2009). Figure 2.1 clearly shows the important areas of the body so the athlete knows which part the questions are related to and then follows a logical order through the area identifying the degree of any possible pain. The questions then follow a logical order down the body covering all the general points of the body. The results can then be passed on to a sports rehabilitation practitioner for further focused functional assessment of the identified problem areas (Dawson et al. 2009).

Assessment by the means of questionnaire is not, of course, a new methodology, but the integration with functional and physical tests to form a holistic process is. The way in which the screening elements are integrated into the injury prevention process can have a large impact on the athletes involved. If athletes feel part of this process then they could take ownership and really fully commit to the measure that is ultimately proposed. If athletes are insufficiently involved then they might see the process and resultant programme as unnecessary and therefore not worthy of expending too much energy on. The other situation is where the athlete could learn too much and become de-motivated about their long-term future in the sport and as a result not commit. The latter two scenarios will mean that the practitioners plans may fail and the problems identified will probably arise with a negative effect on performance.

A more simplistic approach to questionnaires can also produce good results in reducing the amount of tests that an athlete needs to perform. The reduction of testing time is important in large squads as the amount of time the practitioner spends with the athlete reduces the amount of time the coach can spend working on sport-specific training. Therefore, the need to develop a useful tool to identify the members of the squad who do not need any further testing or those who can be dealt with in a reduced fashion is important. The process shown in Figure 2.2 indicates a simple pathway to group the squad into different levels of testing through a simple set of questions. This has been shown to be reliable in identifying conditions and more importantly not missing any problems. The questionnaire again needs to be administered by a practitioner to insure the validity of the answers (Berg-Rice et al. 2007).

When the screening questions shown in Figure 2.2 were completed by a practitioner the potential injuries were correctly found in 92% of athletes screened. When the same process was completed by a non-practitioner only 80% of cases where positively screened. Although the results of the screening process show that it needs to be conducted by the relevant practitioner, the overall impact on the average number of days lost through injury was still similar 23 (non-screened) versus 21 days. This suggests that stringent follow-up tests are still needed to ensure that problems are dealt with effectively (Berg-Rice et al. 2007).

The initial questionnaire used for screening can also access the athlete's psychological state towards injury and the social factors that could affect their own approach to problems. The Orebro Musculoskeletal Pain Screening Questionnaire (OMPSQ)

How to answer the questionnaire:

Please answer by putting a cross in the appropriate box- one cross for each question. Answer every question, even if you have never had trouble in any part of your body. Please answer questions from left to right before going down to the next body region. This picture shows how the body has been divided. Limits are not sharply defined and certain parts overlap. You should decide for yourself which part (if any) is or has been affected.

	Have you ever had trouble (ache, pain or discomfort) in:	If 'No', go on to the next body region. If 'Yes', please continue	At the time of initial onset of the trouble, what was your age?	Have you ever been hospitalised because of the trouble?	Have you ever had to change jobs or duties (even temporarily) because of the trouble?	Have you had trouble (ache, pain, discomfort) at anytime during the best 12 months?	If 'No', go on to the next body region. If 'Yes', please continue	Have you had trouble (ache, pain, discomfort) at anytime during the last month (4 weeks)?	Have you had trouble (ache, pain, discomfort) today?	During the **best 12 months** have you at anytime:			
										been prevented from doing your normal work (at home or away from home) because of the trouble?	seen a doctor, physio- therapist, chiropractor or other such person because of the trouble?	taken medication because of the trouble?	taken sick leave from work/studies because of the trouble?
NECK	☐ No ☐ Yes		___ years	☐ No ☐ Yes	☐ No ☐ Yes	☐ No ☐ Yes		☐ No ☐ Yes	☐ No ☐ Yes	☐ No ☐ Yes	☐ No ☐ Yes	☐ No ☐ Yes	☐ No ☐ Yes
SHOULDERS	☐ No ☐ Yes		___ years	☐ No ☐ Yes	☐ No ☐ Yes	☐ No ☐ Yes		☐ No ☐ Yes	☐ No ☐ Yes	☐ No ☐ Yes	☐ No ☐ Yes	☐ No ☐ Yes	☐ No ☐ Yes
UPPER BACK	☐ No ☐ Yes		___ years	☐ No ☐ Yes	☐ No ☐ Yes	☐ No ☐ Yes		☐ No ☐ Yes	☐ No ☐ Yes	☐ No ☐ Yes	☐ No ☐ Yes	☐ No ☐ Yes	☐ No ☐ Yes
ELBOWS	☐ No ☐ Yes		___ years	☐ No ☐ Yes	☐ No ☐ Yes	☐ No ☐ Yes		☐ No ☐ Yes	☐ No ☐ Yes	☐ No ☐ Yes	☐ No ☐ Yes	☐ No ☐ Yes	☐ No ☐ Yes
WRISTS/HANDS	☐ No ☐ Yes		___ years	☐ No ☐ Yes	☐ No ☐ Yes	☐ No ☐ Yes		☐ No ☐ Yes	☐ No ☐ Yes	☐ No ☐ Yes	☐ No ☐ Yes	☐ No ☐ Yes	☐ No ☐ Yes
LOW BACK	☐ No ☐ Yes		___ years	☐ No ☐ Yes	☐ No ☐ Yes	☐ No ☐ Yes		☐ No ☐ Yes	☐ No ☐ Yes	☐ No ☐ Yes	☐ No ☐ Yes	☐ No ☐ Yes	☐ No ☐ Yes
HIPS/THIGHS	☐ No ☐ Yes		___ years	☐ No ☐ Yes	☐ No ☐ Yes	☐ No ☐ Yes		☐ No ☐ Yes	☐ No ☐ Yes	☐ No ☐ Yes	☐ No ☐ Yes	☐ No ☐ Yes	☐ No ☐ Yes
KNEES	☐ No ☐ Yes		___ years	☐ No ☐ Yes	☐ No ☐ Yes	☐ No ☐ Yes		☐ No ☐ Yes	☐ No ☐ Yes	☐ No ☐ Yes	☐ No ☐ Yes	☐ No ☐ Yes	☐ No ☐ Yes
ANKLES/FEET	☐ No ☐ Yes		___ years	☐ No ☐ Yes	☐ No ☐ Yes	☐ No ☐ Yes		☐ No ☐ Yes	☐ No ☐ Yes	☐ No ☐ Yes	☐ No ☐ Yes	☐ No ☐ Yes	☐ No ☐ Yes

Figure 2.1 The Extended Nordic Musculoskeletal Questionnaire (NMQ-E) (Dawson et al. 2009). Reproduced, with permission, from Dawson, A.P., Steele, E.J., Hodges, P.W., & Stewart, S. (2009). Development and Test-Retest reliability of an extended version of the nordic musculoskeletal questionnaire (NMQ-E): A Screening instrument for musculoskeletal pain. The journal of Pain, 10 (5), 517–526 © 2009 Elsevier.

18 INJURY PREVENTION AND SCREENING

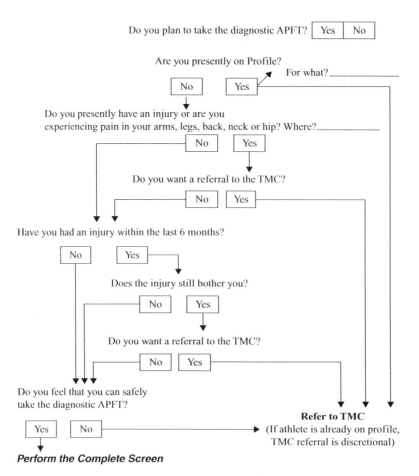

Figure 2.2 Initial screening questions (Berg-Rice et al. 2007). Reproduced, with permission, from Berg-Rice, V.J., Conolly, V.L., Pritchard, A., Bergeron, A., & Mays, M.Z. (2007). Effectiveness of a screening tool to detect injuries furing army health care specialist training. Work, 29, 117–188, © 2007 IOS Press.

has been used in research to look at potential problems and the subject's pain avoidance. The athlete's injury history will have an effect on their ability to deal with injury and how they rate their current musculoskeletal state. The OMPSQ factors into the score: fear avoidance; how well the person perceives they can deal with pain; how distressed they have been in the past about injuries; and the athlete's own rating of their function. These scores were then combined and factors given to the rating to produce three predicting factors of how many days the athlete will miss on average a year. The first predicting factor was the function group of questions, which significantly ($p = 0.001$), predicted the amount of 'Sick' days the athlete would have over the course of a three-year period (Westman et al. 2008). The functional assessment questions looked into how athletes perceived the injury affected them and their ability to perform. The second factor that significantly predicted the amount of the missed training days due to problems was the pain factor. The athletes' pain and injury history was factored into this predictor to significantly predict the amount of days the athlete would miss during the next three years, ($p = 0.0026$) (Westman et al. 2008). The final factor in this questionnaire, which was labelled fear-avoidance and was the pain that the athlete had experienced, did not significantly predict the amount of missed training days. The last factor included the athlete's fear of training due the perceived affect it would have on an injury and

Characteristic of Assessment	Coding	N
Quality of pain studied	Intensity	12
	General pain	8
	Severity	2
	"Axial, peripheral, and global pain"	1
	" Intensity, steady, brief ("shock-like") and skin pain ('pain elicited by nonpainful stimulation of the skin')"	1
	" Maximum, minimum, and current pain. The mean of the 3 measurements provided the pain index."	1
	"Subjects' overall perception of back pain."	1
	Not provided	4
Scale type	100-mm line	12
	10-cm line	9
	0–100	4
	0–90 mm	1
	Not provided	4
Scale anchors	Anchors provided	13
	Anchors not provided	17
Reporting period	Current	5
	Previous week/Last 7 days	5
	Daily	2
	"At rest and during activity"	1
	Hourly	1
	Morning, evening, and mean	1
	Not provided	15

Figure 2.3 VAS characteristics (Laslett et al. 2004). Reproduced, with permission, from Litcher-Kelly, L., Martion, S.A., Broderick, J.E., & Stone, A.A. (2007). A systematic review of measures used to assess chronic musculoskeletal pain in clinical and randomized controlled clinical trials. The journal of pain, 8 (12), 906–913, © 2007 Elsevier.

therefore the length of time away from competition (Westman et al. 2008).

The scale by which athletes are often asked to report pain can be varied but the most reliable and most common is the Visual Analog Scale (VAS). The characteristics of the VAS can be seen in Figure 2.3. The scale can have a variety of anchor points and reporting periods. The way in which it categorises pain can also be varied with global, intense and general terms, all being used to interact with the patient to help them identify their pain. There is a need for a 'golden measure' of assessment of patient pain, and the VAS seems to be the most commonly used method at present (Litcher-Kelly et al. 2007).

Functional assessments can follow a screening questionnaire or be the injury prevention process in its entirety. The test used to functionally assess an athlete can be varied and differ depending upon the area of the body and the activity the athlete competes in. The back and knee are two of the major areas that the screening process needs to assess due to their importance to locomotion and therefore the resultant sporting performance. The flexibility and posture can be governed extensively by the back and in particular the lower back strength of the athlete. If the athletes suffer from lower back pain (LBP) then they can often miss training sessions due to the lack of mobility or work at a reduced level. The importance to screen this area of the body properly and plan for any potential problems is essential in an athlete's sporting life.

Revels model can be used as screening test for LBP, with no specific conditions. Athletes with specific LBP would need to be further screened to assess the full extent of their problems. These findings were shown in research to be applicable for 11% of the population, as the testing procedure lacked specificity (Laslett et al. 2004). The Revels testing

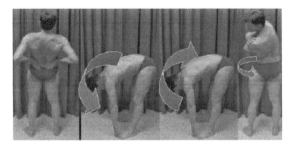

Figure 2.4 The Revels model physical examination (Laslett et al. 2004). Reproduced from Laslett, M., Oberg, B., Aprill, C.N., & McDonald, B. (2004). Zygapophysial joint blocks in chronice low back pain: A test of revel's model as a screening test. BMC Musculoskeletal Disorders, 5 (43), 1–6, public domain information courtesy of BioMed Central Ltd.

procedure involves the patient being asked to assess the back pain currently experienced using a VAS, then the highest amount of pain and then the lowest amount of pain experienced. Subjects were then asked to complete the exercises shown in Figure 2.4. On completion of the exercises the subjects were then asked to re-evaluate their pain scores on the VAS. Using the Revels model the subjects were categorised into groups that had met the criteria and those that had not. The groups that had met the criteria had a further assessment on the lower back joints. As previously stated this only produced a positive result in those subjects with very general problems, but was significantly linked ($p = 0.04$) to the patients amount of 'sick days', i.e. those with a positive Revels test had a greater number of 'sick days' (Laslett et al. 2004).

The safety of tests like the Revels model has been questioned due to the danger imposed by possible further damage to existing injuries. The condition of most concern is vertebral artery dissection (VAD). When assessing the back, practitioners must ensure that the patient's history is fully recorded before examination, ruling out any symptoms for VAD. The patient should be assessed initially by means of non-provocative manipulation. If the area of the cervical spine needs to be manipulated then the practitioner should ensure that all symptoms for VAD have been eliminated and decide whether the provocative manipulation results are necessary or conclusions can be drawn through other means (Thiel and Rix 2005).

Another area on the body that is of paramount importance to the practitioner is the knee. An injury to the anterior cruciate (ACL) is one of the most serious problems that can happen to an athlete and have a serious effect on their career (Bonci, 1999). The prevention of injury to the ACL and the knee is very important and is dependent on identifying a series of risk factors. The major risk factors associated with ACL injuries include lower extremity malalignments, ligamentous laxity, lower extremity muscular strength considerations, neuromuscular control, hormonal influences, intercondylar notch width and the biomechanics of the athletes' sporting techniques (Bonci, 1999). The rehabilitator can only improve and modify the strength and neuromuscular control risk factors, which is why these are key areas of a screening process (Bonci, 1999). The misalignment of the body can increase risk of injury, particularly in the back and knee. and this is very important in the lower limbs in athletes. The locomotion part of any sport technique means that any problem with the lower limb will be subjected to extreme forces. The knee in particular can only absorb a certain amount of these forces and through the nature of sport the athlete will at some point suffer injury as a result. The degrees of misalignment that will cause sufficient risk to cause potential knee problems is varied, depending on several assessments (Bonci 1999). (For greater detail regarding ACL injuries see Chapter 21, The Knee.) In order to gain a full assessment of these risks, the athlete's foot pronation, knee recurvatum, tibial torsion and posture need to be measured. Neuromuscular control tests for the lower limb often involve a variety of movements but with common goals: to obtain an objective assessment of function and to challenge dynamic knee stability during landing and deceleration (Bonci 1999). A range of tests can be used to achieve these goals, with the single-leg hop, one-leg vertical jump, timed hop, figure of eight running, side stepping and stair running being a few of the commonly used ones (Lephart et al. 1991). These tests can also be combined with isokinetic testing to give a rounded picture of the knee (Barber et al. 1990; Lephart et al. 1991). The way the athlete's foot lands during their gait has a large impact on the distribution of the force created. If the level of pronation is too high then one of the body's natural shock absorbers, the heel, will be unable to prevent force from travelling up to the knee. The contact phase is where the level of pronation in the foot is critical. If the correct amount is present the movement will be inefficient and shock absorption will not occur adequately in the heel.

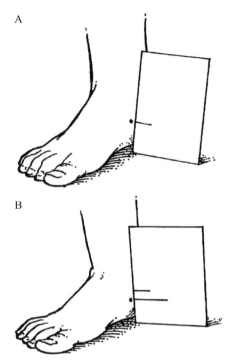

Figure 2.5 Navicular drop test (Bonci, 1999). Reproduced, with permission, from Bonci, C.M. (1999). Assessment and evaluation of presisposing fators to anterior cruciate ligament injury. Journal of athletic training, 34 (2), 155–164. © 1999 National Athletic Trainers' Association, Inc., www.nata.org/jat.

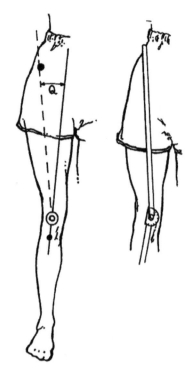

Figure 2.6 The Q angle (Bonci, 1999). Reproduced, with permission, from Bonci, C.M. (1999). Assessment and evaluation of presisposing fators to anterior cruciate ligament injury. Journal of athletic training, 34 (2), 155–164. © 1999 National Athletic Trainers' Association, Inc., www.nata.org/jat.

Pronation is important to the support phase, but not as stated in the contact to the shock absorption problems (Bonci 1999). The navicular drop test is commonly used to assess the pronation of the foot, and identifies the difference between the subtalar joint (STJ) in a seated neutral position (position A in Figure 2.5) and a weight bearing position (position B in Figure 2.5). Athletes who have had ACL problems generally have a difference of 13mm when compared to healthy individuals who will have a difference of just 8mm (Bonci 1999). Genu recurvatum is also known as knee hyperextension and is normally due to problems in the athlete's skeleton and/or movement pattern abnormalities and proprioception. The problem is often increased due to soft tissue laxities, which cause occasional hyperextension of the knee. Over the course of the athlete's career the extent and occurrences of the hyperextensions will begin to place extra strain on the ACL. This measurement is normally assessed through the extent of laxity of the thumb, fingers, elbows and hyperextension of the knee (Bonci 1999). An athlete with two or more postural problems that can be interlinked can have an increased risk of knee injuries. The increase of the Q angle can be an indicator of a problem occurring in the knee and of increased strain being placed on the ACL. The Q angle is the angle between the force line of the quadriceps and the line of pull of the patella tendon (see Figure 2.6) (Merchant, et al. 2008). The Q angle in a normal athlete is between 10 and 15 degree (Bonci, 1999). The Q angle is a useful measure of potential knee problems but the validity of this is under discussion (Merchant, et al. 2008). The varity of methods used to assess the Q angle has caused problems with the standardisation of the results found. Therefore, the use of an athlete's Q angle must be used in conjunction with other tests to complete the screening picture (Smith et al. 2008).

The strength of the muscles around the knee has an important role to play in stabilising the joint and therefore reducing the chances of injury. If, for example, the hamstring is more than 15% weaker than the other lower limb muscles than the athlete is 2.6 times more likely to suffer lower limb injury (Knapik et al. 1991; Gabbe et al. 2009). Muscle imbalances between limbs as well as within are important in assessing the potential for injury in the knee, particularly in females where such imbalances are found in 20–30% of athletes (Bonci 1999; Gabbe et al. 2004). The ratio of flexion and extension of the joint is also a measure of potential injury, where an athlete with a ratio (eccentric to concentric) of less than 0.75 are 1.6 times more likely to suffer knee injuries (Moul 1998). Isokinetic testing at 60, 180 and 300 degrees can help accurately establish these ratios and therefore identify potential problems, but this is very sensitive to velocity changes in the testing protocol (Moore and Wade 1989). Traditional functional assessments can also provide this information, albeit, at a lower degree of accuracy, but still establish the athletes' readiness for competition. The neuromuscular control of the athlete can be measured through a series of jumps and hops to assess the stability and awareness the athlete has of their knee movement. This is normally assessed on recovery from an injury and the most reliable test results are found using two or more jump types (Bonci 1999).

Isokinetic testing can be used in a variety of ways to profile an athlete to identify potential areas of injury concern. The simplest method is to evaluate the power output of the movements associated with the sport played and compare these against the competitive requirements. If there is a deficit then the athlete is more likely to be at risk to injury when performing that movement and can be discouraged from doing it until their power output is improved (Rosenblum and Shankar 2006). When this is complete it has been found to offer about a 10% reduction in injuries (Rosenblum and Shankar 2006) (see Chapter 3, assessing performance for greater detail regarding isokinetic assessment).

The flexibility of the muscles in the lower limbs plays an important role in the prevention of injuries to the area and should go hand in hand with a muscle development programme. The role of equipment used in the athletes' chosen sport can also have a major effect on the potential injuries that they could suffer. For example, the height of the saddle in cycling can have implication for a range of conditions in the knee of cyclists. If the saddle is too high the

Figure 2.7 Metatarsal head position (Callaghan 2005). Reproduced, with permission, from Callaghan, M.J. (2005). Lower body problems and injury in cycling. Journal of bodywork and movement therapies, 9, 226–236, © 2005 Elsevier.

athlete could suffer from illotibial band (ITB) pain and potentially suffer from ACL strains (Callaghan 2005). If conversely the saddle is too low then the athlete could suffer from patellofemoral pain, LBP and anterior knee pain (Callaghan 2005).The technique of the athlete needs to be addressed to ensure that there are limited biomechanical faults, which could also lead to injury. The way in which cyclists position their feet on the pedal (see Figure 2.7) is important for two main reasons. If the position is incorrect then the cyclist's cadence and resultant power will be affected and performance will be hindered. Secondly if the foot is incorrectly positioned it can cause knee and back problems depending on whether it is either too far forward or too far back. The ability to rotate the heel using the toe as a fixed pivot is also important to reduce stress on the knee and also increase efficiency of movement (Callaghan and Jarvis 1996). Through kinematic analysis using pressure plates in the clips the optimum reduction of pain in the knee whilst minimising the impact on the power phase of cycle can be found. The clip system now recommended to cyclists bearing in mind these two factors is the 'floating clip' system, due to the amount of rotation given (Callaghan 2005). The role of the practitioner in assessing the athlete for potential injury should not just involve functional anatomical test but should look at the sport as a whole (Callaghan 2005; Callaghan and Jarvis 1996).

The problems that an athlete could potentially have are not just confined to the major joint of the back and the knee, the musculoskeletal system also needs to be assessed for any potential abnormalities. A simple Gait Arms Legs Spine (GALS) test can be used to help identify abnormalities (Beattie et al. 2008). The testing procedure has been found to be 95% accurate at helping to identify musculoskeletal abnormalities. Figure 2.8 shows the characteristics that the athlete will be assessed with, with an overall score worked out at the end. The Gait is simply classified as abnormal or normal, and then the arms, legs and spine are identified by appearance and movement. The appearance of the limb will first be assessed for abnormality and then the movement of the limb will follow using the same process. The subjects are also asked general questions about any pain or stiffness in their muscles or joints (Beattie et al. 2008).

The GALS offers a useful screening of the muscles and with very little deviation between practitioners, up to 95% agreement in this research feature (Beattie et al. 2008). GALS does, however, offer different results depending on the condition, with it providing 53% of positive results with acute conditions versus 95% in chronic conditions. This highlights the potential problems with diagnosing and screening musculoskeletal problems. The length of time that athletes have had the issue will affect the way they answer the questions and how it will appear to the practitioner. The variability of the issues can mean that there is no 'golden test' for screening and a holistic approach to screening is the best approach to increase accuracy of the results and therefore increase injury prevention (Beattie et al. 2008).

The assessment of an athlete's gait can often lead to the identification of potential and current musculoskeletal problems. This can be completed by the practitioner 'eye balling' athletes as they walk or run in front of them or through the use of a biomechanics. The athletes gait can be assessed through the use of force plates and video analysis. The history of gait analysis can be traced as far back as Leonardo da Vinci (Paul 2005). The aims are relatively still the same although with many more varied applications and not merely in the interest of locomotion. An athlete's gait can be looked at extensively in terms of performance improvements, especially in elite sport where even the small percentage improvement of efficiency could lead to success. Gait can also be used in the prevention of injury through the identification of where the forces are travelling during locomotion and their possible impact (Paul 2005). The use of a force plate can greatly enhance the accuracy of the screening process by being able to easily identify imbalances between the left and right limbs. The exact forces can be measured and their direction so a picture of how much of an impact a gait abnormality will have on the athlete can be drawn up. Figure 2.9 indicates a normal foot trace and resultant hip forces generated by the athlete. The forces can be expressed in terms of percentage body mass or any unit of force the practitioner wishes to work in (Paul 2005). The biomechanical approach can be used in conjunction with functional assessments to add detail and identify the extent of an identified problem, forming a complete screen (Paul 2005).

The use high-speed video analysis can add further depth to a screening process and in particular the assessment of the knee for potential anterior cruciate knee ligament injuries. The kinematic model used for linking the moments of force calculated through video analyses with the appropriate screening problem is shown in Figure 2.10 (McClean, Walker, Ford, Myer, Hewett, & Van den Bogert, 2005). It is important when using video analysis to have an established model of reference to relate the analysis to in order to accurately identify the errors (McClean et al. 2005).

Figure 2.11 shows the correlation of results between 2D analysis and 3D analysis, when looking at knee valgus, during a side step motion. The correlations found where moderate to good, with the best correlation displayed when the valgus angles where greater. You could interpret this as a lack of sensitivity in the 2D analysis methodology, which is indicated in the results. The peak rotation force was later when using 2D analysis, as seen in Figure 2.11 (McClean et al. 2005). The lack of movement in the trace supports the lower sensitivity of the 2D analysis. However, the correlation does exist, so the use of field base 2D analysis is a valid tool in the screening process (McClean et al. 2005). 3D analysis is the 'gold standard' measurement for lower limb angles

GAIT
- Symmetry & smoothness of movement
- Stride length & mechanics
- Ability to turn normally & quickly

ARMS (Hands)
- Wrist/finger swelling/deformity
- Squeeze across 2nd to 5th metacarpals for tenderness (Indicates synovltls)
- Turn hands over, Inspect muscle wasting & forearm pronation/suplnation

ARMS (Grip Strength)
- Power grip (tight fist)
- Precision grip (oppose each finger to thumb)

ARMS (Elbows)
- Full extension

ARMS (Shoulders)
- Abduction & external rotation of shoulders

LEGS (Feet)
- Squeeze across metatarsals for tenderness (indicates synovitis)
- Calluses

LEGS (Knees)
- Knee swelling/deformity, effusion
- Quadriceps muscle bulk
- Crepitus during passive knee flexion

LEGS (Hips)
- Check internal rotation of hips

SPINE (Inspection from behind)
- Shoulders & iliac crest height symmetry
- Scoliosis
- Paraspinal, shoulder, buttocks, thighs & calves muscles normal
- Popliteal or hind foot swelling or deformity

SPINE (Inspection from front)
- Quadriceps normal in bulk & symmetry
- Swelling or at Varus or valgus deformity at knee
- Forefoot of midfoot deformity action normal
- Ear against shoulder on either side to check lateral cervical spine flexion
- Hands behind head with elbows back (check rotator cuff muscles, acromioclavicular, sternoclavicular & elbow joints)

SPINE (Inspection from side)
- Normal thoracic & lumbar lordosis
- Normal cervical kyphosis
- Normal flexion (lumbosacral rhythm from lumbar lordosis to kyphosis) while touching toes

SPINE (Trigger point tenderness)
- Supraspinatus muscle tenderness (exaggerated response)

Figure 2.8 GALS testing characteristics (Beattie et al. 2008). Reproduced, with permission, from McClean, S.G., Walker, K., Ford, K.R., Myer, G.D., Hewett, T.E., & Van den Bogert, A.J. (2005). Evaluation of a Two dimensional analysis method as a screening and evaluation tool for anterior ligament injury. British Journal of Sport Medicine, 39, 355–362 © 2005 BMJ Publishing Group Ltd.

SCREENING METHODS   25

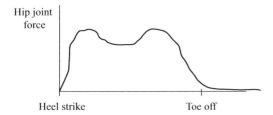

Figure 2.9 Typical foot contact and hip force trace. Reproduced, with permission, from McClean, S.G., Walker, K., Ford, K.R., Myer, G.D., Hewett, T.E., & Van den Bogert, A.J. (2005). Evaluation of a Two dimensional analysis method as a screening and evaluation tool for anterior ligament injury. British Journal of Sport Medicine, 39, 355–362 © 2005 BMJ Publishing Group Ltd.

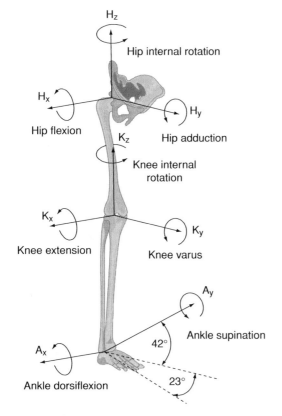

Figure 2.10 Kinematic model used in motion analysis (McClean et al. 2005). Reproduced from Beattie, K.A., Bobba, R., Bayoumi, I., Chan, D., Schabort, S., Boulos, P., et al. (2008). Validation of the GALS musculoskeletal screening exam for use in primary care: a pilot study. BMC Musculoskeletal Disorders, 9 (115), 1–8. public domain information courtesy of BioMed Central Ltd.

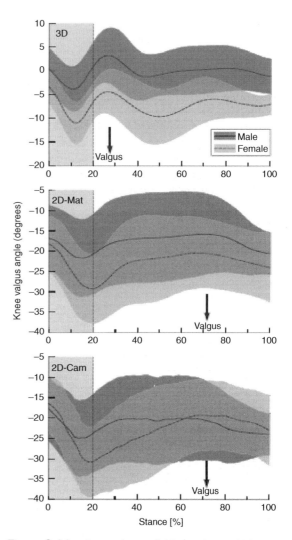

Figure 2.11 Comparison of kinematic analysis 3D versus 2D (McClean et al. 2005). Reproduced, with permission, from Paul, J.P. (2005). The history of Musculoskeletal modelling in human gait. Theoretical Issues in Ergonomics Science, 6 (3–4), 217–224 © 2005 Taylor & Francis Ltd.

and velocity variables. The use of frontal plane video camera can produce analysis that correlates well with the use of a 3D system when looking at a side step or side jump movement, which are also standard neuromuscular assessment tests (McClean et al. 2005).

The effect of injury on an athlete's gait can be extensive and still be apparent long after the injury appears to have been treated and the damage repaired. This can be due to the muscles becoming deactivated and then not becoming reactivated through the rehabilitation process. When an injury occurs the body adapts by ensuring extra force is exerted by the surrounding tissues, for example the ligaments and other supporting muscles. When the muscle has recovered from damage the extra force continues to be exerted by those tissues and therefore the muscle becomes deactivated and untrained (Komura and Nagano 2004). Deactivation can affect other muscles causing an overall reduction in force output. For example, if this occurs in the iliopsoas then there can be a resultant reduction of power in the soleus. This can have a large affect on the athlete's technique and therefore performance, so ideally you need the other tissues to compensate for the injured muscle without having any long-term effects on technique (Komura and Nagano 2004). The muscles which seem to have the greatest impact on the gait of an athlete are the gluteus medialis (GMED), gluteus minimus (GMIN), hamstrings (HAM), adductor longus, adductor magnus, tensor fasciae latae, gluteus maximus (GMAX), iliopsoas (ILIPSO), rectus femoris (RF), vastus (VAS), gastrocnemius (GAS), soleus (SOL) and the tibialis anterior (TIBANT) (Komura and Nagano 2004). If the RF is damaged then the knee movement is created by the VAS so there is not a loss of motion. When the GMAX is deactivated hip extension is caused by the GMED and the Ham. When looking at knee extension when the HAM is deactivated the VAS created the movement to allow locomotion to continue (Komura and Nagano 2004). See Figure 2.12 for further muscle adaptations. Knowledge of the movement and which muscle can perform the movement can help in correctly assessing injury or potential problems. Rather than assuming that there are no issues surrounding the GMAX because the athlete can successfully perform hip flexion, the practitioner needs to identify whether the movement is completed by the correct muscle. If this process is not completed then an injury could go undiagnosed during the screening process leading to problems in the surrounding areas and technique at a later date (Komura and Nagano 2004). The potential problems will not only be limited to the surrounding muscles and ligaments but also the bones. During locomotion bone on bone contact force is apparent, but limited when the correct muscles are activated. If those muscles become deactivated then bone on bone force increases and can result in injury.

The movement of force is also increased as the bone on bone forces have been shown to increase in other joint in the limb. If the HAM and ILIPSO are deactivated then not only are the bone to bone forces increased around the hip but also in the ankle due to the resultant muscles, the VAS, having to create the missing force. Deactivation of the GMAX increased the bone on bone force more at the knee rather than the hip as would be expected with the GMED and HAM creating the force instead of the GMAX, (Komura and Nagano 2004). Refer to Figure 2.13 for a more detailed analysis of the movement of bone to bone forces. The body's ability to maintain locomotion is undoubtedly a positive aspect but in a sporting context the hidden damage that could be caused is a problem. The need for a screening process to highlight fully and in depth which muscles are damaged and so could become deactivated as a result is required by the practitioner to ensure the long-term health of the athlete (Komura and Nagano 2004).

The need for sporting bodies and teams to adopt a proactive approach to injury prevention is very apparent as is the need to increase education of coaches and athletes so as to improve performance. An example where this is being completed is in cricket (Dennis et al. 2008). Functional tests will probably be the backbone of a sports team's screening process due to the ease of setup and accessibility to trained staff to administer the tests. The screening process usually consists of a battery of functional tests which the members of the squad will complete en masse. The knee extension test is usually used to assess the athlete's hamstring muscle length. The athlete will lie in a supine position with their hip flex on the tested knee at an angle of 90 degrees. The athlete will then be asked to flex the knee to its maximum position and the angle recorded. The final value will be 90 minus the recorded angle. If the athlete can extend fully, the hip will be moved to a 30 degree position and the recorded angle will be subtracted from 120 degrees (Dennis et al. 2008).

The Modified Thomas Test (MTT) can be used to assess hip abduction range and hip extension. The test is performed over the edge of a plinth with the athlete lying in a supine position (Figure 2.14). The abduction and hip flexion movements are then carried out with assistance from the practitioner and a goniometer is used to record the angle achieved (Dennis et al. 2008).

SCREENING METHODS 27

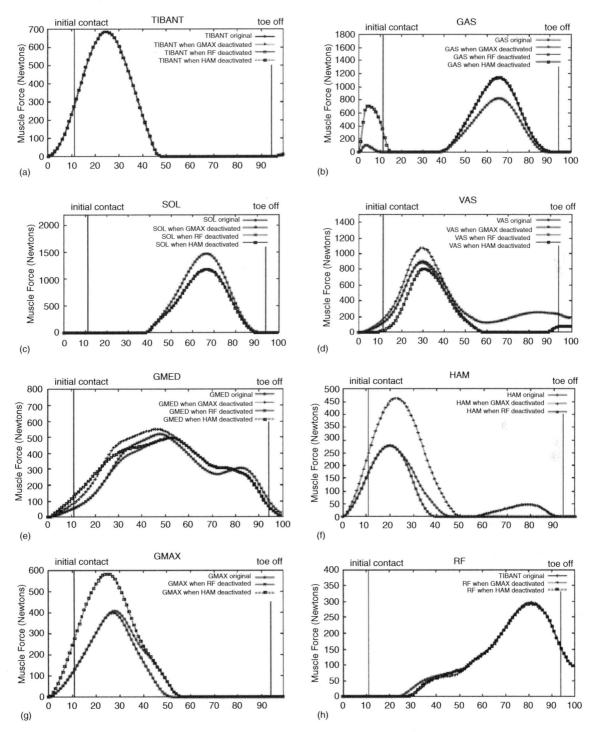

Figure 2.12 Muscle deactivation force production movement (Komura and Nagano 2004). Reproduced, with permission, from Komura, T., & Nagano, A. (2004). Evaluation of the infulence of muscle deactivation on other muscles and joints during gait motion. Journal of biomechanics, 37, 425–436 © 2004 Elsevier.

28 INJURY PREVENTION AND SCREENING

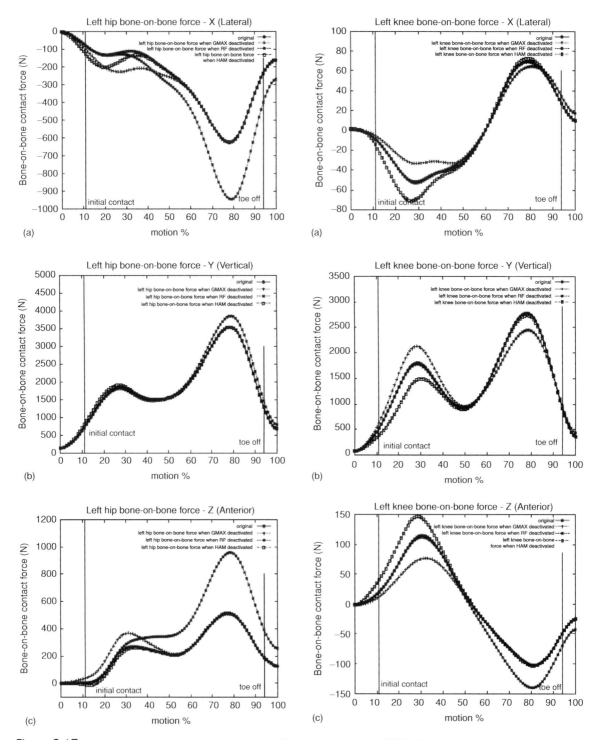

Figure 2.13 Bone to bone contact force movements (Komura and Nagano 2004). Reproduced, with permission, from Komura, T., & Nagano, A. (2004). Evaluation of the infulence of muscle deactivation on other muscles and joints during gait motion. Journal of biomechanics, 37, 425–436 © 2004 Elsevier.

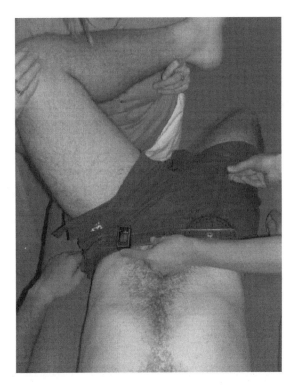

Figure 2.14 The Modified Thomas Test (MTT) (Dennis et al. 2008). Reproduced, with permission, from Dennis, R.J., Finch, C.F., Elliott, B.C., & Farhart, P.J. (2008). The reliability of musculoskeletal screening tests used in cricket. Physical Therapy in sport, 9, 25–33 © 2008 Elsevier.

The athlete's hip internal and external rotation can also be measured in the prone position and then with the testing leg fully straightened the practitioner can move the limb to as far as possible and measure the angle from the tibia line to the relative vertical line (Dennis et al. 2008).

The combined elevation test measures the distance from the base of the thumb to the floor once the arms are fully elevated and the subject is in a prone position with each arm fully extended. This is used to measure thoracic extension strength and range of movement. The rest of the body remains in contact with the ground. Whilst holding their breath, athletes need to hold this position for the duration of the test (Dennis et al. 2008).

The prone four-point hold test is used to measure the strength of the core muscles by assessing the length of time they can hold a neutral lumbopelvic position (see Figure 2.15). The test is stopped if the

Figure 2.15 The prone four-point hold test (Dennis et al. 2008). Reproduced, with permission, from Dennis, R.J., Finch, C.F., Elliott, B.C., & Farhart, P.J. (2008). The reliability of musculoskeletal screening tests used in cricket. Physical Therapy in sport, 9, 25–33 © 2008 Elsevier.

athlete starts to experience any back pain to reduce the chances of injury (Dennis et al. 2008). The athlete's time is recorded as a measure for this test with normative non-back pain sufferers aiming for 72.5 +/−32.6 s (Schellenberg et al. 2007).

The athletes' calf heel raises and ankle dorsiflexion ability are also important in cricket especially in bowlers. The athlete is asked to lung towards a wall and the maximum distance they could keep their feet planted on the ground whilst still touching the wall with their knee is recorded for their dorsiflexion ability. The athlete needs to repeatedly perform calf raises to full height until failure, with the amount and duration they complete used to calculate their cycle per second figure (Dennis et al. 2008).

The bridging hold (see Figure 2.16) assesses the gluteal musculature endurance. The athlete is asked to support the lower back to prevent arching whilst their leg position is being confirmed and the length of time they can hold this position is recorded. A healthy athlete with no back pain should aim for a time of 170.4 $\pm$ 42.5s (Schellenberg et al. 2007). If the athlete experiences back or hamstring pain then the test is aborted to reduce the chances of injury. The leg must remain fully elevated and fully extended throughout the test, and then repeated on the other side using the same process (Dennis et al. 2008). This battery of tests gives the practitioner an extensive amount of data on the functional elements associated, in this case, to cricket but they are relevant to most sport movements. The muscular isometric ability of certain parts of the body is also measured to give an insight into not only the

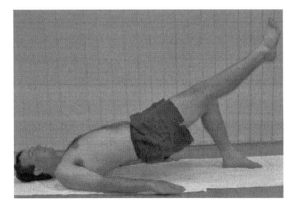

Figure 2.16 Bridging hold position (Dennis et al. 2008). Reproduced, with permission, from Dennis, R.J., Finch, C.F., Elliott, B.C., & Farhart, P.J. (2008). The reliability of musculoskeletal screening tests used in cricket. Physical Therapy in sport, 9, 25–33 © 2008 Elsevier.

power but the endurance of certain muscle groups on the athlete. From this data the practitioner can assess areas of weakness and potentially plan a preventative injury programme to ultimately improve performance levels through a reduction in injuries (Dennis et al. 2008).

The implementation of injury prevention within sport is varied, dependent on the sport and possibly linked to the financial ability of the team or individual to employ the correct practitioners to implement such support. In the Australian football league, studies have shown that coaches have a limited education in preparation of the squad practices when it comes to injury prevention. The general trend in football is for coaches to focus on maintaining core temperature and to work on flexibility. Although these elements are important to injury prevention in a contact sport, procedures should be implemented in training sessions to reduce lower limb injuries (LLI). The prevention of LLIs can be completed through the use of lateral movement practices and foot work drills. The coaches were also lacking in knowledge about how to treat old injuries and how to avoid re-injury. The importance of a cool down is also an area that needs improvement with the traditional approach of concentrating on the warm up before practice being the dominant injury prevention measure and nothing planned post practice. The coaches agree that this lack of knowledge needs to be addressed through coach education to improve performance (Twomey et al. 2009).

Validity and reliability of screening methods

When screening athletes for injury prevention it is important for the practitioner to remain objective towards the client and ensure that the recorded results are valid and therefore reliable. If the measurements are being administered by a team of practitioners then the inter-practitioner reliability needs to high. Good inter-practitioner reliability is achieved by each test being performed using the same protocol and by having the failure points standardised. Figure 2.17 indicates the reliability of a series of well-used musculoskeletal screening tests and the difference between two practitioner's results. The practitioners were asked to perform the series of tests shown in Figure 2.17, which were repeated using the same subjects (Gabbe et al. 2004).

In the majority of tests only the MTT of the quads indicated a significant difference between the practitioners, but this difference was within the error range for this test as it was within 95% of the standard error measurement (SEM). The other tests, the sit and reach, the lumber extension, the slump test, the active hip internal rotation range of movement (IR ROM) test, and the active knee extension (AKE) test, indicated excellent inter practitioner reliability (Gabbe et al. 2004). The sets of tests assessed are well used in the field of musculoskeletal screening so the reliability of the testing protocols needs to be good, as does the education of the practitioners in the use of the tests, as indicated in this research (Gabbe et al. 2004).

The ability of the practitioner to produce the same measurement between clients and on the same client over a series of visits is paramount to the test. The practitioner has to provide reliable measures to ensure that progress can be monitored and the problems are correctly diagnosed to increase injury prevention. The second part of the research indicates that again, overall, the application of the tests is excellent with only two retests being significantly different from the first test. These were the sit and reach test and the passive straight leg raise (PSLR) test; both were within the SEM for these tests, indicting the practitioners' techniques at completing the protocols were

Test	Rater A. mean (SD)	Rater B. mean (SD)	p Value (t test)	ICC (2.1)	95% CI ICC	SEM	95% CI SEM
Sit and Reach (cm)	3.0 (8.5)	3.7 (8.6)	0.22	0.97	0.91–0.99	2	3
Lumbar extension (cm)	9.7(4.5)	9.4(5.3)	0.47	0.95	0.85–0.98	1	2
Slump (°)	19.8 (11.5)	22.2(10.3)	0.05	0.92	0.77–0.97	3	6
Active hip IR[a] (°)	27.1(6.2)	26.0(7.3)	0.10	0.94	0.82–0.98	2	3
Active hip ER[b] (°)	22.2(5.6)	21.1 (5.6)	0.15	0.88	0.67–0.96	2	4
PSLR[c](°)	70.2(14.2)	68.7 (14.6)	0.32	0.93	0.80–0.97	4	7
AKE[d] (°)	29.8 (14.7)	30.5 (14.5)	0.65	0.93	0.80–0.98	4	8
MTT[e] (hip flexor) (°)	1.5(8.8)	1.9(9.7)	0.67	0.92	0.79–0.97	3	5
MTT (quadriceps) (°)	68.9(8.1)	65.7 (10.8)	0.005	0.90	0.72–0.96	3	5

[a]IR, internal rotation.
[b]ER, external rotation.
[c]PSLR, Passive Straight Leg Raise.
[d]AKE, Active Knee Extension.
[e]MTT, Modified Thomas Test.

Figure 2.17 Inter-practitioner reliability scores (Gabbe et al. 2004). Reproduced, with permission, from Gabbe, B., Bennell, K., Wajswelner, H., & Finch, C.F. (2004). Reliability of common lower extremity musculoskeletal screening tests. Physcial Therapy in Sport, 5, 90–97 © 2004 Elsevier.

excellent. The problems arising from the two protocols occur due to the use of very similar movements to complete the test, and, although these are very small errors, this is something that needs to be monitored. The ability to maintain objectiveness is important and, as the results show in Figure 2.18, this is the case when using these screening tests (Gabbe et al. 2004).

The use of video analysis can enhance the reliability of the test results and can be used in the field as well as in a clinic or laboratory situation. Many of the tests that have been discussed are functional assessments but practitioners can also perform a series of field-based movements that will enable strength screening to be completed. The field tests can be recorded and software used post completion to complete the measurements and increase the accuracy of the test. Figure 2.19 outlines a series of field tests that give an indication of muscular strength and any possible imbalances that might exist within the athlete. The results indicate excellent SEM values for the entire test batch displayed, demonstrating the validity of the protocols (Miller and Callister 2009).

The video analysis of the same test can then be used to gain the functional assessments required for screening post collection. This is shown in Figure 2.18 which displays the reliability of the test again showing excellent SEM figures. The higher values indicated in the video analysis of these movements are indicative of stability in the joints (Miller and Callister 2009). The use of this type of test battery could be implemented on large squads where time is limited to complete the screening session. The use of video analysis and the simplistic nature of the tests do not decrease accuracy, as the values in Figures 2.18 and 2.19 indicate, but also help to increase objectivity (Miller and Callister 2009). The use of video analysis would also allow the practitioner to use more dynamic tests in the screening process to ensure that the athletes place similar forces on the body that they would experience in their sport. The static measurements are important but they need to be supported with dynamic movements to ensure that the screening process is sports specific (McClean et al. 2005).

The need to have objective measures and not let personal views interfere with the screening process is important and vital to the success of the injury prevention plans and the reputation of the practitioner. The need to ensure that practitioners have common core objectives and common protocols is also apparent in the increasing multi-cultural and cosmopolitan nature of today's sport. When athletes move around the globe they need to be sure that

Test	Rater	Session 1 mean (SD)	Session 2 mean (SD)	p Value (t test)	ICC (3.1)	95% CI ICC	SEM	95% CI SEM
Sit and Reach (cm)	A	3.0(8.5)	2.7(8.7)	0.31	0.99	0.98–1.00	1	2
	B	3.7(8.6)	2.6(8.7)	0.04	0.98	0.94–0.99	1	2
Lumbar extension (cm)	A	9.7(4.5)	10.1 (4.3)	0.22	0.86	0.89–0.99	1	2
	B	9.4(5.3)	9.6(5.7)	0.75	0.89	0.79–0.96	2	4
Slump(°)	A	19.8 (11.5)	21.1 (11.7)	0.19	0.95	0.85–0.98	3	5
	B	22.3(10.3)	24.1 (12.0)	0.31	0.80	0.51–0.93	5	9
Active hip IR[a] (°)	A	27.1 (6.2)	27.9 (6.7)	0.35	0.83	0.57–0.94	3	5
	B	26.0(7.3)	26.4(7.7)	0.62	0.92	0.78–0.97	2	4
Active hip ER[b] (°)	A	22.2(5.6)	22.8 (5.0)	0.37	0.90	0.73–0.96	2	3
	B	21.1(5.6)	20.7 (6.3)	0.69	0.83	0.57–0.94	2	3
PSLR[c] (°)	A	70.2(142)	65.7 (12.8)	0.01	0.91	0.75–0.97	4	8
	B	68.70	68.07	0.69	0.91	0.74–0.97	4	8
AKE[d] (°)	A	29.8(14.7)	30.2 (14.2)	0.76	0.96	0.88–0.96	3	6
	B	30.5 (14.5)	31.1 (13.2)	0.66	0.94	0.82–0.98	3	7
MTT[e] (iliopsoas) (°)	A	1.5(8.8)	−0.4 (8.6)	0.36	0.63	0.20–0.86	5	10
	B	1.9(9.7)	3.0 (9.6)	0.54	0.75	0.41–0.95	5	9
MTT (quadriceps) (°)	A	68.9(8.1)	69.4 (10.9)	0.80	0.69	0.30–0.88	5	10
	B	65.7(8.3)	66.4 (10.8)	0.74	0.69	0.29–0.88	5	10

[a]IR, internal rotation.
[b]ER, external rotation.
[c]PSLR, Passive Straight Leg Raise.
[d]AKE, Active Knee Extension.
[e]MTT, Modified Thomas Test.

Figure 2.18 Musculoskeletal test-retest reliability figures (Gabbe et al. 2004). Reproduced, with permission, from Gabbe, B., Bennell, K., Wajswelner, H., & Finch, C.F. (2004). Reliability of common lower extremity musculoskeletal screening tests. Physcial Therapy in Sport, 5, 90–97 © 2004 Elsevier.

the scores they receive in one country and testing facility will be conducted through the same methods as in another country. These objectives can only be achieved through maintaining high training standards and ensuring continued professional training is compulsory amongst practitioners (Coady et al. 2003).

Risk assessment in injury prevention

Injury prevention and musculoskeletal screening is a form of assessment of risk on an athlete. The process can be as detailed as the practitioner desires but the loss of accuracy will suffer with a lower detailed screening process. Rather than viewing this process as a collection of singular tests, it should be viewed as building a picture of the athlete's functional capabilities, a performance matrix. Once this matrix is established the 'weak links' can be worked on to improve the overall matrix and therefore the resultant performances of the athlete. The sites used in the proposed matrix are upper neck (UP), lower neck (LN), upper back (UB), shoulder blade (SB), shoulder joint (SJ), low back/pelvis (LB/P), hip (H) and lower leg (LG). These have a series of tests to score which have an upper and lower threshold and cover all the directions indicated in Figure 2.20. The results are plotted in a 3D space which enables the athlete's weak links to be identified easily, with the ideal result being a complete block on the high threshold side of the cube (Mottram and Comerford 2008).

A series of tests are completed by the athlete that will cover a series of directions and sites. The tests can be easily scored and then classified by threshold limit and then marked on the matrix. The weak links that need to be identified are usually areas of instability in the tested movements which occur when the athlete is not fatigued. Figure 2.21

	Mean				Change in mean (%)	95% CI	Typical error as a CV(%)	95% CI	ICC(r)	95% CI	Minimum-raw-change required(°)
	Trial 1	Trial 2	Trial 3								
Step up ANQ[c]	162.0	162.8	161.5	Trial 2–1 3–2	0.43 −0.79	−0.82 to 1.70 −2.11 to 0.55	3.10 3.31	2.57–3.90 2.75–4.16	0.751 0.759	0.594–0.853 0.606–0.858	6.31
Thigh to horizontal[c]	69.4	69.7	69.1	Trial 2–1 3–2	−0.07 −0.79	−2.02 to 1.93 −2.68 to 1.14	4.82 4.80	3.98–6.11 3.98–6.05	0.721 0.767	0.545–0.836 0.617–0.863	4.24
Single-leg vertical jump											
ANQ[c]	167.2	171.5	171.2	Trial 2–1 3–2	2.48 −0.09	0.78–4.21 −1.92 to 1.78	4.16 4.61	3.45–5.24 3.82–5.81	0.684 0.608	0.497–0.811 0.392–0.761	8.76
Thigh to horizontal	73.7	76.1	75.9	Trial 2–1 3–2	3.15 −0.14	1.21–5.13 −2.04 to 1.80	4.74 4.78	3.93–5.97 3.96–6.02	0.681 0.679	0.492–0.809 0.489–0.807	4.40
Single-leg drop vert jump											
ANQ[c]	172.1	174.7	175.4	Trial 2–1 3–2	1.62 0.41	−0.01 to 3.29 −1.30 to 2.16	4.04 4.29	3.35–5.08 3.56–5.40	0.642 0.436	0.438–0.784 0.173–0.641	8.75
Thigh to horizontal[c]	77.2	78.9	78.8	Trial 2–1 3–2	2.32 −0.16	0.58–4.09 −2.15 to 1.87	4.26 5.01	3.54–5.37 4.16–6.32	0.717 0.482	0.543–0.831 0.230–0.674	4.15
Side spring											
ANQ[c]	158.7	159.2	158.0	Trial 2–1 3–2	0.16 −0.74	−1.63 to 1.99 −2.32 to 0.87	4.51 3.99	3.74–5.68 3.31–5.03	0.705 0.799	0.526–0.824 0.666–0.883	9.01
Thigh to horizontal[c]	92.8	94.3	93.2	Trial 2–1 3–2	1.65 −1.17	0.10–3.23 −2.63 to 0.32	3.82 3.71	3.17–4.81 3.08–4.67	0.718 0.742	0.546–0.833 0.580–0.848	4.46

Figure 2.19 Field-based test video analysis reliability (Miller and Callister 2009). Reproduced, with permission, from Miller, A., & Callister, R. (2009). Reliable lower limb musculoskeletal profiling using easily operated portable equipment. Physical therapy in sport, 10, 30–37 © 2009 Elsevier.

34 INJURY PREVENTION AND SCREENING

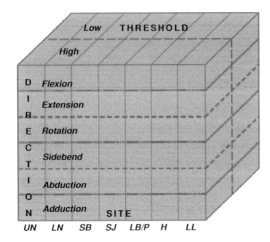

Figure 2.20 The performance matrix (Mottram and Comerford 2008). Reproduced, with permission, from Miller, A., & Callister, R. (2009). Reliable lower limb musculoskeletal profiling using easily operated portable equipment. Physical therapy in sport, 10, 30–37 © 2009 Elsevier.

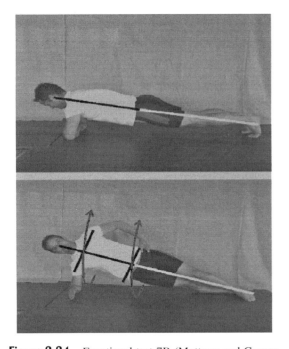

Figure 2.21 Functional test 7B (Mottram and Comerford 2008). Reproduced, with permission, from Mottram, S., & Comerford, M. (2008). A new perspective on risk assessment. Physical therapy in sport, 9, 40–51 © 2008 Elsevier.

7B	Elbows push up + twist to side support		Results	
'Weak link'			Fail[×]	
Load	Site	Direction	L	R
High	Neck	Rotation	☐	☐
	Shoulder blade (WB)	Hitch	☐	☐
		Drop	☐	☐
		Winging	☐	☐
	Shoulder joint (WB)	Forward glide	☐	☐
		Rotation (medial)	☐	☐
	Low back	Extension	×	×
		Rotation	☐	☐
		Sidebend	☐	☐
	Hip (WB)	Flexion	☐	☐
		Adduction	☐	☐

Figure 2.22 Functional test 7B mark sheet (Mottram and Comerford 2008). Reproduced, with permission, from Mottram, S., & Comerford, M. (2008). A new perspective on risk assessment. Physical therapy in sport, 9, 40–51 © 2008 Elsevier.

indicates a familiar starting position for a functional test but with an additional twist to add further analysis. The movement is then scored in detail using the scheme indicated in Figures 2.22 and 2.23. This one movement effects scoring for five sites and seven directions, giving an indication of the ease at which a complex picture can be built up of the athlete (Mottram and Comerford 2008).

The athlete's weak links can be improved through targeted training which is possible as the site and direction of movement is clear. The use of core stability training is often used to help improve the weak links. Where motor control needs to be improved low threshold training is found to rectify both global and local problems effectively. Once the motor control abnormalities have been rectified at a lower threshold level they can then be built on at a higher level to ensure the weak link is improved. Training an athlete to ensure that they are prepared for injury needs a detailed profile of the athlete's strength and weaknesses (Mottram and Comerford 2008). If the weak links are not isolated correctly then they could be improved on with stronger areas thus not actually improving the area as both the strong and weak are training at the same rate. The use of the type of testing schedule is not in the individual tests but with the links that can be drawn between them. The use of trunk bridging has been shown to indicate lower back problems (Schellenberg et al. 2007), which although may not produce a definite indicator

Test 7B	Elbows push up + twist to side support
Start position	○ Lie face down propped on elbows with hands pointing to opposite elbow ○ Knees and feet together ○ Shoulders midway between hitched and dropped ○ Taking weight through the arms, lift hips and knees of floor pushing off the toes ○ Make a straight line with legs and trunk and head
Test movement	● Keeping the pelvis neutral in a straight line with the legs and trunk, shift the upper body weight onto one elbow ● As the weight shifts, turn the whole body 900 from the shoulder so that the whole body is side on with the pelvis and knees unsupported and in a straight line with the legs and trunk ● The forearm and feet are the only contact points ● The weight bearing upper arm should be vertical

Performance Matrix analysis					Weak link	
	L	R	Load	Site	Direction
Can you prevent the back from side bending as the turn is initiated?	Yes ☐ No ☐ →	Yes ☐ No ☐ →	High	Low back (lumbo-pelvic)	Sidebend
Can you prevent the pelvis from leading the twist? (keep the back and pelvis turning together)	Yes ☐ No ☐ →	Yes ☐ No ☐ →	High	Low back (lumbo-pelvic)	Rotation
Can you prevent the back from arching?	Yes ☐ No ☐ →	Yes ☐ No ☐ →	High	Low back (fumbo-pelvic)	Extension
Can you prevent the pelvis and bottom hip from dropping towards the floor in the side position?	Yes ☐ No ☐ →	Yes ☐ No ☐ →	High	Hip (bottom leg)	Adduction
Can you prevent the hips from flexing? (keep the legs and trunk in a straight line)	Yes ☐ No ☐ →	Yes ☐ No ☐ →	High	Hip	Flexion
Can you prevent the weight-bearing (WB) shoulder blade winging?	Yes ☐ No ☐ →	Yes ☐ No ☐ →	High	Shoulder blade (WB) (scapula)	Winging
Can you prevent the weight-bearing (WB) shoulder blade hitching?	Yes ☐ No ☐ →	Yes ☐ No ☐ →	High	Shoulder blade (WB) (scapula)	Hitch (elevation)
Can you prevent the weight-bearing (WB) shoulder blade dropping?	Yes ☐ No ☐ →	Yes ☐ No ☐ →	High	Shoulder blade (WB) (scapula)	Drop (downward rotation/depression)
Can you prevent forward protrusion of the head of the weight-bearing (WB) shoulder joint?	Yes ☐ No ☐ →	Yes ☐ No ☐ →	High	Shoulder Joint (gleno-humeral)	Forward glide
Can you prevent the weight-bearing (WB) forearm from turning towards the feet (medial rotation) as the body twists?	Yes ☐ No ☐ →	Yes ☐ No ☐ →	Low	Shoulder joint (WB) (gleno-humeral)	Rotation (medial)
Can you prevent the head from turning or tilting?	Yes ☐ No ☐ →	Yes ☐ No ☐ →	High	Neck	Rotation

Figure 2.23 Mark scheme for functional test 7B (Mottram and Comerford 2008). Reproduced, with permission, from Mottram, S., & Comerford, M. (2008). A new perspective on risk assessment. Physical therapy in sport, 9, 40–51 © 2008 Elsevier.

of problems particularly in an athletic population, could help identify a weakness which the body will try to accommodate by over-compensating in another area to absorb the forces (see Figures 2.10 and 2.11).

With the correct identification of weak links, injury prevention programmes can become more specialised and therefore meet the needs of an individual athlete to eradicate potential problems occurring due to the body overcompensating due to a weakness (Mottram and Comerford 2008). If the athlete is not prescribed appropriate conditioning to reduce the risk then problems will still occur. The conditioning should be functional and appropriate to the athlete's sport. The conditioning must match the sporting movements, associated muscle actions, directional changes, velocities and loads experienced during the sport. If these conditions are adhered to then the injury risk will be reduced (Mottram and Comerford 2008).

Conclusions

Injury prevention and musculoskeletal screening are important aspects of modern sport. They need to be approached in the most complete and holistic manner in order to correctly identify the apparent weakness in the athlete and implement changes to correct them in order to fulfil the goal of improving athletic performance. The process needs to be detailed and not just include the traditional functional assessments. The need for pre-screening through the use of a questionnaire like the NIMQ-E (Dawson et al. 2009) will often eliminate athletes or areas from testing so that valuable time is not taken up with unnecessary tests. The athlete's ability to deal with pain can also be used to gauge how well they might deal with an injury and whether they could potentially mask an injury (Westman et al. 2008). The range of functional tests is vast but the practitioner needs to be objective in testing so that valid and reliable results are achieved and the whole process is not undermined (Gabbe et al. 2004). The use of biomechanics to add further detail to the analysis should not be overlooked as a history of injury can often be confusing and the details gained from traditional screening methods alone might not solve the problem (Whiting and Zernicke 1998). Finally, injury prevention is the assessment of risk as to when an athlete is going to break down (Mottram and Comerford 2008). A new approach to screening through the use of a performance matrix can help target weak links and improve these potential breakdowns, therefore enabling sporting performance to continue uninterrupted (Mottram and Comerford 2008).

Recommendations

For a detailed musculoskeletal screening process the following stages are recommended:

1. The screening process should start with a detailed examination of their perception of where they might have pain, using the examples shown in Figures 2.1 and 2.8.

2. This can also be support with the use of the VAS to gain an understanding of the athlete's ability to deal with pain and their pain history.

3. The physical examination should take the form of the functional assessments indicated in Figures 12–16 (Dennis et al. 2008).

4. The athelte should then have a more dynamic assessment completed as detailed in research by Mottram and Comerford (2008) with the possible use of video anaylsis to add a more detailed and objective analysis (McClean et al. 2005).

5. On completion of these stages a matrix of the athlete's strengths and weakness can be built to ensure that a programme is put together with a strength and conidioning coach to ensure that problem areas are improved on.

6. The coach and athelte need to be made aware of recovery techniques to help erradicate any problems that could be occuring due to training intensity, such as in proper cool down and rehydration stratigies.

7. For more details on the athlete's problem areas, the use of a biomechanist will add depth to the analysis and could offer an insight into technique error as a possible cause of injury (Callaghan and Jarvis 1996; Paul 2005).

References

Barber, S.D., Noyes, F.R., Mangine, R.E., McCloskey, J.W. and Hartman, W. (1990) Quatnitiative assessment of functional limiatations in normal and anterior cruciate ligament deficient knees. *Clinical Orthopaedics and Related Research*, 255, 204–214.

Beattie, K.A., Bobba, R., Bayoumi, I., Chan, D., Schabort, S., Boulos, P. et al. (2008) Validation of the GALS musculoskeletal screening exam for use in primary care: a pilot study. *BMC Musculoskeletal Disorders*, 9 (115), 1–8.

Berg-Rice, V.J., Conolly, V.L., Pritchard, A., Bergeron, A. and Mays, M.Z. (2007) Effectiveness of a screening tool to detect injuries furing army health care specialist training. *Work*, 29, 117–188.

Bonci, C.M. (1999) Assessment and evaluation of presisposing fators to anterior cruciate ligament injury. *Journal of Athletic Trajning*, 34 (2), 155–164.

Callaghan, M.J. (2005) Lower body problems and injury in cycling. *Journal of Bodywork and Movement Therapies*, 9, 226–236.

Callaghan, M.J. and Jarvis, C. (1996) Evaluation of elite British cyclists: The role of the squad medical. *British Journal of Sports Medicine*, 30, 349–353.

Coady, D., Walker, D. and Kay, L. (2003) The attitudes and beliefs of clinicians invloved in teaching undergraduate musculoskeletal clinical examination skills. *Medical Teacher*, 25 (6), 617–620.

Dawson, A.P., Steele, E.J., Hodges, P.W. and Stewart, S. (2009) Development and Test-Retest reliablity of an extended version of the nordic musculoskeletal questionnaire (NMQ-E): A screening instrument for musculoskeletal pain. *The Journal of Pain*, 10 (5), 517–526.

Dennis, R.J., Finch, C.F., Elliott, B.C. and Farhart, P.J. (2008) the reliability of musculoskeletal screening tests used in cricket. *Physical Therapy in Sport*, 9, 25–33.

Gabbe, B., Bennell, K., Wajswelner, H. and Finch, C.F. (2004) Reliability of common lower extremity musculoskeletal screening tests. *Physcial Therapy in Sport*, 5, 90–97.

Gabbe, B., Bennell, K., Finch, C. and Wajswelner, H. (2009) Predictors of hamstring injury in elite Australian football. *Physcial Therapy in Sport*, 87, 87–88.

Knapik, J.J., Bauman, C.L., Jones, B.H., Harries, J.M., & Vaughan, L. (1991) Preseason strength and flexability imbalances associated with athletic injuries in female collegiate athletes. *American Journal Sport Medicine*, 1, 76–81.

Komura, T. and Nagano, A. (2004) Evaluation of the infulence of muscle deactivation on other muscles and joints during gait motion. *Journal of Biomechanics*, 37, 425–436.

Laslett, M., Oberg, B., Aprill, C.N. and McDonald, B. (2004) Zygapophysial joint blocks in chronice low back pain: A test of revel's model as a screening test. *BMC Musculoskeletal Disorders*, 5 (43), 1–6.

Lephart, S.M., Perrin, D.H., Minger, K. and Fu, F. (1991) Functional performance test for the enterior crucitate ligament insufficient athlete. *Athleteic Trainer*, 26, 44–50.

Litcher-Kelly, L., Martion, S.A., Broderick, J.E. and Stone, A.A. (2007) A systematic review of measures used to assess chronic musculoskeletal pain in clinical and randomized controlled clinical trials. *The Journal of Pain*, 8 (12), 906–913.

McClean, S.G., Walker, K., Ford, K.R., Myer, G.D., Hewett, T.E. and Van den Bogert, A.J. (2005) Evaluation of a Two dimensional analysis method as a screening and evaluation tool for anterior ligament injury. *British Journal of Sport Medicine*, 39, 355–362.

Merchant, A.C., Arendt, E.A., Dye, S.F., Fredericson, M., Grelsamer, R.P., Leadbetter, W.P. et al. (2008) The female knee: anatomic variations and the female-specific total knee design. *Clinical Orthopaedics And Related Research*, 466, 3059–3065.

Miller, A. and Callister, R. (2009) Reliable lower limb musculoskeletal profiling using easily operated portable equipment. *Physical Therapy in Sport*, 10, 30–37.

Moore, J.R. and Wade, G. (1989) Prevention of anterior cruciate ligament injuries. *Natonal Strength and Conditioning Association Journal*, 11, 35–40.

Mottram, S. and Comerford, M. (2008) A new perspective on risk assessment. *Physical Therapy in Sport*, 9, 40–51.

Moul, J. (1998) Differences in selected predictors of anterior cruciate ligament tears between male and female NCAA division I Collegiate Baskeball players. *Journal of Athletic Trainnier*, 33, 118–121.

Paul, J.P. (2005) The history of musculoskeletal modelling in human gait. *Theorectical Issues in Ergonomics Science*, 6 (3–4), 217–224.

Rosenblum, K.E. and Shankar, A. (2006) A Study of the efffects of isokinetic pre-emplyment physical capabilityscreening in the reduction of musculoskeletal disorders in a labor intensive work environment. *Work*, 26, 215–228.

Schellenberg, K.L., Lang, M.J., Cahn, M.K. and Burnham, R.S. (2007) A clinical tool for office assessment of lumbar spine stabilization endurance: prone and supine bridge maneuvers. *American Journal of Physical Medicine & Rehabilitation*, 86 (5), 380–386.

Smith, T.O., Hunt, N.J. and Donell, S.T. (2008) The reliability and validity of the Q-angle: a systematic review. *Knee Surgery, Sports Traumatology, Arthroscopy: Official Journal of the ESSKA*, 16 (12), 1068–1079.

Thiel, H. and Rix, G. (2005) Is it time to stop functional pre-manipulation testing of the cervical spine? *Manual Therapy*, 10, 154–158.

Twomey, D., Finch, C., Roediger, E. and Lloyd, D. (2009) Preventing lower limb injuries: is the latest evidence being translated into the football field? *Journal of Science and Medicine in Sport*, 12, 452–456.

Westman, A., Linton, S.J., Ohrvik, J., Wahlen, P. and Leppert, J. (2008) Do psychosocial factors predict disability and health at a 3-year follow-up for patients with non-acute musculoskeletal pain? A validation of the orebro musculoskeletal pain screening questionaire. *European Journal of Pain*, 12, 641–649.

Whiting, C.W. and Zernicke, R.F. (1998) *Biomechanics of Musculoskeletal Injury*. Champaign, IL: Human Kinetics.

3

Assessment and Needs Analysis

Paul Comfort and Martyn Matthews
University of Salford, Greater Manchester

To condition athletes effectively, training must reflect the conditions encountered in sport. To achieve this, programme designers must: (1) analyse the demands of the sport; (2) identify the individual characteristics of the athlete (strengths and weaknesses; training history); (3) tailor and prioritise training to allow each individual athlete to meet these specific demands.

To effectively test an athlete's fitness, and therefore develop an appropriate training and rehabilitation regime, sports rehabilitators and strength and conditioning coaches must identify the essential components of the sport/activity in question. This 'needs analysis' requires the gathering of accurate, precise and reliable data, ideally from the published literature, combined with detailed observation of training and competition. Appropriate fitness tests can then be selected and conducted, with comparisons made to determine individual strengths and weaknesses that inform the implementation of appropriate training that focuses on the sport, the athlete and any identified injury risk. It is essential that the sports rehabilitator develop an applied awareness of such methods of assessing an athlete in order to complement their clinical assessments/skills when determining an athlete's readiness to return to sport. A more detailed understanding of the demands of the sports will also be invaluable in terms of implementing effective and evidence-based injury prevention programmes.

The primary aim of this chapter is, therefore, to explore how to conduct an appropriate needs analysis. The chapter begins with an exploration of the different components required in order to undertake an effective needs analysis, including the metabolic and mechanical demands of the sport/activity. This leads the reader on to a section on fitness testing, which discusses the variety of testing modalities available, and includes some discussion of the validity and reliability of these methods. The chapter then provides two detailed summaries of needs analyses as examples (football and rugby league) of the process as a whole.

Analysing the demands of sport

To analyse the specific demands of the sport, sports rehabilitators must consider the demands placed on the muscular, nervous, endocrine, cardiovascular, respiratory and skeletal systems. They must consider the length of the event; the type, speed and frequency of movement involved; the pattern of play and work–rest ratios; the nature of contact with other players or opponents; and the competition structure. They must consider the combined demands of both training and performance and the injury risk that arises in each. They must consider the nutritional and psychological demands of the sport, and the effect that these may have on performance, recovery and rehabilitation.

Sports Rehabilitation and Injury Prevention   Edited by Paul Comfort and Earle Abrahamson
© 2010 John Wiley & Sons, Ltd

In general, the demands of sport can be categorised as those with a *metabolic* emphasis and those with a *mechanical* emphasis.

Metabolic demands

The metabolic demands of a sport are determined by the biochemical pathways used for energy production in that sport. These energy pathways produce adenosine tri-phosphate (ATP), the only source of energy for muscular contraction. There are three systems responsible for energy production, the *phosphagen* system, the *glycolytic* system and the *aerobic* system.

The *phosphagen* system is a very powerful system producing large quantities of ATP from phosphocreatine (PCr). Unfortunately, the combined stores of ATP and PCr (the phosphagen system) only provide enough energy for 6–8 seconds of intense muscular contraction. The phosphagen system therefore predominates in sports requiring short duration, maximal intensity bouts of effort. These include weight lifting, high jump and short sprints (60, 100 and 200m). The phosphagen system therefore plays a crucial role in team sports requiring multiple short bursts of activity.

The *glycolytic* system is another powerful system for producing ATP for high-intensity activity. It is less powerful than the phosphagen system but can produce ATP for longer. Like the phosphagen system, the glycolytic system starts working immediately exercise starts but, unlike the phosphagen system, it takes about 5 seconds to reach maximal production capacity and lasts for 30–40 seconds at maximal intensity. The glycolytic system therefore predominates in longer duration, high intensity activities (e.g. 400m), and during the sprint finish of longer events. It therefore plays a crucial role in the performance of all athletes. Together the phosphagen and glycolytic systems are *anaerobic*, or oxygen independent.

The *aerobic* system is oxygen dependent and can produce vast quantities of ATP very efficiently, but at a much slower rate. Again, the aerobic system starts producing ATP as soon as exercise begins but does not reach full capacity for several minutes. The aerobic system therefore predominates in events requiring high levels of endurance. These events include continuous, long duration, moderate intensity events such as rowing, cycling and distance running, but also team events, where high intensity bursts of activity (phosphagen system dominated) have to be repeated throughout the course of a long game (up to 90 minutes). During these latter events, the aerobic system plays a crucial role in replenishing phosphocreatine and therefore dominates during the recovery between bursts of activity. Although it is tempting to categorise sports into phosphagen, glycolytic and aerobic dominated, it is essential to remember that most sports require a combination of all three systems. For example, football utilises the phosphagen system most during the repeated high intensity bouts of activity, the glycolytic system most during extended periods of high intensity play, and the aerobic system during recovery phases (including half-time); a two-hour training session, for an elite sprinter, will focus each individual activity on the phosphagen and possibly glycolytic system, however, each bout of recovery will be aerobic. This must be accounted for in training.

Training should reflect the specific nature of energy system usage by targeting the dominant pathways, but must also recognise the contributing role of the other systems. Within each sport there is clearly room for variety and a change of focus as the training year progresses; however the ultimate goal is maximal performance in a specific task. For team sports, however, it is essential to determine whether it is an athlete's aerobic performance, or sprint performance that is the limiting factor, and therefore form the primary focus of training.

Mechanical demands

Movement specificity

The mechanical demands of sport determine the movements that athletes should train. Exercises that are similar to the actual movements encountered in sport should be prioritised; for example, there is a triple extension of the ankles, knees and hips during a vertical jump and therefore exercises that involve a rapid extension of these joints, such as squat jumps, the clean or snatch, should be targeted. By focusing on *movement pattern specificity,* athletes can reinforce and condition the motor programmes used in skilled performance. These programmes control the precise order, timing and force application to enable the muscles to produce a predetermined movement (Enoka 2002). The more practiced and efficient these programmes, the better the performance of the skill. For example, a rugby player who focuses practice on the foot patterns required to side step an opponent can enhance side-stepping performance by executing quicker, more efficient motor programmes during the game. However, these are generally trained and refined during skill training.

The training methods that transfer best to actual sporting performance usually involve coordinated movements across multiple joints rather than strict isolation exercises. In sport, no muscle works in isolation. Isolating specific muscles, then, is non-functional; gains in strength, power or endurance occur only in the trained muscle and fail to integrate with the whole movements required for sporting performance. Athletes must consider how to train *movements, not muscles*. Training that focuses on the whole movement enhances sports performance more effectively than the training of isolated joint movements (Bompa 1999; McGill 2006). Moreover, closed kinetic chain (CKC) exercises have a greater effect on functional performance when compared to open kinetic chain (OKC) exercises (Augustsson et al., 1998; Blackburn and Morrissey 1998). *Integrate, don't isolate.*

Muscle action

As well as training to *produce force,* it is also essential to develop an athlete's ability to *accept force*. Sport requires athletes to reduce and absorb external forces, often at high speeds, in three dimensions, and in an unpredictable environment. Athletes must train for deceleration and force-acceptance as well as force production. Training tends to focus on concentric force development; however, many sports require *heavy load eccentric muscle actions*, which can also be *high velocity*, especially during rapid decelerations and changes of direction. Eccentric training, jump landings, *dynamic control and stabilisation* training, along with *specific jumping and agility drills,* will increase an athlete's capacity to control and manage the specific forces encountered during sport.

For example: athletes typically train hamstrings by using knee flexion exercises (leg curls) and hip extension exercises (stiff-legged deadlifts). In function, however the hamstrings also act to control and decelerate the limb (as in kicking a football), act antagonistically to the rectus femoris to prevent hip flexion (as in squatting or jumping), and act as an ACL agonist by preventing anterior tibial translation (Li et al. 1999; Ebben and Leigh 2000). Movements that target these attributes may include Nordic hamstring lowers (Fig 3.8a, 3.8b) (Askling et al. 2003; Mjolsnes et al. 2004; Clark et al. 2005; Arnason et al. 2007), drop jump landings (including correct landing during plyometric activity) and lunges (Jonhagen et al. 2009), increasing progressively in terms of both velocity and amplitude. Eccentric and plyometric training has also been shown to decrease the risk/incidence of both ACL and hamstring strain injuries (Hewett et al. 1996; Heidt et al. 2000; Clark et al. 2005; Mjolsnes et al. 2004; Wilkerson et al. 2004; Hewett et al. 2005; Mandelbaum et al. 2005; Myer et al. 2005).

Direction and velocity of force

Sporting movements often require athletes to produce and accept forces in multiple planes, at various speeds, all in a fluid and ever-changing environment. Sports rehabilitators, athletes and coaches therefore need to identify what these movement patterns are and, where safe to do so, tailor training and rehabilitation to mimic these. For example, functional movement patterns are enhanced by training at greater speeds, in multiple directions, and under varied and unpredictable conditions. This challenges an athlete's balance and proprioception, enhancing their ability to stabilise joints and maintain posture, allowing the transfer of forces efficiently from one body section to another. Exercises that incorporate fast eccentric loading in the initial phases and place a high demand on an athlete's ability to dynamically stabilise their joints under varying conditions also allows them to develop greater control and accept higher forces quickly. For example, plyometric training should begin with primarily vertical movements, followed by forward momentum, then lateral movements finally progressing to multidirectional movements, which progressively become more sport specific (Dugan 2005).

Sporting movements occur quickly, often between 30 and 260 ms. For example at the beginning of the race, when a sprinter drives out of the blocks and accelerates up to full speed, the ground contact time can be greater than 200ms (Mero 1988; McKenna and Riches 2007), whereas when the athlete reaches peak running velocity, the contact time is nearer to 70–125ms (Kunz and Kaufmann 1981; Mann and Herman 1985; Moravec et al. 1987; Chu and Korchemny 1993; Weyland et al. 2000; McKenna & Riches 2007).

Once the demands of the sport have been analysed it is essential to determine to what extent each athlete can meet those demands. This is achieved through fitness testing.

Fitness testing

The importance of fitness testing

To determine the current status of an athlete's fitness, and monitor the progress made during both training and rehabilitation, specific components of fitness must be assessed. Fitness testing allows coaches and rehabilitators to identify an athlete's strengths and weaknesses, enabling them to tailor and adjust training and rehabilitation according to the athlete's greatest need(s). This optimises the use of training time and resources, helping to achieve maximal performance gains and enhance rehabilitation as efficiently as possible. Regular fitness testing provides vital information to athletes and their support teams and should therefore form part of any athlete's development programme. Moreover, fitness testing is also used to monitor the effectiveness of training programmes, establish a baseline that may be used to monitor the progress of rehabilitation post-injury, provide a motivational tool for athletes (particularly those that train independently) and enable coaches identify future talent. Without fitness testing, it is impossible to accurately and objectively monitor an athlete's progress or assess readiness and readiness to return to sport post-injury.

For tests to be effective and reflect the changes in an athlete's fitness they must be valid and reliable, repeated at regular intervals using carefully controlled procedures, and be understood by athletes and coaches. To ensure this, the principles of validity, reliability and objectivity must be taken into account.

Validity, reliability and objectivity

Validity refers to what is actually measured. Some tests directly measure that which is required; what you see is what you get. For example, in a 40m sprint test, time is recorded and, if electronic timing equipment is used, the time recorded is an accurate reflection of the time taken to complete the test; in a maximum strength test the highest weight lifted is recorded (1 repetition maximum – 1RM); and in a skin-fold calliper test, skin-fold thickness is recorded. In the latter case, the measurement of skin-fold thickness is only a measure of skin-fold thickness, not body fat. The calculation of body fat is an estimation based on prediction equations (Jackson et al. 1980; Pollock and Jackson 1984).

The next consideration is *reliability* (see Table 3.1 for specific reliability data). This is the extent to which scores are consistent and repeatable across time or between testers, and therefore reflects the ability to detect actual changes with time. Even direct measurements may be subject to errors. For example, if a 40m sprint is timed by hand, the timekeeper must determine when the athlete started and finished. There may be a delay as the brain processes this information and also a delay before the button is pressed and the clock stopped. Hand timing, then, relies on an individual interpreting what they see and deciding when to press the button. This inbuilt inaccuracy may be greater than any differences in actual time between testing sessions, and so may not be sensitive enough to reflect legitimate training improvements. Other challenges to the reliability of tests include variations with equipment, test environment, the weather, warm-up procedures, or subject motivation between testing sessions. In terms of assessing body composition via skin fold measurements, errors in testing can include improper site selection and measurement, use of different callipers, and intra- and inter-tester variation (Pollock and Jackson 1984).

Objectivity refers to any bias that originates with either the tester or the athlete. Tester bias may include subtle differences in testing protocols (such as the positioning and timing of skin-fold measurements), or interpretation of test performance (stopping the clock when the tester perceives that the athlete crossed the line).

Subject bias may include situations where the athlete being tested aims to manipulate the results. For

Table 3.1 Reliability of common methods of assessing performance

Test	Reliability (ICC)	Author
Yo-yo	Related to VO_2 max ($r = 0.75$, $p<0.001$) via treadmill direct gas analysis in adult male soccer players	Castagna et al. (2006)
T-test (agility)	0.98	Pauole et al. (2000)
Vertical jump	Jump mat: $r = 0.967$	Leard et al. (2007)
(Squat jump)	Jump mat: $r = 0.97$	Markovic et al. (2004)
	Jump and reach: $r = 0.906$	Leard et al. (2007)
	Force Plate: $r = 0.75$–0.99*	Moir et al. (2005)
Single-leg vertical jump	0.86 dominant leg, 0.82 non-dominant leg	Maulder and Cronin (2005)
Standing long jump	0.95	Markovic et al. (2004)
Single-leg horizontal jump	0.90 dominant leg, 0.89 non-dominant leg	Maulder and Cronin (2005)
Standing triple jump	0.93	Markovic et al. (2004)
Star excursion balance test	0.82–0.87	Plisky et al. (2006)
(SEBT)	≥ 0.86	Kinzey and Armstrong (1998)
Hop tests:		
Single hop for distance	0.92 (SEM = 4.61cm)	Ross et al. (2002)
	0.96 (SEM = 4.56cm)	Bolgla and Keskula (1997)
Triple hop for distance	0.97 (SEM = 11.17cm)	Ross et al. (2002)
	0.95 (SEM = 15.44cm)	Bolgla and Keskula (1997)
Cross-over hop for distance	0.93 (SEM = 17.74cm)	Ross et al. (2002)
	0.96 (SEM = 15.95cm)	Bolgla and Keskula (1997)
6m hop for time	0.92 (SEM = 0.06s)	Ross et al. (2002)
	0.66 (SEM = 0.13s)	Bolgla and Keskula (1997)
Bench trunk curl	0.94 (females) and 0.88 (males)	
Isokinetic knee flexion and extension (peak torque)	>0.90 at 60°/s	Sole et al., (2007)
	>0.97 at 60°/s	Maffiuletti et al., (2007)
	>0.98 at 120°/s and 180°/s	
	0.82 at 60°/s	Li et al. (1996)
	0.83 at 120°/s	
	0.80 at 60°/s	Impellizzeri et al. (2008)
	0.89 at 60°/s	Lund et al. (2005)
Yo-yo endurance test	Related to VO_2 max ($r = 0.75$, $p < 0.00002$) via treadmill direct gas analysis, related to peak treadmill speed at VO_2 max ($r = 0.87$, $p < 0.0003$) in male soccer players.	Castagna et al. (2006)
	Related to peak treadmill speed at VO_2 max ($r = 0.71$, $p = 0.0001$) in adult male basketball players	Castagna, Imperellizzeri, Rampinini, et al (2007)
Yo-yo recovery test	Related to peak treadmill speed at VO_2 max ($r = 0.71$, $p < 0.0003$) in adult male soccer players	Castagna, Imperellizzeri, Chamari et al., (2006)

SEM = standard error of measurement
*Unloaded, 30% & 60% 1RM squat

example, many teams expect an athlete's fitness to improve throughout the course of pre-season training and sometimes impose fines on those athletes that do not improve. There is therefore a temptation for some athletes to intentionally under-perform in the first battery of tests so that they are more likely to record an improvement as pre-season training progresses, thereby avoiding a fine.

Table 3.2 Advantages/disadvantages of field and lab based tests?

Laboratory tests		Field tests	
Advantages	Disadvantages	Advantages	Disadvantages
Lots of information	Expensive	Widely available	Changes in testing environment
Reproducible	Non-functional	Functional	Equipment must be accurate
Precise and direct measurements	Can take a lot of time	Time efficient for teams	Some errors in prediction

Laboratory versus field testing

Despite the lack of direct transfer to competition, laboratory tests do have several advantages. A laboratory allows the same test to be reproduced under similar conditions on separate occasions. This increases the sensitivity of the test and allows subtle changes in fitness to be monitored over time without interference from varying environmental conditions (Table 3.2).

Field-testing, in contrast, has a number of advantages over laboratory-based tests. Tests can be devised that more closely mimic the requirements of the particular sports. For example, the changes of direction that occur in the Multi Stage Fitness Test (Bleep Test), and particularly the yo-yo test, closely resemble some of the movement characteristics and work–rest ratios of team sports. It could be argued that performance in the yo-yo test gives a better indication of a soccer player's ability to perform in an intermittent multi-directional activity like soccer, than their performance in a laboratory-based treadmill test. Castagna et al. (2006) found that VO_2 max data collected via the yo-yo test was strongly and significantly related to VO_2 max ($r = 0.75$, $p<0.001$) via treadmill direct gas analysis in adult male soccer players. Metaxas et al. (2005) found a 10.5–13.3% variation, in VO_2 max measurements, assessed via the yo-yo endurance test, the yo-yo intermittent test and continuous and intermittent treadmill tests. However, the authors concluded that yo-yo field tests should be used to monitor aerobic fitness in team sports, as they are easy to administer and incorporate into training sessions during the competitive season. In a review of literature, Bangsbo et al., (2008) concluded that the yo-yo intermittent recovery test is a simple and valid method of assessing an individual's capacity to perform repeated intense exercise bouts, and monitoring changes in performance capacity.

Additional considerations, when selecting testing methods, are time and reproducibility. In terms of the use of the yo-yo test for team sports, it is far more time efficient than individually assessing the whole team, and it is also highly reproducible if the testing conditions (clothing, equipment, time of day, warm-up) remain constant. If working with a marathon runner, however, the most valid and reliable method of assessing aerobic capacity would be via direct gas analysis.

Field tests also have the advantage that they are simpler, easier to set up and administer, and can often be applied to several athletes at once. As each athlete has minimal disruption to their routine, field tests may therefore be the most practical choice for regular monitoring of training gains.

Where available, it is sometimes possible to take traditional laboratory equipment into the field. Mobile gas analysis devices, such as the Metamax™ (Fig 3.1) allow sophisticated tests to be conducted in an ecologically valid setting.

Test order

When conducting several tests, the order in which the individual tests are performed is vital. The performance of a previous test can impact the performance of a subsequent one. For example, a test to assess anaerobic endurance requires a high intensity of effort for an extended duration, causing considerable fatigue and leaving the athlete below normal capacity for some time. Any subsequent tests, performed whilst the athlete is still fatigued, will be severely impaired. If that test is agility, and on a re-test the athlete scores higher than before, then athletes and coaches cannot conclude that agility has actually improved. Several factors could have contributed to a better agility result. These include: an improved anaerobic endurance, leading to less induced fatigue; improved

Figure 3.1 The Metamax testing system.

cardiovascular fitness, leading to quicker recovery; and lower motivation levels during the anaerobic test, leading to less fatigue. All of these may allow the athlete to start the agility test in a better state of readiness and therefore record improvements that are not necessarily down to improvements in agility.

To ensure that any changes recorded actually reflect genuine improvements, the impact of previous tests must be minimised. The National Strength and Conditioning Association (NSCA) (Baechle & Earle, 2008) suggest the following test order (Table 3.3).

Agility and *speed* are tested first as they are short, relatively non-fatiguing activities that require only a few minutes for full recovery. Testing these early will not impact later results. *Strength* should be tested next, again because the actual tests require little recovery (5 minutes) and have little impact on subsequent muscular or cardiovascular endurance. *Muscular/Anaerobic endurance* should be tested after strength as these tests can induce fatigue in the muscles and has a major impact on subsequent strength, skill and speed performance. Depending on the requirements of the sport, *cardiovascular fitness* (usually assessed via a maximal test that also stresses the anaerobic energy systems) may also need to be addressed. In this case it should either replace the anaerobic endurance test, or be tested on a separate occasion.

The tests

When designing a battery of fitness tests, several factors must be considered. These include: selection of tests (which ones and how many), order of tests, recovery period between tests, what equipment and any changes to account for different playing positions (Table 3.4).

The choice of tests should reflect the characteristics of both the sport and individual player position. For example, a cycle based test will have almost no transfer to sports that require athletes to run; a strength assessment using a bench press will not necessarily transfer to throwing or punching activities; a constant-pace running-based endurance test for a goalkeeper will be less functional than a test of repetitive explosiveness or agility.

The order of tests should be chosen to minimise interference between tests (Baechle and Earle 2008).

Table 3.3 Test order

1 Agility
2 Speed
3 Muscular strength
4 Muscular endurance
5 Aerobic capacity (ideally on a separate day)

Table 3.4 Factors to consider when designing a battery of tests

How many tests?	Maximum of seven
What tests?	Decision based on functional characteristics of the sport. Will an improvement in the test result in an improved performance in the sport?
What order?	Agility, speed, strength/muscular endurance, anaerobic/aerobic endurance
How long in between?	Enough for full recovery from previous test
What equipment?	Timing gates; tape measure; cones; sports hall, etc.
Changes to account for different positions	Different tests required to gain useful information about different positions. For example: goal keeper versus midfield player

The length of rest between tests should be long enough to allow complete recovery so that interference is minimised. The equipment should allow tests to be accurately reproduced and limit tester objectivity (timing gates versus hand timing).

Depending on where tests originated, distances may be expressed in metres or yards. In reality, a 40m test is no better or worse than a 40 yard test. The important issues to consider are consistency across time (when repeating tests) and an awareness of any differences when comparing test results to normative data.

Assessing agility

Shuttle runs are widely used as a test of agility, providing information on explosiveness, acceleration, deceleration, turning ability, functional lower body strength and body control. No agility test should last longer than about 10 seconds. Beyond this, performance is determined far more by an athlete's speed-endurance capabilities than agility.

With all agility runs, careful placing of extra timing gates (before and after the turn) can help establish whether improvements in overall performance are the result of improved speed or improved turning ability.

T-test

The T-test is a 40m (or 40 yard) agility test (Figure 3.2) that incorporates forward sprinting, side-to-side shuffling and backwards running. It is particularly suited to both team and racquet sports.

Intraclass reliability for the T-test has also been shown to be as high as 0.98 (Pauole et al. 2000), therefore small changes in performance should be a direct result of adaptations to training. As with assessing sprint ability, it is essential that timing gates are used to assess performance in this test.

Illinois agility test

The Illinois agility test (Figure 3.3) is designed to test acceleration, deceleration, cutting and turning ability. Performance is determined by the time that is taken to complete the course. The use of timing gates is essential.

Speed

The best way to test an athlete's speed is to assess their performance over sport specific distances. For example, short sprints, lasting 10–40 m, are used to mimic the sprint distances typically observed during

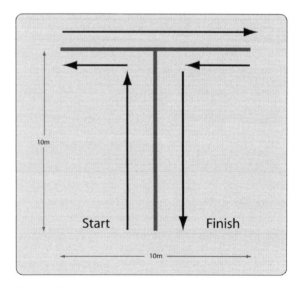

Figure 3.2 T-test.

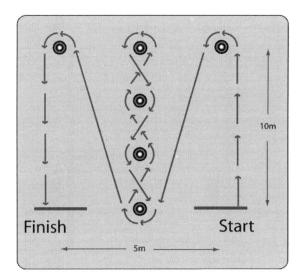

Figure 3.3 Illinois agility test.

team sports; 6s or 10s sprints on a cycle ergometer are used to test speed in track cyclists; 10, 20, 30 or 40 m sprints in multiple sprint sports; and 100m sprints on a rowing ergometer are used to test speed in rowers.

These short sprints are the best way of assessing functional speed as they actually do measure short sprint ability. Because they all start from standing, they also give an indication of explosiveness and acceleration ability. As with all such tests (including agility tests) timing must be performed with timing gates or similar electronic method. (Hand timing is not an acceptable method of timing, due to the level of accuracy/reliability.)

Sprint tests

Tests consist of three sprints from a standing start, with full recovery between runs. The best time is always taken. Some researchers take the average of three; however it is usually not appropriate to use this method as a single poor result can skew the recorded figure.

Starting position and height of the timing gates also needs to be standardised. Small differences in times as a result of starting position are not interchangeable (Duthie et al. 2006), with starting positions with feet parallel resulting in slower times than those where the dominant foot is placed forwards (Cronin et al. 2007). Cronin and Templeton (2008) also observed an error of ≤1.3% (equal to 0.7s) between the times achieved with the gates positioned at hip and shoulder height. This was attributed to the legs breaking the beam earlier with the gates positioned at hip height.

A 40m-sprint test, with an additional timing gate placed at the 20m-ine, gives useful extra information on an athlete's initial acceleration. The distances can be altered for different sports to make the test sport specific.

Vertical jump tests

Vertical jump tests are widely used to assess single- and double-legged vertical jumping ability. They focus on a particular performance parameter (height jumped) and in this respect are highly functional to those sports requiring vertical jumping ability, such as basketball and volleyball. Variations of the test can include, a two-legged take off, a one-legged take off, a step and one-legged take off (tennis, soccer), or measuring the height jumped after a drop off of a small platform.

Two common methods are employed to determine height jumped. The first uses contact mats to determine flight time. The second is a simple jump and reach method. Both methods appear to be valid and reliable tests ($r = 0.967$ and 0.906 respectively) when compared to a three-camera motion analysis system (Leard et al. 2007). Another method is to attach a linear position transducer to a belt/harness that the athlete wears. On jumping, the device records the linear displacement during each jump, and can provide reliable data regarding jump height, peak force (intraclass correlation coefficient (ICC) = 0.977–0.982), mean force (ICC = 0.924–0.975), and time to peak force (0.721–0.964) (Cronin et al. 2004).

Although vertical jump tests do not measure power (they measure height jumped, unless performed on a force plate), the ability to jump high is closely correlated with this parameter. This relationship has permitted the development of prediction equations for peak power output during the vertical jump, which removes the need for force plates to monitor athletes' peak power during the vertical jump. Keir et al.(2003) found that peak power (W) can be predicted (within 2%) via height jumped (cm) and body mass (kg), with a very high level of reliability ($r > 0.9999$, CV $<0.2\%$) (see Table 3.1, and refer to Keir et al. (2003) *Journal of Strength and Conditioning Research* for the Nomogram).

Vertical jump assessments can also be performed with additional load (also referred to as squat jumps), depending on the requirements of the sport, however these do require the use of a force plate to determine peak power output, peak force and rate of force development. In trained individuals, high test–retest correlations (ICC range: 0.75–0.99) are observed during unloaded and loaded jump squats (30 and 60% 1RM) with familiarisation not necessary due to low individual variation (CV range: 1–2–7.6%) between tests (Moir et al. 2005).

Maulder and Cronin (2005) adapted the vertical jump from a bilateral to a unilateral test and found that test–retest reliability remained high (ICC = 0.86 dominant leg, 0.82 non-dominant leg). Performance of unilateral assessments can highlight limb asymmetries and therefore act as a potential marker of injury risk.

Horizontal jump tests

Horizontal jump tests are useful to assess power in a horizontal direction. Tests, such as the standing long jump and the standing triple jump, where the athlete attempts to jump for distance from a standing start, give useful information, not only about an athlete's horizontal hopping ability but also about an athlete's ability to control their landing. Both of these tests are highly reliable with ICC of 0.95 and 0.93 and coefficient of variation of 2.4% and 2.9% respectively (Markovic et al., 2004).

Maulder and Cronin (2005) adapted the standing long jump from a bilateral to a unilateral test and found that test–retest reliability remained high (ICC = 0.90 dominant leg, 0.89 non-dominant leg). The study also found that the performance in the horizontal jumps was a better predictor of 20m sprint ability than vertical jumps (r = 0.73 and 0.66 respectively).

Hop tests

There are a number of hop tests available (hop for distance, triple hop for distance, cross-over hop for distance, and 6m hop for time, see Figure 3.4), which were primarily developed to assess power based performances during horizontal movements, but which are now commonly used in clinical environments

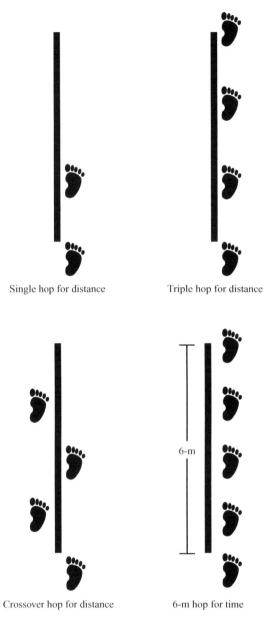

Figure 3.4 Hop tests.

to assess/monitor rehabilitation from lower limb injury. Along with the performance measures obtained during these tests, kinematic evaluation of the performances of these tasks may also highlight additional risk factors for knee injury if poor lower limb control and landing mechanisms are identified (Fitzgerald et al. 2001). Additional benefits of the hop tests are the ability to determine bilateral asymmetry, and monitor progress in performance and neuromuscular control post injury (Reid et al., 2007).

Single hop for distance

The athlete starts with their toe on a start line, hands on hips and their non-involved leg held in front at 90° of hip flexion to prevent any countermovement. The subject then hops as far as possible with the tester measuring from the initial toe position (start line) to the heel strike. Participants must be stable on landing. Research has demonstrated a high level of reliability (ICC r = 0.92–0.96) with a standard error of measurement of 4.61–4.56cm (Bolgla and Keskula 1997; Ross et al. 2002).

Triple hop for distance

Performed in the same way as the single hop for distance, only three consecutive hops are performed with the total distance measured from the start line to the final heel strike. Again participants must be stable on landing. Research has demonstrated a high level of reliability (ICC r = 0.95–0.97) with a standard error of measurement of 11.17–15.55cm (Bolgla and Keskula 1997; Ross et al. 2002).

Cross-over hop for distance

The athlete starts in the same position as for the other hops, only this time it is essential that a tape (≥ 8 m) measure is stuck to the ground leading away from start line. If the athlete begins on their right leg, they need to start to the right of the tape. Each hop is performed as in the triple hop for distance, with the only difference being that the athlete must cross the measuring tape during each hop. Research has demonstrated a high level of reliability (ICC r = 0.93–0.96) with a standard error of measurement of 15.95–17.74cm (Bolgla and Keskula 1997; Ross et al. 2002).

Six-meter hop for time

The athlete begins on the start line and hops as quickly as possible on the appropriate leg to the finish line (6m away). This has been conducted using a stop-watch, but due to the range of human error (as discussed elsewhere in this chapter) it would be much more accurate and reliable if timing gates are positioned at 6m intervals. Research has demonstrated a varying level of reliability (ICC r = 0.66–0.92) and a standard error of measurement of 0.06–0.13s (Bolgla and Keskula 1997; Ross et al. 2002). The more varied range of reliability and standard error of measurement may be due to the fact that stopwatches were used for timing.

Star excursion balance test

The star excursion balance test (SEBT) is commonly performed to assess dynamic balance and stability across multiple planes of movement. Poor performance in one limb compared to the other is a good indicator of bilateral instability and possible imbalance. Especially when combined with the results of assessments such as the hop tests.

The SEBT is performed with the athlete standing on the middle of the grid (see Figure 3.5) on the leg to be tested. The grid consists of lines extending out from the centre at 45 degrees (see Figure 3.6). Foot position should be standardised with the heel in the centre of the grid and the big toe on the anteriorly projected line. With the other leg, the individual reaches as far as possible along each line, in turn, returning the leg to the start position between each attempt (*the reaching leg must not be used for support*) (Figure 3.5). The distance reached in each direction is recorded. To standardise the test, and provide familiarisation the participants should be provided 4–6 practice attempts followed by a

Figure 3.5 SEBT.

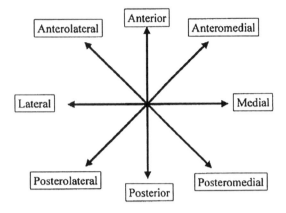
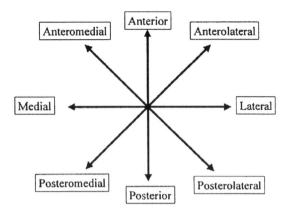

Figure 3.6 SEBT set up.

5-minute rest prior to the actual testing to eliminate any fatigue. Kinzey and Armstrong (1998) demonstrated that practice attempts raised the reliability of testing from r = 0.67–0.87, depending on the direction, to r>0.86.

This test has been shown to be a good indicator of progress of dynamic postural and lower limb control during rehabilitation from knee (Herrington et al. 2009) and ankle injury and chronic ankle instability (Gribble et al. 2004; Munn et al. 2009) via comparison of the performances of the injured to the non-injured leg. Plisky et al. (2006) also found the SEBT to be a good predictor of lower extremity injury and recommended its inclusion into pre-participation screening.

It is essential that the SEBT and similar dynamic balance/stability tests are performed prior to any possible fatiguing activity as Gribble et al. (2004) found that fatigue resulted in a noticeable decrease in performance in the SEBT.

Wingate Anaerobic Test

The Wingate anaerobic test (**WAnT**) is a 30-second cycle test providing information on peak power, mean power, muscle endurance and muscle fatigability. It is a relatively simple and inexpensive test that is reliable and repeatable (especially in motivated subjects), is sensitive to change over time, and truly reflects a person's anaerobic performance capacity. Although the test is well established and widely used in an educational setting, its application to most sports is limited. Highly specific for sprint cyclists, but not functional for other sports; however a number of studies, summarised by Inbar et al. (1996) have found strong associations with some field tests for assessing power, including 40m sprint speed (r = 0.84) and vertical jump (r = 0.70), but weak associations with others (Sargeant anaerobic skating test, r = 0.32).

There are also a number of modified versions of the WAnT, with varying loads and varying durations. When monitoring peak power output, these are all highly reliable, but it is essential to use the same loads if comparisons of peak power are going to be made at a later date. Using modified protocols to predict fatigue index and minimum power also appear to be highly reliable (R^2 = 0.84 and 0.91 respectively) (Stickley et al. 2008).

When testing athletes, consistency between tests is essential. For the WAnT, the equipment used (pedal crank length, toe clips, seat height, cycle geometry, resistance setting) can be adjusted to exact settings, and therefore made both highly specific to the individual cyclist's actual bike, and also reproducible across time. Care must also be taken to ensure consistency of environment (controlled laboratory conditions) as well as motivation of the athlete. Familiarisation with the test is also essential, as Barfield et al. (2002) demonstrated that subjects exhibit an increase in peak power output (14%) and mean power output (5%) due to a practice/familiarisation effect. It

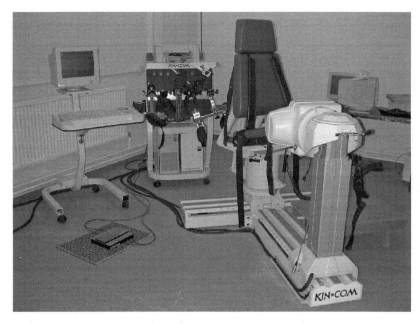

Figure 3.7 Isokinetic Dynamometer and balance assessment platform by Chatanooga.

is therefore recommended that individuals perform a familiarisation test several days prior to any baseline measurements.

It is important to consider the validity of using such laboratory tests to predict on-the-field performance, especially when the mode of laboratory testing (e.g. cycling) is mechanically different from the requirements of the sport (e.g. sprinting and jumping). This was highlighted by Baker and Davis (2002), who found that peak power output during a modified WAnT was unrelated to sprinting (10, 30 and 40m) and jumping (vertical and horizontal) performance, however, jumping and sprinting test performance were highly related (r = 0.80–0.91; p<0.01).

Assessing strength and muscular endurance

Strength is defined as the ability of a muscle group to develop maximal contractile force against a resistance in a single contraction. *Muscular endurance* is the ability of a muscle group to exert sub-maximal force over repeated contractions or for extended periods.

When assessing strength and muscular endurance we must therefore take account of the specific strength requirements of the sport. These include range of movement, joint angle, body position, type of muscle action (concentric, eccentric, isometric), and also velocity of movement and the speeds at which strength must be applied. It should therefore be possible to both train and assess strength in a functional manner.

There are three major areas of strength assessment: isotonic, isokinetic and isometric.

Isotonic measurement

Isotonic tests use free weights and are used to determine both *maximal strength* (via the one repetition maximum (1RM) test), which gives an indication of the maximum weight that can be lifted by a muscle group, and *local muscular endurance*, where an athlete's ability to perform multiple repetitions of a sub-maximal load is assessed.

One repetition maximum test

To record an accurate test score for 1RM the following procedure should be followed (adapted from Kraemer and Fry 1995):

1. Warm up with 5–10 repetitions of the test exercise at a load of 40–60% of estimated 1RM. Rest for 1–2 minutes

2. Perform 3–5 repetitions at 60–80% of estimated 1RM. Rest for 4–5 minutes

3. Increase the weight by between 5 and 10% and attempt a 1RM lift

4. If successful, rest for a further 4–5 minutes, increase weight by approximately 5% and repeat

5. Repeat until a 1RM is achieved.

Note, a novice may require 6–7 attempts to reach their 1RM. A more experienced lifter, with a more accurate perception of their true strength, may need 3–5 attempts.

Because 1RM testing requires maximal effort, it is usually confined to the more familiar free-weight squat, deadlift and bench press exercises. Wisloff et al. (2004) found that 1RM squat had a strong correlation with 10m sprint time ($r = 0.94$, $p<0.001$), 30m sprint time ($r = 0.71$, $p<0.01$) and vertical jump height ($r = 0.78$, $p<0.02$). Moreover, Baker (2001) found a strong correlation ($r = 0.87$) between maximal strength, during the bench press, and peak power output during bench throws using a variety of loads. 1RM testing, however, may not always be functional, and any interpretation of results, or predictions of future performance based on these results, should take this into account.

Furthermore testers should be aware that athletes unfamiliar with the test will under-perform when compared with more experienced lifters. As such, tests should always incorporate a familiarisation period prior to testing to account for any learning effect.

Three or five repetition maximum test

Many athletes, particularly if they do not lift free weights regularly as part of their conditioning programme, are not always comfortable with the maximal effort required for 1RM testing. As such it may be appropriate to conduct a 3RM or a 5RM test instead. Strength is still being assessed (there is a high correlation between 3RM and 5RM scores and 1RM scores), but the athletes may be happier with a sub-maximal (but still very heavy) load.

It is important with all strength and local muscular endurance assessment to ensure that technique remains constant throughout the test and between testing sessions. A rigid and repeatable set of instructions should be used to ensure that the tests remain reflective of actual changes in performance.

Isokinetic measurement

Isokinetic dynamometry has been shown to be a valid and reliable method of assessing strength that is sensitive to changes over time and as such is widely used by professional clubs to assess relative strengths across opposing muscle groups and identify potential for injury (see Table 3.1) (Li et al. 1996; Lund et al. 2005; Maffiuletti et al. 2007; Sole et al. 2007; Impellizzeri et al. 2008).

Thigh muscle imbalance appears to be an indicator of increased risk of injury (Engstrom and Renstrom 1998), with an imbalance between the hamstrings and quadriceps (eccentric hamstring to concentric quadriceps ratio of $\leq 1:1$, or a conventional concentric hamstring: quadriceps ratio of $<2:3$) indicating increased risk of hamstring injuries (Knapik et al. 1991; Yamamoto 1993; Croisier et al. 2002; Cameron et al. 2003; Foreman et al. 2006), and ACL injuries (Holcomb et al. 2007). This is due to hamstrings co-contraction during knee flexion, which minimises both anterior and lateral tibial translation (Ahmed et al. 2006; Escamilla et al. 2001; Kingma et al. 2004), decreases shear forces (Li et al. 1999) increases knee stability (Aagaard et al. 1998).

The major disadvantage, however, is that the movements are restricted to single joints and the velocity is *usually* limited to slow, non-functional, speeds. Furthermore it is useful to know the angles at which these peak forces occur (or at least what the relationship is when the quadriceps are producing peak torque) and how this relationship changes as the velocity of movement increases towards functional speeds.

Isometric measurement

Isometric strength assessment requires measurement of the static force produced by muscles using dynamometers (usually handgrip or back-lift). This type of assessment may be appropriate for sports requiring high levels of isometric strength or isometric strength endurance. Examples include the maximal grip strength and isometric grip strength endurance of a rock climber.

AEROBIC ENDURANCE

The most widely available tests for aerobic endurance are those performed with the minimum of equipment. Such tests include those that can be performed 'in the field' such as the multi-stage fitness test (MFST) or 'bleep' test, yo-yo test, 3-kilometre run, 5-kilometre run, 1.5-mile run, 12-minute run. Other more sport specific tests may include time-trials for cycling or rowing. For example, 10000m row or 20km cycle performed on a cycle or rowing ergometer.

Yo-yo endurance test

The yo-yo endurance test appears to have a good level of reliability (Castagna et al. 2006, 2007) (see Table 3.1). Metaxas et al. (2005) also concluded that yo-yo field tests should be used to monitor aerobic fitness in team sports, as they are easy to administer and incorporate into training sessions during the competitive season.

Yo-yo intermittent recovery test

Castagna, et al., (2006) found peak VO_2, assessed via yo-yo *intermittent* endurance test and an incremental treadmill test, had no significant relationship ($p>0.05$) indicating that the yo-yo intermittent endurance test is a weak predictor of aerobic fitness in moderately trained youth soccer players, compared to laboratory assessment. Similarly, Metaxas et al. (2005) found a 10.5–13.3% variation, in VO_2 max measurements, assessed via the yo-yo *endurance* test, yo-yo *intermittent* test, and continuous and intermittent treadmill tests. However, a review of literature, Bangsbo et al. (2008) concluded that the yo-yo intermittent recovery test is a simple and valid method of assessing an individual's capacity to perform repeated intense exercise bouts, and monitoring changes in performance capacity. Additional considerations when selecting testing methods are time and reproducibility. For team sports, the yo-yo test, may be a far more time efficient method than individually assessing the whole team, and also remains highly reproducible if the testing conditions (clothing, equipment, time of day, warm up) remain constant. If working with a marathon runner, however, the most valid and reliable method of assessing aerobic capacity would be via direct gas analysis. For one individual testing duration would be similar whether using direct gas analysis or a maximal prediction method, such as the yo-yo test.

Graded exercise test

A graded exercise test (GXT) in a human performance laboratory can reveal greater detail in terms of maximum oxygen uptake and anaerobic threshold but it is often the length of time that an athlete lasts in such a test that is the best (and simplest) predictor of functional performance ability. Graded exercise tests involve a progressive increase in exercise intensity and are designed such that the athlete lasts between 10 and 12 minutes before the point of exhaustion. These tests must be specific to the athlete's sport and use large muscle groups.

In a laboratory, the usual criteria for determining VO_2max include: a respiratory exchange ratio >1.15; heart rate (HR) in last stage ± 10 beats.min^{-1} of HRmax; and/or a plateau in VO_2 with increasing work rate.

Needs analysis for different sports

In certain sports, particularly those well supported by the literature, it is possible to formulate a detailed needs analysis. Below are example needs analyses for football and rugby league.

For sports where the published literature is lacking it is essential that all of the demands of the sport are considered in terms of the demands of the sport, as identified earlier in this chapter. Initially, this needs to focus on the dominant energy system required, followed by the velocity of movements, force generation and force acceptance, direction of force application and development. Once these concepts have been decided the rehabilitator can determine the methods of assessment that are essential to test, and also which components of fitness/performance need to be prioritised.

Needs analysis for football

Match analysis reveals that distances covered during a 90-minute soccer game range from 9,845–11,527 metres (Hoff, 2005), although distances as great as 13km have been reported (Shephard and Astrand 2000; Bangsbo et al. 2006), with average intensity exceeding 70% VO_2max (Bangsbo et al. 2006). The distance covered during the game is generally divided between walking (25%), jogging (37%), submaximal cruising (20%), sprinting (11%) and moving backwards (7%) (Shephard and Astrand 2000). The multiple sprints average 10–15 metres (2–4s in duration, approximately every 90s), interspersed with jogging, walking, running backwards, and rapid changes in direction. There are approximately 50 rapid, high-velocity movements in amateurs (Withers et al. 1982) and 150–250 rapid, high-velocity movements in elite athletes (Bangsbo et al. 2006). The rapid, high-velocity movements involve high-force eccentric and concentric muscle actions, while maintaining balance and control of the ball (Hoff 2005). Fatigue usually occurs, during a game due to glycogen depletion, however temporary fatigue between multiple short sprints may result from temporary depletion of intramuscular phosphocreatine concentrations (Bangsbo et al. 2006).

Average VO_2max measurements of football players appear to be above 60ml.kg.min^{-1}, with individual measures sometimes exceeding 70ml.kg.min^{-1}; average body mass and percentage body fat are approximately 77kg and 10% respectively (Reilly et al. 2000a,b; Arnason et al. 2004; Hoff 2005). More specific data is presented in Table 3.5.

It is worth noting that the data in Table 3.5 is 'normative' and not necessarily optimal. For example research has shown that increasing VO_2max leads to an increase in performance on the pitch. Helgerud et al. (2001) found that increasing VO_2 max by 10% (58.1 ± 4.5 to 64.3 ± 3.9ml.kg.min^{-1}) improved running economy (6.7%), increased distance covered by 20%, increased the number of sprints by 100%, and resulted in a 24% increase in the number of involvements with the ball.

It is also essential to acknowledge that energy expenditure, during training sessions and competition, has been estimated between 1400 and 1800kcal, dependant on playing position (Reilly et al. 2000a; Bangsbo et al. 2006). In relation to the energy expenditure during training, it is also worth noting that elite squads regularly train twice per day making rapid and efficient replenishment of energy substrates essential.

Football injuries

When conducting a needs analysis it is essential to identify the common injuries, and their causes as this may inform the choice of screening tests used to identify injury risk.

During two seasons of professional English football, it has been shown that the most common types of injuries were hamstring strains and anterior cruciate (ACL) injuries, accounting for up to 21% and 8% of all injuries respectively (Hawkins et al. 2001; Woods et al. 2002, 2004). The majority of hamstring injuries in football are non-contact in nature (Hawkins et al. 2001; Woods et al. 2002, 2004; Hagglund et al. 2005; Walden et al. 2005), highlighting that the risk of injury is intrinsic and may be offset through appropriate conditioning. The literature suggests that there are two non-contact mechanisms responsible for hamstring strain; one resulting from high speed running (Yamamoto 1993; Woods et al. 2004), and the other during stretching movements carried out by extreme range of motion (ROM) (Askling et al. 2002), both resulting in high velocity eccentric loading (Kujala et al. 1997; Cameron et al. 2003; Brockett et al. 2004). The strain is most likely to occur during two stages of the running cycle; late forward swing and toe off (Stanton 1989) as, at this stage, the hamstrings decelerate hip flexion and knee extension (Hoskins and Pollard 2005) resulting in large eccentric loads.

In terms of injury prevention, these common mechanisms of injury have implications for conditioning. It is essential that the hamstrings are conditioned via not only the 'normal' concentric emphasised exercises, but also via eccentric muscle actions with exercises such as 'Nordic hamstring lowers' (see Figure 3.8a. b), which have been shown to decrease the risk of hamstring injury (Askling et al. 2003; Mjolsnes et al. 2004; Clark et al. 2005; Arnason et al. 2007). It is also essential to progress on to higher velocity eccentric exercise such as plyometrics (deceleration training), which has also been shown to have a beneficial effect in preventing and rehabilitating hamstring strain injuries (Kaminski et al. 1998; Brockett et al. 2001, 2004; Proske &

Table 3.5 Characteristics of elite football players

Height (cm)	Weight (kg)	Body fat%	VO$_2$ max	1RM squat	Vertical jump	Sprint (s)	Level	Age	Subjects	Reference
180.9±4.9	76.9±7.0		63.7±5.0 ml.kg.min^{-1}	164.6±21.8 kg 135.0±16.2 kg	56.7±6.6 cm 53.1±4.0 cm		Norwegian elite	23.8±3.8	n = 14 n = 15	Wisloff et al. 1998
177.6±6.4	77.6±8.6	12.8±5.2	60.9±3.4 ml.kg.min^{-1}		63.0±8.0 cm	10yrd = 1.7±0.1 40yrd = 5.0±0.2	NCAA	19.9±1.3	n = 25	Silvestre et al. 2006a
177.6±6.3	77.5±9.2	13.9±5.8	59.4±4.2 ml.kg.min^{-1}		61.6±7.1 cm	10yrd = 1.7±0.1 40yrd = 4.9±0.2	NCAA	19.9±1.3	n = 27	Silvestre et al. 2006b
181.7±0.5 179.6±0.5	77.0±0.7 75.7±0.7	9.9±0.5 11.2±0.5	63.2±0.4 61.9±0.7		39.4±0.4 cm 38.8±0.7 cm		Elite-Division 1 Division 1 (Iceland)	24.2±0.2 23.6±0.4	n = 306	Arnason et al. 2004
169.1±5.7 168.6±4.8 168.8±4.6	68.17±6.9 67.74±4.8 69.87±4.6				23.6±3.5 cm 21.4±4.5 cm 20.3±4.3 cm	10m = 1.95±0.34 10m = 2.14±0.41 10m = 2.21±0.45	Youth Elite Youth Sub-elite Youth Recreational	16.3±1.26 16.4±1.32 16.2±1.29	n = 18 n = 18 n = 18	Gissis et al. 2006
171.0±0.05	63.1±1.1	11.3±2.1	59.0±1.7 ml.kg.min^{-1}		55.80±5.82cm	5m = 1.04±0.03 15m = 2.44±0.07 30m = 4.31±0.14	Youth Elite	16.4	n = 16	Reilly et al. 2000b
175.0±0.06	66.4±2.5	13.9±3.8	55.5±3.8 ml.kg.min^{-1}		50.21±7.58cm	5m = 1.07±0.06 15m = 2.56±0.12 30m = 4.46±0.21	Youth Sub-elite	16.4	n = 15	

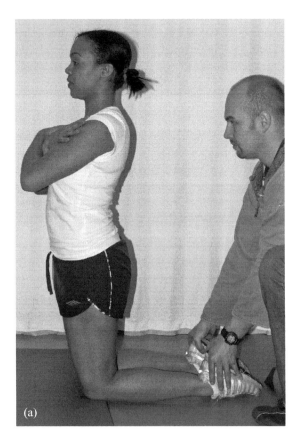

Figure 3.8 (a) Nordic hamstring lowers (curls) – start position. (b) Nordic hamstring lowers (curls) – descent.

Morgan 2001; Proske et al. 2004; Clark et al. 2005; Comfort et al. 2009).

It is also worth noting that the incidence of injury during training has been reported to be higher pre season (4.2±2.9 per 1000 hours) compared to during the competitive season (2.1±2.2 per 1000 hours) (Hagglund et al. 2005; Walden et al. 2005).

In summary, soccer is a sport dominated by short (10–15m) intermittent (every 90s on average) sprints, and high-speed changes of direction, over an extended period of 90 minutes. Normative fitness data appear in Table 3.2, which also provides a guide for suitable methods of assessment of the different components of fitness. Training should be targetted to those attributes that enhance sprint and agility performance (including speed, strength, power and correct deceleration mechanics), cardiovascular endurance to enhance repeated sprint ability, and injury prevention, with special emphasis on the areas highlighted as deficient following an appropriate battery of fitness tests.

Needs analysis for rugby league

During a game of rugby league (80 mins), distances covered can be as great as 10,000 metres (Meir et al. 2001), with the majority of the activity being low intensity activities such as jogging and walking, interspersed with high intensity short duration sprints that include periods of high force generation and force acceptance during cutting and turning. In contrast to soccer, rugby league also includes multiple bouts of high force generation and acceptance due to the number of tackles – up to 40 tackles per game (Brewer and Davis 1995). Average levels of aerobic capacity have been reported as high as 56ml/kg/min (Brewer and Davis 1995), with no significant differences between forwards and backs, excluding body mass, however, more specific data is presented in Table 3.3). In a review of literature, Gabbett (2005a) found that average heart rates during competition were 78%, 84% and 93% of maximum heart rate for amateur, semi-professional and professional athletes respectively.

It is worth noting that athletes with a higher lean body mass and more playing experience appear to be preferentially included during team selection (Gabbett 2002b).

To fuel training and performance, athletes consume on average 4230kcal.day^{-1} (range

Table 3.6 Characteristics of rugby league players

Position	Weight (kg)	Body fat%	VO$_2$max (ml/kg/min)	1RM squat (kg)	Vertical jump (cm)	Sprint (s)	Level	Age (years)	Subjects	Reference
Forwards	49.0–82.1		29.6–45.7		21.7–41.6	10m = 2.15–2.67 20m = 3.51–4.34	Sub-elite Junior	12.2–15.7	n=88	Gabbett 2002a
Backs	41.5–69.4		33.8–52.6		28.2–44.7	10m = 2.10–2.54 20m = 3.46–4.16				
Forwards	78.6–105.5		40.3–52.4		33.1–55.3	10m = 1.97–2.28 20m = 3.28–3.68	Sub-elite Senior (Australia)	17.2–27.2	n=71	
Backs	78.8–92.8		42.3–52.6		35.1–54.3	10m = 1.93–2.19 20m = 3.21–3.65				
Forwards	97±10.0		45.8±4.4		40.7±7.9	10m = 2.19±0.16 20m = 3.56±0.17 30m = 4.94±0.10	Semi-professional (Australia)	24±4	n=66	Gabbett 2002b
Backs	88±7.0		48.0±3.6		46.7±10.4	10m = 2.09±0.11 20m = 3.38±0.17 30m = 4.68±0.09				
Forwards	86.2–95.4	18.2–21.6	35.4–40.8		33.7–40.5	10m = 2.57–2.67 40m = 6.69–6.89	Amateur (Australia)	28.6		Gabbett, 2000
Backs	74.7–84.7 80.7–91.7	15.0–20.0	37.8–32.2 47.5–51.7		37.8–42.2 53.6–57.4	10m = 2.43–2.63 10m = 1.77–1.83 20m = 3.06–3.14 40m = 5.46–5.61	Amateur (Australia)	24.2 (18+)	n=52	Gabbett, 2005b
Forwards	98.4±7.7	13.5±2.9					Elite Professional (Australia)	25.4±3.5	n=44	Lundy et al. 2006
Backs	85.5±6.7	11.1±2.7						25.0±3.7	n=30	
	92.2±11.4		46.9±5.8		50.7±9.8	10m = 2.06±0.18 20m = 3.36±0.23 40m = 5.83±0.31	Elite Professional (Australia)	23.7±4.3	n=26	Gabbett et al. 2007

2671–6917kcal.day^{-1}), consisting of 6g.kg.day^{-1} of carbohydrate, 2g.kg.day^{-1} of protein (Lundy, et al., 2006), with very little variation in nutrient intake preceding competition. Due to the volume of training usually performed by these athletes, they may benefit from slightly higher (7–8g.kg.day^{-1}) carbohydrate intakes (Jeukendrup and Gleeson, 2004), along with slightly lower protein intakes ($\leq$1.6g.kg.day^{-1}) (Lemon et al., 1994; Campbell et al., 2007).

Rugby injuries

The most common injuries in rugby league are musculotendinous injuries to the lower limbs (Hoskins et al. 2006). Knee injuries appear to range from 8.0 to 27.7%, hamstring and groin injuries from 8.0 to 19.7%, and ankle injuries from 6.0 to 12.4% (Seward et al. 1993; Gibbs 1994; Orchard 2004; Gabbett and Domrow 2005). Between 38.8 and 91% of rugby injuries occur due to collisions and tackles and therefore may not be preventable (Hoskins et al. 2006, Gabbett 2005b).

Gabbett (2008) found that the most common site of injuries in junior (under 19 years) rugby league was the shoulder (15.6 per 1000 playing hours), followed by the knee (10.1 per 1000 playing hours). The most common injury type was a sprain (24.7 per 1000 playing hours). Injuries were most commonly sustained while being tackled (19.2 per 1000 playing hours) and during tackling (10.1 per 1000 playing hours).

In summary, rugby league requires athletes to be capable of covering distances as great as 10,000 metres during the 80 minutes of game play; consisting of low intensity activities such as jogging and walking, interspersed with high intensity short duration sprints, cutting and turning and high levels of impact during the average of 40 tackles per player per game. Normative data of a range of method of assessment is presented in Table 3.6, which may also provide a good indication of suitable tests for assessing performance in this sport. Training should be targetted at those attributes that enhance sprint and agility performance (including strength, power and correct deceleration mechanics), cardiovascular endurance to enhance repeated sprint ability and injury prevention. Special emphasis on the areas highlighted as deficient following an appropriate battery of fitness tests. In contrast to soccer, a strong upper body is also required to cope with the forces exerted during the high impact tackles (although this should not be the main focus of training).

References

Aagaard, P., Simonsen, E.B., Magnusson, S.P., Larsson, B. and Dyhre-Poulsen, P. (1998) A new concept for isokinetic hamstring: quadriceps muscle strength ratio. *American Journal of Sports Medicine*, 26 (2): 231–237.

Ahmed, C.S., Clark, A.M., Heilmann, N., Schoeb, J.S., Gardner, T.R. and Levine, W.N. (2006) Effect of gender and maturity on quadriceps to hamstring ratio and anterior cruciate ligament laxity. *American Journal of Sports Medicine*, 34 (3), 370–374.

Arnason, A., Sigurdsson, S.B., Gudmundsson, A., Holme, I., Engebretsen, L. and Bahr, R. (2004) Physical fitness, injuries and team performance in soccer. *Medicine and Science in Sports and Exercise*, 36 (2), 278–285.

Arnason, A., Anderson, T.E., Holme, I., Engebretsen, L. and Bahr, R. (2007) Prevention of hamstring strains in elite soccer: an intervention study. *Scandinavian Journal of Medicine and Science in Sports*, 18 (1), 40–48.

Askling, C. M., Lund, H.,Saartok, T. and Thorstensson, A. (2002) Self-reported hamstring injuries in student dancers. *Scandinavian Journal of Medicine and Science in Sports*, 12 (4), 230–235.

Askling, C., Karlsson, J. and Thorstensson, A. (2003) Hamstring injury occurrence in elite soccer players after preseason strength training with eccentric overload. *Scandinavian Journal of Medicine and Science in Sports*, 13, 244–250.

Augustsson, J., Esko, A., Thomee, R. and Svantesson, U. (1998) Weight training of the thigh muscles using closed vs. open kinetic chain exercises: a comparison of performance enhancement. *Journal of Orthopaedic Sports Physical Therapy*, 27 (1), 3–8.

Baechle, T.R. and Earle, R.W. (2008) *Essentials of Strength and Conditioning Training*, 3rd edn. Champaign, IL: Human Kinetics.

Baker, D. (2001) A series of studies on the training of high-intensity muscle power in rugby league football players. *Journal of Strength and Conditioning Research*, 15 (2), 198–209.

Baker, J.S. and Davies, B. (2002) High intensity exercise assessment: Relationships between laboratory and field measures of performance. *Journal of Science and Medicine in Sports*, 5 (4), 341–347.

Bangsbo, J., Mohr, M. and Krustrup, P. (2006) Physical and metabolic demands of training and match-play in the elite football player. *Journal of Sports Science, 24* (7), 665–674.

Bangsbo, J., Iaia, F.M. and Krustrup, P. (2008) The Yo-yo intermittent recovery test: A useful tool for evaluation of physical performance in intermittent sport. *Sports Medicine, 38* (1), 37–51.

Barfield, J., Sells, P.D., Rowe, D.A. and Hannigan-Downs, K. (2002) Practice effect of the wingate anaerobic test. *Journal of Strength and Conditioning Research, 16* (3), 472–473.

Bolgla, L.A. and Keskula, D.R. (1997) Reliability of lower extremity functional performance tests. *Journal of Orthopaedic and Sports Physical Therapy, 26* (3), 138–142.

Bompa, T.O. (1999) *Periodization: Theory and Methodology of Training*, 4th edn. Champaign, IL: Human Kinetics.

Brewer, J. and Davis, J. (1995) Applied physiology of rugby league. *Sports Medicine, 20* (3), 129–135.

Brockett, C.L., Morgan, D.L. and Proske, U. (2001) Human hamstring muscles adapt to eccentric exercise by changing optimum length. *Medicine and Science in Sports and Exercise, 33* (5) 783–790.

Brockett, C.L., Morgan, D.L. and Proske, U. (2004) Predicting hamstring strain injury in elite athletes. *Medicine and Science in Sports and Exercise, 36* (3) 379–387.

Cameron, M., Adams, R. and Maher, C. (2003) Motor control and strength as predictors of hamstring injury in elite players of Australian football. *Physical Therapy in Sport, 4*, 159–166.

Campbell, B., Kreider, R.B., Ziegenfuss, T., Bounty, P.L., Roberts, M., Burke, D., Landis, J., Lopez, H. and Antonio, J. (2007) International Society of Sports Nutrition position stand: protein and exercise. *Journal of the International Society of Sports Nutrition, 4*, 8.

Castagna, C., Impellizzeri, F.M., Bellardinelli, R., Abt, G., Coutts, A., Chamari, K. and D'Ottavio, S. (2006) Cardiorespiratory responses to Yo-yo intermittent endurance test in non-elite youth soccer players. *Journal of Strength and Conditioning Research, 20* (2), 326–330.

Chu, D., & Korchemny, R., (1993). Sprinting stride actions: analysis and evaluation. *National Strength and Conditioning Association Journal, 15* (1), 48–53.

Clark, R., Bryant, A., Culgan, J. and Hartley, B. (2005) The effects of eccentric hamstring strength training on dynamic jumping performance and isokinetic strength parameters: a pilot study on the implications for the prevention of hamstring injuries. *Physical Therapy in Sport, 6*, 67–73.

Comfort, P., Green, C.M. and Matthews, M.J. (2009) Training considerations for athletes post hamstring strain injury. *Strength and Conditioning Journal, 31* (1), 68–74.

Cronin, J.B. and Hansen, K.T. (2005) Strength and power predictors of sports speed. *Journal of Strength and Conditioning Research, 19* (2), 349–357.

Cronin, J.B. and Templeton, R.L. (2008) Timing light height affects sprint times. *Journal of Strength and Conditioning Research, 22* (1), 318–320.

Cronin, J.B., Hing, R.D. and McNair, P.J. (2004) Reliability and validity of a linear position transducer for measuring jump performance. *Journal of Strength and Conditioning Research, 18* (3), 590–593.

Cronin, J.B., Green, J.P., Levin, G.T., Brughelli, M.E. and Frost, D.M. (2007) Effect of starting stance on initial sprint performance. *Journal of Strength and Conditioning Research, 21* (3), 990–992.

Croisier, J.L., Forthomme, B., Namurois, M.H., Vanderthommen, M. and Crielaard, J.M. (2002) Hamstring muscle strain recurrence and strength performance disorders. *American Journal of Sports Medicine, 30* (2), 199–203.

Dugan, S.A. (2005) Sports-related knee injuries in female athletes. *American Journal of Physical Medicine and Rehabilitation, 84* (2), 122–130.

Duthie, G.M., Pyne, D.B., Ross, A.A. Livingstone, S.G., and Hooper, S.L. (2006) The reliability of ten-meter sprint time using different start techniques. *Journal of Strength and Conditioning Research, 20* (2), 246–251.

Ebben, W.P. and Leigh, D.H. (2000) The role of the back squat as a hamstring training stimulus. *Journal and Strength Conditioning, 22* (5), 15–17.

Engstrom, B.K. and Renstrom, P.A. (1998) How can injuries be prevented in the World Cup soccer athlete? *Clinical Sports Medicine, 17* (4), 755–768.

Enoka, R.M. (2002) *Neuromechanics of Human Movement*, 3rd Ed. Champaign,IL: Human Kinetics.

Escamilla, R.F., Fleisig, G.S., Zheng, N., Lander, J.E., Barrentine, S.W., Andrews, J.R. Bergemann, B.W., Moorman, C.T. (2001) Effects of technique variations on knee biomechanics during the squat and leg press. *Medicine and Science in Sports Exercise, 33* (9), 1552–1566.

Fitzgerald, G.K., Lephart, S.M., Hwang, J. H. and Wainner, M.R.S. (2001) Hop tests as predictors of dynamic knee stability. *Journal of Orthopaedic Sports Physical Therapy, 31* (10), 588–597.

Foreman, T.K., Addy, T., Baker, S., Burns, J., Hill, N. and Madden, T. (2006) Prospective studies into the causation of hamstring injuries in sport: A systematic review. *Physical Therapy in Sport, 7*, 101–109.

Gabbett, T.J. (2000) Physiological and anthropometric characteristics of amateur rugby league players. *British Journal of Sports Medicine*, 34, 303–307.

Gabbett, T.J. (2002a) Physiological characteristics of junior and senior rugby league players. *British Journal of Sports Medicine*, 36, 334–339.

Gabbett, T.J. (2002b) Influence of physiological characteristics on selection in a semi-professional first grade rugby league team: a case study. *Journal of Sports Science*, 20, 399–405.

Gabbett, T.J. (2005a) Science of rugby league football: A review. *Journal of Sports Science*, 23 (9), 961–976.

Gabbett, T.J. (2005b) Changes in physiological and anthropometric characteristics of rugby league players during a competetive season. *Journal of Strength and Conditioning Research*, 19 (2), 400–408. 2005b..

Gabbett, T.J. (2008) Incidence of injury in junior rugby league players over four competetive seasons. *Journal of Science and Medicine in Sport*, 11, 323–328.

Gabbett, T.J., Domrow, N. (2005) Risk factors for injury in subelite rugby league players. *American Journal of Sports Medicine*, 33 (3),428–434.

Gabbett, T.J., Kelly, J. and Pezet, T. (2007) Relationship between physical fitness and playing ability in rugby league players. *Journal of Strength and Conditioning Research*, 21 (4), 1126–1133.

Gibbs, N. (1994) Common rugby league injuries. Recommendations for treatment and preventative measures. *Sports Medicine*, 18 (6), 438–450.

Gissis, I., Papadopoulos, C., Kalapotharakos, V. I., Sotiropoulos, A., Komsis, G. and Manolopoulos, E. (2006). Strength and speed characteristics of elite, subelite, and recreational young soccer players. *Research in Sports Medicine*, 14, 205–214.

Gribble, P.A., Hertel, J., Denegar, C.R. and Buckley, W.E. (2004) The effects of fatigue and chronic ankle instability on dynamic postural control. *Journal of Athletic Training*, 39 (4), 321–329.

Hagglund, M., Walden, M. and Ekstrand, J. (2005). Injury incidence and distribution in elite football – a prospective study of the Danish and the Swedish top divisions. *Scandinavian Journal of Sports Medicine*, 15, 21–28.

Hawkins, R.D., Hulse, M.A., Wilkinson, C., Hodson, A. and Gibson, M. (2001) The association football medical research programme: an audit of injuries in professional football. *British Journal of Sports Medicine*, 35 (1), 43–47.

Heidt, R.S., Sweeterman, L.M., Carlonas, R.L., Traub, J.A. and Tekulve, F.X. (2000) Avoidance of soccer injuries with pre-season conditioning. *American Journal of Sports Medicine*, 28 (5), 659–662.

Helgerud, J., Engen, L.C., Wisloff, U. and Hoff, J. (2001). Aerobic endurance training improves soccer performance. *Medicine and Science in Sport and Exercise*, 33 (11), 1925–1931.

Herrington, L., Hatcher, J., Hatcher, A. and McNicholas, M. (2009) A comparison of star excursion balance test reach distances between ACL deficient patients and asymptomatic controls. *The Knee*, 16, 149–152.

Hewett, T.E., Stroupe, A.L., Nance, T.A. and Noyes, F.R. (1996) Plyometric training in female athletes. Decreased impact forces and increased hamsrting torques. *American Journal of Sports Medicine*, 24 (6), 765–773.

Hewett, T.E., Myer, G.D., Ford, K.R. and Palumbo, J.P. (2005) Neuromuscular training improves performance and lower-extremity biomechanics in female athletes. *Journal of Strength and Conditioning Research*, 19 (1), 51–60.

Hoff, J. (2005). Training and testing physical capacities for elite soccer players. *Journal of Sports Science*, 23 (6), 573–582.

Holcomb, W. R., Rubley, M. D., Lee, H. J. and Guadagnoli, M. A. (2007) Effect of hamstring emphasised resistance training on hamstring quadriceps strength ratios. *Journal of Strength and Conditioning Research*, 21 (1), 41–47.

Hoskins, W. and Pollard, H. (2005) The management of hamstring injury – Part 1: Issues in diagnosis. *Journal of Manipulative and Physiological Therapeutics*, 10 (2), 96–107.

Hoskins, W., Pollard, H., Hough, K.,and Tully, C. (2006) Injury in rugby league. *Journal of Science and Medicine in Sport*, 9, 46–56.

Impellizzeri, F.M., Bizzini, N., Rampinini, E., Cereda, F. and Maffiuletti, N.A. (2008) Reliability of isokinetic imbalance ratios measured using the Cybex NORM dynamometer. *Clinical Physiology and Functional Imaging*, 28, 113–119.

Inbar, O., BarOr, O. and Skinner, J.S. (1996) *The Wingate Anaerobic Test*. Champaign, Il: Human Kinetics.

Jackson, A. S., Pollock, M. L. and Ward, A. (1980) Generalized equations for predicting body density of women. *Medicine and Science in Sports and Exercise*, 12 (3), 175–81.

Jeukendrup, A. and Gleeson, M. (2004) Sport Nutrition. An Introduction to Energy Production and Performance. Champaign, Il: Human Kinetics.

Jonhagen, S., Ackermann, P., and Saartok, T. (2009) Forward lunge: a training study of eccentric exercises of the lower limbs. *Journal of Strength and Conditioning Research*, 23 (3), 972–978.

Kaminski, T.W., Webberson, C.V. and Murphy, R.M. (1998) Concentric versus enhanced eccentric hamstring strength training: Clinical implications. *Journal of Athletic Training*, 33 (3), 216–221.

Keir, P.J., Jamnik, V.K. and Gledhill, N. (2003) Technical-Methodological Report: A nomogram for peak leg power output in the vertical jump. *Journal of Strength and Conditioning Research*, 17 (4), 701–703.

Kingma, I., Aalbersberg, S. and van Dieen, J.H. (2004) Are hamstrings activated to counteract shear forces during isometric knee extension efforts in healthy subjects? *Journal of Electromyography and Kinesiology*, 14 (3), 307–15.

Kinzey, S.J. and Armstrong, C.W. (1998) The reliability of the star-excursion test in assessing dynamic balance. *Journal of Orthopaedic and Sports Physical Therapy*, 27 (5), 356–360.

Knapik, J.J., Bauman, C.L., Jones, B.H., Harris, J.M. and Vaughan, L. (1991) Preseason strength and flexibility imbalances associated with athletic injuries in female collegiate athletes. *American Journal of Sports Medicine*, 19 (1), 76–81.

Kraemer, W. and Fry, A. (1995) Strength testing development and evaluation of methodology. In Maud, P.J. and Foster, C. (Eds) *Physiological Assessment of Human Fitness*. Champaign, IL: Human Kinetics.

Kujala, U.M., Orava, S. and Jarvinen, M. (1997) Hamstring injuries: current trends in treatment and prevention. *Sports Medicine*, 23 (6), 397–404.

Kunz, H. and Kaufmann, D.A. (1981) Biomechanical analysis of sprinting: decathletes versus champions. *British Journal of Sports Medicine*, 15, 177–181.

Leard, J.S., Cirillo, M.A., Katsnelson, E., Kimiatek, D.A., Miller, T.W., Trebincevic, K. and Garbolosa, J.C. (2007) Validity of two alternative systems for measuring vertical jump height. *Journal of Strength and Conditioning Research*, 21 (4), 1296–1299.

Li, R.C.T., Wu, Y., Mafulli, N., Chan, K.M. and Chan, J.L.C. (1996) Eccentric and concentric isokinetic knee flexion and extension: a reliability study using the Cybex 6000 dynamometer. *British Journal of Sports Medicine*, 30, 156–160.

Li, G., Rudy, T.W., Sakane, M., Kanamori, A., Ma, C.B. and Woo, S.L.Y. (1999) The importance of quadriceps and hamstring muscle loading on knee kinematics and in-situ forces in the ACL. *Journal of Biomechanics*, 32, 395–400.

Lund, H., Sondergaard, K., Christensen, R., Bulow, P., Henriksen, M., Bartels, E.M., Danneskiold-Samsoe, B. and Bliddal, H. (2005) Learning effect of isokinetic measurements in healthy subjects and reliability and comparability of Biodex and Lido dynamometers. *Clinical Physiology and Functional Imaging*, 25, 75–82.

Lundy, B., O'Connor, H., Pelly, F., and Caterson, I. (2006). Antrhopometric characsitics and competition dietary intakes of professional rugby league players. *International Journal of Sports Nutrition and Exercise Metabolism*, 16, 199–213.

Maffiuletti, N.A., Bizzini, M., Desbrosses, K., Babault, N. and Munzinger, U. (2007) Reliability of knee extension and flexion measurements using the Con-Trex isokinetic dynamometer. *Clinical Physiology and Functional Imaging*, 27, 346–353.

Mandelbaum, B.R., Silvers, H.J., Watanabe, D.S., Knarr, J.F., Thomas, S.D., Griffin, L.Y., Kirkandall, D.T. and Garrett, W. (2005) Effectiveness of a neuromuscular and proprioceptive training program in preventing anterior cruciate ligament injuries in female athletes. *American Journal of Sports Medicine*, 33 (7), 1003–1010.

Mann, R. and Herman, J. (1985) Kinematic analysis of olympic sprint performance: Men's 200 meters. *Journal of Applied Biomechanics*, 1 (2), 175–180.

Markovic, G., Dizdar, D., Jukic, I. and Cardinale, M. (2004) Reliability and factorial validity of squat and countermovement jump tests. *Journal of Strength and Conditioning Research*, 18, (3), 551–555.

Maulder, P. and Cronin, J. (2005) Horizontal and vertical jump assessment: reliability and symmetry, discriminative and predictive ability. *Physical Therapy in Sport*, 6, 74–82.

McGill, S. (2006) *Ultimate Back Fitness and Performance*, 3rd edn. Waterloo, ON: Backfitpro.

McKenna, M. and Riches, P.E. (2007) A comparison of sprinting kinematics on two types of treadmill and overground. *Scandinavian Journal of Medicine and Science in Sports*, 17 (6), 649–55.

Meir, R, Newton, R, Curtis, E, Fardell, M. and Butler, B. (2001) Physical fitness qualities of professional rugby league football players: Determination of positional differences. *Journal of Strength Conditioning Research*, 15, 450–458.

Mero, A. (1988) Force-time characteristics and running velocity of male sprinters during the acceleration phase of sprinting. *Research Quarterly for Exercise and Sport*, 94 (2), 94–98.

Metaxas, T.I., Koutlianos, N.A., Kouidi, E.J. and Deligiannis, A.P. (2005) Comparative study of field and laboratory tests for the evaluation of aerobic capacity in soccer players. *Journal of Strength and Conditioning Research*, 19 (1), 79–84.

Mjolsnes, R., Arnason, A., Osthagen, T., Raastad, T. and Bahr, R. (2004) A 10-week randomized trial comparing eccentric vs. concentric hamstring strength training in well trained soccer players. *Scandinavian Journal of Medicine and Science in Sports, 14* (5), 311–317.

Moir, G., Sanders, R., Button, C. and Glaister, M. (2005) The influence of familiarization on the reliability of force variables measured during unloaded and loaded vertical jumps. *Journal of Strength and Conditioning Research, 19* (1), 140–145.

Moravec, P., Ružička, J., Sušanka, P., Dostal, E., Kodejs, M., and Nosek, M. (1987). The 1987 International Athletic Foundation/IAAF Scientific Project Report: time analysis of the 100 meters events at the II World Championships in Athletics. *New Studies in Athletics, 3* (3), 61–96.

Munn, J., Sullivan, S. J. and Schneiders, A. G. (2009) Evidence of sensorimotor deficits in functional ankle instability: a systematic review with meta-analysis. *Journal of Science and Medicine in Sports*. (E-pub ahead of print).

Myer, G.D., Hewett, T.E. and Ford, K.R. (2005) Reducing knee and anterior cruciate ligament injuries among female athletes: A systematic review of neuromuscular training interventions. *Journal of Knee Surgery, 18* (1), 82–88.

Orchard, J. (2004) Missed time through injury and injury management at an NRL club. *Sport Health, 22* (1),11–19.

Pauole, K., Madole, K., Garhammer, J., Lacourse, M. and Rozenek. (2000) Reliability and validity of the T-test as a measure of agility, leg power, and speed in college-aged men and women. *Journal of Strength and Conditioning Research, 14* (4), 443–450.

Plisky, P.J., Rauh, M.J., Kaminski, T.W. and Underwood, F.B. (2006) Star excursion balance test as a predictor of lower extremity injury in high school basketball players. *Journal of Orthopaedic Sports Physical Therapy, 36* (12), 911–919.

Pollock, M.L. and Jackson, A.S. (1984) Research progress in validation of clinical methods of assessing body composition. *Medicine and Science in Sports and Exercise, 16* (6), 606–615.

Proske, U. and Morgan, D.L. (2001) Muscle damage from eccentric exercise: mechanism, mechanical signs, adaptation and clinical applications. *Journal of Physiology, 537* (2), 333–45.

Proske, U., Morgan, D.L., Brockett, D.L. and Percival, P. (2004) Identifying athletes at risk of hamstring strains and how to protect them. *Clinical and Experimental Pharmacology and Physiology, 31* (8), 546–550.

Reid, A., Birmingham, T.B., Stratford, B.W., Alcock, G.K. and Giffin, J.R. (2007) Hop testing provides a reliable and valid outcome measure during rehabilitation after anterior cruciate ligament reconstruction. *Physical Therapy, 87* (3), 337–349.

Reilly, T., Bangsbo, J. and Franks, A. (2000a) Anthropometric and Physiological Predispositions for Elite Soccer. *Journal of Sports Science, 18* (9), 669–683.

Reilly, T., Williams, A. M., Nevill, A., and Franks, A. (2000b). A multidisciplinary approach to talent identification in soccer. *Journal of Sports Science, 18,* 695–702.

Ross, M.D., Langford, B., and Whelan, P.J. (2002) Test-retest reliability of 4 single-leg horizontal hop tests. *Journal of Strength and Conditioning Research, 16* (4), 617–22.

Seward, H., Orchard, J., Hazard, H. *et al.* (1993) Football injuries in Australia at the elite level. *Medical Journal of Australia, 159* (5), 298–301.

Shephard, R. and Astrand, P. (2000) *Endurance in Sport*. Oxford: Blackwell Scientific.

Silvestre, R., Kraemer, W.J., West, C., Judelson, D.A., Spiering, B.A., Vingren, J.L., Hatfield, D. L., Anderson, J.M. and Maresh, C.M. (2006a) Body composition and physical performance during a National Collegiate Athletic Association Division I men's soccer season. *Journal of Strength and Conditioning Research, 20* (4), 962–970.

Silvestre, R., West, C., Maresh, C.M. and Kraemer, W.J. (2006b) Body composition and physical performance in men's soccer: A study of a National Collegiate Athletic Association division I team. *Journal of Strength and Conditioning Research, 20* (1), 177–183.

Sole, G., Hamren, J., Milosavljevic, S., Nicholson, H. and Sullivan, S.J. (2007) Test–retest reliability of isokinetic knee extension and flexion. *Archives of Physical Medicine and Rehabilitation, 88,* 626–631.

Stanton, P.E. (1989) Hamstring injuries in sprinting – the role of eccentric exercise. *Journal of Orthopaedic and Sports Physical Therapy, 10,* 343–9.

Stickley, C.D., Hetzler, R.K. and Kimura, I.F. (2008) Prediction of anaerobic power values from an abbreviated WAnT protocols. *Journal of Strength and Conditioning Research, 22* (3), 958–965.

Walden, M., Hagglund, M. and Ekstrand, J. (2005) Injuries in Swedish elite football – a prospective study on injury definitions, risk for injury and injury pattern during 2001. *Scandinavian Journal of Medicine and Science in Sports, 15,* 118–125

Weyand, P. G., Sternlight, D. B., Bellizzi, M. J. and Wright, S. Faster top running speeds are achieved with greater ground forces not more rapid leg movements. *Journal of Applied Physiology, 89,* 1991–1999.

Wilkerson, G.B., Colston, M.A., Short, N.I., Neal, K.L., Hoewischer, P.E. and Pixley, J.J. (2004)

Neuromuscular changes in female collegiate athletes resulting from a plyometric jump-training program. *Journal of Athletic Training*, *39* (1), 71–23.

Wisløff, U., Helgerud, J. and Hoff, J. (1998) Strength and endurance of elite soccer players. *Medicine and Science in Sports Exercise*, *30* (3), 462–467.

Wisløff, U., Castagna, C., Helgerud, J., Jones, R. and Hoff, J. (2004) Strong correlation of maximal squat strength with sprint performance and vertical jump height in elite soccer players. *British Journal of Sports Medicine*, *38*, 285–288.

Withers, R.T., Maricic, Z., Wasilewski, S. and Kelly, L. (1982) Match analysis of Australian professional soccer players. *Journal of Human Movement Studies*, *8*, 159–176.

Woods, C., Hawkins, R., Hulse, M. and Hodson, A. (2002) The Football Association medical research programme: an audit of injuries in professional football – analysis of preseason injuries. *British Journal of Sports Medicine*, *36* (6), 436–441.

Woods, C., Hawkins, R.D., Maltby, S., Thomas, A. and Hodson, A. (2004) The Football Association Medical Research Programme: an audit of injuries in professional football – analysis of hamstring injuries. *British Journal of Sports Medicine*, *38* (1)**,** 36–41.

Yamamoto, T. (1993) Relationship between hamstring strains and leg muscle strength. *Journal of Sports Medicine and Physical Fitness*, *33* (2), 194–199.

Part 3

Pathophysiology of musculoskeletal injuries

4

Pathophysiology of skeletal muscle injuries

Dr Lee Herrington and Paul Comfort
University of Salford, Greater Manchester

Skeletal muscle injuries are a common occurrence, especially during sporting activities. For example, hamstring strains accounted for 21% of injuries reported during two seasons of professional English football (Hawkins et al. 2001; Wood et al. 2002, 2004). The most common injuries reported in rugby league are also musculotendinous injuries to the lower limbs (Hoskins et al. 2006), with hamstring and groin injuries making up 8.0–19.7% of all reported injuries (Seward et al. 1993; Gibbs 1994; Orchard 2004; Gabbett and Domrow, 2005). An understanding of the pathophysiology of muscle injury and repair is essential in order to provide the optimum treatment interventions and to limit injury or re-injury to a muscle. Reducing the risk of re-injury is an essential component of the rehabilitative process, as musculoskeletal injuries such as hamstring strains are compounded by a high recurrence rate of 12–31% with approximately 1 in 13 athletes re-injuring within the first year of return to sport (Dadebo et al. 2004; Petersen and Holmich 2005; Orchard and Best 2002).

The aim of this chapter is to introduce the reader to the structure and function of skeletal muscle, progressing on to the pathophysiology of skeletal muscle injury and repair. Developing a detailed understanding of this process will help the sports rehabilitator to treat such injuries effectively and efficiently, thereby reducing time to return to sport and recurrence of injury. The chapter will culminate in a case study that will provide the reader with an applied example of this process.

Anatomy

Skeletal muscle consists of both contractile and non-contractile elements. The non-contractile elements provide a framework for individual muscle cells and encase the nerves and blood vessels (Jarvinen et al. 2005). The connective tissue is composed of three levels of sheath: the epimysium, a dense fibrous sheath surrounding the entire muscle belly; perimysium, which binds together muscle fibres to form fascicles; and the endomysium, which surrounds each muscle fibre (see Figure 4.1). The connective tissue sheaths are continuous and extend beyond the muscle to form myotendinous junctions. This means when a muscle contracts it pulls on the sheaths, which in turn transmit the force to the bone to be moved (Kossmann and Huxley 1961; Huxley 1975; Jarvinen et al. 2005).

An individual muscle fibre is a long cylindrical cell with multiple nuclei bound by its sarcolemma (Lutz and Lieber 1999). The sarcolemma invaginates

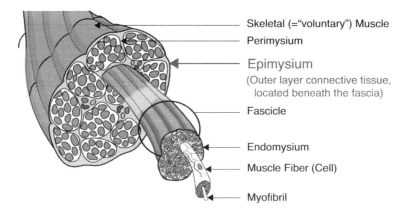

Figure 4.1 Anatomy of Skeletal Muscle.

around the muscle fibre forming T-tubules. Each muscle fibre contains a large number of myofibrils, which are the contractile elements of the muscle and consist of protein myofilaments. These are actin (thin filament, isotropic, I bands) and myosin (dark filament, anisotropic, A bands) and it is these bands that give muscle its striated appearance. These myofilaments are organised into repeating functional units called sarcomeres. In a resting muscle the actin filaments overlap the myosin to a certain extent. Each myofibril is surrounded by the sarcoplasmic reticulum, which is a store for release of calcium (Kossmann and Huxley 1961; Huxley 1975; Lutz and Lieber 1999) (see Figure 4.2).

Physiology

Skeletal muscles are designed to produce voluntary movement by applying forces to bones and joints via a muscle contraction. When a muscle contracts the sarcomeres shorten. This is due to the 'Sliding

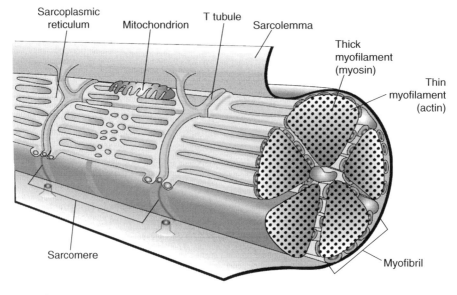

Figure 4.2 A myofibril.

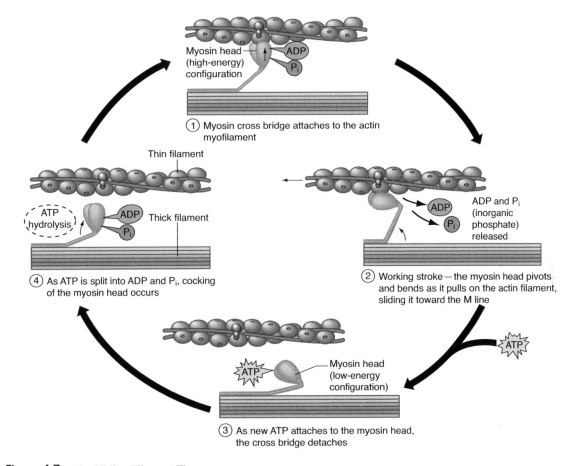

Figure 4.3 The Sliding Filament Theory.

Filament Theory' proposed by Huxley in 1954. The myosin filaments remain static and the actin filaments slide in and out producing force (the A band remains a constant length; the I band becomes shorter). The force is generated by crossbridges forming between the myosin (via the globular myosin heads) and actin binding sites (Kossmann and Huxley 1961; Huxley 1975) (see Figure 4.3).

For this to occur, the muscle fibre is stimulated by a nerve impulse creating an action potential across the sarcolemma. The action potential is propagated down the T-tubules triggering the release of calcium from the sarcoplasmic reticulum. The calcium binds to troponin, a regulatory protein on the surface of actin, exposing the actin binding site. A crossbridge is then formed between actin and myosin, resulting in a contraction cycle powered by ATP. Each muscle fibre is part of a motor unit so a number of muscle fibres are stimulated at the same time (Kossmann and Huxley 1961; Huxley 1975).

Pathophysiology

There are a number of types of muscle injury that can occur: laceration, contusion and strain (Garrett 1996; Huard et al. 2002; Jarvinen et al. 2005). A laceration occurs when the muscle is cut by an external object, this usually occurs during traumatic accidents such as road traffic or industrial accidents. A contusion occurs when there is a compressive force to the muscle and usually occurs in contact sports (Jarvinen et al. 2005), for example in football when two players collide, knee to thigh in a tackle. Strain injuries occur when muscle fibres cannot withstand excessive tensile forces placed on them and are therefore generally associated with eccentric muscle action

(Mair et al. 1996; Pull and Ranson, 2007). Strains most commonly occur in muscles working across two joints e.g. hamstrings, gastrocnemius (Jarvinen et al. 2005) during periods of rapid acceleration and deceleration, by placing the muscle in a lengthened state over two joints and contracting forcefully (Stanton 1989; Farber and Buckwalter 2002; Brockett et al. 2004; Hoskins and Pollard 2005; Askling et al. 2006, 2007a, 2007b).

When the muscle is strained the initial injury is usually associated with disruption of the distal myotendinous junction and fibres distal to this but still near the myotendinous junction. Injuries to the muscle belly only occur with the application of very high forces. The contractile elements are the first tissues to be disrupted; with the surrounding connective tissue not being damage until high forces are applied (Hasselman, et al. 1995). The contractile elements are relative stiff in comparison to the surrounding connective tissue and hence become disrupted at lower forces than the surrounding connective tissue.

Muscles heal by a repair process that can be divided into two phases: (1) the destruction/injury phase resulting in rupture and necrosis of the muscle fibres; and (2) repair and regeneration.

Destruction/injury phase

This phase results in damage to the vascular supply and as oxygen can no longer reach the cells, they die and release lysosomes (Schiaffino and Partridge 2008). Excessive force to a muscle fibre results in tearing of the sarcoplasm and the cells respond by forming a contraction band (condensation of cytoskeletal material) creating a protective barrier. The lysosomes are pivotal in this vital process keeping the necrosis to a local area and preventing it from spreading along the whole length of the cell (Jarvinen et al. 2005).

Within 15 minutes of injury the damaged tissue consists of disrupted extracellular tissue and dead cells, platelets and plasma, which themselves release powerful enzymes such as thrombin thereby setting off an inflammatory cascade (Schiaffino and Partridge 2008). A haematoma is formed to fill the gap between ruptured muscle fibres.

Clinically tissue inflammation presents as redness, heat, swelling and pain of the tissues. The redness, heat and swelling are due to an increased blood flow and so blood within vascular beds in the area, with the swelling developing as a result of the increased local tissue pressure due to inflammatory exudates (local capillaries become more permeable) leaking into the interstitial space (Schiaffino and Partridge 2008). The pain is due to the initial damage to local nerves and irritation of nerves in the area from the inflammatory chemicals release by the damaged tissue (Evans 1980).

Repair and regeneration

After the initial injury, the inflammatory chemicals released from the injured tissues attract lymphocytes and macrophages to the area (Jarvinen et al. 2005). Macrophage phagocytosis of the necrotic material then occurs removing the debris (Tidball and Wehling-Henricks 2007). The regeneration process starts within 3–6 days following injury, reaching a peak between day 7 and 14. The regenerative capacity of skeletal muscle is provided by satellite cells, specialised cells underneath the basal lamina of each muscle fibre. In response to injury they proliferate and differentiate into myoblasts and then become multinucleated myotubes (Jarvinen et al. 2005). These then fuse with parts of the muscle fibre that have survived the initial injury and attempt to breech the gap in the muscle (Lieber and Friden 2002). Recent findings have shown that bone marrow stem cells may play a role in contributing to regenerating muscle fibres and replenishing the satellite cells, however it is debatable as to how big a role this is (Gates and Huard 2005; Jarvinen et al. 2005).

At the same time as myotubes are attempting to cross the injury site, fibrin is being laid down creating an irregular meshwork of short fibres by the fibroblasts, which have differentiated from the macrophages drawn to the area earlier in the process, this acts as a connective tissue scaffold (Schiaffino and Partridge 2008). This irregular fibrin network obviously reduces the ability of the myotubes to cross the rupture and make good an effective contractile unit. It is possible to minimise the irregular nature of this connective tissue mesh and also the amount present by applying suitable tensile loads (i.e. not high enough to cause re-rupture) (Lehto et al. 1985).

This regeneration process can occur rapidly, within 10 days post injury the injured muscle can regain much of its contractile ability with progressively applied loading, set at levels below the pain threshold (Nikolaou et al. 1987).

Treatment of muscle injuries

It is frequently cited within the sports medicine literature that the initial treatment of musculoskeletal injuries should be rest, ice, compression and elevation (RICE). However, this acrimony is likely to be more valid in the less metabolically active tissues such as ligament and bone (Gates and Huard 2005). Muscle with its capacity for rapid regeneration due to the nature of its constituent tissues requires a modified approach (Peterson and Holmich 2005; Thorsson et al. 2007). This modification is based around the balance between absolute rest and the absolute level of loading to stimulate appropriately the rapidly developing tissues.

Rest and mobilisation

Early mobilisation, rather than immobilisation or complete rest, has been advocated (Thorsson et al. 2007). Studies have shown early mobilisation aids with regeneration of muscle fibres (stimulation of satellite cells and myotube formation improves capillary growth into the area and aids with more parallel orientation of collagen and muscle fibres (Nikolaou et al. 1987, Taylor et al. 1993; Goldspink 1999). The exact level of loading is difficult to judge, it must be sufficient to stimulate and challenge the developing tissues, but not so great that it causes tissue breakdown (Kannus et al. 2002). This includes early weight bearing to help promote scar tissue re-alignment (Croisier 2004; Jarvinen et al. 2005) and controlled running to reduce muscle inhibition (Herrington, 2000)

Immobilisation and poorly controlled early loading (low or high), have been shown to lead to the development of contracted scar tissue, which blocks the linking of myotubes across the injury, thereby stopping the formation of a functional contractile element, and the surrounding areas then become more susceptible to further injury (Beiner and Jokl 2001; Lehto et al. 1985).

Ice

Studies have shown that ice results in a significantly smaller haematoma and less inflammation in the initial stages of the injury (Jarvinen et al. 2005). Reducing tissue temperature results in vasoconstriction, thereby limiting the amount of bleeding in the area (Beiner and Jokl 20012). It also reduces the metabolic rate of the tissue and therefore reduces the demand for oxygen (Hubbard et al. 2004) decreasing the hypoxic damage.

Clinical studies have shown the optimal duration to apply ice is for 5–10 minutes in the initial stages and repeat every 60 minutes (Croisier 2004; Hubbard et al. 2004) within the first 24–48hours to reduce the inflammatory effects.

Compression and elevation

Compression is an area where research is lacking, it has been stated that it results in a reduction in the severity of bleeding and swelling following an injury (Kannus et al. 2002; Jarvinen et al. 2005), though the only evidence often present in support of this theory is from studies involving ligament injuries. A recent clinical study by Thorsson et al. (2007) utilising 40 athletes with calf injuries found compression resulted in no significant difference in reducing muscle haematoma, or speed of recovery of the injury using compression

Elevation is still one of the preferred and easiest methods of immediate management used in sports medicine for muscular injuries (Hergenroeder 1998). It simply relies on the use of gravity to promote venous return and lymphatic flow to drive swelling/oedema from the area (Hergenroeder 1998). Again research is lacking in this area, with most authors when citing research, using studies on ankle ligament injuries and even here the actual positive benefits when the limb is dependent again are very limited (Tsang et al. 2003). It would appear that the greatest benefit that comes from compression and elevation is that it ensures that the athlete rests during the acute inflammatory phase of the first 72 hours following injury.

Strengthening exercise

Isometric exercise can begin after 2–5 days and should be performed within the limits of pain (Jarvinen et al. 2005). Frequency, duration and intensity are limited by the patients' pain. Some therapists advocate three sets of 10 repetitions using 5–10 second holds to begin with at intensity within pain tolerance (Pull and Ranson 2007). These then are undertaken at multiple angles, beginning in mid range then progressing to inner range (shortened position) then outer range (lengthened position). Once these can be undertaken in a pain-free manner throughout the available range, then isotonic exercises can commence. Dynamic movement and isotonic contraction is then incorporated again starting in the strongest position (mid range, close to a 90° joint angle) progressing to and finishing in the functionally most relevant (outer range with eccentric and concentric contractions most often).

There is preferential atrophy of type 1 muscle fibres with disuse (Stockmar et al. 2006), and high loads and rates of force development are most likely to over-stress the healing tissue. Therefore, initially an endurance based programme should be used, (three sets of 15 repetitions at 40–60% of one repetition maximum) this would be progressed to strength (4–6 sets of 3–6 repetitions at 85–95% of one repetition maximum and then power training (3–5 sets of 3-5 repetitions at 75–85% of one repetition maximum) (Kraemer et al. 2002), or plyometrics depending on the specific requirements of the muscle. The exact nature of the progressions, directions and velocity of movement will depend on which muscle has been injured and the requirements of the sport, as will whether or not the progressions have a major stability/proprioception element.

Within the body certain muscles can be regarded as having a role that is predominately about the generation of force/power and they tend to work mostly in a single plane (usually the sagittal) these are often called mobiliser muscles. Other muscles can be regarded as having a stability role, they control motion of the body and often have the ability to contract in multiple planes; they are often called stabiliser muscles (see Table 4.1).

The nature of the role of the muscles affects the choice of rehabilitation exercise we will choose towards the end stage of rehabilitation. Power/mobiliser muscles do not have a stability role and so will not require exercise progressions that involve a proprioceptive challenge, whereas, the stabiliser muscles will. Similarly, those muscles that predominately work in an open kinetic chain manner will not require exercise progressions involving closed kinetic chain exercises.

Stretching

Passive stretching (at the end of available range) should be avoided for the first 72 hours as a minimal period, possibly the athlete should not stretch for the first 7–10 days following injury (Neidlinger-Wilke et al. 2002). The reasons for this are twofold. Firstly, the healing tissue is weak and intolerant of tensile loading and so is likely to be damaged by uncontrolled stretching. The second reason is there is no physiological need in the early stages to stretch, as the scar does not beginning to shrink until around the tenth day post injury when fibroblasts begin to be converted to myofibroblasts and contract and draw the wound ends together. Prior to the tenth day post injury it would be more appropriate to take the muscle through its full available pain-free range without any attempt to force the muscle beyond this point.

Once it is appropriate to begin stretching the muscle; that is, elongating the tissue beyond its available range, then careful passive stretching can be performed. Each stretch should be held at the end of available range within the limits of pain. Time, frequency, duration and intensity of stretch remain debateable in the literature. Some research suggests passive stretching should be held for a minimum of 15 seconds with 6–8 sets per day (Roberts and Wilson 1999), however, Bandy et al. (1997) reviewed stretches at 15 seconds, 30 seconds and 60 seconds. They determined that 30 seconds was the optimum duration, also stating that longer periods after 30 seconds was ineffective in promoting additional stretch.

As with strengthening exercises stretching exercises need to be progressed in order that the tissue adapts to the different types of load once the athlete is comfortable (pain free) with passive stretching (Bandy et al. 1997).

Electrotherapy

Pulsed shortwave diathermy is particularly helpful for enabling re-absorption of the muscular haematoma as it is particularly effective in more

Table 4.1 Examples of different muscle's roles

Mobiliser muscles	Stabiliser muscles
Hamstrings	Gluteus medius/minimus
Quadriceps	Adductor longus
Gastrocnemius	Tibialis posterior
Pectoralis major	Infraspinatus
Latissimus dorsi	Subscapularis

vascularised tissues such as skeletal muscle (Robertson and Baker 2001). It is thought to work at the cell membrane level, resulting in an 'up regulation' of cellular behaviour. This results in an improved rate of oedema dispersion, resolution of the inflammatory process and promotes a more rapid rate of fibrin fibre orientation and deposition of collagen (Robertson and Baker 2001).

Another modality that can be used to help speed up a recovery muscle injury is therapeutic ultrasound. This is the use of high intensity sounds waves, which research has been shown can help recovery of tissues at a cellular level by increasing ion transport across cells and increasing metabolism within the cell (Wilkin et al. 2003) and increasing fibroblastic and angiogenic activity (ter-Haar et al. 1978). However, evidence-based research remains limited in the effectiveness and reliability of therapeutic ultrasound treatment with the majority of supporting evidence coming from in-vitro cell culture studies (Robertson 2002) even though many clinicians still use it as their main electrotherapy treatment.

Muscle stimulation may prove a further useful electrotherapeutic adjunct for the treatment of muscle injuries. It has been shown to decrease oedema, muscle inhibition and the rate of strength loss with inactivity (Thornton et al. 1998). However, muscle stimulation has not been shown to be useful in the regaining of strength in the injured athlete (Snyder-Mackler et al. 1995).

Other factors

As discussed earlier, management of muscle injuries includes prevention of re-injury. For example, with hamstring injuries, many factors have been cited in literature as potential causes of re-injury. These include: previous injury (Verall et al. 2001; al. Crossier et al. 2002, 2003; Arnason et al. 2004; Foreman et al. 2006); lack of flexibility (Knapik et al. 1991; Hennessey and Watson 1993; Jonhagen et al. 1994; Bennell et al. 1998; Cross and Worrell 1999; Witvrouw et al. 2000; Funk et al. 2001; Brockett et al. 2004; Foreman et al. 2006); inadequate warm up (Worrell 1994; Worrell et al. 1994); fatigue (Worrell 1994; Worrell et al. 1994); muscle strength imbalance (Knapik et al. 1991; Yamamoto, 1993; Cameron et al. 2003; Crossier et al. 2003; Foreman et al. 2006); and poor coordination (Cameron et al. 2003; Brockett et al. 2004; Foreman et al. 2006).

Hamstring strains have also been associated with eccentric loading (Kujala et al. 1997; Cameron et al. 2003; Brockett et al. 2004), such as during rapid deceleration.

An understanding of the mechanics of the sport can be helpful in analysing the cause of the injury in the first place. This information can normally be revealed during the initial subjective assessment of the athlete, using simple questioning such as 'how did the injury happen, can you demonstrate (without re-injuring of course)'.

Example of muscle injury (hamstring)

Hamstring strains are one of the most common injuries in sport and can result in a lengthy period out of the game if not treated effectively (Clark et al. 2005).

The hamstring is a two-joint muscle and is most susceptible to injury in sports involving sprinting and kicking (Stanton 1989; Brockett et al. 2004; Hoskins and Pollard 2005; Askling et al. 2006). The majority of injuries occur in the biceps femoris and at the musculotendinous junction, although they can also occur to the semimembranosus during stretching (Askling et al. 2007b).

Mechanism of injury

As stated earlier, injury often occurs during sprinting and the point of failure has been shown to occur in the terminal swing phase just prior to foot strike. This is when the hamstrings have to work eccentrically to decelerate the tibia and control knee extension (Clark et al. 2005). Hamstring strains are commonly reported in sprinters when speed is maximal or close to maximal (Askling et al. 2006) and during powerful eccentric muscle actions (Brockett et al. 2004). The strain is most likely to occur during two phases of the running cycle; late forward swing and toe off (Stanton 1989) as during this phase the hamstrings decelerate hip flexion and knee extension (Hoskins and Pollard 2005) resulting in large eccentric loads. It has also been found that whilst sprinters sustain their injuries during high-speed running, dancers sustain injuries whilst performing slow stretching type exercises (Askling et al. 2006). In activities such as dancing most hamstring injuries occur during stretching (hip flexion with knee extension) (Askling et al. 2006, 2007a, 2007b), resulting in an eccentric load, with the proximal end of the semimembranosus as the site of injury (Askling et al. 2007b).

Predisposing factors

The predisposing factors of hamstring strain injury are multifactorial, including poor lumbar posture (Hennessey and Watson 1993), previous injury (Crossier et al. 2003; Arnason et al. 2004; Foreman et al. 2006), lack of flexibility (Knapik et al. 1991; Hennessey and Watson 1993; Bennell et al. 1998; Kaminski et al. 1998; Cross and Worrell 1999; Funk et al. 2001; Brockett et al. 2004; Foreman et al. 2006), inadequate warm up (Worrell 1994; Worrell and Smith 1994), fatigue (Worrell 1994; Worrell and Smith 1994), strength imbalance and inadequate quadriceps to hamstring ratio (Knapik et al. 1991; Cameron et al. 2003; Crossier et al. 2003; Foreman et al. 2006), and poor coordination (Cameron et al. 2003; Brockett et al. 2004; Foreman et al. 2006).

The length of the muscle when peak torque is produced has also been postulated as a predisposing factor, with the hypothesis that the greater the knee extension angle at which peak torque is produced the lower the risk of injury (Clark et al. 2005).

The literature suggests that there are two types of hamstring strain; one resulting from high-speed running, as in football (Woods et al. 2004) and athletics (Yamamoto 1993), and the other during stretching movements carried out at extreme range of motion (ROM) (Askling et al. 2002). Reported causes of hamstring strains include poor lumbar posture (Hennessey and Watson 1993), previous injury (Verall et al. 2001; Crossier et al. 2002; Arnason et al. 2004; Foreman et al. 2006), lack of flexibility (Hennessey and Watson 1993; Funk et al. 2001; Brockett et al. 2004; Jonhagen et al. 1994; Cross et al. 1999; Witvrouw et al. 2000; Bennell et al. 1998; Foreman et al. 2006; Knapik et al. 1991), inadequate warm up (Worrell, 1994; Worrell et al. 1994), fatigue (Worrell, 1994; Worrell et al. 1994), strength imbalance and inadequate Quadriceps to Hamstring ratio (Knapik et al. 1991; Crossier et al. 2003; Yamamoto, 1993; Cameron et al. 2003; Foreman et al. 2006), and poor coordination (Cameron et al. 2003; Brockett et al. 2004; Foreman et al. 2006). Hamstring strains have also been associated with eccentric loading (Kujala et al. 1997; Cameron et al. 2003; Brockett et al. 2004), such as during rapid deceleration.

Symptoms

Hamstring strains can often be diagnosed by the mechanism of injury resulting in sudden onset of pain. The patient presents with reduced hamstrings contraction against resistance and reduced stretch. Local bruising/haematoma is often present with pain on palpation. They may also find it difficult to walk and are unable to run or sprint (Croisier et al. 2002).

Treatment

Acute management is as stated previously in that relative rest, ice, compression and elevation are indicated to reduce inflammation and provide optimal environment for repair (Thorsson et al. 2007).

Gentle stretches and early weight bearing can also be commenced within the pain-free range to assist with correct fibre orientation (Bandy et al. 1994, 1997, 1998; Goldspink 1999; Sherry and Best 2004; Peterson and Holmich 2005). Localised soft tissue techniques, including cryotherapy in the acute phase (Herrington 2000; Hubbard et al. 2004) and electrotherapy once the acute phase has settled, aid the reduction of any muscle spasm and helps soft tissue repair (Herrington 2000; Robertson and Baker 2001). In particular, ultrasound may be used to assist in the breakdown of scar tissue and promotion of tissue healing, and is interferential in the reduction of swelling and inflammation (Wilkin et al. 2004).

Strengthening exercises are vital to try to prevent further injury. Initially starting with isometric and progressing onto isotonic as pain allows (Jarvinen et al. 2005). As injury is most likely to occur during eccentric activity, it is vital that eccentric exercise is incorporated into the programme in the later stages. The Nordic eccentric exercise has been shown to improve the torque angle at the knee, and rugby union teams that incorporated it into their training programmes found a reduced number of hamstring strains (Kujala et al. 1997; Clark et al. 2005; Brooks et al. 2006; Gabbe et al. 2006; Arnason et al. 2008). Eccentric training programmes should be closely monitored as they can lead to delayed onset muscle soreness; a low volume (3–5 sets of three repetitions) high frequency (3–4 times a week) may be most appropriate.

Exercises are then progressed to functional activities (e.g. running/sprint training), a return to sport once there is full strength and pain-free movement, and to completion of progressive running programmes and functional tests (Herrington 2000). As discussed earlier, sports specific rehabilitation is vital to return the athlete back to their functional sport. Not only can this return the athlete back to sport quicker, but help prevent further injury later on.

Functional fitness must also be maintained throughout the rehabilitation process without aggravating the injury. Examples could include cycling, walking, upper body weights and swimming, so long as these are pain free (Croisier et al. 2002). This would not be functional for most sports, but may be useful in reducing the detraining effect regularly associated with a reduction in training volume and intensity following injury.

Prevention

As stated in the example given there are a number of factors that can predispose to skeletal muscle injury and if these can be controlled/prevented then the risk of injury will be reduced.

Examples may include adequate warm up and stretches prior to sports participation; a conditioning programme consisting of eccentric, plyometric, sports specific and cardiovascular exercise and optimum treatment following previous injury (Kujala et al. 1997; Herrington, 2000; Clark et al. 2005; Brooks et al. 2006; Gabbe et al. 2006; Arnason et al. 2008). As an example, a hamstring in a football can be caused by a running/sprint deceleration activity which incorporates eccentric control of the hamstring muscle at higher speeds. This eccentric control of muscle must be re-implemented within the rehabilitation programme once the initial injury is managed in order to rehabilitate the fibre reorientation and to help reorganise the neuromuscular pathways which control the activity at such speed (Marqueste et al. 2004). Therefore, as a simple example sports specific rehabilitation may involve sprinting over short distances with sudden stopping and re-sprinting in another direction over many repetitions in order to mimic the control required for football. To begin with, this may require slower timed sessions to begin with, and as time progresses, this process is speeded up to match speed. This not only helps recruit the right muscle fibre type, but also helps restore proprioception and regain functional fitness.

Fatigue has also been shown to be a precursor to injury as a fatigued muscle is not able to absorb as much energy prior to failure due to a reduced ability to generate force (Mair et al. 1996). Therefore, rehabilitation should also be geared towards increasing the duration of rehabilitation sessions and trying to maintain fitness even with injury. Using the example above, the footballer with the hamstring strain can still cycle, swim, aqua-jog to help maintain a certain level of fitness without re-injuring the hamstring during treatment. Similarly, once the injury is repaired, emphasis should be placed on increasing duration of sessions, and hence helping to ensure endurance is not compromised.

It can be seen that rehabilitation therapists have a vital role to play in preventing and treating injuries to skeletal muscle.

Summary key points of muscle healing and rehabilitation

- RICE should be implement as soon as possible following acute injury

- Early mobilisation and weight bearing should also be encouraged

- Stretching and strength exercises can start within pain-free range as soon as possible

- Fitness and conditioning of the athlete should be incorporated within the early rehabilitation programme without compromising the injury

- Specificity, and functional fitness are imperative to help return the athlete back to sport without recurrence.

References

Arnason, A., Sigurdsson, S.B., Gudmundsson, A., Holme, I., Engebretsen, L. and Bahr, R. (2004) Risk factors for injuries in football. *American Journal Sports Medicine*, *32* (1), S4–16.

Arnason, A., Anderson, T.E., Holme, I., Engebretsen, L. and Bahr, R. (2008) Prevention of hamstring strains in elite soccer: an intervention study. *Scandinavian Journal of Medincine and Science in Sports*, *18* (1), 40–48.

Askling, C. M., Lund, H., and Saartok, T. and Thorstensson, A. (2002) Self reported hamstring injuries in student dancers. *Scandinavian Journal of Medincine and Science in Sports*, *12* (4), 230–235.

Askling, C. M., Saartok, T. and Thorstensson, A. (2006) Type of acute hamstring strain affects flexibility, strength, and time to return to pre-injury level. *British Journal of Sports Medicine*, 40 (1), 40–44.

Askling, C.M., Tengvar, M., Saartok, T. and Thorstensson, A. (2007a) Acute first-time hamstring strains during high-speed running: a longitudinal study including clinical and magnetic resonance imaging findings. *American Journal of Sports Medicine*, 35 (2), 197–206.

Askling, C.M, Tengvar, M., Saartok, T. and Thorstensson, A. (2007b) Acute first-time hamstring strainsduring slow-speed stretching: clinical, magnetic resonance imaging, and recovery characteristics. *American Journal of Sports Medicine*, 35 (10), 1716–24.

Bandy, W.D. and Irion, J.M. (1994) The effect of time on static stretch on the flexibility of the hamstring muscles. *Physical Therapy*, 74 (9), 845–852.

Bandy, W.D., Irion, J.M. and Briggler, M. (1997) The effect of time and frequency of static stretching on the flexibility of the hamstring muscles. *Physical Therapy*, 77 (10), 1090–1096.

Bandy, W.D., Irion, J.M. and Briggler, M. (1998) The effect of static stretch and dynamic range of motion training on the flexibility of the hamstring muscles. *Journal of Orthopaedic Sports and Physical Therapy*, 27 (4), 295–300.

Beiner, J. M. and Jokl, P. (2001) Muscle contusion injuries: Current treatment options. *Journal of American Academy of Orthopaedic Surgery*, 9 (4), 227–237.

Bennell, K., Wajswelner, H., Lew, P., Schall-Riaucour, A., Leslie, S., Plant, D. and Cirone, J. (1998) Isokintic strength does not predict hamstring injury in Australian Rules footballers. *British Journal of Sports Medicine*, 32 (4), 309–314.

Brockett, C.L., Morgan, D.L. and Proske, U. (2004) Predicting hamstring strain injury in elite athletes. *Medicine and Science in Sports and Exercise*, 36 (3), 379–387.

Brooks, J.H.M., Fuller C.W., Kemp S.P.T. and Reddin, D.B. (2006) Incidence, risk and prevention of hamstring muscle injuries in professional Rugby Union. *American Journal of Sports Medicine*, 34, 1297–1306.

Cameron, M., Adams, R. and Maher, C. (2003) Motor control and strength as predictors of hamstring injury in elite players of Australian football. *Physical Therapy in Sport*, 4, 159–166.

Clark, R., Bryant, A., Culgan, J.P., and Hartley, B. (2005) The effects of eccentric hamstring strength training on dynamic jumping performance and isokinetic strength parameters: a pilot study on the implications for the prevention of hamstring injuries. *Physical Therapy in Sport*, 6, 67–73.

Croisier, J.L. (2004) Factors associated with recurrent hamstring injuries. *Sports Medicine*, 34 (10), 681–695.

Croisier J.L., Forthomme, B., Namurois, M-H., Vanderthommen, M. and Crielaard J.M. (2002) Hamstring muscle strain recurrence and strength performance disorders. *American Journal of Sports Medicine*, 30, 199–203.

Croisier, J.L., Forthomme, B., Namurois, M.H., Vanderthommen, M., and Crielaard, J.M. (2003) Hamstring muscle strain recurrence and strength performance disorders. *American Journal of Sports Medicine*, 30 (2), 199–203.

Cross, K.M. and Worrell, T.W. (1999) Effects of a static stretching program on the incidence of lower extremity musculotendinous strains. *Journal of Athletic Training*, 34 (1), 11–14.

Dadebo, B., White, J. and George, K.P. (2004) A survey of flexibility training protocols and hamstring strains in professional football clubs in England. *British Journal of Sports Medicine*, 38 (4), 388–394.

Evans, P. (1980) The Healing Process at Cellular Level: A review. *Physiotherapy*, 66 (8), 256–259.

Farber, J., and Buckwalter, K. (2002) Magnetic resonance imaging in non-neoplastic muscle disorders of the lower extremity. *Radiologic Clinics of North America*, 40 (5), 1013–1031.

Foreman, T.K., Addy, T., Baker, S., Burns, J., Hill, N. and Madden, T. (2006) Prospective studies into the causation of hamstring injuries in sport: A systematic review. *Physical Therapy in Sport*, 7, 101–109.

Funk, D., Swank, A.M., Adams, K.J. and Treolo, D. (2001) Efficacy of moist heat pack application over static stretching on hamstring flexibility. *Journal of Strength and Conditioning Research*, 15 (1), 123–126.

Gabbe, B.J., Branson, R. and Bennell, K.L. (2006) A pilot randomised controlled trial of eccentric exercise to prevent hamstring injuries in community-level Australian Football. *Journal of Science and Medicine in Sport*, 9 (1–2), 103–109.

Gabbett, T.J. and Domrow, N. (2005) Risk factors for injury in subelite rugby league players. *American Journal of Sports Medicine*, 33 (3), 428–434.

Garrett, W.E. (1996) Muscle strain injuries. *American Journal of Sports Medicine*, 24 (6), S2–8.

Gates, C. and Huard, D. (2005) Management of skeletal muscle injuries in military personnel. *Operative Techniques in Sports Medicine*, 13, 247–256.

Gibbs, N. (1994) Common rugby league injuries. Recommendations for treatment and preventative measures. *Sports Medicine*, 18 (6), 438–450.

Goldspink, G. (1999) Changes in muscle mass and phenotype and the expression of autocrine and systemic growth factors by muscle response to stretch and overload. *Journal of Anatomy*, *194* (3), 323–334.

Hasselman, C.T., Best, T.M., Seaber, A.V., Garrett, W.E. (1995) A threshold and continuum of injury during active stretch of rabbit skeletal muscle. *American Journal of Sports Medicine*, *23*, 65–73.

Hawkins, R.D., Hulse, M.A., Wilkinson, C., Hodson, A. and Gibson, M. (2001) The association football medical research programme: an audit of injuries in professional football. *British Journal of Sports Medicine*, *35* (1), 43–47.

Hennessey, L. and Watson, A.W.S. (1993) Flexibility and posture assessment in relation to hamstring injury. *British Journal of Sports Medicine*, *27* (4), 243–246.

Hergenroeder, A. (1998) Prevention of sports Injuries. *American Journal of Paediatrics*, *101* (6), 1057–1063.

Herrington, L. (2000) Patients with hamstring muscle strains returning to sport in less than fourteen days, a report of the treatment used. *Physical Therapy in Sport*, *1* (4), 137–138.

Hoskins, W. and Pollard, H. (2005) The management of hamstring injury – Part 1: Issues in diagnosis. *Manual Therapy*, *10* (2), 96–107.

Hoskins, W., Pollard, H., Hough, K.,and Tully, C. (2006) Injury in rugby league. *Journal of Science and Medicine in Sport*, *9*, 46–56.

Huard, J., Li, Y., Fu, F.H. (2002) Muscle injuries and repair: Current trends in research. *Journal of Bone and Joint Surgery*, *84*, 822–832.

Hubbard. T.J., Aronson, S.L. and Denegar, C.R. (2004) Does cryotherapy hasten the return to participation? A systematic review. *Journal of Athletic Training 39* (1), 88–94.

Huxley, A.F. (1975) The origin of force in skeletal muscle. *Ciba Foundation Symposium*, *31*, 271–90.

Jarvinen, T.A., Jarvinen, T.L., Kaariainen, M., Kalimo, H., and Jarvinen, M. Muscle injuries: Biology and treatment. *American Journal of Sports Medicine*, *33*(5), 745–764.

Jonhagen, S., Nemeth, G. and Eriksson, E. (1994) Hamstring injuries in sprinters. The role of concentric and eccentric hamstring muscle strength and flexibility. *American Journal of Sports Medicine*, *22* (2), 262–266.

Kaminski, T.W., Webberson, C.V. and Murphy, R.M. (1998) Concentric versus enhanced eccentric hamstring strength training: Clinical implications. *Journal of Athletic Training*, *33* (3), 216–221.

Kannus, P., Parkkari, J., Järvinen, T.A. and Järvinen, A.H. (2002) Basic science and clinical studies coincide: active treatment approach is needed after a sports injury. *Scandinavian Journal of Medicine and Science in Sports*. *13* (3), 150–154.

Kossmann, C.E. and Huxley, H.E. (1961) The contactile structure of cardiac and skeletal muscle. *Circulation*, *24*, 328–335.

Knapik, J.J., Bauman, C.L., Jones, B.H., Harris, J.M. and Vaughan, L. (1991) Preseason strength and flexibility imbalances associated with athletic injuries in female collegiate athletes. *American Journal of Sports Medicine*, *19* (1), 76–81.

Kraemer, W.J., Adams, K., Cafarelli, E., Dudley, G.A., Dooly, C., Feigenbaum, M.S., Fleck, S.J., Franklin, B., Fry, A.C., Hoffman, J.R., Newton, R.U., Potteiger, J., Stone, M.H., Ratamess, N.A. and Triplett-McBride, T. (2002) American College of Sports Medicine, Position Stand. Progression models in resistance training for healthy adults. *Medicine and Science in Sports and Exercise*, *34* (2), 364–380.

Kujala, U.M., Orava, S. and Jarvinen, M. (1997) Hamstring injuries: current trends in treatment and prevention. *Sports Medicine*, *23* (6), 397–404.

Lehto, M., Duance, V.C., and Restall, D. (1985) Collagen and fibronectin in a healing skeletal muscle injury: an immunohistological study of the effects of physical activity on the repair of injured gastrocnemius muscles in rats. *Journal of Bone and Joint Surgery*, *67B* (5), 820–828.

Lieber, R.L. and Friden, J. (2002) Mechanisms of muscle injury gleaned from animal models. *American Journal of Physical Medical Rehabilitation*, *81*, S70–79.

Lutz, G.J. and Lieber, R.L. (1999) Skeletal muscle myosin II structure and function. *Exercise and Sports Science Reviews*, *27*, 63–77.

Mair, S.D., Seaber, A.V., Glisson, R.R. and Garrett, W.E. (1996) The role of fatigue in susceptibility to acute muscle strain injury. *American Journal of Sports Medicine*, *24* (2), 137–143.

Marqueste, T., Alliez, J., Alluin, O., Jammes, Y. and Decherchi, P. (2004) Neuromuscular rehabilitation by treadmill running on electrical stimulation after peripheral nerve injury and repair. *Journal of applied Physiology*, *96*, 1988–1995.

Neidlinger-Wilke, C., Grood, E., Claes, L. and Brand, R. (2002) Fibroblast orientation to stretch begins within three hours. *Journal Orthopaedic Research*, *20*, 953–956.

Nikolaou, P.K., Macdonald, B.L., Glisson, R.R., Seaber, A.V. and Garrett, W.E. (1987) Biomechanical and histological evaluation of muscle after controlled strain injury. *American Journal of Sports Medicine*, *15*, 9–14.

Orchard, J. (2004) Missed time through injury and injury management at an NRL club. *Sport Health*, *22* (1) 11–19.

Peterson, J. and Holmich, P. (2005) Evidence based research of hamstring injuries in sport. *British Journal of Sports Medicine*, 29 (6), 319–323.

Pull, M. R., and Ranson, C. (2007) Eccentric muscle actions: Implications for injury prevention and rehabilitation. *Physical Therapy in Sport*, 8, 88–97.

Roberts, J. and Wilson, K. (1999). Effect of stretching duration on active and passive range of motion in the lower extremity. *British Journal of Sports Medicine*. 33 (4), 259–263.

Robertson, V. (2002) Dosage and treatment response in a randomised control trial of therapeutic ultrasound. *Physical Therapy in Sport*, 3 (3), 124–133.

Robertson, V.J. and Baker, K.G. (2001) A review of therapeutic ultrasound: effectiveness studies. *J Physical Therapy*, 81 (7), 1339–50.

Schiaffino, S. and Partridge, T. (2008) *Skeletal Muscle Repair and Regeneration*. London: Springer.

Seward, H., Orchard, J., Hazard, H., et al. (1993) Football injuries in Australia at the elite level. *Medical Journal of Australia*, 159 (5), 298–301.

Sherry, M.A. and Best, T.M. (2004) A comparison of 2 rehabilitation programs in the treatment of acute hamstring strains. *Journal of Orthopaedic and Sports Physical Therapy*, 34 (3), 116–125.

Snyder-Mackler, L., Delitto, A. and Bailey S. (1995) Strength of the quadriceps femoris muscle and functional recovery after reconstruction of the anterior cruciate ligament. *Journal of Bone and Joint Surgery*, 77A, 1166–1173.

Stanton, P.E. (1989) Hamstring injuries in sprinting – the role of eccentric exercise. *Journal of Orthopaedic and Sports Physical Therapy*, 10, 343–9.

Stockmar, C., Lill, H., Trapp, A., Josten, C. and Punkt, K. (2006) Fibre type related changes in the metabolic profile and fibre diameter of human vastus medialis muscle after anterior cruciate ligament rupture. *Acta Histochemica*, 108 (5), 335–42.

Taylor, D.C., Dalton, J.D., Seaber, A.V., Garrett, W.E. (1993) Experimental muscle strain injury. Early functional and structural deficits and the increased risk for reinjury. *American Journal of Sports Medicine*, 21, 190–194.

Ter-Haar, G., Dyson, M. and Tlabert, D. (1978) Ultrasonically induced contraction of mouse uterine smooth muscle in-vivo. *Ultrasonics*, 16, 275–276.

Thornton, R., Mendel, F. and Fish, D. (1998) Effects of electrical stimulation on edema formation in different strains of rats. *Physical Therapy*, 78, 386–394.

Thorsson, O., Lilja, B., Nilsson, P. and Westlin, N. (2007) Immediate external compression in the management of acute muscle injury. *Scandinavian Journal of Medicine and Science in Sports*, 7 (3), 182–190.

Tidball, J. G., and Wehling-Henricks, M. (2007) Macrophages promote muscle membrane repair and muscle fibre growth and regeneration during modified muscle loading in mice in vivo. *Journal of Physiology*, 578 (1), 327–336.

Tsang, K.K., Hertel, J., Denegar, C. (2003) Volume decreases after elevation and intermittent compression of postacute ankle sprains are negated by gravity-dependent positioning. *Journal of Athletic Training*, 38, 320–324.

Verrall, G.M., Slavotinek, J.P., Barnes, P.G., Fon, G.T. and Spriggins, A.J. (2001) Clinical risk factors for hamstring muscle strain injury: a prospective study with correlation of injury by magnetic resonance imaging. *British Journal of Sports Medicine*, 35, 435–439.

Wilkin, L.D., Merrick, M.A., Kirby, T.E. and Devor, S.T. (2004) Influence of therapeutic ultrasound on skeletal muscle regeneration following blunt contusion. *International Journal of Sports Medicine*, 25 (1), 73–77.

Witvrouw, E., Bellemans, J., Lysens, R., Danneels, L., and Cambier, D. (2001) Intrinsic risk factors for the development of patellar tendonitis in the athletic population. A two year prospective study. *American Journal of Sports Medicine*, 29 (2), 190–195.

Woods, C., Hawkins, R., Hulse, M. and Hodson, A. (2002) The Football Association medical research programme: an audit of injuries in professional football – analysis of preseason injuries. *British Journal of Sports Medicine*, 36 (6), 436–441.

Woods, C., Hawkins, R. D., Maltby, S., Hulse, M., Thomas, A. and Hodson, A. (2004) Football Association Medical Research Programme. The Football Association Medical Research Programme: an audit of injuries in professional football – analysis of hamstring injuries. *British Journal of Sports Medicine*, 38 (1), 36–41.

Worrell, T.W. (1994) Factors associated with hamstring injuries: an approach to treatment and preventative measures. *Journal of Sports Medicine*, 17 (5), 338–345.

Worrell, T.W., Smith, T.L. and Winegardner, J. (1994) Effect of hamstring stretching on hamstring muscle performance. *Journal of Orthopaedic and Sports Physical Therapy*, 20 (3), 154–159.

Yamamoto, T. (1993) Relationship between hamstring strains and leg muscle strength. A follow up study of collegiate track and field athletes. *Journal of Sports Medicine and Physical Fitness*, 33 (2), 194–199.

5

Tendons

Dr. Stephen Pearson
University of Salford, Greater Manchester

Introduction

Tendons are an important link in the locomotor system, enabling muscle forces to be utilised to perform complex movements and actions. As such, their characteristics can affect the ability to generate appropriate forces when attempting, for example, fine motor tasks, or indeed those tasks requiring high rates of force development. In order to be able to respond to these demands, the tendon is designed to be both rigid enough to enable efficient transfer of forces from muscle to bone, but also compliant enough to allow storage of energy for later use.

It is because of these diverse requirements on the tendon that they are frequently at risk of injury, both acutely where perhaps force has been applied in an unusual or inappropriate fashion, or chronically where micro damage has accumulated over time leading to trauma.

This chapter will outline the tendon physiology from a molecular level to gross tissue level and how this relates to its function and its subsequent mechanical properties. The tendon characteristics are then put in context of the action of muscle and ultimately the ability to produce functional force. Next, tendon injury will be discussed in context of its use both from an acute and also chronic perspective. The ability of the tendon to regenerate and optimal or appropriate healing strategies will also be put into a context and illustrated by appropriate by case studies.

Basic tendon anatomy and physiology

Tendon gross structure can be described as either sheet like or rounded. Some tendinous structures may be a combination of both, with the tendon being more rounded in the mid portion and tending to resemble sheet-like fascia at their attachments (aponeuroses) with the adjoining muscle. Within the tendon there is a structure consisting of closely packed collagen molecules, which are assembled into fibrils, fibres, fibre bundles or sub fascicles, fascicles and tertiary fibre bundles (see Figure 5.1). This arrangement of collagens into a hierarchical structure enables the tendon to resist tensile loading and ensures that there is minimal risk of catastrophic rupture. Each fibre bundle is wrapped with a layer of connective tissue called the endotenon; blood vessels, nerves and lymphatic elements can pass through this layer to supply the tendon. The tendon is covered by an outer layer of connective tissue called the epitenon; this is contiguous with the layer below (endotenon). Most tendons are then covered with a looser outer layer paratenon; this tissue is lined with synovial cells and allows for tendon movement or gliding.

Tendon consists of tendon cells (fibroblasts) that lie in longitudinal rows and are elongated cells extending within the tendon structure and communicate via gap junctions within the three-dimensional space of the tendon. Also there is the extracellular matrix (ECM) that forms the scaffold for the tendon

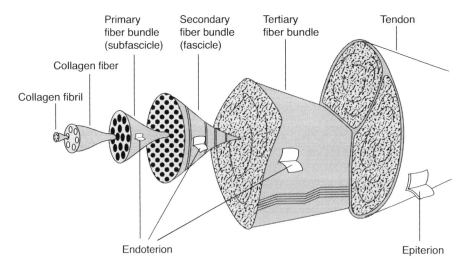

Figure 5.1 Arrangement of collagen fibres and covering layers in the tendon structure. Reproduced, with permission, from Kannus, P. (2000). Structure of the tendon connective tissue. Scand. J. Med. Sci. Sports. 10:312–320. (p. 313) © 2000 John Wiley & Sons Ltd.

and which consists of a range of collagens (I, II, III, V, VI, IX and XI), proteoglycans and water (see Table 5.1 for relative proportions of components).

The largest proportion of collagen contained in tendon is type I, this lends strength to the tendon structure, whereas, the proteoglycans gives the tendon its viscoelastic properties.

It is now well known that tendon is a metabolically active tissue and adapts readily to influences such as mechanical loading and unloading (Heinemeier et al. 2009). This can be understood as a dynamic mechanism responding to feedback signalling via mechanotransduction pathways in an attempt to optimise the tissue characteristics to its function. Previous work has shown that tendon adapts to loading by modifying its mechanical properties and dimensions. Similarly it has been observed that tendon mechanical properties such as compliance are increased with unloading due to regulation of structural proteins, i.e. collagens. In line with this is the observation that tendon elastic properties are associated with the attached muscle (Muraoka et al. 2005) in that a muscle that is able to generate large force will generally be attached to a tendon that has adapted to cope with these forces.

The mechanisms of how the tendon responds to mechanical signals at a cellular level are still not well understood but are thought to involve the transduction of mechanical stressors via the ECM. The ECM forms the substrate for the anchoring of cells and enables the tendon to maintain its structural integrity. It has been suggested that forces generated end to end along the tendon are transferred along it by the matrix in shear connecting along the length of the discontinuous collagen fibres (Ker 1999).

Components called integrins, which are described as heterodimeric transmembrane proteins, form cell surface receptors that connect the tendon ECM to the cytoskeleton. Thus these are part of the adhesion complex, forming adhesion foci complexes at the cell membrane and are thought to be involved in the complex signalling processes resulting in modifications of structural proteins. These molecules when activated for example via mechanical stress, can initiate chemical signalling cascade processes, such as tyrosine phosphorylation, nuclear factor transcription factor (NF-$_k$B), mitogen activated protein kinase (MAPK) pathway activation, either directly or as part of the process involving other growth factors (see Figure 5.2). Another ECM molecule that is receiving some attention is tenascin C; this is also seen to be up regulated with mechanical stimulus and is a ligand for integrin. The actual role of tenascin in tendon tissue is as yet not clear, but may involve actions related to tissue remodelling and general maintenance of the intercellular structures enabling gliding of bordering collagen bundles (Riley et al. 1996).

How integrins sense strain is not clear, in that although conformational changes are seen when

Table 5.1 Components of tendon tissue and relative constituent proportion (taken from Ker 2002). Reprinted, with permission, from Burgess KE, Pearson SJ, Breen L, Onambélé GN. (2009). Tendon structural and mechanical properties do not differ between genders in a healthy community-dwelling elderly population. J Orthop Res. 2009 Jun;27(6):820–5. (p. 823) © Elsevier

	Wet mass (%)	Dry mass (%)
Water	65[b], 53–70[e], 75[g], 60[j]	
Collagen: total		50[a], 76[b,c], 99.7[d], 60–85[e], 65–80[f]
Type I		67[g], 65[h]
Type III		10.0[b], 3.5[c], 1.8[g], 0[h]
Type VI		2.0[d]
Elastic fibers		2[e], 1–2[f]
Hyaluronan		0.01[f], 0.09[g], 0.10[h]
Other proteins: total		4.6[j]
Sulfated proteoglycans		
Decorin		1.0[j]
Biglycan		'Half the decorin amount'[d]
Aggrecan		'Little'[i]
Other glycoprotein		
COMP		1.0[j]
Anorganic		<0.2[f]

This list is indented only to give an indication of the main extracellular components. Other proteins which might be relevant, but for which I have not found quantitative data include tenascin-C, the small proteoglycans biglycan, fibromodulin, and lumican and the glycoproteins undulin and PRELP (Proline arginine-rich end leucine-rich repeat protein). Abbreviation: COMP = cartilage oligomeric matrix protein.
[a]Bank et al. (1999): human supraspinatus and biceps brachii.
[b]Birch et al. (1999): horse forelimb superficial digital flexor.
[c]Birch et al. (1999): horse forelimb deep digital flexor.
[d]Derwin et al. (2001): mouse tail.
[e]Elliott (1965): general review.
[f]Kannus (2000): general review, giving ranges.
[g]Riley et al. (1994): human supraspinatus.
[h]Riley et al. (1994): human biceps.
[i]Smith et al. (1997): horse.
[j]Vogel and Meyers (1999): bovine forelimb deep digital flexor.

forces are applied and ligands bind, other possible strain sensing mechanisms may involve a link between the calcium channel responses to stretch (reviewed in Chiquet 1999). Calcium channels are observed to open transiently for a relatively short period with stretch, this response is seen to be down regulated by interfering with the integrins (Chen and Grinnell 1995).

Functional aspects of tendon

Tendon is sited between the muscle and bone and acts to transmit forces generated by the muscle to the bone to enable movement. The forces experienced by tendons are usually tensile, although due to their arrangements with the bony anatomy, tendons may also experience compressive and shear forces. For example, tendon can glide over bony areas to allow moment arms to be utilised, potentially increasing torque transmitted via the tendon. The tensile properties of tendon allow for its capacity to resist rupture under normal loading conditions. Tendon is described as having viscoelastic qualities. These properties enable the tendon to act both as a damper to reduce potentially damaging high forces, but also as a storage medium for energy generated during movement.

It is usual to describe the tendon response to loading in terms of strain (percentage change of length from rest) and stress (force per unit area). A typical stress–strain curve for human tendon can be seen in Figure 5.3.

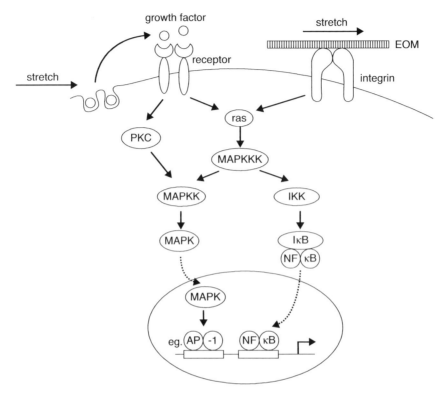

Figure 5.2 Schematic diagram showing possible signalling pathways for the up regulation of collagen synthesis via mechanosensitive pathways (from Chiquet 1999). Reprinted, with permission, from Chiquet, M. (1999). Regulation of extracellular matrix gene expression by mechanical stress. Matrix Biology. 18:417–426. © Elsevier (p. 422)

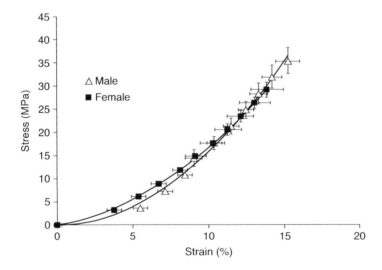

Figure 5.3 Stress–strain curve for elderly males and females (mean age 72 yrs) (Burgess et al. 2009b). Reproduced, with permission, from Burgess KE, Pearson SJ, Breen L, Onambélé GN. (2009). Tendon structural and mechanical properties do not differ between genders in a healthy community-dwelling elderly population. J Orthop Res. 2009 Jun;27(6):820–5. (p. 823) (C) 2009 John Wiley & Sons Ltd.

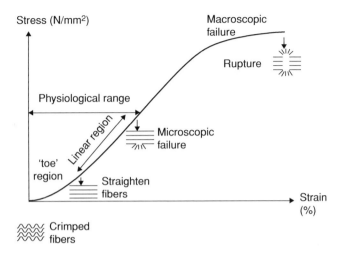

Figure 5.4 Diagram showing elongation properties of tendon with progressive loading. Low load region (toe) shows characteristic greater strain than linear region, with subsequent plastic region leading to rupture.

With respect to the strain, tendon has the characteristics of larger strain at low load, due in part to the crimp structure of the collagen fibrils, and a linear region where strain is proportional to stress, beyond this point the tendon starts to become plastic in that it takes on a new resting length when the load is removed. If stress is continued to be applied then ultimately tendon rupture will result (see Figure 5.4). Tendons with less crimp have been observed to fail before those with a more pronounced tendon crimp pattern (Wilmink et al. 1992). This may be reflective of the mechanical connections between the collagen fibrils and the ECM or some qualitative differences in the collagen structure.

Regarding the ultimate tensile stress for tendon tissue, a conservative limit has been suggested of 100 MPa (Bennett et al. 1986). However Johnson et al. (1994) examined human patellar tendons and estimated that the ultimate tensile strength was 65 ± 15 MPa for young subjects.

Ker (2002) discusses 'stress for life', where a functional stress is determined from the ratios of the tendon to the muscle cross-sectional area (csa) and the product of the maximal isometric stress developed by the muscle (0.3 MPa, Wells 1965), with ranges from 8–100 MPa, but more normally approximating 13 MPa. It can be seen then that in certain circumstances a tendon may be subjected to stress which may be nearing its rupture limit, for example the patellar tendon has been reported to be subjected to approximately 12.2 kN during actions such as competitive weightlifting (Zernicke et al.1977), but under normal circumstances there is a relatively large margin of safety. This is perhaps best illustrated by considering the tendons involved in locomotion, such as the Achilles; it has been determined that this structure may experience forces in the order of 7 kN during running (Komi et al. 1992), this equates to approximately 8.5 times body mass for an individual of mass 80 kg. This may be even greater during actions such as long jump or triple jumping and has been reported to be in the order of 9 kN during jumping (Komi 1990).

However, with regard to tendinous damage, it is more likely generally that tendons suffer what is termed fatigue damage resulting from repeated loading cycles (see section on Chronic tendon injury).

Tendon has been shown to become damaged at strain values approximating 15% (Stäubli et al. 1999), although these values have been previously reported to be as high as 30% (Haut and Pawlison 1990). A functional design aspect of tendon to help limit damage is that of the ability of the tendon fascicles to slide with respect to each other whilst simultaneously transmitting forces along the tendon length. With respect to strain, it has previously been shown that there are qualitative differences between male ands female tendon, with females having greater compliance in comparison to males (Onambele et al. 2007). Similarly previous work has reported that in response to acute stretch females shown larger decreases in tendon stiffness in contrast to males, this may have implications where stretching is used as part of a warm up (Burgess et al. 2009a). This may reflect differences in the collagen content and/or composition between males and females. These gender differences do not seem to be retained in elderly tendon, however (Burgess et al. 2009b), possibly reflecting the changing hormonal influences on tendon being less disparate between the genders with age and more similarities in lifestyle.

At the levels of loading seen routinely, tendon may suffer micro damage at levels of strain seen normally, that is below 10%, leading to accumulated deterioration in tendon (see section on Chronic tendon injury). Normal ranges of strain under conditions of maximal isometric contraction for human tendon structures are within the ranges of 6–14%. The range of normalised elasticity (Young's modulus) for tendon is approximately 0.8–2 GPa. This property represents the normalised (corrected for differences in tendon length and cross section) stiffness of the tissue and is the gradient of the stress-strain curve.

As tendons are also elastic they tend to stretch when muscle contraction takes place. If, for example, the external load is very heavy, then as the muscle contracts the tendon may stretch initially, the action of the tendon here could be thought of as similar to a spring being stretched and energy stored. This may be functionally important, an example of this could be where an athlete is sprinting or jumping and the tendon is being subjected to stretch-shortening cycles. This enables the movement to be carried out with higher efficiency than if the tendon did not stretch and then shorten, in that the tendon is initially storing energy during the stretch, to be released during the shortening or recoil period. It has been reported that tendon has the ability to return ~93% of its stored energy during the recoil phase, the rest being lost as heat energy (Alexander 2002). This particular property of tendon is perhaps useful where for example cyclical movements are being carried out with changes in kinetic and potential energy or to increase the power component as in jumping. The tendon can act to help smooth out these changes and store some of the energy as elastic strain energy to be returned to the system when required, reducing the overall work required for the muscles (i.e. running). However, where high rates of

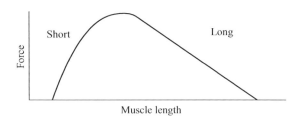

Figure 5.5 Length–tension relationship for skeletal muscle.

force development are required, tendon compliance could be seen as potentially detrimental to optimal performance.

The mechanical properties of the tendon, in particular the stiffness, can also be seen to potentially affect the characteristics of muscle contraction. A muscle's contractile ability in part is described by its length tension curve, by which the number of elements that generate muscle force (cross bridges) are increased or decreased dependant on the muscle length (see Figure 5.5). Here it can be seen that if the tendon is compliant and stretches under load, the muscle will shorten, and this could affect its ability to generate force. As can be seen from Figure 5.5, if the muscle is starting from a position near maximal force and the tendon stretches, the muscle will shorten moving down the force generating curve.

The position where the tendon attaches to bone can have functional implications. A muscle can only extend or shorten over a given range; the points at which tendon attaches to the bone can thus affect the movement of the limb to which the muscle–tendon complex is attached. This can be seen to be useful from a functional standpoint where for example force is preferred over range of movement and visa versa (see Figure 5.6 for example). Here, where the tendon attaches closer to the joint centre of rotation as in Figure 5.6A, the limb being affected can move further for a given muscle shortening than where the tendon attaches further from the rotation centre (Figure 5.6B). This is at a cost of reduced torque generating ability.

Tendon injury and its management

Tendon injury can be debilitating in so far as a damaged tendon will likely reduce the capacity for transfer of forces via the muscle to the bone. This has obvious implications for a range of activities ranging

ACUTE TENDON INJURY

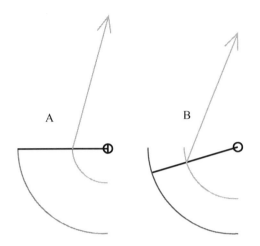

Figure 5.6 Differences in limb range of movement with tendon attachment. A: Tendon attaches closer to joint centre of rotation. B: Tendon attaches further from centre of rotation than A.

from normal everyday movement to elite level athletic performances where perhaps high levels of force and rates of force development are prerequisite.

Injury to the tendon can be characterised as acute resulting in either a catastrophic rupture or a less major tear of the tendon tissue. Alternatively the tendon may suffer chronic pathological changes which may be inflammatory or not even detectable. These factors can affect the tendon at numerous sites along its length. There appears to be no set consensus on description for the degenerative or pathological changes to tendon (see Table 5.2 for descriptors, taken from Wang et al. 2006).

With regard to acute tendon injuries, although a number of these could be identified due to extrinsic factors, such as direct contact with external body, or accidental unaccustomed loading such as extreme shear, the majority of acute tendon injuries show prior predisposing factors. These potentially contributory factors can be thought of as chronic inflammatory or accumulative degenerative conditions and may be undetected prior to the acute trauma.

Acute tendon injury

Immediate tendon injury

This can be perhaps best described as either complete rupture of the tendinous structure or partial rupture. Complete rupture will be self-evident and accompanied by loss of function, pain and swelling usually following the incident. On inspection it may be obvious that rupture has occurred in that a gap may be seen where the tendon would normally be. More specific diagnosis of rupture can be carried out using imaging techniques, such as ultrasound (see Figure 5.7) or magnetic resonance imaging. Tendon injuries of this nature tend to be unilateral and of those reported for the achilles, 75% were related specifically to sport (Jozsa et al. 1989 cited in Leppilahti and Orava 1998). The tendons mostly seen to be suffering rupture are those bearing higher functional loads, such as the achilles and patellar tendons. Site of

Table 5.2 Description for degeneration or pathological changes in tendon

Term	Description	Reference
Tendinitis or tendonitis	Inflammation of the tendon	Clancy et al. (1976); Almekinders and Temple (1998); Curwin and Stanish (1984); Schepsis and Leach (1987)
Tendinosis	Asymptomatic tendon degeneration	Puddu et al. (1976); Järvinen et al. (1997)
Tendinopathy	Generic description for tendon disorders	Khan et al. (2002); Maffulli et al. (1998)
Paratenonitis	Inflammation of the paratenon	Maffulli et al. (1998); Jozsa and Kannus (1997)
Peritendinitis	Inflammation of the peritendon	Maffulli et al. (1998); Jozsa and Kannus (1997)
Spontaneous tendon rupture	Tendon rupture without any predisposing symptoms	Kannus and Jozsa (1991)
Partial rupture	Incomplete tendon tear	Karlsson et al. (1992); Karlsson et al. (1991)
Enthesopathy	Tendon-bone junction disorders	Maffulli et al. (1998); Kvist (1991)

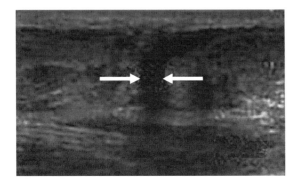

Figure 5.7 Saggital view of achilles tendon rupture indicated by arrows (from imaging consultant.com).

tendon rupture has also been previously suggested to be associated with hypovascular regions of the tendon. For example, achilles tendon ruptures typically occur 2–6 cm proximal to the insertion site at the calcaneus. This area along with the insertion site is also known to be poorly supplied with blood (Carr and Norris 1989; Niculescu and Matusz 1988).

Causes

There are a number of factors which are thought to be responsible for acute tendon injury. These factors can be separated into intrinsic and extrinsic factors.

A sudden, large force or torque applied through the tendon, perhaps at an oblique angle could result in a partial or complete tear. How this force or torque presents may be due to either environmental factors – incorrect or accidental placement of the limbs in context of the surface, abnormal surface conditions – and/or neural control mechanism failure, which may be accompanied by inappropriate muscle balance. But it is also suggested that in a number of acute injuries to the tendon, prior predisposing pathological factors may have been present. Here, when strenuous activity is routinely carried out it is possible that cumulative damage may occur leading to a catastrophic failure of the tissue (Järvinen et al. 2001). The mechanisms underlying this will be discussed in more detail in section on Chronic tendon injury.

It has been reported that males present with tendon injuries more often than females, however, it is not clear if this is due to more participation in sport by males and thus more exposure to risk of injury (Clayton and Court-Brown 2008). Onambele et al. (2007) demonstrated qualitative differences between male and female tendons in young adults with females having more compliant structures. Whether this would result in a tendon more likely to be injured is not clear, however, if the two tendon structures have similar strain limits it might be suggested that for a given force the female tendon would be more likely to undergo permanent deformation and subsequent damage.

More recently it has been reported that certain classes of drugs, including antibiotics (flouroquinolines) and corticosteroids, can affect the collagen tissue directly, leading to weakened structures that may rupture under high loading conditions (Sode et al. 2007). The mechanism by which flouroquinolines are thought to increase the risk of tendon injury is linked to the increased activation of metalloproteinase, a regulator of collagen degradation.

There has been suggested association between anabolic steroids use and tendon injury. Of the studies carried out in animal models there have been reports of morphological changes to the collagen structures. In humans, however, this has not been shown. Evans et al. (1998) examined tendon tissue from individuals who had self-administered anabolic steroids for between 1 and 10 years. Subsequent light and electron microscopy analysis of the tendon indicated no differences from normal. There is some initial case study evidence to suggest there is an association between steroid use and tendon rupture. One such case involved an individual presenting with a relatively rare case of bilateral quadriceps tendon rupture, which was subsequently suggested to be associated with the earlier intake of anabolic steroids (Liow and Tavares 1995). Regarding the mechanism responsible for injury with the intake of steroids, it may be that as the ability to generate high forces increases due to the anabolic effects on muscle, the tendon simply cannot keep pace with this increased ability and therefore succumbs to injury.

Previously, Unverferth and Olix (1973) showed, in a number of cases involving athletes who had received localised corticoid steroid injection of the tendon, that there was an association with the use of corticosteroid and subsequent tendon rupture. A series of experiments on animal tendons led them to suggest this may have been due to resultant disruption of the tendon collagen matrix, and this was aligned with a reduced modulus of elasticity for those tendons.

Recent evidence has indicated that there may be a link between certain specific gene polymorphisms and tendon injury. Mokone et al. (2005, 2006), showed a relationship between the COL5A1 and tenascin C gene and achilles tendon injury rates. These genes can be seen to be important to the structural integrity of tendon as the COL5A1 gene is known to code for collagen structural components, whereas the tenascin C gene encodes for the extracellular matrix component tenascin C. This component is understood to be involved with the transmission of forces throughout the tendon and is a component that is upregulated in response to mechanical loading, suggesting a role in tendon remodelling.

With respect to the optimal method for repair and subsequent rehabilitation treatment of tendon ruptures, no consensus exists. Although there is increasing evidence to suggest that where the individuals are relatively active and able, surgery may be advantageous, providing better functional capacity over non-operative options (Schepsis et al. 2002); whereas, for those who are more sedentary, less invasive non-operative treatments may be advocated. The following case study illustrates the use of a recent methodology to aid in the repair of a ruptured rotator cuff tendon. Here a patient underwent surgery involving the use of a "cascade membrane" to knit the tendon and accelerate healing. The membrane consisted of a thin layer of autologous fibre saturated in platelets, the idea being that platelets are responsible for the production of growth factors responsible for promoting healing. Here the membrane provided both mechanical strength to allow connection of the tendon ends and an environment for optimal healing. After surgery the patient was immobilised for a period of four weeks and a standard rotator cuff rehabilitation protocol carried out. MRI examination of the repair after six months indicated a complete and robust repair of the ruptured tendon (Maniscalco et al. 2008).

Chronic tendon injury

Definitions

There are a number of definitions for chronic tendon pathological changes generally associated with increased tendon usage, these range from terms to describe degeneration and the failing of the healing response, to those used to describe inflammation and morphological adaptation of the tendon tissue (see Table 5.3 for characteristics of tendon pathologies). A more general and suitable clinical term that does not attempt to describe the pathologies associated with the condition has been suggested as "tendinopathy" (Mafulli et al. 1998).

Causes

Tendinopathy can be directly associated with the volume or intensity of tendon loading and this can be modified by other factors such as age, gender (Aström et al. 1998; Riley et al. 2001), body mass, disease and oral contraceptives in females (Holmes and Lin 2006). Where loading is inappropriate the tendon may suffer continual minor damage from which the tendon is not able to recover for any number of reasons, leading to eventual clinical signs of injury or damage. As a mechanism for 'non-coping' tendons, the vasculature supplying the tendon may have a part to play. Here the blood supply is required to provide requisite healing components as part of the process. If this system is not able to carry out this task optimally the tendon may not recover from the trauma. Previously poorly perfused tendon tissue has been linked with spontaneous tendon ruptures (Yepes et al. 2008). Interestingly though, tendinopathy shows hypervascularisation of the tendon without an associated healing response, this contrast is difficult to explain and may be indicative of breakdown of the cascade system responsible for normal healing. This increased vascularisation has been reported to be responsible for the pain associated with tendinopathy (Alfredson et al. 2003). Recently, researchers have suggested that insertional overuse injuries may be characterised by compressive forces rather than tensile loading. Here the suggestion is that the change from tensile to compressive loading at the tendon origin during knee flexion could result in adaptations of the tendon structure making it more susceptible to injury (Johnson et al. 1996; Hamilton and Purdam 2004). However, recently a study investigating the aetiology of patellar jumper's knee reported that forces present at the patellar origin site were in fact tensile in nature, showing that patellar strain increased with loading but also that tendon strain increased with decreased patellar-patellar tendon angle (Lavagnino et al., 2008). The minimal patellar-patellar tendon angle was also shown to coincide with predictions of knee angle for maximal

Table 5.3 Characteristics of tendon pathologies (from Xu and Murrell 2008). Reprinted, with permission, from Xu, Y and Murrell, G. (2008). The basic science of tendinopathy. Clin. Orthop. Relat. Res. 466:1528–1538. (p. 1529). © The Association of Bone and Joint Surgeons

Tendon category	Macroscopic	Ultrasound/doppler imaging	Light microscope	Electron microscope
Normal	• Brilliant white • Firm texture	• Parallel hyperechoic • Regular fibre structure	• Organised parallel collagen fibres • Spindle shaped tenocytes • Parallel nucleus arrangement	• Dense collagen fibre structure • Uniform fibre diameter and alignment
Tendinopathic tendon	• Grey/brown • Thin disorganised	• Localised widening • Hypoechoic areas • Irregular structures • Neovascularisation	• Disorganised collagen fibres • Loss of aligned nuclei • Increased vascular ingrowth • Increases in proteoglycans and glycosaminoglycans	• Angulation of collagen fibres • Heterogeneity of collagen diameters and

knee loading (60°) (Pflum et al. 2004). Hence, here it can be seen that either deep knee flexion, high loading or a combination of both could predispose the tendon to injury where the architecture of the tendon–bone connection could cause increased localised tendon strain.

A common response to injury or damage is inflammation; hence the molecular markers of inflammation can be used to identify potential damage. Tendon responds differentially to cyclical loading or strain. A previous study examined the effects of strain on tendon expression of inflammatory mediators. Where strains were relatively low, that is approximating 4%, no deleterious effects on tendon were shown; however, with strains of 8% up-regulation of genes associated with inflammation (COX-2, MMP-1 and prostaglandin E_2) were noted (Yang et al. 2005). This study suggests then that moderate load or strain may be appropriate for tendon homeostasis, whereas, higher loading can lead to damage, which may accumulate if frequent.

Although inflammation is common to injury, it is generally thought that tendinopathy is not associated with inflammation. In that tendons generally present as painful and histological degenerative changes are evident such as increased cellularity and rounded cells rather than elongated or striated cells, along with ECM remodelling, which includes reductions in total collagen (Riley et al. 1994).

Diagnosis of tendinopathy would include the taking of patient history to determine if pain was present and for how long. The characteristics of pain are also considered, that is present only after activity, present during but not sufficient to interfere with the activity, present during activity lasting into period after activity ceases and sufficient to interfere with activity, or ultimately total tendon failure may present. A visual inspection can be made of the kinematics of the segments to identify any irregular aspects. Assessment of the affected site may reveal swelling, tenderness and nodularity or crepitus. Other valid tools used are the Victorian Institute of Sport Assessment Questionnaire (VISA) for the patellar or a variant of this for the achilles (Visentini et al. 1998; Robinson et al. 2001). Imaging techniques allow for the quantification of the tendinopathologies, and typically ultrasound or magnetic resonance imaging is used to determine tendon anomalies. Linear ultrasound probes with a frequency range of between 7.5 and 15 MHz are able to spatially resolve superficial tissue to less than 0.1 mm (Martinoli et al. 1993). Ultrasound imaging has been utilised to describe three different levels of tendinopathy: (1) normal tendon image, (2) enlarged tendon area, and (3) hypoechoic area to tendon (Archambault et al. 1998). MRI is a very useful imaging methodology, in part because of its high spatial and contrast resolution, and is able to generate 3D images of the tendon structures; however, it is still relatively expensive and operator intensive in contrast to ultrasound imaging.

Treatment

Treatment of tendinopathy can be by a number of different methods.

As tendinopathy has been associated with increased vascularisation and also pain, a previous study examined the therapeutic value of utilising a sclerosing agent (Polidocanol) in order to try and reduce the localised vascularisation. Öhberg and Alfredson (2002) reported that the use of Polidocanol in 10 patients presenting with achilles tendinosis resulted in 8 out of 10 patients showing improvements in the condition and reductions in vascularisation of the affected areas.

A case study is presented here for a 35-year-old tennis player complaining of pain at the flexor carpi ulnaris tendon site. The use of a sclerosing agent was applied based on the finding of neovascularisation of the injury site. Polidocanol (1.5 ml) was injected to the site of neovascularisation, guided by power and laser doppler spectrophotometry. This initial treatment reduced the neovascularisation by approximately 25%. Further treatment after a short rest was in the form of eccentric resistance loading (Therabands) of the forearm for a period of 12 weeks (6 × 15 repetitions day^{-1}) after which the patient reported no incidence of wrist pain (Knobloch et al. 2007).

Tendon healing

Tendons by nature have a much lower metabolic rate than associated tissue such as skeletal muscle, with tendon having an oxygen consumption of approximately 13% of skeletal tissue (Zernicke et al. 1977). This results in a much slower healing process than would be evident with higher metabolic rates. This less adequate metabolic process is partly due to the limited blood supply to the tendon tissue. Subsequent to injury there are distinct but overlapping healing responses, the early phase is delineated by inflammation, next the secondary phase involves synthesis and is termed remodelling, followed by the final stage where refinement of the tissue takes place and is appropriately termed the modelling phase.

Cytokines, in particular neutrophils are associated with initiation of the inflammatory phase, vascular permeability is increased and damaged material is removed via phagocytosis, with monocytes and macrophages invading the damage site within the first 24 hours. These mechanistic responses also bring about initiation of angiogenesis and tenocyte proliferation and migration to the damage site, which results in increased type III collagen synthesis. During the remodelling stage, which is evident after the first two days or so, type III collagen synthesis continues to increase, with the peak occurring during this phase. Next, after a number of weeks there is a down regulation of collagen production, during weeks 6–10 there is a gradual change in the new tissue from cellular to fibrous. It is during this 'consolidation' stage that the collagen fibres align to the direction of stress. With time there is a gradual reduction in both the tenocyte activity and vascularisation of the tendon tissue injury site. In addition, during this period there is a changeover from predominantly type III collagen production to type I. After a period of approximately 10 weeks the remaining fibrous tissue begins to be modified into tendon tissue that is scar-like, this process continues for up to one year and is sometimes referred to as the maturation phase.

Key points and summary

Tendon is a highly adaptive structure capable of coping with varying demands put on it.

The viscoelastic nature of tendons modifies the muscle output potential and enables the efficient storage and release of energy.

The mechanisms by which tendons "sense" mechanical loading are still not well understood but may involve proteins that connect the tendon ECM to the cytoskeleton. These proteins may somehow be connected via stretch sensitive ion channels to enable signalling events leading to collagen synthesis.

Injury to tendon can be categorised as acute or chronic. Where acute injury occurs it is suggested that predisposing factors are present in many cases.

Overuse injuries can be described as chronic tendon injuries and fall into the description for tendinopathies.

Clinical diagnosis of tendinopathies involves manual and visual elements. Pain may or may not be present, but changes to the tendon may include swelling, crepitus, tenderness, and or nodularity.

Imaging has been utilised to help diagnose tendinopathy, both US and MRI are able to identify stages associated with tendinopathic changes, such as enlargement of the tendon and hypoechogenic areas to the tendon.

Treatments of acute and/or chronic tendon injuries are dependant on site of injury, activity levels or expectations of patient, the specific tendon injured and the grade of injury.

Many treatments involving pharmacological intervention are available, for example NSAIDS have been used with some degree of success to reduce swelling and symptoms of tendinopathy. More recently, sclerosing injections have been utilised with good results to reduce the neovascularisation seen in some cases (described in earlier section). It remains, however, that no one treatment is a best option where non-invasive conservative treatments are advocated.

Where surgery is advocated there are again many different methodologies present to repair tendon, this chapter has presented a recent development whereby growth factors are encouraged within the repair by use of a patch saturated with platelets. Rehabilitation of the tendon following either acute or chronic injury involves a period of rest or immobilisation followed by a gradual, progressive development of strengthening and mobilising exercises. Recently eccentric exercise has been advocated as a particularly useful adjunct to successful rehabilitation of tendon injuries; suggested mechanisms for this include high frequency oscillation of force during loading (Rees et al. 2008).

In conclusion, tendon injury is a complex, multifaceted issue. Injury can be cumulative or sudden and can appear suddenly without any precognition of a problem. Repair of tendon can take the form of surgery, which tends to have a better outcome in terms of functionality and time taken to get back to pre-injury levels of activity, or conservative, which usually results in fewer complications such as infections. There is currently no consensus on the best form of treatment, either repair or rehabilitation. However, there are a number of recent developments that show promise.

References

Alexander, R.M. (2002) Tendon elasticity and muscle function. *Comparative Biochemistry and Physiology, Part A, Molecular and Integrative Physiology 133* (4), 1001–11.

Alfredson, H., Ohberg, L. and Forsgren, S. (2003) Is vasculo-neural ingrowth the cause of pain in chronic Achilles tendinosis? An investigation using ultrasonography and colour Doppler, immunohistochemistry, and diagnostic injections. *Knee Surgery Sports Traumatology and Arthroscopy. 11* (5), 334–338.

Almekinders, L.C. and Temple, J.D. (1998) Etiology, diagnosis, and treatment of tendonitis: an analysis of the literature. *Medicine and Science in Sports and Exercise, 30* (8), 1183–1190.

Archambault, J.M., Wiley, J.P., Bray, R.C., Verhoef, M., Wiseman, D.A. and Elliott, P.D. (1998) Can sonography predict the outcome in patients with achillodynia? *Journal of Clinical Ultrasound, 26* (7), 335–339.

Aström, M. (1998) Partial rupture in chronic achilles tendinopathy. A retrospective analysis of 342 cases. *Acta Orthopaedica Scandinavica 69* (4), 404–407.

Bennett, M.B., Ker, R.F., Dimery, N.J. and Alexander, R.M. (1986) Mechanical properties of various mammalian tendons. *Journal of Zoology, 209*, 537–548.

Burgess, K.E., Graham-Smith, P. and Pearson, S.J. (2009) Effect of acute tensile loading on gender-specific tendon structural and mechanical properties. *Journal of Orthopaedic Research, 27* (4), 510–516.

Burgess, K.E., Pearson, S.J., Breen, L. and Onambélé, G.N. (2009) Tendon structural and mechanical properties do not differ between genders in a healthy community-dwelling elderly population. *Journal of Orthopaedic Research, 27* (6), 820–825.

Carr, A.J. and Norris, S.H. (1989) The blood supply of the calcaneal tendon. *The Journal of Bone and Joint Surgery, 71* (1), 100–101.

Chen, B.M. and Grinnell, A.D. (1995) Integrins and modulation of transmitter release from motor nerve terminals by stretch. *Science 269* (5230), 1578–1580.

Chiquet, M. (1999) Regulation of extracellular matrix gene expression by mechanical stress. *Matrix Biology*, *18*, 417–426.

Clancy, W.G. Jr., Neidhart, D. and Brand, R.L. (1976) Achilles tendonitis in runners: a report of five cases. *The American Journal of Sports Medicine*, *4* (2), 46–57.

Clayton, R.A. and Court-Brown, C.M. (2008) The epidemiology of musculoskeletal tendinous and ligamentous injuries. *Injury*, *39* (12), 1338–1344.

Curwin, S. and Standish, W.D. (1984) *Tendinitis: Its etiology and treatment*. Lexington, MA: The Collamore Press.

Evans, N.A., Bowrey, D.J. and Newman, G.R. (1998) Ultrastructural analysis of ruptured tendon from anabolic steroid users. *Injury*, *29* (10), 769–773.

Hamilton, B. and Purdam, C. (2004) Patellar tendinosis as an adaptive process: a new hypothesis. *British Journal of Sports Medicine*, *38* (6), 758–761.

Haut, R.C. and Powlison, A.C. (1990) The effects of test environment and cyclic stretching on the failure properties of human patellar tendons. *Journal of Orthopaedic Research*, *8* (4), 532–540.

Heinemeier, K.M., Olesen, J.L., Haddad, F., Schjerling, P., Baldwin, K.M. and Kjaer, M. (2009) Effect of unloading followed by reloading on expression of collagen and related growth factors in rat tendon and muscle. *Journal of Applied Physiology*, *106* (1), 178–186.

Holmes, G.B. and Lin, J. (2006) Etiologic factors associated with symptomatic achilles tendinopathy. *Foot and Ankle International*, *27* (11), 952–959.

Järvinen, T.A., Kannus, P., Paavola, M., Järvinen, T.L., Józsa, L. and Järvinen, M. (2001) Achilles tendon injuries. *Current Opinions in Rheumatology*, *13* (2), 150–155.

Johnson, G.A., Tramaglini, D.M., Levine, R.E., Ohno, K., Choi, N.Y. and Woo, S.L. (1994) Tensile and viscoelastic properties of human patellar tendon. *Journal of Orthopaedic Research*, *12* (6), 796–803.

Johnson, D.P., Wakeley, C.J. and Watt, I. (1996) Magnetic resonance imaging of patellar tendonitis. *The Journal of Bone and Joint Surgery*, *78* (3), 452–457.

Jozsa, L. and Kannus, P. (1997) Overuse injuries in tendons. In Jozsa, L. and Kannus, P. (Eds) *Human Tendons: anatomy, physiology and pathology*, pp 164–253. Champaign IL, Human Kinetics.

Kannus, P. and Józsa, L. (1991) Histopathological changes preceding spontaneous rupture of a tendon. A controlled study of 891 patients. *The Journal of Bone and Joint Surgery (American)*, *73* (10), 1507–1525.

Karlsson, J., Lundin, O., Lossing, I.W. and Peterson, L. (1991) Partial rupture of the patellar ligament. Results after operative treatment. *The American Journal of Sports Medicine*, *19* (4), 403–408.

Karlsson, J., Kälebo, P., Goksör, L.A., Thomée, R. and Swärd, L. (1992) Partial rupture of the patellar ligament. *The American Journal of Sports Medicine*, *20* (4), 390–395.

Ker, R.F. (1999) The design of soft collagenous load-bearing tissues. *The Journal of Experimental Biology*, *202*, 3315–3324.

Ker, R.F. (2002) The implications of the adaptable fatigue quality of tendons for their construction, repair and function. *Comparative Biochemistry and Physiology Part A*, *133*, 987–1000.

Khan, K.M., Cook, J.L., Kannus, P., Maffulli, N. and Bonar, S.F. (2002) Time to abandon the "tendinitis" myth. *British Medical Journal*, *324* (7338), 626–627.

Knobloch, K., Spies, M., Busch, K.H. and Vogt, P.M. (2007) Sclerosing therapy and eccentric training in flexor carpi radialis tendinopathy in a tennis player. *British Journal of Sports Medicine*, *41* (12), 920–921.

Komi, P.V. (1990) Relevance of in vivo force measurements to human biomechanics. *Journal of Biomechanics.*, *23* (Suppl 1), 23–34.

Komi, P.V., Fukashiro, S. and Järvinen, M. (1992) Biomechanical loading of Achilles tendon during normal locomotion. *Clinics in Sports Medicine*, *11* (3), 521–531.

Kvist, M. (1991) Achilles tendon injuries in athletes. *Annales Chirurgiae et Gynaecologiae*, *80* (2), 188–201.

Lavagnino, M., Arnoczky, S.P., Elvin, N. and Dodds, J. (2008) Patellar tendon strain is increased at the site of the jumper's knee lesion during knee flexion and tendon loading: results and cadaveric testing of a computational model. *The American Journal of Sports Medicine*, *36* (11), 2110–2118.

Leppilahti, J. and Orava, S. (1998) Total achilles tendon rupture. *Sports Medicine*, *25* (2), 79–100.

Liow, R.Y. and Tavares, S. (1995) Bilateral rupture of the quadriceps tendon associated with anabolic steroids. *British Journal of Sports Medicine*, *29* (2), 77–79.

Maffulli, N., Khan, K.M. and Puddu, G. (1998) Overuse tendon conditions: time to change a confusing terminology. *Arthroscopy*, *14* (8), 840–843.

Maniscalco, P., Gambera, D., Lunati, A., Vox, G., Fossombroni, V., Beretta, R. and Crainz, E. (2009) The "Cascade" membrane: a new PRP device for tendon ruptures. Description and case report on rotator cuff tendon. *Acta Biomedica*, *79* (3), 223–226.

Martinoli, C., Derchi, L.E., Pastorino, C., Bertolotto, M. and Silvestri, E. (1993) Analysis of echotexture of tendons with US. *Radiology 186* (3), 839–843.

Mokone, G.G., Gajjar, M., September, A.V., Schwellnus, M.P., Greenberg, J., Noakes, T.D. and Collins, M. (2005) The guanine-thymine dinucleotide repeat polymorphism within the tenascin-C gene is associated with achilles tendon injuries. *The American Journal of Sports Medicine*, 33 (7), 1016–1021.

Mokone, G.G., Schwellnus, M.P., Noakes, T.D. and Collins, M. (2006) The COL5A1 gene and Achilles tendon pathology. *Scandinavian Journal of Medicine and Science in Sports*, 16 (1), 19–26.

Muraoka, T., Muramatsu, T., Fukunaga, T. and Kanehisa, H. (2005) Elastic properties of human Achilles tendon are correlated to muscle strength. *Journal of Applied Physiology*, 99 (2), 665–669.

Niculescu, V. and Matusz, P. (1988) The clinical importance of the calcaneal tendon vasculature (tendo calcaneus). *Morphologie et Embryology*, 34 (1), 5–8.

Ohberg, L. and Alfredson, H. (2002) Ultrasound guided sclerosis of neovessels in painful chronic Achilles tendinosis: pilot study of a new treatment. *British Journal of Sports Medicine*, 36 (3), 173–177.

Onambélé, G.N., Burgess, K. and Pearson, S.J. (2007) Gender-specific in vivo measurement of the structural and mechanical properties of the human patellar tendon. *Journal of Orthopaedic Research*, 25 (12), 1635–1642.

Pflum, M.A., Shelburne, K.B., Torry, M.R., Decker, M.J. and Pandy, M.G. (2004) Model prediction of anterior cruciate ligament force during drop-landings. *Medicine and Science in Sports and Exercise*, 36 (11), 1949–1958.

Puddu, G., Ippolito, E. and Postacchini, F. (1976) A classification of Achilles tendon disease. *The American Journal of Sports Medicine*, 4 (4), 145–150.

Rees, J.D., Lichtwark, G.A., Wolman, R.L. and Wilson, A.M. (2008) The mechanism for efficacy of eccentric loading in Achilles tendon injury; an in vivo study in humans. *Rheumatology*, 47 (10), 1493–1497.

Riley, G.P., Harrall, R.L., Constant, C.R., Chard, M.D., Cawston, T.E. and Hazleman, B.L. (1994) Tendon degeneration and chronic shoulder pain: changes in the collagen composition of the human rotator cuff tendons in rotator cuff tendinitis. *Annals of the Rheumatic Diseases*, 53 (6), 359–366.

Riley, G.P., Harrall, R.L., Cawston, T.E., Hazleman, B.L. and Mackie, E.J. (1996) Tenascin-C and human tendon degeneration. *American Journal of Pathology*, 149 (3), 933–943.

Riley, G.P., Goddard, M.J. and Hazleman, B.L. (2001) Histopathological assessment and pathological significance of matrix degeneration in supraspinatus tendons. *Rheumatology*, 40 (2), 229–230.

Robinson, J.M., Cook, J.L., Purdam, C., Visentini, P.J., Ross, J., Maffulli, N., Taunton, J.E. and Khan, K.M. (2001) Victorian Institute of Sport Tendon Study Group. The VISA-A questionnaire: a valid and reliable index of the clinical severity of Achilles tendinopathy. *British Journal of Sports Medicine*, 35 (5), 335–341.

Schepsis, A.A. and Leach, R.E. (1987) Surgical management of Achilles tendinitis. *The American Journal of Sports Medicine* 15 (4), 308–315.

Schepsis, A.A., Jones, H. and Haas, A.L. (2002) Achilles tendon disorders in athletes. *The American Journal of Sports Medicine*, 30 (2), 287–305.

Sode, J., Obel, N., Hallas, J. and Lassen, A. (2007) Use of fluroquinolone and risk of Achilles tendon rupture: a population-based cohort study. *European Journal of Clinical Pharmacology*, 63 (5), 499–503.

Stäubli, H.U., Schatzmann, L., Brunner, P., Rincón, L. and Nolte, L.P. (1999) Mechanical tensile properties of the quadriceps tendon and patellar ligament in young adults. *The American Journal of Sports Medicine*, 27 (1), 27–34.

Unverferth, L.J. and Olix, M.L. (1973) The effect of local steroid injections on tendon. *The Journal of Sports Medicine*, 1 (4), 31–37.

Visentini, P.J., Khan, K.M., Cook, J.L., Kiss, Z.S., Harcourt, P.R. and Wark, J.D. (1998) The VISA score: an index of severity of symptoms in patients with jumper's knee (patellar tendinosis). Victorian Institute of Sport Tendon Study Group. *Journal of Science and Medicine in Sport*, 1 (1), 22–28.

Wang, J.H-C., Iosifidis, M.I. and Fu, F.H. (2006) Biomechanical basis for tendinopathy. *Clinical Orthopaedics and Related Research*, 443, 329–332.

Wells, J.B. (1965) Comparisons of mechanical properties between slow and fast mammalian muscles. *Journal of Physiology*, 178, 252–269.

Wilmink, J., Wilson, A.M. and Goodship, A.E. (1992) Functional significance of the morphology and micromechanics of collagen fibres in relation to partial rupture of the superficial digital flexor tendon in racehorses. *Research in Veterinary Science*, 53 (3), 354–359.

Xu, Y. and Murrell, G.A. (2008) The basic science of tendinopathy. *Clinical Orthopaedics and Related Research*, 466 (7), 1528–1538.

Yang, G., Im, H.J. and Wang, J.H. (2005) Repetitive mechanical stretching modulates IL-1beta induced

COX-2, MMP-1 expression, and PGE2 production in human patellar tendon fibroblasts. *Gene, 363,* 166–172.

Yepes, H., Tang, M., Morris, S.F. and Stanish, W.D. (2008) Relationship between hypovascular zones and patterns of ruptures of the quadriceps tendon. *The Journal of Bone and Joint Surgery (American), 90* (10), 2135–2141.

Zernicke, R.F., Garhammer, J. and Jobe, F.W. (1997) Human patellar-tendon rupture. *The Journal of Bone and Joint Surgery (American), 59* (2), 179–183.

Józsa, L., Kvist, M., Bálint, B.J., Reffy, A., Järvinen, M., Lehto, M., Barzo, M. The role of recreational sport activity in Achilles tendon rupture. A clinical, pathoanatomical, and sociological study of 292 cases. (1989). *American Journal of Sports Medicine. 17* (3), 338–43.

6

Pathophysiology of ligament injuries

Dror Steiner
Chartered Osteopath

As ligaments injuries are common both in athletes and the general public, this chapter's aim is to familiarise the reader with the anatomy, physiology and pathology of ligamentous structure and function, and then progress on to the healing process of ligamentous structures. Later on, this chapter will illustrate this process by introducing two of the most common ligamentous injuries and their available treatments.

Introduction

Ligaments of the skeletal system are dense connective tissues that attach bones across a joint. Ligaments play an important role in the neuro-muscular system by providing joints' stability and sending sensory feedback to the central nervous system with information about stress, tension, joints' motion, stretch and pain (Riemann and Lephart 2002a).

Anatomy

Fibroblasts are the cells that reside in ligaments producing both the matrix and the collagen fibre. Fibroblasts are arranged in longitudinal rows parallel to the fibres and communicate with each other by a gap junction, a mechanism that is poorly understood (Chi et al. 2005).

The matrix is composed of proteoglycans and collagen fibres. In ligaments the most common proteoglycan is decorin, which strengthens the links between collagen fibres (Ilic et al. 2005).

Ligaments are made up primarily of type I collagen, with normally 1–2% oelastic fibre (Frank 1996). During ligaments' healing, type III collagen levels increase and this is the reason for the weakening of ligaments at that point (Frank et al. 1987). Ligaments, such as ligamentum flavum and ligament nuchae, have a different structure with a higher content of elastic fibre, which allows recoil and saves muscular energy moving the spine back to the anatomical position from flexion (Yong-Hing and Reilly 1976).

Like other connective tissues, ligaments have a hierarchal structure (Figure 6.1).

Each ligament is built by multiple fascicle units. Each fascicle carries the collagen fibrils and fibroblasts cells in longitudinal lines and encapsulates them in a loose connective tissue called endoligament. Fascicles are separate enough to allow shearing movements occurring at different joint's position (Frank 1996). The entire ligament is covered with a vascular epiligament, a loose connective tissue that does not undergo much tension during the ligament's tightness (Ian et al. 2002).

The enthesis is the insertion of a ligament to the nearby bone and has a special arrangement (Figure 6.2). This is where the collagen fibres fan-out and are attached diagonally to bones via the periosteum. This arrangement helps in dissipate the ligament stress more evenly to the bone during different joint's position (Benjamin et al. 2006).

PATHOPHYSIOLOGY OF LIGAMENT INJURIES

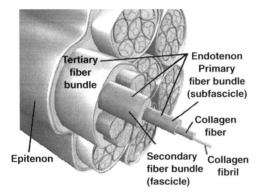

Figure 6.1 Ligament structure.

One character of the fascicle is the "crimps" – a waviness of the fibrils. The crimps give the ligaments its pseudo-elasticity and disappear once the ligament is stretched (Scott 2003).

Many ligaments are attached to the joint capsule, tendons and other connective tissue and cannot be separated anatomically and functionally, hence the common injuries where more than one tissue is damaged.

Blood supply

The blood supply to ligaments deserves a special examination as in most cases it is the limiting factor in healing injured ligaments. Compared with other tissue, ligaments are hypovascular and their blood supply is better closer to the bones' attachments; the middle section is poorly supplied (Bray et al. 1996).

The blood supply to ligaments may arrive from three places. The epiligament carries blood along the ligament, where blood vessels branch out to the endoligament and inside the fascicles. A build up of tension in the ligament will reduce the amount of blood circulating, but will recover at rest (Bray et al. 1996). The periosteal blood supply supplies mostly the enthesis region of the ligament. The surrounding connective tissues such as fat, joint capsule or muscles may carry some blood that collaterally supplies ligaments (Arnoczky 1985; Bray et al. 1996).

Following injury, the vascularity increases for about 40 weeks while the ligament's fibres become disorganised. The increased vascularity allows healing. The final post-injury state is a reduced vascularity (compare to the pre-injury level) with poor vessel organisation in the scar tissue. Ligaments probably do not have the ability to keep the original vascular organisation, causing reduced vascularity in the chronic healing stage, which may be the causes of higher level of reinjures.

Nerve supply

Ligaments carry two types of sensory impulses to the central nervous system: mechanoreceptor and pain.

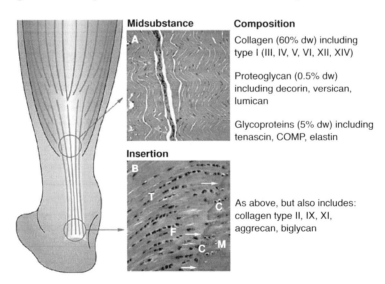

Figure 6.2 The enthesis. Reprinted, with permission, from Nature Clinical Practice in Rheumatology (2008) 4, 82–89 doi: 10.1038/ncprheum0700 © Nature Publishing (2008)

Mechanoreceptors signal mechanical events occurring in the tissue and have an important role in the coordinated motion pattern. Ruffini receptors are the most common mechanoreceptors in ligaments and joint capsule, whilst the others are Pacinian, Golgi tendon organ-like and free nerve endings. All of these receptors allowing the central nervous system to assess the amount of stress joints undergo and execute patterns of muscle contraction to help protect joints over stretching (Riemann and Lephart 2002b).

Following an injury both the peripheral and central nervous system undergo modifications.

One study found that the post-injury instability of the anterior crutiate ligament is due, in part, to remodelling of the central nervous system. The remodelling is probably due to the habitual reduction of usage of that area (Valeriani et al. 1996). As ligaments become strained and torn so the nerves supply loses a certain amount of receptors. This by itself can damage the peripheral nervous system and prevent it from accurately suppling the central nervous system with a real-time sensation of what happened to the joint.

In any case, following an injury it is probably both the central and peripheral nervous systems that are damaged and so supply corrupted sensory input to the central nervous system. This, in turn corrupts the muscular contraction pattern (Riemann and Lephart 2002b). One author's hypothesis is that some cases of back pain, for example, may originate from chronic ligamentous tension, which sends a corrupt sensory message to spinal muscles to contract continuously, causing those muscle to be in chronic tension and fatigue (Panjabi 2007).

Physiology

Ligaments have similar mechanical properties to other connective tissues: viscoelasticity and stress-strain (Norking and Levangie 2005).

The first mechanical property of a ligament is its non-linear stress-strain relationship (see Figure 6.3).

The stress-strain graph (Figure 6.3) demonstrates three regions:

1. Toe region - this is where the deformity (ligament length) is high while the force applied is low. Anatomically, this is where the fibrils' crimps slack begins to tighten. The toe region is where most ligaments are at rest.

2. Linear region - this is where a stress build up creates a linear build up of strain or stretch to the ligament. This region demonstrates ligaments during a normal joint's movement.

3. Failure region - this is where even a mild increased stress to the ligament creates a large deformation as the ligament is overstretched or torn.

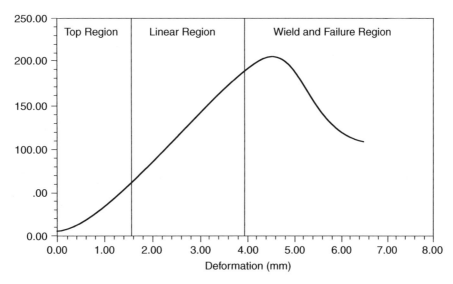

Figure 6.3 Stress-Strain graph. Reprinted from Mow, V.C. and Hayes, W.C. 'Basic Orthopaedic Biomechanics' (ISBN: 9780881677966) © 1991, Lippencott, Williams and Wilkins

98 PATHOPHYSIOLOGY OF LIGAMENT INJURIES

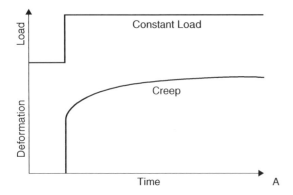

Figure 6.4 Ligament creep. Reprinted from Mow, V.C. and Hayes, W.C. 'Basic Orthopaedic Biomechanics' (ISBN: 9780881677966) © 1991, Lippencott, Williams and Wilkins

The stress-strain graph never perfectly occurs as the ligament also exhibits viscoelasticity property. Viscoelasticity is a tendency of the tissue to stretch and return slowly to its normal form whilst dampening the shearing force. There are three viscoelasticity tissue behaviours: creep, stress-relaxation and hysteresis.

Creep is the tendency of slowly increasing the ligament's deformation under a load and the return to normal shape once the load is taken away (Figure 6.4). For example, under a load the ligament will slowly elongate, and then return to its original length once the load is taken away. The rate of creep increases at high temperature.

The second viscoelasticity property is stress relaxation (Figure 6.5). This means that under a constant deformation the stress will be reduced. For example, if a ligament will elongate during load, the stress will be reduced compared to a situation where the ligament will stay at a constant length.

The third viscoelastistic property is hysteresis (Figure 6.6). Under load, ligaments undergo deformation that does not immediately return to the original length after unloading the joint. The difference between the length before loading and after unloading represents the amount of energy lost in the process. If, for example, the joint is loaded and unloaded many times, the stress-strain curve would move to the right, showing an increase strain to the same stress.

Figure 6.6 illustrates how loading a ligament many times in a short period can cause a strain or a tear if there is not enough recovery time between the cycles.

Normal changes to ligament through life

Ligaments exhibit adaptation to external and internal changes.

The effect of exercise and long-term immobility to ligaments are known and directly relate to the property of collagen fibrils to increase its thickness in response to exercise and reduce thickness in response to immobility (Kannus et al. 1992).

There are no differences between pre-puberty male and female ligamentous structure. However, following puberty females show an increased joint laxity (Quatman et al 2008). Before puberty the rate

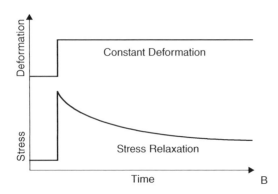

Figure 6.5 Stress relaxation relationship in ligaments. Reprinted from Mow, V.C. and Hayes, W.C. 'Basic Orthopaedic Biomechanics' (ISBN: 9780881677966) © 1991, Lippencott, Williams and Wilkins

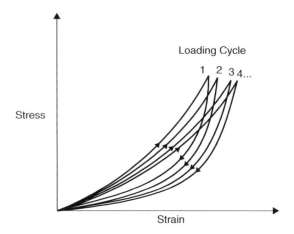

Figure 6.6 Hysteresis. Reprinted from Mow, V.C. and Hayes, W.C. 'Basic Orthopaedic Biomechanics' (ISBN: 9780881677966) © 1991, Lippencott, Williams and Wilkins

of anterior crutiate ligament (ACL) injury is equal male to female, but the post-puberty injury rate is at least 3-fold higher in adolescent females (Agel et al. 2005).

The knee crutiate ligaments express the sex hormone receptors for oestrogen, testosterone and relaxin (Faryniarz et al. 2006). There is an increased risk of ACL strain during the pre-ovulation stages of menstrual cycle, where hormone levels change from a low to high level (Slauterbeck et al. 2002). Due to different individual hormonal profiles during menstrual cycle there is a significant difference to the amount of ligamentous laxity and risk of injury (Shultz et al. 2006).

In the elderly there is an increase in joint capsule and ligaments laxity, which may be one of the reasons for the increase in incidents of osteoarthritis (Rudolph et al. 2007).

Pathology

Pathological changes in ligaments may occur due to structural and functional failure. Any strain to a ligament may cause a long-term joint instability. There are agreed degrees of ligaments strain: 1st degree is mild, 2nd degree moderate and 3rd degree is a complete tear (Chen et al. 2008).

Following a strain, ligaments do not heal by producing an identical tissue; instead, a scar tissue is formed. The scar tissue presents with an uneven matrix, smaller in diameter collagen fibers, weaker collagen crosslinking and a limited creep (Figure 6.7).

The blood circulation is also affected with a long-term reduction in blood circulation to the scar tissue (Bray et al. 1996). Although scar tissue is formed relatively fast, there are still mechanical and chemical changes months and years after the injury, though it never has a normal appearanceor function again, resulting in a high level of reoccurrences (Frank et al. 1999). Ankle sprains, for example, reoccur in 73% of all athletes (Yeung et al. 1994).

As ligaments provide the proprioceptive sensory nerve supply to the central nervous system, any ligament injury may cause further problems in the neuromuscular system.

Treatment and the healing process

Following a strain injury, the gap within the ligament fills with blood to start the inflammation phase. Fibroblasts proliferate, the ligaments revascularise

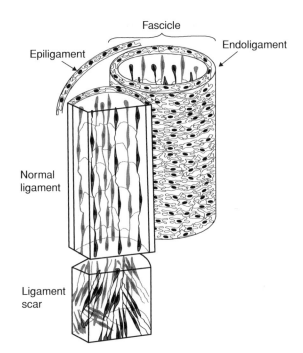

Figure 6.7 The healing ligament and creating of a scar tissue. Reprinted from Mow, V.C. and Hayes, W.C. 'Basic Orthopaedic Biomechanics' (ISBN: 9780881677966) © 1991, Lippencott, Williams and Wilkins

and the gap becomes filled with scar tissue (Frank et al. 1992). Fibroblasts start producing scar tissue made by type III collagen. This collagen-type material creates a rapid, disorganised structure with weaker cross links to fill the gap between the ligament's edges quickly (Frank et al. 1987).

Followed is the long remodelling period, which can take months or even years as the matrix and collagen fibres are rearranged to have stronger bonds.

The healing process depends on few parameters: an isolated strain heals better than when it combines with other tissue's injury; the degree of the injury; certain ligaments heal better than others (such as the MCL compared with the ACL); and failure of the repair process can appear months post injury (Frank et al. 1992). The viewpoints of the main two treatments are the immobility, long rest and braces versus the sooner-than-later active treatment (see below).

A systematic review finds no data to support any benefits from using braces in sprained ligament (Pietrosimone et al. 2008).

Another study found that early activity, rather than long immobility, would make the healing time shorter, more complete and that the ligaments would appear stronger (Woo et al. 1987; Frank et al. 1992). Immobility affects the ligament on a cellular level with the fibroblasts appearing uneven.

For grades I–II a non-operative active treatment approach is the most common (Chen et al. 2008) and usually comprises rest (for a limited period), ice, compression and elevation (RICE), isometric and isotonic exercises and proprioceptive training. The rationale of using the RICE treatment is to reduce the inflammation during the acute phase.

Proprioceptive training concentrates mostly on the lower limb, in particularly post sprain in the ankle joint. The idea behind it is to maintain the proprioceptive neural function by continuously using the joint in a challenged position. It has been found that even the use of unsupervised, home proprioceptive exercises reduces the rate of reoccurance in atheletes (Hupperets et al. 2009).

Another treatment method is exercise, which aims to improve the proprioceptive property of the damaged ligament and to strengthen the muscles surrounding the affected joint (Olsen et al. 2005).

Examples of ligament injuries
Lateral ankle sprain

The ankle joint is one of the most common sites of ligament injury with a higher reoccurrence rate compared to other ligament strains (Hubbard and Hicks-Little 2008). It occupies 0.6% of all cases in Accident and Emergency in the U.K., with higher rate of 1.3% in young girls (Bridgman et al. 2003). At 14%, ankle sprains are the most common acute sports injury (Fong et al. 2009) with 59% of reoccurrence (Yeung et al., 1994).

The most common ligamentous injury is to the lateral ankle, and to the anterior talofibular ligament in particular (Daniel et al. 2009).

The anterior talofibular ligament is the weakest ankle joint ligament while limiting a relatively large inversion at the subtar joint (Attarian et al. 1985).

Lateral ankle injury happens mostly by an excessive supination due to mechanical and/or functional instability. Mechanical instability can arise due to certain muscular weaknesses, knee or other joint problems, or degenerative changes. Functional instability can arise from a lack of proprioception or coordination (Hertel 2002).

Increased factors for sprain injuries are previous sprains, wearing shoes with air cells, being overweight, not stretching before sports activities, increased foot width and weak active eversion and dorsiflexion (Daniel et al. 2009). Immobilisation of the sprained joint was found to be less effective than active treatment (Kerkhoffs et al. 2001).

The most common treatment for grade I and II during an acute sprained ankle is RICE (Cooke et al. 2003). Early exercise, which includes dorsiflexion and plantarflexion range-of-motion exercises, and isometric and isotonic strength-training exercises are recommended by many authors (Trevino et al. 1994; Lynch and Renstrom 1999; Safran et al. 1999).

Another treatment is balanced and coordination training performed on the wobble board, both as a treatment and preventative measurements (Mattacola and Dwyer 2002). It appears to halve the reoccurrence of a sprained ankle ligament. This exercise enhances the proprioceptive neural function of the central nervous system and reduce the response time of the peroneal and other muscles around the ankle joint to protect the ligaments from reinjury (McKeon and Hertel 2008).

A meta-analysis found that for grade III sprain ankle the possibility of surgical ligament reconstruction is controversial (Lynch and Renstrom 1999). This research found that only late ligament reconstruction appears to be better than a conservative treatment after all other treatment modalities fail to produce joint stability.

Anterior crutiate ligament sprain

ACL resists anterior tibial translation on the femur and rotational motion at the knee joint. ACL can withstand multiple stresses and varying tensile strains at the same time, a property that is not yet achieved with any ACL artificial reconstruction (Duthon et al. 2006). Non-contact ACL injuries can arise during deceleration and acceleration motions, an excessive quadriceps contraction together with reduced hamstrings co-contraction, where the knee internally rotated at or near full knee extension (Shimokochi and Shultz 2008). In all of those position there is a higher force applies on the ACL. ACL injury is more common in contact sports, with the highest rate in footballers (Hootman et al. 2007). The exact damage to the ligament will determine the length of the healing process. For example, if the blood supply pathway from the patellar fat pad of the synovial membrane is also damaged, the healing time will take longer (Toy et al. 1995).

A procedure of warm up to improve running, changing running direction quickly, landing from jumping technique, balance and strength was found to significantly reduce knee injury rate (Olsen et al. 2005).

For a complete ACL rupture, ligament reconstruction is one of the treatment options (Woo et al. 2006). The new ligament is taken from the hamstring or quadriceps tendons and can be construct from a single or double bundle.

Summary key points of ligaments

- Ligaments offer joints stability with sensory feedback to the central nervous system

- Ligaments' blood supply is the limiting factor in many healing process of sprain

- Ligaments heal by the process of laying down scar tissue, which exhibits structurally different tissue organisation and is weaker compared to the original ligament

- Most frequent treatment for grade I–II sprains are RICE, early mobilisation, isometric and isotonic strengthening exercise, heuromuscular rehabilitation and return to normal function as soon as possible. For grade III injury, reconstructive surgery is an option.

References

Agel, J., Arendt, E.A. and Bershadsky, B. (2005) Anterior cruciate ligament injury in national collegiate athletic association basketball and soccer: a 13-year review. *American Journal of Sports Medicine*, *33* (4),524–530.

Arnoczky, S.P. (1985) Blood supply to the anterior cruciate ligament and supporting structures. *Orthopaedic Clinic of North America*, *16* (1), 15–28.

Attarian, D.E., McCrackin, H.J., DeVito, D.P., McElhaney, J.E. and Garrett, W.E. (1985) A biomechanical study of human ankle ligaments and autogenous reconstructive grafts. *American Journal of Sports Medicine*, *13*, 377–381.

Benjamin, M., Toumi, H., Ralphs, J.R., Bydder, G. Best, T.M. and Milz, S. (2006) Where tendons and ligaments meet bone: attachment sites ('entheses') in relation to exercise and/or mechanical load. *Journal of Anatomy*, *208* (4), 471–490.

Bray R.C., Rangayyan R. and Frank, C.B. (1996) Normal and healing ligament vascularity: a quantitative histological assessment in the adult rabbit medial collateral ligament. *Anatomy*, *188*, 87–95,

Bridgman, S.A., Clement, D., Downing, A., Walley, G., Phair, I. and Maffulli, N. (2003) Population based epidemiology of ankle sprains attending accident and emergency units in the West Midlands of England, and a survey of UK practice for severe ankle sprains. *Emergency Medicine Journal*, *20*, 508–510.

Chen, L., Kim, P.D., Ahmad, C.S. and Levine, W.L. (2008). Medial collateral ligament injuries of the knee: current treatment concepts. *Current Reviews in Musculoskeletal Medicine*, *1* (2), 108–113.

Chi, S.S., Rattner, J.B., Sciore, P., Boorman, R. and Lo, I.K.Y. (2005) Gap junctions of the medial collateral ligament: structure, distribution, associations and function. *Journal of Anatomy*, *207* (2), 145–154.

Cooke, M.W., Lamb, S.E., Marsh, J. and Dale, J. (2003) A survey of current consultant practice of treatment of severe ankle sprains in emergency departments in the United Kingdom. *Emergency Medicine Journal*, *20*, 505–507.

Duthon, V.B., Barea, C., Abrassart, S., Fasel, J.H., Fritschy, D. and Menetrey, J. (2006) Anatomy of the anterior cruciate ligament. *Knee Surgery, Sports Traumatology, Arthroscopy*, *14*, 204–213.

Faryniarz, D.A., Bhargava, M., Lajam, C., Attia, E.T. and Hannafin, J.A. (2006) Quantitation of estrogen receptors and relaxin bindingin human anterior cruciate ligament fibroblasts. *In Vitro Cellular and Developmental Biology - Animal*, *42* (7), 176–181.

Fong, D.T.P., Chan, Y-Y., Mok, K-M., Yung, P.S.H. and Chan, K-M. (2009) Understanding acute ankle ligamentous sprain injury in sports. *Sports Medicine, Arthroscopy, Rehabilitation, Therapy and Technology*, *1*, 14.

Frank, C.B. (1996) Ligament injuries? Pathophysiology and healing. In Zachazewski, J.E., Magee, D.J. and Quillen, W.S. (Eds), *Athletic Injuries, and Rehabilitation*. Philadelphi, PA: WB Saunders.

Frank C., Woo, S., Andriacchi, T., Brand, R., Oakes, B., Dahners, L., DeHaven, K., Lewis, J. and Sabiston, P. (1987) Normal ligament: structure, function and composition. In Woo, S.B.J.A. (Ed.), Injury and Repair of the Musculoskeletal Soft Tissues, pp 45–101. American Academy of Orthopaedic Surgeons.

Frank, C., Hart, D.A. and Shrive, N.G. (1999) Molecular biology and biomechanics of normal and healing ligaments – a review. *Osteoarthritis Cartilage 7*, 130–140.

Fong, D.T.P., Chan,Y.Y., Mok, K.M., Yung, P.S.H. and Chan, K.M. (2009) Understanding acute ankle ligamentous sprain injury in sports. *Sports Medicine Arthroscopy Rehabilitation Therapy Technology*, *1*, 14.

Hertel, J. (2002) Functional anatomy, pathomechanics and pathophysiology of lateral ankle instability. *Journal of Athletic Training*, *37* (4), 364–375.

Hootman, J., Randall, D. and Agel, J. (2007) Epidemiology of collegiate injuries for 15 sports: Summary and recommendations for injury prevention initiatives. *Journal of Athletic Training*, *42* (2), 311–319.

Hubbard, T.J. and Hicks-Little, C.A. (2008) Ankle ligament healing after an acute ankle sprain: An evidence-based approach. *Journal of Athletic Training*, *43* (5), 523–529.

Hupperets, M.D.W., Verhagen, E.A.L.M. and van Mechelen, W. (2009) Effect of unsupervised home based proprioceptive training on recurrences of ankle sprain: randomised controlled trial. *British Medical Journal*, *339*, 2684.

Ilic, M.Z., Carter, P., Tyndall, A., Dudhia, J. and Handley, C.J. (2005) Proteoglycans and catabolic products of proteoglycans present in ligament. *Biochemistry Journal*, *15*, 385 (Pt 2), 381–388.

Kannus, R., Jòzsa, L., Renström, R., Järvtoen, M., Kvist, M., Lento, M., Oja, P. and Vuorl, I. (1992) The effects of training, immobilization and remobilization on musculoskeletal tissue. *Scandinavian Journal of Medicine & Science in Sports*, *2* (3) 100–118.

Kerkhoffs, G.M., Rowe, B.H., Assendelft, W.J., Kelly, K.D., Struijs, P.A. and van Dijk, C.N. (2001) Immobilisation for acute ankle sprain. A systematic review. *Archives of Orthopaedic and Trauma Surgery*, *121*, 462–471.

Lynch, S.A. and Renstrom, P.A. (1999) Treatment of acute lateral ankle ligament rupture in the athlete. Conservative versus surgical treatment. *Sports Medicine*, *27*, 61–71.

Mattacola, C.G. and Dwyer, M.K. (2002) Rehabilitation of the ankle after acute sprain or chronic instability. *Journal of Athletic Training*, *37* (4), 413–429.

McKeon, P.O. and Hertel, J. (2008) Systematic review of postural control and lateral ankle instability, part II: Is balance training clinically effective. *Journal of Athletic Training*, *43* (3), 305–315.

Norking, C. and Levangie, P. (2005) *Joint Structure And Function: A Comprehensive Analysis*, 3rd edn. F.A. Davis.

Olsen, O-E., Myklebust, G., Engebretsen, L., Holme, I. and Bahr, R. (2005) Exercises to prevent lower limb injuries in youth sports: cluster randomised controlled trial. *British Medical Journal*, *330* (7489), 449.

Panjabi, M. (2007) Letter to the Editor concerning "A hypothesis of chronic back pain: ligament subfailure injuries lead to muscle control dysfunction". *European Spine Journal*, *16*, 1733—1735.

Pietrosimone, B.G., Grindstaff, T.L., Linens, S.W., Uczekaj, E. and Hertel, J. (2008) A systematic review of prophylactic braces in the prevention of knee ligament injuries in collegiate football players. *Journal of Athletic Training*, *43* (4), 409–415.

Quatman, C.E., Ford, K.R., Myer, G.D., Paterno, M.V. and Hewett, T.E. (2008) The effects of gender and maturational status on generalized joint laxity in young athletes. *Journal of Science and Medicine in Sport*, *11* (3), 257–263.

Riemann, B.L. and Lephart, S.M. (2002a) The sensorimotor system, Part I: The physiologic basis of functional joint stability. *Journal of Athletic Training*, *37* (1), 71–79.

Riemann, B.L. and Lephart, S.M. (2002b) The sensorimotor system, part II: The role of proprioception in motor control and functional joint stability. *Journal of Athletic Training*, *37* (1), 80–84.

Rudolph, K.S., Schmitt, L.C. and Lewek, M.D. (2007) Age-related changes in strength, joint laxity, and walking patterns: Are they related to knee osteoarthritis? *Physical Therapy 87* (11), 1422–1432.

Safran, M.R., Zachazewski, J.E., Benedetti, R.S., Bartolozzi, A.R., 3rd and Mandelbaum, R. (1999) Lateral ankle sprains: a comprehensive review: part 2: treatment and rehabilitation with an emphasis on the athlete. *Medicine and Science in Sports and Exercise*, *31*, 438–447.

Scott, J.E. (2003) Elasticity in extracellular matrix 'shape modules' of tendon, cartilage, etc. A sliding proteoglycan-filament model. *Journal of Physiology*, *553* (2), 335–343.

Shimokochi, Y. and Shultz, S. (2008) Mechanisms of non-contact anterior cruciate ligament injury. *Journal of Athletic Training*, *43* (4), 396–408.

Shultz, S.J., Gansneder, B.G., Sander, T.C., Kirk, S.E. and Perrin, D.H. (2006) Absolute hormone levels predict the magnitude of change in knee laxity across the menstrual cycle. *Journal of Orthopaedic Research*, *24* (2), 124–131.

Slauterbeck, J.R., Ruzie, S.F., Smith, M.P., et al. (2002) The menstrual cycle, sex hormones, and anterior cruciate ligament injury. *Journal of Athletic Training*, *37* (3), 275–280.

Toy, B.J., Yeasting, R.A., Morse, D.E. and McCann, P. (1995) Arterial supply to the human anterior crutiate ligament. *Journal of Athletic Training*, *30* (2), 149–152.

Trevino, S.G., Davis, P. and Hecht, P.J. (1994) Management of acute and chronic lateral ligament injuries of the ankle. *Orthopedic Clinics of North America*, *25*, 1–16.

Valeriani, M., Restuccia, D., DiLazzaro, V., Franceschi, F., Fabbriciani, C. and Tonali, P. (1996) Central nervous system modifications in patients with lesion of the anterior cruciate ligament of the knee. *Brain*, *119* (Pt 5), 1751–1762.

Woo, S.L-Y., Wu, C., Dede, O., Vercillo, F. and Noorani, S. (2006) Biomechanics and anterior cruciate ligament reconstruction. *Journal of Orthopaedic Surgery*, *1*, 2.

Yeung, M.S., Chan, K.M., So, C.H. and Yuan, W.Y. (1994) An epidemiological survey on ankle sprain. *British Journal of Sports Medicine*, *28*, 112–116.

Yong-Hing, K, Reilly, J. and Kirkaldy-Willis, W.H. (1976) The ligamentum flavum. *Spine*, *1* (4), 226–234.

Woo, S.L., Inoue, M., McGurk-Burleson, E. and Gomez, M.A. (1987) Treatment of the medial collateral ligament injury. II: Structure and function of canine knees in response to differing treatment regimens. *American Journal of Sports Medicine*, *15* (1), 22–29.

7
Pathophysiology of skeletal injuries

Sarah Catlow
University College Plymouth St Mark & St John, Plymouth

Introduction

Skeletal injuries are common in sport, especially in contact sports, such as football and rugby, and in individual sports such as skiing and gymnastics. This chapter provides an overview of the skeleton and its component parts. Special attention is paid to the process of bone formation or ossification and its unique implications for healing following skeletal injuries. The chapter further considers principles of rehabilitation in relation to skeletal injury and pathology. The use of annotated diagrams and schema, within the chapter, further emphasise important processes in skeletal organisation (Figure 7.1).

The skeleton can be divided into two subgroups:

1. Axial skeleton – bones of the skull, vertebral column, ribs and sternum

2. Appendicular skeleton – bones of the upper and lower limbs.

Bone is a living, well-organised, vascular form of connective tissue. It is largely composed of an organic protein, collagen, and an inorganic mineral, hydroxyapatite, which combine to provide a mechanical and supportive role in the body (Smith et al. 1983). Bone is a dynamic tissue that requires stress for normal development. The capacity of bone to adapt its structure to imposed loads has become known as Wolff's Law:

> **Wolff's law:**
>
> "states that bone responds to the stresses that are imposed upon it by rearranging its initial architecture in the best way to withstand stress (Porter 2008)

Bones function as:

- Protectors of vital organs – provides mechanical protection for most of the body's internal organs, thereby reducing the risk of injury to them

- Supportive structures – the skeleton is the framework of the body it provides attachment for skeletal muscles

- Levers – the skeleton assists with movement

- Reservoirs for calcium and phosphorus – storage for minerals (calcium and phosphorus), which are released when needed into the blood

- Blood producing cells – develops red blood cells in the bone marrow.

106 PATHOPHYSIOLOGY OF SKELETAL INJURIES

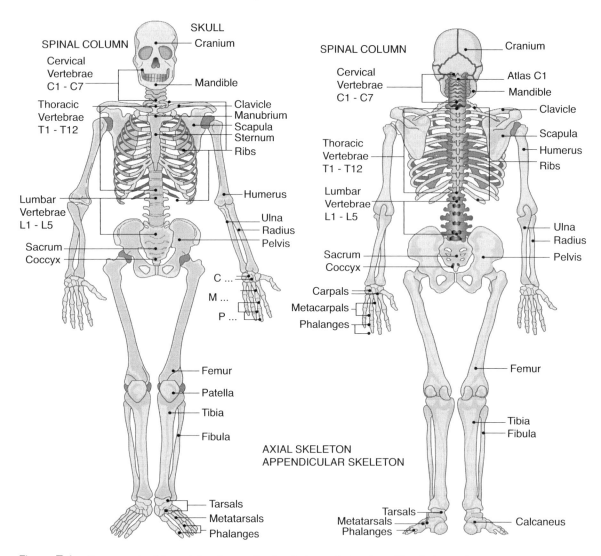

Figure 7.1 Components of the axial and appendicular skeletons (taken from Google images).

Bone structure

There are primarily three types of bone, namely: woven, compact and cancellous (Figure 7.2).

Woven bone

Woven bone is normally remodelled and replaced with either compact or cancellous bone. Woven bone is found during embryonic development, during fracture healing (callus formation), and in some pathological states, such as hyperparathyroidism and Padget Disease (Recker et al. 1992).

Compact bone

Compact bone is the outer structure and provides mechanical strength, while cancellous bone forms the inner structure and its function is the metabolic unit of the bone (Figure 7.3). Compact bone is dense bone and surrounds the cancellous bone. The primary structural unit of compact bone is an osteon, which is also known as a Haversian system. Osteons consist of cylindrical shaped lamellar bone that surrounds longitudinally oriented vascular channels called Haverisan canals; horizontally

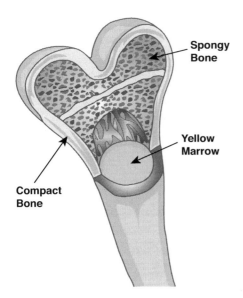

Figure 7.2 Illustrate the structure of bone (taken from Google images).

oriented vascular channels, which are known as Volkmann canals, connect adjacent osteons.

Cancellous bone

Cancellous (spongy) bone consists of spicules of bone enclosing cavities containing marrow (blood-forming cells). This type of bone is strong but the spaces make it light and flexible. Cancellous bone is always covered and therefore protected by compact bone. In long bones, cancellous bone is found in the epiphysis (at the end of the long bone); in some it also extends down inside into the shaft. In all other bones, it forms the central mass of bone within a compact bone lining.

In adults, 80% of the skeleton is compact bone. However, the relative proportions of compact and cancellous bone vary in different parts of the skeleton. For example, in the lumbar spine, cancellous bone accounts for about 70% of the total bone tissue, whereas in the femoral neck and radial diaphysis, it accounts for about 50% and 5%, respectively (Kanis 1994; Einhorn 1996; Fleisch 1997).

Bone covering

The periosteum is a membrane that lines the outer surface of all bones, (Netter 1987) except in the area of a joint where articular cartilage is present. It consists of dense irregular connective tissue. The periosteum is divided into a fibrous layer (outer) and an oseogenic layer (inner). The fibrous layer contains fibroblasts, while the oseogenic layer contains progenitor cells that develop into osteoblasts, which are cells that form and repair bones. The periosteum has nociceptors nerve endings present and is sensitive to injury. It also provides nourishment by providing the blood supply. It is attached to bone by strong collagenous fibres called Sharpey's fibres, which extend to the outer circumferential and interstitial lamellae. It also provides an attachment for muscles and tendons. Bone, including the marrow, periosteum, metaphysic, diaphysis and epiphysis are richly supplied with blood vessels. Studies by Shim et al. (1967) and Tothill and MacPherson (1986) reported that approximately 7% of the cardiac output is sent to the skeleton.

Classification of bones

Bones are usually classified according to their shape. Table 7.1 summarises the main categories of bones.

Bone formation and growth

Ossification is the name given to the formation of bone. There are three cells involved in bone metabolism:

1. Osteoblasts (bone forming cells) – mononuclear cells of mesenchymal origin.

2. Osteoclasts (bone eating cells).

3. Osteocytes (cells of the matrix) – found in mature adult bone.

Signalling pathways between these cells help regulate the balance between bone formation and bone reabsorption.

Osteoblasts are the bone-forming cells and they originate from local mesenchymal stem cells (bone marrow stroma or connective tissue mesenchyme). These stem cells undergo proliferation and differentiate to preosteoblasts and then to mature osteoblasts (Triffitt 1996). The plasma membrane of osteoblasts is rich in alkaline phosphatase, which enters the systemic circulation. The plasma concentration of this

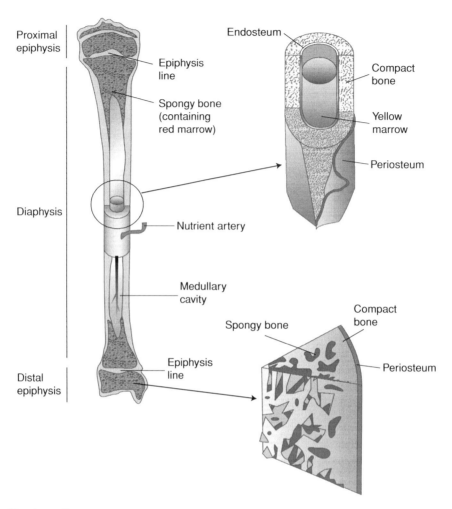

Figure 7.3 Structure of bone.

enzyme is used as a biochemical marker of bone formation. Osteocytes originate from osteoblasts embedded in the organic bone matrix, which subsequently become mineralised. They have numerous cell processes forming a network of thin canaliculi that connects them with active osteoblasts and flat lining cells. Osteocytes probably play a role in the homeostasis of this extracellular fluid and in the local activation of bone formation and/or resorption in response to mechanical loads (Nijweide et al. 1996).

Osteoclasts are giant cells containing 4–20 nuclei that reabsorb bone. Osteoclastic reabsorption takes place at the cell/bone interface in a sealed-off microenvironment (Baron 1996; Teitelbaum et al., 1996). The mechanism of bone resorption involves the secretion of hydrogen ions and proteolytic enzymes into the sub-osteoclastic resorbing compartment. The hydrogen ions dissolve the bone minerals, thereby exposing the organic matrix to the proteolytic enzymes (Baron 1996; Teitelbaum et al., 1996). These enzymes, which include collagenases and cathepsins, are responsible for the breakdown of the organic matrix.

Bone metabolism is under constant regulation by a host of hormonal and local factors. Three of the calcitropic hormones that most effect bone metabolism are parathyroid hormone, vitamin D and calcitonin (Table 7.2).

Table 7.1 Categorisation of bones

Type of bone	Function
Long bones	This type of bone functions as a lever
	They have greater length than width and consist of a shaft and a variable number of endings (extremities)
	They are usually somewhat curved for strength
	Examples include femur, tibia, fibula, humerus, ulna and radius
Flat bones	This type of bone has a broad surface for muscle attachment and is used for the protection of underlying organs
	Examples include cranial bones (protecting the brain), the sternum and ribs (protecting the organs in the thorax), and the scapulae.
Sesamoid bones	This type of bone develops in some tendons in locations where there is considerable friction, tension and physical stress. However their presence and quantity varies considerably from person to person
	Examples include patellae and under 1st metatarsal
Irregular bones	This type of bone has complicated shapes and so cannot be classified into any of the above (shape-based) categories. Their shapes are due to the functions they fulfil within the body e.g. providing major mechanical support for the body yet also protecting the spinal cord (in the case of the vertebrae)
	Examples include the vertebrae and some facial bones
Short bones	These types of bone are roughly cube-shaped and have approximately equal length and width
	Examples include ankle and wrist bones

Cartilage

Cartilage is a non-vascular connective tissue that is divided, according to its minute structure, into:

- hyaline cartilage (articular) – covers joint surfaces

- fibrocartilage – knee meniscus, vertebral discs

- elastic cartilage – outer ear

These different forms of cartilage are distinguished by their structure, elasticity, and strength.

In general, cartilage is a tough, fibrous and blood vessel-free connective tissue that forms flexible linkages, supporting structures and acts as a shock absorber in joints such as the knee.

Hyaline (articular) cartilage is the most common type of cartilage. In addition to being found in articulated joints, hyaline cartilage forms the majority of the skeleton of a fetus. Later in fetal development,

Table 7.2 Hormones that most effect bone metabolism

Hormone	Description
Parathyroid hormone	Produced via the parathyroid glands, which are small endocrine glands in the neck
	Humans have four parathyroid glands, which are usually located behind the thyroid gland
	Increases the flow of calcium into the calcium pool
	Maintains body's extracellular calcium pool level at a relatively constant level
	Osteoblasts are the only cells that have parathyroid hormone receptors
	Parathyroid hormone has antagonistic effects to those of calcitonin.
Vitamin D	Fat soluble molecule
	Stimulates intestinal and renal calcium binding proteins and facilitates active calcium transport
	Inhibits parathyroid hormone secretion
Calcitonin	Serves to inhibit calcium dependent cellular metabolic activity

it is replaced by bone. The free surfaces of most hyaline cartilage (except that found in joints) are covered by a layer of fibrous connective tissue known as perichondrium. The perichondrium is rich in a type of cell known as the fibroblast. Compositionally, hyaline cartilage is made of water (75% by weight), collagen (10% by weight, mainly collagen type II), with the remainder being non-fibrous material, such as chondroitin sulphate and keratan sulphate. The collagen provides strength and makes hyaline cartilage resistant to compression. Also, the collagen provides a means by which the cartilage can be anchored to bone.

Cartilage is a metabolically active tissue that under normal conditions is maintained in a relatively slow state of turnover by a sparse population of chondrocytes distributed throughout the tissue (Naujok et al. 2008).

Common cartilage injuries

It is well known that lesions which are confined to the hyaline cartilage alone have little or no capacity to heal (Naujoks et al. 2008). In general, the individual becomes symptomatic and a significant progression to osteoarthritis is possible (Lohmander 2003).

Osteoarthritis (OA) is the most common form of arthritis, which is a leading cause of physical disability, increased healthcare usage and impaired quality of life (Felson 1990; Guccione et al. 1994). The term OA also applies particularly to the degeneration and excessive wear of cartilage. This condition develops and progresses with an increase in age. Epidemiological studies have demonstrated that participation in certain competitive sports increase the risk for OA (Kujala et al. 1994; Buckwalter and Lane 1997). Moderate regular running has low, if any risk leading to OA (Lane et al. 1993; Newton et al. 1997). Sport activities that appear to increase the risk for OA include those that demand high-intensity, acute, direct joint impact as a result of contact with other participants, playing surfaces or equipment (Buckwalter and Lane 1997).

Common skeletal injuries and their manifestation

As previously stated bone provides structural support and protection, facilitating movement and mineral storage. Therefore, injuries to bone can compromise any of these functions and interrupt daily functions.

Bone fractures

Fractures are potentially serious injuries, damaging not only the bone but also the soft tissue in the surrounding area (diagram of types of fractures are illustrated in Figure 7.4) (Table 7.3). Although the bone tissue itself contains no nociceptors, bone fracture can be very painful, due to (1) the breaking in the continuity of the periosteum; (2) oedema of nearby soft tissues, caused by bleeding of torn periosteal blood vessels, evoking pressure pain; and (3) spasms in muscles trying to hold bone fragments in place.

The severity of a fracture depends on its location and the damage done to the bone and tissue near it. Serious fractures can have dangerous complications if not treated promptly; possible complications include damage to blood vessels or nerves and infection of the bone (osteomyelitis) or surrounding tissue.

Clinical features of a fracture

The features of a fracture are many and varied, depending on the cause and nature of the injury. The clinical features are listed below:

- pain
- deformity
- oedema
- muscle spasm
- abnormal movements
- loss of function
- shock
- limitation of joint movement.

Stress fractures

Stress related bone injuries have become commonplace amongst the members of our increasingly active society and account for up to 10% of cases in a typical sport medicine practice (Jones et al. 1989)

Table 7.3 Types of bone fracture

Type of fracture	Description of fracture
Compound (open)	Occurs when the sharp ends of the broken bone protrude through the individual's skin
Closed	Skin remains intact
Depressed or fissured	Occurs when a sharp localised blow depresses a segment of cortical bone below the level of surrounding bone – example – fractured skull
Greenstick	Seen in children – fracture is on one side of the bone but does not tear the periosteum of the opposite side
Spiral	This is caused by opposite rotator forces pulling on the bone
Oblique	The fracture is oriented at an angle of ≥ 30 degrees to the axis of the bone
Transverse	The fracture is oriented at a right angle to the axis of the bone
Avulsion	The fracture is caused by a sudden muscle contraction, with the muscle pulling off the portion of the bone
Comminuted	The fracture involves multiple fracture fragments
Stress	The fracture results from stresses repeated with excessive frequency to the bone

Stress fractures occur in normal and abnormal bones that have been subjected to repeated traumas (Figure 7.5). Other terms used to describe this injury include crack fracture, pseudofracture, spontaneous fracture and exhaustion fracture (Belkin 1980; Jones et al. 1989). These fractures occur in weight-bearing and non-weight bearing bones. There are two general types of stress fractures, insufficiency fracture and a fatigue fracture.

Insufficiency fracture results from normal stress applied to abnormal bone (Pentecost et al. 1964) that is weakened by an underlying disorder such as osteoporosis, rheumatoid arthritis, osteomalacia or Paget's disease (Stafford et al. 1986)

Fatigue fractures occur when normal bones are subjected to increased loads and repetitive stresses. None of these stresses is individually capable of producing a fracture, but combined they will lead to

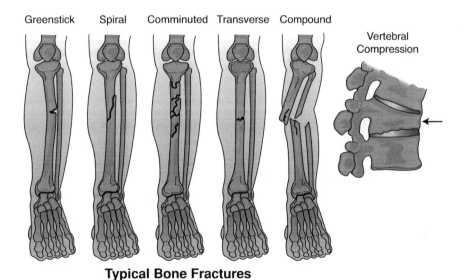

Figure 7.4 Diagram illustrating common fractures.

112 PATHOPHYSIOLOGY OF SKELETAL INJURIES

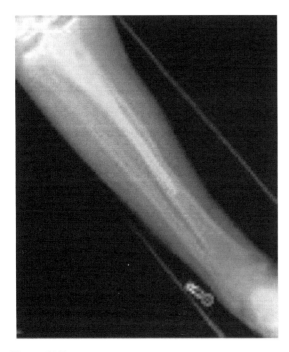

Figure 7.5 X ray of a stress fracture.

mechanical failure over time (Belkin 1980). It is not known or clear whether compressive, gravitational or muscular forces are most responsible for fatigue fractures. Several mechanisms have been proposed:

- Weight bearing
  - Weight bearing plays a role in some stress fractures.
  - It is unlikely this mechanism is solely responsible for the development of a stress fracture. This is supported by the fact that stress fractures occur in both weight-bearing and non-weight bearing bones.

- Muscle actions
  - Muscles may provide enough repetitive force to create a stress fracture. This is a likely mechanism for upper limb stress fractures.
  - With training muscle strengthens quicker than bone causing a mis-match that may lead to osseous fatigue failure (Daffner 1978)

- Muscle fatigue
  - Muscles provide shock absorption by taking the force away from the bone, thereby protecting it from fracture. As the muscle fatigues the stress is then transferred to the bone.

Running is responsible for the greatest number of stress fractures (fatigue fractures). This occurs especially when an individual:

- demonstrates a poor training technique (changing intensity or duration)

- anatomic and biomechanical factors, such as leg length discrepancy, external rotation of the hip, excessive pronation (Sullivan et al. 1984; Giladi et al., 1991)

- poor footwear

Osteoporosis

Osteopenia is not an injury but if left untreated can lead to the development of osteoporosis. Osteopenia is when the Bone mineral density (BMD) is lower than normal but not low enough to be classed as osteoporosis.

Osteoporosis is an established and well-defined disease that affects more than 75 million people in Europe, Japan and the USA, and causes more than 2.3 million fractures annually in Europe and the USA alone (WHO 2000). It is a systemic skeletal disease characterised by low bone density and micro-architectural deterioration of bone tissue with a consequent increase in bone fragility (WHO 2000). Early osteoporosis is not usually diagnosed and remains asymptomatic; it does not become clinically evident until fractures occur. Loss of bone density occurs with advancing age and rates of fracture increase markedly with age, giving rise to significant morbidity and some mortality (WHO 1994).

Osteoporosis is three times more common in women than in men, partly because women have a lower peak bone mass and partly because of the hormonal changes that occur at the menopause. Just after menopause the rate of bone mass loss in females is up to 10 times faster than in men of the same age (Reginster et al. 2006).

Oestrogens have an important function in preserving bone mass during adulthood, and bone loss occurs as levels decline, usually from about the age of 50 years. Also women live longer than men (the 1994 revision of the United Nations (1995) global population estimates and projections) and therefore have greater reductions in bone mass.

According to the WHO diagnostic criteria, women with bone density levels more than 2.5 standard deviations below the young adult reference mean are considered to have osteoporosis (Kanis et al. 1994). Individuals with bone density below this threshold who also sustain a fracture meet the definition of "established or severe osteoporosis". Among British women aged 50–59 years, for example, the prevalence of osteoporosis (as defined by a WHO Study Group) at the femoral neck of the hip is 4% and at any site is 15%. These figures rise to 48% and 70%, respectively, in women aged 80 years and over (WHO 2000)

One mode of athletic activity, gymnastic training, invokes high impact loading strains on bone, which many have powerful osteogenic effects (Taaffe et al. 1997). It has been reported that regional and total body density in competitive collegiate gymnasts exceeds that of runners, swimmers and non-athletic women regardless of menstrual cycle status (Myburgh et al. 1993).

The Female Athlete

The female triad is defined as a serious syndrome (Micklesfiled et al. 2007) consisting of three interrelated components:

1. disordered eating

2. amenorrhea

3. osteoporosis.

The athletes most at risk are those participating in sports in which success is determined by thinness and aesthetics (Micklesfiled et al. 2007). The link between current and past menstrual dysfunction and potential deleterious effects on bone mineral density has been thoroughly investigated in female athletes (Drinkwater et al. 1984; Marcus et al. 1985; Micklesfiled et al. 1998; Miller et al. 2006)

Results of treatments for reversing bone loss in athletes with menstrual dysfunction, such as hormone replacement, have shown equivocal outcomes (Gibson et al. 1999). However the effectiveness of a reduced training load (Warren et al. 2003) and improved nutrition on the resumption of regular menses, and the consequent increase in BMD, suggests that energy balance is also implicated in the relationship between compromised bone health and menstrual dysfunction (Micklesfield et al. 2007). Extensive reports confirm that female athletes who present with menstrual dysfunction have a lower BMD than amenorrhoeic athletes (Marcus et al. 1985). Traditionally it has been accepted that the reason for the decrease in BMD that accompanies menstrual dysfunction is chronic hypo-oestrogenism (Micklesfield et al. 2007). However, the use of oral contraceptives and the restoration of regular menses in these athletes have not been successful in reversing this bone loss (Mazess et al. 1991; Keen 1997; Micklesfield et al. 1998). Another mechanism may be responsible for the bone loss seen in athletes with menstrual dysfunction. This is confirmed by Micklesfield et al. (2007) in a study that showed a relationship between the occurrence of bone stress injuries and disordered eating patterns, as well as a high training load, this mechanism maybe more related to energy balance than to hypo-oestrogenism.

The osteoporosis risk factors are summarised below:

- female sex

- premature menopause

- age

- primary or secondary amenorrhoea

- slight body build

- primary and secondary hypogonadism in men

- Asian or Caucasian race

- previous fragility fracture

- glucocorticoid therapy

- maternal history of hip fracture

- low body weight

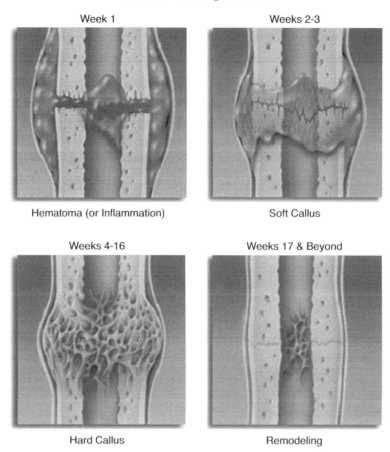

Figure 7.6 Fracture healing process.

- cigarette smoking
- excessive alcohol consumption
- prolonged immobilisation
- vitamin D deficiency
- low dietary calcium intake.

Healing re-modelling process during injury/rehabilitation

Healthy bone remodelling occurs at many simultaneous sites throughout the body where bone is experiencing growth, mechanical stress, microfractures or breaks. About 20% of all bone tissue is replaced annually by the remodelling process. The total process takes about 4–8 months, and occurs continually throughout our lives. The healing potential of bone is influenced by a variety of biochemical, biomechanical, cellular, hormonal and pathological mechanisms (Kalfas 2001).

The first stage of bone healing is referred to as the inflammatory phase (also known as the granulation stage, fracture stage or clot phase). This stage has two parts to it: during the first part of this stage the surviving cells are sensitised to chemical messengers that are involved in the healing process, this stage is completed within seven days. The second part, which lasts for about two weeks, is the development of a clot around the fracture site; this is not seen within stress fractures. After the clot has been formed, granulation tissue forms within the space between the fracture fragments. This granulation tissue then activates macrophages.

The second stage is known as the reparative phase (callous stage), and can be divided into the soft callous and hard callous stages. During the soft callous stage the osteoblasts and chondrocytes within the granulation tissue begins to make cartilage and woven bone matrices. This newly formed callus begins to mineralise after approximately a week. This mineralisation concludes with the formation of a fracture/hard callus, this callus is detectable on X-rays due to the calcium it contains. The creation and mineralisation of the callus can take 4–16 weeks to complete.

The third stage is called the remodelling phase (consolidation phase), In the remodelling phase the process may occur over months to years and consists of restoring the fractured bone to its normal size, shape and strength (Kalfas 2001). Adequate strength usually develops by six months.

Effects on the bone healing process

Nutrition

Calcium is far the most abundant mineral found within the body and contributes to the structure of teeth and bone (Thatcher et al. 2009) Calcium plays an important role in helping attain peak bone mass during bone development and preventing fractures in later life. The daily recommended allowance of calcium intake is 800–1200 mg (Thatcher et al. 2009). Multiple factors can affect the bioactivity of calcium:

- high-fat or high fibre diets can interfere with or decrease the activity of calcium

- large doses of zinc supplementation or mega doses of vitamin A can lower calcium bioactivity

- high protein diets can decrease calcium reserves by increasing urinary excretion of calcium

(Placzek and Boyce 2006)

In addition, alcohol consumption can decrease the absorption of calcium and various medications (heparin, glucocorticoids) can affect calcium activity. Vitamin D regulates calcium absorption and excretion, especially when calcium intake is low.

When calcium levels in the blood drop, parathyroid hormone (PTH) is released. PTH causes calcium to be released from the bones; this then raises the low calcium levels in the blood. Osteoporosis may result from chronically high levels of PTH (Groff and Gropper 2000).

Wolff's law and the effect it has on bone healing

When mechanical stresses are put on bone, the bone has the ability to adapt by changing size, shape and structure. When optimal stress is placed on bone there is a greater bone deposition than bone resorption. This results in an increased bone density and a hypertrophy of perosteal bone.

Effect of NSAID's on fracture healing

Bone repair is a complex process initiated by injury and an inflammatory response. Prostaglandins mediate inflammation, influence the balance of bone formation and resorption; processes that are essential for new bone formation. NSAIDs inhibit cyclooxygenases, which are essential for prostaglandin production (Dumont et al. 2000). It has been shown that long-term excessive use of these medications may reduce normal bone healing (Placzek and Boyce 2006).

Conditions that have a negative effect

The following list gives a range of conditions that have a negative effect on the bone healing process:

- infection

- poor reduction (poor realignment of fracture)

- loss of local blood supply due to injury

- vascular injury

- failure to make callus (metabolic abnormalities)

- formation of scar and fat tissue instead of callus

- poor nutrition

- alcohol abuse

- smoking.

Summary

Bone is a dynamic fully functional organ of the human body (Gormley and Hussey 2005). It has a strong network of blood vessels and nerves; it constantly remodels itself; it can support extreme loads during exercise; and it can repair itself. The presentation of optimal bone health should be a priority of all health professionals (Gormley and Hussey 2005).

References

Baron, R. (1996) Molecular mechanisms of bone resorption: therapeuticimplications. *Revue du Rhumatisme (English Edition)*, 63, 633–638.

Belkin, S.C. (1980) Stress fractures in athletes. *Orthopaedic Clinics North America*, 11, 735–742.

Buckwalter, J. and Lane, L. (1997) Athletics and osteoarthritis. American Journal of Sports Medicine, 25, 873–881.

Daffner, R.H. (1978) Stress fractures: current concepts. *Skeletal Radiology*, 2, 221–229.

Drinkwater, B., Nilson, K., Chesnut, C., Bremner, W., Shainholt, S. and Southworth, M. (1984) Bone mineral contentof ammenorrheic and eumenorrheic athletes. *New England Journal of Medicine*, 311, 277–281.

Dumont, A.S., Verma, S., Dumont, R. and Hurlbert, R. (2000) Nonsteroidal anti-inflammatory drugs and bone metabolism in spinal fusion surgery. A pharmacological quandary. *Journal of Pharmacological and Toxicological Methods*, 43, 31–39.

Einhorn, T.A. (1996) The bone organ system: form and function. In Marcus, R., Feldman, D. and Kelsey, J. (Eds), *Osteoporosis*, pp 3–22. San Diego, CA, Academic Press..

Fleisch, H. (1997) Bone and mineral metabolism. In *Bisphosphonates in Bone Disease. From the laboratory to the patient*, 3rd edn, pp 11–31. London: Parthenon.

Gibson, H., Mitchell, A., Reeve, J. and Harries, M. (1999) Treatment of reduced bone mineral density in athleteics amenorrhea; a pilot study. *Osteoporos International*, 10, 284–289.

Giladi, M., Milgrom, C., Simkin, A. and Danon, Y. (1991) Stress fractures. Identifiable risk factors. *American Journal of Sports Medicine*, 19, 647–652.

Gormley, J. and Hussey, J. (2005) *Exercise Therapy Prevention and Treatment of Disease*. Oxford: Blackwell.

Groff, J. and Gropper, S. (2000) *Advanced Nutrition and Human Metabolism*, 3rd edn. Florence, KY: Wadsworth.

Jones, B.H., Harris, J.M.C.S., Vinh, T.N. and Rubin, C. (1989) Exercise-induced stress fractures and stress reactions of bone: epidemiology, etiology, and classification. *Exerc ise and Sport Science Research*, 17, 379–422.

Kalfas, I.H. (2001) Principles of bone healing. *Neurosurgery Focus*, 10 (4), Article 1

Kanis, J. (1994) Pathogenesis of osteoporosis and fracture. In *Osteoporosis*, pp 22–55. Oxford: Blackwell Science.

Kanis, J.A., et al (1994) Assessment of fracture risk and its application to screening for postmenopausal osteoporosis. Synopsis of a WHO Report. *Osteoporosis International*, 4, 368–381.

Keen, A. and Drinkwater, B. (1997) Irreversible bone loss in former amenorrheic athletes. *Osteoporosis International*, 7, 311–315.

Kujala, U., Kaprio, K. and Sarna, S. (1994) Osteoarthrisits of the weight bearing joints of the lower limbs in former elite male athletes. *British Medical Journal*, 308, 231–234.

Lane, N., Michel, B., Bjorkengren, A., Oehlert, J., Shi, H. and Block, D. (1993) The risk of osteoarthritis with running and aging: a 5 year longitudinal study. *Journal of Rheumatology*, 20, 461–468.

Lohmander, L.S. (2003) Tissue engineering of cartilage: do we need it, can we do it, is it good and can we prove it? *Novartis Foundation Symposium*, 249, 2–10.

Marcus, R., Cann, C., Madvig, P., Minkoff, J., Goddard, M., Bayer, M., Martin, M., Gaudiani, L., Haskell, W. and Genant, H. (1985) Menstrual function and bone mass in elite women distance runners. Endocrine and metabolic features. *Annals of Internal Medicine*, 102, 158–163.

Mazess R.B., Trempe, J.A. and Bisek, J.P. (1991) Calibration of dual-energy x-ray absorptiometry for bone density. *Journal of Bone Mineral Research*, 6, 799–806.

Micklesfield, L., Reyneke, L., Fataar, A. and Myburgh, K. (1998) Long-term restoration of deficits in bone mineral density is inadequate in premenopausal women with prior menstrual irregularity. *Clinical Journal of Sport Medicine*, 8, 155–163.

Micklesfield, L., Hugo, J., Johnson, C., Noakes, T. and Lambert, E. (2007) Factors associated with menstrual dysfunction and self-reported bone stress injuries in female runners in the ultraand half marathons of the two oceans. *British Journal of Sports Medicine*, 41, 679–683.

Myburgh, K., Bachrach, L., Lewis, S., Kent, K. and Marcus, R. (1993) Low bone mineral density at axial and appendicular sites in amenorrheic athletes. *Medicine and Science in Sports and Exercise*, 25, 1197–1202

Naujok, C., Meyer, U., Wiesmann, H., Jasche-Meyer, J., Hohoff, A., Depprich, R. and Handschel, J. (2008) Principles of cartilage tissue engineering in TMJ reconstruction. *Head Face Medicine*, 4, 1–7.

Netter, F.H. (1987) Musculoskeletal System: Anatomy, Physiology, and Metabolic Disorders. Summit, NJ: Ciba-Geigy.

Newton, P., Mow, V., Gardner, T., Buckwalter, J. and Albright, J. (1997) Winner of the 1996 Cabaud Award. The effect of lifelong exercise on canine articular cartilage. *American Journal of Sports Medicine*, 25, 282–287.

Nijweide, P.J. et al. (1996) The osteocyte. In Bilezikian, J.P., Raisz, L.G. and Rodan, G.A. (Eds), *Principles of Bone Biology*, pp 115–126. San Diego, CA: Academic Press.

Pentecost, R.L., Murray, R.A. and Brindley, H.H. (1964) Fatigue, insufficiency, and pathologic fractures. *JAMA*, 187, 1001–1004.

Placzek, J.D. and Boyce, D.A. (2006) *Orthopaedic Physical Therapy Secrets*, 2nd edn. St Louis, MI: Mosby Elsevier.

Porter, S.B. (2008) *Tidy's Physiotherapy*, 14th edn. Amsterdam: Elsevier Health Sciences.

Recker, R.R., Davies, K.M., Hinders, S.M., Heaney, R.P., Stegman, M.R. and Kimmel, D.B. (1992) Bone gain in young adult women. *Journal of the American Medical Association*, 268, 2403–2408.

Reginster, J., Adami, S., Lakatos, P., Greenwald, M., Stepan, J., Silverman, S., Christiansen, C., Rowell, L., Mairon, N., Bonvoisin, B., Drezner, M., Emkey, R., Felsenberg, D., Cooper, C., Delmas, P. and Miller, P. (2006) Efficacy and tolerability of once-monthly oral ibandronate in postmenopausal osteoporosis: 2 year results from the MOBILE study. *Annals of Rheumatic Disease*, 65, 654–661.

Shim, S.S., Copp, D.H. and Patterson, F.P. (1967) An indirect method of bone-blood flow measurement based on the bone clearance of a circulating bone-seeking radioisotope. Journal of Bone and Joint Surgery,. 49–A, 693–702.

Smith, E.L., Hill, R.L., Lehman, I.R., Lefkowitz, R.J., Handler, P. and White, A. (1983) *Principles of Biochemistry: Mammalian Biochemistry*, 7th edn. New York, NY: McGraw-Hill.

Stafford, S.A., Rosenthal, D.I., Gebhardt, M.C., Brady, T.J. and Scott, J.A. (1986) MRI in stress fracture. *AJR*, 147, 553–556.

Sullivan, D., Warren, R.F., Pavlov, H. and Kelman, G. (1984) Stress fractures in 51 runners. *Clinical Orthopaedic Relation Research*, 187, 188–192.

Taaffee, D., Robinson, T., Snow, C. and Marcus, R. (1997) High impact exercise promotes bone gain in well trained female athletes. *Journal of Bone and Mineral Research*, 12 (2), 255–260.

Teitelbaum, S.L., Tondravi, M.M. and Ross, F.P. (1996) Osteoclast biology. In Marcus, R., Feldman, D. and Kelsey, J. (Eds), *Osteoporosis*, pp 61–94. San Diego, CA: Academic Press.

Thatcher, J., Thatcher, R., Day, M., Portas, M. and Hood, S. (2009) *Sport and Exercise Science*. Exeter: Learning Matters.

Tothill, P. and MacPherson, J.N. (1986) The distribution of blood flow to the whole skeleton in dogs, rabbits and rats measured with microspheres. *Clinical and Physical Physiology Measures*, 7 (2), 117–123.

Triffitt, J.T. (1996) The stem cell of the osteoblast. In Bilezikian, J.P., Raisz, L.G. and Rodan, G.A. (Eds), *Principles of Bone Biology*, pp 39–50. San Diego, CA: Academic Press.

United Nations (1995) *The sex and age distributions of population. The 1994 revision of the United Nations global population estimates and projections*. New York, NY: United Nations.

Warren, M.P., Fox, R.P., Holderness, C.C., DeRogatis, A.J., Hamilton, W.G. and Hamilton, L. (2003) Persistent osteopenia in women with amenorrhea and delayed menarche despite hormone replacement therapy: A longitudinal study. *Fertility and Sterility*, 80 (2), 398–404.

World Health Organization (1994) Assessment of fracture risk and its application to screening for postmenopausal osteoporosis. Report of a WHO Study Group. *WHO Technical Report Series, No. 843*. Geneva~:WHO.

WHO (2000) WHO Scientific Group Meeting on Prevention and Management of Osteoporosis, April. Geneva: WHO.

8

Peripheral nerve injuries

Elizabeth Fowler
University of Salford, Greater Manchester

Introduction

This chapter aims to introduce the structure and function of nerves and the neurological system, and the pathophysiology of common nerve injuries. The chapter also reviews some common nerve injuries, their assessment and evidence based treatment.

Anatomy of nerves

The nervous system comprises of the central, autonomic and peripheral nervous systems (Gallant 1998) the latter of which consists of cranial nerves, spinal nerves, peripheral nerves and peripheral components of the autonomic nervous system (Gardner and Bunge 1984). The neurone is the structural unit of the nervous system and is responsible for conducting messages from one part of the body to another (Marieb 1998). The basic unit of the neurone is the axon, more commonly known as the nerve fibre (Butler 1991) (Figure 8.1) and can be either mylenated or non-mylenated. Mylenated axons are surrounded by a myelin sheath and have one Schwann cell per axon. The Schwann cell is frequently interrupted by nodes of Ranvier; a structure that allows impulses to be conducted from one node to the next (Topp and Boyd 2006; Campbell 2008). Non-mylenated fibres however, have several axons associated with only one Schwann cell (Topp and Boyd 2006; Campbell 2008). Individual mylenated fibres and groups of non-mylenated fibres are surrounded by three connective tissue layers; the endoneurium, the perineurium and the epineurium (Topp and Boyd, 2006; Campbell 2008).

The innermost connective tissue layer is the endoneurium, which is composed of longitudinally aligned collagen fibres and therefore, plays an important role in protecting the axon from tensile forces (Butler 1991). Encompassing the endoneurial components, axon and Schwann cells is the perineurium, the second layer of connective tissue, whose primary responsibility is to act as a primary barrier to external forces (Lundborg 1988). Collectively, all the contents surrounded by the perineurium form what is known as a nerve fascicle (Topp and Boyd 2006). Collections of nerve fascicles then form a nerve and these are surrounded by the outermost layer of connective tissue, the epineurium (Campbell 2008). The epineurium is regarded as the most resistant connective layer to tensile forces (Sunderland 1978) as it surrounds, protects and cushions the nerve fascicles (Butler 1991). Collectively all three connective tissue layers not only protect the axon, but are also structurally developed to cope with the tensile stresses and compressive forces nerves typically have to endure (Butler 1991). In everyday movements and postures, nerves are subjected to various mechanical stresses (Topp and Boyd 2006), which can elongate, compress or increase strain within the nerve (Shacklock 1995).

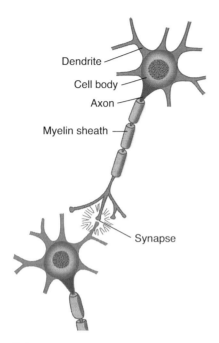

Figure 8.1 Structure of nerves.

The ability of the nervous system to withstand and adapt to the mechanical stresses placed on it, is essential to prevent injury (Shacklock 1995). Alterations in the physical stress to which a biological tissue is subjected, causes predictable responses within the tissue according to Mueller and Maluf's (2002) Physical Stress Theory. Injury to tissues is caused by excessive physical stress via any of the following mechanisms:

1. High magnitude stress applied to the tissue for a brief duration; spinal cord injury is a typical outcome from this mechanism of injury.

2. Low magnitude stress applied for long duration or repetitively; an example of nerve injury from this particular mechanism of injury is Carpal Tunnel Syndrome at the wrist.

3. Moderate stress applied to a tissue many times. Cubital Tunnel Syndrome at the elbow, for example in a javelin thrower, whereby repetitive high load forces are exerted through the elbow and consequently the ulnar nerve, is an example (Mueller and Maluf 2002).

Biological tissues, such as nerves, have an ideal physical stress range that they can tolerate to maintain homeostasis. This is called the "maintenance stress range" (Mueller and Maluf 2002). Stress levels lower than the maintenance range decreases a tissue's tolerance to physical stress; for example during immobilisation of a limb, muscle atrophy is a typical bi-product of being in a cast for a prolonged period of time. Contrastingly, when physical stresses are exceedingly higher than the homeostatic stress range, the associated tissues are unable to tolerate these and any subsequent stresses and ultimately become injured (Mueller and Maluf 2002).

The mechanical stresses a nerve is subjected to are not the only factor which must be taken into account with nerve injury as the nature of the surrounding structures a nerve traverses or passes through must also be considered (Butler 1991). The nervous system is surrounded by and comes into contact with many different anatomical structures which may be firm and unyielding, such as the radial nerve in the spiral groove of the humerus, or soft, such as the tibial nerve in the posterior thigh musculature (Butler 1991). Additionally, anatomically narrow passages through which the nerve must pass may also predispose individual nerves to entrapment neuropathies, such as the ulnar nerve in the cubital fossa of the elbow (Bencardino and Rosenberg 2006). The extent of an injury to a nerve is dependant on the mechanism of injury, as traumatic injuries, such as a gun-shot wound (i.e. high magnitude stress), will significantly damage a nerve's integrity, whilst a low magnitude stress, such as prolonged intermittent compression over a long duration of time, will have less of an impact on the nerve. Nerve injuries were initially classified by Seddon (1943) based on the severity of the injury and the potential for reversibility of the condition (Bencardino and Rosenberg 2006).

Classification of nerve injury

Neurapraxia, axonotmesis and neurotmesis are the three categories of nerve injuries classified by Seddon (1943) each describing different degrees of injury to the nerve's anatomical structures (Table 8.1). Neurapraxia is the most benign injury and typically results from compression and/or traction (Perlmutter and Apruzzese 1998; Bencardino and Rosenberg 2006). Pathologically, the nerve is

Table 8.1 Classification and characteristics of nerve injury

Classification	Anatomical presentation	Clinical presentation	Prognosis
Neuropraxia	Segmental demyelenation	Slight motor loss Minimal sensory involvement	Excellent Up to 12 week recovery
Axontmesis	Loss of axonal continuity Connective tissue intact	Substantial loss of motor, sensory and autonomic function	Good Surgery may be required however
Neurotmesis	Complete disruption of nerve fibre and connective tissues	Significant muscle fibre atrophy Sensory loss	Poor Surgery is a necessity

Adapted from Seddon (1943).

intact but is temporarily unable to transmit signals (Campbell 2008) due to injury to the myelin about the nodes of Ranvier (Bencardino and Rosenberg 2006). Consequently, the clinical presentation is that of motor loss, with little sensory involvement and little disturbance of sympathetic innervation (Perlmutter and Apruzzese 1998). With remyelenation of the nerve, motor and sensory dysfunction will gradually be restored, usually within 12 weeks (Novak and Mackinnon 2005; Bencardino and Rosenberg 2006). The prognosis for neuropraxia is excellent (Perlmutter and Apruzzese 1998) and complete recovery is expected (Novak and Mackinnon 2005).

When there is loss of continuity of the axon and myelin sheath, but the majority of the connective tissue structure is maintained, the injury is classified as axonotmesis; the second classification (Seddon 1943). The endoneurium and Schwann sheath are preserved, but injury to the axon is evident, accompanied with secondary Wallerian degeneration (Bencardino and Rosenberg 2006). A clinical finding of significant loss of motor, sensory and autonomic function distal to the injury site is evident (Perlmutter and Apruzzese 1998). Neural regeneration will occur, but at a rate of approximately 1mm per day and over-all the prognosis for recovery is good (Novak and Mackinnon 2005). Stinger Syndrome is also an example of axontmesis, particularly should the injury recur within a short period of time.

The most severe peripheral nerve injury is referred to as neurontmesis and involves complete disruption of the nerve trunk and connective tissue (axons, endoneurial tube, Schwann sheath and epineurium) (Campbell 2008). The prognosis for recovery is poor as there is no potential for nerve regeneration (Bencardino and Rosenberg 2006) and therefore surgical intervention is often required in an attempt to re-establish the continuity of the peripheral nerve (Perlmutter and Apruzzese 1998). Avulsion of the nerve, via stab wounds or gunshot, is a typical example of neurontmesis; it is a rare occurrence in sport.

The majority of nerve injuries in sport will typically involve neurapraxia or axontmesis and therefore prognosis for recovery is generally good.

Assessment of nerve injury

Neural and non-neural tissues should be assessed by the clinician in all patients presenting with pain (Nee and Butler 2006) by means of a comprehensive subjective and physical examination. All the physical examination findings should complement and support the information obtained from the patient via subjective examination (Hall and Elvey 1999). Nee and Butler (2006) proposed numerous clinical features that may be evident in the subjective and physical assessment of patients presenting with peripheral neuropathic pain, such as complaints of tingling, burning and parasthesia in addition to antalgic postures and movement impairments being present. Motor or sensory losses, or both, may also be evident in the assessment of patients presenting with nerve pain. However, when nerve injury is suspected, the assessment of the integrity of the nerve is vital.

Evaluation of the function of the muscles innervated by the injured nerve should be assessed via manual resistance testing to determine the motor function of the nerve (Figure 8.2 and Table 8.2). Likewise, the areas of innervation should be assessed via touch to assess the sensory function of the

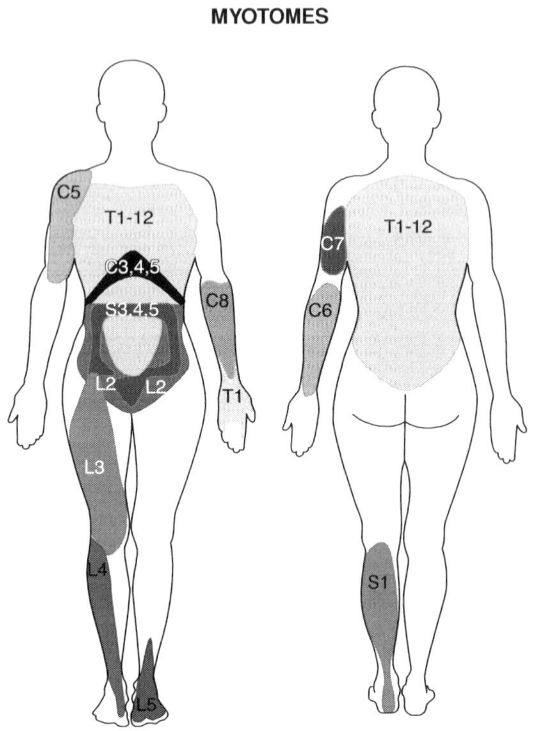

MYOTOMES

Figure 8.2 Myotomes.

Table 8.2 Myotomes of the upper and lower limb

Myotomes of the Upper limb	
Nerve root/nerve	
C1	Cervical flexion
C2	Cervical flexion
C3	Cervical lateral flexion
C4	Shoulder elevation
C5	Shoulder abduction
C6	Elbow flexion
C7	Elbow extension
C8	Thumb extension
T1	Finger abduction
Myotomes of the Lower limb	
Nerve root/nerve	
L1	Hip flexion
L2	Hip adduction
L3	Knee extension
L4	Ankle dorsi-flexion
L5	Great toe extension
S1	Ankle plantar flexion
S2	Knee flexion

injured nerve (Dahlin 2008) (Figure 8.3). Knowledge of the myotomes and dermatomes of the upper and lower extremities is important to conduct a thorough assessment of the peripheral nervous system. Nee and Butler (2006) also recommend incorporating neurodynamic tests into the physical assessment to determine the mechanosensitivity of the nervous system.

Neurodynamic testing

Nerves slide and stretch during limb movements to allow for changes in nerve bed length (Babbage et al. 2007) and whilst healthy nerves can tolerate strain and compression, injured or inflamed nerves become sensitive to mechanical stimuli and can inflict pain on movement (Bove et al. 2005).

Neurodynamic tests were developed to evaluate peripheral nerve sensitivity to movement and to infer underlying pathomechanics (Topp and Boyd 2006). To determine if neural tissues contribute to the patient's symptoms, it is important to move the neural structure in the area in question, without moving the musculoskeletal tissues in the same region; a task attempted by structural differentiation during neurodynamic tests (Shacklock 1995). The concept behind structural differentiation is to move a joint remote to the area where the patient experiences symptoms; should this decrease or increase symptoms, neural involvement into patient symptoms should be suspected. For example, in a patient presenting with posterior thigh pain, moving the head from cervical flexion to extension during the slump test should have no effect on patient symptoms if the pain is of a non-neural origin. However, should the clinician suspect the sciatic nerve is a contributing factor to patient symptoms, moving the cervical spine from flexion to extension should decrease the pain and/or permit a further range of movement at the knee joint.

Neurodynamic tests for the upper and lower extremities have been developed; each one intent on identifying whether a specific nerve is contributing to the symptomatic patient. The termination of the movement during a neurodynamic test is dependant on whether the clinician wants to stop at the point

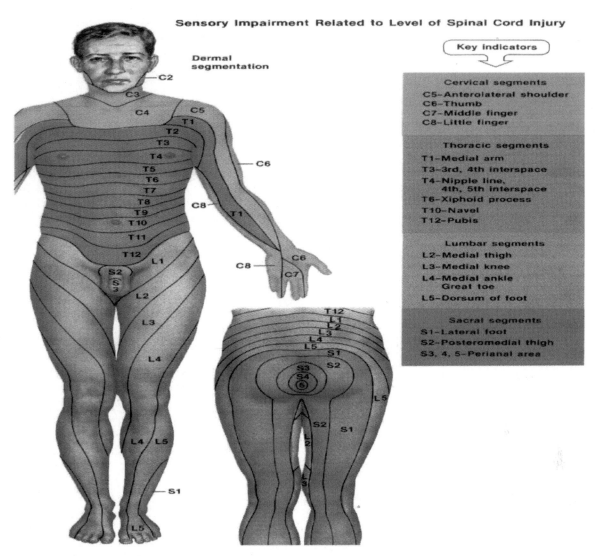

Figure 8.3 Dermatomes.

where resistance is experienced by the clinician (Shacklock 1995; Nee and Butler 2006) or to the point where the patient reports onset of pain or submaximal pain (Coppieters et al. 2002). Three tests for the upper limb are described below, each focused on examining the integrity of the three primary nerves in the upper extremity; the median, radial and ulnar nerves. For the lower limb, the slump test and straight leg raise tests will be illustrated. In-depth and more detailed clinical neurodynamic tests have been presented by Shacklock (2005).

Upper limb neurodynamic test with median nerve bias

In patients presenting with suspected median nerve pathology whereby symptoms may be localised to the median nerve, such as carpal tunnel syndrome or pronator tunnel syndrome, incorporating a neurodynamic test into the physical examination is essential to determine if the nerve contributes to patient symptoms. To conduct this test, the patient lies supine on the plinth with arms relaxed by the side of the body.

Keeping the head in a neutral position, the clinician places a hand over the superior aspect of the shoulder with the aim of depressing the shoulder as the test commences. The clinician then holds the patient's ipsilateral hand and places the glenohumeral joint in 90° abduction and 90° lateral rotation, the forearm in full supination and the wrist in extension. The clinician then supports the patient's arm on their thigh with the aim of preventing adduction of the shoulder during the test. Once in this position, the clinician slowly extends the elbow to either onset of resistance or pain (Figure 8.4). To structurally differentiate, the clinician can either ask the patient to execute contralateral side flexion of the head for distal symptoms or the wrist can be moved from extension into flexion for proximal symptoms (Shacklock 2005).

Upper limb neurodynamic test with radial nerve bias

In patients presenting with radial neuropathy such as radial tunnel syndrome, the clinician requires a physical test which can be conducted to determine the sensitivity of the nerve to stretch. To do this, a clinical test was developed to focus specifically on the radial nerve and is conducted as follows: the patient lies supine with arms resting by the side of the body but in a diagonal direction across the plinth. This allows the clinician to apply shoulder depression using the anterior aspect of their hip whilst conducting the test. The patient's starting position involves placing the elbow

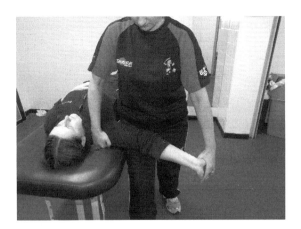

Figure 8.4 Upper limb neurodynamic test with median nerve bias.

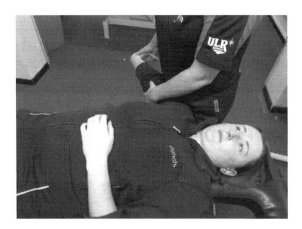

Figure 8.5 Upper limb neurodynamic test with radial nerve bias.

at 90° and the wrist and fingers in neutral. The clinician then applies shoulder depression before moving the elbow into extension and the wrist and fingers into flexion whilst internally rotating the glenohumeral joint. Finally, glenohumeral abduction is applied and again the clinician ceases the movement at the point of resistance or pain (Figure 8.5). To structurally differentiate, shoulder depression can be removed for distal symptoms, whilst moving the wrist from flexion to neutral may identify proximal lesions (Shacklock 2005).

Upper limb neurodynamic test with ulnar nerve bias

When symptoms are evident in the anatomical pathway of the ulnar nerve or the lower trunk of the brachial nerve (Shacklock 2005), neurodynamic testing specific to this nerve is important. The testing sequence for an upper limb neurodynamic test with ulnar nerve bias requires the patient to lie supine on the plinth, with the head in a neutral position and the arms relaxed by the side of the body. The clinician prepares for depression of the shoulder as for the median nerve test. The patient's shoulder joint is abducted slightly whilst the elbow is maintained in extension and the forearm slightly pronated. The wrist and hand of the patient remains in a neutral position, following which, the clinician then commences the test by depressing the shoulder before applying extension to the wrist and fingers and pronation to the forearm. The elbow is then brought

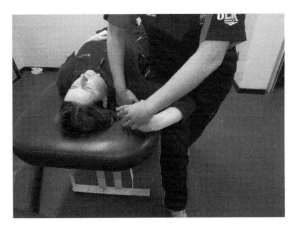

Figure 8.6 Upper limb neurodynamic test with ulnar nerve bias.

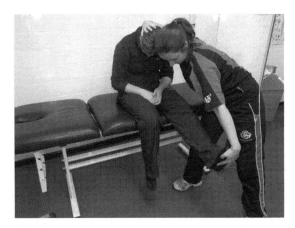

Figure 8.7 Lower limb neurodynamic test: The slump test.

into a flexed position; following which, the shoulder is gradually moved into an abducted position (Figure 8.6). The clinician again uses a leg to support the moving limb; thereby making the movement more fluid. Again once resistance or pain is experienced by the clinician or patient respectively, shoulder abduction ceases and the structural differentiation manoeuvre is applied; either contralateral cervical side flexion for distal symptoms or radial deviation for proximal symptoms (Shacklock 2005).

Lower limb neurodynamic test: The slump test

The slump test is used to evaluate the dynamics of the central and peripheral nervous systems from the head, along the spinal cord and sciatic nerve tract and its extensions in the foot (Shacklock 2005). The slump test should be a component of the clinical examination in patients who present with spinal symptoms (Butler 1991) and/or spinal, pelvic and lower limb conditions whereby pain is experienced in the neural distribution of the sciatic nerve and its extensions (Butler 1991; Shacklock 2005). To conduct the slump test, the patient is requested to side on the plinth with the back of the knees to the edge of the plinth and to "sag" the upper body, or bring their shoulder towards their hips, thereby resulting in thoracic flexion. The clinician then applies overpressure at C7 spinous process with the medial aspect of the forearm (Figure 8.7). The patient is instructed to bring their chin to chest, following which the clinician applies dorsi-flexion to the ankle before slowly extending the knee to the point of resistance or pain. To structurally differentiate, the release of cervical flexion and ankle dorsi-flexion are used for distal and proximal lesions respectively.

Lower limb neurodynamic test: the straight leg raise

The straight leg raise (SLR) is a similar neurodynamic test to the slump test, intent in evaluating the integrity of the lumbosacral trunk and plexus, sciatic nerve and its expansion in the leg and foot (Shacklock 2005). The format for the SLR is as follows: the patient lies supine on the plinth with arms resting by the side of the body. The clinician uses one hand to support the posterior calf, just proximal to the ankle joint; whilst the other hand is placed on the anterior aspect of the knee joint to ensure knee extension is maintained throughout. Maintaining this position, the clinician then begins to lift the leg off the bed, thereby conducting hip flexion, moving to the point of resistance or patient pain. Following this the clinician then applies dorsi-flexion to the ankle by way of structurally differentiating (Figure 8.8). An increase in symptoms at this point indicates a neural involvement in the patient's condition. To ascertain whether the peroneal division of the sciatic nerve is sensitive to stretch, whilst conducting the SLR, the clinician can plantar-flex and invert the ankle whilst moving the ankle into dorsi-flexion and eversion emphasises the tibial nerve (Shacklock 2005).

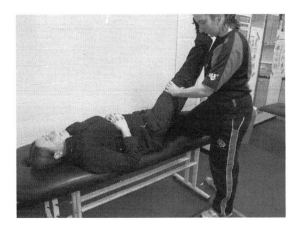

Figure 8.8 Lower limb neurodynamic test: the straight leg raise.

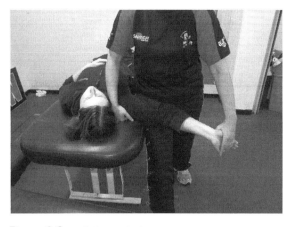

Figure 8.9 Sliding technique.

Neurodynamic tests as treatment tools

A positive neurodynamic test is constituted by a reproduction or increase in symptoms during the test, which is subsequently decreased with the removal of the structural differentiating manoeuvre (Maitland 1985); difference in responses between limbs (Nee and Butler 2006); differences in available range of motion (Coppieters et al. 2002; Nee and Butler 2006); or where there is structural differentiation supporting a neurogenic source (Butler 2000). Whilst all the aforementioned tests have been described from a diagnostic point of view, each of these can be modified and incorporated into treatment plans, should their inclusion as a treatment modality be warranted. Modified versions of the neurodynamic tests can be used in clinical practice in the treatment plan for peripheral nerve injuries and can be specific to each patient. The purpose of utilising neurodynamic tests as treatment tools is to minimise scarring and stretching of the nerve, and maintain or restore normal nerve excursion and function (Wehbé and Schlegel 2004).

Two neural treatment techniques will be considered in this chapter; sliding and tensioning techniques, the principle of which can be applied to any neurodynamic test and considered for use in the treatment plan for neuropathy patients. A "sliding" technique was defined by Coppieters et al. (2009), as a combination of movements that elongate the nerve bed at one joint, whilst simultaneously reducing nerve bed length at an adjacent joint, to counterbalance the elongation. For example, during the neurodynamic test with median nerve bias, instead of having the patient placed in the final testing position, with contralateral side flexion of the cervical spine, the patient would move the elbow into extension whilst simultaneously moving the cervical spine into ipsilateral side flexion (Figure 8.9). The opposing technique is the "tensioning" technique, which the same authors define as the elongation of the nerve bed at two adjacent joints (Coppieters et al. 2009) and again, referring back to the median nerve test, would involve the patient actively extending the elbow whilst simultaneously side flexing the cervical spine into the contralateral direction (Figure 8.10).

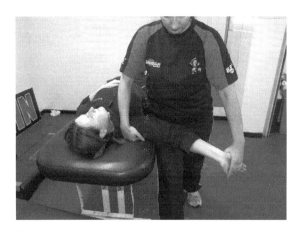

Figure 8.10 Tensioning technique.

The sliding technique causes more longitudinal excursion of a nerve than the tensioning technique (Coppieters et al. 2009), and therefore, from a clinical viewpoint, may be more suitable for highly irritable or acute neural conditions as it still encourages movement of the inflamed nerve, whilst limiting the strain placed on the nerve. A tensioning technique, however, is proposed to induce higher tension within the nerve and therefore less excursion (Coppieters et al. 2009) than a sliding technique and can therefore be considered more aggressive and possibly suitable for less irritable conditions. The clinician should therefore consider the implementation of sliding or tensioning techniques into a treatment plan for the patient presenting with neural symptoms, with the former to be used in acute, irritable conditions and the latter technique for use in the end stage of rehabilitation.

Ultimately, neurodynamic tests can be modified and used in the treatment plan for the neuropathic patient and can also be incorporated into a home exercise programme (Kostopoulos 2004). Typical patient responses to neurodynamic tests as both a diagnostic and treatment tool, of which the patient should be aware, range from reporting feelings of "stretching", tissue tension, light numbness or a slight increase of pain during the technique; all which should reduce or dissipate on cessation of the test or treatment (Kostopoulos 2004).

Treatment plans for nerve injury

Immediately post-injury, inflammation occurs, thereby rendering injured tissues less capable of tolerating stress compared to their pre-morbid level (Mueller and Maluf 2002). Nerve fibres become sensitive to stretch and low intensity pressure following injury (Dilley et al. 2005). Therefore the importance of protecting the tissues during this acute inflammation stage from subsequent stress cannot be underestimated (Mueller and Maluf 2002). Conservative treatment is the initial recommended strategy in numerous neuropathies not requiring urgent medical attention. Within this category of treatment, several authors suggest incorporating non-steroidal anti-inflammatory drugs (NSAIDs), ice, rest, elimination of the aggravating activity, physical and manual therapy into the conservative treatment plan (McKean 2009; Shapiro and Preston 2009); the primary aim of which is to decrease inflammation and restore nerve function. Brief treatment options will be provided in this section, following the discussion of each specific injury.

Peripheral nerve injury

Neurological conditions are common in athletes (Dimberg and Burns 2005) often dependant on the nature and intensity of the sporting activity (Toth 2009). A clinician may be confronted with signs and symptoms of neurological injuries affecting various aspects of the nervous system, such as spinal nerve roots, peripheral nerve injuries or plexopathies. Possessing the ability to recognise and diagnose injuries specific to sporting activities and then subsequently to treat the injury, is vital to a clinician in sport (Toth 2009) to ensure a rapid return to play for the athlete.

The upper extremity is particularly vulnerable to nerve injury (Dahlin 2008) due to its high mobility (Aldridge et al. 2001) and it is therefore unsurprising that the brachial plexus is the most commonly injured plexus in the body (Wilbourn 2007). The brachial plexus is formed by the ventral rami of the spinal nerves C5 to T1, which enter the posterior cervical triangle between the scalene anterior and medius muscles (Pratt 2005). Superior, middle and inferior nerve trunks of the brachial plexus are then formed by the C5–C6, C7 and C8–T1 nerve roots respectively (Reid and Trent 2002; Pratt, 2005). These trunks divide into anterior and posterior portions behind the clavicle whereby these divisions then unite to form lateral, medial and posterior cords which enter the axilla (Reid and Trent 2002). The end result of the brachial plexus is the formation of peripheral nerves which supply the upper limb (Pratt 2005) (Figure 8.11). The primary peripheral nerves which are subsequently discussed in this chapter are the axillary nerve, long thoracic nerve, suprascapular nerve, ulnar nerve, median nerve and radial nerve; the anatomical locations of which can be viewed in Figure 8.12.

Brachial plexus neuropathy

Stinger Syndrome

One of the most common brachial plexus neuropathy is Stinger Syndrome (Hershman 1990), which is more common in young adults who participate in sport (Unlu et al. 2007) and particularly in contact

128 PERIPHERAL NERVE INJURIES

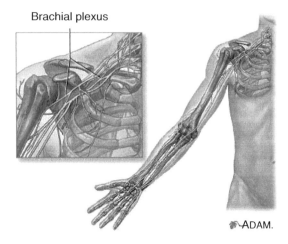

Figure 8.11 The brachial plexus and its pathways.

sports (Wilbourn 2007). The superior trunk of the brachial plexus (C5 and C6) is thought to be the structure injured in this syndrome (Aldridge et al. 2001; Dimberg and Burns 2005; Wilbourn 2007). Tensile overload is one mechanism of injury for the brachial plexus (Hershman 1990), whereby the head and shoulder of the symptomatic side have been forced in opposite directions to each other (Weinstein 1998; Dimberg and Burns 2005) thereby causing stretch and traction of the nerve. Alternatively, another mechanism of injury in Stinger Syndrome is where a compressive force forces the head and neck into the posterolateral corner, causing nerve root compression and thereby injuring the nerve (Weinstein 1998). Finally, injury to the brachial plexus can occur as the result of a direct blow to

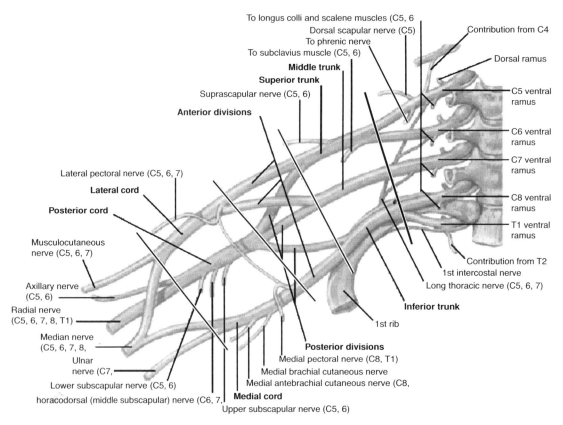

Figure 8.12 Pathways of the nerves for the upper limb

the plexus region (Weinstein 1998; Reid and Trent 2002).

Symptoms reported for Stinger Syndrome generally involve a single upper extremity (Dimberg and Burns 2005; Wilbourn 2007) with sharp, burning pain and numbness accompanied by short-lived weakness down the arm (Weinstein 1998; Aldridge et al. 2001). Patients who report bilateral or lower extremity symptoms should be treated with caution as spinal cord injury is a potential injury (Dimberg and Burns 2005). An important finding with Stinger Syndrome is that symptoms are unilateral (Wilbourn 2007) and short-lived. Neuropraxia and axontmesis are typically associated with this type of injury.

If the symptoms resolve within minutes and there is no evidence of reduced strength or disrupted movement of the neck, the athlete can be considered for return to play in the same competition (Dimberg and Burns 2005). However, the player should be continuously monitored during the remainder of the competition and re-assessed over the subsequent weeks. Cervical spine rehabilitation should be undertaken in the Stinger Syndrome patient, aiming to protect the injured structures and control pain and inflammation in the initial stages, following which flexibility and strength imbalances and deficits should be addressed (Weinstein 1998). Postural abnormalities should also be considered over the course of treating patients with cervical spine and nerve injuries. A treatment strategy for Stinger Syndrome can include NSAIDs, modification of activity, neurodynamic sliding techniques in the acute stage, manual therapy of the cervical spine, range of movement activities for the shoulder and strengthening of the shoulder and neck muscles.

Thoracic outlet syndrome

The thoracic outlet region in the upper extremity contains three structures; the subclavian artery, subclavian vein and the brachial plexus, all of which pass under the clavicle and subclavius muscle, before then travelling beneath the pectoralis minor muscle close to its insertion point at the coracoid process, whereafter, the structures enter the axilla space (Atasoy 2004). These neurovascular structures pass through the interscalene triangle; comprising of the anterior and middle scalene muscles and the first rib, whilst en-route to the axilla (Brantigan and Roos 2004) and are considered vulnerable to compression within this area. Restriction of space for the neurovascular structures to pass through within the thoracic outlet can cause compression of the neural and/or vascular structures, subsequently resulting in symptoms (Crosby and Wehbé 2004). Compression of one or more of the structures can lead to the diagnosis of thoracic outlet syndrome (TOS). The clinical presentation of TOS can be variable, presenting as either a neurological or a vascular condition, or both. However, with Brantigan and Roos (2004) reporting 98% of cases to be neurological in origin and Crosby and Wehbé (2004) agreeing that the majority of patients have neurogenic symptoms, for the purpose of this section, TOS will be discussed in relation to the brachial plexus only.

TOS can develop insidiously, or as a consequence of a previous traumatic injury, such as whiplash or a similar such trauma, which ultimately causes chronic muscle spasm within the neck or shoulder (Brantigan and Roos 2004). Whilst the initial injury may be considered to be relatively minor, the resultant effect of the injury on the brachial plexus may be delayed and the condition may take considerable time to develop. The patient may complain of arm elevation exacerbating symptoms, such as shoulder press or serving in tennis, whilst exercise can induce a similar response, but only following cessation of the exercise, rather than during it (Brantigan and Roos 2004). Neck stiffness may also be evident in patients with suspected TOS, and therefore the physical examination should include a comprehensive assessment of the neck, shoulder and the upper extremity. Typically evident in patients with neurogenic TOS, is the reproduction of symptoms when the clinician places mild to moderate pressure over the brachial plexus in the supraclavicular fossa of the symptomatic side (Brantigan and Roos 2004). Brantigan and Roos (2004) recommend incorporating the EAST (elevated arm stress test) into the clinical examination to diagnose TOS as few individuals with this condition are actually able to complete the test.

In the majority of cases, conservative treatment will alleviate symptoms and address the cause of TOS unless there is significant neural loss or vascular compression (Crosby and Wehbé 2004). Incorporation of NSAIDs, postural control including during sleep, manual therapy such as thoracic and first rib mobilisation and muscle and nerve gliding exercises into the treatment plan should be undertaken for patients with TOS (Crosby and Wehbé

2004). Exercise therapy such as neck side flexion, neck rotation, chin tucks, shoulder shrugs and shoulder circles to name but a few, can also be beneficial for this condition (Crosby and Wehbé 2004). Crosby and Wehbé (2004) present a comprehensive overview of treatment strategies for TOS.

Axillary nerve

The axillary nerve is the terminal branch of the posterior cord of the brachial plexus, receiving its contributions from C5 and C6 nerve roots (Bencardino and Rosenberg 2006). Axillary nerve injury is a well-known complication of anterior glenohumeral dislocation (Perlmutter and Apruzzese 1998; Goslin and Krivickas 1999) evident in 9–18% of this injury (Perlmutter and Apruzzese 1998). The close proximity of the anatomical course of the axillary nerve to the shoulder joint increases its susceptibility to injury, secondary to shoulder dislocation. The nerve travels inferior to the humeral head before passing through the quadrangular space, where it then wraps horizontally around the posterior aspect of the surgical neck of the humerus (Pratt 2005). It then enters the deltoid muscle which it innervates, along with teres minor (Goslin and Krivickas 1999).

Glenohumeral dislocation typically occurs in an inferior-anterior direction (Owens and Itamura 2000) whereby on dislocation, the humeral head moves into the infra-articular fossa (Pratt 2005). As a consequence of the forced, abrupt movement of the humeral head, a traction force is placed on the axillary nerve (Perlmutter and Apruzzese 1998) resulting in stretching of the nerve around the humeral head and tension propagating to the infraclavicular brachial plexus (Pratt 2005); the overall outcome resulting in nerve injury (Pratt 2005). Blunt trauma to the anterior/lateral aspect of the shoulder, typically seen in tackling during rugby, whereby a compressive force is applied to the axillary nerve is another common mechanism of injury for this structure (Perlmutter and Apruzzese 1998).

Injury to the axillary nerve with shoulder dislocations may go unnoticed initially as the bony injury may dominate the clinical picture (Perlmutter and Apruzzese 1998) and consequently for an athlete participating in contact sports, a thorough neurological examination with shoulder dislocation should be undertaken (Perlmutter and Apruzzese, 1998). A well-defined patch of sensory loss over the lateral shoulder, in addition to weakness in shoulder abduction (deltoid) and external rotation (teres minor) are typical clinical reporting for axillary neuropathy (Perlmutter and Apruzzese 1998; Goslin and Krivickas 1999). Neurological complications following shoulder dislocation, of the brachial plexus and axillary nerve are enhanced in patients >50 years or if the shoulder remains dislocated for greater than 12 hours (Pratt 2005).

Restoring full range of motion at the shoulder is a priority in the conservative treatment for the axillary nerve (Perlmutter and Apruzzese 1998; Goslin and Krivickas 1999. NSAIDs and analgesics will assist with pain and inflammation control during the acute stage. Strengthening exercises for the rotator cuff and periscapular muscles should also be undertaken (Aldridge et al. 2001) in addition to nerve gliding exercises.

Long thoracic nerve

The long thoracic nerve arises from the ventral rami of C5–C7 spinal nerves (Goslin and Krivickas 1999) and innervates exclusively the serratus anterior muscle (Goslin and Krivickas 1999; Aldridge et al. 2001; Pratt 2005). The relative immobility of the long thoracic nerve is a primary factor in the occurrence of injury to this structure, as it is anchored at regular, short intervals throughout it's anatomical course (Pratt 2005), particularly at the scalene medius and serratus anterior muscles (Goslin and Krivickas 1999). Consequently, traction or stretch of the long thoracic nerve, such as with extreme excursion of the shoulder girdle, is a common mechanism of injury for this structure (Goslin and Krivickas 1999; Pratt 2005). Palsy of this nerve has been associated with repetitive motion or muscular hypertrophy in sport (Toth et al. 2005). One case reported an elite marksman presenting with long thoracic neuropathy, due, according to the author, to the positional stress imposed during repetitive shooting postures whilst holding the gun (Woodhead 1985). Maintaining static postures that keep the shoulder girdle elevated for long periods is an alternate way in which injury to the long thoracic nerve occurs (Goslin and Krivickas 1999; Pratt 2005). This is evident in patients with Saturday night palsy (Pratt 2005), which originates from patients falling asleep on a chair with the shoulder draped over the back of a chair, thereby subjecting the long thoracic nerve to prolonged stretching. Due to the relative superficiality of the nerve, it is also susceptible to external compression (Pratt 2005).

Scapula winging is the most prevalent clinical finding associated with injury to the long thoracic nerve (Aldridge et al. 2001; Goslin and Krivickas 1999) as a result of weakness of the serratus anterior muscle (Owens and Itamura 2002). This clinical finding is best demonstrated by having the patient perform a push-up against a wall (Owens and Itamura 2002). An ache or burning pain in the posterior shoulder with associated weakness in arm elevation and abduction is an early finding in patients with long thoracic nerve pathology (Owens and Itamura 2002), in addition to them having difficulty performing overhead activities (Aldridge et al. 2001), possibly due to serratus anterior being unable to efficiently laterally rotate the scapula during shoulder elevation.

Treatment strategies for long thoracic neuropathy involve rest and avoidance or modification of the aggravating activity (Goslin and Krivickas 1999), NSAIDs and physical therapy. Exercises aiming to fix the scapula against the thorax are advised to prevent further overstretching of the nerve or scapula winging (Goslin and Krivickas 1999). Maintaining shoulder range of motion (Aldridge et al. 2001) and increasing the strength of trapezius and the rhomboids (Goslin and Krivickas 1999) should be additional aims for the treatment plan.

Suprascapular nerve

The suprascapular nerve originates from the superior nerve trunk (C5–C6 nerve roots) at Erbs' point (Aldridge et al. 2001; Pratt 2005; Bencardino and Rosenberg 2006) and is responsible for innervating supraspinatous and infraspinatous muscles, which abduct and laterally rotate the shoulder respectively (Goslin and Krivickas 1999; Pratt 2005). The suprascapular nerve is vulnerable to entrapment (Goslin and Krivickas 1999) as it passes through the suprascapular foramen before curving around the spinoglenoid notch; both anatomical points of nerve entrapment (Pratt 2005).

Injury to this structure can occur as a result of acute stretching, a blow to the superior aspect of the shoulder (Owens and Itamura 2001) whereby the shoulder is forcibly depressed (Goslin and Krivickas 1999), or can be of insidious onset (Aldridge et al. 2001). Repetitive scapular motion, such as that which occurs during overhead activities in tennis or badminton can stretch and compress the suprascapular nerve and induce entrapment neuropathy (Bencadino and Rosenberg 2006). Patients with suprascapular neuropathy may complain of pain at the superior border of the scapula (Goslin and Krivickas 1999), which may co-exist with impingement signs and symptoms (Owens and Itamura 2002) or weakness (Goslin and Krivickas 1999). Weakness and pain during shoulder abduction and lateral rotation may be evident, due to dennervation of the supraspinatous and infraspinatous muscles respectively (Goslin and Krivickas 1999; Aldridge et al. 2001). Point tenderness over the area of nerve compression is evident in patients with entrapment of the nerve whilst horizontal adduction of the shoulder may elicit patient symptoms as the tension on the nerve is increased with this manoeuvre (Aldridge et al. 2001).

The management for injury to the suprascapular nerve should involve modifying or ceasing the aggravating activity (Goslin and Krivickas 1999; Aldridge et al. 2001), whilst incorporating NSAIDs and scapular stabilisation exercises into the rehabilitation programme (Aldridge et al. 2001). Restoring full range of motion at the shoulder, increasing strength of the scapular stabilisers and rotator cuff muscles (Goslin and Krivickas 1999) are additional aims for a treatment programme for this neuropathy.

Ulnar nerve

The ulnar nerve is the end point of the medial cord of the brachial plexus, composed of fibres from C8 and T1 nerve roots (Rokito et al. 1996; Aldridge et al. 2001; Bencardino and Rosenberg 2006) and is extremely mobile at the elbow joint. It is therefore susceptible to direct trauma, compression and traction; to name but a few causes of injury to this structure (Izzi et al. 2001).

Cubital tunnel syndrome

Cubital tunnel syndrome is the second most common neuropathy in the upper extremity (Bencardino and Rosenberg 2006) and the commonest entrapment neuropathy at the elbow (Salama and Stanley 2008), being responsible for entrapment of the ulnar nerve. The potential for ulnar nerve compression at the cubital tunnel is unsurprising considering its anatomical arrangement (Pratt 2005). The arcuate ligament and medial collateral ligament of the elbow form the roof and floor aspect of the tunnel respectively (Pratt 2005; Bencardino and Rosenberg 2006). During elbow flexion the points of attachments for the structures of the cubital tunnel are pulled further apart, resulting in tightening of both the floor and roof of the tunnel. In addition, during forearm flexion, the trochlea of the humerus rotates under the medial collateral ligament and elevates the floor of the tunnel, culminating in decreased space for the ulnar nerve to glide and move (Pratt 2005). Therefore, it is unsurprising that this syndrome is prevalent in throwing athletes (Izzi et al. 2001), many of whom undertake repetitive elbow joint movements, with forced extension, such as seen in pitching a baseball. During repetitive, high-velocity throwing, athletes, particularly in those who demonstrate excessive laxity at the elbow (Badia and Stennett; Rokito et al. 1996), increased traction is placed on the nerve during the movement, all of which culminates in an increased susceptibility to ulnar neuropathy (Rokito et al. 1996).

Clinically, patients with cubital tunnel syndrome may present with diminished sensation in the ulnar aspect of the fourth finger and all of the fifth finger (Izzi et al. 2001; Salama and Stanley 2008) and elbow pain radiating to the hand with sensory symptoms, particularly at night, when the elbow is maintained in flexion for long periods of time during sleep (Bencardino and Rosenberg 2006). Weakness in the finger abductors, and thumb adductor or struggling to maintain a powerful grip may be additional clinical complaints (Izzi et al. 2001; Shapiro and Preston 2009). In the overhead athlete, pain along the medial joint line may be the first clinical sign of ulnar neuropathy or complaints of a popping or snapping sensation at the elbow during flexion and extension of the forearm, due to subluxation of the ulnar nerve, may also be reported (Aldridge et al. 2001).

In the early stages of nerve compression, provocative testing may be the only positive finding (Novak and Mackinnon 2005). A positive Tinel's sign should be performed at the entrapment site and along the course of the nerve whereby reproduction of patient symptoms in the relevant neural distribution constitutes a positive test. Tapping the ulnar nerve behind the medial epicondyle is a Tinel's test for the ulnar nerve at the elbow (Salama and Stanley 2008). Positional and provocation tests should also be undertaken in potential entrapment neuropathy, whereby each test is held for one minute, and are deemed positive if there is alteration in sensations in the correct neural distribution (Table 8.3) (Novak and Mackinnon 2005). It is important in cubital tunnel syndrome to differentially diagnose the possibility of thoracic outlet syndrome, lower cervical neuropathy or compressive ulnar neuropathy at the wrist (Bencardino and Rosenberg 2006). Medial epicondylitis, medial ligament laxity and ulnar neuritis all require differential diagnosis by the clinician (Rokito et al. 1996).

Conservative treatment of cubital tunnel syndrome can include avoidance of the aggravating activity (Badia and Stennett 2006), NSAIDs, altering throwing technique, manual therapy, nerve gliding (Kostopoulos 2004) and exercise therapy such as a progressive strengthening exercise programme (Badia and Stennett 2006). A night splint may be indicated in early cases, particularly if the patient sleeps prone, with the elbow tucked under a pillow whereby local pressure can be placed on the ulnar nerve. In this position, traction can also be placed on the nerve, if the elbow remains flexed and shoulder abducted during the sleeping posture (Izzi et al. 2001; Salama and Stanley 2008). Night splints should only be utilised if the patient complains of symptoms whilst sleeping.

Table 8.3 Positive tests for nerve compression in the upper extremity

Nerve	Site of entrapment	Provocation test
Radial nerve	Distal forearm	Forearm pronation with wrist ulnar deviation. Pressure over tendinous junction of extensor carpi radialis and brachioradialis
Ulnar nerve	Cubital tunnel	Elbow flexion and pressure on ulnar nerve at the cubital tunnel region
Median Nerve	Proximal forearm	Forearm supination with pressure in the region of pronator teres
	Carpal tunnel	Wrist flexion and/or extension with pressure proximal to carpal tunnel

Adapted from Novak and Mackinnon (2005).

Radial nerve

The radial nerve is the larger branch of the posterior cord of the brachial plexus (Bencardino and Rosenberg 2006), with part of its anatomical course running obliquely around the posterior shaft of the humerus in the spiral groove (Pratt 2005), making it therefore very vulnerable to injury following fracture of the humerus. Radial nerve injury can occur secondary to humeral shaft fracture, as a result of the inappropriate use of axillary crutches or due to entrapment at the elbow (Bencardino and Rosenberg 2006). The radial nerve also has the potential to become entrapped in callus formation during the healing process following humeral fracture (Pratt 2005).

Radial tunnel syndrome

Compression of the radial nerve at the elbow is referred to as radial tunnel syndrome and more commonly affects the posterior interosseus nerve (Bencardino and Rosenberg 2006). It is commonly seen in racquet sport athletes or swimmers where repetitive pronation and supination occurs (Izzi et al. 2001; Bencardino and Rosenberg 2006). The posterior interosseus nerve is vulnerable to compression at various sites within the radial tunnel (Bencardino and Rosenberg 2006), of which the arcade of Frohse is one of the hazardous anatomical structures that can compress the nerve at the elbow (Pratt 2005). The arcade of Frohse is a fibrous arch formed by the proximal portion of the superficial head of the supinator muscle (Pratt 2005; Bencardino and Rosenberg 2006) and under which the radial nerve passes (Bencardino and Rosenberg 2006).

Clinically, a patient with compression of the radial nerve at the elbow will present with poorly localised pain to the antero-lateral aspect of the elbow, which can be provoked by manoeuvers that stretch or compress the nerve (Izzi et al. 2001). Tenderness over the radial nerve along the radial tunnel, pain on resisted supination and a positive Tinels' sign over the radial forearm are factors that may be present on physical examination (Bencardino and Rosenberg 2006). Differential diagnosis for lateral epicondylitis of the elbow is vital as radial tunnel syndrome can masquerade as or co-exist with this condition (Bencardino and Rosenberg 2006; Salama and Stanley 2008).

Management of radial tunnel syndrome should incorporate pain-relieving modalities such as NSAIDs (Izzi et al. 2001; Salama and Stanley 2008), nerve gliding exercises (Badia and Stennett 2006), activity modification and splinting (Izzi et al. 2001; Salama and Stanley 2008).

Median nerve

The median nerve is formed by contributions from the medial and lateral cords of the brachial plexus (Pratt 2005; Bencardino and Rosenberg 2006) containing motor and sensory fibres from the C5–T1 nerve roots (Bencardino and Rosenberg 2006). Whilst carpal tunnel syndrome is the most common nerve entrapment of the upper extremity (Bencardio and Rosenberg 2006), it is more common in the general population as opposed to athletes, as the latter typically suffer from nerve compression at the elbow (Izzi et al. 2001).

Pronator teres syndrome

Pronator teres syndrome is compression of the median nerve (Izzi et al. 2001; Kostopoulos 2004) and is the most common cause of median nerve entrapment at the elbow (Bencardino and Rosenberg 2006). It is more prevalent in athletes as opposed to the general population (Badia and Stennett 2006) with hypertrophy of the pronator teres cited as one of the potential inflictors of compression on the nerve (Izzi et al. 2001). External compression on the forearm muscles is another source of medial neuropathy discussed by Badia and Stennett (2006) whereby one of the authors reported medial nerve compression in an acrobat who wrapped a curtain of material around the forearm as part of an acrobatic routine during numerous performances, thereby repetitively compressing the median nerve and ultimately inflicting injury.

The clinical features of pronator teres syndrome are often vague (Kostopolous 2004). However, patients with this syndrome may complain of volar arm pain, with or without hand pain, which may be exacerbated by repeated forearm pronation and wrist flexion (Izzi et al. 2001). Numbness in the median nerve distribution with repetitive pronation and supination, but not elbow flexion and extension, is another indicator of this neuropathy (Bencardino and Rosenberg 2006). Weakness in resisted forearm pronation, wrist flexion and radial deviation are additional clinical presentations with this condition, in additional to thenar atrophy and an inability to oppose or flex the thumb (Bencardino and Rosenberg 2006). Tenderness on palpation can be evident over the pronator muscle (Izzi et al. 2001).

Patients with pronator teres syndrome are advised to abstain from the aggravating activity or modify their choice of equipment or technique (Badia and Stennett; Salama and Stanley 2008). NSAIDs, rest (Salama and Stanley 2008), nerve gliding exercises and stretching of the pronator teres muscle are some of the treatment strategies available to clinicians for this neuropathy. Splinting is another option presented by Salama and Stanley (2008) for treatment of this condition.

Carpal tunnel syndrome

Carpal tunnel syndrome (CTS) is the most common entrapment neuropathy in the upper body (Shapiro and Preston 2009), involving compression of the median nerve as it traverses through the wrist at the carpal tunnel (Rempel and Diao 2004) and presents relatively frequent in athletes (Aldridge et al. 2001). The carpal tunnel is composed of the stiff carpal bones of the wrist, which make up the floor and walls of the tunnel and the flexor retinaculum, which acts as the roof of the tunnel (Kostopoulos 2004). Nine tendons, the medial nerve, synovium and radial and ulnar bursae occupy the carpal tunnel (Rempel and Diao 2004). It is therefore quite apparent how this particular area may be subjected to entrapment neuropathy due to the limited space within the tunnel, and the high volume of structures occupying this space.

CTS is considered to develop as the result of repetitive wrist use and commonly presents in gripping athletes, such as archery, racquet and throwing athletes (Izzi et al. 2001). The syndrome, according to Kostopoulos (2004) is defined by the signs and symptoms of the median nerve at the wrist, as sensory loss and parasthesia are commonly present in the distribution of the median nerve with this condition (Shapiro and Preston 2009). A thorough investigation of the wrist and cervical spine should be undertaken to eliminate the potential for double crush syndrome (Kostopoulos 2004). Clinically, patients complain of pain in the wrist and hand with parasthesia evident in the lateral three and a half fingers (Izzi et al. 2001) in addition to reporting nocturnal pain (Aldridge et al. 2001; Izzi et al. 2001). In chronic conditions, symptoms may be reported above the carpal region, as far distal as the cervical spine (Kostopoulos 2004). Abductor pollicis weakness is the most common motor weakness associated with CTS (Izzi et al. 2001) and a diminished grip strength may be evident when compared to the asymptomatic side (Aldridge et al. 2001). A positive Tinel's sign or Phalens test is indicative of CTS (Aldridge et al. 2001; Shapiro and Preston 2009) and thenar atrophy may be observed in advanced cases of CTS (Aldridge et al. 2001). Symptoms can be exacerbated during sleep if the wrist is maintained in a flexed or extended position (Shapiro and Preston 2009).

Conservative treatment of CTS can include NSAIDs, active rest, modification of the aggravating activity, tendon and nerve gliding (Aldridge et al. 2001; Izzi et al. 2001; Kostopoulos 2004) and exercise therapy. Splinting of the wrist is another treatment option for clinicians for CTS management.

Lower limb nerve injuries

Comprising of the lumbar and sacral plexuses, the lumbosacral plexus originates from L1, L2, L3 and, in part, L4 anterior primary rami (APR) (Wilbourn 2007). The remaining APR of L4 fuses with L5 APR to form the lumbosacral trunk, which subsequently contributes to the formation of the sacral plexus; an entity responsible for providing sensation to the majority of the lower limb (Wilbourn 2007). The lower branch of L2, all of L3 and the upper branch of L4 terminate by dividing into anterior and posterior divisions; the former forming the obturator nerve and the latter forming the femoral nerve. The superior and inferior gluteal nerves are formed by the posterior divisions of L4, L5 and S1, and the posterior divisions of L5, S1 and S2, respectively (Figure 8.13). The sacral plexus provides sensation to the gluteal region and, with the exception of the anterolateral thigh and a lengthy strip of medial leg, the lower limb. It innervates the pelvic floor muscles, gluteals and tensor fascia latae, hamstrings and all the muscles of the leg and foot (Wilbourn 2007). Lumbosacral plexopathies are infrequent, compared to brachial plexus injury, possibly due in part to the protective anatomical arrangement of the pelvis and surrounding musculature, in addition to the nerves being associated with less mobile structures (Wilbourn

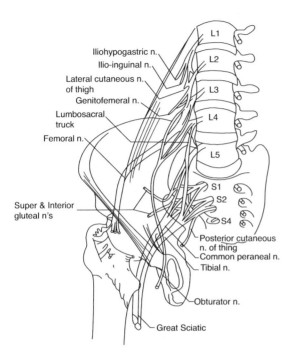

Figure 8.13 Lumbosacral plexus.

2007). The anatomical arrangement of the sciatic, tibial and peroneal nerves can be viewed from Figure 8.14 and are subsequently discussed in relation to injury and treatment, particularly in sporting populations.

Sciatic nerve

Originating from L4, L5, S1 and S2 nerve roots and consisting of a peroneal and tibial division, which are encased in a common sheath, the sciatic nerve is formed (Katirji 1999). The sciatic nerve exits the pelvis via the greater sciatic notch before traversing into the posterior thigh where approximately mid-thigh; it splits into the tibial nerve and common peroneal nerve (Yuen and So 1999). The tibial nerve is responsible for innervating the majority of the hamstrings; the short head of biceps femoris being the exception as it is innervated by the common peroneal nerve (Katirji 1999; Yuen and So 1999).

Piriformis syndrome

The piriformis muscle originates from the anterior surface of the sacrum, inserting into the greater trochanter of the femur (Papadopoulos and Khan 2004). It traverses to its insertion, via the sciatic notch; a point where all neurovascular structures which enter the buttock from the pelvis descend, either superior or inferior to piriformis; the sciatic nerve, being one such structure, exits the pelvis below the muscle (Papadapoulous and Khan 2004). Piriformis syndrome arises from the belief that a hypertrophied piriformis muscle compresses the sciatic nerve causing pain in the nerve's distribution (Tiel 2008), a theory which is plausible considering the intimate anatomical arrangement between these two structures. The resultant effect of this nerve compression is symptoms indicative of proximal sciatic nerve dysfunction, a neuropathy that has become known as piriformis syndrome (Shapiro and Preston 2009).

A patient presenting with suspected piriformis syndrome generally complains of pain and tenderness in the buttock region at the sciatic notch, particularly during prolonged sitting on hard surfaces (Shapiro and Preston 2009). Complaints of buttock pain, with or without accompanying ipsilateral radiating pain in the sciatic nerve's distribution, is common; symptoms may be exacerbated by stretching the piriformis via adduction and medial rotation of the hip joint (Papadapoulous and Khan 2004; Shapiro and Preston 2009). The clinical assessment should include a comprehensive neurological examination inclusive of motor, sensory and reflex tests (Papadapoulous and Khan 2004). Palpation of the sciatic notch should also be undertaken as this can reproduce symptoms (Papadapoulous and Khan 2004; Shapiro and Preston 2009). However, the intolerance to prolonged sitting on hard surfaces is probably the most prominent sign of this condition (Papadapoulous and Khan 2004); and therefore this condition may be commonly seen in rowers. The noticeable difference between piriformis syndrome and L5 radiculopathy is the latter presents with back pain, altered reflexes, sensory loss and muscle weakness of the hamstrings and gastrocnemius (Shapiro and Preston 2009).

Treatment options for piriformis syndrome should include NSAIDs, physical therapy, neurodynamic treatment techniques and stretching of the muscle (Papadapoulous and Khan 2004; Shapiro and Preston 2009). Should conservative treatment fail, injection or surgical release are more advanced options according to the same authors.

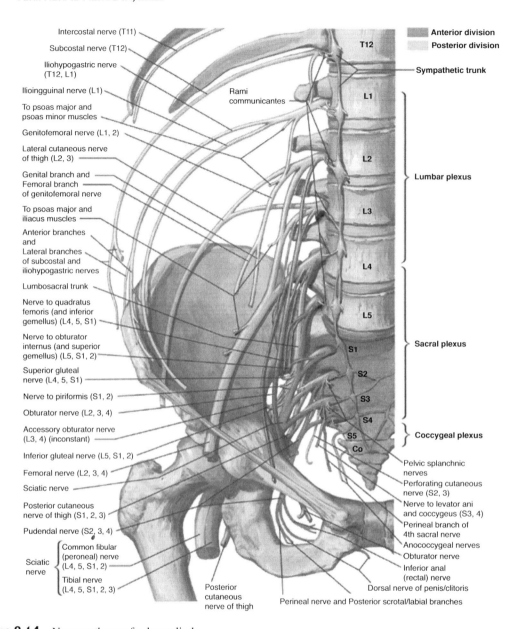

Figure 8.14 Nerve pathways for lower limb.

Posterior thigh injury

In athletes presenting with an injury to the posterior thigh there is not automatically an injury to the hamstrings (Woods et al. 2004). For example, almost 20% of Australian Football League (AFL) players had a normal Magnetic Resonance Imaging (MRI) scan of the hamstrings, despite presenting with pain in the posterior thigh (Verrall et al. 2001). The consequences of such findings for clinicians is such that, in a patient presenting with posterior thigh pain, where there is difficulty identifying by palpation the painful site, lack of bruising, a vague mechanism of injury or complaints of spinal pain, mechanosensitivity of the neural system should be suspected (Butler 1991). The nervous system may incur injury as a direct or indirect result of hamstring muscle injury; the latter of which may occur as a consequence of the natural formation of scar tissue during the healing process (Shacklock 1995). Fibrosis, lesions or intermuscular adhesions following muscle injury can reduce the fluidness of movement between the muscle and the nerve, and thereby lead to secondary nerve damage (Turl and George 1998; Shacklock 2005). Considerable overlap in terms of clinical features exist between Grade 1 or minor hamstring muscle strains and injuries involving referred pain (Verrall et al. 2003). Considering that the clinical signs associated with Grade 2 and 3 muscle strains, such as swelling and bruising, are typically absent from Grade 1 strains (Kornberg and Lew 1989; Turl and George 1998), differentiation of Grade I hamstring injury from referred pain is difficult (Kornberg and Lew 1989).

Suspicion of sciatic nerve involvement in hamstring strains is warranted, particularly considering how one in five athletes will have an absence of muscle pathology on MRI scan (Verrall et al. 2001), despite presenting clinically as a muscle strain. As an MRI scan is not an easily accessible diagnostic tool to many clinicians working in sport, neurodynamic testing, via the slump test is often the clinical test used to differentially diagnose posterior thigh pain.

Whilst the treatment of posterior thigh pain, if considered to be muscular, should focus on returning the muscle to pre-morbid abilities, the sciatic nerve should not be ignored. Whilst slump stretching, used in conjunction with traditional treatment (not mobilisation) modalities, has been shown to significantly decrease the recovery period from hamstring injury in professional AFL players (Kornberg and Lew 1989); no research has evaluated the effect sliding or tensioning techniques has on the return to play. However, as with all the aforementioned neuropathies, sliding or tensioning neurodynamic techniques should be included in the treatment of posterior thigh injury; the selected technique dependant on the irritability of the patient's condition. Sliding techniques should be considered in the acute stage management, progressing to the more aggressive tensioning techniques in the latter stages of rehabilitation.

Peroneal nerve injury

The common peroneal nerve (CPN) originating from the sciatic nerve, descends through the posterior thigh before traversing the popliteal fossa, whereby it then wraps around the fibular neck before passing through the fibular tunnel (Katirji 1999), following which the nerve divides into the superficial and deep peroneal nerve (McKean 2009; Shapiro and Preston 2009). The fibular tunnel is a tendinous tunnel between the peroneus longus muscle and the fibula (Katirji 1999) and it is therefore a common site of compression for this nerve.

Peroneal neuropathy is prevalent in individuals who repetitively squat (Shapiro and Preston 2009), such as a catcher in baseball, as this position can induce prolonged stretching of the nerve. However, it also occurs in individuals with minimal body fat, due to the minimal amount of adipose tissue available to surround and protect the nerve from external forces or compression (Shapiro and Preston 2009). An acute injury to the CPN may occur in runners, whereby the nerve is forcibly stretched as the result of a severe ankle inversion injury (Shapiro and Preston 2009), whilst fracture to the fibular head or knee dislocation can also injury this nerve (Katirji 1999).

Typically patients complain of pain, burning or numbness down the anterolateral aspect of the lower leg with a loss in sensation evident in the dorsum of the foot (Katirji 1999; McKean 2009; Shapiro and Preston 2009). Clinical examination can reveal weakness in the ankle evertor muscles (McKean 2009), the ankle and toe dorsi-flexors (Shapiro and Preston 2009), the resultant effect being foot-drop (Katirji 1999). As a consequence of foot-drop, susceptibility to falls is increased in patients as the foot may get trapped due to the inability to dorsi-flex the foot when walking (Katirji 1999). A positive Tinel's sign may be present at the fibular neck (Katirji 1999; McKean 2009; Shapiro and Preston 2009) and pain can also be reported at the lateral fibular neck (Shapiro and Preston 2009).

Conservative treatment for non-traumatic peroneal neuropathy should include NSAIDs, active rest, physical therapy and neurodynamic sliding or tensioning techniques, such as the SLR with plantar-flexion and inversion and manual therapy. In thin athletes, padding on the fibular head may be recommended to protect the peroneal nerve from external compressive forces. Avoidance of crossing the legs is also advised and ankle bracing should be considered in patients presenting with foot-drop to prevent ankle inversion sprains (Katirji 1999; Shapiro and Preston 2009).

Tibial nerve injury

The tibial nerve originates from the ventral divisions of the L5, S1 and S2 nerve roots and is the largest of the two major divisions of the sciatic nerve (Oh and Meyer 1999), traversing through the popliteal fossa, entering the lower leg between the medial and lateral head of gastrocnemius (Franson and Baravarian 2006), before passing on the medial aspect of the ankle and entering the foot via the tarsal tunnel (Oh and Meyer 1999). The tarsal tunnel is a fibro-osseous tunnel comprising of the flexor retinaculum as the roof of the tunnel and the medial wall of the talus and calcaneus and distal medial aspect of the tibia as the floor of the tunnel (Oh and Meyer 1999; Franson and Baravarian 2006; McKean 2009). Numerous anatomical structures course through the tarsal tunnel, such as the tibialis posterior tendon, posterior tibial nerve and flexor hallucis longus and flexor digitorum tendons, along with numerous blood vessels (Franson and Baravarian 2006). One of the most common entrapment neuropathies of the lower limb is tarsal tunnel syndrome involving the tibial nerve (Oh and Meyer 1999).

Tarsal tunnel syndrome

Tarsal tunnel syndrome (TTS) is a compression neuropathy of the posterior tibial nerve as it passes through the tarsal tunnel at the ankle on the medial side (Franson and Baravarian 2006). The diagnosis of TTS is based on the subjective history and objective examination of a patient (Franson and Baravarian 2006). Patients can present with numerous symptoms such as tingling, numbness or pain at the toes, through the arch of the foot or the heel. Prolonged standing or walking may exacerbate symptoms (Franson and Baravarian 2006) and therefore this is an injury that may be prevalent in hikers or runners, the latter of which may be vulnerable to TTS (Shapiro and Preston 2009) due to the repetitive ankle dorsi-flexion and plantar flexion which occurs whilst running (McKean 2009).

An insidious onset of TTS is often reported (Shapiro and Preston 2009) and symptoms can vary depending on the location of the tibial nerve and its branches (McKean 2009). Patients may complain of numbness or tingling and pain in the medial heel, arch of the foot or the sole, particularly on the medial aspect (McKean 2009; Shapiro and Preston 2009). These symptoms may increase during running, subsequently decreasing with rest (KcKean, 2009). The two most prominent clinical signs are a positive Tinel's test and sensory impairment of the terminal branches of the plantar nerve. Percussion over the tibial nerve proximal to the upper border of the tarsal tunnel behind the medial malleolus is the best location for the Tinel's test. Evaluating sensory impairment, via pin-pricking, and comparing to the non-involved foot is recommended by Oh and Meyer (1999). It is vital to differentially diagnose TTS from L5 or S1 radiculopathy and sensory peripheral neuropathy. Lumbosacral radiculopathy generally presents with a positive straight leg raise (SLR), weakness of the calf muscles, pain radiating from the lower back and a negative Tinel's sign, whereas distal sensory peripheral neuropathy, has an objective stocking distribution sensory loss in the foot, including the dorsum and lateral aspects, and an absent Tinel's sign (Oh and Meyer 1999; Oh 2007). In TTS there should be no objective sensory loss over the dorsum of the foot, although it may be evident on the dorsum of the toes (Oh 2007). Bilateral TTS rarely occurs; however, if suspected, the presence of a distal sensory neuropathy must be considered (Oh 2007). Plantar fasciitis is another condition requiring consideration during the diagnosing process in patients with suspected TTS. Plantar fasciitis generally presents with localised pain, which is absent of sensory abnormalities; two factors not evident in TTS.

As with all the previously mentioned neuropathies, conservative treatment can include NSAIDs, active rest, manual therapy, neurodynamic siding or tensioning techniques and exercise therapy. An additional treatment option to consider is using orthotics or modify footwear to correct any excessive pronation, to minimise the stress placed on the tibial nerve (McKean 2009; Shapiro and Preston 2009).

Summary

The upper limb, being more mobile than the lower extremity is more susceptible to nerve injury and therefore the clinicians working in sport should expect a higher incidence of upper limb neuropathies than lower limb, particularly in athletes participating in contact sports such as rugby union. Not all neural injuries in the sporting population will be traumatic in nature, particularly for those semi-professional or recreational athletes, who are in employment and may be susceptible to chronic neuropathies of the upper limb. In this particular cohort, whereby diversity exists in the type of employment, the clinician needs to be aware of neuropathies that can present as a consequence of occupation; such as CTS and TOS. Rather than being blinkered by neuropathies that typically occur in sport, the clinician who has a vast array of knowledge of various conditions will not only aid the athletes in their sporting activity, but also their activities of daily living.

Case study

A 30-year-old male football player, who competed once a week, reported sustaining a left hamstring injury during a game, three days prior to initial assessment, whereby he felt an "electric" pulse shoot into the ischial tuberosity region and had to stop playing immediately. The patient reported lumbar spine stiffness and a feeling of being unable to straighten the knee whilst walking. He first noticed discomfort in the gluteal area approximately three weeks prior to assessment, which was exacerbated whilst driving, particularly whilst resting the left foot on the foot pedal. However, over the three-week period the symptoms gradually increased, whereby the patient reported a feeling of tightness in the buttock and hamstring during competition and training.

During the physical assessment, the patient reported stiffness in the lumbar spine and hamstring during lumbar flexion. No abnormalities were reported in the sacroiliac joint. All passive and resisted hip movements were negative. Resisted knee flexion was weak but pain free, whilst all other resisted movements for the knee were normal. No bruising was apparent in the gluteal or posterior thigh region. The patient reported reproduction of symptoms during the Straight Leg Raise ankle dorsi-flexion, which was relieved with the release of dorsi-flexion. Likewise, a positive slump test, in terms of symptom reproduction and decreased knee ankle, was also observed in both the right and left limb. The patient experienced pain on palpation, with central posterior-anterior pressure on L4, L5 and S1 vertebrae, which were found on palpation to be hypomobile.

It was apparent the hamstring strain occurred as a bi-product of lumbar spine pathology. A core treatment strategy utilised over the course of rehabilitation was the use of neural sliding techniques during the slump test. Considering the acuteness and irritability of the condition and the fact neural symptoms had been experienced for three weeks prior to hamstring injury onset, sliding techniques were deemed the most appropriate neurodynamic treatment tool. The patient was instructed to execute the sliders as frequently as possible, particularly if he had been driving for a significant period of time. Manual therapy and a home exercise programme of stabilisation exercises were also implemented into the patient's treatment plan to address the underlying lumbar spine problem. Finally, a progressive rehabilitation programme for the hamstrings was also implemented. Three weeks following injury, the patient returned to competition.

References

Aldridge, J.W., Bruno, R.J., Strauch, R.J. and Rosenwasser, M.P. (2001) Nerve entrapment in athletes. *Clinics in Sports Medicine*, *20*, 95–122.

Atasoy, E. (2004) Thoracic outlet syndrome: anatomy. *Hand Clinics*, *20*, 7–14.

Babbage, C.S., Coppieters, M.W. and McGowan, C.M. (2007) Strain and excursion of the sciatic nerve in the dog: Biomechanical considerations in the development of a clinical test for increased neural mechano sensitivity. *Veterinary Journal*, *174*, 330–336.

Badia, A. and Stennett, C. (2006) Sports-related Injuries of the Elbow. *Journal of Hand Therapy*, *19*, 206–227.

Bencardino, J.T. and Rosenberg, Z.S. (2006) Entrapment neuropathies of the shoulder and elbow in the athlete. *Clinics in Sports Medicine*, *25*, 465–487.

Bove, G.M., Zaheem, A. and Bajwa, Z.H. (2005) Subjective nature of lower limb radicular pain. *Journal of Manipulative and Physiological Therapeutics*, *28*, 12–14.

Brantigan, C.O. and Roos, D.B. (2004) Etiology of neurogenic thoracic outlet syndrome. *Hand Clinics*, *20*, 17–22.

Butler, D. (1991) *Functional Anatomy and Physiology*. Melbourne: Churchill Livingstone.

Butler, D. (2000) *The Sensitive Nervous System*. Adelaide: NOIGroup Publications.

Campbell, W.W. (2008) Evaluation and management of peripheral nerve injury. *Clinical Neurophysiology*, *119*, 1951–1965.

Coppieters, M., Stappaerts, K. and Janssens, K. (2002) Reliability of detecting "onset of pain" and "submaximal pain" during neural provocation testing of the upper quadrant. *Physiotherapy Research International*, 7, 146–156.

Coppieters, M.W., Hough, A.D. and Dilley, A. (2009) Different nerve-gliding exercises induce different magnitudes of median nerve longitudinal excursion: An in vivo study using dynamic ultrasound imaging. *Journal of Orthopaedic and Sports Physical Therapy*, 39, 164–171.

Crosby, C.A. and Wehbé, M.A. (2004) Conservative treatment for thoracic outlet syndrome. *Hand Clinics*, 20, 43–49.

Dahlin, L. B. (2008) Nerve injuries. *Current Orthopaedics*, 22, 9–16.

Dilley, A., Lynn, B. and Pang, S.J. (2005) Pressure and stretch mechanosensitivity of peripheral nerve fibres following local inflammation of the nerve trunk. *Pain*, 117, 462–472.

Dimberg, E.L. and Burns, T.M. (2005) Management of common neurologic conditions in sports. *Clinics in Sports Medicine*, 24, 637–662.

Franson, J. and Baravarian, B. (2006) Tarsal tunnel syndrome: A compression neuropathy involving four distinct tunnels. *Clinics in Podiatric Medicine and Surgery*, 23, 597–609.

Gallant, S. (1998) Assessing adverse neural tension in athletes. *Journal of Sport Rehabilitation*, 7, 128–139.

Gardner, E. and Bunge, R. (1984) *Gross Anatomy of the pPeripheral Nervous System*. Philadelphia, PA: Saunders.

Goslin, K.L. and Krivickas, L.S. (1999) Proximal neuropathies of the upper extremity. *Neurologic Clinics*, 17, 525–548.

Hall, T.M. and Elvey, R.L. (1999) Nerve trunk pain: physical diagnosis and treatment. *Manual Therapy*, 4, 63–73.

Hershman, E. (1990) Brachial plexus injuries. *Clinics in Sports Medicine*, 9, 11–29.

Izzi, J., Dennison, D., Noerdlinger, M., Dasilva, M. and Akelman, E. (2001) Nerve injuries of the elbow, wrist, and hand in athletes. *Clinics in Sports Medicine*, 20, 203–217.

Katirji, B. (1999) Peroneal neuropathy. *Neurologic Clinics*, 17, 567–591.

Kornberg, C. and Lew, P. (1989) The effect of stretching neural structures on grade one hamstring injuries. *Journal of Orthopaedic and Sports Physical Therapy*, June, 481–487.

Kostopoulos, D. (2004) Treatment of carpal tunnel syndrome: a review of the non-surgical approaches with emphasis in neural mobilization. *Journal of Bodywork and Movement Therapies*, 8, 2–8.

Lundborg, G. (1988) *Nerve Injury and Repair*. Edinburgh: Churchill Livingstone.

Maitland, G. (1985) The slump test: examination and treatment. *Australian Journal of Physiotherapy*, 31, 215–219.

Marieb, E. (1998) *Human Anatomy and Physiology*., San Francisco, CA: Benjamin/Cummings Science.

McKean, K.A. (2009) Neurologic running injuries. *Physical Medicine and Rehabilitation Clinics of North America*, 20, 249–262.

Mueller, M. and Maluf, K. (2002) Tissue adaptation to physical stress: A proposed "Physical Stress Theory" to guide physical therapists practice, eductation and research. *Physical Therapy*, 82, 383–403.

Nee, R. J. and Butler, D. (2006) Management of peripheral neuropathic pain: Integrating neurobiology, neurodynamics, and clinical evidence. *Physical Therapy in Sport*, 7, 36–49.

Novak, C.B. and Mackinnon, S.E. (2005) Evaluation of nerve injury and nerve compression in the upper quadrant. *Journal of Hand Therapy*, 18, 230–240.

Oh, S.J. and Meyer, R.D. (1999) Entrapment neuropathies of the tibial (posterior tibial) nerve. *Neurologic Clinics*, 17, 593–615.

Owens, S. and Itamura, J.M. (2000) Differential diagnosis of shoulder injuries in sports. *Operative Techniques in Sports Medicine*, 8, 253–257.

Papadopoulos, E.C. and Khan, S.N. (2004) Piriformis syndrome and low back pain: a new classification and review of the literature. *Orthopedic Clinics of North America*, 35, 65–71.

Perlmutter, G. and Apruzzes, W. (1998) Axillary nerve injuries in contact sports; recommendations for treatment and rehabilitation. *Sports Medicine*, 26, 351–361.

Pratt, N. (2005) Anatomy of nerve entrapment sites in the upper quarter. *Journal of Hand Therapy*, 18, 216–229.

Reid, S. and Trent, V. (2002) Brachial plexus injuries – report of two cases presenting to a sports medicine practice. *Physical Therapy in Sport*, 3, 175–182.

Rempel, D.M. and Diao, E. (2004) Entrapment neuropathies: pathophysiology and pathogenesis. *Journal of Electromyography and Kinesiology*, 14, 71–75.

Rokito, A.S., Iviciviahon, P.J. and Jobe, F.W. (1996) Cubital tunnel syndrome. *Operative Techniques in Sports Medicine*, 4, 15–20.

Salama, A. and Stanley, D. (2008) (i) Nerve compression syndromes around the elbow. *Current Orthopaedics*, 22, 75–79.

Seddon, H. (1943) Three types of nerve injury. *Brain*, 66, 237–288.

Shacklock, M. (1995) Neurodynamics. *Physiotherapy*, *81*, 9–16.

Shacklock, M. (2005) *Clinical Neurodynamics: A new System of Musculoskeletal Treatment*. Oxford: Elsevier Health.

Shapiro, B.E. and Preston, D.C. (2009) Entrapment and compressive neuropathies. *Medical Clinics of North America*, *93*, 285–315.

Sunderland, S. (1978) *Nerves and Nerve Injuries*. Edinburgh: Churchill Livingstone.

Tiel, R.L. (2008) Piriformis and related entrapment syndromes: Myth and fallacy. *Neurosurgery Clinics of North America*, *19*, 623–627.

Topp, K.S. and Boyd, B.S. (2006) Structure and biomechanics of peripheral nerves: Nerve responses to physical stresses and implications for physical therapist practice. *Physical Therapy*, *86*, 92–109.

Toth, C. (2009) Peripheral nerve injuries attributable to sport and recreation. *Physical Medicine and Rehabilitation Clinics of North America*, *20*, 77–100.

Toth, C., McNeil, S. and Feasby, T. (2005) Peripheral nervous system injuries in sport and recreation. A systematic review. *Sports Medicine*, *35*, 717–738.

Turl, S.E. and George, K.P. (1998) Adverse neural tension: A factor in repetitive hamstring strain? *Journal of Orthopaedic and Sports Physical Therapy*, *27*, 16–21.

Unlu, M., Kesmezacar, H. and Akgun, I. (2007) Brachial plexus neuropathy (stinger syndrome) occurring in a patient with shoulder laxity. *Acta Orthopaedica et Tramatologica Turcica*, *41*, 74–79.

Verrall, G.M., Slavotinek, J.P., Barnes, P.G. and Fon, G.T. (2003) Diagnostic and prognostic value of clinical findings in 83 athletes with posterior thigh injury – Comparison of clinical findings with magnetic resonance imaging documentation of hamstring muscle strain. *American Journal of Sports Medicine*, *31*, 969–973.

Verrall, G. M., Slavotinek, J.P., Barnes, P.G., Fon, G.T. and Spriggins, A.J. (2001) Clinical risk factors for hamstring muscle strain injury: a prospective study with correlation of injury by magnetic resonance imaging. *British Journal of Sports Medicine*, *35*, 435–439.

Wehbé, M.A. and Schlegel, J.M. (2004) Nerve gliding exercises for thoracic outlet syndrome. *Hand Clinics*, *20*, 51–55.

Weinstein, S.M. (1998) Assessment and rehabilitation of the athlete with a "stinger": A model for the management of noncatastrophic athletic cervical spine injury. *Clinics in Sports Medicine*, *17*, 127–135.

Wilbourn, A.J. (2007) Plexopathies. *Neurologic Clinics*, *25*, 139–171.

Woodhead, A. (1985) Paralysis of the serratus anterior in a world class marksman. *American Journal of Sports Medicine*, *13*, 359–362.

Woods, C., Hawkins, R.D., Maltby, S., Hulse, M., Thomas, A. and Hodson, A. (2004) The Football Association medical research programme: an audit of injuries in professional football - analysis of hamstring injuries. *British Journal of Sports Medicine*, *38*, 36–41.

Yuen, E.C. and So, Y.T. (1999) Sciatic Neuropathy. *Neurologic Clinics*, *17*, 617–631.

Part 4
Effective clinical decision making

9

An introduction to periodisation

Paul Comfort and Martyn Matthews
University of Salford, Greater Manchester

Periodisation

Probably the most influential factor behind recent advances in sports performance is the greater understanding shown by coaches, athletes and sports scientists of Training Theory. Central to this is the concept of *periodisation*, or the structured and sequential planning of training to allow the athlete to make optimal gains in sports performance and produce their best performances in key competitions, with minimal risk of overtraining or injury. Previous research has also demonstrated that compared to non-periodised training programmes, periodised training programmes result in greater increases in strength, power and sports performance (Fleck 1999; Fleck 2002; Rhea and Alderman 2004).

Specifically, periodisation involves the planned and structured variation in training type, volume (intensity and duration), and rest to achieve sport-specific goals. This is achieved via the periodical progression of specific aspects of fitness within a specified time frame. This style of programme allows for gradual and progressive overload, ensuring an optimal stimulus for adaptation whilst allowing adequate time for recovery and adaptation.

Progressive overload is essential in all aspects of physical training, whether the aim is to optimise performance, improve body image, or rehabilitate an injury. In a periodised model, the athlete progresses through phases (meso-cycles), each of which targets a specific fitness attribute, but which is planned such that the intensity and specificity advances from phase to phase and culminates in the athlete reaching their peak of sport-specific fitness at the exact time of key competitions. For example, subsequent phases may focus on basic anatomical and functional adaptations (increased range of motion, increased tendon strength, improved co-ordination and increased stability), strength endurance, maximal strength, power and, ultimately, sport-specific speed and power. Such a progression requires appropriate adjustments in the frequency, intensity and duration of individual training sessions and the appropriate selection of exercises, sets and repetitions within each session. (For more detail on recommended sets, repetitions and loads see Chapter 13 Strength and Conditioning.) With this in mind, periods of training are divided into smaller, more manageable chunks that allow the athlete to maintain both physical and mental input throughout.

This chapter will focus on training cycles and their specific goals, and will consider how best to integrate these into performance, injury prevention and rehabilitation goals, with specific emphasis on ensuring the athlete's readiness to perform.

By the end of this chapter the reader should have a better knowledge of periodisation, and its different phases, an understanding of the application of periodisation to injury prevention and rehabilitation

146 AN INTRODUCTION TO PERIODISATION

Table 9.1 Training phase descriptions

Cycle	Size	Example
Macro-cycle	Large block. Represents several months or even years	The training year; a whole Olympiad
Meso-cycle	Medium block. Usually consisting of a number of weeks or months of training	Often with a specific focus such as strength endurance, maximal strength, power, peaking, active rest, aerobic endurance, or anaerobic threshold
Micro-cycles	Smaller blocks of time	Individual training day/week

programmes, and the integration of this into the athlete's strength and conditioning programme.

Training cycles

Most effective training programmes break training down into cycles. This allows athletes to focus on different attributes at different times, helps with the prevention of overtraining, and is convenient as training can be made to coincide with our own cycles (for example, weeks, seasons, years and Olympiads). Training cycles can last from days to years and, depending on their length, are termed *macro-cycles*, *meso-cycles* and *micro-cycles* (Table 9.1). A macro-cycle represents several months or years, for example, the general conditioning period, the whole training year, or the four years of an Olympiad. A meso-cycle is a smaller block, usually consisting of a number of weeks or months of training (usually prioritised towards a particular training outcome, such as strength endurance, maximal strength, power, peaking, active rest, aerobic endurance, or anaerobic threshold), interspersed with periods of reduced training to facilitate rejuvenation. In this way, meso-cycles can be structured to maximise the recuperative effects of the recovery periods and minimise the risk of overtraining. Micro-cycles are smaller blocks such as an individual training day/week.

Many coaches use a 4-week cycle where the training load increases up to week three, followed by an easier (unloading) week to allow full recovery (usually a decrease in volume but not intensity). This is a basic form of shock training with the four weeks termed ordinary, development, shock and recovery (Matveyev 1981). A meso-cycle typically consists of one or two full 4-week cycles. A 4-week meso-cycle may look like Figure 9.1. An 8-week meso-cycle may look like Figure 9.2.

Meso-cycles that use such a 4-week training block place a great deal of short-term accumulative stress on the body. It is possible that full recovery between sessions or even micro-cycles is not achieved until week four. Prior to this the athlete will accumulate fatigue as the overall training load increases. This in

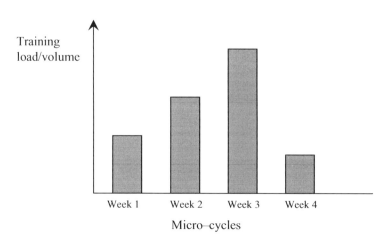

Figure 9.1 An example of a 4-week meso-cycle.

TRAINING CYCLES

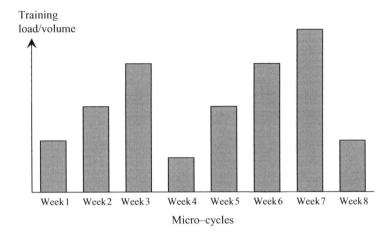

Figure 9.2 An example of a 8-week meso-cycle.

turn can have implications for the trainability of particular fitness attributes (Balyi 2001). For example the training of speed, maximal strength, technique and the learning of new skills should be emphasised during the early weeks of each four-week cycle, when the athlete is relatively fresh and receptive to such methods. During later weeks the most trainable attributes will be cardiovascular endurance, anaerobic power, local muscular endurance and speed endurance. The improvements in fitness from such a pattern of training are illustrated in Figure 9.3.

The training year

The most common macro-cycle is the training year, which in turn consists of a number of meso-cycles and micro-cycles. Every training year consists of a preparation phase, a performance phase and a recuperation phase (Figure 9.4), which vary in length depending on the competition schedule of each particular sport.

The preparation phase itself may be considered in greater detail and further divided into a general preparation period, a specific preparation period and a pre-competition period (Matveyev 1965). This gives a training year with five distinct phases: *general-conditioning period*; *specific-conditioning period*; *pre-competition period*; *competition* (or performance) *period*; *transition* (or recuperation) *period* (see Table 9.2). Figure 9.5 shows Bompa's (1999) representation of the annual training plan.

General conditioning period

The aim of the general conditioning period is to develop a broad base of fitness that acts as a foundation

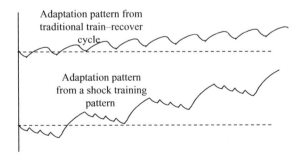

Figure 9.3 A comparison of traditional and shock training adaptations.

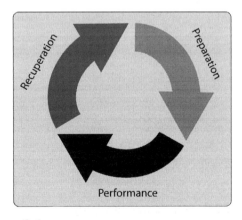

Figure 9.4 Example of macrocycle phases.

148 AN INTRODUCTION TO PERIODISATION

Table 9.2 Training phases

Phases of the training year (Matveyev 1965)	Aim
General conditioning period	To develop a broad base of fitness involving strength, power, local muscular endurance, cardiovascular fitness and flexibility
Specific conditioning period and pre-competition period	To provide a progressive transition from the broad general fitness towards the highly specific fitness required by competition
Competition (or performance) period	To maintain their hard earned fitness, remain injury free and peak for key competitions
Transition (or recuperation) period	To achieve full recuperation in readiness for the general preparation period ahead

for the high-intensity, sport specific training that follows. This period predominantly focuses on strength endurance (8–20 repetitions, 3–5 sets, 3–4 days per week) (Plisk and Stone 2003), but also includes some power exercises, cardiovascular fitness, flexibility training and technique work. This phase is characterised by a progressive increase in both volume (primary aim) and intensity (secondary aim) of training.

Note: Although this period is about general preparation, athletes should not neglect their practice of sport specific skills. These skills should be incorporated into the training via dedicated, short duration, high quality practices, or included as part of a functional dynamic warm-up. This will help ensure that athletes maintain the skills and coordination required in competition. For example, during this period sprinters may be focusing many of their track sessions on speed endurance, rather than pure speed. This, however, does not preclude the athletes performing short bursts of maximum speed running, or quick feet coordination drills, as part of their warm-up.

Specific conditioning and pre-competition period

The aim of this phase is to provide a progressive transition from the broad general fitness towards the highly specific fitness required by competition. It can last from a few weeks (English Premiership football) through to two or three months (sprinters) and consists of a progressive increase in intensity and decrease in volume of training with a greater emphasis of sports specific activities and skills, functional speeds and actual sports practices. Strength training focuses on basic strength (4–6 repetitions, 3–5 sets, 3–5 days per week) (Plisk and Stone 2003).

Competition period

The length of the competition period varies for different sports (soccer — 9 months, athletics — 3–4 months, cricket — 5 months); however the aims remain the same. Athletes need to maintain their hard earned fitness, optimise performance, remain injury free, and peak for key competitions. This phase

	The annual plan				
Phases of training	Preparatory		Competitive		Transition
Aim	General preparation	Specific preparation	Pre-competitive	Competitive	Transition
Meso cycles					
Micro cycles					

Figure 9.5 Annual training plan (after Bompa 1999).

usually focuses on lower volume strength and power based activities (2–3 repetitions, 3–5 sets, 3–6 days per week depending on frequency of competition) (Plisk and Stone 2003).

Transition/recuperation period/active rest

This period usually lasts around a month, depending on the sport, the stresses of the previous competitive season and the timing of the next one. The aims are to achieve full recuperation in readiness for the general preparation period ahead. This period allows any injuries to heal and a period of rehabilitation to occur so that the athlete is ready to resume full training. It is also a time of relaxation and enjoyment. The stresses of competitive sport are far more than just physical. Athletes need time to un-wind and relax. Rest should be taken from the hard training and competition but athletes must also remain active and partake in activities unrelated to their sport. This helps with weight management but also enables the athlete to maintain a degree of general fitness going into the general preparation phase. The activities should be novel and fresh and not too demanding (low volume: 1–3 repetitions, 1–3 sets, ~3 days per week) (Plisk and Stone 2003). A similar strategy can be adopted when peaking/tapering for competition.

There are specific goals for each meso-cycle of training (usually completed in the order below) as proposed by Bompa (1999). These phases include:

- Anatomical adaptation – the development of technique and preparation of the musculoskeletal system (including connective tissues) for the heavier loads used in the following stages (these are usually addressed in the general conditioning phase).

- Hypertrophy (strength endurance) – high volume (repetitions of 8–20 at 50–80% –1RM, for 4–6 sets).

- Strength – which can be subdivided into maximal strength (repetitions of 3–6 at 80–90% –1RM, for 4–8 sets), speed strength (repetitions of 8–15 at <70% –1RM, for 3–4 sets) and strength endurance (repetitions of 20–30 at 30–40% –1RM, for 2–3 sets) (Plisk and Stone 2003).

- Power – can be trained via two primary methods; plyometrics and variations of the 'Olympic lifts'. However, it is worth noting that few athletes 'naturally' have the attributes to perform 'Olympic lifts' safely and effectively and therefore this needs to be developed in the early stages of training (McGill 2006). It is also important to select the component of the lift that is most suitable for the training goal, for example lower limb biomechanics during the second pull is similar to the mechanics of a vertical jump. During 'Olympic lifts' and their variations it is not necessary to use maximal load as Kawamori et al. (2005) showed that lighter loads (50–70% –1RM) resulted in a greater power output due to a higher velocity. Cormie et al. (2007a, 2007b) also demonstrated that when performing activities (such as jump squats) optimal loading, to achieve peak power, is achieved using only body mass, with no additional external resistance.

These phases, but especially the anatomical adaptation phase, need to take into account the 'laws of strength training' and the principles of 'functional training' (Table 9.3) to ensure that the individual is appropriately conditioned to perform the more

Table 9.3

Laws of Strength Training (Bompa 1999)	Functional Training (McGill 2006)
Develop joint flexibility (active ROM)	Develop intra-muscular co-ordination of fibres within a muscle (fast and slow twitch)
Develop tendon strength	Develop inter-muscular co-ordination between muscle groups (efficiency of movement)
Develop core strength	
Develop stabilisers and fixators (balance between agonists and antagonists)	Develop facilitatory and inhibitory reflex pathways (optimal efficiency through the kinetic chain, affected by balance and posture)
Train movements not muscles	Motor learning (optimal efficiency of specific movement)

intense phases of training that follow. This approach will not only reduce the risk of injury, but also ensure that each phase of training is as productive as possible/intended.

Differences between sports

The length of the preparation, competition, transition periods and associated meso-cycles, changes from sport to sport and athlete to athlete. For example, a track and field athlete may have two or three key competitions throughout the course of a three-month competitive season. The rest of the year is effectively preparation for these key competitions. In contrast, an English Premiership football team may only have a six- or seven-week preparation period preceding a nine-month competitive season, during which players are expected to compete on a weekly or a twice-weekly basis. Swimmers often have major competitions arranged on 13- or 14-week cycles. Professional boxers may have between three and four fights a year, with their training always geared towards preparation for the next fight. Despite the different conditions encountered during the preparation for these sports, the following general principles still apply. A gradual build-up in volume (and intensity) early on; as the competition approaches the intensity or quality increases further, whilst the volume tapers off.

Application to sports performance

For athletes with limited training experience the incorporation of a periodisation plan provides an excellent framework to develop an athlete's lifting skills, as well as their strength and power. For example, if an individual wishes to compete in Olympic weightlifting but has never actually completed a structured resistance training programme, they need to start with the basic components of the 'Olympic' lifts, and progress to the point where they are as strong as possible in each of these components, while they learn the complete 'Olympic' lifts. Each meso-cycle should teach progressively more complex components of the lifts to ensure correct technique and to allow for the appropriate anatomical and functional adaptations (neuromuscular control, active range of motion, joint stability, tendon strength) to occur until the complete lifts (snatch and clean and jerk) can be performed (Table 9.4). See Appendix 1 and 2 for illustrations of the clean and snatch.

By the end of the first macro-cycle, the athlete should be confident/competent at performing each of the primary lifts, but should have also developed a considerable level of strength in the key exercises (back squat, front squat, overhead squat, deadlift, Romanian deadlift, see Figures 9.6–9.10). During the next macro-cycle, the athlete can focus on developing additional strength in each of the key exercises of the individual lifts, while progressively increasing load on the 'Olympic lifts'. This allows the athlete to perform the primary lifts with additional weight whilst maintaining good technique at high velocities to ensure the development of power.

The next macro-cycle should place greater emphasis on the loads lifted during the clean and jerk and the snatch. The previous phase should have ensured adequate anatomical adaptations during each of the components of the lifts to ensure adequate conditioning for the lifts to be performed at higher loads. To ensure that the client can decelerate during the catch phase of the snatch, initial deceleration training begins with drop jumps and then progresses to snatch balances, which mimic the catch phase of the snatch (Figure 9.11) and develop adequate trunk strength to decelerate the torso and bar during the rapid decent to the catch phase; a similar progression can also be seen for the clean (drop jumps, clean balance) (Figure 9.12) and the split jerk (split squat, overhead split squat), (Figure 9.13). It is also essential that the velocity of each movement is trained in a progressive manner, as can be seen in the progression from military press to push press.

Olympic lifts are commonly used within strength and conditioning program, due to the fact that increases in performance in these lifts has been shown to increase athletic performance (Hori et al. 2005; Tricoli et al. 2005; Hori et al. 2005; Channell and Barfield, 2008). Appropriate conditioning and preparation for these athletes is essential to ensure that these athletes are adequately prepared to perform these types of exercises.

As mentioned above, each phase has a specific role, which ensures that the athlete is appropriately conditioned for the subsequent phase. Anatomical adaptation focuses on the development of

Table 9.4 Example of a macro-cycle 1: weight lifting

Meso-cycle 1	Meso-cycle 2	Meso-cycle 3	Meso-cycle 4	Meso-cycle 5
Training day 1				
Overhead squat	Overhead squat	Overhead squat	Overhead squat	Overhead squat
Back squat	Back squat	Back squat	Back squat	Back squat
Romanian deadlift*	Romanian seadlift*	Romanian seadlift*	Romanian deadlift*	Romanian deadlift*
Barbell shrug*	Power shrug*	Power snatch	Hang snatch	Snatch
Drop jump landings	Snatch balance	Snatch balance	Snatch balance	
Training day 2				
Front squat	Front squat	Front squat	Front squat	Front squat
Drop jump landings	Clean balance (Drop and candatch)	Clean balance (Drop and candatch)	Clean balance (Drop and catch)	
Deadlift#	Deadlift#	Deadlift#	Deadlift#	Deadlift#
Split squat	Split squat	Overhead split squat	Overhead split squat	
Barbell shrug#	Power shrug#	Power clean	Hang clean	Clean
Military press	Push press	Jerk	Split jerk	Split jerk
Leg flexion/glute ham raise	Leg flexion/glute ham raise	Leg flexion/glute ham raise	Leg flexion/glute ham raise	Leg flexion/glute ham raise

*Using snatch grip
#Using clean grip

Key:
- 15 × 2–3 @ 60%1RM — Anatomical Adaptations / Technique
- 8–12 × 3–4 @ 65–80% 1RM — Hypertrophy / Strength Endurance
- 3–6 × 4–6 @ 85–95% 1RM — Strength
- 1–3 × 4–6 @ 95–100% 1RM — Maximal Strength
- 3v–5 × 3–5 @ 75–85% 1RM — Power
- 85–95% 1RM — Maintenance

technique and preparation of the musculoskeletal system (including connective tissues) for the heavier loads used in the following stages in the strength phase. Hypertrophy (strength endurance) in this example is not a specific goal; therefore this is used as a transition phase between anatomical adaptations and strength. The focus of the strength training phase is to maximise the loads that the athlete can use during the power phase. The strength phase is subdivided into two phases: strength and maximal strength, which allows appropriate progression of loads through to near maximal loads (>90% –1RM). The power phase incorporates two primary methods: plyometrics and variations of the 'Olympic lifts', which provide the opportunity for the athlete to develop near maximal force at high velocity (Olympic variations), and near maximal velocity high force production (plyometrics) that assist with initial acceleration and maximal velocity sprinting, respectively. Table 9.5 illustrates a generic periodised programme for a Sprinter. This only describes the gym based conditioning programme, which would require amendments based on additional sports specific/technique training and manipulation of volume and intensity near periods of competition.

Application to rehabilitation

An understanding of periodisation is essential for complete and effective rehabilitation. This applies both to the phases of progression through rehabilitation, but also to an understanding of the training year of the particular sport to which the athlete is returning. For example, a multiple sprint sport athlete (football, rugby, hockey) has had their ACL

Figure 9.6 Back Squat.

Figure 9.8 Overhead squat (Snatch Catch Position).

Figure 9.7 Front Squat (Clean Catch Position).

Figure 9.9 Deadlift & Clean Start.

Figure 9.10 Romanian (Stiff Leg) Deadlift.

Figure 9.11 Overhead Squat (Snatch Catch Position).

reconstructed, has regained 'normal' ROM and can perform activities of daily living without any problems or pain. The athlete wants to begin training with the squad and playing in matches. However, most non-contact ACL injuries occur during rapid deceleration and cutting manoeuvres (Agel 2005) or during landing (Boden et al. 2000a, 2000b), which both involve high velocity, high force eccentric loading. 'Normal' ACL rehabilitation does not appropriately condition the athlete for such demands, therefore it is essential that the rehabilitation process is specific to the demands of the sport. To achieve this, it is essential to understand the mechanisms of ACL injury. Recent evidence suggests that decreased activation and possible inadequate strength in the gluteus medius, especially in female athletes, can lead to increased knee valgus (Zazulak et al. 2005; Hanson et al. 2008), which is a common factor in acute ACL injury. Other causative factors are poor/inappropriate technique, with limited knee flexion during deceleration; muscle imbalance between hamstring and quadriceps (h:q ratio <0.6) (Li et al. 1999; Ahmed 2006) or an eccentric hamstring to concentric quadriceps ratio <1.0 (Holcomb et al. 2007); delayed activation of the hamstrings, especially in females (Hewett et al. 1996; Malinzak et al. 2001; Zazulak et al. 2005), both of which can result in an increased knee valgus during eccentric loading (Markolf et al. 1995; Malinzak et al. 2001; Ford et al. 2003; McLean 2004; Hewett et al. 2005a).

Therefore, to reduce the risk of non-contact ACL injuries and their recurrence, it is essential that the athlete is appropriately conditioned to cope with rapid deceleration (high velocity eccentric loading) and changes in direction (high velocity eccentric loading followed immediately by high velocity concentric muscle actions – plyometric movements). As females, generally, appear to be quadriceps dominant during deceleration in landing and cutting manoeuvres (Hanson et al. 2008), it is essential that adequate gluteus medius and maximus strength is developed (this must be developed in a functional

Figure 9.12 Catch position for the clean, and bottom of the front squat and front squat drop jumps.

manner, not via isolation of the gluteal muscles), along with adequate hamstring strength to reduce anterior translation of the tibia during knee flexion.

Eccentric hamstring exercises, such as Nordic hamstring lowers, are more effective than concentric training (Kaminski et al. 1998; Mjolsnes et al. 2004; Kilgallon et al. 2007), improving the hamstring: quadriceps ratio, especially at higher velocities (Mjolsnes et al. 2004; Holcomb et al. 2007). This type of activity will also condition the muscles for the specific eccentric component required during deceleration, albeit at a slower velocity.

Based on the periodisation model, and taking into account the current research and guidelines on plyometric training, there are a number of phases (mesocycles) required to train the athlete to a point where they are appropriately conditioned for return to their sport.

The end stage for the athlete is to be able to perform high-intensity plyometric movements; however, in order for them to be able to perform high-intensity plyometrics safely it is recommended that the athlete is sufficiently conditioned

Figure 9.13 Split Jerk Catch Position & Overhead Lunge.

Table 9.5 Example 2: Periodised programme for a sprinter

	General preparation	Hypertrophy (Transition)	Strength	Maxi Strength	Specific preparation	Pre-Competition
	Anatomical adaptation/endurance				Power	Speed
	Meso-cycle 1	Meso-cycle 3	Meso-cycle 4	Meso-cycle 5	Meso-cycle 6	Meso-cycle 7

Training day 1

Back Squat (10–20 × 3 @ 50–60%)	Back Squat (8–12 × 4 @ 65–80%)	Back Squat (4–6 × 4–5 @ 80–90%)	Back Squat (2–4 × 5–6 @ 90–95%)	Back Squat Complex (5 reps, 1 set, @ 90%–1RM) 4 mins rest THEN:	Back Squat Complex (5 reps, 1 set, @ 90%–1RM); 4 mins rest THEN:
Jump Squats (5–6 × 3 @ Body Mass)	Jump Squats (5–6 × 4 @ Body Mass)	Jump Squats (5–6 × 4–5 @ Body Mass ⇓ 5–10%)	Jump Squats (5–6 × 5–6 @ Body Mass ⇓ 10–20%)	Jump Squats (3–5 × 5 @ Body Mass)*1	Jump Squats (2–3 × 4 @ Body Mass)*1.
			Linear Box Jumps (5–6 × 3 @ Body Mass)	Linear Box Jumps (8–10 × 3 @ Body Mass ⇓ 5–10%)	
Stiff Leg Deadlift (8–10 × 3 @ 50–60%)	Stiff Leg Deadlift (8–12 × 3 @ 65–80%)	Stiff Leg Deadlift (6–6 × 3–4 @ 80–90%)	Stiff Leg Deadlift (2–4 × 4–5 @ 90–95%)	Stiff Leg Deadlift (4 reps, 2 sets, @ 90% –1RM)	
		Hang-Clean (4–6 × 3 Technique, low resistance ~40%)	Hang-Clean (6–10 × 3 medium resistance ~60%)	Hang Power-Clean (2–3 × 3 @ 70%)*3	
Mid-Thigh Power Shrug (Snatch-grip) (10 × 3 @ 60–70%)	Mid-Thigh Power Shrug (6–10 × 3 @ 70–80%)	Mid-Thigh Power Shrug (4–6 × 3 @ 80–90%)	Mid-Thigh Power Snatch (6–10 × 3 low/medium resistance 50–60%; emphasise speed)	Mid-Thigh Power Snatch (2–3 × 3 @ 60–70%)	Mid-Thigh Power Snatch (2–3 × 3 Low resistance 60–70%; emphasise speed); 4 mins rest THEN:
					Linear Box Jumps (10–15 × 3 @ Body Mass*4

(Continued)

Table 9.5 (*Continued*)

	General preparation		Specific preparation	Pre-Competition
	Anatomical adaptation/endurance	Hypertrophy (Transition)	Strength	Max' Strength	Power	Speed
	Meso-cycle 1	Meso-cycle 3	Meso-cycle 4	Meso-cycle 5	Meso-cycle 6	Meso-cycle 7

Training Day 2

Deadlift (15 × 2–3 @ 50–60%)	Deadlift (8-12 × 3–4 @ 65–75%)	Deadlift (4 @ 80–90%)	Deadlift (2–4 × 4 @ 90–95%)	Deadlift (4 reps, 2 sets, @ 90% –1RM)	Deadlift (4 reps, 1 set, @ 90% –1RM)
Mid-Thigh Power Shrug (2nd Pull) (15 × 2–3 @ 50–60%)	Mid-Thigh Power Shrug (8–12 × 3–4 @ 65–75%)	Mid-Thigh Power Shrug (4 × 4 @ 80–90%)	Mid-Thigh Power Shrug (2–4 × 4–6 @ 90–95%)	Mid-Thigh Power Shrug (6 × 6 @ 80%)*2	Mid-Thigh Power Shrug (3 × 4 @80%)*2
Walking Lunges	Walking Lunges	Walking Lunges	Walking Lunges (4–6 × 4 @ 85–90% –1RM)	Walking Lunges (5 reps/leg, 2 sets, @ 85–90% –1RM), 4 mins rest THEN:	Walking Lunges (5 reps/leg, 2 sets, @ 85–90% –1RM) 4 mins rest THEN:
Split Jump Lunge (8–12 × 2 @ Body Mass	Alternate Split Jump Lunge (6 – 8 /leg × 3 @ Body Mass ⇓ Weighted Vest)	Alternate Split Jump Lunge (4 – 6/leg × 3 @ Body Mass ⇓ Weighted Vest)	Alternate Leg Bounds (10/leg × 2 @ Body Mass)	Alternate Leg Bounds (15/leg × 2 @ Body Mass ⇓ Weighted Vest)	Alternate Leg Bounds (10/leg × 2 @ Body Mass)
Nordic Hamstring Lowers (4 × 3 @ Body Mass)	Nordic Hamstring Lowers (4 × 4 @ Body Mass)	Nordic Hamstring Lowers (6 × 3 @ Body Mass)	Nordic Hamstring Lowers (6 × 3 @ Body Mass ⇓ Weighted Vest)	Nordic Hamstring Lowers (6 × 3 @ Body Mass ⇓ Weighted Vest)	

Calf Raises (15 × 2–3 @ 50–60%)	Calf Raises (8–12 × 3–4 @ 65–75%)	Calf Raises (4 @ 80–90%)	Calf Raises (2–4 × 4 @ 90–95%)	Calf Raises (4 reps, 2 sets, @ 90% –1RM) 4–5 mins rest THEN: Single Leg Hops (15 × 2 @ Body Mass)
		Double Leg Hops (30 × 2 @ Body Mass)	Double Leg Hops (50 × 2 @ Body Mass)	Single Leg Hops (10 × 2 @ Body Mass)

Key :

General Preparation	Anatomical Adaptations / Technique / Muscular Endurance
	Hypertrophy / Transition
	Strength
Specific Preparation	Maximal Strength
	Power
Pre-Competition	Peaking/CompetitionPhase
Maintenance	Maintenance (Prevention of Detraining)

Notes:

*1 Cormie et al. (2007a, 2007b) demonstrated that peak power was achieved during jump squats when body mass alone was used as resistance, when taking body mass into account. Stone et al. (2003) had previously reported peak power to be achieved using 10% 1–RM squat, during a jump squat, however, they did not assess peak power during an unloaded jump squat. Their results demonstrated a decrease in peak power in direct relationship with an increase in load.

*2 Cormie et al. (2007a, 2007b) also demonstrated that peak power, during power cleans, is achieved at 80% 1–RM.

*3 Kawamori et al. (2005) demonstrated that a load of 70% or 1–RM during a hang power – clean results in the greatest peak power output. Hori et al. (2008) have also shown a significant relationship (p<0.05) between hang power clean performance and 20m sprint performance, therefore increasing performance in the hang power clean may result in improved acceleration over the initial 20m.

*4 Rest for 5–10 seconds between each repetition to prevent fatigue and maintain velocity.

Cormie et al. (2007a) demonstrated that combining lower body strength and power training is more effective than power training alone, at increasing power especially against additional resistance. Therefore the example programme above combines both strength and power based activities, but alters the focus from maximising strength, in the general preparation phases, to optimising power during the pre-competition phase.

Peak power during the jump squat (50–70% 1–RM) and split jump squat (30–60% 1–RM) are related initial acceleration during sprints, as assessed by 5-m sprint times (Sleivert and Taingahue 2004). Thomas et al. (2007) showed a gender effect on optimal power load during the jump squat, with 30–40% 1–RM as the optimal load for females and 30–50% 1–RM as the optimal load for males.

Table 9.6 Example macro-cycle for ACL rehabilitation

Meso-cycle 1 10–15 reps, 2–3 Sets, 50% –1RM	Meso-cycle 2 15–20 reps, 2 Sets 55–60% –1RM	Meso-cycle 3 8–12 reps, 4 sets 65–75% –1RM	Meso-cycle 4 4–6 reps, 5–6 Sets 80–90% –1RM	Meso-cycle 5 2–4 reps, 6 sets 90–95% –1RM	Meso-cycle 6 10 reps, 4 sets 40% –1RM
Training Day 1					
Split Squats (10 reps, 3 sets)	Bulgarian Split Squats (10 reps, 3 sets)	Single Leg Squats*1 (5 reps, 3 sets)	Single Leg Squats (8 reps, 3 sets)	Single Leg Squats (10 reps, 3 sets)	Single Leg Squats (12–15 reps, 3 sets)
Back Squat*1 Romanian Deadlift	Back Squat*1 Romanian Deadlift	Back Squat Romanian Deadlift	Back Squat Romanian Deadlift	Back Squat Romanian Deadlift	Back Squat Romanian Deadlift
	Drop Jumps*3 (5 reps, 2 sets)	Drop Jumps (5 reps, 3 sets)	Drop Jumps (8 reps, 2 sets)	Drop Jumps (10 reps, 2 sets)	Drop Jumps (10 reps, 3 sets)
		Nordic Hamstring Lowers (3 reps, 2 sets)	Nordic Hamstring Lowers (3 reps, 3 sets)	Nordic Hamstring Lowers (5 reps, 3 sets)	Nordic Hamstring Lowers (5 reps, 3 sets)
		Double Leg Hops (10 reps, 2 sets)	Double Leg Hops (15 reps, 3 sets)	Single Leg Hops (10 reps, 2 sets)	Single Leg Hops (10 reps, 3 sets)
Training Day 2					
Deadlift*2 Leg Flexion / Glute Ham Raise	Deadlift Leg Flexion / Glute Ham Raise	Deadlift Leg Flexion / Glute Ham Raise	Deadlift Leg Flexion / Glute Ham Raise	Deadlift Leg Flexion / Glute Ham Raise	Deadlift Leg Flexion / Glute Ham Raise
		Nordic Hamstring Lowers (3 reps, 2 sets)	Nordic Hamstring Lowers (3 reps, 3 sets)	Nordic Hamstring Lowers (5 reps, 3 sets)	Nordic Hamstring Lowers (5 reps, 3 sets)
	Drop Jumps*3 (5 reps, 2 sets)	Drop Jumps (5 reps, 3 sets)	Drop Jumps (8 reps, 2 sets)	Drop Jumps (10 reps, 2 sets)	Drop Jumps (10 reps, 3 sets)
Split Squats (10 reps, 3 sets)	Bulgarian Split Squats (10 reps, 3 sets)	Single Leg Squats (5 reps, 3 sets)	Single Leg Squats (8 reps, 3 sets)	Single Leg Squats (10 reps, 3 sets)	Single Leg Squats (12–15 reps, 3 sets)
		Double Leg Hops (10 reps, 2 sets)	Double Leg Hops (15 reps, 3 sets)	Single Leg Hops (10 reps, 2 sets)	Single Leg Hops (15 reps, 3 sets)
Calf Raises	Calf Raises	Calf Raises	Calf Raises	Calf Raises	Calf Raises

*1 During the initial introduction of the single leg squat and back squat it is essential that the exercise is performed within a pain free ROM, which may initially be limited. If this is the case it is essential to continue to develop active ROM in the joint.

*2 During the initial introduction of the Deadlift it is essential that the exercise is performed within a pain free ROM, which may initially be limited and require the bar to be elevated on to blocks to initially decrease the level of knee flexion required. If this is the case it is essential to continue to develop active ROM in the joint.

*3 Drop jump landings should focus on force acceptance (deceleration), while focussing on the correct landing technique. This can be progressed in terms of intensity from increasing the height of the step / platform from which the exercise is performed. Alternatively the intensity and difficulty of the movement can be increased by progressing to single leg drop jump landings.

Table 9.7 Progression of example programme

Meso-Cycle	Aim
Meso-cycle 1: Anatomical Adaptations	Develop ROM and technique in each exercise, including split squats as a unilateral exercise, with low level deceleration
Meso-cycle 2: Strength Endurance	Increase muscular endurance, improve technique, introduce correct landing technique and higher speed deceleration training
Meso-cycle 3: Hypertrophy	Transitional phase from strength endurance to strength phase
Meso-cycle 4: Strength	Increase strength introduce focussed eccentric loading of hamstrings to further improve H:Q ratio
Meso-cycle 5: Maximal Strength	Optimize strength prior to speed strength development, in preparation for plyometric training
Meso-cycle 6: Speed Strength (High Velocity)	Progressively decrease loading from 90–95% –1RM down to 40% –1RM while progressively increasing velocity of movements, in preparation for full integration of plyometric training

to achieve the following stages (see also Table 9.6 and 9.7):

- Developing lower limb control during bilateral and then unilateral exercises

- Developing pain-free ROM for a full-depth squat, while addressing the hamstring to quadriceps ratio

- Developing strength in squatting movements, progressing to the point that the athlete can correctly squat 1.5 × body mass, pain free, through a full ROM (Holcomb et al. 1998; Bompa 1999), *prior to high intensity plyometrics* (depth jumps). It is essential that the hamstrings are not neglected at this stage as squatting is not a sufficient training stimulus to ensure their development (McCaw and Melrose 1999; Escamilla et al. 1998, 2001; Anderson et al. 2006)

- Low intensity plyometric movements, such as low-level drop jumps should be practiced to ensure that the athlete has the appropriate foot placement, limb movement, balance and posture (McGill 2006), *while developing strength*

- Plyometric movements should be integrated beginning with primarily vertical movements, followed by forward momentum, then lateral movements finally progressing to multidirectional movements which gradually become more sport specific.

It is essential that during a conditioning programme (such as the previous example Tables 9.6 and 9.7) the athlete still addresses other issues, such as ensuring restoration of adequate range of motion for their sport, whilst prioritising the training of correct landing, cutting and deceleration strategies to minimise the risk of recurrence of ACL injury during their skill training (Ford et al. 2003, Hewett et al. 2005a; Hanson et al. 2008).

As long as the athlete exhibits all of the prerequisites for plyometric training identified above, by the end of micro-cycle 5, they should be able to progress to full plyometric training programme, after micro-cycle 6.

Plyometric training should also be progressive and periodised to allow improvement in terms of increasing velocity of movement or deceleration forces, depending on the aims of the athlete and the requirements of the sport. At the same time, it is essential to continue with *some* strength training to prevent a detraining effect and therefore a loss of strength. Progression for sport specific activities should develop from unidirectional, to bidirectional, to multidirectional movements (Heidt et al. 2000).

Summary

Although periodisation has previously been used primarily by strength and conditioning coaches to optimise performance for specific competitions, the same approach can be used in the development of both rehabilitation programmes and injury prevention programmes. This approach ensures that athletes are appropriately conditioned prior to progressing on to the next, more demanding phase of their rehabilitation/training programme.

References

Ahmed, C.S., Clark, A.M., Heilmann, N., Schoeb, J.S., Gardner, T.R. and Levine, W.N. (2006) Effect of gender and maturity on quadriceps to hamstring ratio and anterior cruciate ligament laxity. *American Journal of Sports Medicine*, 34 (3), 370–4.

Agel, J., Arendt, E.A. and Bershadsky, B. (2005) Anterior cruciate ligament injury in National Collegiate Athletic Association basketball and soccer: A 13-year review. *American Journal of Sports Medicine*, 33 (4), 524–530.

Andersen, L.L., Magnussun, S. P., Nielson, M., Haleem, J., Poulsen, K. and Aagaard, P. (2006) Neuromuscular activation in conventional therapeutic exercises and heavy resistance exercises: Implications for rehabilitation. *Physical Therapy*, 86, 683–697.

Balyi, I. (2001) Long term athlete development. Presentation to National Coaching Foundation – High Performance Coaching Workshops. Leeds Metropolitan University.

Boden, B.P., Dean, G.D., Feagin, J.A. and Garrett, W.E. (2000a) Mechanisms of anterior cruciate ligament injury. *Orthopedics*, 23 (6), 573–578.

Boden, B.P., Griffin, L.Y. and Garrett, W.E. (2000b) Etiology and prevention of noncontact ACL injury. *Physical Sports Medicine*, 28 (4), 53–60

Bompa, T.O. (1999) *Periodization;* Theory and Methodology of Training, 4th edn. Champaign, IL: Human Kinetics.

Channell, B.T. and Barfield, J.P. (2008) Effect of Olympic and traditional resistance training on vertical jump improvement in highschool boys. *Journal of Strength Conditioning Research*, 22 (5), 1522–1527.

Cormie, P., McCaulley, G.O. and McBride, J.M. (2007a) Power versus strength-power jump squat training: influence on the load-power relationship. *Medicine and Science in Sports and Exercise*, 39 (6), 996–1003.

Cormie, P., McCaulley, G.O., Triplett, N.T. and McBride, J.M. (2007b) Optimal loading for maximal power output during lower body resistance exercises. *Medicine and Science in Sports and Exercise*, 39 (2), 340–349.

Escamilla, R.F., Fleisig, G.S., Zheng, N., Barrentine, S.W., Wilke, K.E. and Andrews, J.R. (1998) Biomechanics of the knee during closed kinetic chain and open kinetic chain exercises. *Medicine and Science in Sports and Exercise*, 30 (4), 556–569.

Escamilla, R.F., Fleisig, G.S., Zheng, N., Lander, J.E., Barrentine, S.W., Andrews, J.R., Bergemann, B.W. and Moorman, C.T. (2001) Effects of technique variations on knee biomechanics during the squat and leg press. *Medicine and Science in Sports and Exercise*, 33 (9), 1552–1566.

Fleck, S.J. (1999) Periodized strength training: A critical review. *Journal of Strength Conditioning Research*, 8, 235–242.

Fleck, S.J. (2002) Periodization of training. In Kraemer, W.J. and Hakkinen. K. (Eds), *Strength Training for Sport*, pp. 55–56. Oxford: Blackwell Science.

Ford, K.R., Myer, G.D. and Hewett, T.E. (2003) Valgus Knee Motion during Landing in High School Female and Male Basketball Players. *Medicine and Science in Sports and Exercise*, 35, 1745–1750.

Hanson, A.M., Padua, D.A., Blackburn, T.J., Prentice, W.E. and Hirth, C.J. (2008) Muscle activation during side-step cutting manoeuvres in male and female soccer athletes. *Journal of Athletic Training*, 43 (2), 133–143.

Heidt, R.S., Sweeterman, L.M., Carlonas, R.L., Traub, J.A. and Tekulve, F.X. (2000) Avoidance of soccer injuries with pre-season conditioning. *American Journal of Sports Medicine*, 28 (5), 659–662.

Hewett, T.E., Stroupe, A.L., Nance, T.A. and Noyes, F.R. (1996) Plyometric training in female athletes: Decreasing impact forces and increasing hamstring torques. American Journal of Sports Medicine, 24 (6), 765–773.

Hewett, T.E., Myer, G.D., Ford, K.R. and Palumbo, J.P. (2005a) Neuromuscular training improves performance and lower-extremity biomechanics in female athletes. *Journal of Strength Conditioning Research*, 19 (1), 51–60.

Hewett, T.E., Myer, G.D., Ford, K.R., Heidt, R.S., Colosimo, A.J., Mclean, S.G., Van den Bogert, A.J., Paterno, M.V. and Succop, P. (2005b) Biomechanical measures of neuromuscular control and valgus loading of the knee predict anterior cruciate ligament injury risk in female athletes. *American Journal of Sports Medicine*, 33, 492–501.

Holcomb, W.R., Kleiner, D.M. and Chu, D.A. (1998) Plyometrics: Considerations for safe and effective training. *Strength and Conditioning Journal*, 20 (3), 36–39.

Holcomb, W.R., Rubley, M.D., Lee, H.J. and Guadagnoli, M.A. (2007) Effect of hamstring emphasised resistance training on hamstring quadriceps strength ratios. *Journal of Strength Conditioning Research*, 21 (1), 41–47.

Hori, N., Newton, R.U., Andrews, W.A., Kawamori, N., McGuigan, M.R., and Nosaka, K. (2008) Does performance of hang power clean differentiate performance of jumping, sprinting, and changing of direction. *Journal of Strength Conditioning Research,* 22 (2), 412–418.

Hori, N., Newton, R.U., Nosaka, K. and Stone, M.H. (2005) Weightlifting exercises enhance athletic performance that requires high load speed strength. *Strength and Conditioning Journal,* 27 (4), 50–55.

Kaminski, T.W., Webberson, C.V. and Murphy, R.M. (1998) Concentric versus enhanced eccentric hamstring strength training: Clinical implications. *Journal of Athletic Training,* 33 (3), 216–221.

Kawamori, N., Crum, A.J., Blumert, P.A., Kulik, J.R., Childers, J.T., Wood, J.A., Stone, M.H. and Haff, G.G. (2005) Influence of different relative intensities on power output during the hang power clean: Identification of the optimal load. *Journal of Strength Conditioning Research,* 19 (3), 698–708.

Kilgallon, M., Donnelly, A.E. and Shafat, A. (2007) Progressive resistance training temporarily alters hamstring torque angle relationship. *Scandinavian Journal of Medicine and Science inSports,* 17 (1), 18–24.

Li, G., Rudy, T.W., Sakane, M., Kanamori, A., Ma, C.B. and Woo, S.L.Y. (1999) The importance of quadriceps and hamstring muscle loading on knee kinematics and in-situ forces in the ACL. *Journal of Biomechanical,* 32, 395–400.

Malinzak, R.A., Colby, S.M., Kirkendall, D.T., Yu, B. and Garret, W.E. (2001) A comparison of knee joint motion patterns between men and women in selected athletic tasks. *Clinical Biomechanics,* 16, 438–445

Markolf, K.L., Burchfield, D.M., Shapiro, M.M., Shepard, M.F., Finerman, G.A. and Slauterbeck, J.L. (1995) Combined knee loading states that generate high anterior cruciate ligament forces. *Journal of Orthopaedic Research,* 13 (6), 930–935.

McCaw, S.T. and Melrose, D.R. (1999) Stance width and bar load effects on leg muscle activity during the parallel squat. *Medicine and Science in Sports and Exercise,* 31 (3), 428–436.

Matveyev, L.P. (1965) *Periodization of Sport Training.* Moscow: Fizkultura I Sport.

Matveyev L.P. (1981) *Fundamentals of Sports Training.* Moscow: Progress.

McGill, S. (2006) *Ultimate Back Fitness and Performance,* 3rd edn. Waterloo, ON: Backfitpro.

McLean, S.G., Huang, X., Su A. and Van Den Bogert, A.J. (2004) Sagittal plane biomechanics cannot injure the ACL during sidestep cutting. *Clinical Biomechanical,* 19 (8), 828–838.

Mjolsnes, R., Arnason, A., Osthagen, T., Raastad, T. and Bahr, R. (2004) A 10-week randomized trial comparing eccentric vs. concentric hamstring strength training in well trained soccer players. *Scandinavian Journal of Medicine and Science in Sports,* 14 (5), 311–317.

Plisk, S.S. and Stone, M.H. (2003) Periodization strategies. *Strength and Conditioning Journal,* 25(6), 19–37.

Rhea, M.R. and Alderman, B.L. (2004). A meta-analysis of preiodized versus non-periodized strength and power training programs. *Research Quarterly for Exercise and Sport,* 75, 413–422.

Sleivert, G., and Taingahue, M. (2004) The relationship between maximal jump-squat power and sprint acceleration in athletes. *European Journal of Applied Physiology,* 91 (1), 46–52.

Stone, M.H., O'Bryant, H.S., McCoy, L., Coglianese, R., Lehmkuhl, M. and Schilling, B. (2003) Power and maximum strength relationships during performance of dynamic and static weighted jumps. *Journal of Strength and Conditioning Research,* 17 (1), 140–147.

Thomas, G.A., Kraemer, W.J., Spiering, B.A., Volek, J.S., Anderson, J.M. and Maresh, C.M. (2007) Maximal power at different percentages of one repetition maximum: Influence of resistance and gender. *Journal of Strength and Conditioning Research,* 12 (2), 336–342.

Zazulak, B.T., Ponce, P.L., Straub, S.J., Medvecky, M.J., Avedisian, L. and Hewett, T.E. (2005) Gender comparisons of hip muscle activity during single-leg landings. *Journal of Orthopaedic Sports Physical Therapy,* 35 (5), 292–299.

10

Management of acute sport injury

Jeffrey A. Russell
University of California, Irvine

Introduction and aims

The experience of acute injury can be devastating to an athlete. Catastrophic injury is an obvious example, but even less severe occurrences can adversely affect an athlete because of fears and thoughts of the unknown: How severe is it? Can I carry on playing? What if I have to sit out – for how long? Will I lose my place on the team? All of these and many more can negatively impact a situation that is already antithetical to sport "activity." Thus the sport rehabilitator and the rest of the healthcare team must ensure that the management of acute injury creates the best environment possible for a successful return of the athlete to full function. With this as the ultimate goal, the aims of the chapter are to describe the nature and occurrence patterns of acute sport injury, to present a general plan for assessing acute sport injury, to recommend appropriate methods for managing acute sport injury; and to provide a foundation on which to base more detailed presentations of specific injuries in various body regions.

Common acute injuries in sport

Injury epidemiology

Intuitively it is not possible to eliminate injuries in sport, however, the incidence of injuries can be reduced. For the year 2002 the Royal Society for the Prevention of Injuries reported 710,018 injuries in the United Kingdom in activities classified as sport (excluding in educational venues), running, cycling or riding (Royal Society for the Prevention of Accidents 2009). The UK's Leisure Accident Surveillance System (LASS) estimated that the total number of sport injuries occurring in the United Kingdom in 2002 (the last year for which data are available) was 824,182 (Department of Trade and Industry 2003). More than 525,000 of these occurred in ball sports where implements (e.g., sticks, bats) were not used. This reflects the immense popularity – and the high injury rates – of football and rugby. Football injury rates have been reported to be 25.9 per 1000 player-match hours (i.e. injuries occurring in matches) and 3.4 per 1000 player-training hours (i.e. injuries occurring in training) in England's Premiership and First and Second professional divisions (Hawkins and Fuller 1999). Rugby's injury rate is even higher: in preparation for and participation in the Rugby World Cup by the England national team, injury rates were reported as 218 per 1000 player-match hours and 6.1 injuries per 1000 player-training hours, with an overall rate of 17 injuries per 1000 hours (Brooks et al. 2005). In the 2007 Rugby World Cup epidemiology study, injury rates for all teams taken together were reported at 83.9 injuries per 1000 player-match hours and 3.5 injuries per 1000 player-training hours (Fuller et al. 2008).

Sadly LASS only offer raw statistics without epidemiological interpretation; research studies add clarity to the data. However, it is clear that all sports carry inherent risk of injury for participants. It is therefore important to understand the epidemiological terms related to injury. Incidence is the number of new occurrences of an injury or illness in a given time period and prevalence is the number of injuries that are present in a given population at a specified time (Anonymous 2005; Caine et al. 2006). Three specific terms have been presented in the context of analysing sport injury incidence (Knowles et al. 2006): incidence proportion, incidence rate and clinical incidence. Incidence proportion is the number of injured athletes divided by the number of athletes at risk and represents the average risk of injury for an athlete (Knowles et al. 2006). Incidence rate refers to the number of injuries in a given quantity of time. Clinical incidence is the overall number of injuries that occur in a given group of athletes, a number important to those responsible for decisions about how to allot clinical resources to healthcare settings (Knowles et al. 2006).

One of the key factors of any systematic injury survey that provides data about acute injury epidemiology is the definition of injury and related parameters (van Mechelen et al. 1996; Ekstrand and Karlsson 2003; Brooks and Fuller 2006; Hodgson et al. 2007). The National Collegiate Athletic Association's Injury Surveillance System (ISS) has recorded injuries in university athletes in America since 1982 (National Collegiate Athletic Association 2008). A reportable injury is defined in this database as one that:

> Occurs as a result of participation in an organized intercollegiate [i.e. inter-university] practice or competition; requires medical attention by a [healthcare professional]; and results in restriction of the student-athlete's participation or performance for one or more days beyond the day of injury. (p. 109)

In other words, an episode is not considered an injury until, at a minimum; it compromises an athlete's ability to participate in their sport on the day following the episode.

This definition is not universal, however. Hodgson et al. (2007) recommend a much more encompassing definition. Contrast of the foregoing classification with the following underscores the difficulty in comparing injury rates among research articles. King et al. (2009) recommended this definition of injury for the entirety of rugby league:

> Any pain or disability that occurs during participation in rugby league match or training activities that is sustained by a player, irrespective of the need for match or training time loss or for first aid or medical attention. An injury that results in a player requiring first aid or medical attention is referred to as a "medical attention injury" and an injury that results in the player being unable to partake in full part of future training and/or match activities is referred to as a "time loss" injury. (p. 13)

The dissimilarity with the ISS definition is obvious and substantial. Analogous variation exists, as well, between the ISS and King et al. definitions of exposure. An "athlete exposure" is the ISS's basic measure of risk, being defined as one athlete participating in one training session or one competition in which they are exposed to the potential for being injured (National Collegiate Athletic Association 2008). The definition provided by King et al. (2009) is:

> [For matches] The product of the number of players, the number of matches between two different teams, and the match duration for the period that the study is being conducted. The recommended recording of exposure time is the duration of the matches being studied in hours. (p.14)

> [For training] The product of the number of players, the number of training sessions, and the training duration for the period that the study is being conducted. The recommended recording of exposure time is the duration of the training sessions being studied in hours. (p. 14)

In a study of the ISS for the years 1988–2004 (Hootman et al. 2007), injuries are statistically more likely to occur during competitions than during training (13.8 versus 4.0 injuries per 1,000 athlete exposures). Also, injuries during pre-season training were significantly higher (6.6 injuries per 1,000 athlete exposures) than injuries occurring during a sport's season (2.3 injuries per 1,000 athlete exposures) and after the season (1.4 injuries per 1,000 athlete exposures). The general tendencies of these results – if not the injury rates – are similar to observations made about rugby injuries in England (Brooks et al. 2008) and Australia (Gabbett 2004). Another interesting finding is that across all sports, university male athletes are 3.5 times more likely to suffer a non-time-loss injury than a time-loss injury (Powell and Dompier 2004). This ratio is even higher in

Table 10.1 Injury risk data for selected sports (National Athletic Trainers' Association 2007)

Sport	Time-loss injuries per 1,000 athlete exposures	Injury risk[1]	Health Care Index[2]
Basketball, men	5.5	7	2.4
Basketball, women	5.1	5	4.0
Football, men	7.8	4	2.8
Football, women	7.4	5	3.6
Gymnastics, men	6.8	8	3.7
Gymnastics, women	7.8	8	4.0
Hockey, women	4.7	4	2.8
Ice hockey, men	7.3	6	1.8
Ice hockey, women	4.9	5	1.0
North American football, men	8.7	8	3.1
Outdoor athletics, men	3.7	2	1.1
Outdoor athletics, women	3.1	2	1.1
Volleyball, men	7.5	5	4.0
Volleyball, women	4.4	5	3.5

[1] Catastrophic and non-catastrophic risks combined; higher number indicates greater relative risk. 8 is maximum
[2] Higher number indicates more healthcare required; 4.0 is maximum

female athletes: they are over 5 times more likely to suffer a non-time-loss injury.

The relative need for healthcare delivery to participants of a given sport is a pertinent topic in any discussion about sport injury incidence. A very useful tool for determining the healthcare requirements of a variety of sports has been proposed and is based on injury incidence rate and the relative personnel load of providing care for injuries incurred in university sports (National Athletic Trainers' Association 2007). Injury rate index is based on risk of injury combined with risk of catastrophic (brain and spine) injury. Table 10.1 summarises relevant data. A higher Health Care Index suggests that a sport requires more medical attention. While comparable data for cricket and rugby could not be found, the information presented is helpful for the sport rehabilitator's understanding about the nature of injury in sport and the demand for healthcare each sport engenders.

Specific injuries encountered

It is surprisingly difficult to locate summary data on the incidences of specific injuries in sport. Apart from the epidemiological challenges previously outlined, there are some practical reasons for this. Countless articles offer reports about injuries in certain sports, but assessing these systematically is a monstrous task. In addition, the types of injuries sustained by children, adolescents, young adults, middle-aged adults and senior adults likely differ, as do injuries encountered in recreational sport, competitive sport and sport training. Combine these diversities with the variety in sports, rules, equipment, venues, training regimens and healthcare access across the globe and the task of assessing sport injury epidemiology appears overwhelming at best. The USA's National Collegiate Athletic Association maintains excellent data for university athletes that offer some insight into those body regions that are most affected and those injuries that occur most commonly in certain sports (Hootman et al. 2007). Selected data are provided in Tables 10.2 and 10.3.

Initial assessment of acute sport injury

Assessment of all injuries in sport participants should follow the standard history, inspection (observation), palpation, and special testing paradigm commonly applied by healthcare professionals. These steps can be summarised as follows (for greater detail read Chapter 11: Musculoskeletal Assessment).

Table 10.2 Injury incidence by body region for university athletes from 1988 through 2004 (Hootman et al. 2007)

Body region	Percent of all injuries	
	Matches	Training
Lower extremity	53.8	53.7
Upper extremity	18.3	21.4
Trunk/back	13.2	10.0
Head/neck	9.8	12.8
Other	4.9	2.2

History

This first step in injury evaluation is fundamental to success. It entails a series of questions directed to the athlete that are designed to elicit the factors surrounding the injury and any corollary findings (such as previous history of injury) that may be important in arriving at a diagnosis. The exact questions are modified depending on the situation, but some sample queries for athletes with acute injuries are given below.

- When did this happen?
- What were you doing at the time of the injury?
- How did the injury occur?
- Did you hear or feel a pop, crack, snap, or other unusual sensation?
- Have you injured this area previously?
- How does it feel now compared to when the injury first occurred?
- How is the injury limiting your activity?

Inspection (observation)

This step requires a careful visual analysis of the injured area and adjacent regions. Several pieces of information are collected before the examiner touches the injured area. The contralateral limb should be used for comparison as long as it does not exhibit atypical characteristics because of a prior injury or other reason. Examples of observational information to gather about the type and extent of injury include:

- Deformity (e.g. joint incongruity, limb angulations, anatomical landmark displacement)
- Swelling

Table 10.3 Incidence of common injuries in university athletes, by sport, from 1988 through 2004 (Hootman et al. 2007)

Sport	Ankle ligament sprains		Anterior cruciate ligament injuries		Concussions	
	Percent of all injuries	Injury rate per 1000 exposures	Percent of all injuries	Injury rate per 1000 exposures	Percent of all injuries	Injury rate per 1000 exposures
Basketball, men	26.6	1.30	1.4	0.07	3.2	0.16
Basketball, women	24.0	1.15	4.9	0.23	4.7	0.22
Football, men	17.2	1.24	1.3	0.09	3.9	0.28
Football, women	16.7	1.30	3.7	0.28	5.3	0.41
Gymnastics, women	15.4	1.05	4.9	0.33	2.3	0.16
Hockey, women	10.0	0.46	1.6	0.07	3.9	0.18
Ice hockey, men	4.5	0.23	1.2	0.06	7.9	0.41
Ice hockey, women*	2.8	0.14	0.7	0.03	18.3	0.91
North American football, men	13.6	0.83	3.0	0.18	6.0	0.37
Volleyball, women	23.8	1.01	2.0	0.09	2.0	0.09

*Data collection began in the academic year 2000–2001

- Discolouration (e.g. pallor, ecchymosis, erythaema)

- Gait discrepancies or difficulties

- Difficulties climbing to the examination couch

- Reduction in joint range of motion.

Palpation

This is the first instance when the examiner actually touches the injured athlete. A systematic probing of the injured area and surrounding tissues gain important details and may help the clinician identify pathologies that accompany the primary injury. Once again the contralateral limb is used as a reference. Examples of findings from effective palpation include:

- Localisation of pain and tenderness (including deciding which structures are painful and which are not as part of a differential diagnostic process)

- Atypical anatomical contours

- Crepitus

- Effusion and oedema.

Special testing

Following completion of the history, inspection and palpation steps the examiner conducts a series of assessments to gather further insight into the nature and severity of the injury. These vary widely depending on the body region being evaluated and the diagnostic clues gained in the first portion of the examination. (Magee (2008) provides an excellent and comprehensive handbook of musculoskeletal examination that the sport rehabilitator will find invaluable as a reference in addition to the present text.) They may include manual tests performed by the examiner to identify joint instability or certain functional activities that the examiner asks the athlete to perform. The results of this analysis add to the diagnostic process and help broaden the information on which clinical decision making is based.

Once again, testing the contralateral extremity is important as a comparison. However, there are three important caveats for the examiner. First, the uninjured limb should be tested before the injured one in order to familiarise the athlete with the evaluation procedures and to allay their apprehensions. If the athlete cannot relax, muscle guarding will compromise the testing. Second, a previously injured contralateral joint may not be a suitable "normal" standard against which to judge the currently injured joint. Third, a systematic method for evaluating each region of the body is most effective and efficient for completing the examination process without missing important diagnostic details.

Serving as one common example that will be encountered by virtually all sport rehabilitators, the evaluation of an acute knee injury includes special tests such as:

- Apprehension sign for patellar dislocation/subluxation

- Lachman's test for anterior cruciate ligament integrity

- Anterior drawer test for anterior cruciate ligament integrity

- Posterior gravity sign for posterior cruciate ligament integrity

- Pivot shift test for anterolateral rotary instability

- McMurray's test for meniscal damage

- Apley's grind test for meniscal damage

- Ottawa Knee Rules (Stiell et al. 1995, 1996; Tigges et al. 1999) to rule out fracture.

Assessment case study: ankle sprain

One of the most common injuries in sport is ankle sprain (Garrick 1977; Yeung et al. 1994; Fong et al. 2007; Hootman et al, 2007), so it will serve as an example of how acute injury evaluation is applied. The sport rehabilitator confronted with this type of ankle injury will assign a severity grade to the sprain based on the results of the clinical examination (see Table 10.4). A partially torn ligament – categorised as a grade 2 sprain – can encompass a wide range of damage, whilst a grade 3 sprain is a complete rupture

Table 10.4 Summary of ligament injury severity grading

Injury grade	Ligament damage	Sensation to examiner with stress exam	Associated pain for patient
Grade 1	None to minimal (stretched)	No instability	Mild to moderate
Grade 2	Partially torn	Mild to moderate instability; endpoint is present	Moderate
Grade 3	Completely torn (ruptured)	Moderate to severe instability; soft or indistinct endpoint	Moderate to severe

of a ligament. The determination by the examiner of a limit – or an endpoint – to their stress exam of the joint is important. The endpoint is created by any remaining intact ligament restricting the movement. A grade 3 sprain does not exhibit such a limit; rather, an indistinct, soft endpoint is present because the torn ligament cannot oppose the stress the examiner places on the ankle.

However, the examiner must remember that many subjective subtleties exist in evaluating ligament damage because there are virtually infinite ways that ligaments supporting a joint can be injured. The amount of joint instability may be graded instead of the degree of injury to a specific ligament (Hildebrand et al., 2007). For instance, a grade 2 instability may be present when one ligament is completely torn and the others around a joint are not. But, partial tears to two or more ligaments about the joint may also result in the same type of instability. This underscores the need for the sport rehabilitator to practise a systematic method of ligament injury evaluation.

The following section outlines the stages of injury assessment carried out by a sport rehabilitator. An injured basketball player presents to the sport injury clinic to be evaluated by a member of staff. He is wearing shorts and walks with a limp. He climbs onto the examination couch and the sport rehabilitator begins the assessment.

Acute ankle injury: history

There are certain customary questions to ask an injured athlete, and some of these depend on the body region involved. A framework of "What, When, Where, How?" may be helpful. It is ideal to develop a routine examination methodology in order to facilitate the process and ensure one does not miss important diagnostic clues.

Examiner: What happened to your ankle?
Player: I jumped up for a rebound and when I came down, my foot landed on top of Ian's and it rolled over.
Examiner: When did this occur?
Player: Yesterday afternoon during training.
Examiner: What did you feel when it happened?
Player: It hurt a lot, and I felt it pop.
Examiner: Has it been painful since the injury occurred?
Player: It hurt loads when it happened, but I just walked it off. Later it felt better and I've been able to carry on with life OK.
Examiner: How does it feel now?
Player: It still hurts, but not as much.
Examiner: Have you ever hurt this ankle before?
Player: No.
Examiner: How about the other one?
Player: No, not at all.
Examiner: OK, I'm going to examine it now.

Acute ankle injury: inspection

At this point the examiner looks carefully at both the injured and uninjured ankles to assess swelling, ecchymosis (bluish discolouration), and other visible signs that may add information to the assessment. The athlete's ankle exhibits an area of swelling anterior to the lateral malleolus that is approximately the size of a table tennis ball.

Examiner: Where would you say it hurts the most? Can you place one finger at that spot?

The player points to a location about 1.5 cm anterior to the lateral malleolus; this is an area over the anterior talofibular ligament and the spot where the swelling is located. There are no other areas of swelling nor are there other outward signs of injury.

Examiner: Can you show me the way your ankle and foot turned when you landed on Ian's foot?

The player demonstrates an inversion movement of the foot with slight plantar flexion.

Acute ankle injury: palpation

First the examiner palpates the uninjured extremity in order, again, to familiarise the player with the process and to reduce his apprehension. Then the examiner begins to gently palpate various places about the foot, ankle and leg of the player's injured limb, starting with areas that are not likely to be painful in order to keep the player at ease. It is crucial that the clinician be well versed in clinical anatomy for this portion of the injury examination. Locations and structures the examiner palpates are:

- Ligaments (to investigate possible sprains)
  - deltoid ligament
  - anterior talofibular ligament
  - calcaneofibular ligament
  - anterior tibiofibular ligament
  - calcaneocuboid ligament
- Bones (to investigate the possibility of associated fractures)
  - base of the 5th metatarsal, to assess an avulsion fracture
  - metaphysis and diaphysis of the 5th metatarsal, to assess a shaft fracture or Jones (proximal diaphyseal/metaphyseal) fracture
  - lateral malleolus
  - fibula, extending up the leg to the fibular neck in order to assess a Maisonneuve fracture
  - dorsal talus
  - navicular
  - calcaneus
- Tendons and muscles (to investigate possible strains)
  - Achilles tendon
  - peroneal tendons, extending up the leg to their muscles
  - tibialis anterior tendon
  - tibialis posterior tendon.

The examiner does not note any pain or crepitus on bony structures, so the suspicion for fracture is low. Palpating the soft tissue structures does not elicit pain except for the anterior talofibular ligament; this is accompanied by a sensation of ballottement as the examiner presses over the swollen tissue.

Acute ankle injury: special testing

Now the examiner proceeds to the final phase of the injury evaluation. Once again, the contralateral side receives attention first; the player reported no history of injury to this side so it will serve as "normal" for comparison purposes. The following tests are performed there and then repeated on the injured limb. (For more specific detail regarding special testing for the ankle please see Chapter 22)

- Eversion (varus) stress test to evaluate the deltoid ligament
- Forcing the foot into dorsiflexion to evaluate the anterior tibiofibular ligament (this forces slight separation of the tibiofibular syndesmosis)
- Twisting the foot in the ankle mortise to evaluate the anterior tibiofibular ligament (this also separates the syndesmosis)
- Inversion (valgus) stress test in anatomical position (ankle neutral) to evaluate the calcaneofibular ligament
- Inversion stress test in plantar flexion to evaluate the anterior talofibular ligament
- Anterior drawer test with the ankle in neutral position to evaluate the anterior talofibular ligament.

The examiner must be careful to not fall into the trap of finalising an injury diagnosis without gathering all potentially useful information with a thorough history, inspection, palpation and special testing protocol. Many inexperienced clinicians have been

Table 10.5 Ottawa Ankle Rules (see Figure 10.1 also) (Stiell et al. 1992, 1994; Leddy et al., 2002; Nugent 2004)

Ankle should be X-rayed if…	…the posterior half of the distal 6 cm of the fibula or tibia or tip of the lateral malleolus is painful to palpation…	AND	…the patient cannot bear weight for 4 steps on the injured limb at the time of injury and at the time of the evaluation (limping is irrelevant)
Midfoot should be X-rayed if…	…the base of the 5th metatarsal or the navicular bone is painful to palpation…	AND	…the patient cannot bear weight for 4 steps on the injured limb at the time of injury and at the time of the evaluation (limping is irrelevant)

distracted from a proper diagnostic conclusion because they did not assess an injury carefully enough. That being said, however, there are certainly natural tendencies that are anticipated when the mechanism of an injury is known. In the case of our basketball player here, we are aware that he inverted and plantar flexed his ankle when he landed on another player's foot. This mechanism is very typical of an anterior talofibular ligament sprain (Garrick 1977), but completing the battery of tests listed above will help identify other potentially injured structures.

Our basketball player exhibits the following results of ligament testing:

- Eversion stress: not painful and no instability (firm endpoint) compared to contralateral
  - Impression: deltoid ligament intact

- Forced dorsiflexion: no pain elicited
  - Impression: anterior tibiofibular ligament intact

- Twisting foot in the mortise: no pain elicited
  - Impression: anterior tibiofibular ligament intact

- Inversion stress test in anatomical position: very slightly painful anterior to the lateral malleolus at end of range and no instability compared to contralateral
  - Impression: calcaneofibular ligament intact; based on its location, the mild pain is likely emanating from the anterior talofibular ligament

- Inversion stress test in plantar flexion: moderate pain that worsens at end of range of movement and questionable slight instability compared to contralateral
  - Impression: grade 1 or 2 anterior talofibular ligament sprain

- Anterior drawer test: moderate pain that worsens at end of range of movement and slight instability compared to contralateral
  - Impression: grade 2 anterior talofibular ligament sprain.

Parenthetically, as part of an acute ankle assessment the Ottawa Ankle Rules (Stiell et al. 1992, 1994) should be applied. These rules (Table 10.5 and Figure 10.1) have greatly reduced the unnecessary X-rays previously associated with ankle injury management in the A&E department and other healthcare

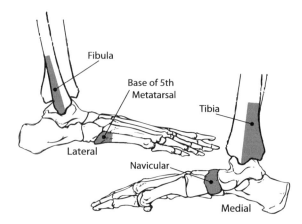

Figure 10.1 Pictorial representation of the Ottawa Ankle Rules showing the locations – the shaded areas – where bony tenderness indicates the advisability of obtaining ankle X-rays. Reproduced, with permission, from Figure 1 from Russell, J.A., 'Acute Ankle Sprain in Dancers' *Journal of Dance Medicine and Science*, 14, 2010 © J. Michael Ryan Publishing, Inc.

settings (Milne 1996; Leddy et al., 1998; Bachmann et al., 2003; Nugent 2004). The rules provide guidance for clinical personnel in determining when ankle injuries require radiography for suspected fracture, and they are appropriate and effective for sport medicine settings (Leddy et al., 1998; Papacostas et al., 2001; Leddy et al., 2002). Implementing the rules for sport ankle injuries prevented missed fractures, decreased X-ray exposure, saved unnecessary healthcare expenditures and fostered patient satisfaction (Leddy et al., 2002). The Ottawa Ankle Rules appear to be most useful when applied by healthcare workers, as patients may not be able to apply them accurately to their own ankle injuries (Blackham et al., 2008). Sport rehabilitators and other allied health professionals should familiarise themselves with these important adjuncts to ankle injury evaluation.

Acute ankle injury: information collating and decision making

After executing the history, inspection, palpation and special testing steps, the sport rehabilitator arrives at the clinical impression of a grade 2 sprain of the anterior talofibular ligament. Following is a summary of the examination details that led to this decision.

- Mechanism of injury: inversion with plantar flexion, a movement that places the anterior talofibular ligament under tension

- Location of swelling and pain: anterior to the lateral malleolus, which is the anatomical location of the anterior talofibular ligament

- Pain and slight laxity in the ankle: occurs during inversion and plantar flexion testing that place tension on the injured ligament, suggesting that a portion of the ligament is disrupted; note that in this case there is an equivocal impression when considering only the inversion exam with plantar flexion – the anterior drawer test result adds clinical information

- Associated fracture likelihood: not likely, as the player arrived at the clinic with full weight bearing ambulation and palpation of the bony structures did not reveal pain; thus, the Ottawa Ankle Rules suggest no fracture

Certainly injury examination is very subjective, although a systematic evaluation scheme helps reduce variability in the assessment process. Nonetheless, there is no substitute for the experience of examining as many injuries as possible in order to gain the ability to discern subtleties present in the wide variety of cases. Both proper clinical education and clinical experience are essential for effective injury care.

First aid and initial therapeutic measures

The acronym PRICE is the standard for acute care of sport injury. This is a reminder of five steps: application of Protection, Rest, Ice, Compression and Elevation (Flegel 2008). An alternative, RICES – Rest, Ice, Compression, Elevation, Stabilisation – has been proposed (Knight 2008; Knight and Draper 2008), but portrays the same basic meaning. The generally accepted period for the acute treatment described below is the first 48–72 hours after the injury. Knight (2008) suggests for clarity's sake that acute care be subdivided into three phases: emergency (encompassing CPR or urgent transport to A&E), immediate (from time of injury to 12 hours post-injury) and transition care (from 12 hours to 4 days post-injury). No matter how the initial period of time following an injury is apportioned, the success of follow-up treatment depends substantially on the initial treatment that is applied.

Protection by bracing, splinting, or non-weight bearing transport of an injured athlete is undertaken when necessary to minimise the risk of further trauma to an injured area. Proper first aid techniques at this initial stage are crucial to the athlete's well-being. Rest from the activity that caused the injury, or similar activities, is warranted when the opportunity exists for reinjury or further injury.

In the last several years, the "gold standard" of cold – or cryotherapy – as an injury treatment has been analysed for its efficacy in evidence-based practice (Ernst and Fialka 1994; Lessard et al. 1997; Bleakley et al. 2004; Hubbard and Denegar 2004; Collins 2008). In spite of questions raised about the therapeutic effects of cold treatment, it is helpful for pain reduction (Hubbard and Denegar 2004; Algafly and George 2007). Other reasons for applying cold to acute sport injuries include minimising oedema and effusion (Merrick 2007) and controlling secondary cellular hypoxia (Dale et al., 2004) and the broader secondary metabolic injury (Merrick et al.

1999; Knight and Draper 2008) in the injured region. Whilst cold slows oedema formation, it does not reduce oedema that is already present (Knight and Draper 2008).

A typical application of therapeutic cold should last no longer than 20 minutes (Bleakley et al. 2006; Flegel 2008); approximately one hour should elapse before another cold treatment in order to allow the tissue to rewarm. As with all treatment procedures there are important caveats. More body tissue heat is given off (resulting in lower tissue temperature) to a 10° cold pack than to a 20° cold pack (Knight and Draper 2008) and by body regions with a thinner insulating fat layer (Otte et al. 2002) or a larger surface area covered by the treatment modality (Knight and Draper 2008). Therefore, longer treatment times or more intense forms of treatment further increase heat loss and can instigate tissue damage (Knight 1995; Knight and Draper 2008). Moreover, research suggests that muscular power and functional performance are reduced even after a 10 minute ice application (Fischer et al. 2009).

Circumferential compression in an injured extremity enhances control of oedema by increasing the external tissue pressure to promote lymphatic drainage in the region of the injury and by assisting venous return (Wilkerson 1991; Wilkerson and Horn-Kingery 1993; Delis et al. 2000; Mora et al. 2002; Vanscheidt et al. 2009). Application of an elastic or crepe bandage provides some compression and a general reassurance to the athlete, although a pneumatic stirrup-type of brace has been shown to be more effective in treating ankle sprains (Boyce et al. 2005). Intermittently applied compression may be beneficial, as well, especially when applied in conjunction with cryotherapy (Starkey 1976; Quillen and Rouillier 1982; Mora et al. 2002). Research in this area related to patients with chronic oedematous pathology is instructive (Delis et al. 2000; Vanscheidt et al. 2009) as significant improvements in swelling have been shown and a dose-response relationship has been reported (Vanscheidt et al. 2009).

Of concern in the common sport injury of ankle sprain are the contours of the medial and lateral malleoli. This topology about the ankle may prevent the areas anterior and posterior to the malleoli from receiving adequate compression by an elastic or crepe bandage. This is because the bandage bridges across from the anterior and posterior ankle surfaces to the malleoli, thus leaving the spaces

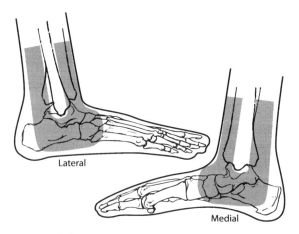

Figure 10.2 The shaded areas denote the shape and location for placement of a foam or chiropody felt horseshoe pad as a compressive element in treating acute ankle sprain. Reproduced, with permission, from Figure 1 from Russell, J.A., 'Acute Ankle Sprain in Dancers' *Journal of Dance Medicine and Science*, 14, 2010 © J. Michael Ryan Publishing, Inc.

adjacent to the malleoli without sufficient external pressure. Horseshoe shaped pads cut from 1 cm (3/8 inch) foam or chiropody felt and positioned underneath the bandage during wrapping – as shown in Figure 10.2 – may improve elimination of oedema in these areas (Wilkerson and Horn-Kingery 1993; Merrick 2004).

Elevation of the injured extremity so the ankle is positioned above the level of the heart is another technique that promotes a decrease in swelling (Rucinkski et al. 1991). Gravity assists venous return, as well as lymphatic drainage of fluid collected in the interstitial spaces (Dale et al. 2004). However, a return of the limb to the non-elevated, or gravity-dependent, position may reverse the oedema reduction gained by elevation (Tsang et al. 2003). Thus, elevation for as much time as possible is important during the period of acute injury management.

Pain management in acute musculoskeletal sport injury

A number of therapeutic modalities are available today for treatment of various aspects of sport injury. The presentation of these will be confined to those categorised as physical agents (Merrick 2007) or pharmacological agents that are designed to control pain in the acute stage of injury. It is important in applying any type of modality to a patient or client that the clinician has an appropriate aim for such application as well as an understanding about the physiology of the modality and its effects on the body.

Physical agents

The two traditional physical agents best suited to successful management of acute injury pain are cryotherapy and electrotherapy (Merrick 2004). There are many methods of cryotherapy (Swenson et al. 1996; Bleakley, et al. 2004), though not all are equivalent in their efficacy (Bleakley et al. 2004). As previously mentioned, cold treatment is effective for ameliorating pain among other therapeutic effects like slowing development of haematoma and oedema, decreasing nerve conduction velocity, reducing muscle spasm (Swenson et al. 1996), reducing local blood flow in the injured area (Knight and Londeree 1980) and minimising secondary metabolic injury (Merrick et al. 1999; Knight and Draper 2008).

In terms of method of application, it is somewhat difficult to make a recommendation because techniques and published findings are so varied (Bleakley et al. 2004, 2007). A simple ice bag is usually the most expeditious and inexpensive means of applying cold. It has many proponents (McMaster et al. 1978; Merrick 2004; Bleakley, et al. 2006; Knight and Draper 2008), although it has been shown that a 20 minute cold whirlpool bath procured a longer tissue temperature reduction than did a 20 minute crushed ice pack treatment for the leg (Myrer et al. 1998). A protocol of a 10 minute ice pack application, 10 minutes without ice and then another 10 minutes of ice pack lowered tissue temperature better than a single 20 minute ice pack treatment (Bleakley et al. 2006). The effective cooling by ice is related to the phase change that it undergoes from a solid to a liquid as it melts (Merrick et al. 2003). The combination of therapeutic cold and exercise, or cryokinetics (Hayden 1964; Knight 1995; Bleakley et al. 2007; Knight and Draper 2008), is especially beneficial because of the way early motion promotes subsequent rehabilitation. In this technique pain is managed by the cold treatment and then the athlete undertakes a controlled exercise regimen, followed by another cycle of cold.

Transcutaneous electrical nerve stimulation (TENS) is an electrotherapy modality shown to be effective for musculoskeletal pain (Hsueh et al. 1997; Bertalanffy et al. 2005). It electrically stimulates sensory nerves around an injured body region to interfere with pain stimuli. The gate control theory (Melzack and Wall 1965; Dickenson 2002) defines how high-frequency TENS mediates the sensation of pain. Pain impulses on α-δ and C nerve fibres in the spinal cord are interrupted by sensory information transmitted via α-β afferent nerves (Merrick 2004). Thus, if the TENS signal stimulates the α-β fibres to a great enough degree, those nerve signals will close a "gate" in the spinal cord for the pain stimuli and be perceived instead of the pain.

Although it has been applied to acute musculoskeletal pain scenarios, systematic review of the research is inconclusive about the efficacy of TENS (Walsh et al. 2009), and a lack of randomisation in clinical trials leads to over-estimation of the positive effects of the treatment (Carroll et al. 1996). In acute low back pain patients, therapeutic exercise with TENS was not assistive to patient recovery in comparison to exercise alone (Herman et al. 1994). Neither was it a benefit for patients with acute postoperative pain following lumbar surgery (McCallum et al. 1988). However, in a randomised, double-blind study of trauma patients at A&E TENS was as effective as oral medication for analgesia (Ordog 1987). There are several different TENS units on the market, a myriad of musculoskeletal injuries and numerous protocols with which TENS can be applied to these injuries. Discussing all of these variations is beyond the scope of this chapter; but, because TENS is a relatively safe modality, there is minimal risk to utilising it for management of acute pain in sport injuries and discontinuing it if it proves ineffective for a given athlete.

High-voltage, low-frequency electrical stimulation also can be useful for managing acute pain (Merrick 2004). The intention for this modality is to elicit release of an opioid substance called beta-endorphin that naturally occurs in the body (Hughes et al. 1984; Bender et al. 2007); beta-endorphin is a powerful mediator of pain (Bender et al. 2007). The treatment may be somewhat uncomfortable to the patient because of the waveform and frequency, but this stimulation is what makes the technique effective in bringing forth release of beta-endorphin and consequent pain relief after treatment (Merrick 2004).

A final type of electrotherapy that often is effective for acute pain management is interferential current (Merrick 2004; Jorge et al. 2006). This technique's name is derived from its two electromagnetic fields that cross each other; at their point of intersection is a localised field of interference caused by their competing current phases. The current in this field continually changes, a property that precludes accommodation by the nerves in the treatment area and which yields analgesia via the gate control mechanism (Merrick 2004). Whilst interferential current has been shown to be similarly effective compared to TENS, research about it is scant (Johnson and Tabasam 2003). Certainly its usage is confined to a clinical setting, perhaps making TENS preferable simply because of easier device portability and the ability of patients to operate the equipment.

Pharmacological agents

Topical analgesics

Topical methods of analgesia are enjoying renewed interest in healthcare because of the challenges inherent with administering pain medications via other routes (Stanos 2007). Among the non-prescription compounds available, counterirritants are one of the oldest and most widely used. They are marketed under such brand names as Sports Muscle Rub, Muellergesic, BioFreeze, Tiger Balm and Deep Heat Cream. These preparations contain capsaicin, menthol, camphor, garlic and other ingredients that possess both distinct aromatic properties and an ability to act on the nociceptors of the skin. Their characteristic "counterirritation" reduces the perception of the musculoskeletal pain "irritation."

NSAID topical ointments, especially those containing ibuprofen or diclofenac, show promise as satisfactory alternatives to oral NSAIDs. A double-blind study of the effects of oral versus topical gel ibuprofen reported no difference between the two delivery methods for all therapeutic measures (Whitefield et al. 2002). Other studies corroborate the efficacy of topical NSAIDs (Dominkus et al. 1996; Rovenský et al. 2001; Banning 2008); but, this depends on the relative depth (e.g. superficial versus intra-articular) of the pain's origin (Dominkus et al. 1996; Miyatake et al. 2009).

Over-the-counter oral analgesics

Paracetamol (acetominophen) is the most widely used analgesic compound in the world (Jalan et al. 2007). It is generally thought to be a safe and effective medication (Graham et al. 2003; Kehlet and Werner 2003), although some concern exists that liver toxicity is possible even at commonly used non-prescription doses (Jalan et al. 2007). Paracetamol also has been shown to be as efficacious in acute musculoskeletal pain as non-steroidal anti-inflammatory drugs (Woo et al. 2005; Gøtzsche 2006). Nonetheless the drug is not without side-effects and controversy (Moynihan 2002; Ahmad 2007); and, assuredly it must not be offered nor ingested indiscriminately. The sport rehabilitator has a role and responsibility in properly advising athletes, patients and clients in the approved use of this and other medicinal preparations.

Compared to Paracetamol, NSAIDs are much more likely to elicit unfavourable reactions. They are well known for adversely affecting the gastrointestinal system (McCarthy 2001), and more recent evidence has associated adverse cardiovascular effects with cyclo-oxygenase 2 (COX-2) inhibitor NSAIDs (Jüni et al. 2004; Vardeny and Solomon 2008). Assessing the risks and benefits of this class of pharmaceuticals can be complex (Patrono and Rocca 2009). As mentioned above, some studies suggest that for typical pain from acute musculoskeletal injury, NSAIDs and Paracetamol are similarly effective. Unquestionably, administration of prescription NSAIDs or over-the-counter NSAIDs (e.g. ibuprofen) at prescription strength doses is unethical without the care of a GP or consultant. The sport rehabilitator must adhere to approved applications of NSAIDs and discourage improper NSAID use when advising individuals under their care.

Concussion

An understanding of concussion as an acute injury is crucial to the sport rehabilitator. Whilst it is not the typical musculoskeletal injury, it can be considerably more serious and troublesome to manage. Concussion is defined as "a trauma-induced alteration in mental status that may or may not involve loss of consciousness" (American Academy of Neurology 1997, p. 582). It has gained increasing attention in recent years because – in addition to its trauma-induced

Table 10.6 Neurological injuries to participants in a variety of sports (NA = not available) (adapted from Toth 2008)

Sport	Acute neurological injury rate per 100 athlete-exposures	Incidence of mild traumatic brain injury per 1,000 athlete-exposures
Basketball, males 18–23 yrs	1.0	0.3
Basketball, males 22–39 yrs	1.9–6.4	NA
Basketball, females 18–23 yrs	0.4	0.5
Basketball, females 22–39 yrs	2.5–6.7	NA
Boxing, amateur	14–20	11–77
Boxing, professional	21–45	186–251
Cricket	<0.1	<0.1
Football, males 18–23 yrs	18.8	1.1
Football, males 22–35 yrs	105	NA
Football, females 18–23 yrs	16.4	1.4
Football, females 22–35 yrs	109	NA
Football, North American, 18–23 yrs	1.5–4.0	2.3–6.1
Hockey, males 17–23 yrs	12.6	1.1
Hockey, females 17–23 yrs	7.9	0.5–0.7
Ice hockey, males 18–23 yrs	0.5	1.5–4.2
Ice hockey, females 18–23 yrs	1.3	2.7
Ice hockey, males 20–36 yrs	11.9	6.6
Snowboarding	0.4	6.1
Taekwondo	6.3	NA
Wrestling, males 17–23 yrs	7.3	1.3

metabolic compromise of the brain (Katayama et al. 1990; Giza and Hovda 2001) – insidious sequelae often are associated with it in the form of associated pathologies such as subdural hematoma (Kersey 1998; Mori et al. 2006) or post-concussion syndrome (Fazio et al. 2007; Yang et al. 2009).

Whilst the basic steps of acute injury evaluation presented earlier apply to concussion, this injury is being treated separately herein because of its importance and the special techniques necessary for successfully handling it. Several medical associations have published position papers outlining proper management of concussion (American Academy of Neurology 1997; Guskiewicz et al. 2004; American College of Sports Medicine 2006; McCrory et al., 2009) in order to help decrease the morbidity and mortality of this injury. The sport rehabilitator who attends pitchside must be well versed in management of concussion and related conditions as catastrophic injury can be more alarming and potentially deadly than most other situations that confront sport medicine workers.

Toth (2008) comprehensively reviewed the literature to identify neurological injury rates sustained in sport and recreation. Table 10.6 summarises his analysis. Collision sports (e.g. rugby, North American football), sports where the participants or projectiles move at high velocity (e.g. alpine skiing, motorsports, skateboarding, cricket, baseball, ice hockey) and pugilistic sports (e.g. boxing, martial arts) are particularly prone to injuries of the head and neck. Certainly protective headgear are helpful against direct blows in some sports, but they do not prevent all head injuries and concussion can occur with or without helmets (American College of Sports Medicine 2006). In football, headgear does not offer protection against ball contact head injuries, but it does appear to reduce the severity of head-to-head contact injuries (Withnall et al. 2005). Of particular consequence in sport-related brain injuries are rapid acceleration or deceleration of the head or high velocity angular head motion (Holbourn 1945; Ommaya and Gennarelli 1974). Concussions are usually graded 1 through 3 according to severity (American Academy of Neurology 1997; Randolph 2001), as outlined in Table 10.7.

Athletes who experience concussion can safely resume sport participation if the return to play

Table 10.7 Characteristics of concussion grades (American Academy of Neurology 1997; Randolph 2001)

Concussion grade	Identifying characteristics
1	Transient confusion
	No loss of consciousness
	Concussion symptoms or mental status abnormalities resolve in less than 15 minutes
2	Transient confusion
	No loss of consciousness
	Concussion symptoms or mental status abnormalities last more than 15 minutes
3	Any loss of consciousness

Table 10.8 University of Pittsburgh's signs and symptoms of concussion (Lovell et al. 2004)

Signs observed by medical staff	Appears to be dazed or stunned
	Is confused about assignment
	Forgets plays
	Is unsure of game, score, or opponent
	Moves clumsily
	Answers questions slowly
	Loses consciousness
	Shows behaviour or personality change
	Forgets events prior to play (retrograde)
	Forgets events after hit (post-traumatic)
Symptoms reported by athlete	Headache
	Nausea
	Balance problems or dizziness
	Double or fuzzy/blurry vision
	Sensitivity to light or noise
	Feeling sluggish or slowed down
	Feeling "foggy" or groggy
	Concentration or memory problems
	Change in sleep pattern (appears later)
	Feeling fatigued

decisions are carefully made by healthcare professionals using appropriate individualised clinical guidelines (Randolph 2001; Lovell et al. 2004; American College of Sports Medicine 2006; Alla et al. 2009; Makdissi et al. 2009). However, such evaluations are complex. Evidence suggests that athletes who sustain multiple concussions exhibit cumulative neurological effects (Collins et al. 2002; Iverson et al. 2004), experience a substantial reduction in memory ability (Iverson et al. 2004) and demonstrate several pitchside neurological examination abnormalities (Collins et al. 2002) when compared to athletes with no history of concussion. Moreover, return to participation prior to resolution of head injury symptoms heightens susceptibility to further concussive episodes, even if the subsequent trauma is relatively minor (Kersey 1998; Lovell et al. 2004; Mori et al. 2006).

Acute evaluation of concussion

Concussion is not an easy diagnosis in the sport environment. The precipitating event may be associated with neither direct head trauma nor loss of consciousness (Lovell et al. 2004). Players usually are anxious to return to the match/competition; thus, they may not report a head injury or they may attempt to evade examination in favour of resuming play. A wide variety of signs and symptoms must be assessed, many of which are presented in Tables 10.8 and 10.9. A consultant neurologist is invaluable in this instance, and conservative management is paramount at all times. When in doubt, exclude the athlete from participation until concussion is disproven or completely resolved because there is a risk of cumulative neurological deficits in athletes who sustain subsequent concussions, especially before resolution of the antecedent episode (Cantu 1988; McCrory and Berkovic 1998). Prudence indicates that the sport rehabilitator and other healthcare professionals maintain an emergency action plan that can be implemented as required when they are confronted with a head injured athlete. Consciousness, airway integrity, breathing and circulation are fundamental diagnostic signs to assess and act on.

Presuming a non-emergency episode of potential concussion, the sport rehabilitator should embark on the history, inspection, palpation and special testing paradigm described previously. Appropriate modification of these is warranted. For example, an athlete's mental status may preclude their ability to offer satisfactory answers to injury history questions. (This substantiates the value of qualified pitchside healthcare practitioners for firsthand observation of an injury episode that provides key information to the evaluation.) The observation step coincides heavily with the special testing phase of the examination because so many of the special tests require visual

Table 10.9 American Academy of Neurology's (1997) signs and symptoms of concussion. Reproduced, with permission, from Table 1 and 2 from 'Practice Parameter: the management of concussion in sports' © American Academy of Neurology.

Features of concussion frequently observed	Vacant stare (befuddled facial expression)
	Delayed verbal and motor responses (slow to answer questions or follow instructions)
	Confusion and inability to focus attention (easily distracted and unable to follow through with normal activities)
	Disorientation (walking in the wrong direction, unaware of time, date and place)
	Slurred or incoherent speech (making disjointed or incomprehensible statements)
	Gross observable incoordination (stumbling, inability to walk tandem/straight line)
	Emotions out of proportion to circumstances (distraught, crying for no apparent reason)
	Memory deficits (exhibited by the athlete repeatedly asking the same question that has already been answered, or inability to memorise and recall 3 of 3 words or 3 of 3 objects in 5 minutes)
	Any period of loss of consciousness (paralytic coma, unresponsiveness to arousal)
Symptoms of concussion	Early (minutes to hours):
	Headache
	Dizziness or vertigo
	Lack of awareness of surroundings
	Nausea or vomiting
	Late (days to weeks):
	Persistent low grade headache
	Light-headedness
	Poor attention and concentration
	Memory dysfunction
	Easy fatigability
	Irritability and low frustration tolerance
	Intolerance of bright lights or difficulty focusing vision
	Intolerance of loud noises, sometimes ringing in the ears
	Anxiety and/or depressed mood
	Sleep disturbance

interpretation by the examiner. Palpation in this instance is usually utilised to determine cranial or facial tenderness associated directly with the offending blow or muscular tenderness that results from reflexive contraction of the neck musculature when the head is struck. Table 10.10 outlines diagnostic criteria for assessing concussion; this information provides an excellent pitchside guide for the sport rehabilitator.

There are important cautions for the sport rehabilitator when evaluating the athlete with a closed head injury. In his report of neuropsychological testing algorithms for suspected concussion in high school, university and professional athletes, Randolph (2001) identifies the increased sophistication necessary for each subsequent sport proficiency level because of the higher stakes for participants (e.g. stature, salary, importance of a given competition). He offers a model for neuropsychological evaluation that is customised for each level and also recommends baseline evaluations be collected from every player, something which is clearly more feasible for the professional clubs.

Concussion case study

An 18-year-old North American football player arrived at his university for the beginning of pre-season training. A typical training schedule was followed with the players wearing only T-shirts, shorts and headgear for two days and then wearing headgear and shoulder pads for the next few days in order

Table 10.10 Pitch side evaluation of suspected concussion (adapted from the American Academy of Neurology 1997)

Mental status testing	Orientation	Report time, place, person and situation (circumstances of injury)
	Concentration	Count presented digits backward (e.g. 3-1-7, 4-6-8-2, 5-3-0-7-4)
		Name months of the year in reverse order
	Memory	Name the teams in prior match
		Recall 3 words and 3 objects at 0 and 5 minutes
		Recall recent newsworthy events
		Recall details of the match (plays, moves, strategies, etc.)
External provocative tests	Any appearance of associated symptoms is abnormal, e.g. headaches, dizziness, nausea, unsteadiness, photophobia, blurred or double vision, emotional liability or mental status changes	Complete 35 metre sprint
		Complete 5 push-ups
		Complete 5 sit-ups
		Complete 5 knee bends
Neurologic tests	Pupils	Observe symmetry and reaction
	Coordination	Touch finger to nose
		Touch right finger to left finger
		Observe tandem gait
	Sensation	Touch finger to nose (eyes closed)
		Balance in Romberg test

to acclimatise to the workload and environmental conditions. On the second day of training with the full complement of equipment the player participated in a tackling drill and received a blow to the head that was hard, but not extreme. He was obviously shaken by the play, and he came off the field to be evaluated by the medical staff.

Concussion: history

The player denied loss of consciousness – which was corroborated by the medical staff who witnessed his collision and the events immediately subsequent – but he did report a mild instance of "seeing stars." He was oriented to his surroundings and able to accurately relate the circumstances of the injury. On questioning he indicated that his head felt "a little foggy." He was slightly disoriented and his vision was unfocused for a few seconds. These symptoms resolved in less than five minutes. A mild headache was the player's one enduring symptom.

Concussion: inspection

The player's pupils were of appropriate and equal size and responsive to light. He did not exhibit any abnormal signs; otorrhea, rhinorrhea and Battle's sign were all absent. No signs of serious injury were outwardly apparent.

Concussion: palpation

The player's skull, cervical spine and neck musculature were systematically palpated for areas of point tenderness that could indicate pathology. There was some tenderness in his right trapezius muscle along the neck, but all other areas were non-tender.

Concussion: special testing

The player's vital signs were within normal limits at baseline and continued so. Evaluations of the cranial nerves were normal. Ocular tracking was normal, although the player mentioned needing to concentrate to complete the task. His visual acuity was normal. He completed the Romberg test for balance successfully with his eyes closed. Motor coordination testing was normal. The player's neck soreness was accentuated slightly by active resistive muscle testing for right lateral flexion and hyperextension of the cervical spine.

Concussion: information collating and decision making

Based on the above described information that was gathered during the acute injury examination the player was diagnosed with a mild grade 2 concussion. This was due primarily to his descriptions of how he felt during and immediately following the injury. Although his symptoms generally resolved quickly, he continued to report a headache that lasted beyond 15 minutes and slight neck musculature tenderness. Thus, the player was excluded from further participation pending complete symptom resolution. The player's headgear was collected in order to prevent an attempt to return to training without medical clearance. The coach was informed that the player would be ineligible for participation until his symptoms subsided. The player was instructed about the signs and symptoms of head injuries and their follow-up care and was dismissed with a plan for his teammates to monitor him during the next 24 hours. Specific directions about reporting new or worsening symptoms were provided, and the medical staff made arrangements for his follow-up care.

Concussion: follow-up for return to sport

The player presented himself to the sport medicine clinic on a daily basis. In close communication with the team physician, the sport rehabilitator evaluated the player at each visit using standard concussion evaluation techniques. A fortnight later the player's headache was completely resolved and all other concussion tests were normal. However, he continued to exhibit mild myofascial neck pain with resisted cervical motions. This was presumed to be a residual sign of reflexive contraction of the neck musculature during the injury episode.

At two weeks post-injury the player was referred to the radiology department for a CT scan of his neck in order to investigate the persistent pain. Review of the CT images revealed the player suffered from a previously undetected congenital tripartite C1 vertebra. On the basis of this finding the player was excluded from further participation in collision sports and counselled about alternative sporting outlets that would not expose his cervical spine to catastrophic risk. It was obviously fortunate that during his preceding several years of North American football involvement he did not sustain a disastrous neck injury.

This case exemplifies the need for alertness to symptoms that may be unrelated to the primary pathology. A conservative approach is particularly appropriate for injuries to the central nervous system.

Conclusion

Acute management entails evaluation of the injury and application of appropriate treatments based on decisions informed by the evaluation's results. This process, especially the evaluation, is enhanced by the clinician's presence at training sessions and competitions where they can witness the mechanism of injuries. Following the injury evaluation and categorisation of the injury's nature and severity, prompt administration of proper on-site care should ensue. In non-emergent cases this will include judicious application of the PRICE method: protection, rest, ice, compression and elevation.

Success in managing the acute stage of injury generally leads to success in the post-acute stage and onward as the athlete is prepared by healthcare professionals and others on the sport healthcare team for re-entry to participation. Insofar as possible, it is incumbent on the sport rehabilitator to ensure this process goes smoothly, to offer physical and psychological support and to keep the athlete's best interests foremost during the progression back to full activity.

References

Ahmad, S.R. (2007) Safety of recommended doses of paracetamol. *Lancet*, *369* (9560), 462–463.

Algafly, A.A. and George, K.P. (2007) The effect of cryotherapy on nerve conduction velocity, pain threshold and pain tolerance. *British Journal of Sports Medicine*, *41* (6), 365–369.

Alla, S., Sullivan, S.J., Hale, L. and McCrory, P. (2009) Self-report scales/checklists for the measurement of concussion symptoms: a systematic review. *British Journal of Sports Medicine*, *43* (Suppl I), i3–i12.

American Academy of Neurology (1997) Practice parameter: the management of concussion in sports (summary statement). *Neurology*, *48*, 581–585.

American College of Sports Medicine (2006) Concussion (mild traumatic brain injury) and the team physician: a consensus statement. *Medicine and Science in Sports and Exercise*, *38* (2), 395–399.

Anonymous. (2005) *Stedman's Medical Dictionary for the Health Professions and Nursing*. Philadelphia, PA: Lippincott Williams and Wilkins.

Bachmann, L.M., Kolb, E., Koller, M.T., Steurer, J. and ter Riet, G. (2003) Accuracy of Ottawa ankle rules to exclude fractures of the ankle and mid-foot: systematic review. *British Medical Journal*, 326 (7386), 417–423.

Banning, M. (2008) Topical diclofenac: clinical effectiveness and current uses in osteoarthritis of the knee and soft tissue injuries. *Expert Opinion on Pharmacotherapy*, 9 (16), 2921–2929.

Bender, T., Nagy, G., Barna, I., Tefner, I., Kádas, É. and Géher, P. (2007) The effect of physical therapy on beta-endorphin levels. *European Journal of Applied Physiology*, 100 (4), 371–382.

Bertalanffy, A., Kober, A., Bertalanffy, P., Gustorff, B., Gore, O., Adel, S. and Hoerauf, K. (2005) Transcutaneous electrical nerve stimulation reduces acute low back pain during emergency transport. *Academic Emergency Medicine*, 12 (7), 607–611.

Blackham, J.E.J., Claridge, T. and Benger, J.R. (2008) Can patients apply the Ottawa ankle rules to themselves? *Emergency Medicine Journal*, 25 (11), 750–751.

Bleakley, C., McDonough, S. and MacAuley, D. (2004) The use of ice in the treatment of acute soft-tissue injury: a systematic review of randomized controlled trials. *American Journal of Sports Medicine*, 32 (1), 251–261.

Bleakley, C.M., McDonough, S.M., MacAuley, D.C. and Bjordal, J. (2006) Cryotherapy for acute ankle sprains: a randomised controlled study of two different icing protocols. *British Journal of Sports Medicine*, 40 (8), 700–705.

Bleakley, C., O'Connor, S., Tully, M., Rocke, L., MacAuley, D. and McDonough, S. (2007) The PRICE study (Protection Rest Ice Compression Elevation): design of a randomised controlled trial comparing standard versus cryokinetic ice applications in the management of acute ankle sprain [ISRCTN13903946]. *BMC Musculoskeletal Disorders*, 8 (1), 125.

Boyce, S.H., Quigley, M.A. and Campbell, S. (2005) Management of ankle sprains: a randomised controlled trial of the treatment of inversion injuries using an elastic support bandage or an Aircast ankle brace. *British Journal of Sports Medicine*, 39 (2), 91–96.

Brooks, J.H.M. and Fuller, C.W. (2006) The influence of methodological issues on the results and conclusions from epidemiological studies of sports injuries: illustrative examples. *Sports Medicine*, 36 (6), 459–472.

Brooks, J.H.M., Fuller, C.W., Kemp, S.P.T. and Reddin, D.B. (2005) A prospective study of injuries and training amongst the England 2003 Rugby World Cup squad. *British Journal of Sports Medicine*, 39 (5), 288–293.

Brooks, J.H.M., Fuller, C.W., Kemp, S.P.T. and Reddin, D.B. (2008) An assessment of training volume in professional rugby union and its impact on the incidence, severity, and nature of match and training injuries. *Journal of Sports Sciences*, 26 (8), 863–873.

Caine, D., Caine, C. and Maffulli, N. (2006) Incidence and distribution of pediatric sport-related injuries. *Clinical Journal of Sport Medicine*, 16 (6), 500–513.

Cantu, R.C. (1988) Second-impact syndrome. *Clinics in Sports Medicine*, 17, 37–44.

Carroll, D., Tramer, M., McQuay, H., Nye, B. and Moore, A. (1996) Randomization is important in studies with pain outcomes: systematic review of transcutaneous electrical nerve stimulation in acute postoperative pain. *British Journal of Anaesthesia*, (6), 798–803.

Collins, M.W., Lovell, M.R., Iverson, G.L., Cantu, R.C., Maroon, J.C. and Field, M. (2002) Cumulative effects of concussion in high school athletes. *Neurosurgery*, 51 (5), 1175–1181.

Collins, N.C. (2008) Is ice right? Does cryotherapy improve outcome for acute soft tissue injury? *Emergency Medicine Journal*, 25 (2), 65–68.

Dale, R.B., Harrelson, G.L. and Leaver-Dunn, D. (2004) Principles of rehabilitation. In Andrews, J.R., Harrelson, G.L. and Wilk, K.E. (Eds) *Physical Rehabilitation of the Injured Athlete*, 3rd edn. Philadelphia, PA: Saunders, pp. 157–188.

Delis, K.T., Azizi, Z.A., Stevens, R.J.G., Wolfe, J.H.N. and Nicolaides, A.N. (2000) Optimum intermittent pneumatic compression stimulus for lower-limb venous emptying. *European Journal of Vascular and Endovascular Surgery*, 19 (3), 261–269.

Department of Trade and Industry (2003) *24th (Final) Report of the Home and Leisure Accident Surveillance System*, London: Department of Trade and Industry.

Dickenson, A.H. (2002) Gate Control Theory of pain stands the test of time. *British Journal of Anaesthesia*, 88 (6), 755–757.

Dominkus, M., Nicolakis, M., Kotz, R., Wilkinson, F.E., Kaiser, R.R. and Chlud, K. (1996) Comparison of tissue and plasma levels of ibuprofen after oral and topical administration. *Arzneimittel-Forschung*, 46 (12), 1138–1143.

Ekstrand, J. and Karlsson, J. (2003) The risk for injury in football. There is a need for a consensus about definition of the injury and the design of studies. *Scandinavian Journal of Medicine and Science in Sports*, 13 (3), 147–149.

Ernst, E. and Fialka, V. (1994) Ice freezes pain? A review of the clinical effectiveness of analgesic cold therapy. *Journal of Pain and Symptom Management*, 9 (1), 56–59.

Fazio, V.C., Lovell, M.R., Pardini, J.E. and Collins, M.W. (2007) The relation between post concussion symptoms and neurocognitive performance in concussed athletes. *NeuroRehabilitation*, 22 (3), 207–216.

Fischer, J., Van Lunen, B.L., Branch, J.D. and Pirone, J.L. (2009) Functional performance following an ice bag application to the hamstrings. *Journal of Strength and Conditioning Research*, 23 (1), 44–50.

Flegel, M.J. (2008) *Sport First Aid*, Champaign, IL: Human Kinetics.

Fong, D.T.-P., Hong, Y., Chan, L.-K., Yung, P.S.-H. and Chan, K.-M. (2007) A systematic review on ankle injury and ankle sprain in sports. *Sports Medicine*, 37 (1), 73–94.

Fuller, C.W., Laborde, F., Leather, R.J. and Molloy, M.G. (2008) International Rugby Board Rugby World Cup 2007 injury surveillance study. *British Journal of Sports Medicine*, 42 (6), 452–459.

Gabbett, T.J. (2004) Incidence of injury in junior and senior Rugby League players. *Sports Medicine*, 34 (12), 849–859.

Garrick, J.G. (1977) The frequency of injury, mechanism of injury, and epidemiology of ankle sprains. *American Journal of Sports Medicine*, 5 (6), 241–242.

Giza, C.C. and Hovda, D.A. (2001) The neurometabolic cascade of concussion. *Journal of Athletic Training*, 36 (3), 228–235.

Gøtzsche, P.C. (2006) [Paracetamol has the same effect as non-steroidal anti-inflammatory agents in acute musculoskeletal injuries]. *Ugeskrift for Laeger*, 168 (20), 1981–1982.

Graham, G.G., Scott, K.F. and Day, R.O. (2003) [Tolerability of paracetamol]. *Drugs*, 63 (Spec. No. 2), 43–46.

Guskiewicz, K.M., Bruce, S.L., Cantu, R.C., Ferrara, M.S., Kelly, J.P., McCrea, M., Putukian, M. and Valovich McLeod, T.C. (2004) National Athletic Trainers' Association position statement: management of sport-related concussion. *Journal of Athletic Training*, 39 (3), 280–297.

Hawkins, R.D. and Fuller, C.W. (1999) A prospective epidemiological study of injuries in four English professional football clubs. *British Journal of Sports Medicine*, 33 (3), 196–203.

Hayden, C.A. (1964) Cryokinetics in an early treatment program. *Physical Therapy*, 44, 990–993.

Herman, E., Williams, R., Stratford, P., Fargas-Babjak, A. and Trott, M. (1994) A randomized controlled trial of transcutaneous electrical nerve stimulation (CODETRON) to determine its benefits in a rehabilitation program for acute occupational low back pain. *Spine*, 19 (5), 561–568.

Hildebrand, K.A., Hart, D.A., Rattner, J.B., Marchuk, L.L. and Frank, C.B. (2007) Ligament injuries: pathophysiology, healing, and treatment considerations. *in* Magee, D.J., Zachazewski, J.E. and Quillen, W.S. (Eds) *Scientific Foundations and Principles of Practice in Musculoskeletal Rehabilitation*. St. Louis, MO: Saunders, pp. 23–47.

Hodgson, L., Gissane, C., Gabbett, T.J. and King, D.A. (2007) For debate: consensus injury definitions in team sports should focus on encompassing all injuries. *Clinical Journal of Sport Medicine*, 17 (3), 188–191.

Holbourn, A.H.S. (1945) The mechanics of brain injuries. *British Medical Bulletin*, 3 (6), 147–149.

Hootman, J.M., Dick, R. and Agel, J. (2007) Epidemiology of collegiate injuries for 15 sports: summary and recommendations for injury prevention initiatives. *Journal of Athletic Training*, 42 (2), 311–319.

Hsueh, T.-C., Cheng, P.-T., Kuan, T.-S. and Hong, C.-Z. (1997) The immediate effectiveness of electrical nerve stimulation and electrical muscle stimulation on myofascial trigger points. *American Journal of Physical Medicine and Rehabilitation*, 76 (6), 471–476.

Hubbard, T.J. and Denegar, C.R. (2004) Does cryotherapy improve outcomes with soft tissue injury? *Journal of Athletic Training*, 39 (3), 278–279.

Hughes, G.S., Lichstein, P.R., Whitlock, D. and Harker, C. (1984) Response of plasma beta-endorphins to transcutaneous electrical nerve stimulation in healthy subjects. *Physical Therapy*, 64 (7), 1062–1066.

Iverson, G.L., Gaetz, M., Lovell, M.R. and Collins, M.W. (2004) Cumulative effects of concussion in amateur athletes. *Brain Injury*, 18 (5), 433–443.

Jalan, R., Williams, R. and Bernuau, J. (2007) Paracetamol: are therapeutic doses entirely safe? *Lancet*, 368 (9554), 2195–2196.

Johnson, M.I. and Tabasam, G. (2003) An investigation into the analgesic effects of interferential currents and transcutaneous electrical nerve stimulation on experimentally induced ischemic pain in otherwise pain-free volunteers. *Physical Therapy*, 83 (3), 208–223.

Jorge, S., Parada, C.A., Ferreira, S.H. and Tambeli, C.H. (2006) Interferential therapy produces antinociception during application in various models of inflammatory pain. *Physical Therapy*, 86 (6), 800–808.

Jüni, P., Nartey, L., Reichenbach, S., Sterchi, R., Dieppe, P.A. and Egger, M. (2004) Risk of cardiovascular events and rofecoxib: cumulative meta-analysis. *The Lancet*, 364 (9450), 2021–2029.

Katayama, Y., Becker, D.P., Tamura, T. and Hovda, D.A. (1990) Massive increases in extracellular potassium and the indiscriminate release of glutamate following concussive brain injury. *Journal of Neurosurgery*, 73 (6), 889–900.

Kehlet, H. and Werner, M.U. (2003) [Role of paracetamol in the acute pain management]. *Drugs, 63* (Spec. No. 2), 15–22.

Kersey, R.D. (1998) Acute subdural hematoma after a reported mild concussion: a case report. *Journal of Athletic Training, 33* (3), 264–268.

King, D.A., Gabbett, T.J., Gissane, C. and Hodgson, L. (2009) Epidemiological studies of injuries in rugby league: suggestions for definitions, data collection and reporting methods. *Journal of Science and Medicine in Sport, 12* (1), 12–19.

Knight, K.L. (1995) *Cryotherapy in Sport Injury Management*. Champaign, IL: Human Kinetics.

Knight, K.L. (2008) More precise classification of orthopaedic injury types and treatment will improve patient care. *Journal of Athletic Training, 43* (2), 117–118.

Knight, K.L. and Draper, D.O. (2008) *Therapeutic Modalities: The Art and Science*. Baltimore, MD: Lippincott Williams and Wilkins.

Knight, K.L. and Londeree, B.R. (1980) Comparison of blood flow in the ankle of uninjured subjects during therapeutic applications of heat, cold, and exercise. *Medicine and Science in Sports and Exercise, 12* (1), 76–80.

Knowles, S.B., Marshall, S.W. and Guskiewicz, K.M. (2006) Issues in estimating risks and rates in sports injury research. *Journal of Athletic Training, 41* (2), 207–215.

Leddy, J.J., Smolinski, R.J., Lawrence, J., Snyder, J.L. and Priore, R.L. (1998) Prospective evaluation of the Ottawa Ankle Rules in a university sports medicine center: with a modification to increase specificity for identifying malleolar fractures. *American Journal of Sports Medicine, 26* (2), 158–165.

Leddy, J.J., Kesari, A. and Smolinski, R.J. (2002) Implementation of the Ottawa ankle rule in a university sports medicine center. *Medicine and Science in Sports and Exercise, 34* (1), 57–62.

Lessard, L.A., Scudds, R.A., Amendola, A. and Vaz, M.D. (1997) The efficacy of cryotherapy following arthroscopic knee surgery. *Journal of Orthopaedic and Sports Physical Therapy, 26* (1), 14–22.

Lovell, M.R., Collins, M. and Bradley, J. (2004) Return to play following sports-related concussion. *Clinics in Sports Medicine, 23* (3), 421–441.

Magee, D.J. (2008) *Orthopedic Physical Assessment*. Philadelphia, PA: Saunders Elsevier.

Makdissi, M., McCrory, P., Ugoni, A., Darby, D. and Brukner, P. (2009) A prospective study of postconcussive outcomes after return to play in Australian football. *American Journal of Sports Medicine, 37* (5), 877–883.

McCallum, M.I.D., Glynn, C.J., Moore, R.A., Lammer, P. and Phillips, A.M. (1988) Transcutaneous electrical nerve stimulation in the management of acute postoperative pain. *British Journal of Anaesthesia, 61* (3), 308–312.

McCarthy, D.M. (2001) Prevention and treatment of gastrointestinal symptoms and complications due to NSAIDs. *Best Practice and Research Clinical Gastroenterology, 15* (5), 755–773.

McCrory, P.R. and Berkovic, S.F. (1998) Second impact syndrome. *Neurology, 50*, 677–683.

McCrory, P., Meeuwisse, W., Johnston, K., Dvorak, J., Aubry, M., Molloy, M. and Cantu, R. (2009) Consensus Statement on Concussion in Sport: the 3rd International Conference on Concussion in Sport held in Zurich, November 2008. *Journal of Athletic Training, 44* (4), 434–448.

McMaster, W.C., Liddle, S. and Waugh, T.R. (1978) Laboratory evaluation of various cold therapy modalities. *American Journal of Sports Medicine, 6* (5), 291–294.

Melzack, R. and Wall, P.D. (1965) Pain mechanisms: a new theory. *Science, 150* (3699), 971–979.

Merrick, M.A. (2004) Therapeutic modalities as an adjunct to rehabilitation. in Andrews, J.R., Harrelson, G.L. and Wilk, K.E. (Eds) *Physical Rehabilitation of the Injured Athlete*, 3rd edn. Philadelphia, PA: Saunders, pp. 51–98.

Merrick, M.A. (2007) Physiological bases of physical agents. In Magee, D.J., Zachazewski, J.E. and Quillen, W.S. (Eds) *Scientific Foundations and Principles of Practice in Musculoskeletal Rehabilitation*. St. Louis, MO: Saunders, pp. 237–254.

Merrick, M.A., Rankin, J.M., Andres, F.A. and Hinman, C.L. (1999) A preliminary examination of cryotherapy and secondary injury in skeletal muscle. *Medicine and Science in Sports and Exercise, 31* (11), 1516–1521.

Merrick, M.A., Jutte, L.S. and Smith, M.E. (2003) Cold modalities with different thermodynamic properties produce different surface and intramuscular temperatures. *Journal of Athletic Training, 38* (1), 28–33.

Milne, L. (1996) Ottawa ankle decision rules. *Western Journal of Medicine, 164* (1), p. 67.

Miyatake, S., Ichiyama, H., Kondo, E. and Yasuda, K. (2009) Randomized clinical comparisons of diclofenac concentration in the soft tissues and blood plasma between topical and oral applications. *British Journal of Clinical Pharmacology, 67* (1), 125–129.

Mora, S., Zalavras, C.G., Wang, L. and Thordarson, D.B. (2002) The role of pulsatile cold compression in edema resolution following ankle fractures: a randomized clinical trial. *Foot and Ankle International, 23* (11), 999–1002.

Mori, T., Katayama, Y. and Kawamata, T. (2006) Acute hemispheric swelling associated with thin subdural hematomas: pathophysiology of repetitive head injury in sports. *Acta Neurochirurgica. Supplement*, 96, 40–43.

Moynihan, R. (2002) FDA fails to reduce accessibility of paracetamol despite 450 deaths a year. *British Medical Journal*, 325 (7366), p. 678.

Myer, J.W., Measom, G. and Fellingham, G.W. (1998) Temperature changes in the human leg during and after two methods of cryotherapy. *Journal of Athletic Training*, 33 (1), 25–29.

National Athletic Trainers' Association (2007) *Recommendations and Guidelines for Appropriate Medical Coverage of Intercollegiate Athletics*. Dallas, TX: National Athletic Trainers' Association.

National Collegiate Athletic Association (2008) *2008–09 NCAA Sports Medicine Handbook*, Indianapolis, IA: National Collegiate Athletic Association.

Nugent, P.L. (2004) Ottawa ankle rules accurately assess injuries and reduce reliance on radiographs. *Journal of Family Practice*, 53 (10), 785–788.

Ommaya, A.K. and Gennarelli, T.A. (1974) Cerebral concussion and traumatic unconsciousness: correlation of experimental and clinical observations on blunt head injuries. *Brain*, 97 (1), 633–654.

Ordog, G.J. (1987) Transcutaneous electrical nerve stimulation versus oral analgesic: A randomized double-blind controlled study in acute traumatic pain. *American Journal of Emergency Medicine*, 5 (1), 6–10.

Otte, J.W., Merrick, M.A., Ingersoll, C.D. and Cordova, M.L. (2002) Subcutaneous adipose tissue thickness alters cooling time during cryotherapy. *Archives of Physical Medicine and Rehabilitation*, 83 (11), 1501–1505.

Papacostas, E., Malliaropoulos, N., Papadopoulos, A. and Liouliakis, C. (2001) Validation of Ottawa ankle rules protocol in Greek athletes: study in the emergency departments of a district general hospital and a sports injuries clinic. *British Journal of Sports Medicine*, 35 (6), 445–447.

Patrono, C. and Rocca, B. (2009) Nonsteroidal antiinflammatory drugs: past, present and future. *Pharmacological Research*, 59 (5), 285–289.

Powell, J.W. and Dompier, T.P. (2004) Analysis of injury rates and treatment patterns for time-loss and non-time-loss injuries among collegiate student-athletes. *Journal of Athletic Training*, 39 (1), 56–70.

Quillen, W.S. and Rouillier, L.H. (1982) Initial management of acute ankle sprains with rapid pulsed pneumatic compression and cold. *Journal of Orthopaedic and Sports Physical Therapy*, 4 (1), 39–43.

Randolph, C. (2001) Implementation of neuropsychological testing models for the high school, collegiate, and professional sport settings. *Journal of Athletic Training*, 36 (3), 288–296.

Rovenský, J., Miceková, D., Gubzová, Z., Fimmers, R., Lenhard, G., Vögtle-Junkert, U. and Screyger, F. (2001) Treatment of knee osteoarthritis with a topical non-steroidal antiinflammatory drug: results of a randomized, double-blind, placebo-controlled study on the efficacy and safety of a 5% ibuprofen cream. *Drugs Under Experimental and Clinical Research*, 27 (5–6), 209–221.

Royal Society for the Prevention of Accidents. (2009) *Home and leisure accident statistics*. [online]. Birmingham, UK. [accessed 14th March 2009]. <http://www.hassandlass.org.uk/query/index.htm>.

Rucinkski, T.J., Hooker, D.N., Prentice, W.E., Shields, E.W. and Cote-Murray, D.J. (1991) The effects of intermittent compression on edema in postacute ankle sprains. *Journal of Orthopaedic and Sports Physical Therapy*, 14 (2), 65–69.

Stanos, S.P. (2007) Topical agents for the management of musculoskeletal pain. *Journal of Pain and Symptom Management*, 33 (3), 342–355.

Starkey, J.A. (1976) Treatment of ankle sprains by simultaneous use of intermittent compression and ice packs. *American Journal of Sports Medicine*, 4 (4), 142–144.

Stiell, I.G., Greenberg, G.H., McKnight, R.D., Nair, R.C., McDowell, I. and Worthington, J.R. (1992) A study to develop clinical decision rules for the use of radiography in acute ankle injuries. *Annals of Emergency Medicine*, 21 (4), 384–390.

Stiell, I.G., McKnight, R.D., Greenberg, G.H., McDowell, I., Nair, R.C., Wells, G.A., Johns, C. and Worthington, J.R. (1994) Implementation of the Ottawa ankle rules. *Journal of the American Medical Association*, 271 (11), 827–832.

Stiell, I.G., Greenberg, G.H., Wells, G.A., McKnight, R.D., Cwinn, A.A., Cacciotti, T., McDowell, I. and Smith, N.A. (1995) Derivation of a decision rule for the use of radiography in acute knee injuries. *Annals of Emergency Medicine*, 26 (4), 405–413.

Stiell, I.G., Greenberg, G.H., Wells, G.A., McDowell, I., Cwinn, A.A., Smith, N.A., Cacciotti, T. and Sivilotti, M.L. (1996) Prospective validation of a decision rule for the use of radiography in acute knee injuries. *Journal of the American Medical Association*, 275 (8), 611–615.

Swenson, C., Swärd, L. and Karlsson, J. (1996) Cryotherapy in sports medicine. *Scandinavian Journal of Medicine and Science in Sports*, 6 (4), 193–200.

Tigges, S., Pitts, S., Mukundan, S., Jr., Morrison, D., Olson, M. and Shahriara, A. (1999) External validation of the Ottawa knee rules in an urban trauma center in the United States. *American Journal of Roentgenology*, *172* (4), 1069–1071.

Toth, C. (2008) The epidemiology of injuries to the nervous system resulting from sport and recreation. *Neurologic Clinics*, *26* (1), 1–31.

Tsang, K.K., Hertel, J. and Denegar, C.R. (2003) Volume decreases after elevation and intermittent compression of postacute ankle sprains are negated by gravity-dependent positioning. *Journal of Athletic Training*, *38* (4), 320–324.

van Mechelen, W., Hlobil, H. and Kemper, H.C. (1996) Incidence, severity, aetiology and prevention of sports injuries: a review of concepts. *Sports Medicine*, *14* (2), 82–99.

Vanscheidt, W., Ukat, A. and Partsch, H. (2009) Dose-response of compression therapy for chronic venous edema – higher pressures are associated with greater volume reduction: Two randomized clinical studies. *Journal of Vascular Surgery*, *49* (2), 395–402.

Vardeny, O. and Solomon, S.D. (2008) Cyclooxygenase-2 inhibitors, nonsteroidal anti-inflammatory drugs, and cardiovascular risk. *Cardiology Clinics*, *26* (4), 589–601.

Walsh, D.M., Howe, T.E., Johnson, M.I. and Sluka, K.A. (2009) Transcutaneous electrical nerve stimulation for acute pain. *Cochrane Database of Systematic Reviews*, *15* (2), p.CD006142.

Whitefield, M., O'Kane, C.J.A. and Anderson, S. (2002) Comparative efficacy of a proprietary topical ibuprofen gel and oral ibuprofen in acute soft tissue injuries: a randomized, double-blind study. *Journal of Clinical Pharmacy and Therapeutics*, *27* (6), 409–417.

Wilkerson, G.B. (1991) Treatment of the inversion ankle sprain through synchronous application of focal compression and cold. *Journal of Athletic Training*, *26*, 220–237.

Wilkerson, G.B. and Horn-Kingery, H.M. (1993) Treatment of the inversion ankle sprain: comparison of different modes of compression and cryotherapy. *Journal of Orthopaedic and Sports Physical Therapy*, *17* (5), 240–246.

Withnall, C., Shewchenko, N., Wonnacott, M. and Dvorak, J. (2005) Effectiveness of headgear in football. *British Journal of Sports Medicine*, *39* (Suppl 1), i40–i48.

Woo, W.W.K., Man, S.-Y., Lam, P.K.W. and Rainer, T.H. (2005) Randomized double-blind trial comparing oral paracetamol and oral nonsteroidal antiinflammatory drugs for treating pain after musculoskeletal injury. *Annals of Emergency Medicine*, *46* (4), 352–361.

Yang, C.-C., Hua, M.-S., Tu, Y.-K. and Huang, S.-J. (2009) Early clinical characteristics of patients with persistent post-concussion symptoms: A prospective study. *Brain Injury*, *23* (4), 299–306.

Yeung, M.S., Chan, K.M., So, C.H. and Yuan, W.Y. (1994) An epidemiological survey on ankle sprain. *British Journal of Sports Medicine*, *28* (2), 112–116.

11

Musculoskeletal assessment

Julian Hatcher
University of Salford, Greater Manchester

Clinical assessment procedures

A major skill that a competent Sports Rehabilitator must have is the ability to assess conditions that are presented. Needless to say, these come in various clinical presentations. In addition to this variety of signs and symptoms, there are numerous ways of assessing these conditions, and the various tests and questions to be asked are many. Life can, however, be made a little simpler by using the logical systems first advocated by the late Dr James Cyriax (Cyriax 1985a). For many therapists, particularly physiotherapists, training in assessment has been based on conceptual foundations of renowned therapists such as Geoffrey Maitland (Maitland et al. 2005). Although there are no problems with assessment methods based around Maitland's philosophy, there are some fundamental differences behind his philosophy and that of Cyriax. These differences are based around the fact that Cyriax was a physician; a doctor who's job it was to make a diagnosis in order to prescribe some treatment to the patient. This was his role during the 1950s and 1960s. This is very different from Maitland, who was an Australian physiotherapist, had no clinical autonomy back in the 1970s the like of which therapists enjoy today; in fact he would provide treatments based on a doctor's diagnosis. Maitland did, however, accept the fact that sometimes the diagnosis was not always accurate, and his methods of assessment would at the very least attempt to identify whether or not the presenting condition was of musculoskeletal origin, rather than some kind of systemic origin. He could treat the former cases, but would refer the latter cases back to the referring medic. Since both of these major assessment philosophies were introduced, medical and therapeutic knowledge has advanced and both methods of assessment have adapted accordingly. It is because of the strong emphasis in the original conceptual foundations that this chapter will concentrate on the fundamentals of assessment based on Cyriax philosophy. It must be stressed, however, that this system of assessment has been adapted from his methods, by both the author, and by orthopaedic medicine clinical groups:

- Orthopaedic Medicine Seminars

- Society of Orthopaedic Medicine

- Association of Chartered Physiotherapist in Orthopaedic Medicine.

It is by no means meant to be a definitive method of assessment; however, it will help the manual therapist to reach a diagnosis in over 90% of cases. It must also be stated that this is only the basic assessment; there are many additional tests that can be added to end of each of these assessments.

This chapter has three main aims. Firstly, it aims to give sports rehabilitators the conceptual foundations of musculoskeletal assessment skills, using a logical system of questioning and physical testing, in order to diagnose musculoskeletal disorders. This will enable rehabilitators to assess and diagnose many of the common disorders that affect physically active people; more detailed assessment techniques and tests relevant to specific regions of the body will be presented in the relevant chapters of this book. Secondly, this chapter aims to give a simple but effective guideline for pitch-side assessment that can be utilised in most immediate clinical sporting scenarios. Thirdly, this chapter aims to provide a framework for recording assessment and treatment details, using a commonly used system for recording medical notes of this kind. The information within this chapter is designed to be an accompaniment to learning, and is no substitute for the practical learning and role play that has been shown to significantly improve the diagnostic skills of practitioners (Smith et al. 2005).

Fundamentals of assessment

Before progressing, let us look at some fundamentals of assessment. Firstly, how useful are X-rays (radiographs) when it comes to assessment and diagnosis? From a doctor's perspective (certainly back in the 1950s), this was a major method of assessing the musculoskeletal system. Cyriax realised in his early professional career that there were many conditions where pain was present, yet the radiographic findings were unequivocal.

The problem with X-rays is that they only show dense body tissues such as the bones (Figure 11.1). All soft moving tissues are radio-translucent. If pain arises from a soft tissue, then X-rays reveal only one of two things:

1. NAD – bones appear normal; that is, negative X-ray, or

2. X-ray shows some symptomless abnormality, for example cervical spondylosis

In addition to this, it often requires two radiographs in order to interpret meaningful information as an X-Ray is a two-dimensional image of a three-dimensional structure. Figures 11.2 and 11.3 show a loose body within the knee joint, which

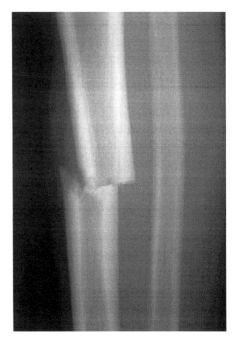

Figure 11.1 X-Rays only show dense tissues like bone.

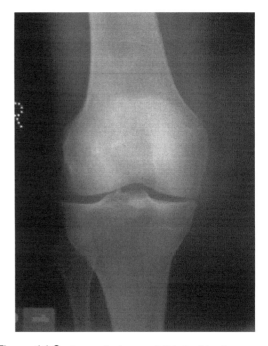

Figure 11.2 Loose body not visible in this view.

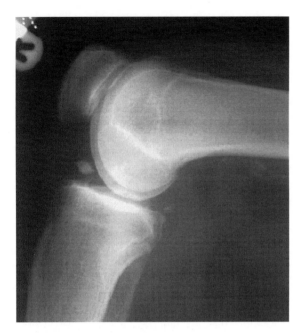

Figure 11.3 Loose body becomes visible in lateral view.

only becomes truly apparent in the lateral view (Figure 11.3).

Secondly, how useful is palpation in assessment and diagnosis? There is always a great temptation in assessing painful problems to go ahead and put a finger to the sore area and palpate. The problem here is that not all pain is felt at the site of origin; hence palpation can very often deceive. Soft tissues have the habit of referring pain to other areas; often they refer pain on a segmental basis. Referred pain is not too confusing if following the simple "rules" that appear towards the end of this chapter.

Following on from that, it should be obvious that any treatment needs to be directed at the source of, and not necessarily the site of, symptoms. There are, however, certain circumstances when the latter may be appropriate, but it cannot be assumed to be enough. For example, a client suffering with referred pain from tennis elbow (lateral epicondylitis) may require treatment to the source of problem, such as the bone-tendon junction, but may also benefit from pain relieving modalities such as TENS (transcutaneous electrical nerve stimulation). In this case, the treatment electrodes may be placed around the area of referred pain, not just the causal site. Treatment should have a beneficial effect on the particular tissues.

Again this sounds obvious, but too often therapists have been guilty of performing treatments without really taking this into full consideration. The aims of treatment, therefore, should be to influence the cause of symptoms, not just relieve the symptoms.

Primary decisions for assessment

In the quest of diagnosis, there are some primary decisions to be made even before the assessment can begin:

- About which joint does the lesion lie?

- In what sort of tissue does the lesion lie? Contractile/inert

- Is the pain reproduced by the test?

The first question should really be answered by the initial subjective history. If the right sorts of questions are asked, then an initial impression may become evident, at least to the extent where the rehabilitator can decide at least which joint assessment to perform. The next two questions are answered by applying the Cyriax principle of *selective tissue tensioning* (Cyriax 1985a). Each tissue is selectively stressed, whilst at the same time not allowing any tension to occur at other tissues. Tissues are conveniently divided into two tissue types:

1. contractile: muscles, tendons and all corresponding junctions (musculo-tendinous and teno-osseous junctions)

2. inert: bones, cartilage, ligaments, capsule, bursae, fasciae, nerve root and dura mater.

One of the inert structures above also displays an additional characteristic that is significant in diagnosis, namely the joint's fibrous capsule. Cyriax observed that joints lost their range of motion in predictable ways, which he termed Capsular Patterns:

> When the capsule of a joint becomes inflamed, whether by trauma, infection or degeneration, it contracts and restricts the available range of movement in a set pattern. This pattern is the same for that joint but may be different for different joints, for example:
> *Shoulders display the same Capsular Pattern as each other, yet this differs from all knees.*

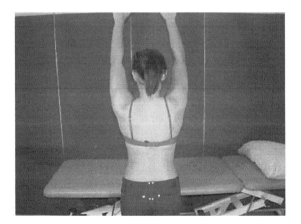

Figure 11.4 Active testing of shoulder flexion.

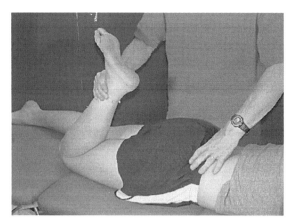

Figure 11.5 Passive testing of knee flexion.

A loss of range that is not in common with the known Capsular Pattern is called a Non-capsular Pattern.

Active, passive and resisted movements

Active movements are often not very helpful in diagnosis as all tissues are under tension simultaneously. However, they can give an indication of *willingness* to move, in addition to onset of pain, available range and end-feel to joint motion. Figure 11.4 shows active testing of shoulder flexion.

Passive movements stress the inert structures mainly, and provide an indication of onset of *pain*, *range* and *end-feel*. Passive movements should be exactly that; movements where the client is relaxed and does not attempt to help or join in with the movement. If this happens, then the client would be recruiting the contractile tissues, which may give a false positive result. Figure 11.5 shows passive testing of knee flexion.

Resisted movements put the contractile components under tension and give an idea of *pain* and *strength*. Resisted tests must be performed to the maximum, and should be isometric contractions in order to ensure that the inert structures are not stressed at the same time. In order to get maximum contraction, the joint needs to be in a relatively neutral position (to allow inert structures to be in a non-stretched position). This should be mid-range in order to allow the angle of force from the contractile unit to act in the most mechanically advantageous position, plus allows the optimum overlap between actin and myosin filaments for strong contraction. If only 90% strength is used, and the muscle lesion lies within the untested 10% of fibres, then a false negative result would be obtained by the test. Figure 11.6 shows resisted testing of the quadriceps.

Essentially, it should be remembered that all tests, whether passive or resisted, are only as sensitive and specific as the rehabilitator is at performing them; there is no substitute for practising assessment techniques.

Finally, it may be appropriate to use *palpation* in order to further localise the lesion. Further additional tests may then be carried out to aid this process in some cases; this simple form of assessment will allow the rehabilitator to gain a good clinical

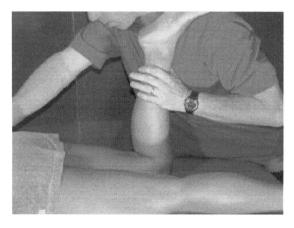

Figure 11.6 Resisted testing of knee extension.

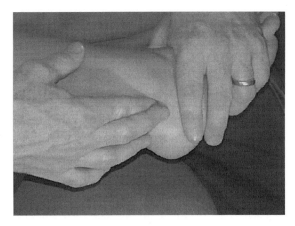

Figure 11.7 Palpation of the peroneus brevis tendon.

impression of diagnosis in 90% of cases. Figure 11.7 shows palpation of the peroneus brevis tendon.

End-feels

The normal feel to the end of passive joint range of motion is often called the end-feel. The following are classed as normal end-feels:

- *Hard* – usually due to bony opposition and feels like a solid immoveable block to movement (e.g. olecranon process meeting the olecranon fossa during full elbow extension).

- *Soft* – usually due to soft-tissue opposition, which feels like a squashing of a sponge (e.g. hamstrings meeting calf musculature during flexion of the knee).

- *Elastic* – usually the most common end-feel for synovial joints and is associated with stretching of the joint capsule and the feeling of elastic recoil when releasing the tension (e.g. lateral rotation of the shoulder joint).

Abnormal end-feels, or pathological end-feels as they indicate the presence of pathology, are as follows:

- *Springy* – usually due to the presence of cartilage being trapped between joint surfaces and feels like a firm resistance to movement that has some give when pressure is increased (e.g. meniscus in knee flexion or extension).

- *Spasm* – this is a hard end-feel caused by sudden activation of muscles in response to the movement through pain, or apprehension. It is different from a normal hard end-feel in that it is not always repeatable at the same point in the range of motion.

- *Empty* – this is when no end-range is actually reached either due to the onset of extreme pain, or because of major joint disruption (i.e. torn ligaments/capsule) which therefore offer no resistance to movement.

Referred pain

Referred pain is described as an error in perception; pain perceived elsewhere than at its true site is termed "referred". The sensation of pain is an extremely complex phenomenon, and very subjective, however, certain aspects of pain perception are known (Butler 1991, 2000).

- site of pain – this is sensed by the body in the sensory cortex of the brain

- memory of pain – this is sensed in the temporal lobes of the brain

- degree of pain – this is sensed in the frontal lobes (amount of tension in these frontal lobes may govern the patient's response to pain).

Referred pain, although complex in nature, can be said to follow some general rules (Cyriax 1985a):

- does not cross the mid-line of the body

- has a tendency to refer distally

- always refers segmentally (within a dermatome)

- may be felt in all or part of a dermatome

- is often felt or perceived as being deep.

The use of simple rules can help to make some sense of referred pain:

- usually, the deeper the site of lesion, the more vague the reference of pain and the greater the spread of reference

- this is also true of the location of the lesion; that is the more proximal, the more vague and increase in spread of reference

- in most cases, the stronger the stimulus, the increase in spread of reference.

Clinical orthopaedic examination

The following subheadings make the whole clinical assessment easier to breakdown for the purpose of greater understanding. Imagine the situation where an unknown client enters your clinic, requiring your professional services. After introducing yourself, you need to sit down with the client and glean a subjective history from them. The subjective history is literally exploring what the client feels has happened to themselves and how it is affecting them now. The objective examination, which comes after, allows the rehabilitator to perform a series of objective tests – namely passive and resisted movements. The process of assessment is attempting to put together several pieces of a puzzle, and to stand back and look at these and attempt to recognise the puzzle; it is rare that assessment yields every piece, so some sound anatomical knowledge and clinical reasoning skills will be needed for more complex patterns of clinical findings, and it is this particular skill that is hardest to gain. The rehabilitator looks for specific key elements (e.g. onset, behaviour and symptoms, etc.) in much the same way that a detective looks for key elements in any investigation (e.g. motive, opportunity, forensic evidence, etc.). Diagnosis is not, therefore, made on the basis of a single positive finding, but on a pattern of clinical features.

Subjective examination

The first stage of the subjective examination is that of observation.

Observation

- Face, posture and gait

It is particularly important to gauge how much pain and discomfort is the client in; one of the most accurate ways of doing this is to observe their face when they walk in the clinic, and when you greet them and introduce yourself to them. At this point, they are unaware that you may be observing them, and will tend to act in a more natural manner. The face can give clear indications of pain, whether severe, or prolonged, and may also give an idea of whether they have had disturbed sleep. It is a good opportunity to observe their posture also, and the way they move and walk. Someone with an ankle problem may walk on their heel with a relatively fixed and extended knee, whilst someone with a knee problem may well walk on their toes in order to prevent full knee extension.

The next stage is the interview stage.

Subjective history

During the subjective history taking, several questions need to be asked in order to help with the diagnostic process. The order of questions is ultimately a personal thing and does not have to follow the format set out below. It is worth noting, however, that it is highly likely that as a rehabilitator you are likely to not only ask someone to get relatively undressed, but you are going to invade their "personal space" by actually physically handling their limbs or body in some manner. In which case, it seems to me to be really important to gain some rapport with the client as quickly as possible, and hence ask questions about themselves of a more general (but no less important) nature early, and follow up with the most personal questions towards the end of the subjective interview. It may also be helpful as an aide memoir to use the keywords listed below as a template around which to ask relevant questions. These are linked together as pairs:

- age and occupation

- site and spread

- onset and duration

- behaviour and symptoms
- previous medical history and medication.

Age

Essentially, one of the first things required to establish is rapport, so to do that, you need to ask someone's age. One of the useful things here is that some musculoskeletal conditions are age related, for example, slipped epiphysis in young adolescent boys, or degenerative conditions in the over 50s.

Occupation

Paired with age is occupation; it is important to establish exactly what the client does for a living. Are they a professional athlete of some kind, or are they a recreational athlete? Either way, you need to establish the activity levels they need to regain as part of your treatment plan. A professional athlete is likely to require more rehabilitation than an office worker who plays a little tennis at weekends. Equally, it needs to be established whether work or recreational activity may have any impact on the current injury – either causative or preventing recovery in some way.

Site

Asking the client to indicate where they feel their problem is can be enlightening in that it may indicate whether the problem is a local one, or indeed referred from elsewhere in the body. If one understands the rules of referred pain, then it can be seen that a client indicating medial knee pain with a vague wave of the hand around the area, is suffering from a different complaint to another client who indicated their medial knee pain with the pointing of one finger. In the latter case, it is likely that local pathology is allowing the client to localise the sensation somewhat more accurately than in the former case, where pain is referred to the knee from some pathology of the hip joint.

Spread

Linked to site of pain or symptoms, is that of spread. Ask the client: does their pain stay localised to the same area, or does it move in anyway? Again, knowledge of referred pain may further add to the evidence that the symptoms being experienced are from a referred source more proximal to the site of perceived pain.

Onset

It is always useful to attempt to find out what factors may have led to the onset of the client's problem. Generally there are three forms of onset: gradual, sudden and insidious. The latter is a term that refers to an onset of unknown origin and can apply itself across both of the former two onsets (i.e. insidious gradual, or insidious sudden). Sudden onsets are usually associated with trauma, and as such a "mechanism of injury" may be established. Using sound anatomical and biomechanical knowledge, it may help to identify the tissues likely to be at fault causing symptoms. Gradual onsets may be the result of overuse, or repeated trauma (e.g. tendonitis). Examples of insidious insets may be pathologically more severe or even sinister; insidious gradual may be the onset of a tumour, insidious sudden may reflect a systemic problem such as gout.

Duration

This is an interesting one in that really it provides an idea of how likely you are to be able to change the client's symptoms. Generally speaking, the longer someone has a condition, the less likely you are to be able to change or cure it. That does not mean to say that one should not attempt to change it; you may be the first therapist to be approached about the condition, or the first to accurately diagnose the problem.

Behaviour

What you are looking for here is: how do the symptoms change during the day, are they constant, and what aggravates and indeed what eases them? Is there any diurnal variation? That is, what are the symptoms like first thing in the morning, later in the day, at night? Conditions affecting the capsule and ligaments tend not to like extensive static periods and feel "stiff" when first moving, though feel better once moved. Tendonitis tends to be worse either during or after activity, whereas muscle lesions hurt on movements that recruit that particular muscle. It is rare for symptoms such as pain from musculoskeletal origin to be constantly present (this is often the manifestation of internal organ problems or systemic disease). Often, musculoskeletal pain is intermittent and movement dependant, and identifying what movements cause the onset of symptoms, and what positions, etc. cause relief, can help identify the tissue at fault. It may also help with provision of treatment in that someone who identifies that movement and application of warmth helps their knee, it would seem unwise not to incorporate some kind of heat treatment and mobilisation or exercise into their treatment/rehabilitation programme.

Symptoms

The most obvious symptom here is pain; however, some clients complain of other sensations, such as locking, giving way, pins and needles (paraesthesia) and even numbness (anaesthesia). They may also complain of stiffness (as a symptom, as opposed to lack of range of motion – which you would measure as objective sign). There are many authors who have written about pain and the types of pain that people experience, and relate this to the various types of tissue that give pain (Butler, 1991, 2000; Travell and Simons 1992; Simons et al. 1999). The fact that the pathophysiological mechanisms behind pain and its perception is so complex is further complicated when psychological influences are included (Butler 2000). This makes clinical reasoning of pain to become rather subjective and may lead to many opportunities for misunderstanding. For example, clients have described the pain of sciatica as "being like a red-hot poker down the back of my leg", or sharp knee pain, "feeling like I have been stabbed". As both the client and the rehabilitator need to fully understand what is meant by such descriptions, and either or both may misinterpret what is really being experienced, diagnosis based on such reasoning may be flawed. It would seem to suffice, therefore, when a client claims they have pain; objective and subjective examination will indicate the rest!

Previous medical history

Many therapists like to ask questions around this subject earlier in the interview; however this may result in digging up matter that is really quite personal, so leaving it to the end allows you first to establish rapport with your client. Again, there are many things to consider here; ideally keep it reasonably simple. Several forms of treatment are contraindicated by the presence of certain conditions, and there are too many to consider at this stage of assessment (as you have no idea what treatments you wish to perform at this stage). Therefore, ask the client whether they have had any previous major operations, accidents or illnesses. The common response to this is a glib, "No". It is useful to check and ask, "Are you sure?" This has the purpose of just getting the client to realise that this is an important question that requires careful consideration. This may well yield the same response; however, you are more likely to be able to trust this response in terms of accuracy. A good example of the reason for asking such questions is this: a female client complains of low back pain of insidious gradual nature, and in her past history reveals she has had a hysterectomy. What needs to be established is why did she have a hysterectomy (was it due to cervical cancer – which can lead to spinal secondary metastases)? This can be ascertained by asking whether the client is still under the medical team for this, or has she been discharged (i.e. clear of the condition). This may help to change the potential worrying diagnoses of spinal cancer to one of remote possibility (note it does not dismiss the possibility).

Medication

Linked closely to past medical history is the question of medication. It may be useful to see if the client is taking any medication for the current condition, and whether or not it is beneficial. However, it is also another way of establishing whether there are any other underlying illnesses or diseases not disclosed in the previous questions. As there are a plethora of possible medications that clients use, we are particularly on the lookout for anticoagulants and long-term corticosteroid use. This is because in the case of the latter, ligaments and joints may become damaged by such medication; in the case of the former medication, potentially violent or high-impact treatment/rehabilitation processes may potentially cause bleeding that is not easy to stop (Grieve 1991).

Objective examination

The next stage is the beginning of the objective examination and should begin with a general inspection of the affected area.

Inspection

This is an opportunity to inspect the site of lesion briefly looking for evidence of the following:

- bony deformity
- colour changes
- wasting
- swelling.

Bony deformity could indicate joint subluxation, dislocation, fracture or postural changes, whilst colour changes may indicate presence of bleeding or signs of inflammation. Musculature associated with the site of lesion may indicate the presence of long-standing problems, or even neurological problems. Swelling would indicate the presence of an inflammatory process. Figure 11.8 shows a discrepancy in the relative scapula position from left to right.

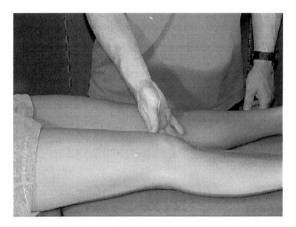

Figure 11.9 Palpation for heat.

This is followed up by an initial brief palpation of the area (see Figure 11.9) specifically looking for:

- heat – indicating presence of inflammation
- swelling – indicating presence of inflammation
- synovial thickening – indicating presence of long standing inflammation
- pulses – indicating blood flow to through the area.

It is only once this initial observation phase is complete that the actual objective testing and measurement should begin. It is always necessary to establish whether the client feels any symptoms whilst at rest, before beginning any testing.

Objective testing and measurement

- active movements (if appropriate) for willingness to move
- passive movements for pain, range and end-feel. (passive stretching/squeezing may occur to inert tissues)
- resisted movements for pain and power (isometric contraction of contractile tissue components without passive stretching).

The number and variety of these tests will be specific to particular joint regions of the body and will be

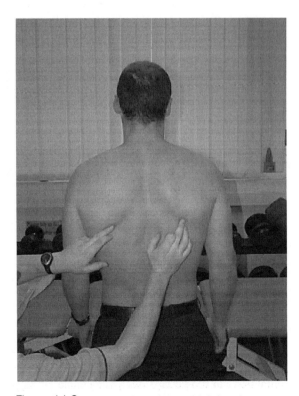

Figure 11.8 Observation of postural deformity.

dealt with in subsequent specific chapters relevant to that region.

It may be appropriate to include some *neurological tests* such as testing of reflexes, and strength tests for myotomes, and sensation loss for dermatomes, particularly in spinal assessments where nerve roots may be involved in the lesion.

It is only after completion of all these tests would *palpation* be performed to localise the lesion further and, in some more difficult conditions, *further additional tests* may be performed, such as X-rays, scans, blood tests, etc.

Emergency pitchside assessment

This is all very well for the clinic scenario when the "walking wounded" literally walk in to the clinic; but what happens on the pitch side, poolside or trackside?

Generally speaking, there are specific programmes of study that must be completed before entering this situation for many professional sports in the UK (Soccer, Rugby Union and Rugby League), and as such there are specific elements of Advance Life Saving that have to be adhered to. Aside from those sports, a common acronym that can be applied is that of SALTAPS:

S: STOP play. If the player has gone down injured get there as quickly as you can. (Check the rules regarding access with the referee beforehand).
A: ASK the player (unless unconscious) what happened. Remember, as detailed a history as possible will be important. Be aware of facial expression, posture adopted, etc. as this may indicate the degree of pain (and hence the severity of injury). Is the player talking sensibly? If the player is unresponsive, further first aid should be administered accordingly.
L: LOOK at the specific limb. Be vigilant for obvious signs, for example bleeding, discolouration/bruising, immediate swelling, bone/joint deformity, muscle spasm.
T: TOUCH the injury site only if the athlete will let you. (Again this can indicate just how serious the injury may be.) Palpate gently to find the site of pain – note the athlete's response. Do not be sadistic in your handling.
A: ACTIVE movement: can the athlete move the limb painlessly through the full range of movement?
P: PASSIVE movement: Only if 'A' above applies attempt to move the joint to the end of its range and note the response. (Techniques for these movements will be dealt with in subsequent chapters.)
S: STAND-UP and play-on: Can the player resume immediately or are they trying to 'run-it-off'. In both instances keep a close eye on them to make sure they recover fully and quickly. Be prepared to replace them.

Obviously, this approach does not give an accurate diagnosis, but allows a quick and easy method for assessing the situation in front of you, and later you can apply the more in-depth approach to diagnosing any remaining problem in more controlled conditions.

Assessment and evaluation notation

When any kind of assessment is performed, accurate notes must be recorded to comply with regulations within the country of work. For example, in the UK, rehabilitators must ensure that records are retained or destroyed in accordance with Department of Health Records Management (NHS Code of Practice – part 2). In other parts of the world this may be different (and certainly may be different within different States within the US). By UK Law, all records should be retained for eight years following conclusion of treatment. The only exceptions to this are for children whose notes must be retained until the client reaches their 25th birthday. That said, there are different ways of recording notes and there are no such rules as to how this should be done; however, notes

CASE REPORT

SUBJECTIVE ASSESSMENT

OBSERVATION
- Tired and exasperated in appearance.

HISTORY
Age: 56 years old Male
Occupation: Warehouse Worker (no absence of leave taken with this present condition). Sedentary leisure interests only, such as watching football.
Site: Superior aspect of left shoulder
Spread: Left deltoid region and lateral aspect of arm radiating to the elbow
Onset: Insidious – unable to relate onset to any particular incident.
Duration: 6 months (Nov '06)
Behaviour: *Constant* ache, although a little less aware when distracted by manageable activities
Intermittent stabbing pains lasting approximately 4-5 minutes.
Aggravated by putting coat on (arm in sleeve), unable to reach above shoulder height.
Eased only mildly with paracetamol and Ibuprofen. Nothing else helps.
Awakes at least three times a night due to pain.
Unable to lie on left shoulder
Diurnal pattern – worse on waking and at rest.
Symptoms: No cervical, Vestibular or neurological deficit.
Medical: *Past Medical History* - Diabetes Mellitus: diagnosed 1990 (currently participating in a trial), Nephrectomy 1980, Hypertension, Currently undergoing investigations for osteoporosis – unaware of results.
Drug History - Metformin, Simvastatin, Paracetamol, Ibuprofen, Aspirin and Xanitide (as part of the Diabetes trial).
Previous Treatment – Two steroid injections administered by GP (1st = Jan 07, 2nd = March 07) No effect. Patient described injection site as being on the superior aspect of his shoulder on both occasions. No other treatment received.

OBJECTIVE EXAMINATION

INSPECTION
No deformity, no swelling, no colour change.
Muscle wasting in upper fibres of trapezius and over deltoid region, left.

EXAMINATION
Nothing abnormal detected on palpation.
At rest, experiences an ache over superior aspect of left shoulder and into deltoid region.

Cervical spine: *Active movements*
- Stretch sensation at the end range of each movement – rotation / side flexion/ flexion/ extension/ retraction/ protraction- cervical spine cleared.

Elbow joint: *Passive joint tests*
- Full and pain free – joint cleared.

Shoulder joint: *Elevation tests*
- Active elevation – 110 degrees, limited by pain and stiffness.
- Passive elevation – minimal increase in range achieved, hard end feel with pain reported.
- Painful arc – Not present, 45 degrees active abduction limited by pain and stiffness.

Figure 11.10 An example of a client's notes.

Shoulder joint: *Passive joint tests*
- Passive lateral rotation – 15 degrees, hard end feel with pain reported early range.
- Passive abduction with fixation of scapula – early movement of scapula at ~40 degrees, hard end feel at 50 degrees with pain reported.
- Passive Medial rotation – Hand to lateral aspect of buttock, hard end feel with pain reported.

Shoulder joint: *Resisted tests*
- Rotator cuff, biceps and triceps generalised weakness, grade 4 (Oxford scale) but no increase in pain.

ADDITIONAL TESTS
Not applicable

ASSESSMENT & EVALUATION

DIAGNOSIS
Adhesive Capsulitis, left Shoulder - Secondary to Diabetes. (Stage 2 of the condition).

AIMS
- Reduce Pain
- Increase available Range of Motion (ROM)
- Increase functional strength within Shoulder muscles – particularly rotator cuff muscles

PLAN

TREATMENT
- Advised patient to see G.P. for a review of his pain control
- Grade A longitudinal distraction, distraction in flexion and lateral distraction mobilisations in supine.
- Grade A passive stretches into elevation.
- 'Hold-relax-techniques' for medial and lateral rotation.
- Treatment session to end with interferential therapy.
- The patient was seen twice weekly for physiotherapy in which time he received
- Home exercise programme comprising of capsular stretches, repeated every hour, including setting up own pulley system at home x 3 per day
- If no better within three weeks, plan letter to his G.P. recommending a referral to an Orthopaedic Consultant for possible injection therapy.

PLAN

- Continue with therapy as above with progression to Grade B mobilisations and passive stretches, as pain allows.
- Continue patient's home exercise programme as above.
- Discharge planned for when function has improved to allow overhead should movements with minimal pain or stiffness.

Figure 11.10 (*Continued*)

must be legible and understandable to you and others long after they have been written. One commonly used format for notes is the "SOAP" format, where this is divided into four sections:

- subjective history
- objective examination
- assessment and analysis
- plan

The subjective section records the information regarding age and occupation, site and spread, onset and duration, behaviour and symptoms, and past medical history, whilst the objective section records evidence of deformity, colour changes, wasting, swelling, heat and findings from objective passive and resistive tests. Recordings of the outcome of any special tests used would also be recorded here. It is always worth reporting negative outcomes to test, to show that they have been performed; however, highlight in someway, all the highly significant findings from the subjective and objective examination, as this allows one to quickly identify the salient points when reviewing the same client in the future.

The assessment section would allow the rehabilitator to consider the diagnosis (or diagnoses in the case of multiple problems), and attempt to identify the status of the client and the probable prognosis. Indications of the severity and irritability would be useful here too. It may also be appropriate to consider short-term and long-term aims of treatment. For example, an athlete with an acutely sprained ankle may require taping or strapping and swelling control as short-term goals, whilst requiring gait re-education, proprioception and sport-specific training as long-term goals. By creating a clear intention of how and why to treat, allows the rehabilitator quickly to review their aims of treatment. Together with the previously highlighted findings from the subjective and objective assessment, this allows the rehabilitator to establish a treatment plan that is logical and effective. This plan should also consider how often the client should receive treatment, the frequency and intensity of any exercise or training programme given for home (self-administered) use, and would also be useful to record any criteria for discharge.

An example of notes recorded from a particular client is given in Figure 11.10. Additional information would normally need to be included here (e.g. name, date of birth, name and contact details of GP), however these have been omitted here for the sake of confidentiality. It may be pertinent at this point to remind rehabilitators that is unethical and a breach of professional conduct to disclose information regarding a client to anyone other than another medical or paramedical practitioner, and as such, all records of assessments and treatments should be securely filed away in order to ensure anonymity and confidentiality at all times.

References

Anderson MK, Parr GP and Hall SJ, (2009) Fundamentals of Sports Injury Management. Lippincott, Williams and Wilkins

Boyling J and Jull G (2005) Grieve's Modern Manual Therapy: the Vertebral Column, 3rd Ed. Churchill Livingstone. London

Brotzman SB, (1996) Clinical Orthopaedic Rehabilitation. Mosby

Butler, D.S. (1991) *Mobilisation of the Nervous System*. Melbourne: Churchill Livingstone

Butler, D.S. (2000) *The Sensitive Nervous System*. Adelaide: Noigroup Publications.

Cyriax, J. (1985a) *Textbook of Orthopaedic Medicine, vol. 1. Diagnosis of Soft Tissue Lesions*. London: Balliere Tindall.

Cyriax, J. (1985b) *Textbook of Orthopaedic Medicine, vol. 2. Treatment by Manipulation Massage and Injection*. London: Balliere Tindall.

Maitland, G.D., Hengeveld, E., Banks, K. and English, K. (2005) *Vertebral Manipulation*, 7th edn. London: Butterworth-Heinemann.

Simons, D.G., Travell, J.G. and Simons, L.S. (1999) *Myofascial Pain and Dysfunction, The Trigger Point Manual*, vol 1. Baltimore: Wolters Kluwer Health.

Smith, C.C., Newman, L., Davis, R., Yang, J. and Ramanan, R. (2005) A comprehensive new curriculum to teach and assess resident knowledge and diagnostic evaluation of musculoskeletal complaints. *Medical Teacher* 27(6), 553–558.

Travell, J.G. and Simons, D.G. (1992) *Myofascial Pain and Dysfunction, The Trigger Point Manual*, vol 2. Baltimore: Wolters Kluwer Health.

12

Progressive systematic functional rehabilitation

Earle Abrahamson, Victoria Hyland, Sebastian Hicks, and Christo Koukoullis

London Sport Institute, Middlesex University

Background

Rehabilitation is a complex process demanding the attention and knowledge of a range of issues and applications. Successful rehabilitation of sports injuries is dependent on a progressive plan that addresses and accommodates the injury/pathological issues and conditions. In so doing, the rehabilitation plan incorporates a multitude of decisions and actions, often underpinned by research evidence.

This chapter will explore these issues in relation to progressive rehabilitation by examining the concepts and principles of sports rehabilitation. The main issue and consideration specific to the development of a progressive rehabilitation plan is the careful and logical inclusion of objective criteria. These criteria allow the practitioner to consider how and when, to either progress or regress the rehabilitation in relation to the athlete's needs and goals. Figure 12.1 details the components for progressive rehabilitation planning by considering questions necessary for the successful development of the rehabilitation programme and the logical sequencing of the progressive rehabilitation exercise regime.

Understanding progressive rehabilitation, assessment and observation

On initial observation of an injury it is essential to perform an injury screening (refer to Chapter 2 for a detailed overview of screening procedures). SINS (severity, irritability, nature and stage) and SOAP notes (subjective, objective, assessment and plan) are two acronyms that provide an effective method of gathering information relating to possible trauma (Sleszynski et al. 1999). Table 12.1 describes the notation and how these acronyms are used to organisation injury information and assessment.

Components of rehabilitation including flexibility, the restoration of muscular strength and endurance and enhancing proprioceptive control are concepts that must be managed effectively to ensure optimal injury management (Beam 2002), within the context of physical activity and return to sport. Systematic, functional and progressive rehabilitation involves the process of carefully considering the key components of musculoskeletal trauma. This involves adequate management of the injury in order to prevent further soft tissue trauma and enable a progressive treatment protocol to be implemented effectively. This is

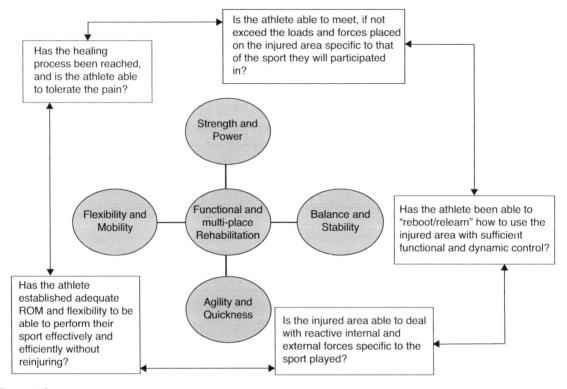

Figure 12.1 Systematic overview of the rehabilitation process.

of particular importance, as previous history of injury can lead directly to an increased risk of trauma following the implementation of a rehabilitation programme (O'Sullivan et al. 2009).

Goal planning is key to the successful implementation of a progressive rehabilitation programme. For example, if an athlete has excessive valgus movement of the knee during a single leg functional activity, it is the role of the rehabilitator to provide effective stabilisation and strengthening exercises for the gluteal muscles in order to provide the lumbo-pelvic hip complex with the stability it requires to maximise total power output throughout the entire kinetic chain (Liebenson 2006; Boudreau et al. 2009). However, during the rehabilitative process it is imperative to ensure strengthening is not focused solely on a specific muscle, this is because during closed kinetic chain exercises, which simulate functional

Table 12.1 Overview of SINS and SOAP notation, adapted from (Borcherding and Morreale 2007)

SINS	Example	SOAP	Example
Severity	The extent to which physical activity is impaired	Subjective	Patients analysis of history/injury concern
Irritability	When do the symptoms arise/subside?	Objective	Range of movement (ROM), MMT, special tests, neural assessment
Nature	Mechanism of injury and factors influencing injury rehabilitation	Assessment	Diagnosis of injury
Stage	Acute, sub-acute, chronic	Plan	Injury management, follow-up rehabilitation sessions

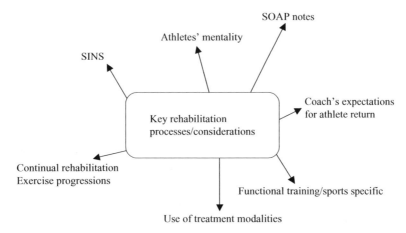

Figure 12.2 Rehabilitation progressions including sport specific progression. (Adapted from: Beam (2002) Rehabilitation including sport specific functional progression for the competitive athlete. *Journal of Bodywork and Movement Therapies* 6 (4), 205–219).

movement, it is inaccurate to make judgement stating a specific muscle is responsible for poor neuromuscular control. Previous research linked to isokinetic muscle testing, which isolates a target muscle, has stated hip musculature is responsible for optimal functional single leg activity; however, fatigue does not occur in a single muscle group when performing closed kinetic chain exercises, which simulate movement patterns more likely to be carried out during everyday actives and athletic performance (Reimer and Wikstrom 2009). Similarly, deficits in muscular strength have been shown to increase the risk of injury. Hollman et al. (2006) suggest that reduced isometric strength of the hip abductors in relation to the hip adductors is associated with malalignment of the lower extremity such as foot pronation and knee valgus. This may result in inadequate neuromuscular control during functional movement increasing the risk of anterior cruciate ligament injury (ACL) (Hewett et al. 2006). Therefore, when prescribing exercises during the rehabilitation of an athlete, it is important to take into account not only strengthening of the effected region of the body, but also to continue strengthening the entire extremity, which can prevent muscle imbalances and potential injury concerns.

It is the synthesis of knowledge coupled with clinical practice that best defines how and when to either progress or regress exercise patterns and routines.

It would be useful to read the chapters on Clinical Reasoning (Chapter 16) and An Introduction to Periodisation (Chapter 9) along with this chapter to better aid understanding of exercise design and decision-making processes.

Consideration and application of the concept of continual progression is useful as it provides the clinician with the ability to justify why specific exercises have been prescribed and how best to regress or progress the exercises depending on the athlete's development through the rehabilitation process. During the final stages of a rehabilitation programme it is important that the athlete performs exercises within their functional range, this involves performing exercises that initiate abnormal motor control and allow minimal mechanical sensitivity (Liebenson 2006). By ensuring exercises are performed within a functional range, this limits the possibility of re-injury (Stracciolini et al. 2007). In addition to introducing the athlete to functional exercise it is important to ensure that movements simulate the patterns of movement the athlete will eventually perform in their sport, this needs to include the relevant muscle actions (concentric, eccentric, isometric), velocity of movement, force generation, power output and rate of force development.

Muscle actions that relate to effective movement patterns in the athlete's sport must be implemented during final stages of rehabilitation. For example, the hamstring muscle complex is a commonly injured region of the body in relation to sports injury (Brockett et al. 2004; Hoskins and Pollard 2005; Askling et al. 2006; Verrall et al. 2006). It is important, however, not to generalise when assessing and initiating injury management for this muscle complex (see the example in Chapter 4, Pathophysiology of Skeletal Muscle Injuries). The demands of various sports often lead to trauma developing in different regions of the hamstring complex. For instance, in sprinters it has been shown that injury consistently occurred in the long head of the biceps femoris; in contrast dancers suffered common trauma at the proximal tendon of the semimembranosus. Return to pre-injury level was shown to be shorter for high-speed trauma as experienced by the sprinters, in contrast to trauma resulting from overstretching from which the dance group demonstrated (Askling et al. 2008). This evidence provides information that shows rehabilitation from hamstring trauma particularly, should consider both the type of sport in which the athlete participates and the common movement patterns in preparation for implementation of a sports-specific functional rehabilitation programme designed to prevent re-injury. This protocol should apply directly to any injury, to ensure successful management and long-term protection against chronic and ongoing injury concerns.

Inflammation and pain management

Liebenson (2006) found that the application of an early stage rehabilitation programme with restoring full pain-free functional range of motion as a main objective, is beneficial and can result in a rapid attainment of pre-set functional tasks. Table 12.2 below details how for example cryotherapy can be used in the acute management of soft tissue injuries.

MacAuley (2001) and Bleakly et al. (2006) suggest that intermittent application of ice may enhance the therapeutic effect of ice in pain relief after acute soft tissue injury. They prescribe the following protocol: "10 minutes ice/water submersion (approx 0°) on 10 minutes off (room temperature) 10 minutes on ice/water submersion (approx 0°)". The submersion may also help with increased pressure and in turn mimic compression effects. The sport rehabilitator is advised to adhere to prescribed guidelines for the use of cryotherapy within progressive injury rehabilitation. These could include the responsible use of cryotherapy, to prevent ice burns that may be caused by direct application or prolonged exposure to cryotherapy. The use of a wet cotton towel similar to that of a dish cloth may act as a suitable barrier between ice and skin (Bleakly et al. 2007). Early intervention of the use of cryotherapy has also been suggested to help speed recovery and return to sport (Hubbard and Denegar 2004), therefore highlighting the importance of early cryotherapy use within the management of an injury.

According to Hubbard and Denegar (2004), this field of applied study and practice would benefit from evidence-based research into the use and barriers of cryotherapy application. This has further been supported by Collins (2008), who claims that the evidence behind cryotherapy is insufficient to the outcomes to improve the outcomes of clinical management of soft tissue injuries.

Range of motion and flexibility

Range of motion (ROM) and flexibility are central to rehabilitation and used not only as markers of assessment, but more importantly as techniques for rehabilitation practices (Small et al. 2008). ROM testing is an essential component of athlete evaluation and provides the practitioner with the acquired information including active and passive ROM, contractile ability of the musculature and occasionally severity of injury (Stracciolini et al. 2007). This can then be used to develop a specific stretching programme with the goal to enable the athlete restore full ROM to allow progression through the appropriate rehabilitative exercises and eventually to return to full sports participation (Beam 2002).

It is important to ensure that prior to progressing rehabilitative exercises and return sport the athlete has sufficient ROM at the joint enabling full capacity of movement to perform the exercises and to perform the sport-specific movements. The athlete must have pain-free active ROM in the frontal, saggital and transverse planes when performing basic movements and have the ability to contract the affected muscle in synergy with the antagonist muscles (Liebenson 2006). Only after an athlete can carry out these movements can a sport specific, functional rehabilitation programme be implemented (Small et al. 2008).

During an initial musculoskeletal evaluation the clinician may have to take the athlete through active and passive range of motion. Active range of motion is when the athlete moves the affected joint through its range without any external influence. This is done as a measure of pain through a functional range and also provides the sports rehabilitator with the

Table 12.2 Summary and suggestions from primary literature for the use of cryotherapy, compression and elevation in the early stages of acute management of a soft tissue injury

Author	Population/clinical question	Exposure	Outcome	Conclusions	Limitation	Practical application
Airaksinen et al. 2003	74 patients with sports-related soft tissue injuries <48 h old	Prospective randomised double-blind study. Application of cold gel vs placebo gel 4 times/day for 14 days	In cold group, measurements at day 7, 14, 28 showed: reduced pain at rest (p < 0.001): reduced pain on movement (p < 0.0001): reduced function disability (p < 0.001)	Shows supporting evidence for using cooling gel	Dose not state which medical service the patients attended	Cooling gel may be used anywhere. Whereas Ice melts and is cumbersome
Sloan et al. 1989	143 patients age 16–50 attending emergency department with acute ankle sprains. 116 patients followed	Prospective randomised trail. "cooling anklet" with elevation for 30 minute in test group and dummy anklet with no elevation in control. Standard therapy for both groups: bandage, NSAIDs, rest, elevation	Cold therapy group showed trent for improvement but did not reach statistical significance. Reduced soft tissue odema (p = 0.07), improvement on injury severity score (p = 0.15), increased weight bearing of function (p = 0.64)	Did not reaching statistical significance after patients used a cooling anklet with elevation for a period of 30 minutes	Single application of ice therapy in otherwise well-designed study; however is somewhat a relatively old study	May be limited to practitioners with funding
Bleakly et al. 2004	To explore the clinical evidence base for the use of cryotherapy. Five sub-questions posed.	Human studied up to 2002. Broad inclusion criteria – trauma and postoperative RCTs. Four outcome measures; pain, swelling, ROM and function	There was a mean PEDro score of 3.4 out of 10. There was marginal evidence that ice plus exercise is most effective, after ankle sprain and postsurgery. There was little evidence to suggest that the addition of ice to compression had any significant effect, but this was restricted to treatment of hospital inpatients.	Many more high-quality studies are needed to ensure adequate evidence-based practice	Clinical question was not focused, being very broad and thorough review of literature therefore reducing the study's validity and applicability	There was little evidence to suggest that the addition of ice to compression had any significant effect

(*Continued*)

Table 12.2 (*Continued*).

Author	Population/clinical question	Exposure	Outcome	Conclusions	Limitation	Practical application
Hubbard and Denegar 2004	Dose cryotherapy hastens return to participation?	Clear and relevant. Focused clinical question	English langue journals 1976–2003. Limits included RCTs only. Four trials found	Used PEDro scoring, range 2–4/10.	Was well-designed systematic review with focus on the clinical question "what is the clinical evidence base for cryotherapy? However, major limitation is paucity of papers obtained for the review	Cryotherapy seems to be effective in decreasing pain if instituted soon after injury and may be effective in speeding return to work or sport.
Bleakly et al. 2006	To compare the efficacy of an intermittent cryotherapy treatment. 44 sports men and 45 members of the public with mild/moderate acute ankle sprains	Randomly allocated, under strict controlled double blind conditions to one of two treatment groups: standard ice application or intermittent ice application	Subjects treated with the intermittent protocol (applied ice for 10 minutes, then removed pack and the iced area was rested at room temperature for 10 minutes; ice then reapplied for a further 10 minutes) had significant less ankle pain on activity than those using a standard 20 minute protocol	Intermittent application may enhance the therapeutic effect of ice in pain relief after acute soft tissue injury. (ice/ice water (approx 0°C)	One week after ankle injury, there were no significant differences between groups in terms of function, swelling, or pain at rest	Intermittent application may be useful for client management of an injury and easy for them to use this protocol for acute pain management.
MacAuley 2001	Ice therpy: how good is the evidence?	The purpose of this systematic review is to identify the original literature on cryotherapy in acute soft tissue injury and produce evidence-based guidance on treatment.	The target temperature is reduction of 10–15°C. Using repeated, rather than continuous, ice applications helps sustain reduced muscle temperature without compromising the skin and allows the superficial skin temperature to return to normal while deeper muscle temperature remains low	Ice is effective in acute soft tissue treatment, but should be applied in repeated applications of 10 minutes to be most effective, avoid side effects, and prevent possible further injury	Reflex activity and motor function are impaired following ice treatment so patients may be more susceptible to injury for up to 30 minutes following treatment	Following intermittent application similar to that stated by Bleakly et al 2006

information about whether to perform a passive range of motion assessment at that joint (Jarvinen et al. 2007). Passive range of motion occurs when the clinician takes the affected limb through its range in order to provide information regarding the integrity of joint or 'end feel'. This usually provides important information regarding the client's condition and informs the clinician about future action, assessment and intervention. In considering normal joint range of motion, resultant injury risk factors and how to relate the importance of full functional range of motion to the athlete's movement patterns in their sport, the clinician can appreciate the relationship between the latter stages of the rehabilitation process and the return of an athlete to a competitive environment (Pizzari et al. 2008). The athlete should not return to full activity until full pain-free functional range has been established (Stracciolini et al. 2007). As flexibility is an expression of range of motion, knowing when and how to incorporate flexibility training into range of motion rehabilitation is important. The clinician must also be able to determine the effectiveness of flexibility training in relation to final stage rehabilitation and the athlete's sport. It is more beneficial to fully and accurately assess movement dynamics in relation to the common muscular movement patterns and then decide whether or not increased range of motion is beneficial to improving the performance of that athlete (Jarvinen et al. 2007).

Flexibility training

The aim of a flexibility programme should be to achieve and maintain an optimum ROM at each joint that is specific to the athletes sport.

Three modalities that are universally employed during a flexibility programme to obtain an increase in joint ROM are static stretching, proprioceptive neuromuscular facilitation (PNF) (Roberts and Wilson 1999; Babault et al. 2009) and self myofascial release (SMR) (Curran et al. 2009)

Static stretching

Static stretching involves taking the limb to a position that produces increases in muscular tension and allows elongation of a muscle with the aim of enhancing tissue extensibility. The proposed neuro-

Figure 12.3 A static stretch for the gastrocnemius.

physiological process by which this occurs is termed autogenic inhibition (Olivo and Magee 2006). This is performed in a slow and controlled manner (consequently the possibility of exciting the muscle spindles and inducing a stretch reflex is minimal) and held for ≥ 30 seconds (Bacurau et al. 2009; Yuktasir and Kaya 2009) and repeated 3–4 times (Costa et al. 2009; Yuktasir and Kaya 2009). As a result of the prolonged period under increased muscular tension the golgi tendon organs (GTOs) may be stimulated and allow for autogenic inhibition to occur, thereby resulting in a decrease in tension and therefore lengthening of the agonist which enables a prolonged and increased stretch of the muscles (Olivo and Magee 2006).

Figure 12.3 showa an example of a static stretch for the gastrocnemius. To initiate the stretch follow these guidelines:

1. facing a wall, place hands out in front of the body

2. place one foot in front of the other ensuring heels stay in contact with the ground

3. maintaining an upright posture, push forwards from the hips keeping the head upright and facing forwards until a stretch is felt in the gastrocnemius of the back leg.

Passive stretching (at the end of available range) should be avoided for the first 72 hours as a minimal period and possibly for the first 7–10 days following injury if more severe (Neidlinger-Wilke et al. 2002).

The reasons for this is that healing tissue is weak and intolerant to tensile loading and is likely to be damaged by *uncontrolled* stretching. Prior to the 10th day post injury it would be more appropriate to take the muscle through its *full available pain free range* without any attempt to force the muscle beyond this point.

Once it is appropriate to begin stretching the muscle, that is, elongating the tissue beyond its available range then careful passive stretching can be performed. Each stretch should be held at the end of available range within the limits of pain. Time, frequency, duration and intensity of stretch remain debateable in the literature. Some research suggests passive stretching should be held for a minimum of 15 seconds with 6–8 sets per day (Roberts and Wilson 1999), however, Bandy et al. (1994, 1997) reviewed stretches at 15 seconds, 30 seconds and 60 seconds. They determined that 30 seconds was the optimum duration (ideally repeated 3 times and performed 3 times per day), also stating that longer periods after 30 seconds was ineffective in promoting additional stretch.

As with strengthening exercises stretching exercises need to be progressed in order that the tissue adapts to the different types of load once the athlete is comfortable (pain free) with passive stretching (Bandy et al. 1994, 1997).

Although implementing a stretching protocol may elicit improvements in performance as a result of increased ROM at a joint (Herda et al. 2008), research has suggested the use of static stretching may have a detrimental effect on performance, immediately post stretching, in relation to sprinting, strength endurance and jumping, all power based activities (Bacurau et al. 2009; O'Sullivan et al. 2009; Hough et al. 2009). PNF has also been linked to a decrease in power production immediately post stretching (Mareket al. 2005). Therefore it is essential that ROM is restored as early in the rehabilitation process as possible to ensure that this does not have a detrimental effect on subsequent strength or power performance.

Proprioceptive neuromuscular facilitation

Proprioceptive neuromuscular facilitation is a form of muscle energy technique (MET) performed passively with a partner involving a voluntary isometric contraction followed by a static stretch phase (Smith and Fryer 2008). There are two main PNF techniques; contract-relax and contract relax-antagonist contract techniques (Olivo and Magee 2006).

The contract relax method requires a limb to be move in to a stretched position and then the agonist undergoes an isometric contraction for 7–10 seconds. As the limb does not move, but the muscle contracts and therefore shortens, the change in length is accommodated by a lengthening of the tendons. This lengthening of the tendon stimulated the GTO and thereby inhibits the agonist. Once the isometric contraction ends the ROM can be increased and the procedure repeated.

The contract relax-antagonist contract technique has been suggested as the most effective in increasing muscle length due to the concept of reciprocol inhibition, which produces increased suppression of the motor pool (Etnyre and Abraham 1986). Reciprocal inhibition is the relaxation that occurs in the opposing muscle that is experiencing increased muscular tension (Olivo and Magee 2006). This is performed with a voluntary isometric contraction of the desired muscle to be stretched lasting approximately 7–10 seconds at a range of 25–40% of maximal voluntary contraction (MVC) followed by an antagonist assisted static stretch phase (Olivo and Magee 2006; Smith and Fryer 2008). The incorporation of a passive stretch supported by the antagonist is believed to produce greater joint flexibility than static stretching (Sady et al. 1982; Guissard 1988; Magnusson et al., 1996; Handel et al. 1997) and produce the longest duration of maintained flexibility than any other form of stretching (Spernoga 2001).

Figure 12.4 illustrates how to perform hamstring PNF, along with the guidelines for correct implementation.

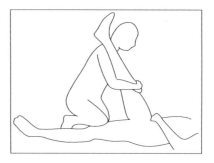

Figure 12.4 Hamstring PNF.

1. The leg is taken to the initial tissue barrier, with the opposing leg straight along the floor. Ensure your hand position is not over the joint.

2. Hold for initial period, stretching the raised leg.

3. Induce an isometric contraction from the athlete in the hamstring musculature.

4. Rest the hamstring muscles and induce an isometric contraction of the antagonist muscles (quadriceps).

5. Rest and repeat the process.

6. Follow the guidelines as stated above for advised muscle contraction force and length of time contractions should be held.

Self myofascial release (SMR)

Self myofascial release (SMR) is a soft tissue technique centred on enhancing ROM through the breakdown of myofascial restrictions in the body's fascial system (Curran et al. 2009). Fascia is a three-dimensional web of connective tissue that envelopes the body's soft tissue from head to toe providing stability and flexibility (Barnes 1997; Myers 1997). SMR is performed by slowly applying a force with the use of a foam roller to tender spots of the muscle (Curran et al. 2009), which are indicative of fascially restricted areas termed "trigger points" (Travell and Simons 1992; Simons et al. 1998). The force is maintained for a time period ranging from 20 to 90 seconds or until a significant reduction in discomfort is attained. This encourages the transformation of a 'knotted' hypertoned bundle of fibrous tissue to a more pliable tissue that is parellel in formation to the fascia (Travell and Simons 1992; Simons et al., 1998). The subsequent increased ROM achieved using SMR has been attributed to the neurophysiological effects on the fascia (Johansson 1962; Schleip 2003) and histological changes in the fascias cellular content (Cantu and Grodin 2001; Sefton 2004; Barnes 2005).

When attempting to enhance ROM at a joint with flexibility training it is important to not only address the muscles that are displaying reduced extensibility but also the synergistic muscles that are underactive or weak and therefore reducing active ROM. In this case, strengthening at the ends of the ROM is required.

Proprioception/neuromuscular control

As suggested by Myer et al. (2005) neuromuscular control training is important to the improvements of athletic performance and biomechanical stability, which in turn reduces the risk of injury.

Proprioception is a term used frequently during rehabilitation and can be defined as a specialised variation of the sensory system of touch that encompasses the sensation of joint movement (kinesthesia) and joint position (joint position sense) (Lephart et al. 1997). These signals are transmitted to the spinal cord via afferent (sensory) pathways (Prentice 2004). The term refers specifically to conscious and subconscious appreciation of a joint potion in space. The efferent (motor) response to sensory information is termed neuromuscular control (Jonsson et al. 1989). Two motor control mechanisms are involved with interpreting afferent information and coordinating efferent responses (Dunn et al. 1986; Prentice 2004), feed-forwards and feedback.

Feed-forward neuromuscular control involves planning movements based on sensory information from past experience (La Coix 1981; Dunn et al. 1986). Feedback process continuously regulates muscle activity through reflex pathways. Feed-forward mechanisms are for preparatory muscle activity; feedback processes are associated with reactive muscle activity. The level of muscle activation, whether it is preparatory or reactive, greatly modifies its stiffness properties (Prentice 2004). From a mechanical perspective, muscle stiffness is the ratio in the change in force to change in length (Dietz et al. 1981; Bach et al. 1983; Dyhre-Poulsen et al. 1991). Muscles that are stiffer resist stretching more effectively and provide more effective dynamic restraint to joint displacement (Bach et al. 1983; Mc Nair et al. 1992); for example, in an ACL deficient knee the increase in hamstring muscle activation increases hamstring stiffness and therefore the functional ability of the knee to reduce anterior translation (Bach et al. 1983; Mc Nair et al. 1992; Myer et al. 2005). The dynamic restraint system is mediated by specialised nerve endings called mechanoreceptors (Grigg 1994), which function by transducing mechanical deformation of tissue into modulated neural signals (Grigg 1994).

Increase tissue deformation is coded by increase afferent discharge rate or increase in mechanoreceptors activated (Grigg 1994; Prentice 2004). The signals provide sensory information concerning internal and external forces action on a joint. Mechanoreceptors (Pacinian corpuscles, Meissner corpuscles and free nerve endings) can be classified as either quick adapting (QA) or slow adapting (SA). Quick adapting receptors cease discharging shortly after the onset of a stimulus, whereas SA continue to discharge as long as the stimulus is present (Clark and Burgess 1975; Schultz et al. 1984; Katonis et al. 1991; Grigg 1994).

In a healthy joint these QA mechanoreceptors are believed to provide conscious and subconscious kinaesthetic sensation in response to a joint movement or acceleration, whilst SA mechanoreceptors provide continuous feedback (Freeman and Wyke 1966; Clark and Burgess 1975) and therefore proprioceptive information relative to the joint position (Freeman and Wyke 1966; Clark and Burgess 1975; Schultz et al., 1984; Katonis et al., 1991; Grigg 1994).

The pre-activation theory suggests that prior sensory feedback (experience) concerning the task is utilised to pre-programme muscle activity patterns. Rehabilitation programmes should be designed to include a proprioceptive component that addresses the following three levels of motor control: spinal reflexes, cognitive programming, and brainstem activity. Such a programme is highly recommended to promote dynamic joint and functional stability (Lephart et al. 1997). There are different proprioceptive states that are important to consider in the design of a progressive rehabilitation programme.

Static proprioceptive development involves exercises with the maintenance of a stable base, while allowing for only minimal movement. During this stage of neuromuscular development the athlete should focus on control of posture and be able to perform a number of modifications of static proprioceptive training before progressing on to a dynamic, more functional proprioceptive training programme (Liebenson 2006).

Dynamic proprioceptive training should only be introduced when the athlete has demonstrated a sufficient level of balance and coordination during the static proprioceptive exercise phase of the neuromuscular control program (Beam 2002). Since dynamic proprioceptive exercises involve greater levels of instability and require a higher demand for accuracy, strength and speed of motion (Myer et al. 2006), balance and coordination are important to ensure the athlete can progress safely without hindering their development. A simple objective test to determine the athlete's progression from static to dynamic exercises can include the Romberg test which assesses the ability to balance. The athlete stands with their feet together and eyes closed (Thuan-Lee and Kapoula 2007). A more advanced test is the stork stand, which involves the athlete standing on one leg usually the effected limb and maintaining the position for a period of at least 30 seconds without touching the floor with the opposing leg or supporting themselves with their other limbs (Melorose et al. 2007). This can then be progressed by being performed with eyes shut. If the athlete finds these tests demanding and cannot perform them effectively there is a deficit in balance that needs to be addressed before progressing onto dynamic exercises.

Table 12.3 shows an example of four different balance tests that could be integrated into all stages of a functional rehabilitation programme to enhance

Table 12.3 Balance tests

Test	Variations	Further progression	Regression
Standing	Double leg Single leg	Increase load	Seated isometric holds
Quarter squat	Double leg Single leg	Change in velocity	Assisted ball squat
Half squat	Double leg Single leg	Increase load and change in velocity	Quarter squat
Jump and hold/hop and hold	Double leg Single leg	Increase height of box to jump off, or increase distance jumped	Unilateral to bilateral, or decrease height of box

proprioceptive responses and neural drive to the working muscles and aid in the athlete's development. This example can be implemented as soon as the athlete is able to weight bear on both legs. Once this capability is obtained, the exercises including taking the individual from a standing position to a countermovement jump can be progressed or regressed depending on the ability of the athlete to perform each exercise without excessive biomechanical dysfunction.

Continual adaptation to sports specific, functional balance training is fundamental to enhancing an athlete's proprioceptive awareness. As a result of proprioceptive training, receptors in soft tissue structures and joint complexes are trained to perform coherent actions, able to initiate dynamic, functional movements (Komi 2003).

Dynamic proprioceptive exercises are performed to enhance the ability of the muscles around the affected joint to control joint motion and stabilise the body during movements in multi-planar directions (Myer et al. 2006; Stracciolini et al., 2007; Pasanen et al 2008; Subasi et al. 2008). When an athlete is performing dynamic neuromuscular control exercises to a high level, it is the practitioner who must continue to progress the exercises, in terms of either intensity or difficulty as required, in order to prevent a plateau development.

For further reading and applied examples regarding specific neuromuscular rehabilitation for selected joints refer to the chapters on the knee (Chapter 21) and ankle (Chapter 22), in the latter section of this book.

By continuing to enhance the athletes neuromuscular development the clinician is allowing for adaptive changes to occur within the neuromuscular system. This can lead to increases in the ability of an athlete to perform plyometric based exercises, increased levels of balance and improve leg coordination in sprint drills (Cameron et al. 2007). Therefore, appropriate progressions during neuromuscular control training will prepare the athlete to progress their rehabilitation on to effective plyometric training.

To implement an effective neuromuscular control programme the clinician should have an expansive knowledge of the anatomical structures and entire kinetic chain. Knowledge of the specific demands and common mechanisms of injury associated with the athlete's sport will also aid in reducing the risk of injury and preventing the re-occurrence of a previous trauma.

The clinician should be able to effectively adapt a neuromuscular programme without the risk of the onset and formation of muscle imbalances, which in turn may lead to musculoskeletal dysfunction and subsequent injury.

For example, decreased neuromuscular control of the trunk can increase the risk of knee injury, specifically anterior cruciate ligament (ACL) injury (Zazulak et al. 2007). The reason for this being that athletes with reductions in neuromuscular control are more susceptible to increased valgus moments of the knee during the impact phase of jump and landing activities, increasing the risk of ACL injury (Hewett et al. 2005).

The altered neuromuscular activation of the hamstrings and quadriceps muscles are believed to play an important role in this risk of ACL injury. Since the hamstrings insert onto the posterior aspect of the tibia and fibular head, they provide a posterior draw force on the knee that resist anterior tibial forces (Bryant et al. 2008). However, if there are deficits in muscle activation of the hamstrings there is a limited ability to protect the knee ligaments (Hewett et al. 2006). Consequently, over-dominant quadriceps activation as a result of under active hamstrings is believed to produce anterior displacement of the tibia, increasing the risk of ACL injury (Griffin et al. 2006).

This is a commonly neglected stage of knee rehabilitation that is of particular importance since ACL injury is one of the most common causes of reduced sport participation and is associated with recurrent injury and an increased risk of osteoarthritis of the knee (Walden et al. 2006).

Another example is the shoulder joint, which allows the greatest ROM of all the joints (Wassinger et al. 2007), but as a consequence compromises its stability. As a result, in order to maintain optimal dynamic stability, effective proprioceptive input is necessary to ensure neuromuscular deficits do not impair the athletes' functional ability (Wassinger et al. 2007).

This is achieved by enhancing sensory motor control and ability of the muscles to respond to a neural stimulus (Roig-Pull and Ranson 2007).

In addition to shoulder instability, lateral ankle sprains are a common injury within the sporting arena, which has lead to research being conducted to understand the most effective interventions in relation to prevention and treatment by means of balance and co-ordination training (McKeon and Hertel 2008). The cause of functional instability of the ankle is likely due to altered neuromuscular feedback, which changes the neuromuscular response within the appropriate musculature (Coughlan and Caulfield 2007).

Therefore, great emphasis should be placed on incorporating neuromuscular control/proprioceptive exercises in the athletes' rehabilitation programme in order to restore and improve balance, coordination and agility thus increasing functional ability of the effected limb and the entire kinetic chain (Subasi et al. 2008). Non-injured athletes can also perform neuromuscular control training during their weekly routine training sessions as part of a injury prevention training programme (Paterno et al. 2004; Pasanen et al. 2008)

Basic concepts of application

As with all stages of injury rehabilitation and the planning of exercise conditioning, the development of a neuromuscular control programme too is also dependant on many factors including gender, injury status, type of sport and level of competition. Since the aim of a proprioceptive training programme is to promote balance, coordination and agility accordingly (Subasi et al. 2008), the athlete must begin first with static exercises before progressing to dynamic exercises. The exercises must develop from simple to complex with emphasis placed on precision, accuracy and control. The clinician can alter a range of variables in order to progress proprioceptive exercises. This can involve changing the rate of speed, amount of simultaneous activities performed at one time, limiting the amount of sight during training and adapting exercises that are more functional with sport specific movements (Risberg et al. 2001)

This is important as adapting the proprioceptive training directly to the neuromuscular response of the athlete throughout the rehabilitation process will simulate similar motor patterns in accordance to the athletes sport. Consequently it has been shown that this may not only enhance performance potential in a competitive sporting environment, but further play a significant role in injury prevention (Hewett et al. 2005; Chappell and Limpisvasti 2008; McKeon and Hertel 2008; Twomey et al. 2008; Zazulak et al. 2007).

It may also be more beneficial to the athlete to perform proprioceptive exercises during the beginning of a rehabilitation session or before a training routine, particularly if the proprioceptive exercises are new to the athlete or are part of more advanced progressions. The reason for this is that it has been shown that exercise induces fatigue on the mechanoreceptors that are situated in the musculature surrounding the joint (McLean et al. 2007). Since mechanoreceptors are responsible for their input of proprioceptive feedback, proprioception is affected along with neuromuscular control and therefore negatively impacts the athletes' ability to perform the proprioception exercises (Myers et al. 1999).

Practical implementation of unstable surface training

The importance of unstable surface training (UST) has been shown to be beneficial in enhancing proprioceptive input, enabling neuromuscular adaptations to commence during the early stage rehabilitation process (Cressey et al. 2007). This form of training has been commonly implemented worldwide with a wide range of products. However, the key consideration is whether this form of training should continue throughout the athlete's rehabilitation. In the latter stages when optimisation of force production and power output are required an unstable surface will result in a decrease in force and power output, as illustrated from the results of Cressey et al. 2007).

The majority of sporting movements in the upper extremity occur in an open chain manner; therefore it may be more beneficial to incorporate unstable surface based exercises under these circumstances, rather than in the lower extremity. Most athletic movements in the lower limb occur in a closed chain fashion, in which unstable surface training may prove detrimental to performance in relation to reduced power output. The potential decrease in power and strength capacity is due to elevated activation of the antagonist muscle in relation to the prime move or agonist, which although can assist in maintaining joint stability can compromise complete activation of the functional agonist. Reciprocal inhibition is a physiological response to UST leading to the resulting decrease in force production.

Table 12.4 Examples of UST compared with stable surface training (adapted from Cressey et al. 2007)

Assessment	Pre-test (watts or secs)	Intervention	Post-test (watts or secs)	% change
Drop jump		2–5 sets of 5–15 reps		
Unstable group	387.8 (w)	Dyna discs	384.0 (w)	0.8%
Stable group	642.8 (w)	Squats	602.57 (w)	3.2%
Countermovement jump		Deadlifts Lunges		
Unstable group	390.6 (w)	SL squats	401.1 (w)	0%
Stable group	588.7 (w)	SL balances	545.8 (w)	2.4%
40-yard sprint				
Unstable	5.02 (sec)		4.93 (sec)	−1.8%
Stable	5.06 (sec)		4.87 (sec)	−3.9%
10-yard sprint				
Unstable	1.73 (sec)		1.67 (sec)	−4.0%
Stable	1.75 (sec)		1.63 (sec)	−7.6%
T-Test				
Unstable	8.33 (sec)		8.09 (sec)	−2.9%
Stable	8.42 (sec)		8.06 (sec)	−4.4%

Table 12.4 shows some practical examples of UST and its comparison to stable surface training. Each of the tests demonstrated show movements that are commonly used within the sporting environment; these include sprinting, jumping and quick changes of direction.

Table 12.4 demonstrates that the implementation of a UST programme can negatively impact the performance of subjects in relation to power and speed-based activities.

Test summary

Jumping assessments

The stable group showed significantly greater results than the unstable group in relation to the bounce drop jumps (BDJ) and counter-movement jumps (CMJ). The UST group did not shown significant improvements in performance following the intervention at post testing. This is in contrast to the stable training group which showed significant improvements.

Sprinting assessments

The stable training group demonstrated improvement in results, in comparison with the unstable group. Pre-test results were similar for both groups in both sprint activities; however the stable group showed greater advances in performance in contrast to the unstable training group. See Table 12.4 for the exact figures.

Agility assessment

There were insignificant differences between the two groups prior to the training interventions. Both the stable training group and UST group demonstrated advances in performance in relation to the pre-test, following the training intervention. See Table 12.4 for the data collated from the T-Test drill.

Summary

Continual adaptation to sports specific, functional balance training is fundamental to enhancing an athlete's proprioceptive awareness. As a result of proprioceptive training, receptors in soft tissue structures and joint complexes are trained to perform coherent actions, able to initiate dynamic, functional movements (Komi 2003). Futhermore by creating a more advanced understanding of postural control requirements amongst athletes participating in different sports, the practitioner may be able to prescribe the most effective exercises in relation to the balance and co-ordination demands of the specific sport, enabling more optimal movement for the athlete (Bressel et al. 2007).

The question of when best to implement a dynamic proprioceptive programme remains controversial. Research has yet to fully support arguments surrounding *pre versus post* training sessions. The important factor to consider here is the influence of continuous eccentric muscle contractions on performance caused from over exertion (Lavender and Nosaka 2007). Intermittent high-intensity eccentric training causes micro-trauma within muscle fibres (Clark et al. 2005). The breakdown of fibres within the skeletal muscle causes slight alterations in the ability of proprioceptive responses to interact effectively with the central nervous system to initiate effective movement patterns in multi-directional planes of motion.

Strength endurance and maximal strength

To initiate a systematic rehabilitation approach through the strength phase it is important to understand the reasons behind why a progressive approach is demanded and how the benefits of strength training can relate to improved performance for an athlete on their return to sport.

It is imperative to ensure athletes progress through the rehabilitation process establishing an adequate level of muscular endurance and then strength (Jarvinen et al. 2007) as required by their individual sport. Isometric strength training needs to be considered in the initial rehabilitation phase, and is often implemented 3–7 days post injury (Jarvinen et al. 2007), usually with a focus on preventing muscle atrophy and/or a loss of strength. Frequency, duration and intensity are limited by the patients' pain. Some therapists advocate three sets of 10 repetitions using 5–10 second holds to begin with at intensity within pain tolerance (Pull and Ranson 2007). These then are undertaken at multiple angles, beginning in mid range then progressing to inner range (shortened position) then outer range (lengthened position).

The sports rehabilitator should continue to monitor the progress of the athlete taking into consideration exercise frequency, intensity and duration. Early phase strengthening exercises, involve an isometric contraction of the agonist muscle with no movement at the joint. The contraction should be performed at different joint angles to initiate muscle fibre activation in different planes of motion. The stabilising force surrounding the joint is the antagonist muscle (Middleton and Smith 2007). Pain management and the acknowledgement of pain free training should be considered simultaneously with the implementation of strength training protocols. This is evident within the isometric strength development, where the primary goal is to create an adaptive response in the muscle, without stressing the joint to excess (Liebenson 2006). Once an athlete is able to perform isometric muscle contractions at various joints without the onset of pain, it may then be advisable to progress the athlete onto dynamic exercises (Newberry and Bishop 2006). Sports rehabilitators must ensure that the athlete has full bilateral range of motion, no swelling or pain prior to implementation of progressive, dynamic, functional based activity (Beam 2002).

During isotonic exercise the athlete must provide a force powerful enough to initiate concentric and eccentric muscle actions, whilst coping with a constant external load. There are different isotonic exercises that could be embedded into the rehabilitation process. These can include use of dumbbells, machine weights, and resistance bands (Beam 2002).

Muscle injuries are associated with preferential atrophy of type 1 muscle fibres with disuse (Stockmar et al. 2006), and high loads and rates of force development are most likely to over stress the healing tissue. Therefore, initially an endurance based programme should be used (3 sets of ≥ 15 repetitions at 40–60% of one repetition maximum) this should be progressed to strength (4–6 sets of 3-6 repetitions at 85–95% of one repetition maximum and then power training (3–5 sets of 3–5 repetitions at 75–85% of one repetition maximum) (Kraemer et al. 2002).

A key consideration during the progression of isotonic training is the velocity at which the movement is performed (Wrbaskic and Dowling 2007). Initially the athlete should perform slow and controlled movements in order to allow for increased neural response to the working muscles and the continued development of neuromuscular control. Once the athlete is able to control the exercise effectively the continual progression to sports specific, functional activity coupled with changes in velocity, increased load and use of different planes of motion, should be considered (Liebenson 2006).

Through training, the body's systems and tissues adapt in direct response to the stresses imposed on

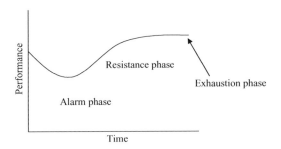

Figure 12.5 Schematic diagram of the general adaptation syndrome.

the body during each training session. The adaptation responses can be explained through the general adaptation syndrome (GAS) first described by Hans Seyle in (1985) (Figure 12.5).

Figure 12.5 illustrates three primary phases of the GAS, which incorporate the initial response or alarm phase, the adaptive response phase and the exhaustion or plateau phase of a resistance or aerobic conditioning programme. The alarm phase often occurs following the introduction of a new stimulus into a training programme and can lead to the onset of delayed onset muscular soreness (DOMS), joint stiffness and a general feeling of discomfort following a training session (Brown 2007). The resistance phase is when training adaptations occur and the muscles are able to respond effectively to the physiological changes within the body (Newton 2006). This phase tests the ability of the body to withstand various mechanical forces, which may ultimately lead to increased neuromuscular function within the trained musculature. If, however, training continues without the appropriate introduction of a new stimulus the exhaustion phase is emphasised (Epley 2004). This can result from too high a volume of training or progressions in intensity that are too large or rapid, resulting in a poorly prescribed and/or implemented training regime. The general adaptation syndrome is useful when designing and deciding on how best to progress or regress exercise intensity. In relation to rehabilitation the practitioner must understand the three stages of the GAS and be able to adapt appropriately, any implemented exercise protocol with regards to injury rehabilitation.

The above diagram illustrates the principles behind the general adaptation syndrome (GAS). GAS considers three primary reactions in relation to a physiological stimulus; these include the alarm, resistance and exhaustion phases.

To adequately and effectively design progressive rehabilitation programmes, it is important to relate biomechanical principles such as force-velocity relationships to the rehabilitation process. Komi (2003) argues that exercise programmes should be devised in accordance with the relative power output requirements of the sport. There are many varied examples, in sport, to highlight the importance of including power specific exercises into the rehabilitation programmes. Sports such as netball, tennis, squash, American football and basketball all require explosive movements in multiplanar directions, and would therefore benefit directly from exercises that incorporate high-velocity actions. However, in relation to progressive rehabilitation it is still imperative to ensure the athlete attains isometric strength prior to continued progression through the strength training

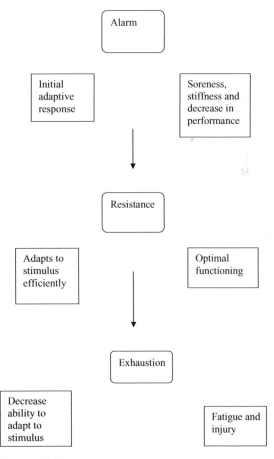

Figure 12.6 Schematic diagram of genaral adaptation syndrome.

214 PROGRESSIVE SYSTEMATIC FUNCTIONAL REHABILITATION

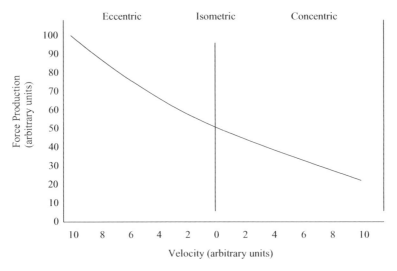

Figure 12.7 Force-velocity curve.

continuum, continually ensuring exercises simulate that of a competitive environment, where the injuries are most likely to occur (Verrall et al. 2005).

Once adequate concentric strength has been developed, it is essential to ensure that both the velocity of the concentric exercises and the emphasis on eccentric strength are developed. Evidence suggests that eccentric training during rehabilitation reduces the risk of recurring injuries, but used prior to injury reduces the occurrence of injury (Askling et al., 2003; Mjolsnes et al., 2004; Clark, et al., 2005; Brooks et al. 2006; Gabbe et al. 2006; Arnason et al. 2008).

Eccentric loading, especially at high velocities such as during plyometric training, may lead to significant muscle damage, especially if there has been limited muscular endurance and strength training (Schache et al. 2008), resulting in delayed onset muscle soreness (DOMS). The rehabilitation process depends on the progress of each individual and therefore must be adapted as required. It is important that the clinician continues to monitor the progress of the athlete and ascertains when to enhance the proprioceptive and strength capacity of the individual (Gokaraju et al. 2008).

Training for muscular strength and endurance must cover all areas of the force velocity curve (Figure 12.7), usually progressing from isometric focused, through to concentric and eccentric, beginning at low velocities and progressing through to high velocity activities.

Plyometric training

In order to appreciate the importance of plyometric training as part of a sport specific rehabilitation programme, the components of athletic movement need to be considered (Epley 2004). All sports and athletic movements require the optimum functioning of active muscles to produce muscular forces at varying speeds through multiple planes of motion. This is known as the force-velocity relationship and can be used to describe the term power.

Power is present in all sports or activities involving rapid force production (Brown et al. 2007). Plyometric training is a form of resistance training that involves high-velocity based exercises characterised by quick eccentric (lengthened) muscle contractions followed by rapid concentric (shortened) muscle contractions. This is achieved through a process termed the stretch-shortening cycle (SSC) (Shiner et al. 2005). These exercises enhance power production by increasing motor-unit recruitment, rate of muscle firing and sensitivity and excitability of the neuromuscular system, thereby promoting optimum neuromuscular efficiency (Newberry and Bishop 2006). Plyometric activities also take advantage of the elastic components of the tissues (tendons, muscles and fascia) along with the increased motor recruitment due to the stretch reflex.

Plyometric training offers a progressive approach to incorporating functional sport specific movements that enhance anaerobic power (Faigenhbaum et al. 2007) strength, (Marques et al. 2008) agility (Thomas et al. 2009) and speed (Myer et al. 2005) in preparation for a full return to sport participation with an emphasis on the prevention of re-injury, through the enhancement of eccentric and neuromuscular control (Gilchrist et al. 2008; Hewett et al. 2006).

Before commencing a plyometric programme the athlete must be assessed for the appropriate components of lumbo-pelvic (core) stability and neuromuscular control (especially during ground contact), including optimum functional movement that will enable the required rate of muscle contraction whilst reducing the risk of potential injury. Constant observational assessments should be conducted, such as postural assessments and lower limb control during landing.

In addition to optimum kinematics, the athlete must also possess appropriate components of strength, proprioception and ROM (Chmielewski et al. 2006). Adequate levels of strength will enable the athlete to perform the exercises with sufficient control minimising the risk of injury. Strength requirements vary depending on the particular exercise, for example Chu (1998) suggests that levels of strength require the athlete to perform a squat with at least 60% of their body weight for five repetitions in no less than five seconds in order to ensure adequate velocity during the concentric phase of loading. The landing or loading phase of plyometric exercise is particularly important as it initiates the stretch-shortening cycle and subsequent force production (Rassier and Herzog 2005). As a result, prior to full plyometric training, emphasis should be placed on perfecting the landing component with the use of technique based drills. This should have already been addressed during earlier stages of rehabilitation that focused on neuromuscular control.

Consequently, it has been suggested that plyometric exercise can be implemented when the athlete can sustain moderate loading during basic strengthening exercises with the ability to perform functional movement patterns efficiently (Chmielewski et al. 2006). Balance, coordination and agility are important components of proprioception that will enable the athlete to control the explosive and intensive movements involved with plyometric training (Chmielewski et al. 2006). An appropriate level of flexibility is also required in order to perform plyometrics. Flexibility is governed by the tissue extensibility at a joint, which in turn dictates the degree of range of motion (ROM) that is available at a joint during active movement (Bradley and Portas 2007). The velocity, force, muscle action and rate of loading can also be manipulated to suit the specific demands of the athlete's sport (Shiner et al. 2005). Frequent functional evaluations will provide the appropriate feedback on whether to regress or progress the athlete's exercises. This will result in a progressive and systematic, goal orientated plyometric exercise programme.

Plyometric training should also be progressive and periodised to allow progression in terms of increasing velocity of movement or deceleration forces, depending on the aims of the athlete and the requirements of the sport. At the same time it is essential to continue with *some* strength training to prevent a detraining effect and therefore a loss of strength. Progression for sport-specific activities should develop from unidirectional, to bidirectional, to multidirectional movements (Heidt et al. 2000).

It is also worth noting that plyometrics can be effectively performed in water (Robinson et al. 2004;

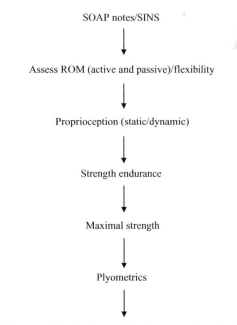

Figure 12.8 The rehabilitation process.

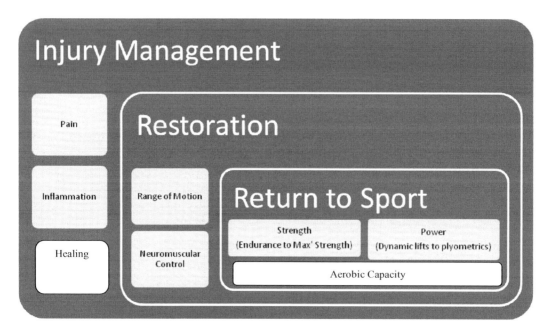

Figure 12.9 Progressive rehabilitation process plan.

Stemm and Jacobson 2007), thereby reducing the impact forces and stress on the muscular system, due to the buoyancy of the water, and allowing plyometric training to be introduced to athletes at an earlier stage in their development/rehabilitation. Burgess et al. (2007) and Kubo et al. (2007) also found that plyometric training has a beneficial effect on tendon and joint stiffness, which may reduce the risk of injuries.

Summary

Figures 12.8 and 12.9 provide a visual analysis of the rehabilitation process and consider the key elements relevant to successful progressive rehabilitation.

References

Airaksinen, O.V., Kyrklund, N., Latvala, K., et al. (2003) Efficay of cold gel for soft tissue injuries. A prosepctive randomised double blined trial. *American Journal of Sports Medicine*, 31, 680–684.

Arnason, A., Anderson, T.E., Holme, I., Engebretsen, L. and Bahr, R. (2008) Prevention of hamstring strains in elite soccer: an intervention study. *Scandinavian Journal of Medicine and Science in Sports*, 18 (1), 40–48.

Askling, C., Karlsson, J. and Thorstensson, A. (2003) Hamstring injury occurrence in elite soccer players after preseason strength training with eccentric overload. *Scandinavian Journal of Medicine and Science in Sports*, 13, 244–250.

Askling, C.M., Saartok, T. and Thorstensson, A. (2006) Type of acute hamstring strain affects flexibility, strength, and time to return to pre-injury level. *British Journal of Sports Medicine*, 40 (1), 40–44.

Askling, C.M, Tengvar, M., Saartok, T. and Thorstensson, A. (2008) Proximal hamstring strains of stretching type in different sports. *The American Journal of Sports Medicine*, 36 (9), 1799–1804.

Babault, N., Kouassi. B.Y.L. and Desbrosses. K (2009) Acute effects of 15 min static or contract-relax stretching modalities on plantar flexors neuromuscular properties. *Journal of Science and Medicine in Sports*, E-pub ahead of print, 425–431.

Bacurau, R.F.P., Monteiro, G.A., Ugrinowitsch, C., Tricoli, V., Cabral, L.F., Ferreira, A. and Aoki, M.S. (2009) Acute effect of a ballistic and a static stretching exercise bout on flexibility and maximal strength. *Journal of Strength and Conditioning Research*, 23 (1), 304–308.

Bandy, W.D. and Irion, J.M. (1994) The effect of time on static stretch on the flexibility of the hamstring muscles. *Physical Therapy*, 74 (9), 845–852.

Bandy, W.D., Irion, J.M. and Briggler, M. (1997) The effect of time and frequency of static stretching on the flexibility of the hamstring muscles. *Physical Therapy*, 77 (10), 1090–1096.

Barnes, M.F. (1997) The basic science of myofascial release: morphological change in connective tissue. *Journal of Bodywork and Movement Therapies*, 1 (4), 231–238.

Barnes, J. (2005) Myofascial release. In Hammer, W.I. (Ed.) *Functional Soft Tissue Examination and Treatment by Manual Methods: New Perspectives*. Gaithersburg, MD: Aspen, pp. 533–548.

Beam, J.W. (2002) Rehabilitation including sport-specific functional progression for the competitive athlete. *Journal of Bodywork and Movement Therapies*, 4, 205–219.

Bleakley, C.M., McDonough, S.M. and MacAuley, D.C. (2004) The use of ice in the treatment of acute soft tissue injuries. A systematic review of randomized controlled trains. *American Journal of Sports and Medicine*, 32, 251–261.

Bleakley, C.M., McDonough, S.M. and Macauley, D.C. (2006) Cryotherapy for acute ankle sprains: a randomised controlled study of two different icing protocols. *British Journal of Sports Medicine*, 40, 700–705.

Bleakley, C.M., O'Connor, S., Tully, M.A., Rocke, L.G., Macauley, D.C. and McDonough, S.M. (2007) The PRICE study (Protection Rest Ice Compression Elevation): design of a randomised controlled trial comparing standard versus cryokinetic ice applications in the management of acute ankle sprain. *BMC Musculoskeletal Disorders*, 19 (8), 125.

Borcherding, S. and Morreale, M.A. (2007) *The OTA's Guide to Writing SOAP Notes*, 2nd edn. Slack Incorporated, p. 39.

Boudreau, S.N., Dwyer, M.K., Mattacola, C.G., Lattermann, C., Uhl, T.L. and Medina Mckeon, J. (2009) Hip-muscle activation during the lunge, single-leg squat, and step-up and-over exercises. *Journal of Sport Rehabilitation*, 18, 91–103.

Bradley, P.S. and Portas, M.D. (2007) The relationship between preseason range of motion and muscle strain injury in elite soccer players. *Journal of Strength and Conditioning Research*, 21 (4), 1155–1159.

Brockett, C.L., Morgan, D.L. and Proske, U. (2004) Predicting hamstring strain injury in elite athletes. *Medicine and Science in Sports and Exercise*, 36 (3), 379–387.

Brooks, J.H., Fuller, C.W., Kemp, S.P. and Reddin, D.B. (2006) Incidence, risk and prevention of hamstring muscle injuries in professional rugby union. *American Journal of Sports Medicine*, 38 (8), 1297–1306.

Brown, L.E. (2007) Strength Training: National Strength and Conditioning Association: NSCA. Champaign, IL: Human Kinetics.

Brown, A.C., Wells, T.J., Schade, M.L., Smith, D.L. and Fehling, P.C. (2007) Effects of plyometric training versus traditional weight training on strength, power, and aesthetic jumping ability in female collegiate dancers. *Journal of Dance Medicine and Science*, 11 (12), 38.

Burgess, K.E., Connick, M.J., Graham-Smith, P. and Pearson, S.J. (2007) Plyometric vs. isometric training influences on tendon properties and muscle output. *Journal of Strength and Conditioning Research*, 21 (3), 986–989.

Cameron, M.L., Adams, R.D., Maher, C.G. and Misson, D. (2007) Effect of the Hamsprint Drills training program on lower limb neuromuscular control in Australian football players. *Journal of Science and Medicine in Sport*, 12, 24–30.

Cantu, R. and Grodin, A. (2001) *Myofascial Manipulation: Theory and Clinical Application*. Baltimore, MD: Aspen.

Chappell, J.D. and Limpisvasti, O. (2008) Effects of a neuromuscular training program on the kinetics and kinematics of jumping tasks. *American Journal of Sports Medicine*, 36 (6), 1081–1086.

Chmielewski, T.L., Myer, G.D., Kauffman, D. and Tillman, S.M. (2006) Plyometric exercise in the rehabilitation of athletes: Physiological responses and clinical application. *Journal of Orthopaedic and Sports Physical Therapy*, 35 (5), 308–319.

Chu, D. (1998) *Jumping into Plyometrics*, 2nd edn. Champaign, IL: Human Kinetics.

Clark, R., Bryant, A., Culgan, J. and Hartley, B. (2005) The effects of eccentric hamstring strength training on dynamic jumping performance and isokinetic strength parameters: a pilot study on the implications for the prevention of hamstring injuries. *Physical Therapy in Sport*, 6, 67–73.

Collins, N.C (2008) Is ice right? Does cryotherapy improve outcome for acute soft tissue injury. *Emergency Medical Journal*, 25 (2), 65–68.

Costa, P.B., Graves, B.S., Whitehurst, M. and Jacobs, P.L. (2009) The acute effects of different durations of static stretching on dynamic balance performance. *Journal of Strength and Conditioning Research*, 23 (1), 141–147

Coughlan, G. and Caulfield, B. (2007) A 4-week neuromuscular training progam and gait patterns at the ankle joint. *Journal of Athletic Training*, 42 (1), 51–59.

Cressey, E.M., West, C.A., Tiberio, D.P., Kraemer, W.J. and Maresh, C.M. (2007) The Effects Of Ten Weeks Of Lower-Body Unstable Surface Training On

Markers of Athletic Performance. *Journal of Strength and Conditioning Research*, 21 (2), 561–567.

Curran, P.F., Fiore, R.D. and Crisco, J.J. (2009) A comparison of the pressure exerted on soft tissue by 2 myofascial rollers. *Journal of Sport Rehabilitation, 17*, 432–442.

Epley, B. (2004) *The Path to Athletic Power: The model conditioning program for championship performance.* Champaign, IL: Human Kinetics.

Etnyre, B.R. and Abraham, L.D. (1986) H-reflex changes during static stretching and two variations of proprioceptive neuromuscular facilitation techniques. *Electroencephalography and Clinical Neurophysiology, 63* (2), 174–179.

Faigenhbaum, A.D., McFarland, J.E., Keiper, F.B., Tevlin, W., Nicholas, A., Kang, R.J. and Hoffman, J.R. (2007) Effects of a short-term plyometric and resistance training program on fitness performance in boys age 12 to 15 years. *Journal of Sports Science and Medicine, 6*, 519–525.

Gabbe, B.J., Branson, R. and Bennell, K.L. (2006) A pilot randomised controlled trial of eccentric exercise to prevent hamstring injuries in community-level Australian Football. *Journal of Science and Medicine in Sport 9* (1–2): 103–109.

Gilchrist, J., Mandelbaum, B.R., Melancon, H., Ryan, G.W., Silvers, H.J., Griffin, L.Y., Watanabe, D.S., Dick, R.W. and Dvorak, J. (2008) A randomized controlled trial to prevent noncontact anterior cruciate ligament injury in female collegiate athletes. *American Journal of Sports Medicine, 36* (8), 1476–1483.

Gokaraju, K., Garikipati, S. and Ashwood, N. (2008) Hamstring Injuries: Trauma. London: Sage.

Griffin, Albohm, Arendt, et al. (2006) Understanding and preventing noncontact anterior cruciate ligament injuries: A review of the Hunt Valley II Meeting. *American Orthopaedic Society of Sports Medicine, 34* (9), 1512–1532.

Guissard, N., Duchateau, J. and Hainaut, K. (1988) Muscle stretching and motoneuron excitability. *European Journal of Applied Physiology, 58* (1–2), 47–52.

Handel, M., Horstmann, D., Dickhuth, H., et al. (1997) Effects of contract-relax stretching training on muscle performance in athletes. *European Journal of Applied Physiology and Occupational Physiology, 76* (5), 400–408.

Heidt, R.S., Sweeterman, L.M., Carlonas, R.L., Traub, J.A. and Tekulve, F.X. (2000) Avoidance of soccer injuries with pre-season conditioning. *American Journal of Sports Medicine, 28*(5), 659–662.

Herda, T.J., Cramer, J.T., Ryan, E.D., McHugh, M.P. and Stout, J.R. (2008) Acute effects of static stretching versus dynamic stretching on isometric peak torque. Electromyography, and mechanomyography of the biceps femoris muscle. *Journal of Strength and Conditioning Research, 22* (3), 809–817.

Hewett, T.E., Ford, K.R. and Myer, G.D. (2006) Anterior Cruciate Ligament Injuries in Female Athletes. *American Journal of Sports Medicine, 34* (3), 490–498.

Hewett, T.E., Myer, G.D. and Ford, K.R. (2006) Anterior Cruciate Ligament Injuries in Female Athletes: Part1, Mechanisms and Risk Factors. *American Journal of Sports Medicine, 34* (2), 299–315.

Hewett, T.E., Myer, G.D., Ford, K.R., Heidt, R.S., Colosimo, A.J. and Mclean, S.G. (2005) Biomechanical Measures of Neuromuscular control and valgus loading of the knee predict anterior cruciate ligament injury risk in female athlete. A prospective study. *American Journal of Sports Medicine, 33* (4), 492–501.

Hollman, J.H., Kolbeck, K.E., Hitchcock, J.L., Koverman, J.W. and Krause, D.A. (2006) Correlations between hip strength and Static foot and knee posture. *Journal of Sports rehabilitation, 15*, 12–23.

Hough, P.A., Ross, E.Z. and Howatson, G. (2009) Effects of dynamic and static stretching on vertical jump performance and electromyographic activity. *Journal of Strength and Conditioning Research, 23* (2), 507–512.

Hoskins, W. and Pollard, H. (2005) The management of hamstring injury – Part 1: Issues in diagnosis. *Manual Therapy, 10* (2), 96–107.

Hubbard T. J., Denegar, C. R., 2004. Does cryotherapy improve outcomes with soft tissue injury? Journal of Athletic Training. *39* (3) 278–279.

Jarvinen, T.A.H., Jarvinen, T.L.N., Kaariainen, M., Aarimaa, V., Vaittinen, S., Kalimo, H. and Jarvinen, M. (2007) Muscle injuries: Optimising recovery. *Best Practice & Research Clinical Rheumatology, 21* (2), 317–331.

Johansson, B. (1962) Circulatory responses to stimulation of somatic afferents with special reference to depressor effects from muscle nerves. *Acta Physiologica Scandinavica, 57* (suppl 198), 1–91.

Jonsson, H., Harrholm, J. and Elmquist, L.G. (1989) Kinematic of active knee extension after tear of anterior cruciate ligament. *American Journal of Sports Medicine, 17*, 796–802.

Katonis, P.G., Assimakopoulos, A.P., Agapitos, M.V. and Exaechou, E.I. (1991) Mechanoreceptors in the posterior cruciate ligament. Acta Orthropedoca Scandanavia, *62* (3), 276–278.

Komi. P.V. (2003) *Strength and Power in Sport: The Encyclopaedia of Sports Medicine, An IOC Medical*

Commision Publication in Collaboration With The International Federation Of Sports Medicine, 2nd edn. Oxford: Blackwell Publishing.

Kraemer, W.J., Adams, K., Cafarelli, E., Dudley, G.A., Dooly, C., Feigenbaum, M. S., Fleck, S.J., Franklin, B., Fry, A.C., Hoffman, J.R., Newton, R.U., Potteiger, J., Stone, M.H., Ratamess, N.A. and Triplett-McBride, T. (2002) American College of Sports Medicine, Position Stand. Progression models in resistance training for healthy adults. *Medicine and Science in Sports and Exercise*, *34* (2), 364–380.

Kubo, K., Morimoto, M., Komuro, T., Yata, H., Tsunoda, N., Kanehisa, H. and Fukunaga, T. (2007) Effects of plyometric and weight training on muscle-tendon complex and jump performance. *Medicine and Science in Sports and Exercise*, *39* (10). 1801–1810.

La Coix, J.M. (1981) The acquisition of autonomic control through biofeedback: The case against an afferent process and a two-process alternative. *Psychophysiology*, *18*, 573–587.

Lavender, A.P. and Nosaka, K. (2007) A light load eccentric confers protection against a subsequent bout of more demanding eccentric exercise. *Journal of Science and Medicine in Sport*, *11*, 291–298.

Liebenson, C. (2006) Functional training for performance enhancement - Part 1: The basics. *Journal of Bodywork and Movement Therapies*, *10*, 154–158.

MacAuley, D. (2001) Ice therapy: how good is the evidence? *International Journal of Sports Medicine*, *22*, 379–384.

Magnusson, P., Simonsen, E., Aagaard, P., Dyhre-Poulsen, P., Malachy, P., McHugh, M. and Kjaer, M. (1996). Mechanical and physiological responses to stretching with and without pre-isometric contraction in human skeletal muscle. *Physical Medicine and Rehabilitation*, *77* (4), 373–378.

Marek, S.M., Cramer, J.T., Fincher, A.L., Massey, L.L., Dangelmaier, S.M., Purkayastha, S., Fitz, K.A. and Culbertson, J.Y. (2005) Acute effects of static and proprioceptive neuromuscular facilitation stretching on muscle strength and power output. *Journal of Athletic Training*, *40*, 94–103.

Marques, M.C., Van den Tillaar, R., Vescovi, J.D. and Gonzales-Badillo, J.J. (2008) Changes in strength and power performance in elite senior female professional volleyball players during the in-season: A case study. *Journal of Strength and Conditioning Research*, *22* (4), 1147–1155.

McKeon, P.O. and Hertel, J. (2008) Systematic review of postural control and lateral ankle instability, Part 2: Is balance training clinically effective? *Journal of Athletic Training*, *42* (3), 305–315.

McLean, S.G., Felin, R.E., Suedekum, N., Calabrese, G., Passerallo, A. and Joy, S. (2007) Impact of fatigue on gender based high risk landing strategies. *Medicine and Science in Sports and Exercise*, *39* (3), 502–514.

McNair, P.J., Wood, G.A. and Marshall R.N. (1992) Stiffness of hamstring muscle and its relationship to function in anterior cruciate deficient individuals. *Clinical Biomechanics*, *7*, 131–173.

Melorose, D.R., Spaniol, F.J., Bohling, M.E. and Bonnette, R.A. (2007) Physiological and performance characteristics of adolescent club volleyball players. *Journal of Strength and Conditioning Research*, *21* (2), 481–487.

Middleton, S.W.F. and Smith, J.E. (2007) Muscle injuries. *Trauma*, *9*, 5–11.

Mjolsnes, R., Arnason, A., Osthagen, T., Raastad, T. and Bahr, R. (2004) A 10-week randomized trial comparing eccentric vs. concentric hamstring strength training in well trained soccer players. *Scandinavian Journal of Medicine and Science in Sports*, *14* (5), 311–317.

Myer, G.D., Ford, K.R., Palumbo, J.P., Hewett, T.E. (2005) Nueromuscular training improves performance and lower-extremity biomechanics in female athletes, Journal of Strength and Conditioning Research Volume *19* (1) page 51–60

Myer, G.D., Ford, K.R., McLean, S.G., Hewett, T.E. (2006) The effects of plyometric vs Dynamic stabilization and balance training on power, balance and landing force in female athletes, *American Journal of Sports Medicine*, Volume *34* (3) pp. 445–455

Myers, T.W. (1997) The anatomy trains. *Journal of Bodywork and Movement Therapies*, *1* (2), 91–101.

Myers, J.B., Guskiewicz, K.M., Schneider, R.A. and Prentice, W.E. (1999) Proprioception and neuromuscular control of the shoulder after muscle fatigue. *Journal of Athletic Training*, *34* (4), 362–367.

Neidlinger-Wilke, C., Grood, E., Claes, L. and Brand, R. (2002) Fibroblast orientation to stretch begins within three hours. *Journal Orthopaedic Research*, *20*, 953–956.

Newberry, L. and Bishop, M.D. (2006) Plyometric and agility training into the regimen of a patient with post-surgical anterior knee pain. *Physical Therapy in Sport*, *7*, 161–167.

Newton, H. (2006) *Explosive Lifting for Sports: Enhanced Edition*. Champaign, IL: Human Kinetics.

Olivo, S.A. and Magee, D.J. (2006) Electromyographic assessment of the activity of the masticatory muscles using the agonist contract antagonist relax technique (AC) and contract relax technique (CR). *Manual Therapy*, *11*, 136–145.

O'Sullivan, K., Murray, E. and Sainsbury, D. (2009) The effect of warm-up, static stretching and dynamic stretching on hamstring flexibility in previously injured subjects. *BioMed Central, 10,* 37.

Pasanen, K., Parkkari, J., Pasanen, M., Hilloskorpi, H., Makinen, T., Jarvinen, M. and Kannus, P. (2008) Neuromuscular training and the risk of leg injuries in female floorball players: Cluster randomised controlled study. *British Journal of Sports Medicine, 42* (10), 802–805.

Paterno, M.V., Myer, G.D., Ford, K.R. and Hewett, T.E. (2004) Neuromuscular training improves single leg stability in young female athletes. *Journal of Orthopaedic Sports Physical Therapy, 34* (6), 305–316.

Pizzari, T., Coburn, P.T. and Crow, J.F. (2008) Prevention and management of osteitis pubis in the Australian Football League: A qualitative ananysis. *Physical Therapy in Sport, 9* (3), 117–125.

Pull, M.R. and Ranson, C. (2007) Eccentric muscle actions: Implications for injury prevention and rehabilitation. *Physical Therapy in Sport, 8,* 88–97.

Rassier, D.E. and Herzog, W. (2005) Considerations on the history dependance of muscle contraction. *Journal of Applied Physiology, 96,* 419–427.

Reimer, R.C. and Wikstrom, E.R. (2009) Functional fatigue of the hip and ankle musculature cause similar alterations in single leg stance postural control. *Journal of Science and Medicine in Sport, 13* (1), 161–166.

Risberg, M.A., Mork, M., Jenssen, H.K. and Holm, I. (2001) Design and implementation of a neuromuscular training program following anterior cruciate ligament reconstruction. *Journal of Orthopaedic Sports Physical Therapy, 31* (11), 620–631.

Roberts, J. and Wilson, K. (1999) Effect of stretching duration on active and passive range of motion in the lower extremity. *British Journal of Sports Medicine, 33* (4), 259–263.

Robinson, L.E., Devor, S.T., Merrick, M.E. and Buckworth, J. (2004) The effect of land vs aquatic plyometrics on power, torque, velocity, and muscle sorenes in women. *Journal of Strength and Conditioning Research, 18* (1), 84–91.

Roig-Pull, M. and Ranson, C. (2007) Eccentric muscle actions: Implications for injury prevention and rehabilitation. *Physical Therapy in Sport, 8,* 88–97.

Sady, S. P., Wortman, M. and Blanke, D. (1982) Flexibility training: ballistic, static or proprioceptive neuromuscular facilitation? *Archives of Physical Medicine and Rehabilitation, 63* (6), 261–263.

Schache, A.G., Wrigley, T.V., Baker, R. and Pandy, M.G. (2008) Biomechanical response to hamstring muscle strain injury. *Gait & Posture, 29,* 332–338.

Schleip, R. (2003) Fascial plasticity—a new neurobiological explanation: part 1. *Journal of Bodywork and Therapies, 7* (1), 11–19.

Schultz, R.A., Miller, D.C., Kerr, C.S. and Miscell, L. (1984) Mechanoreceptors in human cruciate ligaments. *Journals of Bone Joint Surgery, 66-A,* 1072–1076.

Sefton,, J. (2004) Myofascial release for athletic trainers, part 1: theory and session guidelines. *Athletic Therapy Today, 9* (1), 48–49.

Sharman, M.J., Cresswell, A.G., Riek, S. (2006) Proprioceptive Neuromuscular Facilitation Stretching: Mechanisms and Clinical Implications: vol *36* p-929–39; Sports Medicine: School of Human Movement Studies

Shiner, J., Bishop,T. and Cosgarea, A.J. (2005) Integrating low-intensity plyometrics into strength and conditioning programs. *Journal of Strength and Conditioning Research, 27* (2), 10–20.

Simons, D.G., Travell, J.G. and Simons, L.S. (1998) *Travell and Simons Myofascial Pain and Dysfunction: the trigger point manual,* vol. *1.* Baltimore, MD: Williams and Wilkins.

Sloan, J.P., Hain, R. and Pownall, R. (1989) Clinical benefits of early cold therapy in A&E following ankle sprains. *Archives of Emergency Medicine, 6,* 1–6.

Small, K., McNaughton, L., Greig, M. and Lovell, R. (2008) The effects of multidirectional soccer-specific fatigue on markers of hamstring injury risk. *Journal of Science and Medicine in Sport, 1* (1), 120–125.

Smith, M. and Fryer, G. (2008) A comparison of two muscle energy techniques for increasing muscle flexibility of the hamstring muscle group. *Journal of Body Work and Movement Therapies 13,* 312–317.

Spernoga, S.G., Uhl, T.L., Arnold, B.L. and Gansneder, B.M. (2001) Duration of maintained hamstring flexibility after a one-time, modified hold-relax stretching protocol. *Journal of Athletic Training, 36* (1), 44–48.

Stemm, J.D. and Jacobson, B.H. (2007) Comparison of land and aquatic based plyometric training on vertical jump performance. *Journal of Strength and Conditioning Research, 21* (2), 568–571.

Stockmar, C., Lill, H., Trapp, A., Josten, C. and Punkt, K. (2006) Fibre type related changes in the metabolic profile and fibre diameter of human vastus medialis muscle after anterior cruciate ligament rupture. *Acta Histochemica, 108* (5), 335–342.

Stracciolini, A., Meehan, W.P. and d'Hemecourt, P.A. (2007) *Sports Rehabilitation of the Injured Athlete: Clinical Pediatric Emergency Medicine.* Amsterdam: Elsevier.

Subasi, S.S., Gelecek, N. and Aksakoglu, G. (2008) Effects of different warm-up periods on knee proprioception and balance in healthy young individuals. *Journal of Sport Rehabilitation, 17,* 186–205.

Thomas, K., French, D. and Hayes, P.R. (2009) The effect of two plyometric training techniques on muscular power and agility in youth soccer players. *Journal of Strength and Conditioning Research*, 23 (1), 332–335.

Thuan-Lee, T. and Kapoula, Z. (2008) Role of ocular convergence in the Romberg quotient. *Gait and Posture*, 27 (3), 493–500.

Travell, J. and Simons, D. (1992) *Myofascial Pain and Dysfunction: the trigger point manual*, Vol 2. Baltimore, MD: Lippincott Williams and Wilkins.

Verrall, G.M., Slavotinek, J.P. and Barnes, P.G. (2005) The effect of sports specific training on reducing the incidence of hamstring injuries in professional Australian Rules football players. *British Journal of Sports Medicine*, 39, 363–368.

Verrall, G.M., Kalairajah, Y., Slavotinek, J.P. and Spriggins, A.J. (2006) Assessment of performance following return to sport after hamstring muscle strain injury. *Journal of Science and Medicine in Sport*, 9, 87–90.

Walden, M., Hagglund, M. and Ekstrand, J. (2006) High risk of new knee injury in elite footballers with previous anterior cruciate ligament injury. *British Journal of Sports Medicine*, 40, 158–162.

Wassinger, C.A., Myers, J.B., Gatti, J.M., Conley, K.M. and Lephart, S.M. (2007) Proprioception and throwing accurancy in the dominant shoulder after cryotherapy. *Journal of Athletic Training*, 42 (1), 84–89.

Wrbaskic, N. and Dowling, J.J. (2007) The relationship between strength, power and ballistic performance. *Journal of Electromyography and Kinesiology*

Yuktasir, B. and Kaya, F. (2009) Investigation into the long term effects of static and PNF stretching exercises on range of motion and jump performance. *Journal of Bodywork and Movement Therapies*, 13, 11–21.

Zazulak, B.T., Hewett, T.E., Reeves, N.P., Goldberg, B. and Cholewicki, J. (2007) Deficits in neuromuscular control of the trunk predict knee injury risk. A prospective biomechanical epidemiological study. *American Orthopaedic Society of Sports Medicine*, 35 (7), 1123–1130

13

Strength and conditioning

Paul Comfort and Martyn Matthews
University of Salford, Greater Manchester

To enable the appropriate development of rehabilitation programmes that take into account, not only the demands of the athlete's sport, but also the demands of their training regimes Sports Rehabilitator's require a comprehensive understanding of strength and conditioning principles and practices. Collaboration between the strength and conditioning staff and the rehabilitator(s) generally ensures that the athlete is not only appropriately prepared for a return to sport in terms of their injury, but also in terms of their physical conditioning and fitness.

Appropriate conditioning is essential not only to optimise performance, but also to reduce the risk of injury. Research has demonstrated that strength and conditioning training not only improves performance in strength, power and speed related sports and activities (Wilsoff et al. 2004; Cronin and Hansen 2005; Hori et al. 2008), but also in endurance-based sports and activities (Paavolainen et al. 1999; Spurrs et al. 2002; Turner et al. 2003). There is also a large body of evidence that has found that certain methods of strength and conditioning training can reduce injury risk (Ford et al. 2003, Hewett et al. 2005; Kato et al. 2008; Kaminski et al. 1998; Mjolsnes et al. 2004; Kilgallon et al. 2007; Holcomb et al. 2007).

The aim of this chapter is to highlight the different methods of strength and conditioning that are commonly used (emphasised) during different phases of a progressive, periodised training programme, and to summarise the adaptive responses that can be expected from each mode of training. The chapter then progresses on to a summary of how appropriate interventions within a strength and conditioning programme may reduce the risk of injury, and assist in the later stages of a rehabilitation programme, and effectively reduce the risk of re-injury.

In order to fully appreciate the concepts presented in this chapter, this should be studied in conjunction with the chapter on periodisation (Chapter 9).

In most sports, it is not the maximum force produced that determines success; it is the strength that can be produced *explosively*. For example, a sprinter does not have time to produce maximal strength in the short period it takes to leave the blocks at the start of a race. Success in the start depends on another attribute of strength; in this case the magnitude of force that can be produced quickly. The best athletes are not always the strongest but are often the most explosive. Even in highly skilled games like soccer, explosive ability (assessed by short sprint and jumping performance) can differentiate between levels of success (Brewer and Davis 1991; Kollath and Quade 1993). Explosive strength is often referred to as *power*, which is defined as the rate of performing work. The faster any given weight is lifted (or other resistance overcome) the greater the power.

Training for strength

Strength attributes can be enhanced by a number of different training methods including heavy resistance training (Newton and Kraemer 1994; Harris et al. 2008), plyometrics (Verkhoshansky 1986; Markovic 2007; Thomas et al. 2009), complex training (Fleck and Kontor 1986; Duthie et al. 2002; Weber et al. 2008), assisted and resisted training (DeRenne et al. 1990, 1994; Faccioni 1994a, 1994b; Jakalski 1998; Escamilla et al. 2000), explosive isometrics (Olsen and Hopkins 1999; Siff and Verkhoshansky 1999; Kubo et al. 2001; Burgess et al. 2007), eccentric training (Morrissey et al. 1995; Askling et al. 2003; Arnason et al. 2008), and Olympic style lifting (Garhammer and Gregor 1992; Takano 1992; Hoffman et al. 2004; Channell and Barfield 2008). Each of these methods produces adaptations that are specific to the attributes trained. For example, heavy resistance training stimulates the development of maximal strength (power-lifting); power training develops force application at high velocities (throwing and jumping events); explosive isometric training develops rate of force development at low velocities (sprint start) (Kubo et al. 2001); and eccentric emphasised training develops rapid eccentric strength which may decrease the risk of injury during deceleration and agility drills.

Explosive force production

Few sports rely exclusively on maximal strength. Of these, the most notable sport that does rely on strength is Powerlifting, where performance of the bench press, dead lift and squat is exclusively dependent on an athlete's ability to produce force. Powerlifting therefore depends more on strength than power.

In contrast, success in most other sports depends on explosive force production, or power, which in turn depends on both the magnitude of force produced (strength) and the rate at which it is applied (speed). Of these, it is often the *rate of force development* that is more important to sporting success than the magnitude of force developed. For example, in the athletic throwing events it is not the heaviest lifters who throw the discus, shot, hammer, and javelin the furthest, but it is often the most explosive. The most common methods of training for the development of power include Olympic style lifting, plyometrics and complex training.

Whilst there is a relationship between maximal strength and power-based performance such as sprinting and jumping (Wilsoff et al. 2004; Cronin and Hansen 2005), strength alone does not determine success. Rather it is the proportion of that strength that can be utilised quickly and the efficiency with which that strength is integrated into sport-specific movement patterns. For example, a rookie shot putter may possess all the explosiveness necessary for elite performance, but if the power cannot be generated in sport-specific movement patterns and harnessed through correct technique then they will not achieve full potential. Training must be designed such that any gains in strength *transfer* to improved performance.

Rate of force development

Sporting movements occur quickly, often between 30 and 200 ms. The length of the contact time during maximal sprinting is between 70 and 125 ms (Kunz and Kaufmann 1981; Mann and Herman 1985; Moravec et al. 1988), which is not long enough for athletes to produce their maximum force. Therefore, prior to competition, training should focus on maximising the force that can be developed quickly (increasing rate of force development). The quicker a sprinter can generate force then the faster they will be.

Figure 13.1 represents the strength characteristics of two athletes. Athlete B clearly has the highest force production and can be considered the stronger of the two athletes. Athlete A, however, has a greater rate of force development (RFD) and is stronger than athlete B in the initial stages of contraction (up to point C). The arrows D and E represent the differences in force production between athlete A and athlete B at 125ms and 200ms respectively. As most sporting movements require explosive force production within the first 200ms then athlete A can be considered as better conditioned for their sport, and should be more successful. If, however, the athlete was a power lifter, or rugby forward, where production of maximal strength of a longer duration determines success in a lift or scrum respectively, then athlete B can be considered as better conditioned for their sport and should be more successful. The steepness of the curve represents the *rate of*

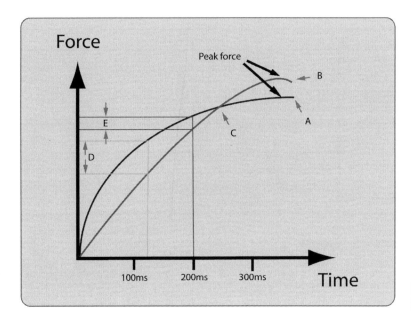

Figure 13.1 Force-time characteristics of two athletes.

force development (RFD). Of course, it is both the magnitude as well as the rate of force development that is essential. For example, Weyand et al. (2000) found that faster running speeds were attributed to greater ground reaction forces, rather than increases in limb speed.

Training to maximize RFD

Rate of force development is best enhanced by training explosive movements where the intention is to accelerate the resistance (either a bar, medicine ball, or the body) as quickly as possible. *Slow* heavy exercises increase strength (and therefore the maximum force athletes can produce), but do not increase the force development within the first 200ms of contraction, required by sport in conditioned athletes (Hakkinen and Komi 1985; Zatsiorsky 1995), although very heavy resistance exercise can increase RFD within the first 200ms, in *individuals who have not previously participated in structured resistance training*, via enhanced neural drive (Aagaard et al. 2002). It is worth noting, however, that all of these types of training play their role within a well developed *periodised* training programme.

Figure 13.2 represents the RFD characteristics of untrained, heavy-resistance trained, and explosive-ballistic trained athletes. Note that the heavy-resistance trained athletes have a higher maximum strength, but take up to 500ms to develop this strength. Explosive-ballistic trained athletes, however, can generate greater force within the 200ms timeframes typically encountered in sport. It is therefore essential that a *complete* rehabilitation programme prepares the athlete for such demands, rather than concentrating on isometric strength, or low velocity concentric strength as with most 'traditional' rehabilitation programmes. In a *periodised model* the heavy training should precede the explosive training.

Although explosive-ballistic training is effective at improving rate of force development and strength in the initial stages of muscle action (up to 200ms), it is still possible to focus training even more accurately. For example, during a sprint, ground contact times, and therefore the time over which an athlete can generate useful force, change throughout the race. At the beginning of the race, when the athlete drives out of the blocks and accelerates for the first few strides, the ground contact time can be greater than 340 and 200ms respectively (Mero 1988), whereas when the athlete reaches peak running velocity, the contact time is nearer to 70–125ms.

This knowledge is crucial to help athletes focus training more specifically on their strengths and

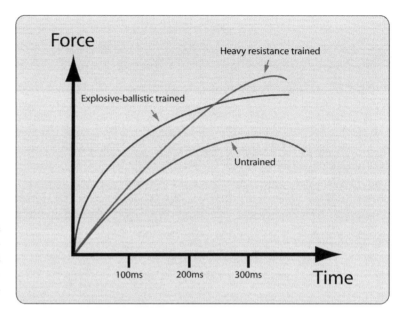

Figure 13.2 Isometric rate of force development characteristics of untrained, heavy resistance trained, and explosive-ballistic trained athletes. Adapted from Hakkinen and Komi, (1985).

weaknesses. For example, if an athlete accelerates well for the first few metres but cannot compete at the faster speeds then they will benefit from training that focuses on shorter ground contact times (*short response* or *reactivity* training) such as plyometrics. If an athlete has a high peak velocity but is slow when accelerating then they should focus on strength and power development over a longer ground contact time (*long response* training) such as weighted jump squats, Olympic lifts, or single leg hopping and bounding. (Single leg hopping has a ground contact time of approximately 230ms according to Aura and Viitasalo (1989).) This has a particular relevance for team sports where the ability to accelerate over the first few metres may be far more important to success than peak running velocity. In soccer, sprints generally last between 12 and 22m (Bangsbo et al. 1991; Drust et al. 1998).

Likewise, long-jumpers and high-jumpers exhibit different ground contact times during the take off (125ms and 200ms respectively). Training for each should therefore focus on different aspects of strength development with the long-jumper placing a greater emphasis on training exercises with a shorter contact time (*short response*) and the high jumper focusing on exercises with a longer contact time (*long response*), especially prior to competition (see Figure 13.3.).

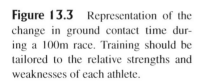

Figure 13.3 Representation of the change in ground contact time during a 100m race. Training should be tailored to the relative strengths and weaknesses of each athlete.

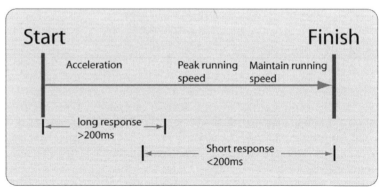

Short response training – reactivity training

Short response (reactivity) training is specifically designed to help athletes accept and produce forces over very short periods (typically 75–200ms). Short response training is therefore fundamental to maximising sprinting speed and jumping ability as these activities require athletes to accept and produce ground reaction forces quickly. The best type of training for reactivity is plyometrics, focusing on those with a short ground contact time.

Plyometric training involves very quick, light, yet powerful activities like quick feet, skips and low amplitude jumps. Emphasis should be placed on speed, maximising the rate of stretch, minimising ground contact time, and good technique, rather than height jumped, the height of the box jumped from and the length of the bound. Although the height of the subsequent jump does increase with increasing drop heights (due to an increase in the rate of stretch at landing) this only occurs up to a threshold height (Bosco and Komi 1979). Above this height, the rate of stretch is too high and the golgi tendon organ (GTO) reflex inhibits muscle contraction (Schmidtbleicher et al. 1988), lessening subsequent jump height. Training aimed at developing the reactive elements of muscle contraction should not use drop heights greater than this threshold. Although Kreighbaum and Katherine (1996) suggest a platform height of no more than 20cm to minimise the risk of injury, for many athletes the threshold occurs at heights much greater than this, although usually no higher than 60cm. It is also worth noting that plyometrics can be effectively performed in water (Robinson et al. 2004; Stemm and Jacobson 2007), therefore reducing the impact forces, due to the buoyancy of the water, and allowing plyometric training to be introduced to athletes at an earlier stage in their development/rehabilitation.

It is also interesting to note that plyometric training has been shown to have a beneficial effect on tendon and joint stiffness (Burgess et al. 2007; Kubo et al. 2007a), which may reduce the risk of injuries.

The type and intensity of exercises incorporated into the training session will determine the number of repetitions performed. For example, mini-hurdle jumps, where the emphasis is on minimising ground contact after a relatively low drop height, can be performed in multiple sets (5) of high quantity (25),

Table 13.1 Short response training

Intensity	Exercise	Reps	Sets
Low	Mini hurdle jumps, cone jumps, line jumps	25	5
Medium	Hexagon drill, line hops	15–20	3
High	Full hurdle jumps, depth jumps	6–10	3–5

high quality (rapid; short contact time) jumps. In contrast, full-hurdle jumps, where the athlete must clear a height of one metre with each jump, elicit a greater impact velocity and, therefore, higher impact forces on landing. To maintain the quality and intensity of movement, the number of repetitions should be kept low (four–eight). Higher intensity exercises (single leg hops, or single leg depth jumps in well-conditioned athletes) may be limited to one to four quality repetitions per set. See Table 13.1 for a summary.

It is essential to remember that plyometric training places a high neurological demand on the athlete and is therefore most effective when athletes are well rested.

Movements such as depth jumps, hopping, and bounding typically have longer ground contact times and come under the heading of *long response* training.

Long response training

Long-response training refers to the training of powerful explosive movements with a contact or application time of more than 200ms. By comparison with short response training long response training is characterised by greater amplitude of movement about the hip, knee, and ankle joints and also by the use of a greater external resistance. Whereas short-response training is dominated by quickness of movement, long response training is dominated more by strength.

Long response movements include activities ranging in speed from faster activities of around 200ms (bounding, tire pulling, single leg hopping and depth jumping) to longer activities of up to 500ms (unweighted squat jumps, one legged jump-squats and weighted jump squats). This type of training can enhance force production, acceleration from a standing start and height jumped. Bosco (1982) found that the

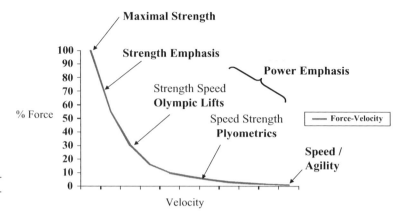

Figure 13.4 Force velocity relationship applied to training methodologies.

inclusion of depth jumps in training resulted in an increased jump height. Research has also demonstrated that including drop jumps (five jumps from a 60cm box) within a dynamic warm up result in an increase in subsequent power based activities (Hilfiker et al. 2007). It is essential that the training programme meets the physiological, biomechanical and metabolic requirements of the sport; these will be introduced later on in the chapter.

Isometric training

In contrast to the beneficial effects of explosive training compared with heavy strength training, research has demonstrated that isometric training can also result in increases in RFD, power and vertical jump height, which is attributed to increases in tendon stiffness (Kubo et al. 2001; Bojsen-Moller et al. 2005; Kubo et al. 2007b; Burgess et al. 2007). It is worth noting, however, that this type of isometric training must be performed with the intention of explosive movement, and should not be the sole method of training. It is generally incorporated into the 'normal' training session as a few additional sets.

A key concept to understand is the force velocity relationship of skeletal muscles, applied to training methodologies and the requirements of sporting activities (see Figures 13.4 and 13.15). This includes both concentric and eccentric muscle actions that are responsible for acceleration and deceleration respectively, which therefore corresponds to improvements in agility. It is essential that these concepts are applied when identifying the mechanics of the sport/activity and integrated into the exercise selection process.

Intensity

Overload

This is usually referred to as 'progressive overload'. More often, however, a form of 'fluctuating overload' (Figure 13.5) is applied as part of a periodised training programme (Siff 2004). Failure to periodise in this manner by attempting an unlimited linear progression (Figure 13.6) will either result in stagnation and a plateau in training (Figure 13.7), or increase risk of injury due to lack of appropriate adaptation. See Chapter 9 for more detail on periodisation.

Overload and super-compensation

For adaptations to occur, the training stimulus must be at a level beyond that normally encountered. Training at a level or intensity that is beyond that normally encountered is called *overload*, and it is this overload that stimulates the adaptations that allow the athlete to tolerate an increased level of training stress.

Training causes physiological, biochemical and mechanical stresses. Immediately post training, these result in tissue damage and metabolic, neural and psychological fatigue, limiting performance and making the athlete (temporarily) less able to perform. (For a period post-training the athlete is actually less fit). Once training has finished, however, recovery begins. This process involves restoration of physiological and biochemical balance (homeostasis), repair of tissue and replenishment of muscle

INTENSITY 229

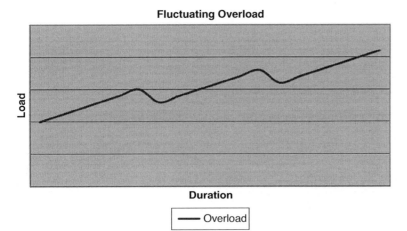

Figure 13.5 Fluctuating Overload.

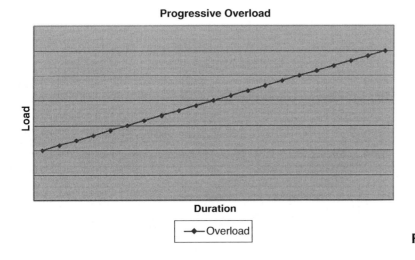

Figure 13.6 Progressive Overload.

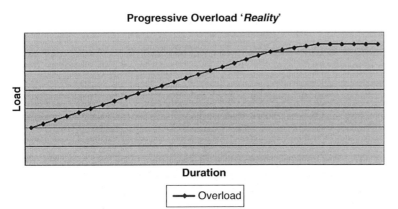

Figure 13.7 Progressive Overload Reality.

230 STRENGTH AND CONDITIONING

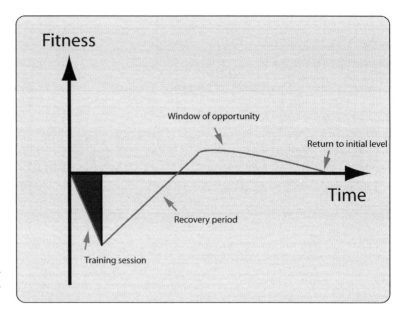

Figure 13.8 The principle of super-compensation from one training session.

and liver fuel stores. Depending on the type, intensity and duration of the training – and the quality of rest and nutrition – full recovery can take from a few hours to several days (or weeks, depending on the intensity and duration of training).

Once pre-exercise levels have been restored, recovery is still not complete. The body continues to adapt, rebuilding and remodelling itself until a higher state of readiness is reached. For a short period of time, an athlete's fitness is enhanced. This is known as *super-compensation* (adaptation). The time period during which super-compensation occurs provides a *window of opportunity* during which the athlete should train again. If the athlete trains during this period of super-compensation, the cycle is repeated and fitness continues to improve. If, however, the athlete does nothing during this key period then the body will re-adapt to the current, lower activity level. Any enhanced state-of-readiness or fitness that occurred as a result of super-compensation will be lost. Figure 13.8 illustrates the principle of super-compensation from one training session.

Timing

The *timing* of the next training session is vital and varies according to the recovery abilities of the athlete and the demands imposed by the previous session. For example, a well-conditioned athlete can recover from a long training run within 24 hours (excluding the skeletal system), whereas the same athlete may take 10 days to recover from a heavy resistance training session involving rapid eccentric activities and supra-maximal loads.

Once an athlete has adapted to a specific training load or stimulus (for example running five miles, three times per week at a constant pace), then further gains can only be made if the training load is increased. This is known as the principle of *progressive overload*. The runner (above) will not improve beyond the fitness level needed to run five miles, three times per week, at that constant pace. Once that level of fitness is achieved it will remain static. In order to improve their running ability further the stimulus must be increased. The runner could run faster, farther, more often, use interval training or run on a more challenging route.

The progressive increases in fitness due to repeated training sessions are illustrated in Figure 13.9 Note that each successive training session coincides with the point during recovery when gains from the previous session have been maximised – the 'window of opportunity'.

Although the stimulus to improve comes from the training session itself, all improvements occur when the athlete rests. If the rest period is too short, and the athlete trains again before full recovery has

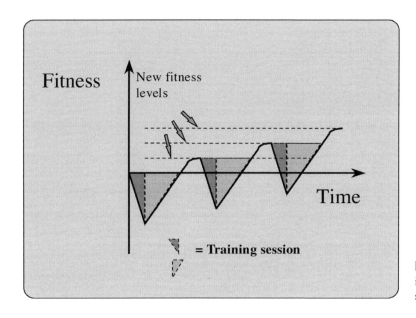

Figure 13.9 Progressive increases in fitness due to successive training sessions.

occurred, then fitness will cease to improve and, if this cycle continues, may even decline (Figure 13.10). During such situations, athletes are more susceptible to overtraining syndrome, injury and illness. Adequate recovery is therefore essential to a successful training programme.

It should be noted that there are times when athletes may perform successive hard training sessions in a block, without allowing full recovery. For a few days, the athletes accumulate fatigue. This shocks the system and gives an opportunity – with sufficient recovery later – to stimulate further adaptations (Figure 13.11), and can also be used when ensuring that an athlete peaks for a specific competition (Pistilli et al. 2008). This is an advanced training method known as *shock training* (planned *Overreaching*) and should only be used by athletes who already have a solid training base. Careful observation and monitoring is required as shock training itself represents the initial stages of overtraining, a condition

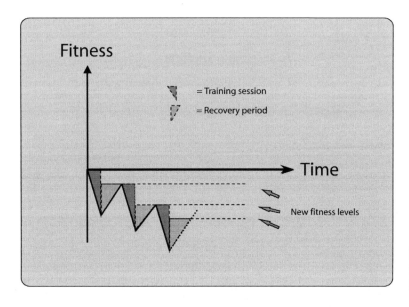

Figure 13.10 Negative adaptation – fitness decrements occurring as a result of inadequate recovery.

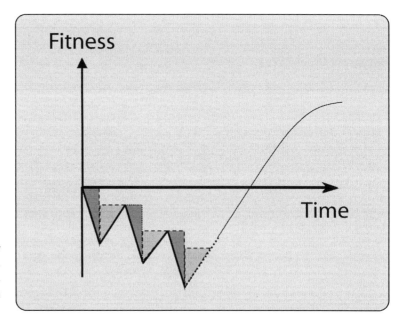

Figure 13.11 Shock training / overreaching – repeated training sessions with inadequate recovery, followed by an extended rest period can lead to a jump in fitness levels.

that is associated with both increased injury and illness risk and well as a reduction in performance. There is often a fine line between providing the optimum stimulus for adaptation and inducing a state of over-training and susceptibility to injury and illness.

Adaptation potential

An athlete's *adaptation potential* (or capacity for improvement) is a function of both their genetic makeup and their current level of fitness. Genetic makeup, an athlete's ultimate potential for success, is fixed and therefore not responsive to training (Brzycki 1989). The major factor determining an athlete's adaptation potential then is their current state of fitness relative to their genetic ceiling.

The primary training aim of all athletes should be to fulfil their genetic potential. This is never easy. As athletes get fitter and more skilled they find it increasingly more difficult to make further improvements – they get diminishing returns for their efforts. As they approach their genetic ceiling, further improvements become almost impossible (Figure 13.12).

Conversely, athletes that are unfit have the greatest capacity to improve. For these athletes, increases in fitness occur rapidly in response to training. These rapid improvements, however, can occur independent of training quality. Even poorly designed training programmes will elicit improvements in the short term. In the medium term, however, such training will plateau, bad habits can form, and the athlete may not fulfil their true potential.

It is crucial; therefore, that athletes and coaches adopt the same, thorough and careful approach to training programme design with all athletes, whatever their age and training status. A well-designed programme in the early stages ultimately lays the foundation for future success.

Training parameters

The training parameters that contribute towards the overall training effect include: *frequency*, *intensity*, and *duration* of training sessions; the quality and quantity of *rest* between and during sessions; and *specificity*. Each of these parameters can be combined and manipulated within a training programme to elicit the desired training effect.(See Tables 13.2 and 13.3 for specific recommendations.)

Rate of adaptation

Different fitness attributes respond to training at different rates. For example, improvements in

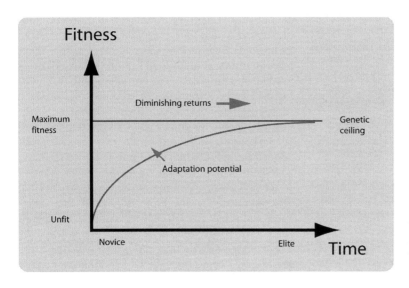

Figure 13.12 Adaptation potential (adapted from Balyi 2001) – Novice athletes have a greater adaptation potential (capacity for improvement) than elite athletes.

Table 13.2 Definitions of training parameters

Training parameters	Definition	Example
Frequency	How often	Training 3x per week; daily
Intensity	How hard	Running at a speed of 10mph; lifting a resistance of 90% of maximum capacity
Duration	How long (in terms of time, distance, or volume)	Cycling for 20 minutes; running for 10 miles; performing a number of sets/repetitions
Rest	What an athlete does to recuperate. (Both quality and quantity are important.)	Two days between each session; 4 minutes between each set. Type: sleep; massage; active rest. Diet: high carbohydrate
Specificity	Type of training	Focus on endurance; speed; sport specific movement patterns

Table 13.3 General recommendations for the development of different fitness attributes

Strength	Endurance	Power		Power (Plyometrics)
1–2 × week	2–3 × week	2–3 × week		2–3 × week
<60 mins	<60 mins	<60 mins		<60 mins
2–6 reps	≥10 reps	Single 1–2	Multi* 3–5	≥10 reps
4–6 sets	3–4 sets	4–8	3–5	4–6 sets
85–95% 1RM	≥75% 1RM	80–90% 1RM	75–85% 1RM	≤30% 1RM

*It is also worth noting that power output can deteriorate during a set of 3–5 repetitions, however this can be minimised, and overload maximised by performing each set as a cluster set (Haff et al. 2003, 2008; Lawton et al. 2006). Each set is performed with a 15–30s rest between each repetition of the set, therefore minimising neurological and metabolic fatigue.

Table 13.4 Adaptation rates following training

Fitness attribute	Rate of adaptation
Flexibility	Days
Strength	Weeks
Sport specific endurance	Weeks to months
Speed	Several months
Work capacity	Months to years

flexibility can occur within days, whereas improvements in speed may take months. In addition, strength improvements can occur within weeks; sport specific endurance may take weeks and months; and overall work capacity will only improve gradually over a period of months and years. Athletes and coaches must also be aware that different body systems adapt at different rates following training (see Table 13.4). For example, with strength training, the rate of adaptation of tendons and bones is slow compared to the rate of adaptation of muscle. The consequences of this are that muscles get stronger quicker and pull on tendons that are yet to adapt to the stresses of training. This leads to the potential for injury.

Detraining

All improvements in fitness are reversible. What an athlete can gain with training, they can lose with detraining. When athletes stop or reduce their training, their body systems adjust accordingly. This presents as a loss of strength and power (Narici et al. 1989; Hakkinen et al. 2000; Weir et al. 1997), a decrease in maximal aerobic capacity (Mujika and Padilla 2001), and a loss of coordination. Although detraining of strength and power is relatively slow (Colliander and Tesch 1992; Hortobagyi et al. 1993), particularly with the inclusion of some eccentric exercise (Housh et al. 1996), the detraining of endurance is more rapid with significant effects noticed in days and weeks (Coyle et al. 1986).

Overall, the de-training effects of a short-term break (1–2 weeks) are relatively small, especially if physical activity levels remain high, however longer or more frequent periods of inactivity have a greater cumulative impact on overall fitness. Thus, when athletes resume training after a layoff or injury, they cannot start at the level (intensity, duration or complexity) at which they left off. Care must therefore be taken when designing training sessions that occur after an injury or other long lay-off. In such cases, a gradual and progressive re-introduction to full training will carry fewer risks of re-injury and produce more sustainable gains than other more intense methods. Many athletes and coaches are keen to make up for lost time and consequently train too-hard, too-soon, risking both injury and possible over-training.

It is far better, therefore, to *minimise fitness losses* in the first place. To minimise fitness losses, injured athletes must examine strategies, within their treatment and rehabilitation, which enable them to continue to exercise. One such strategy may involve *the use of alternative low-risk activities that do not stress the injury*. For example, rowing, stationary cycling and deep water running are all excellent methods of maintaining general cardiovascular fitness when injured; strength training can be targeted towards key movements in non-injured areas; whilst proprioception training of the non-injured limb can provide useful crossover benefits to the injured limb.

Specificity

The most important element of functional conditioning is the *principle of specificity*. The principle of specificity dictates that training should match the demands of the sport. Given that sport consists of many components – with successful athletes needing to start, stop, twist, turn, run, jump, land, shuffle, push, pull, hit, bend, throw, catch, hop, accelerate, decelerate, slide, block, and barge, all within the fluid, dynamic, and unpredictable environment of sport, and all without getting injured – we are presented with a complex problem. It is also worth noting, however, that open kinetic chain exercises, such as leg flexion, have little effect on performance of sport specific movements (Augustsson et al. 1998; Blackburn and Morrissey 1998; Augustsson and Thomee 2000), but may help to address muscle imbalances.

Training that does not relate directly to the movements and activity patterns found in sport, is a waste. Unfortunately, many athletes still employ training methods that are non-sport specific. These methods are often prescriptive, predictable and sterile. Many are based on one-size-fits-all training systems reported to be the secret of another athlete's success.

Many lack sufficient variety, using repetitive schedules that soon lead to a plateau in performance. Most fail to account for the fluid, unpredictable and three-dimensional nature of sport. In essence, they fail to stimulate athletes in a truly sport specific way.

Despite the best intentions of coaches and athletes, too much conditioning time and energy is directed towards superfluous, non-specific activity. For example, when strength training, many athletes monitor their progress by how much they can squat or bench press, and focus training towards these exercises. Sport, however, typically requires strength to be expressed in a faster, more dynamic, and in a three-dimensional fashion. Although a relationship has been demonstrated between squat strength and both sprint speed and vertical jump height (Wilsoff et al. 2004), and heavy bench press appears to be an appropriate pre-load activity for performing complex training on functional movements (Matthews et al. 2009), squatting and bench press *alone* do not prepare an athlete for all movement patterns required in sport. Too much training time devoted to non-functional movements takes training focus away from sports specific activities and causes extra fatigue, forcing the body to recover from training that does have a direct affect on performance. However, the reverse is also true, where athletes spend too much time focusing on sports specific training and lack adequate strength and power development required to produce an increase in sports performance. Less sports specific movements, however, can provide a good solid base (general preparation phase) from which an athlete can progress to more specific conditioning (pre-competition and peaking phases) (see Chapter 9: An Introduction to Periodisation).

To condition athletes effectively, training must mimic the conditions encountered in sport. To achieve this, programme designers must: (1) analyse the demands of the sport; (2) identify the individual characteristics of the athlete (strengths and weaknesses; training history); and (3) tailor and prioritise training to allow each individual athlete to meet these specific demands.

It is essential that the training programme is specific in terms of the metabolic demands (energy systems) (see Chapter 3; Needs Analysis section) and mechanical demands (muscle action, force, velocity, range of motion).

Mechanical demands

The mechanical demands of sport determine the movements that athletes should train. Exercises that mimic the actual movements encountered in sport should be prioritised. By focusing on *movement pattern specificity* athletes can reinforce and condition the actual motor programmes used in skilled performance. These programmes control the precise order, timing, velocity, force application and muscle action to enable the muscles to produce a predetermined movement (Enoka 1994). The more practiced and efficient these programmes, the better the performance of the skill. For example, a rugby player who focuses practice on the foot patterns required to side step an opponent can enhance side-stepping performance by executing quicker, more efficient motor programmes. However, it is also worth noting that if the athlete does not have the eccentric strength to enable them to decelerate, or the power to re-accelerate in another direction practising footwork will have little effect.

The best training for movement pattern specificity is to practice the actual skills involved in sport, however, ensuring progressive overload in these activities is not possible. Repetition of the skill is possible for many sports; however there are sports and situations where it is not practical or safe to perform high volumes of the specific patterns, frequently. A triple-jumper or long-jumper, for example, who only trains by performing the actual jump skills would soon breakdown due to the constant repetition of intense impacts. For these athletes other, less intensive, methods should be used that mimic closely the actual movements involved. These methods focus on the ranges, speeds and forces encountered, allowing the development of fitness attributes in a functional way that may ultimately allow athletes to tolerate more actual (jumping) practice. For example, *heavy force-acceptance strength* training, *fast force-acceptance strength* training, *dynamic control and stabilisation* training, along with *specific jumping drills* will increase an athlete's capacity to tolerate actual skill practice in the highly intensive triple jump.

The training methods that transfer best to actual sporting performance usually involve coordinated movements across multiple joints rather than strict isolation exercises. In sport, *no muscle works in isolation*. Isolating specific muscles, then, is non-functional; gains in strength, power, or endurance occur only in the trained muscle and fail to integrate with the whole movements required for sporting performance. Consideration of how to *train movements, not muscles* is *essential*. Training that focuses on the whole movement enhances functional performance more effectively than the training of isolated joint movements; *integrate, don't isolate*.

For example: A sprint start requires rapid and forceful triple extension of the hip, knee, and ankle, along with rapid flexion at the shoulder on the same side and rapid extension on the opposite side, plus a torso that is rigid enough to control and transfer forces from one body segment to another. Athletes must therefore train movements that target these areas. Multi-joint all-body dynamic movements that are uni-lateral or cyclic (one legged jump squats; split jump lunges) are far more specific than single joint bi-lateral isolation exercises performed slowly (leg extensions; leg curls). Traditional basic strength exercises, like back-squats and squat-jumps, may not be cyclical, but allow the athlete to focus on generating maximal forces as explosively as possible, as required in the initial strides during a sprint start. They are sport specific in terms of being multi-joint triple extension exercises and forces generated, however they are non-sport specific in terms of their velocity and their bi-lateral nature. Typically, sporting movements involve standing or running, producing and accepting forces in multiple directions and planes, and at various speeds, all in a fluid and ever-changing environment. Training should mimic this. Exercises that require athletes to stand on their feet are always more specific to sport than exercises that require them to sit or lie. *Sport is not played lying down*. Equally, exercises that involve multiple joints resemble sporting movements far better than those involving single joints. As previously mentioned it is essential to note that single joint exercises have little effect on performance of sports specific movements (Augustsson et al. 1998; Blackburn and Morrissey, 1998; Augustsson and Thomee, 2000).

Focusing training towards the actual mechanics of the movements (magnitude of force, direction of force, velocity of movement, muscle actions involved, level of stability required, etc.) required by sport is clearly paramount to optimal conditioning. In sport, however, success rarely relies on the execution of one single discreet movement; there are always other movements, either preceding or following the current one. One focus (and often neglected area) of training is training athletes to transfer fluidly and efficiently from one movement to another. In sport, *no movement ever stands alone*. For example, a tennis serve appears to be a one-off stand-alone skill; however this is not the case. Players must be ready to sprint or lunge to any point on the court to intercept the return ball. The follow through from a serve will merge with the next skill, and the more efficiently this occurs the quicker a player can intercept and play the next shot. Equally, if a tennis player can explode laterally from a sideways lunge position (a forehand on the edge of the court) then they stand a better chance of reaching the next shot. If prior to this, however, they exhibit inefficient deceleration mechanics and poor force-acceptance going into the shot, then the whole movement will be slower.

Success in sport, then, requires athletes to train in a sport specific way. By analysing the movement patterns (force, velocity, muscle action, joint angles) involved in sports and replicating those in training, athletes can not only develop strength, speed, power, endurance, and flexibility in a truly functional way, but also integrate and perfect combinations of movements to replicate typical patterns of play.

As well as training to *produce force* we must also focus on the athlete's ability to *accept force*. Sport requires athletes to reduce and absorb external forces often at high speeds, in three dimensions, and in an unpredictable fashion. Athletes must train for deceleration and force-acceptance as well as force production. This will prepare athletes to meet the varied stresses of sport and have a major impact on injury prevention.

For example: The hamstrings are typically trained *concentrically* using knee flexion exercises (leg curls) and hip extension exercises (Romanian deadlifts) (Figure 13.13). In function, however the hamstrings also act *eccentrically* to control and decelerate the limb (as in kicking a football) (Smith, 1999), act antagonistically to the rectus femoris to prevent hip flexion (as in squatting or jumping), and act *eccentrically* as an ACL agonist by preventing anterior tibial translation thereby reducing sheer forces and increasing knee stability (Aagaard et al. 1998; Li et al. 1999; Escamilla et al. 2001; Kingma et al. 2004; Ahmed et al. 2006). Movements that target these attributes may include jumps, one legged squats and squat jumps, and practice kicks increasing progressively in terms of both velocity and amplitude. Prior to this, however, the hamstrings

Figure 13.13 Descent during Romanian deadlift.

Figure 13.14 Descent during Nordic hamstring lowers/curls.

must be appropriately conditioned / strengthened concentrically, and eccentrically at low velocities (for example, using Nordic hamstring lowers) (Figure 13.14).

Sport specific movement patterns are also enhanced by training at greater speeds, in multiple directions, and under varied and unpredictable conditions to challenge an athlete's balance and proprioception, enhancing their ability to stabilise joints and maintain posture, allowing the transfer of forces efficiently from one body section to another. Exercises, such as plyometrics, that incorporate fast eccentric loading in the initial phases and place a high demand on an athlete's ability to dynamically stabilise their joints under varying conditions, also allows them to develop greater control and accept higher forces quickly.

The other advantage is that training the specific movement patterns involved in sport motivates the athlete by helping them see the relevance of the training. This helps the athlete gauge their own progress in a functional way and leads to increased motivation, relaxation, training focus, and general psychological well-being.

Injury prevention

Avoidable versus unavoidable injuries

There are two types of injuries: those that can be avoided through appropriate conditioning; and those that cannot. In situations where a high-velocity, high-energy collision coincides with a vulnerable joint position, injury is usually *unavoidable*. Under such circumstances (vulnerable joint positions; high-velocity, high-energy collisions) there is little that can be done to condition against injury. Although protective clothing, rule changes, and good umpiring can limit such encounters, they can never be completely removed from the dynamic and unpredictable conditions of sport.

In contrast, if an injury occurs during the performance of a common sporting task (twisting away from an opponent; landing from a jump; sprinting for a ball) then it is likely that *appropriate conditioning* could have either prevented the injury, or lessened its severity. In this case a number of researchers have identified that activities that improve lower limb control such as drop landing and plyometric training can decrease knee valgus during deceleration related activities and therefore reduce the risk of non-contact

Table 13.5 Definition of avoidable and unavoidable injuries

Avoidable versus unavoidable injury	
Unavoidable	Characterised by high velocity, high energy collisions and vulnerable joint positions. Cannot be prevented with appropriate conditioning
Avoidable	When the injury occurs during the performance of a common sporting task (twisting away from an opponent; landing from a jump; sprinting for a ball). *Appropriate* conditioning may have either prevented the injury or lessened its severity

ACL injuries (Ford et al. 2003, Hewett et al. 2005; Kato et al. 2008; Hanson et al. 2008)

Of all the factors that may pre-dispose an athlete to injury, *lack of appropriate conditioning* is the one that can be most readily altered. Any injury that occurs without contact with other players may be avoided through correct conditioning. Injuries that occur when landing, twisting, or stopping are often the result of inappropriate muscular activation and inefficient mechanics of landing or deceleration, which may in turn be worsened under conditions of fatigue (see Table 13.5). One example of injury risk being reduced via appropriate conditioning is that of hamstring strains, with a number of authors demonstrating that emphasis on eccentric conditioning of the hamstrings (Nordic Curls) increase hamstring strength and reduce the incidence of injuries (Kaminski et al. 1998; Askling et al. 2003; Mjolsnes et al. 2004; Kilgallon et al. 2007; Holcomb et al. 2007) (see Chapter 3 Assessment and Needs Analysis).

If muscles are activated in an efficient and coordinated manner they can brace the joint under conditions of dynamic load (such as those encountered during sport), and help prevent injuries otherwise attributable to weakness or laxity of ligaments. Appropriate conditioning (involving strength, agility, endurance, force acceptance, control, technique, proprioception and reactive neuromuscular components) is crucial to both the prevention and rehabilitation of injuries.

Female athletes appear more susceptible to non-contact injuries, particularly during twisting and jumping sports, where they sustain four times as many knee injuries than males (Arendt and Dick, 1995). Reasons for this may include joint laxity (Adachi et al. 2008; Beynnon et al. 2006; Hewett et al. 2007), higher Q angle (Tillman et al. 2005), narrow inter-condylar notch (Anderson et al. 2001; Dienst et al. 2007) smaller ACL (Anderson et al. 2001; Chandrashekar et al. 2005; Dienst et al. 2007), and lack of appropriate conditioning (Li et al. 1999; Cowling and Steel, 2001), specifically poor neuromuscular control and gluteal activation. Of these, only lack of conditioning can be addressed. As approximately 80% of all Anterior Cruciate Ligament (ACL) injuries occur without physical contact between players (Noyes et al. 1983; Griffin et al. 2000; Yu and Garrett, 2007), preventative conditioning strategies become paramount, particularly in female athletes (see Table 13.6).

Integrating strength and conditioning into a rehabilitation programme

In the past, the main focus of rehabilitation has been on healing and re-establishing mobility, strength, and endurance about an injured joint. These goals, whilst important components in any rehabilitation programme, are often assessed in ways that are not specific to the individual, taking little account of the specific demands (both on the injured area and the body as a whole) imposed by the sport. For example, strength can been assessed over a single joint (via isokinetic or isometric methods), without account of the specific movement patterns required by

Table 13.6 Possible reasons why female athletes are more susceptible to non-contact injuries

Reason	Can it be altered?
Joint laxity	✗
Higher Q angle	✗
Narrow inter-condylar notch	✗
Smaller ACL	✗
Lack of appropriate conditioning	✓

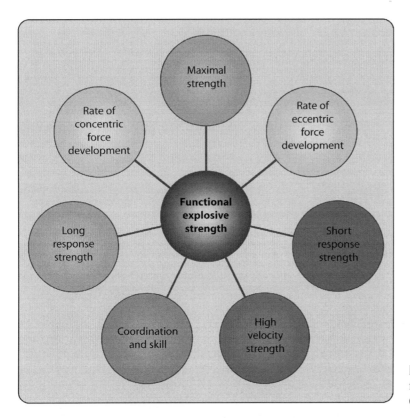

Figure 13.15 The multiple factors used to train functional (Sport-Specific) explosive strength.

sport; ROM has often been assessed passively over single joints and in single planes, rather than actively during sport specific tasks; and endurance has been assessed by the ability to perform a prescribed number of exercise repetitions, rather than the ability to stabilise joints over repeated functional movements and under the conditions of fatigue that mimic those during a game. Athletes can walk out of a clinic with pre-injury levels of strength, endurance and ROM, and, by traditional measures of progress, be considered fit enough to return to sport (see Figure 13.15 and Table 13.7). Unfortunately, achieving clinical and non-functional (non-sport specific) goals means little when it comes to actual sporting performance and prevention of re-injury.

Strength, endurance, and mobility are important factors in rehabilitation but are only effective for performance and injury prevention when combined with adequate *proprioception* and *neuromuscular control*, and integrated into the complex and coordinated movement patterns and skills that characterise sport. These skills typically require precision, occur at high movement velocities in multiple planes, and depend on a constant stream of sensory information

Table 13.7 Traditional versus sport specific (functional) rehabilitation

Traditional rehabilitation	Focuses on the restoration of mobility, strength, and endurance, allowing progress to be monitored via a number of simple tests including isokinetic assessment, ROM assessment, and measures of aerobic fitness
Functional rehabilitation	Bridges the gap between traditional rehabilitation and the functional demands of sport via a sequential progression of ever more challenging and specific exercises

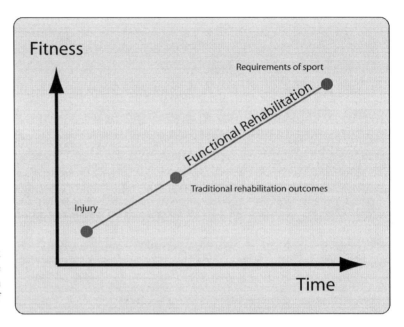

Figure 13.16 Functional (sport specific) rehabilitation bridges the gap between traditional rehabilitation outcomes and the requirements of sport.

to perform properly (see periodised training plan for an athlete post-ACL surgery in Chapter 9, An Introduction to Periodisation).

Sport specific rehabilitation provides a framework for athletes to bridge the gap between these traditional rehabilitation outcomes and the specific demands of sport, and involves a sequential progression of ever more challenging and specific exercises. Athletes begin rehabilitation as soon as possible post-injury and continue until full functional capacity (specific to the demands of their sports) is restored. At each stage, athletes will perform exercises appropriate to their level of healing, and their level of proprioception and control, whilst aiming to incorporate functional elements from their sport. In the early phases post-injury the exercises will be simple and predictable in nature, performed slowly, with minimal loading. In the later stages exercises will be complex, high velocity, high load, multiple plane and unpredictable.

When designing rehabilitation, or conditioning programmes, coaches and rehabilitators must never lose sight of the **specific** demands imposed by the sport, including the ability to exhibit rapid and co-ordinated responses to the ever-changing environments. For a comparison of approaches to rehabilitation see Table 13.7 and Figure 13.16.

References

Aagaard, P., Simonsen, E.B., Magnusson, S.P., Larsson, B. and Dyhre-Poulsen, P. (1998) A new concept for isokinetic hamstring:quadriceps muscle strength ratio. *American Journal of Sports Medicine*, 26 (2), 231–237.

Aagaard, P., Simonsen, E.B., Anderson, J.L., Magnussun, P., and Dyhre-Poulsen, P. Increased rate of force development and neural drive of human skeletal muscle following resistance training. *Journal of Applied Physiology*, 93 (4), 1318–1326.

American Journal of Sports Medicine, Adachi, N., Nawata, K., Maeta, M., and Kurozawa, Y. (2008) Relationship of the menstrual cycle phase to anterior cruciate ligament injuries in teenaged female athletes. *Archives of Orthopaedic Trauma Surgery*, 128 (5), 473–478.

Ahmed, C.S., Clark, A.M., Heilmann, N., Schoeb, J.S. Gardner, T.R. and Levine, W.N. (2006) Effect of gender and maturity on quadriceps to hamstring ratio and anterior cruciate ligament laxity. *American Journal of Sports Medicine*, 34 (3), 370–374.

Anderson, A.F., Dome, D.C., Gautam, S., Awh, M.H., and Rennirt, G.W. (2001) Correlation of anthropometric measurements, strength, anterior cruciate ligament size, and intercondylar notch characteristics to sex differences in anterior cruciate ligament tear rates. *American Journal of Sports Medicine*, 29 (1), 58–66.

Arendt, E. and Dick, R. (1995) Knee injury patterns among men and women in collegiate basketball and soccer. NCAA data and review of literature. *American Journal of Sports Medicine*, 23, 694–701.

Arnason, A., Anderson, T.E., Holme, I., Engebretson, L. and Bahr, R. (2008) Prevention of hamstring strain in elite soccer: an intervention study. *Scandinavian Journal of Medicine and Science in Sports*, 18 (1), 40–48.

Askling, C., Karlsson, J. and Thorstensson, A. (2003) Hamstring injury occurrence in elite soccer players after pre-season strength training with eccentric overload. *Scandinavian Journal of Medicine and Science in Sports*, 13 (4), 244–250.

Augustsson, J. and Thomee, R. (2000) Ability of closed and open kinetic chain tests of muscular strength to assess functional performance. *Scandinavian Journal of Medicine and Science in Sports*, 10, 164–168.

Augustsson, J., Esko, A., Thomee, R., Svantesson, U. (1998) Weight training of the thigh muscles using closed vs open kinetic chain exercises: a comparison of performance enhancement. *Journal of Orthopaedic Sports and Physical Therapy*, 27 (1), 3–8.

Aura, O. and Viitasalo, J.T. (1989) Biomechanical characteristics of jumping. *Journal of Applied Biomechanics*, 5 (1), 89–98.

Balyi, I. (2001) *Long-Term Athlete Development*. High performance coaching workshop. UK National Coaching Foundation, Leeds.

Bangsbo, J., Nørregaard, L. and Thorsø, F. (1991) Activity profile of competition soccer. *Canadian Journal of Sport Sciences*, 16 (2), 110–116.

Beynnon, B.D., Johnson, R.J., Braun, S., Sargent, M., Bernstein, I.M., Skelly, J.M. and Vacek, P.M. (2006) The relationship between menstrual cycle phase and anterior cruciate ligament injury: A case-control study of recreational alpine skiers. *American Journal of Sports Medicine*, 34 (5), 757–764.

Blackburn, J.R. and Morrissey, M.C. (1998) The relationship between open and closed kinetic chain strength of the lower limb and jumping performance. *Journal of Orthopaedic Sports and Physical Therapy*, 27 (6), 430–435.

Bojsen-Moller, J., Magnussen, S.P., Rasmussen, L.R., Kjaer, M., and Aagaard, P. (2005) Muscle performance during maximal isometric and dynamic contractions is influenced by the stiffness of the tendinous structures. *Journal of Applied Physiology*, 99 (3), 986–94.

Bosco, C. and Komi, P.V. (1979) Mechanical characteristics and fiber composition of human leg extensor muscles. *European Journal of Applied Physiology*, 24, 21–32.

Bosco, C., Komi, P.V., Pulli, M., Pittera, C., and H. Montonev, 1982. Considerations of the training of elastic potential of human skeletal muscle. Volleyball Tech. J. 1: 75–80.

Brewer, J. and Davis J. (1991) A physiological comparison of English professional and semi-professional players. Second world congress on science and football. Eindhoven: Routledge

Brzycki, M. (1989) *In A Practical Approach to Strength Training*, revised edn. Indianapolis, IN: Masters Press, pp 13–19.

Burgess, K.E., Connick, M.J., Graham-Smith, P., and Pearson, S.J. (2007) Plyometric vs. Isometric training influences on tendon properties and muscle output. *Journal of Strength and Conditioning Research*, 21 (3), 986–989.

Chandrashekar, N., Slauterbeck, J. and Hashemi, J. (2005) Sex-based differences in the anthropometric characteristics of the anterior cruciate ligament and its relation to intercondylar notch geometry. *American Journal of Sports Medicine*, 33 (10), 1–7.

Channell, B.T. and Barfield, J.P. (2008) Effect of olympic and traditional resistance training on vertical jump performance in high school boys. *Journal of Strength and Conditioning Research*, 22 (5), 1522–7.

Colliander, E.B. and Tesch, P.A. (1992) Effects of detraining following short term resistance training on eccentric and concentric muscle strength. *Acta Physiologica Scandinavica*, 144 (1), 23–29.

Cowling, E.J. and Steel, J.R. (2001) Is lower limb muscle synchrony during landing affected by gender? Implications for variations in ACL injury rates. *Journal of Electromyography Kinesiology*, 11 (4), 263–286.

Coyle, E.F., Hemmert M.K. and Coggan A.R. (1986) Effects of detraining on cardiovascular responses to exercise: role of blood volume. *Journal of Applied Physiology*, 60 (1), 95–99.

Cronin, J.G. and Hansen, K.T. (2005) Strength and power predictors of sports speed. *Journal of Strength and Conditioning Research*, 19 (2), 349–357.

DeRenne, C., Ho, K.W. and Blitzblau, A. (1990) Effect of weighted implement training on throwing velocity. *Journal of Applied Sports Science Research*, 4 (1), 16–19.

DeRenne, C., Buxton, B.P., Hetzler, R.K. and Ho, K.W. (1994) Effect of under- and overweighted implement training on pitching velocity. *Journal of Strength and Conditioning Research* 8 (4), 247–250.

Dienst, N., Schnieder, G., Altmeyer, K., Voelkering, K., Georg, T., Kramann, B. and Kohn, D. (2007) Correlation of of intercondylar notch cross sections to the ACL size: A high resolution MR tomographic in vivo analysis. *Archives of Orthopaedic Trauma Surgery*, *127* (4), 253–60.

Drust, B., Reilly, E. and Rienzi, E. (1998) Analysis of work rate in soccer. *Sports Exercise and Injury 4*, 151–155.

Duthie, G.M., Young, W.B. and Aitken, D.A. (2002) The acute effects of heavy loads on jump squat performance: an evaluation of the complex and contrast methods of power development. *Journal of Strength and Conditioning Research*, *16* (4), 530–538.

Enoka, R.M. (1994) *Neuromechanical Basis of Kinesiology*, 2nd edn. Champaign, IL: Human Kinetics.

Escamilla, R.F., Speer, K.P., Fleisig, G.S. and Barrentine, S.W. (2000) Effect of throwing overweight and underweight baseballs on throwing velocity and accuracy. *Sports Medicine*, *29* (4), 259–272.

Escamilla, R.F., Fleisig, G.S., Zheng, N., Lander, J.E., Barrentine, S.W., Andrews, J.R., Bergemann, B.W. and Moorman, C.T. (2001) Effects of technique variations on knee biomechanics during the squat and leg press. *Medicine and Science in Sports and Exercise*, *33* (9), 1552–1566.

Faccioni, A. (1994a) Assisted and resisted methods for speed development (part I). *Modern Athlete and Coach*, *32* (2), 3–6.

Faccioni, A. (1994b) Assisted and resisted methods for speed development (part II) – resisted speed method. *Modern Athlete and Coach*, *32* (3), 8–12

Fleck, S. and Kontor, K. (1986) Complex Training. *National Strength Conditioning Association Journal*, *8* (5), 66–68.

Ford, K.R., Myer, G.D. and Hewett, T.E. (2003) Valgus knee motion during landing in high school female and male basketball players. *Medicine and Science in Sports and Exercise*, *35* (10), 1745–1750.

Garhammer, J. and Gregor, R. (1992) Propulsion forces as a function of intensity for weightlifting and vertical jumping. *Journal of Applied Sports Science Research 6* (3), 129–134.

Griffin, L.Y., Agel, J., Albohm, M.J., Arendt, E.A., Dick, R.W., Garrett, W.E., Garrick, J.G., Hewett, T.E., Huston, L., Ireland, M.L., Johnson, R.J., Kibler, W.B., Lephart, S., Lewis, J.L., Lindenfeld, T.N., Mandelbaum, B.R., Marchak, P., Teitz, C.C. and Wojtys, E.M. (2000) Noncontact anterior cruciate ligament injuries: Risk factors and prevention strategies. *American Academy of Orthopaedic Surgery*, *8* (3), 141–150.

Haff, G.G., Whitley, A., McCoy, L.B., O'Bryant, H.S., Kilgore, J.L., Haff, E.E., Pierce, K. and Stone, M.H. (2003) Effects of different set configurations on barbell velocity and displacement during a clean pull. *Journal of Strength and Conditioning Research*, *17* (1), 95–103.

Haff, G.G., Hobbs, R.T., Haff, E.E., Sands, W.A., Pierce, K.C. and Stone, M.H. (2008) Cluster Training: A novel method for introducing training program variation. *Strength Conditioning Journal*, *30* (1), 67–76.

Häkkinen, K. and Komi, P.V. (1985) The effect of explosive type strength training on electromyographic and force production characteristic of leg extensor muscles during concentric and various strech-shortening cycle exercises. *Scandinavian Journal of Sports Sciences*, *7*, 65–76.

Häkkinen K., Alen, M., Kallinen, M., Newton, R.U. and Kraemer, W.J. (2000) Neuromuscular adaptation during prolonged strength training, detraining and re-strength-training in middle-aged and elderly people. *European Journal of Applied Physiology*, *83* (1), 51–62.

Harris, N.K., Cronin, J.B., Hopkins, W.G. and Hansen, K.T. Squat jump training at maximal power loads vs. heavy loads: effects on sprint ability. *Journal of Strength and Conditioning Research*, *22* (6), 1742–1749.

Hewett, T.E., Myer, G.D. and Ford, K.R. (2005) Reducing knee and anterior cruciate ligament injuries among female athletes: a systematic review of neuromuscular training interventions. *Journal of Knee Surgery*, *18* (1), 82–88.

Hewett, T.E., Zazulak, B.T. and Myer, G.D. (2007) Effects of the menstrual cycle on anterior cruciate ligament injury: A systematic review. *American Journal of Sports Medicine*, *35* (4), 659–668.

Hilfiker, R., Hubner, K., Lorenz, T. and Marti, B. (2007) Effects of drop jumps added to the warm-up of elite sport athletes with a high capacity for explosive force development. *Journal of Strength and Conditioning Research*, *21* (2), 550–555.

Hoffman, J.R., Cooper, J., Wendell, M. and Kang, J. (2004) Comparison of Olympic vs. traditional power lifting training programs in football players. *Journal of Strength and Conditioning Research*, *18* (1), 129–135.

Holcomb, W.R., Rubley, M.D., Lee, H.J. and Guadagnoli, M.A. (2007) Effect of hamstring-emphasized resistance training on hamstring:quadriceps strength ratios. *Journal of Strength Conditioning Research*, *21* (1), 41–47.

Hori, N., Newton, R.U., Andrews, W.A., Kawamori, N., McGuigan, M.R. and Nosaka, K. (2008) Does performance of hang power clean differentiate performance of jumping, sprinting, and change of direction? *Journal of Strength and Conditioning Research*, *22* (2), 412–418.

Hortobagyi, T., Houmard, J.A., Stevensen, J.R., Fraser, D.D., Johns, R.A. and Israel RG. (1993) The effects of detraining on power athletes. *Medicine and Science in Sports and Exercise 25*, 929–935.

Housh, T.J., Housh, D.J., Weir, J.P. and Weir, L.L. (1996) Effects of eccentric only resistance training and detraining. *International Journal of Sports Medicine, 17*, 145–148.

Jakalski, K. (1998) The pros and cons of using resisted and assisted training methods with high school sprinters: parachutes, tubing and towing. *Track Coach, 144*, 4585–4589.

Kaminski, T.W., Wabberson, C.V. and Murphy, R.M. (1998) Concentric versus enhanced eccentric hamstring strength training: Clinical implications. *Journal of Athletic Training, 33* (3), 216–221.

Kato, S., Urabe, Y. and Kawamura, K. (2008) Alignment control exercise changes lower extremity movement during stop movements in female basketball players. Knee, *15* (4), 299–304.

Kilgallon, M., Donnelly, A.E. and Shafat, A. (2007) Progressive resistance training temporarily alters hamstring torque-angle relationship. *Scandinavian Journal of Medicine of Science and Sports, 17* (1), 18–24.

Kingma, I., Aalbersberg, S. and van Dieen, J. H. (2004) Are hamstrings activated to counteract shear forces during isometric knee extension efforts in healthy subjects? *Journal of Electromyography and Kinesiology, 14* (3), 307–315.

Kollath, E. and Quade, K. (1993) Measurement of sprinting speed of professional and amateur soccer players. In Reilly, T., Clarys, J. and Stibbe, A. (eds) *Science and Football II*. London: Routledge, pp 31–36.

Kreighbaum, E. and Katharine, B.M. (1996) *Biomechanics; A Qualitative Approach for Studying Human Movement*. New Jersey, NJ: Allyn and Bacon, pp 203–204.

Kubo, K., Kanehisa, H., Ito, M. and Fukunaga, T. (2001) Effects of isometric training on the elasticity of human tendon structures in vivo. *Journal of Applied Physiology, 91* (1), 26–32.

Kubo, K., Morimoto, M., Komuro, T., Yata, H., Tsunoda, N., Kanehisa, H. and Fukunaga, T. (2007a) Effects of plyometric and weight training on muscle-tendon complex and jump performance. *Medicine and Science in Sports and Exercise, 39* (10), 1801–1810.

Kubo, K., Morimoto, M., Komuro, T., Tsunoda, N., Kanehisa, H. and Fukunaga, T. (2007b) Influences of tendon stiffness, joint stiffness, and electromyographic activity on jump performances using single joint. *European Journal of Applied Physiology, 99* (3), 235–243.

Kunz, H. and Kaufmann, D.A. (1981) Biomechanical analysis of sprinting: decathletes versus champions. *British Journal of Sports Medicine, 15*, 177–181.

Lawton, T.W., Cronin, J.B. and Lindsell, R.P. (2006) Effect of interrepetition rest intervals on weight training repetition power output. *Journal of Strength and Conditioning Research, 20* (1), 172–176.

Li, G., Rudy, T.W., Sakane, M., Kanamori, A., Ma, C.B. and Woo, S.L.Y. (1999) The importance of quadriceps and hamstring muscle loading on knee kinematics and in-situ forces in the ACL. *Journal of Biomechanics, 32*, 395–400.

Mann, R. and Herman, J. (1985) Kinematic analysis of Olympic sprint performance: men's 200 meters. *Journal of Applied Biomechanics, 1* (2), 151–162.

Markovic, G. (2007) Does plyometric training improve vertical jump height? A meta-analytical review. *British Journal of Sports Medicine, 41* (6), 349–55.

Matthews, M.J., O'Conchuir, C. and Comfort, P. (2009) The acute effects of heavy and light resistances on the flight time of a basketball push pass during upper body complex training. *Journal of Strength and Conditioning Research, 23* (7), 1988–1995.

Mero, A. (1988) Force-time characteristics and running velocity of male sprinters during the acceleration phase of sprinting. *Research Quarterly for Exercise and Sport 94* (2), 94–98.

Mjolsnes, R., Arnason, O., Osthagen, T., Raastad, T. and Bahr, R. (2004) A 10-week randomized trial comparing eccentric vs. concentric hamstring strength training in well-trained soccer players. *Scandinavian Journal of Medicine and Science Sports, 14* (5), 311–317.

Moravec, P., Ružička, J., Sušanka, P., Dostal, E., Kodejs, M. and Nosek, M. (1988) The 1987 International Athletic Foundation/IAAF Scientific Project Report: time analysis of the 100 meters events at the II World Championships in Athletics. *New Studies in Athletics, 3* (3), 61–96.

Morrissey, M.C., Harman, E.A. and Johnson, M.J. (1995) Resistance training modes: Specificity and effectiveness. *Medicine and Science in Sports and Exercise, 27*, 648–660.

Mujika, I. and Padilla, S. (2001) Cardiorespiratory and metabolic characteristics of detraining in humans. *Medicine and Science in Sports and Exercise, 33* (3), 413–421.

Narici, M.V., Roi, G.S., Landoni, L., Minetti, A.E. and Cerretelli, P. (1989) Changes in force, cross-sectional area and neural activation during strength training and detraining of the human quadriceps. *European Journal of Applied Physiology, 59* (4), 310–319.

Newton, R.U. and Kraemer, W.J. (1994) Developing explosive muscular power: Implications for a mixed methods training strategy. *Strength and Conditioning Journal, 16* (5), 21–31.

Noyes, F.R., Mooar, P.A, Matthews, D.S. and Buttler, D.L. (1983) The symptomatic anterior cruciate deficient knee. The long term functional disability in athletically, active individuals. *Journal of Bone and Joint Surgery of American*, 65, 154–162.

Olsen, P.D. and Hopkins, W.G. (1999) The effect of weight training and explosive isometrics on martial-art kicks and palm strikes. *Abstracts Medicine and Science in Sports and Exercise*, 31 (5) Supplement:S177.

Paavolainen, L., Hakkinen, K., Hamalainen, I., Nummella, A. and Rusko, H. (1999) Explosive strength training improves 5-km running time by improving running economy and muscle power. *Journal of Applied Physiology*, 86 (5), 1527–1533.

Pistilli, E.E., Kaminsky, D.E., Totten, L.M. and Miller, D.R. (2008) Incorporating one week of planned overreaching into the training program of weightlifters. *Stength and Conditioning Journal*, 30 (6), 39–44.

Robinson, L.E., Devor, S.T., Merrick, M.E. and Buckworth, J. (2004) The effect of land vs aquatic plyometrics on power, torque, velocity, and muscle soreness in women. *Journal of Strength and Conditioning Research*, 18 (1), 84–91.

Schmidtbleicher, D., Gollhofer, A. and Frick, U. (1988) Effects of a stretch-shortening typed training on the performance capability and innervation characteristics of leg extensor muscles. In de Groot, A.G. et al. (eds) *Biomechanics XI-A*, Vol 7. Amsterdam: Free University Press, pp. 185–189.

Siff, M.C. (2004). *Supertraining*, 6th edn. Denver, CO: Supertraining Institute.

Siff, M.C. and Verkhoshansky, Y.V. (1999) *Supertraining*, 4th edn. Denver, CO: Supertraining Institute.

Smith, A.M. (1999) The coactivation of antagonist muscles. *Canadian Journal of Physiology and Pharmacy*, 59, 733–747.

Spurrs, R.W., Murphy, A.J. and Watsford, M.L. (2002) The effect of plyometric training on distance running performance. *European Journal of Applied Physiology*, 89, 1–7.

Stemm, J.D. and Jacobson, B.H. Comparison of land and aquatic based plyometric training on vertical jump performance. *Journal of Strength and Conditioning Research*, 21 (2), 568–571.

Takano, B. (1992) Resistance exercise: The power clean – perspectives and preparation. *National Strength and Conditioning Journal*, 14 (1), 68–71.

Thomas, K., French, D. and Hayes, P.R. (2009) The effect of two plyometric training techniques on muscular power and agility in youth soccer players. *Journal of Strength and Conditioning Research*, 23 (1), 332–335.

Tillman, M.D., Bauer, J.A., Cauraugh, J.H. and Trimble, M.H. (2005) Differences in lower extremity alignment between males and females. Potential predisposing factors for knee injury. *Journal of Sports Medicine and Physical Fitness*, 45 (3), 355–359.

Turner, A.M., Owings, M. and Schwane, J.A. (2003) Improvement in running economy after 6 weeks of plyometric training. *Journal of Strength and Conditioning Research*, 17 (1), 60–67.

Verkhoshansky, Y.V. (1986) *Fundamentals of Special Strength-Training in Sport* (A. Charniga, Trans.). Livonia, MI: Sportivny Press, p 139).

Weber, K.R., Brown, L.E., Coburn, J.W. and Zinder, S.M. (2008) Acute effects of heavy-load squats on consecutive squat jump performance. *Journal of Strength and Conditioning Research*, 22 (3), 726–730.

Weir, J.P., Housh, D.J., Housh, T.J. and Weil, L.L. (1997) The effect of concentric unilateral weight training and detraining on joint angle specificity, cross training and the bilateral deficit. *Journal of Orthopedic Sports Physiotherapy*, 25, 264–270.

Weyand, P.G., Sternlight, B.D., Bellizzi, M.J. and Wright, S. (2000) Faster top running seeds are achieved with greater ground forces not more rapid leg movements. *Journal of Applied Physiology*, 89 (5), 1991–1999.

Wilsoff, U., Castagna, C., Helgerud, J., Jones, R. and Hoff, J. (2004) Strong correlation of maximal squat strength with sprint performance and vertical jump height in elite soccer players. *British Journal of Sports Medicine*, 38, 285–288.

Yu, B. and Garrett, W.E. (2007) Mechanisms of non-contact ACL injuries. *British Journal of Sports Medicine*, 41 (Supplement 1),47–51.

Zatsiorsky, V.M. (1995) Science and practice of strength training. Champaign, IL: Human Kinetics.

14

Nutritional considerations for performance and rehabilitation

Helen Matthews and Martyn Matthews
University of Salford, Greater Manchester

Introduction

Nutrition is a major consideration for athletes, coaches and rehabilitators and plays a crucial role in training, competition and in the prevention and management of sports injuries. *So, what is sports nutrition?* Sports nutrition encompasses what, when, and how much athletes eat. It takes account of how nutrients are digested and absorbed and how foods are metabolised for energy or assimilated into body tissues.

Correct nutrition, or more specifically the optimal balance of energy and nutrients delivered at the right time, has a range of potential benefits for the athlete. These include: recuperation from and adaptation to training; the maintenance of work-rate throughout a match, race or training session; the maintenance of concentration and coordination; the maintenance of body composition; and provision of an optimum environment for injuries to heal. Without correct nutrition, all of these processes are affected with associated detriments to training, performance, injury risk and injury rehabilitation.

During this chapter you will learn about:

- fundamentals of nutrition
- energy balance
- the nutrients and how to obtain them
- the importance of a balanced diet for health and performance
- energy requirements for specific sports
- nutritional strategies for optimal performance and the evidence to support them
- nutritional strategies for injury prevention
- nutrition for the injured athlete

Fundamentals of nutrition

Energy balance

Food and drink provide energy. Complex chemical processes break down food to provide essential fuels for a range of functions including breathing, blood circulation, chemical reactions, growth, repair, brain function and muscular activity.

Managing energy balance is a nutritional priority for athletes. Food must provide adequate energy for training and competition, but must not provide excess. An excess of energy, over and above that expended, will lead to weight gain. This may be in

---

Sports Rehabilitation and Injury Prevention Edited by Paul Comfort and Earle Abrahamson
© 2010 John Wiley & Sons, Ltd

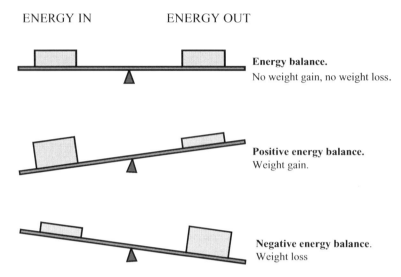

Figure 14.1 Energy balance.

the form of functional muscle tissue in the case of an athlete training for increased muscle hypertrophy, strength and power but, more commonly, may also be in the form of unwanted body fat, for example when an injured athlete is forced to stop training but maintains the same high-energy diet. By contrast, a diet that has insufficient energy will lead to weight loss. This may be the case in sports that require a high volume of training, such as triathlon, where athletes may struggle to consume enough energy to meet the demands of training.

To maintain weight and body composition at the optimal level for sport, athletes must therefore manage both the amount of energy they consume and the amount they expend. This is called managing the *energy balance*.

Energy balance can be summarised by the following: If athletes consume more energy than they use then they will gain weight – hypercalorific diet. If athletes consume as much energy as they use then they will stay the same weight – isocalorific diet. If athletes consume less energy than they use then they will lose weight – hypocalorific diet (Figure 14.1).

Energy and calories

Energy is measured in Joules, however the term most often used in dietetics and amongst athletes is the kcal (one kcal equals 4.2 kJoules). To maintain energy balance, the estimated average energy intakes required in 'healthy' adults are 2550 kcal/day for males and 1940 kcal/day for females (FSA 2008). Additional physical activity needs, however, are not accounted for in these estimated average requirements. Therefore athletes in training will have higher energy requirements and this will increase in proportion to the volume of training that the athlete performs.

Nutrients and where to get them

In addition to energy, food and drink provide nutrients. These are the raw materials, or ingredients, required by the body for optimal health and function.

There are six key nutrient classes. These are:

- carbohydrates
- fats
- proteins
- vitamins
- minerals
- water.

Each of these nutrient classes has specialised roles within the body that are essential for health and optimal performance. An inadequate intake of any of these will result in disease, or impaired physical and cognitive functioning, which can normally only be prevented by the nutrient. Of these nutrients, carbohydrates, fats and proteins provide energy. Water, vitamins and minerals provide no energy but play a vital role in the regulation of body processes. Alcohol is considered by some to be the seventh nutrient as it also provides a source of energy, but it is not essential.

The correct nutrient balance will help athletes remain healthy and perform optimally. Too much or too little of any one nutrient will affect both health and performance.

Carbohydrates

Carbohydrates are the single most important source of energy for athletes. They provide approximately four kcal per gram and are the primary fuel source for high intensity exercise (above 60% VO_2max) (Coggan and Coyle 1988; Carter et al. 2003). Carbohydrates can be classified as simple and complex.

Simple carbohydrates are found in sugary foods such as fruit, fruit juices, sugar and honey and generally provide a quick but short-lasting source of energy. These can be further divided into: monosaccharides, which are simple, one-unit sugars such as glucose, fructose and galactose; and disaccharides, which are formed from two monosaccharides. For example sucrose – or common table sugar – is a combination of glucose and fructose.

Complex carbohydrates are found in starchy foods such as pasta, rice, potatoes, bread, vegetables, cereals, beans and pulses and provide a stable supply of longer-term energy. Complex carbohydrates can be further divided into: polysaccharides (containing more than two monosaccharides), which include starch (from plant sources) and glycogen (from animals), and dietary fibre (non-starch polysaccharide), which is found in plant foods, cereal, fruit and vegetables and is crucial to optimal health.

Foods, and in particular carbohydrates, can also be classified by their glycaemic index. The glycaemic index (GI) is a measure of the extent to which a certain food raises blood glucose (Jenkins et al. 1981). High GI foods, such as white bread, potatoes, cornflakes and jelly beans, cause a sharp rise in blood glucose, which is often short lasting and followed by a rebound drop in blood glucose. Low GI foods, such as porridge, high-fibre cereal, beans, peanuts and apricots, which are high in fibre and/or protein, give a slower, but more sustained release of glucose causing a slower rise in blood glucose and no rebound drop.

For the healthy population, carbohydrates should contribute approximately 50–60% of the energy intake in the diet, with low GI, high-fibre complex carbohydrates providing the majority of this requirement (Salmeron et al. 1997; Foster-Powell et al. 2002). This should increase to approximately 60–70% in an athletic population.

Protein

Protein plays a vital role in the maintenance of all body tissues. It is used as the structural basis of all cells, forms the contractile components of muscle, and is used to synthesise haemoglobin, enzymes, hormones, neurotransmitters and antibodies, helping to maintain the immune system and regulate all essential chemical reactions. Protein, which contains approximately 4 kcal/gram, is also used as an energy source, supplying about 5–10% of energy expenditure (Dohn 1986; Brooks and Mercies 1994).

Proteins are made up of amino acids, and it is these amino acids that are the real building blocks of the body. There are 20 amino acids, 12 of which can be synthesised from other amino acids and are considered *non-essential* and, and eight (nine for children) that are considered *essential* as they cannot be made within the body and therefore must come from the diet.

Protein is found in a variety of foods, however not all foods contain all of the essential amino acids. Those that do contain all the essential amino acids are termed *complete* proteins, which can be found in animal products such as meat, fish, poultry, eggs, milk and yogurt, as well as soya. *Incomplete* proteins have one or more of the essential amino acids missing. These are usually found in plant foods such as beans, pulses, vegetables, grains and rice. In order to get all of the essential amino acids athletes should either aim to eat complete proteins, such as those gained from animal sources, or combine other proteins to ensure that their essential amino acid needs are met.

The recommended intake of protein for a healthy sedentary person is 0.75g/kg.bw/day or approximately 15% of the diet (FSA 2008).

Fat

Fat has many essential functions within the body that include: energy provision, formation of cell membranes and nerve fibres, protection of vital organs, production of hormones, storage and transport of fat soluble vitamins, insulation, suppression of hunger and adding palatability to foods.

The body can store large amounts of energy as fat, mostly in the form of subcutaneous adipose tissue, which can be mobilised and transported to the working muscles as a fuel source during exercise, but also as intramuscular fat where it is more readily utilised within the muscles.

Fat contains approximately 9kcal/gram making it a very efficient way for the body to store large amounts of energy. High fat foods are therefore considered energy dense, making it easy for athletes to inadvertently consume too much energy from these foods. Managing fat intake is therefore a primary concern for athletes to prevent any unwanted increase in body fat.

Athletes should also consider the type of fat in the diet. Naturally occurring dietary fat can be either unsaturated or saturated. *Unsaturated* fatty acids contain one (monounsaturated) or more (polyunsaturated) double bonds between carbon atoms where each double bond replaces two hydrogen atoms. *Saturated* fatty acids contain no double bonds. Unsaturated fatty acids are generally liquid at room temperature and are found in plant sources and fish. Saturated fats tend to be solid at room temperature and come from animal sources. Saturated fats are associated with increased levels of low-density lipoproteins (LDL cholesterol) (Mustad et al. 1997) and an increased risk of coronary heart disease, whereas unsaturated fats are associated with a lower risk by reducing serum LDL (Kratz et al. 2002) and increasing high-density lipoproteins (HDL cholesterol) (Karmally and Goldber 2006).

Trans-fatty acids are formed by a process called hydrogenation. Vegetable oils are chemically changed to give them a higher melting point, effectively making them more solid at room temperature, and are used in the manufacture of margarine and other spreads. When unsaturated fatty acids are altered, and acquire some of the properties of saturated fatty acids through hydrogenation, they are termed trans-fatty acids.

All fats contain a mixture of fatty acid types, but it is the proportion of different fatty acids that makes some fats healthier than others. Fats associated with raised cholesterol levels and heart disease include those high in saturated animal fats (red meat, butter, cream, lard), and trans-fatty acids from hydrogenated vegetable oils (baked goods and some margarines) (Miettinen et al. 1972; Baer et al. 2004). Fats with a greater protective effect include monounsaturated and polyunsaturated fats found in olive oil and oily fish (Bucher 2002).

While most fatty acids can be synthesised within the body, there are two that are essential. The essential fatty acids are linolenic (omega-3) and linoleic (omega-6) fatty acids. Omega-3 fatty acids, in particular, are associated with beneficial effects on the cardio-vascular system (Simopoulos 1999). Linolenic fatty acids are found in oily fish and a range of vegetable oils such as sunflower and sesame. Linoleic fatty acids are found in most plant oils, especially corn and soybean oil.

Current guidelines for fat intake in a healthy diet recommend no more than 30% of energy intake as total fat, and less than 10% as saturated fat (DoH 1991, 1994).

Athletes should aim to avoid or reduce their intake of the following high fat foods: cake, biscuits, chocolate, fat on meat, sausages, pasties, pies, beef burgers, cheese, butter and cream. Moreover, athletes should aim to get the majority of their fat intake from oily fish, white fish, vegetable seeds and oils, soya beans and nuts.

Vitamins

Vitamins are essential organic molecules that cannot be synthesised in the body. Although they are only required in very small quantities, a deficiency can lead to symptoms of disease. Vitamins have a range of specific functions that are essential for health. These include: absorption of nutrients, anti-oxidants and protection of cell membranes, energy metabolism, catalysts for and regulation of chemical reactions and collagen synthesis. There are 13 compounds commonly identified as vitamins that are broadly categorised as either *water-soluble* or *fat-soluble*. A

Table 14.1 Reference nutrient intakes for healthy adults.

Vitamin	RNI (adult male)	RNI (adult female)
A	700µg	600µg
Thiamin (B1)	1.0mg	0.8mg
Riboflavin (B2)	1.3mg	1.1mg
Niacin (B3)	1.7mg	1.3mg
Pyridoxine (B6)	1.4mg	1.2mg
B12	1.5mg	1.5mg
Folic acid	200mg	200mg
C	40mg	40mg
D	10µg (if limited exposure to sunlight)	10µg (if limited exposure to sunlight)
E*	RNI not established	
K*	RNI not established	

Adapted from DoH 1991 – requirements vary for children and pregnant women.

summary of the body's requirements of these nutrients is provided in Table 14.1.

Water-soluble vitamins include vitamin C and the B-complex. They cannot be stored within the body and must therefore be consumed regularly. Any excess intake is normally excreted in the urine.

Vitamin B refers to a host of different vitamins that play an essential role in the release of energy from carbohydrates, fat and protein, and the formation of haemoglobin. They are known as the B-complex and include thiamine (B1), riboflavin (B2), pantothenic acid, niacin (B3), piridoxine (B6), folic acid, cyanocobalamin (B12), and biotin. Deficiencies can lead to fatigue, gastro-intestinal problems, heart failure and nervous disorders (FSA 2008). Vitamin B is found in grains, milk products, eggs, green vegetables, fish, liver, nuts and wholegrain and fortified cereals.

Vitamin C (ascorbic acid) works as an antioxidant and is essential for collagen synthesis, protein metabolism, wound healing, functioning of the immune system and iron absorption. Deficiencies can lead to poor immune function, poor wound healing, and scurvy. Good sources of Vitamin C include citrus fruits, tomatoes and green vegetables.

Fat-soluble vitamins include vitamins A, D, E, and K. These are stored in the body and can accumulate in fatty tissue. Excessive intakes may accumulate to toxic levels.

Vitamin A (retinol and beta-carotene) has many roles. It is an important anti-oxidant, and is essential for vision, immune function, bone health and gene transcription. Deficiency leads to vision impairments, particularly night blindness. Excess can become toxic and lead to numerous problems that could include birth defects, kidney problems, nausea, hair loss, headache, irritability, susceptibility to infection, fissures of the lips, blurred vision, bone and joint pains, muscle pain and weakness (FSA 2003). Whilst it is not possible to establish a Safe Upper Limit for vitamin A, total intakes above 1500 micrograms should be avoided (FSA 2003). Good sources of vitamin A include liver, kidney, milk, and eggs. Good sources of beta-carotene (a precursor to vitamin A) include green leafy vegetables and carrots.

Vitamin D plays a central role in the absorption and regulation of both calcium and phosphorus and is therefore essential for optimum bone health. Deficiency can lead to rickets, osteomalacia, and osteoporosis. Vitamin D is synthesised by the skin in response to UV exposure but can also be gained from foods such as fish, eggs, fortified dairy products and breakfast cereals.

Vitamin E refers to a group of antioxidants that prevent cell-membrane damage by reacting with radicals produced by lipid peroxidation. Vitamin E is found in seeds, nuts, green leafy vegetables, avocado, olives and vegetable oils.

Vitamin K is primarily responsible for blood clotting. A deficiency is associated with severe and uncontrolled bleeding, malformation of developing bone and cartilage calcification. The main dietary sources include green leafy vegetables, avocado, and kiwifruit.

Table 14.2 Summary of the functions and sources of key minerals

Mineral	Function	Foods
Boron	Promotes healthy bones, teeth; metabolism of other minerals	Fruits and vegetables
Calcium	Blood clotting, intracellular signaling, muscle contraction	Milk, yogurt, cheese, other dairy products, sardines, bread, sultanas, vegetables
Chromium	Insulin and glucose tolerance responses	Meat, wholegrain cereals, legumes and nuts
Cobalt	Contained in vitamin B12	Meat, dairy products, eggs
Copper	Formation of haemoglobin; absorption and use of iron; skin, hair pigmentation	Shellfish, liver, meat, cereal products, vegetables
Fluoride	Prevents dental carries; crystalline structure of bones and teeth	Water (variable), tea, seafood,
Iodine	Contained in hormones thyroxine (T4) and triiodothyronine (T3)	Seafood, vegetables and cereals
Iron	Contained in cytochromes, myoglobin, haemoglobin.	Beef, liver, eggs, sardines, apricots, fortified cereals, plain chocolate, bread, vegetables
Magnesium	Mineral present mainly in the bones; maintains electrical potential in nerve and muscle cells	Cheese, milk, chicken, cod, peanuts, bread, marmite, cereal products, potatoes, vegetables
Phosphorus	Contained in bones, teeth; role in energy metabolism	Cheese, eggs, milk, chicken, beef, ham, peanuts, marmite, meat and bread
Potassium	Role in fluid and electrolyte balance; heart muscle activity; metabolism and protein synthesis	Fruit and vegetables, fruit juices, milk, fish, meat
Selenium	Helps heart function; possibly prevents certain cancers	Meat, fish and cereal products
Sodium	Present in extracellular fluid	Table salt, bacon, ham, potato crisps, cereals, marmite, soy sauce
Sulphur	Energy metabolism, enzyme function, and detoxification	Cheese, eggs, nuts, onions, green leafy vegetables, fish, wheat germ
Zinc	Contained in enzymes, transcription factors	Meat and meat products, bread, cheese and eggs, milk, cereal products

For those following a varied and balanced diet, deficiencies in the fat-soluble vitamins are rare.

Minerals

Minerals have several major roles including the formation of bones and teeth, formation of haemoglobin and hormones, muscular contractility, neural conductivity, regulation of acid-base balance, and metabolism. Some minerals occur in the body in relatively large amounts. These are known as the major minerals and include: calcium, phosphorus, sulphur, potassium, chlorine, sodium, magnesium, zinc and iron. Other minerals occur in minute quantities. These are called trace minerals and include: chromium, cobalt, copper, fluorine, iodine, manganese, molybdenum, nickel, selenium, silicon, tin and vanadium (see Table 14.2).

Deficiencies of any mineral can lead to signs and symptoms of disease. Moderate excesses of sodium, potassium, calcium and chlorine are normally excreted by the kidneys, whereas excess intakes of other minerals can be harmful or impair absorption of other micronutrients (e.g. excess calcium inhibits absorption of iron and zinc; excess zinc inhibits absorption of copper) (Fairweather-Tait and Hurrell 1996).

In general, a balanced diet that contains a variety of foodstuffs and includes plenty of fresh fruit and vegetables should provide sufficient quantities of the minerals required for optimal health.

Two of the most important minerals for athletes are iron and calcium.

Iron is an essential element, which is required for oxygen transport (haemoglobin and myoglobin) and electron transport. It is found in the diet in two forms, haem iron and non-haem iron. Haem iron is generally found in animal foods and has greater bioavailability (10–35%). Non-haem iron, generally found in plant foods, has low bioavailability (2–10%) (Zijp et al. 2000).

Low iron levels can be due to inadequate diet, malabsorption or increased iron losses. To ensure an adequate iron intake, athletes should eat a variety of foods including red meat and green leafy vegetables. Iron absorption can be enhanced by the presence of vitamin C. Excessive intakes of iron are associated with gastrointestinal disorders including constipation, nausea, vomiting and diarrhoea, whilst chronic excess may lead to high body iron stores and increased risk of cardiovascular disease or cancer (FSA 2003). The FSA Expert Group on Vitamins and Minerals (2003) suggest that, for most people, an intake of approximately 17 mg/day would not be expected to produce adverse effects.

Calcium is another essential mineral for health and sports performance. It forms an integral part of bone structure, muscle contraction, blood clotting and transmission of nerve impulses. A deficiency can lead to rickets, reduced bone mineral density and osteoporosis, and problems with blood clotting.

The best dietary sources of calcium are milk and milk products, and green leafy vegetables. Overall absorption is generally poor and requires adequate vitamin D (FSA 2003).

Summary of micronutrients

Overall, a diet that is balanced and varied should meet the needs for vitamins and minerals for a healthy population. In athletes, the question of whether vitamins and minerals are required in greater amounts is more complex, and likely to be affected by the interaction between the total volume and intensity of sporting activity and the specific energy and nutrient intakes of each individual (Volpe 2006). It does appear, however, that a deficiency will have a negative impact on performance (Volpe 2006).

Water

Water is by far the most important nutrient. It accounts for 50–75% of body weight, depending on body fat content, with fat tissue containing approximately 20% water, lean tissue approximately 75%, and blood around 95%. Water has many crucial roles including temperature control (sweating), lubrication (brain, eyes, spinal cord and digestive system), a solvent for all biochemical reactions, and the transportation of nutrients and oxygen to, and carbon dioxide and other metabolic waste products away from, the working muscles. For optimal cardiovascular and thermoregulatory function athletes must therefore maintain sufficient body fluid levels and avoid dehydration, which results in a decrease in both cardiovascular (Gonzalez-Alonso et al. 1997) and strength and power performance (Jones et al. 2008). Water is gained from drink, food and metabolic reactions and is lost through urine, faeces, expired air and sweat.

Alcohol

Alcohol is a non-essential nutrient that has a range of negative effects for athletes. It causes intoxication, affects skill and co-ordination, leads to dehydration, de-motivates, inhibits recovery, lowers testosterone and provides an energy source (7kcal/gram) that can only be utilised once it has been deposited as fat. Overall, alcohol is highly detrimental to sporting performance, significantly reducing aerobic performance capacity for several hours after consumption (O'Brien and Lyons 2000; El-Sayed et al. 2005). Alcohol also acts as a diuretic and therefore may result in dehydration.

The importance of a balanced diet

For optimal health and performance it is essential that athletes get the right quantities and balance of all the nutrients. There are several tools and guidelines to ensure the correct quantities and balance of nutrients for health. The most common tool is the reference nutrient intake (RNI), which is the amount of a nutrient required to maintain adequate health for the majority of the population. RNIs are provided for energy, protein, vitamins and minerals. The advantages of using RNIs are that they give individuals a starting point to judge their diet in relation to a

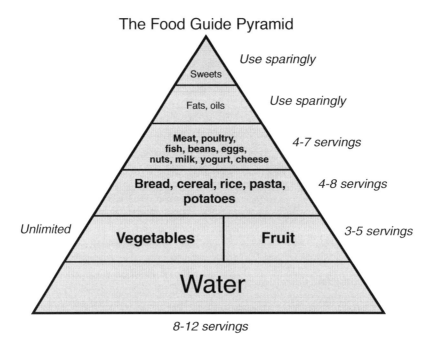

Figure 14.2 The food guide pyramid.

healthy population. RNIs, however, do not take into account the individual needs of athletes determined by the individual, the sport and the level of training. For athletes, a diet that is adequate for health may not be optimal for performance, due primarily to an increased energy expenditure.

The current UK government guidelines for health suggest a balance of approximately 55% carbohydrate, 15% protein and less than 30% fat (of which saturated fat should account for less than 10%) (FSA 2008). Again, these recommendations act as a good starting point for athletes but, as discussed later, there are a number of circumstances where these recommendations are modified. The optimal diet for an athlete will depend on the sport, activity levels, body size, weight gain or weight loss, and the type of training performed at any given time.

The food guide pyramid (Figure 14.2) is a common tool to help individuals achieve a healthy balance of foods within the diet. The pyramid shows the proportion and type of foods, which contribute to a healthy, balanced diet. At the base of the pyramid are carbohydrate-based foods, and requirements for fruit and vegetables. The pyramid reflects the percentage guidelines already outlined for the macronutrients in a healthy diet. Several alternative versions of the pyramid now exist to support specific groups (Painter et al. 2002).

Nutrition for performance

Optimal nutrition for performance is determined by what, when and how much an athlete eats. A diet that is adequate for health is not usually optimal for performance. For example, athletes will require more energy, greater quantities of certain nutrients (carbohydrates, protein) and will need to consume food at key times before, during and after training or competition.

Energy requirements for specific sports/training activities

Energy requirements for athletes are determined by their size, basal metabolic rate, the extra energy that is required for activities performed throughout the day (training and competition), recovery and adaptation from training and competition and whether the athlete is trying to gain or lose weight.

Mean total daily energy expenditures vary between different athletes. For example:

- male boxers (57Kg) expend 2900kcal; male weightlifters (110Kg), 4900kcal; female basketball players (61.4Kg), 3100kcal (Ismail et al. 1997).

- male cross-country skiers (during periods of hard training) use around 8600kcal/day (Sjodin et al. 1994).

- professional road cyclists expend over 6000kcal per day throughout the three-week Tour de France event, and use in excess of 9000kcal per day during hard mountain stages (Saris et al. 1989). This is reflected by high reported energy intakes (in excess of 5450 Kcal) during training and competition (García-Rovés et al. 2000).

In team sports, differences in activity levels will also vary with playing position, resulting in large variations in energy expenditure between individuals on the same team, which must be matched by energy intake. For example, Lundy et al. (2006) report that elite professional rugby league players consume between 2700 and 6900 kcal/day (mean 4230kcal/day) depending on size and position, although it is worth noting that this is 'normative' data and this level of consumption may not necessarily be optimal. In fact, Lundy et al. (2006) recommended that these athletes increase their energy intake through additional carbohydrate consumption.

Macronutrient requirements for performance

Energy needs are met from carbohydrate, fat and protein. The balance of use between these fuels, at any one time, depends on the intensity and duration of exercise. At the exercise intensities normally encountered during training and competition, carbohydrate is usually the primary energy substrate. To optimise performance athletes need to consume the right fuels at the right time.

Carbohydrate requirements for performance

The selection of fuel for muscular work is directly related to the intensity of effort with the higher the intensity, the greater the reliance on carbohydrate as a substrate (see Figure 14.3) (Romjin et al. 1993). As most sports are performed at high intensities carbohydrate becomes the primary source of energy.

In the body, carbohydrate is stored primarily as muscle and liver glycogen. Carbohydrate stores, however, are limited to approximately 450g of muscle glycogen (with a range from 50g after exhaustive exercise to approximately 900g in a large, well-trained, well-rested, and well-fed athlete (Jeukendrup 2003, de Jonge and Smith 2008), 100–120g of liver glycogen, and 5g of circulating blood glucose. Just 2–3 hours of high-intensity activity may be enough to completely deplete these relatively limited glycogen reserves (Coyle et al. 1986). The status of the glycogen stores before activity, therefore, will determine the duration that high intensity exercise can be maintained (Astrand and Rodahl 1986). It is therefore important that, for endurance activities (Coyle et al. 1986) and high-intensity

---

Calculating energy needs

Energy needs can be calculated by the following formula:

Energy EAR (eatimated average requirement) = BMR (basal metabolic rate) × PAL (Physical Activity Level).

PAL refers to the ratio of total energy required over 24 hours to the BMR over 24 hours. For example a PAL of 1.4 represents very low activity levels, 1.6 represents moderate activity levels, and 1.9 represents high activity levels. Most athletes will have a PAL of 1.9 or above.

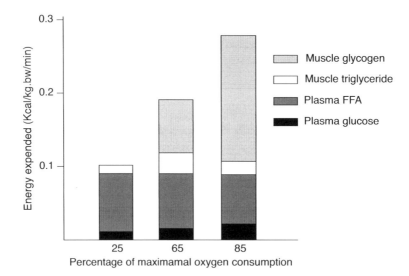

Figure 14.3 Contribution of muscle glycogen, plasma glucose, plasma FFA, and muscle triglyceride to energy expenditure after 30 min of exercise at 25%, 65%, and 85% of maximal oxygen uptake in fasted subjects (from Romjin et al. 1993).

intermittent activities (Balsom et al. 1999), athletes aim to optimise their glycogen stores. Depending on the intensity and volume of training, this is normally achieved with an intake of 5–13g/kg.bw/day(see Table 14.4) (Jeukendrup and Gleeson 2004).

Pre-exercise

To delay fatigue, and therefore enhance performance, athletes should optimise muscle glycogen stores pre-exercise (Astrand and Rodahl 1986). In short, glycogen stores must be sufficient to allow completion of the event at the highest possible intensity. It is important to distinguish between maximal and optimal storage of glycogen. For events where glycogen depletion becomes a limiting factor for performance (e.g. the marathon, a soccer match) it is clearly desirable to maximise stores prior to the race or game. For shorter events, for example, a 10K race, maximising glycogen stores may not be desirable. Every gram of glycogen is stored with approximately 2.7 grams of water so any additional weight from excess glycogen storage has the potential to inhibit performance. Athletes should therefore only seek to 'carbo-load' for sports where glycogen depletion may become a limiting factor in performance.

Carbohydrate-loading

Athletes can maximise muscle glycogen stores through a carbohydrate loading regime (see Table 14.3). There are several protocols for doing this. Early research suggested athletes should perform an exhaustive bout or bouts of exercise followed by 3-4 days of very low carbohydrate intake or very low carbohydrate/high fat intake, followed finally by 3-4 days of extremely high carbohydrate intake (Ahlborg et al. 1967; Bergstrom et al. 1967). Although this did increase muscle glycogen stores, athletes were reluctant to train so hard and felt lethargic and irritable just a few days before competition. A later and more widely adopted protocol involved tapering training over a seven day period and increasing consumption of carbohydrate for three days before the race (Sherman et al. 1981). The advantage of this method was that athletes benefited from increased glycogen stores without the negative side effects. Most glycogen loading protocols used by athletes and researchers involve the Sherman et al. (1981) protocol, or variations thereof (Madsen et al. 1990; Widrick et al., 1993; Burke et al. 2000; James et al. 2001). More recently even higher glycogen levels have been achieved by the inclusion of a 3-minute intense bout of training (150s of cycling at 130% of VO_2 peak followed by 30s of all-out cycling) 24-hours

Table 14.3 Suggested carbo-loading protocols for (1) an endurance event, and (2) a weekly match (for a team sports player)

Endurance event	
Days before the event	Action
7–10 days before the event	Start to taper training
6–4 days before	Consume 7g/kg.bw/day
From 3 days before	Consume 10–13g/kg.bw/day
Weekly match	
From 2 days before the match	Taper training
Days 7–6 before the next match	To facilitate recovery from the previous match – consume 10g/kg.bw/day
Days 5–3 before the next match	To maintain stores during training – consume 7–10/kg.bw/day
From 2 days before the match	To ensure optimal levels of muscle glycogen prior to the next game – consume 10g/kg.bw/day

before the event followed by 10.3g/kg.bw/day of high-glycaemic index carbohydrates (Fairchild et al. 2002). The main advantage of this protocol is not that higher glycogen levels are reached, but that they are achieved in a much shorter time-frame (24 hours) and do not necessarily require an extended period of tapering. Whilst there may be obvious advantages to athletes that compete on a regular basis – for example, team sport athletes – there is, as yet, little evidence to support the use of this method in a practical setting over Sherman et al. (1981), particularly as the potential negative effects of the 3-minute intense bout 24-hours before competition are not yet fully realised.

Liver glycogen stores in particular are sensitive to dietary intake of carbohydrate and can be depleted by the overnight fast. To replenish liver glycogen stores, athletes should eat breakfast and/or consume a pre-competition meal. For example, consume 150–300g of moderate GI carbohydrate three to four hours beforehand. To maintain blood glucose, smaller amounts of carbohydrate (e.g. <50g, of low-moderate GI) can then be consumed in the hour prior to exercise.

Fructose (the sugar often found in fruit, honey and commercial sports drinks) has been suggested as a source of carbohydrate immediately prior to exercise as the insulin response is lower than with glucose (Maughan et al. 1997), however, for many athletes fructose can lead to gastrointestinal discomfort (Murray et al. 1989; Beyer et al. 2005).

During exercise

Whilst the benefits of pre-competition (or pre-training) carbohydrate loading are well established, there is also clear evidence for an ergogenic effect of carbohydrate feeding *during* an event (Coggan and Coyle 1991). Rather than reducing the rate of glycogen utilisation (Bergstrom et al. 1967; Tsintzas et al. 1996), carbohydrate ingestion appears to maintain blood glucose levels late in exercise, thus maintaining carbohydrate oxidation, and therefore performance, during endurance exercise (Coyle et al., 1986, Coggan and Coyle, 1991). Carbohydrate ingestion also improves endurance performance during intermittent high-intensity running in athletes with already high pre-exercise muscle glycogen concentrations (Foskett et al. 2008), and should therefore be considered an essential practice for team sport athletes.

Table 14.4 Summary and practical carbohydrate intake recommendations to replenish muscle glycogen during training

Volume/intensity of training	Approximate duration	Daily carbohydrate requirements*
Moderate training	1–2 hours	5–7g/kg.bw/day
Heavy training	2–3 hours	7–10g/kg.bw/day
Very heavy training	Over 3 hours	10–13/kg.bw/day

*From Jeukendrup and Gleeson (2004).

Carbohydrate feeding during exercise is often best achieved through the use of sports drinks or carbohydrate gels. These are palatable, convenient to take, and generally tolerated well by athletes. Sports drinks typically contain around 60g of carbohydrate per litre, which is close to the maximum amount of exogenous carbohydrate that the body can absorb and oxidise per hour (1-1.1g/min) (Jeukendrup and Jentjens 2000; Wallis et al. 2007; Jeukendrup, 2008). For longer events, athletes can carry small snacks of easily digested high-carbohydrate foods (e.g. sports gels, jelly-beans, jam sandwiches). To identify what they like and what they can tolerate, athletes should practice eating and drinking during training. Depending on the priority (fluid delivery or carbohydrate delivery) athletes can vary the concentration of carbohydrate in the sports drink. For example, a 5% solution will be more appropriate for those athletes wishing to replenish both water and carbohydrate. A 10% solution will be more suited to carbohydrate delivery at the expense of gastric emptying (Maughan 1991) however, during exercise this may result in gastrointestinal distress in some individuals.

It is often difficult for team sport athletes to consume food or fluid during the game. Players and support teams must be organised and proactive during any stoppages, time-outs, injury breaks, and half-times as these breaks provide the perfect opportunity to consume an isotonic carbohydrate-based drink.

Post-exercise

To enhance recovery and provide the fuel for the next match or training session, athletes must restore their muscle glycogen stores as soon as possible post-exercise. If glycogen stores are low at the start of the next training session, muscle glycogen and overall carbohydrate utilisation will be reduced, lowering the exercise intensity that can be maintained.

The rate and extent of glycogen re-synthesis is dependent on the quantity, timing and type of carbohydrate ingestion, and on the nature of recovery (Maughan et al. 1997).

How much?

Glycogen re-synthesis depends on the quantity of dietary carbohydrate consumed. To replenish glycogen stores athletes should consume at least 1g/kg.bw immediately post-exercise (Ivy 1998) or a minimum of 1–1.85g/kg.bw of carbohydrate per hour in the first few hours post-exercise (Jentjens and Jeukendrup 2003) and maintain a daily intake of 5–7g/kg.bw/day during periods of moderate training (1–2 hours per day), 7–10g/kg.bw/day when training load is increased (2–3 hours per day), and 10–13g per kg/day during periods of hard or prolonged training (over 3 hours per day) (Jeukendrup and Gleeson 2004).

When?

Due to elevated enzyme activity (glycogen synthase) (Sherman et al. 1983; Doyle et al. 1993) and cell membrane permeability, the highest rates of muscle glycogen synthesis occur when athletes consume large amounts of carbohydrate (1.0–1.85 g/kg.bw/h) immediately post-exercise, followed by further feeding at 15–60 minute intervals thereafter for up to 6-hours post-exercise (Ivy 1998; Jentjens and Jeukendrup 2003). When carbohydrate feeding is delayed, lower rates of muscle glycogen synthesis will occur (Jentjens and Jeukendrup 2003). Therefore, to facilitate glycogen resynthesis, athletes should consume carbohydrates as soon as possible post-exercise.

What?

High glycaemic index snacks (eg. jaffa cakes, jelly beans, toast and jam) and, in particular, carbohydrate based drinks are convenient, easily and more rapidly absorbed and digested, and will provide sufficient carbohydrate until consumption of a larger meal. In order to maintain the high level of carbohydrate intake and glycogen resynthesis, the post-exercise meal should be based around complex carbohydrates such as pasta, rice or potatoes.

Type of recovery

Passive recovery appears to be a more effective strategy to optimise glycogen resynthesis, particularly in type I muscle fibres (Choi et al. 1994; Fairchild et al. 2003).

Carbohydrates and protein

Although some studies have shown enhanced glycogen resynthesis when carbohydrate feedings are combined with protein, as long as the carbohydrate feedings are adequate, and meet the recommendations, there appears to be no additional benefit on glycogen resynthesis or performance when protein is taken as well (Jentjens et al. 2001; Osterberg et al. 2008). Despite this there may be some additional benefit in terms of protein synthesis and muscle regeneration when both protein (10–20g) and carbohydrate (60–90g) are combined post-exercise (Griewe et al. 2001; Bolster et al. 2004; Dreyer et al. 2008).

Summary/Key points

- Athletes should aim to optimise glycogen levels pre-exercise via breakfast and a pre-exercise meal (low-moderate GI carbohydrates). See table below for examples.

- During exercise consume up to 70g/hour to maintain carbohydrate oxidation and exercise intensity (sports drinks, easily digestible snacks, see the table below for examples).

- Consume high glycaemic index and easily digested carbohydrates immediately post-exercise (1-1.85g/kg.bw).

- Consume 1g/kg.bw high glycaemic index carbohydrates within the next 2 hours post exercise.

Table 14.5 provides examples of carbohydrate-rich foods ideal for pre-, during and post-exercise.

Fat as a fuel for exercise

Whilst high-intensity exercise is fuelled predominantly by carbohydrate, at lower intensities, or late in prolonged exercise, fat becomes increasingly important (Klein et al. 1994; Horowitz and Klein 2000). In contrast to limited carbohydrate stores, fat stores are large, with an average person storing over 100,000 kcal of energy as fat, mainly as subcutaneous adipose tissue. Due to limits in the fat oxidation process, however, these stores cannot be fully utilised for exercise.

There are several steps involved in the oxidation of fat for energy:

1. lipolysis, where stored fat must be broken down to fatty acids for transport around the body

2. transport of fatty acids in the blood stream to the muscle cell

3. transport of fatty acids across the cell membrane and to the mitochondria for oxidation.

The slow rate at which these processes occur is a key limiting factors for athletes and why carbohydrate, which is readily available for oxidation, is the primary fuel for high intensity exercise

Table 14.5 Example carbohydrate rich foods (providing approx. _50g_ carbohydrate _per portion_) for pre-, during-, and post-exercise

Pre	During	Post
1 large bowl of cereal with milk – cornflakes, Frosties, Shreddies, branflakes, 3–4 Weetabix or Shredded Wheat	750ml sports drink	2 slices toast/bread with jam or honey
1 cup of soup and a large bread roll	1–2 gels or sports bars	2–3 crumpets with jam or banana
500ml fruit juice	500ml fruit juice	500ml fruit juice
3/4 can of baked beans on 2 slices of toast	75g jelly babies	3/4 can of baked beans on 2 slices of toast
1 bagel or 2 bread rolls with filling	6 jaffa cakes	250ml milkshake or fruit smoothie
2 slices malt loaf / 2 cereal bars		2 slices malt loaf/2 cereal bars
1 large baked potato		300g mashed potato
1 packet dried fruit		1/3 pizza with topping
1 bowl fruit salad with pot of fruit flavoured yogurt		1 can of rice pudding
2 bananas		6 jaffa cakes/1 mars bar
		1 large currant bun

The relative contribution of fat for fuel depends on both the *intensity* and *duration* of exercise and whether *carbohydrate* has been consumed.

Intensity

At rest the majority of our fuel requirements come from fat. As exercise intensity rises, the rate of lipolysis and oxidation increase to a maximum at approximately 64% VO_2 max (Achten et al. 2002). At higher exercise intensities (approximately 86–89% of VO_2 max) the contribution of fat oxidation to energy expenditure is negligible (Achten et al. 2002; Achten and Jeukendrup 2003), probably due to the restricted transport of fatty acids to the mitochondria via limited blood flow to subcutaneous tissue, and lactate accumulation (Coyle et al. 1997).

Duration

As exercise progresses, fat oxidation plays an increasingly important role. As liver and muscle glycogen stores decrease, the body's ability to oxidise carbohydrate is depleted and fat becomes the dominant fuel, with *an associated drop in exercise intensity* (not ideal during competition).

Carbohydrate intake

Carbohydrate consumption also appears to decrease fat oxidation (Horowitz et al. 1997). Therefore a pre-exercise meal that is high in carbohydrate will reduce fat oxidation. Despite this suppression of fat oxidation, the overall aim is to increase carbohydrate oxidation and therefore pre-exercise carbohydrate intake should be paramount.

Increasing fat oxidation

To enhance the use of fat as an energy source, and therefore save glycogen stores, athletes can adopt several practices. Endurance training is the most effective, increasing the rate of lipid metabolism through increased size and number of muscle mitochondria, increased activity of lipid oxidising enzymes, and increased intra-muscular triacylglycerol stores (Coyle 1995; Phillips et al. 1996). Evidence for dietary manipulation of fat intake to enhance fat oxidation, however, is inconclusive with high-fat diets and supplementation with medium-chain triglycerides failing to show any significant ergogenic benefit (Hawley et al. 1998; Hawley 2002). Whilst caffeine ingestion has been shown to increase endurance performance, usually attributed to an increased level of lipid oxidation (Ivy et al. 1979) it is still unclear whether this is due to enhanced fat metabolism (Graham 2001).

Summary/Key points

- Fat stores in humans are much larger than carbohydrate stores and provide an important source of fuel for exercise.

- Fat oxidation is limited by the steps in the oxidation process.

- Fat oxidation increases with exercise intensity up to 60–65% VO_2 max. At higher intensities fat oxidation is restricted.

- Carbohydrate feeding reduces lipolysis and fatty acid availability, which inhibits fat oxidation and increases carbohydrate usage.

- Manipulation of fat in the diet via high fat diets, or medium chain triglycerides has a limited benefit for performance.

Protein requirements for performance

Do athletes need more protein than non-athletes? Whilst the current UK RNI for protein is 0.75g/kg.body-weight per day (FSA 2008), there are several mechanisms by which athletes could require more. These include: increased oxidation of amino acids during exercise, increased proteolysis as an acute response to exercise, and increased protein synthesis as an adaptation to training.

For endurance and intermittent sprint sports, protein requirements may be increased due to increased content of mitochondrial proteins and increased involvement in oxidative metabolism. To account for this an intake of 1.2–1.4g/kg.bw/day is generally recommended for endurance athletes (Lemon 1995). Consistent levels of high intensity/high volume training where high levels of amino acid oxidation occur may increase protein requirements for endurance athletes to 1.6 g/kg.bw/day (Tarnopolsky 2004; Campbell et al. 2007), with extreme endurance

Table 14.6 Summary of protein requirements for athletes

Population	Protein requirements
Sedentary	0.75–0.8 g/kg.bw/day
Endurance athlete – moderate volume	1.2–1.4 g/kg.bw/day
Endurance athlete – high volume	1.6 g/kg.bw/day
Endurance athlete – extreme	1.6–2.0g/kg.bw/day
Intermittent sport athletes (soccer)	1.4–1.6 g/kg.bw/day
Serious resistance trained athletes	1.7–1.8 g/kg.bw/day
Novice weight trainers in first few weeks	2.0 g/kg.bw/day

From Lemon 1994, 1996; Tarnopolsky 2004; Fink et al. 2005; Campbell et al. 2007.

athletes requiring up to 2.0g/kg.bw/day (Fink et al. 2005) due to athletes' inability to consume adequate levels of carbohydrate throughout the day. Athletes engaged in intermittent sports should aim for an intake of 1.4–1.6g/kg.bw/day (Lemon 1994) (Table 14.6).

For strength and power sports, an increase in muscle mass via increased formation of actin and myosin may increase protein requirements during periods of resistance training (particularly during the initial stages of training) to 1.7–2.0g/kg.bw/day (Lemon 1996, 1997; Campbell et al. 2007) (Table 14.6). There is no mechanism for storing excess dietary proteins in the body so amino acids ingested in excess of the body's immediate requirements are oxidised and the nitrogen excreted.

In general protein intake increases in proportion to energy intake. Those consuming a balanced high-energy diet are likely to meet protein requirements with no need for supplementation. A balanced diet to meet protein needs should include meat, fish and dairy products, cereals, nuts and beans. Alternative protein sources such as Quorn™ are ideal for vegetarian athletes. Athletes who restrict their diet, however, may be at risk of insufficient protein intake, with endurance athletes (via exercise induced appetite suppression), and those competing in weight category or aesthetic sports, at greatest risk.

Timing of protein intake

Pre-exercise. The athlete should aim to create an anti-catabolic environment prior to exercise. For resistance training, Tipton et al. (2001) observed an increase in net amino acid uptake when essential amino acids plus carbohydrates were ingested pre-exercise versus post-exercise. It therefore seems prudent for athletes to consume a protein-based snack (approx. 10g) prior to resistance training. For example milk, yogurt, tuna or turkey sandwich.

Post-exercise. Results appear mixed regarding the ingestion of protein post-exercise (Roy et al. 2000; Rasmussen et al. 2000; Godard et al. 2002; Levenhagen et al. 2002; Rankin et al. 2004; Rowlands et al. 2007). Overall, however, there appears to be some additional benefit of consuming a protein and carbohydrate based snack/drink *immediately* post-exercise. It is unclear whether increases in fat-free soft tissue and strength occur as a result of the additional energy, the presence of amino acids, or both (Roy et al. 2000; Levenhagen et al. 2001, 2002; Rankin et al. 2004).

The athlete should aim to create an anabolic environment post-exercise. This is usually achieved via a protein and carbohydrate snack or drink *immediately* after exercise. For example: milkshake, yogurt and banana, cereals with milk, beans on toast, or a sandwich with a protein filling such as ham, turkey or tuna. Athletes may find that the consumption of a protein-based drink, such as a milk shake, is the most convenient and easily tolerated method to consume and absorb protein quickly post-exercise. It is essential that this drink also contains adequate (75–90g) carbohydrates, as discussed above.

Throughout the day. Increased protein intake leads to increased activity of those enzymes responsible for oxidising protein. Enzyme activity increases quickly in response to a large protein meal, but takes longer to down-regulate (van Hall et al. 1996). A large meal therefore leads to greater levels of protein oxidation (Schauder et al. 1984). To ensure that protein is constantly available for growth and repair, without being preferentially oxidised, athletes should therefore eat small quantities of protein regularly throughout the day.

Summary and practical recommendations

- Protein requirements vary depending on the sport, training, and the individual.

- *Endurance athletes* should consume 1.2–2.0g/kg.bw/day, depending on the volume of training.

- *Intermittent sport athletes* should consume 1.4–1.6g/kg.bw/day.

- *Strength athletes* should consume 1.7–2g/kg.bw/day.

- Eat small quantities at each meal, not all in one go.

- Athletes should consume protein from a variety of sources to ensure they achieve a balance between the essential and non-essential amino acids (complete proteins from animal sources will provide all of the essential amino acids).

- Prior to resistance training consume a protein-based snack.

- Immediately post resistance training consume a protein-based snack.

- Protein intake above those recommended (1.2–2.0g/kg.bw/day) is not necessary for most athletes.

- Ensure adequate carbohydrate intake – the body will use increased levels of protein in a glycogen-depleted state.

Calculating macronutrient requirements for different sports

The tables below show the different nutritional intakes by different athletes.

The examples shown in Tables 14.7 and 14.8 provide a useful reference point for athletes. The practical implications of food intake during training or competition, however, may affect the consumption of each macronutrient and also the overall balance. For example, in practice, a Tour de France cyclist may find it difficult to achieve the 13g/kg.bw/day required to maintain performance. To supply this requires large quantities of easily digestible carbohydrate that is delivered to the cyclist throughout the ride and requires considerable logistical support before, during, and after each stage.

Table 14.7 Example nutritional intake for a Tour de France cyclist

Cyclist – 75Kg – Energy requirements ~ 5500 kcal/day				
Nutrient	Quantity	Total	Energy	% daily intake
Carbohydrate	13g/kg.bw/day	975	3900	69.9
Protein	2.0g/kg.bw/day	150	600	10.8
Fat	120g /day	120	1080	19.3
		Total	5580	100

Fluid for performance

Adequate hydration, and the maintenance of fluid balance, is crucial for performance. During exercise, 75–80% of the energy used by the muscles appears as heat with the greater the intensity of exercise, the greater the heat produced. This heat is dissipated through the evaporation of sweat. It is possible to lose up to 2.5 litres per hour during intense activity

Table 14.8 Example nutritional intake for a rugby league player

Rugby league – 100Kg – Energy Requirements ~ 5000 kcal/day				
Nutrient	Quantity	Total	Energy	% daily intake
Carbohydrate	8g/kg.bw/day	800	3200	64
Protein	1.8g/kg.bw/day	180	720	14.4
Fat	120g /day	120	1080	21.6
		Total	5000	100

(Casa et al. 2000) with losses up to 3.1 litres observed during a 90-minute soccer training session in the heat (Shirreffs et al. 2005), or 2.65 litres in a cool environment (Maughan et al. 2005). The loss of fluid in sweat and associated dehydration contributes to fatigue and hyperthermia during exercise (González-Alonso et al. 1997) with distance runners (5000 and 10,000m) forced to slow their pace by more than 6% following a 2% loss of bodyweight through dehydration (Armstrong et al. 1985). This is further compounded when exercising in hot environments or during events of longer duration. Single bout sprint and power performance, however, does not appear to be negatively affected by dehydration (Watson et al. 2005).

Besides physical performance, dehydration also inhibits co-ordination and increases risk of injury. Because changes in body water content (2% of body weight) can severely impair physical performance, as well as psychomotor, and cognitive performance (Grandjean and Grandjean 2007), potentially increasing risk of injury, it is essential that athletes maintain fluid levels and avoid dehydration. The continued ingestion of fluid therefore becomes a major factor in delaying fatigue during exercise.

Fluid intake during exercise

Fluid intake during exercise has a number of benefits. These include the prevention of dehydration, the maintenance of blood volume, osmolality and viscosity (ensuring cardiac output and the maintenance of performance), and the maintenance of skin blood flow and sweat rate (reducing the risk of hyperthermia and heat stress).

Most athletes, however, do not voluntarily drink sufficient water to prevent dehydration during physical activity (NATA Position Statement: Fluid Replacement for Athletes (2000)). Shirreffs et al. (2005) observed that players only replaced between 9 and 73% (45±16%) of the fluid lost through sweat during a 90-minute training session in the heat. Thirst is a sign of dehydration but, because it is possible to dehydrate by 2% of bodyweight before thirst occurs, athletes must drink before getting thirsty.

Factors affecting fluid intake

There are several factors that affect fluid intake during training and competition. These include availability of fluid, thirst, awareness of sweat losses, opportunity to drink and the palatability of fluid.

Ideal fluid for exercise

Initially fluids should be cool (10–12 °C), palatable, not acidic or gassy and not cause gastrointestinal distress. Water fits this description and is a good place to start, however there are good reasons why drinks should also contain carbohydrate (for energy and to maintain carbohydrate oxidation), salt (0.3–0.7g/l) (to aid fluid retention and stimulate thirst) (Casa et al. 2000), and be isotonic – have an osmolality of 280–300 mOsm/kg (to aid gastric emptying). Sports drinks typically contain a mixture of water, carbohydrates and salt, and benefit athletes through quicker re-hydration after training, quicker refuelling of carbohydrates, stimulating thirst, and being convenient and readily available.

Practical recommendations

What, when, and how much athletes drink will be determined by a range of factors.

What to drink? Athletes must identify whether the priority is to supply fluid or energy. If fluid, then electrolyte solutions with 4–6% carbohydrates will work well (Murray et al. 1999). If carbohydrates are the priority then a more concentrated solution (6–10% carbohydrate) may deliver more carbohydrate. There is, however, some evidence that solutions over 8% may cause gastro-intestinal distress (Shi et al. 2004). Overall, it appears that carbohydrate solutions in the 6–8% range provide the optimal balance of gastric emptying, fluid absorption and carbohydrate delivery. Athletes should experiment with varying concentrations during training to determine what they can tolerate.

When to drink? This will largely depend on the nature of the sport and availability of fluid.

How much to drink? This will depend on whether it is before, during, or after exercise, on losses through sweat, and on how much fluid the athlete can reasonably tolerate.

Prior to exercise: athletes should drink sufficient quantities to ensure they are well hydrated before exercise, with their urine clear for several hours beforehand. As competition approaches, a drink of 500–600ml 2–3 hours before exercise followed by another 200–300 ml 10–20 minutes before exercise begins (Casa et al. 2000) will ensure athletes arrive in a well-hydrated state. Taking regular sips until, the start may be a useful way for athletes to consume these quantities. For longer events, where dehydration may inhibit performance, athletes should drink an additional 400–600ml of water (or carbohydrate solution) *immediately before exercise*. Athletes should experiment in training to ensure they can tolerate both the quantity and the type of drink.

During exercise: drink 100–300ml of water every 15 minutes as tolerated (Rehrer et al. 1990). For team sports, drink at halftime or during breaks in play. Again, athletes should practice drinking during training. Adding carbohydrates and electrolytes, as found in most popular sports drinks, will ensure delivery of carbohydrate for oxidation and glycogen replenishment (during a break), and electrolytes to aid retention.

After exercise: the priority post-exercise is to replenish what was lost (both in terms of fluid and glycogen stores). Fluid intake needs to be about 150% of weight lost during exercise to achieve normal hydration within 6 hours post-exercise (Shirreffs et al. 1996). Ingesting plain water though is largely ineffective as this dilutes plasma and inhibits the secretion of anti-diuretic hormone. Adding sodium (60–80 mmol/litre) will reduce urinary water loss, aiding fluid retention and the recovery of fluid balance (Nose et al. 1988; Sharp 2006). Moreover, adding sodium will trigger thirst and promote drinking. Cool fluids are more palatable. Therefore to promote rapid recovery of fluid balance post exercise, athletes should focus on both volume of fluid (around 150% of weight loss) and sodium content (60–80mmol/litre). The inclusion of carbohydrate (4–6%) will also help to restore glycogen stores.

Hydration for team sports

Team sports such as football, rugby, hockey and netball present another problem. Often athletes cannot consume fluid throughout the match or training session. Athletes therefore need to ensure adequate hydration on days prior to a match, follow the pre-exercise fluid intake guidelines above, aim to consume at least 500ml at half time, and try to drink during any other breaks in play such as during injury breaks. Post-match, athletes should follow the guidelines above for rehydration post-exercise.

Carbohydrate versus fluid delivery

Carbohydrates provide the substrate for glycogen resynthesis and maintenance of blood glucose, but there can be a conflict with fluid absorption at higher carbohydrate concentrations. If an athlete is dehydrated then fluid (and electrolyte) intake will be paramount. A carbohydrate concentration of no more than 4–6% will ensure that gastric emptying is not affected. If carbohydrate is the priority (either for oxidation or glycogen resynthesis) then higher levels up to about 10% may be consumed.

Summary/Key points

- Start well hydrated. Thirst is not an indicator of fluid need but a sign of partial dehydration. Athletes should consume fluids before they are thirsty.

- To avoid dehydration, drink about 500–600 ml in the hours before a race/match and 200–300 ml 10–20 minutes beforehand. Drink regularly throughout exercise (100–300 ml every 10–15 min).

- Carry fluids. This will encourage voluntary fluid consumption.

- Clear (pale yellow) urine is a sign that the athlete is well hydrated; dark urine, that the athlete is under-hydrated.

- Avoid foods and drinks that may have a diuretic effect (alcohol, strong coffee).

- Estimate sweat loss for each athlete by measuring body weight loss during training.

- During exercise, aim to drink sufficient fluids to match sweat loss.

- Combine carbohydrates with fluid ingestion to help replenish glycogen stores.

Vitamin and mineral requirements for athletes

Recommendations for micronutrient intake are largely based on the requirements of healthy, but relatively inactive people. During exercise it is likely that micronutient requirements will increase (Whiting and Barabash 2006).

Micronutrient intake varies widely between individuals and groups, with dietary surveys of athletes showing both high and low reported intakes of some vitamins and minerals, leading to the possibility of a long-term deficient diet or health problems associated with excess intakes. Athlete groups that may be at risk of insufficient micronutrient intake include those on restricted energy intakes (Haymes 1991), vegetarians, female athletes, those involved in endurance or aesthetic sports, and athletes in weight category sports. In these groups, consumption of a multi-vitamin supplement may ensure adequate intakes and avoid deficiency (Beals and Manore 1998).

There is no clear evidence that elevated intakes of vitamins or minerals will increase performance and no evidence that athletes require significantly higher levels of micronutrients than non-athletes. The priority therefore should be to avoid deficiency through the consumption of a diet that is both sufficient and balanced.

There are, however, several key micronutrients that either play a pivotal role during exercise or are particularly prone to deficiency.

Vitamin B

Because vitamin B plays an essential role in the release of energy from carbohydrates, fat and protein, and in the formation of haemoglobin, a deficiency can have serious consequences for the athlete, leading to fatigue and decreases in VO_2 max and power (van der Beek et al. 1994). Athletes at risk of possible deficiency may be those with restricted diets or vegans, whereas athletes with an energy rich diet are unlikely to be deficient. Good sources of the B vitamins include meat, fish, milk, eggs, wholegrain cereals, fortified breakfast cereals and some vegetables. Although, with a balanced and energy rich diet supplementation is generally unnecessary, a multi-vitamin will help meet requirements for athletes who may be unsure of their status.

Vitamin C and antioxidants

Although there is no evidence that athletes need more vitamin C than non-athletes, it is possible that antioxidant supplementation may decrease exercise-induced oxidative stress (Ji 1999; Morillas-Ruiz et al. 2006). Overall, athletes should avoid deficiency and obtain antioxidants via increased consumption of fruit and vegetables. Large doses of single antioxidant compounds are not recommended.

Minerals and exercise

Exercise is associated with increased losses of minerals in sweat and urine. Iron, calcium, magnesium and zinc may be a cause for concern in some athlete groups due to insufficient intakes and increased losses in sweat and urine. Of these, calcium and iron have the biggest impact on health and performance.

Iron

Iron depletion (low iron stores: low serum ferritin) is common in athletes (26% women, 11% men- Malczewska et al. 2001) but does not necessarily affect performance (Risser et al. 1988). Iron deficiency without anaemia may, however, impair adaptation to endurance training (Brownlie et al. 2004) in previously untrained women, but can be corrected with iron supplementation. Iron deficiency with anaemia (low haemoglobin) can impair work capacity and decrease exercise performance (Haas and Brownlie 2001).

Athletes at risk of iron deficiency include young athletes, female athletes (Beard and Tobin 2000; Gropper et al. 2006), athletes on low energy intakes (less than 300kcal/day), athletes in weight category sports, endurance athletes (Spodaryk 1993), vegetarians, and athletes training in hot climates or at altitude.

There are a number of ways to increase iron:

1. Athletes should eat foods rich in haem-iron at least four times per week (e.g. liver, lean red meat) as iron from these foods is readily absorbed.

2. Vegetarians should aim to eat iron-fortified foods (e.g. breakfast cereal) and other non-haem iron food sources (e.g. dried fruit, legumes, green leafy vegetables).

3. Athletes can increase the absorption of iron from non-haem iron foods by consuming them with vitamin C-rich foods (e.g. orange juice) and avoiding tea at meals.

Calcium

Athletes with low energy intakes and who avoid dairy products may not meet their calcium requirements. This is a particular problem for female athletes on low energy intakes (Clarkson 1995) as amenorrhea may further hinder bone development and increase the risk of osteoporosis. The current UK recommended daily intake of calcium is 700mg, with an upper safe limit of 2500mg. Calcium cannot be absorbed without Vitamin D (FSA 2008).

There are a number of ways to increase calcium:

1. Athletes should include three servings per day of low-fat dairy foods. Include these in high carbohydrate meals (e.g. skimmed milk on cereal).

2. Eat fish with bones (e.g. sardines, tinned salmon).

3. If athletes cannot tolerate dairy products then consider calcium-enriched soy products.

4. Eat green leafy vegetables (cabbage, broccoli, spinach).

5. Supplementing calcium to 125% RNI helps maintain bone density when amenorrhea is present.

Summary

Maintaining adequate intakes of vitamins and minerals is essential for health and performance. This can be achieved through a varied and balanced diet. Athletes who consume sufficient energy from a balanced diet are unlikely to have vitamin and mineral deficiencies (Armstrong and Maresh, 1996), however, the use of a multi-vitamin and mineral supplement for groups at risk or on low energy intakes may be appropriate.

Nutrition for injury prevention

Delaying fatigue

The most important nutritional consideration for injury prevention is in delaying the onset of fatigue. If an athlete is fatigued, there is a change in running mechanics (Gerlach et al. 2005; Kellis and Liassou 2009), landing mechanics (King et al. 2005), a decreased ability to maintain joint alignment, control and appropriate muscular activation patterns during potentially risky manoeuvres (Wojtys et al. 1996; Chappell et al. 2005), an increased incidence of high-risk actions (Rahnama et al. 2002), and an actual increase in injury occurrence towards the end of a match or phase of play (Hawkins et al. 2001). To limit fatigue, athletes should consume a diet that allows them to maintain optimal performance throughout the duration of a race, match or training session. The two most important nutrients to prevent fatigue are carbohydrate and water.

Carbohydrate

One of the primary factors linking fatigue and injury is the level of muscle glycogen (Sherman and Costill 1984; Costill and Hargreaves 1992). If muscle glycogen is low, athletes will not be able to maintain exercise intensity, muscles will fatigue and lose strength along with the ability to protect joints, coordination will suffer, protective motor programmes will be replaced by less efficient and more risky movement patterns, awareness of the game and environment will decrease, and reactions will slow (Schlabach 1994). These are the circumstances when injury is most likely to occur.

Therefore the primary nutrient that is required to avoid fatigue is carbohydrate. Athletes who are maintaining high-volume high-intensity exercise are most at risk and must optimise carbohydrate intake. Athletes should aim to consume 5–13g/kg.bw per day (depending on volume and intensity of exercise) by eating pre, during and post exercise (Jeukendrup and Gleeson 2004).

Long-term fatigue

Athletes that are in heavy training for prolonged periods are at risk of progressively depleting glycogen stores leading to a drop off in performance, increased risk of injury, compromised immune system and increased risk of illness, and ultimately overtraining syndrome (Kirwan et al. 1988).

Fluid

Fluid intake is the other primary factor in reducing signs of fatigue. The effects of dehydration largely mimic those of fatigue and can therefore contribute to injury risk. There is also a greater risk of heat injury when dehydrated, as the body is unable to thermo-regulate effectively. Glycogen use increases when dehydrated, further compounding the problems.

Athletes should be aware of and look out for the effects of dehydration. Initially these may include thirst, dark urine, tiredness, lack of concentration, dry skin and headache. Fluid intake should be monitored and matched to sweat losses. Fluid replacement strategies should be in place before, during, and after the game or training (Rehrer et al. 1990; Shirreffs et al. 1996; Casa et al. 2000; Sharp 2006).

Other factors to consider in injury prevention

Iron

Low iron intakes have the potential to affect injury risk through fatigue. In a study on female cross-country runners, over the course of the season, there were 71 injuries that caused a loss of training time. The 34 runners with the lowest ferritin concentrations had twice as many injuries as the 34 runners with the highest ferritin (Loosli et al. 1993). As iron plays a crucial role in the transport of oxygen to muscles, it is likely that athletes with low haemoglobin (caused by iron deficiency) have decreased oxygen delivery to tissues, reducing work capacity (Viteri and Torun 1974), and therefore fatigue more easily.

Bone health

Nutrition can affect bone health in several ways. First, low fat and low energy intakes are associated with an increased risk of stress fractures particularly in physically active women (Frusztajer et al. 1990; Nattiv 2000). Supplementation with calcium (2000mg) and Vitamin D (800IU), however, has been shown to decrease the incidence of stress fractures in female navy recruits by 20% compared with a placebo (Lappe et al. 2008).

In the longer term, a diet that is deficient in energy, fat, calcium or vitamin D may lead to osteoporosis. Moreover other nutrients, such as magnesium and potassium, along with an adequate protein intake, also appear to play a significant role in preventing the loss of bone mineral density (Hannan et al. 2000; Tucker et al. 2001).

Athletes, and in particular female athletes, should therefore ensure they consume a balanced diet containing sufficient energy, protein, fat, calcium and vitamin D with plenty of fresh fruit and vegetables to ensure a balanced micronutrient intake (Tucker et al. 2001).

Nutrition during injury

Injury can lead to a range of complex nutritional issues for some athletes. Body mass management (preventing weight gain during injury, restoration of muscle mass post injury) is crucial for effective rehabilitation. For example, if the athlete has a significant reduction in activity levels as a result of injury, then the diet will need to change to reflect the drop in energy expenditure. Moreover, athletes with poor diets, who have previously avoided weight gain through training, or those who turn to food for comfort, are likely to put on weight when injured, making it harder for them to return to full fitness. If, however, the athlete can maintain energy expenditure through other forms of exercise, then diet may not need to change.

Education is a priority. Athletes must aim for a nutrient rich and healthy diet that is sufficient to maintain energy balance. Athletes should focus on low-fat, low-sugar, high-fibre foods that provide sufficient carbohydrate, protein and fat, and which provide optimal vitamin and mineral intakes.

Despite lower activity levels when injured, athletes who are hospitalised, or subject to long-term incapacity, may still require increased protein (approx 1.4–1.7g/kg/day) to prevent loss of lean tissue, and maintain immune function (Bucci 1994). This can be met through the selection of low-fat protein options such as lean meat, fish and skimmed milk.

Supplements

The use of supplements by athletes requires caution. Whilst there is a substantial body of evidence that some substances found in the diet have an ergogenic or anabolic effect under certain conditions – for example, caffeine for endurance and power performance, and creatine for increasing short-term high-intensity exercise and muscle mass (Birch et al. 1994; Williams and Branch 1998; Greenhaff 2000; Graham 2001; Maughan et al. 2004; Doherty and Smith 2005; Hespel et al. 2006 – many of the elaborate claims made for supplements by manufacturers do not stand up to scientific scrutiny. Moreover, many supplements contain substances not declared on the label and in some cases these substances contravene IOC or WADA doping regulations and would cause an athlete to fail a drugs test (Geyer et al. 2004; Maughan 2005). For example, in an IOC funded study of 634 products labelled as non-hormonal nutritional supplements from 13 countries and 215 different suppliers, 14.8% contained anabolic steroid *precursors* not declared on the label (Schanzer 2002; Geyer et al. 2004). For products purchased in the UK this figure rose to 18.8% (Schanzer 2002; Geyer et al. 2004).

It appears that, for those athletes who may be required to take a drugs test, supplements are another potential source of contamination. It is therefore possible that an athlete could fail a drugs test due to the unintentional ingestion of prohibited substances present in dietary supplements (Maughan 2005). It is simply impossible to know for sure that any given supplement is pure and not contaminated by some substance that may be prohibited.

The principle of strict liability present in the World Anti-Doping Code means that athletes are ultimately responsible for any prohibited substances found in their system (UK Sport (2008) position statement on the use of supplements – July 2008). The unintentional ingestion of prohibited substances is not considered an acceptable excuse and athletes should therefore exhibit extreme caution when deciding on the use of dietary supplements. Moreover, supplements should not be considered a solution to a poor diet and athletes should strive to optimise their nutritional intake before considering the need for supplements.

The full position statement of UK Sport (2008) concerning the use of supplements appears below.

Position statement of UK Sport, Version 5, issued in July 2008.

There is an array of supplements available for athletes to purchase through a range of retail sources that have no prohibited substances listed as ingredients. Despite this there have been several cases whereby supplement products have been contaminated with prohibited substances as defined by the World Anti-Doping Code (WADC) Prohibited List.

UK athletes are advised to be vigilant in their choice to use any supplement. No guarantee can be given that any particular supplement is free from Prohibited Substances.

Athletes should be aware that any product that claims to restore, correct or modify the body's physiological functions should be licensed as a medicine, according to current legislation (for further information visit the Medicines Healthcare products Regulatory Agency website at www.mhra.gov.uk).

Diet, lifestyle and training should all be optimised before considering supplements and athletes should assess the need for supplements by always consulting an accredited sports dietician and/or registered nutritionist with expertise in sports nutrition and a sports and exercise medicine doctor before taking supplements.

An important principle of the World Anti-Doping Code (WADC) is that of strict liability stating athletes are ultimately responsible for any Prohibited Substances found in their system or for the use of any Prohibited Method. Therefore before taking supplements athletes must assess the risk and understand their personal responsibility.

In an attempt to support athletes a number of initiatives have been created globally to identify whether a prohibited substance can be identified within a supplement. As such, supplements may claim to be drug free or safe for drug tested athletes. It is not possible to guarantee that specific supplements will be free of prohibited substances but only to reduce the risk of inadvertent doping by making informed decisions.

In the UK HFL Sports Science has taken the initiative to create a scheme to support athletes in assessing the risk. The Informed-Sport programme is designed to evaluate supplement manufacturers for their process integrity and screening of supplements and ingredients for the presence of prohibited substances that are present on the WADC Prohibited List. The supplements industry has been consulted on this approach and supports its development.

> UK Sport believes this risk minimisation service to be a positive step and welcomes the approach being taken by industry and the HFL owned Informed-Sport programme.
>
> Ultimately we wish to remind athletes that strict liability will still apply and the appropriate sanctions provided to any athlete returning an adverse analytical finding from any supplement product as with all other anti-doping case.

Athletes who are subject to drug testing need to ensure that any commercial sports drink or food is legal. If in doubt, it is very easy to produce a home-made alternative.

For further reading and up to date information on supplements go to: www.uksport.gov.uk and www.100percentme.co.uk

---

Make your own sports drink

For training

1 litre of water

60g table sugar (or ideally powdered glucose/dextrose)

pinch of salt

diet cordial (to taste).

Add contents together. Shake.

For recovery

500ml skimmed milk

banana

2 heaped tablespoons of malted drink powder (Horlicks, Ovaltine, Nesquick)

Add contents together. Blend.

References

Achten, J. and Jeukendrup, A.E. (2003) Maximal fat oxidation during exercise in trained men. *International Journal of Sports Medicine*, 24, 88.

Achten, J.; Gleeson, M. and Jeukendrup, A.E. (2002) Determination of the exercise intensity that elicits maximal fat oxidation *Medicine and Science in Sports and Exercise*, 34 (1), 92–97.

Ahlborg, B., Bergstrom, J., Brohult, J., Ekelund, L.G., Hultman, E. and Maschio, G. (1967) Human muscle glycogen content and capacity for prolonged exercise after different diets. *Forsvarsmedicin*, 3, 85–99.

Armstrong, L.E., Costill, D.L. and Fink, W.J. (1985) Influence of diuretic induced dehydration on competitive running performance. *Medicine and Science in Sports and Exercise*, 17 (4), 456–461.

Armstrong, L.E. and Maresh, C.M. (1996) Vitamin and mineral supplements as nutritional aids to exercise performance and health. *Nutrition Reviews*, 54 (4), S149–S158.

Astrand, P.-O. and Rodahl, K. (1986) Textbook of work physiology, 3rd edn. New York: Mcgraw-Hill.

Baer, D.J., Judd, J.T., Clevidence, B.A. and Tracy, R.P. (2004) Dietary fatty acids affect plasma markers of inflammation in healthy men fed controlled diets: a randomized crossover study. *American Journal of Clinical Nutrition*, 79, 969–973.

Balsom, P.D., Gaitanos, G.C., Söderlund K. and Ekblom, B. (1999) High-intensity exercise and muscle glycogen availability in humans. *Acta Physiologica Scandinavica*, 165 (4), 337–345.

Beals, K.A. and Manore, M.M. (1998) Nutritional status of female athletes with subclinical eating disorders. *Journal of the American Dietetic Association*, 98 (4), 419–425.

Beard, J. and Tobin, B. (2000) Iron status and exercise. *American Journal of Clinical Nutrition*, 72, 594S–597S.

Bergstrom, J., Hermansen, L., Hultman, E. and Saltin, B. (1967) Diet, muscle glycogen and physical performance. *Acta Physiologica Scandinavica*, 71, 140–150.

Beyer, P., Caviar, E. and McCallum, R. (2005) Fructose intake at current levels in the United States may cause gastrointestinal distress in normal adults. *Journal of the American Dental Association*, 105 (10), 1559–1566.

Birch, R., Noble, D. and Greenhaff, P.L. (1994) The influence of dietary creatine supplementation on performance during repeated bouts of maximal isokinetic cycling in man. *European Journal of Applied Physiology and Occupational Physiology*, 69 (3), 268–270.

Bolster, D.R., Jefferson, L.S. and Kimball, S.R. (2004) Regulation of protein synthesis associated with skeletal muscle hypertrophy by insulin-, amino acid- and exercise-induced signalling. *Proceedings of the Nutrition Society*, 63, 351–356.

Brooks, G. and Mercies, J. (1994) Balance of carbohydrate and lipid oxidation during exercise: the 'crossover' concept. *Journal of Applied Physiology*, 76, 2253–2261.

Brownlie, T. 4th, Utermohlen, V., Hinton, P.S. and Haas, J.D. (2004) Tissue iron deficiency without anaemia impairs adaptation in endurance capacity after aerobic training in previously untrained women. *American Journal of Clinical Nutrition*, 79 (3), 437–443.

Bucci, L.R. (1994) Nutrition Applied to Injury Rehabilitation and Sports Medicine. Boca Raton, FL: CRC Press.

Bucher, H. (2002) N-3 polyunsaturated fatty acids in coronary heart disease: a meta-analysis of randomized controlled trials. *The American Journal of Medicine*, 112 (4), 298–304.

Burke, L.M., Hawley, J.A., Schabort, E.J., Gibson, A. Mujika, I. and Noakes, T.D. (2000) Carbohydrate loading failed to improve 100-km cycling performance in a placebo-controlled trial. *Journal of Applied Physiology*, 88, 1284–1290.

Campbell, B., Kreider, R.B., Ziegenfuss, T., Bounty, P.L., Roberts, M., Burke, D., Landis, J., Lopez, H. and Antonio, J. (2007) International Society of Sports Nutrition position stand: protein and exercise. *Journal of the International Society of Sports Nutrition*, 4, 8.

Carter, J., Jeukendrup, A.E., Mundel, T. and Jones, D.A. (2003) Carbohydrate supplementation improves moderate and high-intensity exercise in the heat. *Pflugers Archives*, 446 (2), 211–219.

Casa, D.J., Hillman, S.K., Armstrong, L.E., Montain, S.J., Reiff, R.V., Rich, B.S.E., Roberts, W.O. and Stone, J.A. (2000) National Athletic Trainers' Association Position Statement: Fluid replacement for athletes. *Journal of Athletic Training*, 35 (2), 212–224.

Chappell JD, Herman DC, Knight BS, Kirkendall DT, Garrett WE, Yu B (2005). Effect of fatigue on knee kinetics and kinematics in stop-jump tasks. *Am J Sports Med.* 33 (7): 1022–1029.

Choi, D., Cole, K.J., Goodpaster, B.H., Fink, W.J. and Costill, D.L. (1994) Effect of passive and active recovery on the resynthesis of muscle glycogen. *Medicine and Science in Sports and Exercise*, 26 (8), 992–996.

Clarkson, P.M. (1995) Micronutrients and exercise: Antioxidants and minerals. *Journal of Sports Sciences*, 13 (S1), S11–S24.

Coggan, A.R. and Coyle, E.F. (1988) Effect of carbohydrate feedings during high-intensity exercise. *Journal of Applied Physiology*, 65 (4), 1703–1709.

Coggan, A.R. and Coyle, E.F. (1991) Carbohydrate ingestion during prolonged exercise: Effects on metabolism and performance. *Exercise and Sports Science Reviews*, 19, 1–40.

Costill, D.L. and Hargreaves, M. (1992) Carbohydrate nutrition and fatigue. *Sports Medicine*, 13 (2), 86–92.

Coyle, E.F. (1995) Substrate utilization during exercise in active people. *American Journal of Clinical Nutrition*, 61 (4 Suppl), 968S–979S.

Coyle, E.F., Coggan, A.R., Hemmert, M.K. and Ivy, J.L. (1986) Muscle glycogen utilization during prolonged strenuous exercise when fed carbohydrate. Journal of Applied Physiology, 61 (1) 165-172,

Coyle, E.F., Jeukendrup, A.E., Wagenmakers, A.J. and Saris, W.H. (1997) Fatty acid oxidation is directly regulated by carbohydrate metabolism during exercise. *American Journal of Physiology*, 273 (2), E268–275.

De Jonge, L. and Smith, S.R. (2008) Macronutrients and exercise. *Obesity Management*, 4 (1), 11–13.

Department of Health (1991) *Dietary Reference Values for Food Energy and Nutrients for the United Kingdom: Report of the Panel on Dietary Reference Values of the Committee on Medical Aspects of Food Policy. Report on Health and Social Subjects 41*. London: HMSO.

Department of Health (1994) *Report on Health and Social Subjects No. 46. Nutritional Aspects of Cardiovascular Disease. Report of the Cardiovascular Review Group Committee on Medical Aspects of Food and Nutrition Policy*. London: HMSO.

Doherty, M. and Smith, P.M. (2005) Effects of caffeine ingestion on rating of perceived exertion during and after exercise. A meta-analysis. *Scandanavian Journal of Medicine and Science in Sports*, 15, 69–78.

Dohn, G.L. (1986) Protein as a fuel for endurance exercise. *Exercise and Sports in Science Reviews*, 14, 143–173.

Doyle, J.A., Sherman, W.M. and Strauss, R.L. (1993) Effects of eccentric and concentric exercise on muscle glycogen replenishment. *Journal of Applied Physiology*, 74, 1848–1855.

Dreyer, H.C., Drummond, M.J., Pennings, B., Fujita, S., Glynn, E.L., Chinkes, D.L., Dhanani, S., Volpi, E. and Rasmussen, B.B. (2008) Leucine-enriched essential amino acid and carbohydrate ingestion following resistance exercise enhances mTOR signalling and protein synthesis in human muscle. *American Journal of Physiology and Endocrinology Metabolism*, 294, E392–E400.

El-Sayed, M.S., Ali, N. and El-Sayed, A.Z. (2005) Interaction between alcohol and exercise: Physiological and haematological implications. *Sports Medicine*, 35 (3), 257–269.

Fairchild, T.J., Fletcher, S., Steele, P., Goodman, C., Dawson, B. and Fournier, P.A. (2002) Rapid carbohydrate loading after a short bout of near maximal-intensity exercise. *Medicine and Science in Sports and Exercise*, 34 (6), 980–986.

Fairchild, T. J., Armstrong, A.A., Rao, A., Liu, H., Lawrence, S. and Fournier, P.A. (2003) Glycogen synthesis in muscle fibres during active recovery from intense exercise. *Medicine and Science in Sports and Exercise*, 35 (4), 595–602.

Fairweather-Tait, S. and Hurrell, R.F. (1996) Bioavailability of minerals and trace elements. *Nutrition Research Reviews*, 9, 295–324.

Fink, H.H., Burgoon, L.A. and Mikesky, A.E. (2005) Practical Applications in Sports Nutrition. Sudbury, MA: Jones & Bartlett.

Foskett, A., Williams, C., Boobis, L. and Tsintzas, K. (2008) Carbohydrate availability and muscle energy metabolism during intermittent running. *Medicine and Science in Sports and Exercise*, 40 (1), 96–103.

Foster-Powell, K., Holt, S.H. and Brand-Miller, J.C. (2002) International table of glycemic index and glycemic load values: 2002. *American Journal of Clinical Nutrition*, 76, 5–56.

Frusztajer, N.T., Dhuper, S., Warren, M.P., Brooks-Gunn, J. and Fox, R.P. (1990) Nutrition and the incidence of stress fractures in ballet dancers. *American Journal of Clinical Nutrition*, 51, 779–783.

FSA (2003) *Expert Group on Vitamins and Minerals*. London: Food Standard Agency.

FSA (2008) *Food Standards Agency Manual of Nutrition*, 11th edn. London: TSO (The Stationary Office).

García-Rovés, P.M., Terrados, N., Fernández, S. and Patterson, A.M. (2000) Comparison of dietary intake and eating behavior of professional road cyclists during training and competition. *International Journal of Sport Nutrition and Exercise Medicine*, 10 (1), 82–98.

Gerlach, K.E., White, S.C., Burton, H.W., Dorn, J.M., Leddy, J.J. and Horvath, P.J. (2005) Kinetic changes with fatigue and relationship to injury in female runners. *Medicine and Science in Sports and Exercise*, 37 (4), 657–663.

Geyer, H., Parr, M.K., Mareck, U., Reinhart, U., Schrader, Y. and Schänzer, W. (2004) Analysis of non-hormonal nutritional supplements for anabolic-androgenic steroids: Results of an international study. *International Journal of Sports Medicine*, 25 (2), 124–129.

Godard, M.P., Williamson, D.L. and Trappe, S.W. (2002) Oral amino-acid provision does not affect muscle strength or size gains in older men. *Medicine and Science in Sports and Exercise*, 34, 1126–1131.

González-Alonso, J., Mora-Rodríguez, R., Below, P.R. and Coyle, E.F. (1997) Dehydration markedly impairs cardiovascular function in hyperthermic endurance athletes during exercise. *Journal of Applied Physiology*, 82 (4), 1229–1236.

Graham, T.E. (2001) Caffeine and exercise: Metabolism, endurance and performance review article. *Sports Medicine*, 31 (11), 785–807.

Grandjean, A.C. and Grandjean, N.R. (2007) Dehydration and dognitive performance. *Journal of the American College of Nutrition*, 26 (90005), 549S–554S.

Greenhaff, P.L. (2000) Creatine. In Maughan, R.J. (Ed.) Nutrition in Sport, pp 379–392. Oxford: Blackwell.

Greiwe, J.S., Kwon, G., McDaniel, M.L. and Semenkovich, C.F. (2001) Leucine and insulin active S6 kinase through different pathways in human skeletal muscle. *American Journal of Physiology and Endocrinology Metabolism*, 281, 466–471.

Gropper, S., Blessing, D., Dunham, K. and Barksdale, J. (2006) Iron status of female collegiate athletes involved in different sports. *Biological Trace Element Research*, 109, (1), 1–13.

Haas, J. and Brownlie, T. (2001) Iron deficiency and reduced work capacity: A critical review of the research to determine a causal relationship. *Journal of Nutrition*, 131, 676S–688S.

Hannan, M., Tucker, K., Dawson-Hughes, B., Cupples, L., Felson, D. and Kiel, D. (2000) Effect of dietary protein on bone loss in elderly men and women: The Framingham Osteoporosis Study. *Journal of Bone and Mineral Research*, 15, 2504–2512.

Hawkins, R.D., Hulse, M.A., Wilkinson, C., Hodson, A. and Gibson, M. (2001) The association football medical research programme: an audit of injuries in professional football. *British Journal of Sports Medicine*, 35, 43–47.

Hawley, J.A. (2002) Effect of increased fat availability on metabolism and exercise capacity. *Medicine and Science in Sports and Exercise*, 34 (9), 1485–1491.

Hawley, J.A., Brouns, F. and Jeukendrup, A. (1998) Strategies to enhance fat utilization during exercise. *Sports Medicine*, 25, 241–257.

Haymes, E.M. (1991) Vitamin and mineral supplementation to athletes. *International Journal of Sport Nutrition*, 1 (2), 146–169.

Hespel, P.L., Maughan, R.J. and Greenhaff, P.L. (2006) Dietary supplements for football. *Journal of Sports Sciences*, 24, 749–761.

Horowitz, J.F., Mora-Rodriguez, R., Byerley, L.O. and Coyle, E.F. (1997) Lipolytic suppression following carbohydrate ingestion limits fat oxidation during exercise. *American Journal of Physiology and Endocrinology Metabolism*, 273, E768–E775.

Horowitz, J.F. and Klein, S. (2000) Lipid metabolism during endurance exercise. *American Journal of Clinical Nutrition*, 72 (2 Suppl), 558S–63S.

Ismail, M.N., Wan Nudri, W.D. and Zawiah, H. (1997) Energy expenditure studies to predict requirements of selected national athletes. *Malaysian Journal of Nutrition*, 3 (1), 71–81.

Ivy, J.L. (1998) Glycogen resynthesis after exercise: effect of carbohydrate intake. *International Journal of Sports Medicine*, 19, S142–145.

Ivy, J.L., Costill, D.L., Fink, W.J. and Lower, R.W. (1979) Influence of caffeine and carbohydrate feedings on endurance performance. *Medicine Science in Sports*, 11 (1), 6–11.

James, A.P., Lorraine, M., Cullen, D., et al. (2001) Muscle glycogen supercompensation: absence of a gender-related difference. *European Journal of Applied Physiology*, 85, 533–538.

Jenkins, D.J., Wolever, T.M., Taylor, R.H., Barker, H., Fielden, H., Baldwin, J.M., Bowling, A.C., Newman, H.C., Jenkins, A.L. and Goff, D.V. (1981) Glycemic index of foods: a physiological basis for carbohydrate exchange. *American Journal of Clinical Nutrition*, 34, 362–366.

Jentjens, R. and Jeukendrup, A. (2003) Determinants of post-exercise glycogen synthesis during short-term recovery. *Sports Medicine*, 33 (2), 117–144.

Jentjens, R.L.P.G., Van Loon, L.J.C., Mann, C.H., Wagenmakers, A.J.M. and Jeukendrup, A.E. (2001) Addition of protein and amino acids to carbohydrates does not enhance postexercise muscle glycogen synthesis. *Journal of Applied Physiology*, 91 (2), 839–846.

Jeukendrup, A.E. (2003) Modulation of carbohydrate and fat utilization by diet, exercise, and environment. *Biochemical Society Transactions*, 31 (6), 1270–1273.

Jeukendrup, A. (2008) Carbohydrate feeding during exercise. *European Journal of Sports Science*, 8 (2), 77–86.

Jeukendrup, A. and Gleeson, M. (2004) Sport Nutrition. An Introduction to Energy Production and Performance. Champaign, IL: Human Kinetics.

Jeukendrup, A.E. and Jentjens, R. (2000) Oxidation of carbohydrate feedings during prolonged exercise: Current thoughts, guidelines and directions for future research. *Sports Medicine*, 29 (6), 407–424.

Ji, L.L. (1999) Antioxidants and oxidative stress in exercise. *Proceedings of the Society for Experimental Biology and Medicine*, 222, 283–292.

Jones, L.C., Cleary, M.A., Lopez, R.M., Zuri, R.E. and Lopez, R. (2008) Active dehydration impairs upper and lower body anaerobic muscular power. *Journal of Strength and Conditioning Research*, 2 (2), 455–463.

Karmally, W. and Goldber, I.J. (2006) Can altering carbohydrate, protein and unsaturated fat intake improve patients' blood pressure and lipid profile? *Nature Clinical Practice Cardiovascular Medicine*, 3, 254–255.

Kellis, E. and Liassou, C. (2009) The effect of selective muscle fatigue on sagittal lower limb kinematics and muscle activity during level running. *Journal Orthopaedics Sports Physical Therapy*, 39 (3), 210–220.

King, D., Sigg, J., Belyea, B., Hummel, C. and Buck, M. (2005) *Impact Mechanics during Stop and Go Tasks under Fatigued and Non-fatigued Conditions*. ISB XXth Congress - ASB 29th Annual Meeting, July 31–August 5, Cleveland, Ohio.

Kirwan, J.P., Costill, D.L., Flynn, M.G., Mitchell, J.B., Fink, W.J., Neufer, P.D. and Houmard, J.A. (1988)

Physiological responses to successive days of intense training in competitive swimmers. *Medicine and Science in Sport and Exercise*, 20, 255–259.

Klein, S., Coyle, E.F. and Wolf, R.R. (1994) Fat metabolism during low intensity exercise in endurance trained and un-trained men. *American Journal of Physiology*, 267, E934–940.

Kratz, M., Gülbahçe, E., von Eckardstein, A., Cullen, P., Cignarella, A., Assmann, G., Wahrburg, U. (2002) Dietary mono- and polyunsaturated fatty acids similarly affect LDL size in healthy men and women. *The American Society for Nutritional Sciences Journal of Nutrition*, 132, 715–718.

Lappe, J., Cullen, D., Haynatzki, G., Recker, R., Ahlf, R. and Thompson, K. (2008) Calcium and vitamin D supplementation decreases incidence of stress fractures in female navy recruits. *Journal of Bone and Mineral Research*, 23, 741–749.

Lemon PW. (1997) Dietary protein requirements in athletes. *The Journal of Nutritional Biochemistry*, 8 (2) 52–60.

Lemon, P.W. (1994) Protein requirements of soccer. *Journal of Sports Science*, 12, S17–22.

Lemon, P.W. (1995) Do athletes need more dietary protein and amino acids? *International Journal of Sports Nutrition*, 5, S39–61.

Lemon, P.W. (1996) Is increased dietary protein necessary or beneficial for individuals with a physically active lifestyle? *Nutrition Reviews*, 54 (4 Pt 2), S169–175.

Levenhagen, D.K., Gresham, J.D., Carlson, M.G., Maron, D.J., Borel, M.J. and Flakoll, P.J. (2001) Postexercise nutrient intake timing in humans is critical to recovery of leg glucose and protein homeostasis. *American Journal of Physiology - Endocrinology and Metabolism*, 280 (6), E982–E993.

Levenhagen DK, Carr C, Carlson MG, Maron DJ, Borel MJ, Flakoll PJ. (2002) Post exercise protein intake enhances whole-body and leg protein accretion in human. *Medicine and Science in Sports & Exercise*. 34 (5): 828–37.

Loosli, A.R., Requa, R.K. and Garrick, J.G. (1993) Serum ferritin and injuries in female high school cross country runners. *Medicine and Science in Sports and Exercise*, 25 (5), Supplement abstract 129.

Lundy, B., O'Connor, H., Pelly, F. and Caterson, I. (2006) Anthropometric characteristics and competition dietary intakes of professional rugby league players. *International Journal of Sport Nutrition and Exercise Metabolism*, 16 (2), 199–213.

Madsen, K., Pedersen, P.K., Rose, P. and Richter. E.A. (1990) Carbohydrate supercompensation and muscle glycogen utilisation during exhaustive running in highly trained athletes. *European Journal of Applied Physiology*, 61, 467–472.

Malczewska, J., Szczepańska, B., Stupnicki, R. and Sendecki, W. (2001) The assessment of frequency of iron deficiency in athletes from the transferrin receptor-ferritin index. *International Journal of Sport Nutrition and Exercise Metabolism*, 11 (1), 42–52.

Maughan, R.J. (1991) Fluid and electrolyte loss and replacement in exercise. *Journal of Sports Science*, 9, 117–142.

Maughan, R.J. (2005) Contamination of dietary supplements and positive drug tests in sport. *Journal of Sports Sciences*, 23 (9), 883–889.

Maughan, R.J., Greenhaff, P.L., Leiper, J.B., Ball, D., Lambert, C.P. and Gleeson, M. (1997) Diet composition and the performance of high-intensity exercise. *Journal of Sports Sciences*, 15 (3), 265–75.

Maughan, R.J., King, D.S. and Lea, T. (2004) Dietary supplements. *Journal of Sports Sciences*, 22, 95–113.

Maughan, R.J., Shirreffs, S.M., Merson, S.J. and Horswill, C.A. (2005) Fluid and electrolyte balance in elite male football (soccer) players training in a cool environment. *Journal of Sports Sciences*, 23 (1), 73–79.

Miettinen, M., Turpeinen, O., Karvonen, M.J., Elosuo, R. and Paavilainen, E. (1972) Effect of cholesterol-lowering diet on mortality from coronary heart-disease and other causes. A twelve-year clinical trial in men and women. *Lancet*, 2, 835–838.

Morillas-Ruiz, J.M., Villegas Garcia, J.A., Lopez, F.J., Vidal-Guevara, M.L. and Zafrilla, P. (2006) Effects of polyphenolic antioxidants on exercise-induced oxidative stress. *Clinical Nutrition*, 25 (3), 444–453.

Murray, R., Bartoli, W., Stofan, J., Horn, M. and Eddy, D. (1999) A comparison of the gastric emptying characteristics of selected sports drinks. *International Journal of Sport and Nutrition*, 9 (3), 263–274.

Murray, R., Paul, G.L., Seifert, J.G., Eddy, D.E. and Halaby, G.A. (1989) The effects of glucose, fructose, and sucrose ingestion during exercise. *Medicine and Science in Sports and Exercise*, 21 (3), 275–282.

Mustad, V.A., Etherton, T.D., Cooper, A.D., Mastro, A.M., Pearson, T.A., Jonnalagadda, S.S. and Kris, P.M. (1997) Reducing saturated fat intake is associated with increased levels of LDL receptors on mononuclear cells in healthy men and women. *Journal of Lipid Research*, 38, 459–468.

NATA (2000) Position Statement: Fluid Replacement for Athletes. *Journal of Athletic Training*, 35, 212–224.

Nattiv, A. (2000) Stress fractures and bone health in track and field athletes. *Journal of Science and Medicine in Sport*, 3 (3), 268–279.

Nose, H., Mack, G.W., Shi, X.R., Nadel, E.R. (1988) Role of osmolality and plasma volum during rehydration in humans. *Journal of Applied Physiology*, 65 (1), 325–331.

O'Brien, C. and Lyons, F. (2000) Alcohol and the athlete. *Sports Medicine*, 29 (5), 295–301.

Osterberg, K.L., Zachwieja, J.J. and Smith, J.W. (2008) Carbohydrate and carbohydrate+protein for cycling time-trial performance. *Journal of Sports Sciences*, 26 (3), 227–233.

Painter, J., Rah, J.-H. and Lee, Y.-K. (2002) Comparison of international food guide pictorial representations. *Journal of the American Dietetic Association*, 102 (4), 483–489.

Phillips, S.M., Green, H.J., Tarnopolsky, M.A., Heigenhauser, G.J.F., Hill, R.E. and Grant, S.M. (1996) Effects of training duration on substrate turnover and oxidation during exercise. *Journal of Applied Physiology*, 81, 2182–2191.

Rahnama, N., Reilly, T. and Lees, A. (2002) Injury risk associated with playing actions during competitive soccer. *British Journal of Sports Medicine*, 36 (5), 354–359.

Rankin, J.W., Goldman, L.P., Puglisi, M.J., Nickols-Richardson, S.M., Earthman, C.P. and Gwazdauskas, F.C. (2004) Effect of post-exercise supplement consumption on adaptations to resistance training. *Journal of the American College of Nutrition*, 23 (4), 322–330.

Rasmussen, B., Tipton, K., Miller, S., Wolf, S. and Wolfe, R. (2000) An oral essential amino acid-carbohydrate supplement enhances muscle protein anabolism after resistance exercise. *Journal of Applied Physiology*, 88, 386–392.

Rehrer, N.J., Beckers, E.J., Brouns, F., Ten Hoor, F. and Saris, W.H.M. (1990) Effects of dehydration on gastric emptying and gastrointestinal distress while running. *Medicine and Science in Sports and Exercise*, 22, 790–795.

Risser, W.L., Lee, E.J., Poindexter, H.B.W., West, M.S., Pivarnik, J.M., Risser, J.M.H. and Hickson, J.F. (1988) Iron deficiency in female athletes: its prevalence and impact on performance. *Medicine and Science in Sports and Exercise*, 20 (2), S116–121.

Romjin, J.A., Coyle, E.F., Sidossis, L.S., Gastaldelli, A., Horowitz, J.F., Endert, E. and Wolfe, R.R. (1993) Regulation of endogenous fat and carbohydrate metabolism in relation to exercise intensity and duration. *American Journal of Physiology*, 265 (3 pt 1), E380–391.

Rowlands, D.S., Thorp, R.M., Rossler, K., Graham, D.F. and Rockell, M.J. (2007) Effect of Protein-Rich Feeding on Recovery After Intense Exercise. *International Journal of Sport Nutrition and Exercise Metabolism*, 17 (6).

Roy, B., Fowles, J., Hill, R. and Tarnopolsky, M. (2000) Macronutrient intake and whole body protein metabolism following resistance exercise. *Medicine and Science in Sports and Exercise*, 32, 1412–1418.

Salmeron, J., Manson, J.E., Stampfer, M.J., Colditz, G.A., Wing, A.L. & Willett, W.C. (1997) Dietary fiber, glycemic load, and risk of non-insulin-dependent diabetes mellitus in women. *Journal of American Medical Association*, 277, 472–477.

Saris, W.H.M., van Erp-Baart, M.A., Brouns, F., Westerterp, K.R. and ten Hoor, F. (1989) Study on food intake and energy expenditure during extreme sustained exercise: the Tour de France. *International Journal of Sports Medicine*, 10, S26–S31.

Schanzer, W. (2002) *Analysis of Non-Hormonal Nutritional Supplements for Anabolic-Androgenic Steroids – An International Study. An investigation of the IOC accredited doping laboratory Cologne, Germany*. Available through the official website of the International Olympic Committee.

Schauder, P., Schröder, K. and Langenbeck, U. (1984) Serum branched-chain amino and keto acid response to a protein-rich meal in man. *Annals of Nutrition and Metabolism*, 28 (6), 350–356.

Schlabach, G. (1994) Carbohydrate strategies for injury prevention. *Journal of Athletics Training*, 29 (3), 244–254.

Sharp, R.L. (2006) Role of sodium in fluid homeostasis with exercise. *Journal of the American College of Nutrition*, 25 (3), S61.

Sherman, W.M. and Costill, D.L. (1984) The marathon: dietary manipulation to optimize performance. *American Journal of Sports Medicine*, 12 (1), 44–51.

Sherman, W.M., Costill, D.L., Fink, W.J. and Miller, J.M. (1981) The effect of exercise and diet manipulation on muscle glycogen and its subsequent utilization during performance. *International Journal of Sports Medicine*, 2, 114–118.

Sherman, W.M., Costill, D.L., Fink, W.J., Hagerman, F.C., Armstrong, L.E. and Murray, T.F. (1983) Effect of a 42.2-km footrace and subsequent rest or exercise on muscle glycogen and enzymes. *Journal of Applied Physiology*, 55, 1219–1224.

Shi, X., Horn, M.K., Osterberg, K.L., Stofan, J.R., Zachwieja, J.J., Horswill, C.A., Passe, D.H. and Murray, R. (2004) Gastrointestinal discomfort during intermittent high-intensity exercise: effect of carbohydrate-electrolyte beverage. *International Journal of Sport Nutrition and Exercise Metabolism*, 14 (6), 673–683.

Shirreffs, S.M., Taylor, A.J., Leiper, J.B. and Maughan, R.J. (1996) Post-exercise rehydration in man: Effects of volume consumed and drink sodium content. *Medicine and Science in Sports and Exercise, 28*, 1260–1271.

Shirreffs, S.M., Aragon-Vargas, L.F., Chamorro, M., Maughan, R.J., Serratosa, L., Zachwieja, J.J. (2005) The sweating response of elite professional soccer players to training in the heat. *International Journal of Sports Medicine, 26* (2), 90–95.

Simopoulos, A.P. (1999) Essential fatty acids in health and chronic disease. *American Journal of Clinical Nutrition, 70*, 560S–569S.

Sjodin, A.M., Andersson, A.B., Hogberg, J.M. Westerterp, K.R. (1994) Energy balance in cross-country skiers: a study using doubly labelled water. *Medicine and Science in Sports and Exercise, 26* (6), 720–724.

Spodaryk, K. (1993) Haematological and iron-related parameters of male endurance and strength trained athletes. *European Journal of Applied Physiology, 67* (1), 66–70.

Tarnopolsky, M. (2004) Protein requirements for endurance athletes. *Nutrition, 20* (7–8), 662–668.

Tipton, K.D., Rasmussen, B.B., Miller, S.L. et al. (2001) Timing of amino acid-carbohydrate ingestion alters anabolic response of muscle to resistance exercise. *American Journal of Physiology and Endocrinology Metabolism, 281* (2), 197–206.

Tsintzas, O.-K., Williams, C., Boobis, L. and Greenhaff, P. (1996) Carbohydrate ingestion and single muscle fibre glycogen metabolism during prolonged running in men. *Journal of Applied Physiology, 81*, 801–809.

Tucker, K.L., Hannan, M.T., Kiel, D.P. (2001) The acid-base hypothesis: diet and bone in the Framingham Osteoporosis Study. *European Journal of Nutrition, 40* (5), 231–237.

UK Sport (2008) UK Sports, Nandrolone report and Nandrolone progress report to the UK Sports Council from the expert committee on nandrolone, January 2001 and February 2003. http://www.uksport.gov.uk/pages/supplements_position_statement/

van der Beek, E.J., van Dokkum, W., Wedel, M., Schrijver, J. and van den Berg, H. (1994) Thiamin, riboflavin and vitamin B6: impact of restricted intake on physical performance in man. *Journal of the American College of Nutrition, 13* (6), 629–640.

van Hall, G., MacLean, D.A., Saltin, B. and Wagenmakers, A.J.M. (1996) Mechanisms of activation of muscle branched-chain a-keto acid dehydrogenase during exercise in man. *Journal of Physiology, 494* (3), 899–905.

Viteri, F.E. and Torun, B. (1974) Anemia and physical work capacity. In Garby, L. (Ed.) *Clinics in Hematology*, vol. 3. London: W.B. Saunders, pp. 609–626.

Volpe, S.L. (2006) Micronutrient requirements for athletes. *Clinical Sports and Medicine, 26* (1), 119–130.

Wallis, G.A., Yeo, S.E., Blannin, A.K. and Jeukendrup, A.E. (2007) Dose-response effects of ingested carbohydrate on exercise metabolism in women. *Medicine and Science in Sports and Exercise, 39*, 131–138.

Watson, G., Judelson, D.A., Armstrong, L.E., Yeargin, S.W., Casa, D.J. and Maresh, C.M. (2005) Influence of diuretic-induced dehydration on competitive sprint and power performance. *Medicine and Science in Sports and Exercise, 37* (7), 1168–74.

Whiting, S.J. and Barabash, W.A. (2006) Dietary Reference Intakes for the micronutrients: considerations for physical activity. *Applied Physiology and Nutrition Metabolism, 31*, 80–85.

Widrick, J.J., Costill, D.L., Fink, W.J., Hickey, M.S., McConell, G.K. and Tanaka, H. (1993) Carbohydrate feedings and exercise performance: effect of initial muscle glycogen concentration. *Journal of Applied Physiology, 74*, 2998–3005.

Williams, M.H. and Branch, J.D. (1998) Creatine Supplementation and Exercise Performance: An Update. *Journal of the American College of Nutrition, 17* (3), 216–234.

Wojtys, E.M., Wylie, B.B. and Huston, L.J. (1996) The effects of muscle fatigue on neuromuscular function and anterior tibial translation in healthy knees. *The American Journal of Sports Medicine, 24*, 615–621.

Zijp, I.M., Korver, O. and Tijburg, L.B.M. (2000) Effect of tea and other dietary factors on iron absorption. *Critical Reviews in Food Science and Nutrition, 40* (5), 371–398.

15

Psychology and sports rehabilitation

Rhonda Cohen
London Sport Institute at Middlesex University

Dr Sanna M. Nordin
Dance Science, Trinity Laban

Earle Abrahamson
London Sport Institute at Middlesex University

The aim of this chapter is to illustrate the importance of psychology for a professional working in sports rehabilitation and to provide you with some practical ways of applying psychology in your practice. Psychology has been used with injured athletes in two main ways; firstly in identifying those at risk of injury, and secondly in enhancing recovery. In addition, helping injured athletes to develop psychological skills, such as goal setting or imagery, will ensure that they are not only 'physically fit' following sports rehabilitation but also 'mentally fit' for their subsequent return to sport (Murphy 2005).

The chapter is divided into two sections to help you first focus on theory and issues before examining the application. Section one begins with a brief examination of why psychology is important to you as a sports rehabilitator. It continues on to the importance of the stress model as a basis of identifying players who may be at a higher risk of injury (Williams and Andersen 1998). The section proceeds with a discussion of emotional responses such as the grief-loss model (Kübler-Ross 1969) and its relevance to athletes who are injured. Following on from this is the behavioural issue of adherence. The first section continues with a mention of mental toughness as a way to facilitate adherence and concludes with the SCRAPE model (Hinderliter and Cardinal 2007) as a way of organising the variables of injury recovery. The latter half of the chapter discusses psychological skills training in relation to injury prevention and management.

Why psychology for sports rehabilitators?

Your role as a sports rehabilitator is to prepare the athlete for return to sport or performance as physically fit. However, the reality of the rehabilitative process is that it can be painful, time consuming and uncomfortable. Your patient/client may require techniques for managing pain and relieving stress. Therefore, you will need to address the psychological and emotional issues of your injured athlete. If your athlete is to heal psychologically as well as physically then you need to understand how issues such as stress and adherence affect the rehabilitation process. You may also want to help athletes use psychological skills techniques such as goal setting and self-talk. Yet, the knowledge and application

---

Sports Rehabilitation and Injury Prevention   Edited by Paul Comfort and Earle Abrahamson
© 2010 John Wiley & Sons, Ltd

of psychology to sports rehabilitation is often neglected in the training of sports rehabilitators and other practitioners (Mann et al. 2007). Practitioners often report that they wish they had more training in psychology (Francis, Andersen and Maley 2000; Ninedek and Kolt 2000; Sheppard 2004) as better integration of psychology within a rehabilitation programme will increase a sports rehabilitator's understanding on the psychology of injury (Hemming and Povey 2002). Arvinen-Barrow, Hemmings, Weigand and Becker (2007) pointed out that physiotherapists perceived that athletes suffered psychologically for approximately 83% of the time they were injured. According to Arvinen-Barrow and her colleagues (2007), sports injury professionals utilised some psychological techniques but felt that additional psychological training would be beneficial.

In a recent study, Hamson-Utley, Martin and Walters (2008) found that those working in sports rehabilitation do have positive attitudes on the effectiveness of psychological skills in enhancing recovery. With this positive mind set, you should find that this chapter will help you as a sports rehabilitator to integrate psychological theory and practical suggestions within sport rehabilitation. It aims to assist you in becoming a more skilled practitioner so that you can help your athletes in a more comprehensive way. To begin, the stress–injury relationship will be examined as this is the foundation in understanding why certain athletes may be more prone to injury.

Emotional reponses to sports injury and rehabilitation

Understanding athletes' emotional response to injury has led to a variety of research, which falls into two general categories: cognitive appraisal and the stage model. Cognitive appraisal will be explained through Williams and Andersen's (1998) stress–injury relationship. The stage approach will be examined through Kübler-Ross (1969) research on grief and loss.

Cognitive appraisal: The stress–injury relationship model

There is continued debate regarding stress and destressing. Of course, stress can affect performance, but more importantly to you as a sports rehabilitator, stress raises the injury incidence level, making your athlete more prone to injury. Williams and Andersen (2007) showed that approximately 85% of studies conducted since the 1970s have demonstrated a positive correlation between life-event stress and injury. In fact, stressed athletes run a 2–5 times greater risk of injury than athletes with low life-event stress. This is quite an extraordinary finding and is consistent across numerous sports (Williams and Andersen 2007). Stress as a result of injury is also commonly identified by practitioners as being present during the rehabilitation process. Heaney (2006) noted that stress and anxiety was present in professional footballers 73% of the time that they were injured. For those working with youth teams and academies, stress is also present in children (Nippert and Smith 2008).

Components of the model

Let us examine each of the components of the sport stress–injury relationship, illustrated above by Williams and Andersen (1998), so you can see how an understanding of each of the parts as well as the integration of the model, can be useful to you as a sports rehabilitator.

Personality

Personality is defined as "a dynamic organisation, inside the person, of psychophysical systems that create a person's characteristic patterns of behaviour, thoughts, and feelings" (Carver and Scheier 2000, p.5). The definition above implies that personality is dynamic. We are born with some aspects of a personality (*traits*) and this innate genetic predisposition may be difficult to change or may be unchangeable. For example, some athletes are naturally more anxious and this is known as *trait anxiety*. Your athletes therefore may have certain traits that are integral to their personality and hard to alter. However, other aspects of personality are more environmental oriented and may be changeable. These are referred to as *state* or *situational* characteristics. That is, your athlete may react to certain events in a particular way whether or not they possess a particular trait. For instance, they may become more nervous in a competitive situation, or in your clinic, or before returning back to a game following an injury (e.g. *state anxiety*), even if they are not trait anxious. This can be helped through altering perceptions or implementing coping skills, for example.

It is important to distinguish between trait and state as some athletes may appear only to have problems at certain times and that may be confusing to you as a rehabilitator. Knowing if this is a usual trait (e.g. your athlete is usually anxious) or whether it is the situation (e.g. a specific competitive event or part of the event – such as starting on the blocks) is causing the problem is helpful in helping your athlete to change. If it is a usual trait then you can encourage daily coping strategies for dealing with something that is part of their usual personality. If it is a state then you can help to identify the trigger and promote strategies to assist your athlete in coping with a specific situation.

A specific personality variable that has been related to stress is self-esteem, or how you value yourself. As you might guess, those with lower self-esteem are more likely to feel stressed (Kolt and Roberts 1998). During the sporting season, athletes with low self-esteem and low mood states (e.g. anxiety and depression) were more susceptible to being stressed (Williams et al. 1993), which means that they will be more prone to injury as illustrated by Williams and Andersen (1998). Hence, as a sports rehabilitator, get to know your athlete's personality and remember that even the most confident player on the pitch may suffer from anxiety or low confidence or some sort of a change in state personality when injured.

History of stressors

History of stressors refers to how an athlete feels about previous events and/or experiences. These can be real threats such as recovery from a major injury or a perceived threat such as the worry of recovery from an injury that has returned to complete physical fitness. In addition, history of stressors or previous experience can refer to prior major life events or even the impact of minor occurrences.

Research has demonstrated that experiences can impact on the risk of an athlete getting injured again (Maddison and Prapavessis 2005; Steffen et al. 2008). For example, an athlete who has had previous injuries may be worried about returning to a competitive level of sport. They may be concerned over the severity of previous injury and the potential risk of re-injury and may enter an event apprehensively with elevated levels of stress. The threat of injury may be real or imagined. For example, a study by Chase, Magyar and Drake (2005) found that gymnasts were fearful of injuries especially as they worried about how hard it was to return to competition after suffering from an injury. Therefore, these athletes were more worried about what 'could happen'. Working with athletes on self-confidence can be beneficial in helping support athletes dealing with a history of injuries.

Major life-changing events or minor hassles, can have an impact on the athlete. Major life events can be sporting related, such as returning to sport with a disability, or more generic for example, a relationship breakdown or death of a loved one. This was confirmed by Williams and Andersen (2007) in their review of over 40 studies. Although it is fairly obvious to us that major life events can increase stress levels, it is also possible that much smaller hassles or stressors can have a similar effect. For instance, a study by Fawkner, McMurray and Summer (1999) found that significant increases in minor events can elevate stress. This could be arriving at the competition where all the events are delayed due to rain or arriving late when the minibus you are travelling in breaks down on the way to an event. In all likelihood, it is not only the stress that creates a problem but also because anything that is on your mind, whether small or large, can act as a distraction from the sporting task at hand. A disruption to your athlete's ability to concentrate, attentional disruption, can also increase your athlete's risk of becoming injured.

Coping resources

Hanson et al. (1992) identified coping resources as the best determinants for predicting both severity and number of injuries in an athlete. Therefore a better understanding of coping strategies for you as a sports rehabilitator has the potential to make a significant impact on your practice. A more detailed introduction to coping skills is provided in section two (psychological skills).

If you look at the stress model again you can see how coping strategies feed into the stress response as well as interact with personality. Coping strategies are ways of dealing with problems or situations. Folkman and Lazarus (1984, p141) define it as "constantly changing cognitive and behavioural efforts to manage specific external and /or internal demands

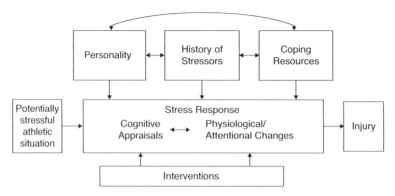

Figure 15.1 The stress and injury model (from Psychosocial antecedents of sport injury: Review and critique of the stress and injury model by J. M. Williams and M. B. Andersen, 1998. Copyright 1998. Reproduced by permission of Taylor and Francis, Inc., http://www.routledge-ny-com).

that appraised as taxing or exceeding the resources of the person". It has been found that a variety of coping strategies are utilised in sport (Nicholls et al. 2007), such as venting, crying, or even the use of alcohol and drugs (Kowalski and Crocker 2001).

There are two types or categories of coping strategies predominantly used: problem-focused and emotion-focused coping (Lazarus and Folkman 1984). Emotion-focused coping is dealing with and managing emotions. Problem-focused coping is focusing on or managing the problem. Within these categories there are a variety of coping strategies (Wethington and Kessler 1991). These include avoidance coping such as dealing with the problem by running away, using alcohol or drugs to cope (Carver et al. 1989). Some athletes turn to religion as a means of support or enlisting help from other people as in social support. So which strategy is the most productive for our athletes in coping with injury?

Research consistently supports the use of various coping strategies for meeting with the demands of injury and life and the necessity for adapting these to meet individual needs. For example, Carson and Polman (2008) cited the benefits of using problem focusing in a case study with a rugby player undergoing ACL rehabilitation, while Gallagher and Gardner (2007) advised that avoidance as a coping technique was detrimental to athletes and is associated with higher levels of negative moods. Social support, on the other hand, is beneficial. Athletes with high levels of social support had fewer injuries regardless of life-event stress, whilst athletes low in social support were prone to more injuries (Hardy et al. 1991; Smith et al. 1990a). Having more social support does help people manage the stresses of life. Therefore, recommend that your injured athletes seek support from friends (inside and outside sport) and family when they are injured – or, ideally, before.

In conclusion, because there is a strong link between having inadequate coping skills and sports injury (Williams et al. 1986), it is important for you as a sports rehabilitator to understand what coping strategies are and how effective strategies may be encouraged.

The next component within the stress–injury model (Figure 15.1) is cognitive appraisal (how an athlete perceives stress) and physiological/attentional changes. These will now be discussed.

Cognitive appraisal

According to the stress model, the way your athlete cognitively interprets a situation is affected by their personality and their coping styles. Cognitive appraisal affects stress levels, which can lead to an increased risk of injury. The word *cognition* pertains to thought patterns and processes. Cognitive appraisal is, therefore, what your athlete *thinks* about a situation which affects their emotional and behavioural responses. Seeing a situation (or injury during the recovery process) as a challenge (facilitative) as opposed to a threatening situation (debilitative) will positively affect behaviour. Perceptions, facilitative or debilitative, can have an effect on how the athlete feels and behaves during competition or the rehabilitative process. Cognitive appraisal is a dynamic process that can change over time and it is up to you as rehabilitator to help your athlete see the positive challenge of the rehabilitative process.

Interventions

Johnson, Ekengren and Andersen (2005), in a study based on the Williams and Andersen (1998) stress–injury model, were able to design a programme for soccer players who they identified as being at risk of injury. An at-risk psychosocial profile was created for each athlete through the use of The Sport Anxiety Scale (Smith et al. 1990b), the Life Event Scale for Collegiate Athletes (Petrie 1992), and the Athletic Coping Skills Inventory-28 (Smith et al.1995). They implemented an intervention scheme which taught athletes skills such as stress management and confidence building. Their study showed that these athletes were injured far less than athletes in a control group. Maddison and Prapavessis (2005) conducted a similar study with rugby players, and found that athletes in the intervention group (who were taught to manage stress) had fewer injuries than athletes in the control group. Such studies powerfully demonstrate how psychological interventions can be used to impact the injury process – something often thought of as purely physical!

Conclusion of the stress model

All the components of the stress model have been described and it should now be clear how an integration of these various components works. An athlete's personality, their previous experiences and ways of dealing with excessive demands feeds into the way they respond to stress. Their stress response is affected by their thoughts, their focus and their bodily reactions. Intervention strategies such as goal setting, self-talk, coping strategies and imagery, which will be discussed in the second part of the chapter, can help reduce levels of stress caused by the stress response and reduction of stress levels can reduce the incidence of injury.

To strengthen your knowledge on the model and coping strategies, refer to: Williams and Andersen (1998) and Lazarus and Folkman (1984).

The next part of the chapter will continue with the emotional response to injury within the stage model (Kübler-Ross, 1969).

Emotional responses to sports injury and rehabilitation: A stage model

Being injured is obviously an emotional experience for your athlete (Heil 1993). Tracey (2003) examined the emotional responses experienced by athletes as a result of injury as well as during rehabilitation and revealed that athletes demonstrate a collection of emotions such as a sense of loss, decreased self-esteem, frustration and anger.

In examining athletic injury, a classic 'stage' model by Kübler-Ross (1969) has been applied to sport and outlines a normal progression of emotions (Figure 15.2). Originally, this model was designed as a framework for understanding the psychological response during the grieving process. However, it was adapted to parallel the stages of emotional response experienced by injured athletes (Crossman 1991). The initial stage is *denial* where athletes find it hard to believe that they are injured. This is usually immediately post-injury. Second is *anger*, which can be directed at the injury, oneself, or even you as medical staff. The third stage is one of bargaining where athletes negotiate a deal, for example, 'I will do all my exercises and be a better person if I can fully recover and return to competition'. Next is a state of depression, followed by acceptance.

The Stages
Denial

There are two aspects of denial. Firstly, the shock state immediately following the injury when the player is in a state of disbelief and may even respond

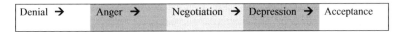

Figure 15.2 Stage model of loss.

with shortness of breath and physical freezing. Secondly, it may progress into a denial period where the athlete still finds it hard to accept their limitations due to injury. You can respond back by going through the reality of the situation though it may take time for the athlete to accept it.

Anger

Secondly, the athlete may express their anger over the situation. They may enter a blame culture in feeling that others have put them here, thinking or saying things such as "The opponent should not have tripped me up ..." Or they may suggest it is your fault for not doing a good enough job with a previous injury. The best way to manage anger is to stay calm, and never respond back with anger. Remember to defuse by allowing the athlete a bit of time to express their feelings and then to acknowledge that you understand that they are feeling angry and that this is quite normal. Can they remember a previous time when they felt angry due to an injury, illness or bad situation? How did they get through it then? What do they feel they could do to get through it this time?

Bargaining

This third stage involves negotiation. The athlete may promise to attend all sessions in exchange for your hard work as a rehabilitator in helping them to return to full recovery. This can include religion where an athlete will promise to be a better person in return for better health.

Depression

Depression is a sense of hopelessness and despair, and athletes may express sadness and apathy through statements such as "my sports career is over". Your athlete may not feel like coming to rehabilitative sessions or working hard at the home exercises you have given them. When athletes feel despair you can try to boost their emotions by examining a positive focus. For example, *what will you do when you complete your rehabilitation? How will you celebrate your return to sport? Can you remember or visualise a time when you felt really successful or happy in your sport?*

Acceptance

Be encouraging and acceptance will be around the corner. Acceptance is acknowledging that an injury has occurred and that the way forward is by working through the rehabilitative process. Your athlete may demonstrate this by saying "I am going to the gym regularly and working on these exercises at home."

Conclusions on the stage model of grief-loss

The Kübler–Ross (1969) stage model has been criticised for being sequential or outlining a set order for this emotional experience. However, psychologists agree that these stages are more like fluid phases and that not all athletes may go through all these emotions in the same order. Some athletes may even not display these emotions at all. So as a professional, just *be aware* of what emotions your client *may* be going through.

Carson and Polman (2008), in working with anterior cruciate ligament patients, identified set stages of shock, depression, relief, encouragement and confidence development. To further your knowledge of the Kübler-Ross grief loss model, read the article entitled 'Psychological responses to injury in competitive sport: a critical review' written by Walker, Thatcher and Lavallee (2007). They conducted a study to help professionals (such as you) who are involved in athletic rehabilitation to understand the impact of psychological factors on injury experience. Their research critically examines models such as grief-loss (e.g. that of Kübler-Ross) and explores the integration of emotional, behavioural and cognitive responses to injury and recovery and what they conclude is a rather complex model.

Behavioural responses to sports injury and rehabilitation

Adherence

Adherence is a type of 'stickability'. For you as sports rehabilitator, it is about getting a patient to stick to a recovery programme which may include adherence to rehabilitation sessions, to a special diet, to home exercises, to attending adapted training programmes and to working on the development of mental strategies. Creating an individualised rehabilitative programme that includes psychological techniques will benefit your patient in facilitating recovery in the most efficient way possible.

Adherence differs from compliance, which assumes that the patient will obediently follow your instruction without question. Adherence is a voluntary, negotiated agreement between the patient and you. Patient involvement is important and is positively related with adherence (Lind et al. 2008). However, not all athletes will follow your rehabilitation programme as agreed, and adherence can range from under-adherence to over-adherence. Under-adherence is doing less than what is prescribed. The cost of under-adherence to the patient may be a slower recovery as well as lower confidence about their progress. Sometimes additional factors can interfere with progress and be the cause of under-adherence. For example, athletes may receive more attention from their coaches and team members when they are injured than when they are fit. They may also want to avoid the pain associated with having to following a regime of rehabilitative exercises. Remember that patients can forget to stick to their programmes once or twice, but forgetting three times is probably no longer an accident. Athletes need to display the same commitment to their rehabilitation as they do to their training, though it is important as sports rehabilitator to remember that today's patients have to balance other commitments as well such as study, friends and family.

Over-adherence, on the other hand, is doing more than one should during recovery. Some athletes will over-adhere to a rehabilitation programme in the hope of recovering quicker. This may be due to not wanting to lose their place in an upcoming competition or a position on a team, perfectionist tendencies, or perhaps pressure from other people such as their coach or teammates.

One way of helping athletes to adhere to their recovery programme is to get them to use a diary. An interesting study by Pizzari, Taylor, McBurney and Feller (2005) examined the relationship between adherence and outcome following ACL surgery. Adherence was measured by a self-reported diary of home exercise and from attendance to appointments. In clinical practice, the use of a diary to track daily and weekly progress is an easy way of both empowering the athlete to track their own progress, and perhaps for you to examine adherence – if they let you see their diary, that is! Of course, willingness to keep a diary in the first place is one indicator that an athlete is keen to be proactive in the rehabilitation process. The results indicated that there was a relationship between adherence to home exercises and outcome, measured by knee function. Home exercise adherence is an important predictor of the rehabilitation adherence and was also demonstrated in a study of patients with wrist fractures (Lyngcoln et al. 2005). The use of a self-reported diary to track home exercise appears to be a useful way of monitoring adherence, you may wish to consider this when working with clients.

Factors that influence adherence

Through interviews with sports physiotherapists, Niven (2007) identified a number of factors that influence adherence in sports rehabilitation programmes. In particular and athlete's personality as well as situational factors impact adherence to rehabilitation programmes. Personality may impact on adherence as athletes who are high in anxiety show reduced adherence to their rehabilitative programmes (Taylor and Taylor 1997). In addition, how your athlete perceives the efficacy of your treatment protocol and the confidence they have in rehabilitation will affect their adherence to your sports injury programme (Brewer et al. 2003; Taylor and May 1996).

To help your athlete to have confidence in you, effective communication and a solid working relationship are essential. In fact, when physiotherapists and athletes have these factors in place, a more effective recovery programme and more positive outcomes result (Crossman 1997; Francis et al., 2000; Ninedek and Kolt 2000).

Determinants of adherence

One way of altering adherence is to identify the factors that determine their adherence to a programme. Crossman (2001) divided the factors into three classifications: *predisposing*, *reinforcing* and *enabling* factors.

Predisposing factors

These are the athlete's own views and thoughts about the recovery process. In order to get an idea from your patient about their perceptions towards rehabilitation you may want to ask about their normal habits and preferences.

For example:

- Do you like working out in the gym or undertaking extra training?

- What do you like most about participating in your sport?

- How will training or home exercises help you to get closer to your goal?

- How are you going to make sure you can achieve your goal of returning fit?

Reinforcing factors

Reinforcing factors are based on the interactions between the athlete and the other significant others such as the sports rehabilitator, coach, the team or other important persons.

For instance, you could ask:

- Are you in touch with the coach and your team?

- Do you feel that there is a good reason to go through the recovery process?

- Has the rehabilitative process been explained to you?

Enabling factors

These are to do with the environment surrounding the rehabilitation treatment. Being able to identify which of these factors is preventing an athlete from attending sessions can help you as sports rehabilitator to improve your client's adherence. For instance, you might ask about:

- Is it easy or hard for you to get to you appointments? What can be done to make attending easier?

- How do you get to these sessions?

- How long does it take you to get home afterwards?

- Are there things (e.g. homework, other appointments, family duties) that make it hard for you to complete your home exercises?

As a rehabilitator it is helpful to identify the predisposing, reinforcing and enabling factors behind adherence in your athlete so that you can work on modifying any unhealthy behaviours or thoughts thus strengthening the rehabilitative process. For further reading please see: Crossman (1997), Francis et al. (2000) and Ninedek and Kolt (2000).

Mental toughness and rehabilitation

Mental toughness is having a psychological advantage in coping with the stressful demands (Fletcher and Fletcher 2005) associated with competition as well as in rehabilitation and return to sport after injury (Jones et al. 2002). Although mental toughness is a relatively new area of research, studies are demonstrating that these factors are important as part of an athlete's mindset. Levy, Polman Clough, Marchant and Earle (2006) found that patients with high mental toughness were more capable of managing pain and displaying a more positive outlook.

Mental toughness is defined by Middleton, Marsh, Martin, Richards and Perry (1997) as "an unshakeable perseverance and conviction towards some goal despite pressure or adversity". Jones, Hanton and Connaughton (2002) state that it is, "the natural or developed psychological edge that enables you to generally cope better than your opponents with the many demands that sport places on a performer" So mental toughness is about being able to cope under pressure and not giving into it.

So does mental toughness help in a rehabilitative setting? Mentally tough individuals have been shown to demonstrate a greater ability to withstand physical pain (Jones et al. 2002) and to recover more quickly from injury (Gucciardi et al 2008). In addition, mentally tough participants perceive their injury to be less severe, feel that they are less susceptible to further injury and focus less on their pain during the course of their rehabilitation (Levy et al. 2006). Perhaps surprisingly, even greater attendance at rehabilitation sessions is demonstrated by those with greater mental toughness (Levy et al. 2005; Marchant and Earle 2006). If you want to explore this area further begin by reading the following: Gucciardi et al. (2008), Jones *et al.* (2002) and Levy et al. 2006.

Implications of emotional and behavioural responses

In summarising section one of this chapter you can see how there are several implications for you as a sports rehabilitator in understanding and incorporating psychological theory into your practice. To start with, you need to be aware of the relationship between stress and injury because you are in a key position to identify and help athletes who are at a greater risk of injury. As a sports rehabilitator it is useful to understand how personality affects the stress process and how an athlete's past can affect how they deal with a stressful situation. As a sports rehabilitator, you are in a good position to encourage athletes to improve their coping strategies in handling stressful major life events as well as in handling daily hassles (Johnson et al. 2005). By catering to the emotional and behavioural needs of your athlete you are more likely to offer a comprehensive treatment that will enable your athlete to be stronger both psychologically and physiologically. Recovery can also be optimised through adherence and mental toughness.

Finally, Tracey (2003) points out that the experience of injury can give an athlete an increased understanding that recovery is a process in which an athlete can reflect and grow emotionally. You, as a rehabilitator, are often in a unique position to become involved in developing intervention programmes for athletes. There have been a variety of psychological interventions identified which facilitate adjustment to injury, such as goal setting, imagery and stress management (Evans et al. 2000; Johnson 2000; Johnson et al. 2005). Interventions in terms of psychological skills training and how they can be used will now be explained.

Psychological skills training in the injury process

The injury process may be simplistically divided into the time before an injury (*pre-injury period*), and the time following injury, lasting until the performer is fully rehabilitated and ready to return to activity (*rehabilitation period*). The role of psychological skills training (PST) in each stage will be outlined in this section. It is important to recognise that while the physical therapist or sports rehabilitator is not a sport psychologist, they are still in a good place to encourage injured athletes to practice mentally, and to give them some basic tips about how to do so (Gordon et al. 1998). This part of the chapter informs you about what psychological skills are, and how they may help prevent injury as well as support injury rehabilitation. You will notice that imagery is mentioned more frequently than other psychological skills; this is only because there is more research into its role and effectiveness. But, one study found that of a number of psychological skills, people actually rate goal setting as their favourite (Brewer et al. 1994). This means that it might be easier to "sell" goal setting to the people you meet; perhaps because it feels like a very practical and not overly 'psychological' technique. Still, the best approach is probably to recommend a variety of psychological skills to injured performers, so that they can choose what suits them best. If you are interested in exploring this topic further, a chapter by Kolt (2000) gives a very interesting example of how physical and psychological treatment following ACL reconstruction can go hand in hand. For additional material, see the books on the psychological aspects of injury by Pargman (2007), Taylor and Taylor (1997), and Heil (1993).

Psychological skills and psychological skills training

Psychological skills (sometimes called *mental skills*) include goal setting, imagery, self-talk and various forms of relaxation, and PST (or *mental skills training, MST*) is simply the systematic training of such skills. For example, a hockey coach who does a bit of informal goal setting with his team in order to prepare for the upcoming season would be said to use the psychological skill of goal setting. If, instead, he designs and implements a programme of goal setting over a number of weeks, perhaps teaching his athletes about how to make goal setting effective and evaluating it at the end, he could be said to be implementing a PST programme. As you might imagine, the effects of the various psychological skills are greater when implemented systematically; therefore, it is a good idea to learn more about, and regularly practice, these skills. You might also be wondering which psychological skill is 'best', or whether you have to study and practise (or recommend) them all. But while that might seem a simple and straightforward question, the answer is that there is no known 'optimal recipe' for how to use PST. Some researchers examine only one skill, and some examine several; however, because of the enormous number of sports, age groups, study purposes, and potential combinations of psychological skills available, there is no easy way in which to answer the question of which psychological skills to practice. But do not worry, sport psychologists often recommend a number of different skills, or combinations of skills, to performers. This way, the performer can choose what works best for them and their particular circumstances, and the chances of success are optimised. The individual psychological skills will now be introduced in turn.

Goal setting

Goal setting is a process of planning ahead for what one wants, and how to get there. This means that there is a huge variety of potential goals both for "normal", healthy sports participation and for rehabilitation. For instance, for one athlete, a rehabilitation goal might be to be back on form by the last game of the season, while others take a more process-oriented approach and set themselves the goal of doing their rehabilitation exercises three times daily. Even if you haven't studied much goal setting, you may still have heard a common acronym for how goal setting is made effective; this is the *SMART* acronym, which indicates that goals should be *specific*, *measurable*, *accurate*, *realistic*, and *timed* (Cox 2001; Weinberg and Gould 2007). It is important that most goals set are focused on individual performance (e.g. re-gaining complete strength of an injured leg) and the processes that contribute to this (e.g. doing three sets of eight reps of the rehabilitation exercises daily for three weeks). Such *process* and *performance goals* are superior to setting mainly *outcome goals* (e.g. returning faster from injury than another injured player, so that you will be the one to make the team; Filby, Maynard and Graydon 1999). This is so because outcome goals depend to some extent on factors outside your control, and may contribute to lowered self-confidence and heightened anxiety. Individually appropriate and relevant goals set jointly between the performer and their rehabilitator and, where appropriate, a coach, will instead promote a sense of control (Taylor and Taylor 1997). In rehabilitation, goal setting is one of the favourite methods for athletes to 'get back on track' (Brewer et al. 1994; Durso Cupal 1998).

Imagery

Imagery is the creation, or re-creation, of experiences in your mind (White and Hardy 1998). It may involve one or more senses, such as *seeing yourself* perform a particular exercise, *feeling* muscles move, and so on. Most imagery work in sport focuses on fairly concrete imagery to do with oneself (e.g. a gymnast imaging herself completing a floor routine), but more abstract, metaphorical images have also been reported, especially in the literature on healing imagery (Korn 1983; Ievleva and Orlick 1991; Green 1992; Evans et al. 2006) and in artistic activities like dance (Hanrahan and Vergeer 2000; Nordin and Cumming 2005). In metaphorical imagery, a performer might imagine the hip joint moving as a wheel or imagine toxins as a black substance that are gradually rinsed out or diluted with the breath – in other words, actions that are not strictly accurate but that may support a person's understanding of a movement, induce relaxation, or similar. The types of imagery that performers may use during rehabilitation include all those experienced in their everyday lives (including *seeing* and *feeling* how they perform certain skills so that they don't forget how to do them while 'out' with an injury). However, performers have also been found to engage in healing-type images when injured (Driediger et al. 2006; Evans et al. 2006; Green 1992; Hanrahan and Vergeer 2000; Ievleva and Orlick 1991; Korn 1983; Milne et al. 2005; Nordin and Cumming 2005; Sordoni et al. 2000, 2002; Short et al. 2004). Healing imagery might include metaphorical images such as those described above, but also more concrete images of physiological processes related to healing, such as tissue repairing itself back to normal. While healing imagery appears to have positive results, be mindful that people sometimes have strong preferences when it comes to imagery types. For example, some people respond really well to images of white blood cells eating local irritants and of cleansing, white air entering the lungs and circulation. For others, such abstract imagery might feel far-fetched or airy-fairy. Some performers value in-depth imagery of anatomical structures and how they function; for others that might be too complex. Thus, be creative in your use and encouragement of imagery, regularly checking that you and the client are 'on the same page'. Offering a menu of images, ranging from strictly anatomical to very abstract, might be a good idea, as is having examples and anecdotes of how such images have been helpful to other clients at the ready.

For those who work with injured performers (e.g. physiotherapists, sports physicians, sports rehabilitators) to encourage their clients to use imagery effectively, they must clearly first learn about how imagery can be used to assist injured athletes (Driediger et al. 2006). In addition, having tried it out yourself first is essential. One useful approach is for the rehabilitator to link imagery in relation to injury with the imagery that an athlete likely already does as part of their training. For example, you could ask them to describe how they rehearse sport skills, strategies, and scenarios in their mind around competition, and then say that imagery in the rehabilitation process is really similar. This should help reassure the athlete that they already have the requisite skills to do rehabilitation imagery. If you inform them that healing imagery has been shown to be effective in well-controlled research studies, it might also help dispel some of the uncertainty or apprehension about a form of imagery that is perhaps less intuitive than simple mental rehearsal (Morris et al. 2005; Wiese et al. 1991).

Self-talk

The cognitive process of talking to oneself is simply referred to as self-talk. Like imagery, it is a basic form of thought and most of us do it all the time – although more or less consciously. Self-talk may be positive, negative, or neutral in nature (Hardy et al. 2001. 2005), and although some athletes report that negative self-talk is motivating for them (Van Raalte et al. 1995; Hardy et al. 2001), mostly the research findings are as you might guess; that is, more positive results come from positive self-talk, and more negative results come from negative self-talk (see Hardy (2006) for a more in-depth discussion). Positive self-talk includes generally motivating statements like "I know I can do this!", whilst negative self-talk might include statements such as "she gave me just too many rehabilitation exercises – I will never be able to do them all every day". Neutral self-talk may be instructional in nature, including statements such as "arm in line with body, shoulders relaxed", perhaps designed to help remember and execute a shoulder rehabilitation exercise correctly.

Relaxation

Whilst a number of PST 'packages' include relaxation, it is rarely investigated on its own in sport psychology; thus, the number of studies examining the effects of relaxation for injured performers are few and far between. Even so, relaxation is a widely used technique that appeals to many performers, as most people intuitively appreciate the feeling of being calm and relaxed. A number of ways of achieving a relaxation response exist, and we will not describe them all here; instead, we will limit ourselves to a brief outline of some methods that are typically employed in PST programmes. Note also that while the sport psychology literature typically refers to relaxation 'techniques' or 'procedures', many people have developed their own ways of relaxing; for instance just by breathing deeply, by listening to music, or by doing yoga or another form of calming exercise.

One of the best known of the established techniques is progressive relaxation, first proposed by Jacobson in 1929. In this technique, the performer first contracts a particular muscle group, and thereafter relaxes it. This way, it is proposed that the athlete will learn to distinguish between the feelings of tension and relaxation. It is notable that imagery seems to play a part in progressive relaxation; that is, a typical progressive relaxation script would guide the participant's attention toward the feeling of their muscles relaxing, in a suggestive fashion. For example, a script used by Gill, Kolt, and Keating (2004, p292) tells participants to "Relax your feet and lower legs. Be aware of the tension being released. Release all the tension. As the tension fades away, focus on the new relaxed feeling in your feet and lower legs. Continue to focus on this feeling." Another form of relaxation is autogenic training, created by psychiatrist Johannes Schultz in 1932. It is somewhat similar to progressive relaxation, with both techniques going through various parts of the body and involving imagery as well as suggestive statements concerning the body being relaxed. Different from progressive relaxation, however, autogenic training focuses on limbs being warm and heavy, and typically also focuses more on the regulation of heart rate, breathing, and temperature (Cox 2002; Noh and Morris 2004).

Combined PST

As noted above, many studies as well as real-life interventions use psychological skills in combination. The following examples give you some idea of how that might be done:

- use *self-talk* to guide yourself through an *imagery* sequence

- *imagine* yourself achieving a *goal*

- setting a *goal* to do *imagery* for 10 minutes daily in order to enhance performance

- use *self-talk* to remind yourself of your *goals*

- and if you have learnt a *relaxation* procedure, simply thinking through that procedure would be a form of *self-talk*, as you would be telling yourself what to do!

Coping skills

Some researchers into PST and injury have looked at the concept of coping skills. Indeed, coping resources are a key component of the Williams and Andersen (1998) model of stress and athletic injury described above. Somewhat similar to psychological skills, coping skills typically include goal setting and/or another form of mental preparation, but also deals with constructs such as coping with adversity, peaking under pressure, concentration, freedom from worry, confidence, achievement motivation, and coachability (Smith et al. 1995). In other words, having good coping skills means that a athlete would handle a variety of situations in a confident, capable way and be resilient to setbacks. Note that while these coping skills should help a performer cope with varying circumstances, the psychological skills literature would typically argue that characteristics such as concentration, self-confidence, and freedom from worry and excessive anxiety, and may be *achieved through* PST. For example, goal setting might be used to improve a footballer's anxiety levels, or imagery might be employed for a runner who would like to improve her self-confidence. So what does all this mean for you in practice? Well, simply that it is important to be clear about *what it is you want to achieve* (e.g. being able to cope with adversity, such as returning from a debilitating, long-term injury), and the *techniques that might be available to help you do so* (e.g. the psychological skills of goal setting, imagery, self-talk and relaxation).

Psychological skills training in the pre-injury period

One of the most fascinating findings that have emerged in recent years is that psychological skills training can help performers avoid injury. The findings related to injury prevention fall into two categories: reduced injury frequency, and reduced injury duration.

Reduced injury frequency

The existing research indicates that a multitude of positive benefits stand to be obtained from PST, including preventing injuries from happening in the first place (Davis 1991; Johnson et al. 2005; Kerr and Goss 1996; Kolt et al. 2004; May and Brown 1989; Perna et al. 2003; Schomer 1990). In one study, a relaxation and imagery intervention reduced injury by as much as 52% (Davis 1991). Intervention studies have used various combinations of psychological skills and stress inoculation training (essentially training performers to handle stress better over time) to reduce injury frequency. And although this means that we do not have a very clear idea of which psychological skill does what, it also suggests that the effect is fairly robust. Moreover, the activities used in the various studies have included soccer (Johnson et al. 2005), alpine skiing (May and Brown 1989), swimming (Davis 1991), gymnastics (Kerr and Goss 1996; Kolt et al. 2004), rowing (Perna et al. 2003) and marathon running (Schomer 1990); this diversity further suggests that the findings are not anomalous, unique or sport-specific. Instead, the studies indicate that the effect is due to PST helping athletes lower their stress levels, feel more confident and optimistic, becoming more aware of their bodies, and building their ability to cope with difficulties.

Reduced injury duration

In a study by Noh, Morris, and Andersen (2005), it was shown that ballet dancers' psychological skills and coping strategies distinguished between not just those with higher versus lower injury *frequencies*, but also between dancers with shorter and longer injury *durations*. Specifically, dancers with shorter injury duration reported less worry and negative stress. They also had greater levels of confidence and achievement motivation. These authors later used their findings to conduct an intervention study, again with ballet dancers (Noh et al. 2007). After teaching one group imagery, self-talk and autogenic training (a form of relaxation), the group improved not only their coping skills, but also spent less time injured than a control group. Another study in rugby yielded similar findings (Maddison and Prapavessis 2005). In their study, rugby players considered to have an "at-risk psychological profile for injury" were identified and undertook a stress management programme. Like the ballet dancers in Noh et al.'s (2007) study, these rugby players gained better coping skills, worried less and spent less time injured than a control group.

Sample PST programme to prevent injury

1. *Set goals regularly.* Make sure they are SMART in nature, and focus more on yourself and your own progress than on how you compare to others. Get regular feedback from knowledgeable people (e.g. coaches) on your progress, and use this as well as your own judgment to keep goals flexible, and to evaluate progress. A training diary is a useful tool to help you keep track of goals and progress.

2. *Use imagery daily.* Explore imagery in all its forms – play around with different types and use it for a range of purposes – improving performance and confidence, reducing stress, and focusing you on your goals. For example, you can imagine yourself performing technical skills in detail, rehearse strategies for how to get out of potential 'trouble' (e.g. being a player down), imagine each component of a game the night before to ensure that you are prepared, see yourself reaching your overall goal, or use metaphorical images to enhance the quality of your performance. Make your images multisensory, because this makes them more effective; so, *see* yourself playing your sport, and *feel* your muscles move efficiently and the emotions you want to feel. You may even want to *hear* the sounds, and *smell* the venue! Regular, deliberate imagery will soon reap benefits for the activity – and the injury prevention will come as a nice bonus.

3. *Use self-talk to your advantage.* Make sure that your inner chatter is beneficial to your performance as well as to your well-being. Having positive self-talk statements 'ready-made' is one of the best ways of banishing negative statements if and when they surface; focus your attention on your positive statement, and repeat it over and over if necessary. With time, you will identify negative self-talk quickly and get rid of it just as quickly. In addition, you can use *instructional* self-talk to focus your attention on the task at hand; for example, a tired runner might feel her mind wander, thereby risking injury due to not focusing on the terrain underfoot. By repeating instructional statements to herself (e.g. feedback from her coach such as 'shoulders loose, arms swinging freely' or metaphors such as 'light as a cheetah'), she keeps her attention in the here-and-now.

4. *Relax – you deserve it!* Many people, and athletes are no exception, experience unwanted stress and tension in their everyday lives. The relaxation strategies described above can be good ways of eliminating some of this tension, as well as providing a pleasant difference for your muscles from their hard work on the court, pitch, dance floor, or ring. You might also want to use relaxation to improve your body awareness – for instance, what parts of your body typically tense up? Why might that be? Is there something you need to change in your exercise, sport, or everyday habits to improve it? It seems likely that being able to relax helps prevent injury through removing excess tension, stress and anxiety, and perhaps through improved body awareness.

Psychological skills training in the rehabilitation period

Benefits that may be obtained from PST in the rehabilitation period

Addressing the psychological side of injury once it has occurred has a multitude of advantages (Heil 1993). For example, rehabilitation may be accelerated with negative emotional experiences minimised and positive emotions maximised; it can activate coping strategies, enhance readiness for return to activity and maintain confidence. These positive outcomes will now be addressed.

Shorter recovery times

Studies have shown that athletes who recover more quickly from an injury use more psychological skills than those who recover more slowly (Ievleva and Orlick 1991). In this particular study, goal setting, positive self-talk and healing imagery were the skills that set the faster healers apart. For example, fast healers spoke to themselves in optimistic terms (e.g. "I can do anything") while those who healed more slowly spoke to themselves more pessimistically (e.g. "What a stupid thing to do"). Fast healers also reported experiencing fewer replay-images of their injury. In other words, it seems that imagery is capable of both slowing recovery down, and speeding it up, suggesting that performers should take care to engage only in facilitative, constructive images. It is possible that the shorter recovery times are a result of better rehabilitation adherence and/or enhanced healing and physical functioning, and these potential benefits are described next.

Greater adherence with rehabilitation schedules

As mentioned above, a key concern for the rehabilitator is how to make performers stick with the recommendations and exercises given to them. Fortunately, low-to-no-cost PST can help you overcome this hurdle. For instance, it has been found that goal setting increases the extent to which athletes rehabilitating from ACL-reconstruction actually do the home exercises given to them (Scherzer et al. 2001). These same authors also found goal setting to improve the work of these athletes while in the clinic; that is, they worked with greater intensity, and adhered to directions better. Positive self-talk was also associated with adherence. Although Scherzer et al. (2001) did not find healing imagery to be related to adherence, other studies have (Gould et al. 1997; Jones and Stuth 1997). Still others have found that management of thoughts and emotions (as is done when using PST) sets those who rehabilitate successfully apart from those who do not (Udry et al. 1997).

Enhanced healing and physical functioning

It is known that athletes can and do use psychological skills to support their physical healing processes (Calmels et al. 2003; Driediger et al. 2006, Evans et al. 2000; Evans et al. 2006; Ievleva and Orlick 1991; Milne et al. 2004; Sordoni et al. 2002). A handful of intervention studies have also examined whether teaching injured athletes' psychological skills can improve aspects of their healing and/or physical functioning, with promising results. For instance, relaxation practice and/or imagery can reduce pain sensations or help increase athletes' pain tolerance (Broucek et al. 1993; Cupal and Brewer 2001; Nicol 1993; Ross and Berger 1996; Sthalekar 1993). Recent work by Law, Driediger, Hall and Forwell (2006) also suggests that athletes who use imagery to cope with pain are more satisfied with their rehabilitation. In a rigorous intervention, Cupal and Brewer (2001) established that relaxation and imagery can help athletes recovering from ACL reconstructions perceive less pain. Participants were also less anxious about re-injuring themselves, and gained greater knee strength than two other groups: one no-intervention control group, and one placebo group who received attention and encouragement, but no PST. This is important, because few studies have a design strong enough to allow us to conclude that PST is truly more effective than simply spending time with an athlete. Cupal and Brewer's (2002) recommendation that more research is required into *how* such effects are obtained is notable, particularly given that another intervention with injured athletes did not observe similar effects (Christakou and Zervas 2007). In this latter study, imagery and relaxation had no effect on athletes' perceptions of pain, or oedema or range of motion following a grade II ankle sprain. As the authors suggested, the role of different imagery types need to be more clearly examined in future research. For now, the simplest recommendation would be for athletes to explore a range of (facilitative) images, including mental rehearsal-type images and healing images, while steering clear of any intrusive, debilitative imagery. In a qualitative interview study with rehabilitating athletes, Evans et al. (2006) provided some support for this idea, and suggested that imagery may change from the early to the late stages of rehabilitation, and as a result of stress. In their study, athletes used imagery of skills to promote self-confidence, while images of healing and pain management helped the rehabilitation process. As an example, consider the following quote from a male semi-professional soccer player with a cartilaginous knee injury:

> I can see the joint itself, I can see the bones, I can see the ligaments, I can imagine where the cartilage is. Once I've actually got that image in my mind I'll actually focus on the point where I feel the pain...I turn it into a color. Now the color I usually see it as originally is red, right, because that's where the pain is...it basically goes from red, to orange, to yellow, to green and through to blue. Once I've got that blue, I just imagine a cold icy feeling right, which I think tends to help the actual pain at that particular moment. (Evans et al. 2006, p. 9).

Another set of psychological skills were used by Beneka and colleagues, who investigated the muscular performance of knee injured athletes during rehabilitation (Beneka et al. 2000). It was found that athletes taught to use goal setting and self-talk did better at this test than those taught to use just one. However, it was more beneficial to use either goal setting or self-talk than to do no PST. Theodorakis and his colleagues have further established the effectiveness of goal setting in enhancing rehabilitation performance. In two separate studies, it was found that teaching participants to set personally relevant goals enhanced strength (Theodorakis et al. 1996, 1997).

Reduced fear and anxiety

When injured, performers often suffer from undesirable fear and anxiety (see earlier section of this chapter on the psychological reactions to injury). Fortunately, PST can be used to decrease these feelings (Suinn, 1975; Heil 1993; Ross and Berger 1996). For example, it has been found that teenage female gymnasts fear getting injured because it is difficult to return from injury, and because being injured prevents them from participating in their sport (Chase et al. 2005). Importantly, this study also established that psychological skills such as imagery and relaxation could help them overcome such fears. A case study with a skier (Suinn 1975) found similar results.

We know from rehabilitation experts (Udry et al. 1997; Karin 2008) that athletes can often return from injury not only physically stronger, but also technically better than they previously were. Highlight to performers, therefore, that injury is an almost inescapable part of sustained high-level participation and it should not be seen as inherently negative. Instead, becoming an expert at injury rehabilitation may be one of the factors that set true high-achievers apart from those who are less able to achieve at a high level. Alerting athletes to this idea, coupled with stories about successfully rehabilitated athletes and PST, may help reduce fear and anxiety.

Maintained well-being

Fortunately PST can support well-being outcomes such as mood and various perceptions of self – perhaps even at the same time as yielding some (or all) of the outcomes outlined above (Evans et al. 2000; Johnson 2000; Driediger et al. 2006). As an example, Johnson (2000) has shown that a combination of relaxation and imagery can enhance mood among competitive athletes with long-term injuries. Athletes who received this intervention were also more ready to return to competition, as judged both by themselves and the physiotherapist treating them. Intriguingly, Johnson's (2000) study indicated that the intervention group receiving relaxation and imagery training was the only one to yield statistically significant improvements: those receiving training in stress management and cognitive control, or in goal-setting, did not. Further support for the role of imagery comes from research by Sordoni, Hall and Forwell (2002). In their study, it was found that healing imagery was positively related to rehabilitation self-efficacy. In other words, those who did more healing imagery felt more capable of rehabilitating successfully. Evans, Hare and Mullen (2006) found rehabilitating athletes to use images of sport skills to enhance their confidence.

Finally, PST may be used to keep motivation up during the drudgery that often accompanies rehabilitation. For example, one injured athlete put it this way:

> Well, for me, one is to motivate yourself to make yourself do all of these dumb little exercises that seem at the time that they are not helping you at all,

but down the road they will. So, to try and keep things in perspective you kind of imagine how they're going to help you get back to competition level. (Driediger et al.,2006, p. 265).

Feeling involved in one's sport, and maintaining sport skills

PST can help an injured performer to feel involved in their sport, rather than left out and unable to do anything to progress athletically. It can also support the maintenance of sport skills, thereby helping athletes return to activity at a good level of proficiency (Ievleva and Orlick 1991; Evans et al. 2006). Consider, for example, the following quote from a male international pole vaulter who had used a lot of imagery during his rehabilitation from a muscular compression injury on his lower back. Without any physical training, he won a National Championship immediately upon his return to activity:

> I've done so much imagery between getting injured and now, it's just, erm, I'm still a lot closer to the real performance than I would be if I had done nothing. I think that that maybe the reason behind the successful performance in the competition last week, in that, I mean, even though I'd been 2 months without any training at all, you know it just kinda came naturally to me, it was amazing, technically I hadn't lost a thing. (Evans et al. 2006, p. 16).

Athletes should be informed that really, rehabilitation is a busy time - if they choose it to be. For example, practicing their rehabilitation exercises and practicing mentally is training that can help them progress as athletes. Importantly, more elite performers are likely to do this automatically, while athletes at lower levels may need more inspiration and support (Sordoni et al. 2002). Athletes can be encouraged to do imagery not only at home but also in the practice setting; by sitting in on practices, helping out (perhaps by support-coaching others), and learning new skills using their imagery skills, an athlete will be in a much better position to return to activity when physically strong enough (White and Hardy 1998; Evans et al. 2000; Kolt 2000). The idea of doing imagery in the "right" setting is one of seven recommendations for how to make imagery effective based on the PETTLEP model (Holmes and Collins 2001). Other important considerations raised by this model is that imagery should be multisensory (kinaesthetic imagery appears especially valuable), performed in real time, and be updated as a performer improves. Emphasise that commitment to PST is an important aspect of commitment to sport in general, and that by doing it, the athletes will return to sport faster.

Sometimes, injured performers experience debilitative images, such as replaying the injury scenario over and over (Ievleva and Orlick 1991; Nordin and Cumming 2005; Evans et al. 2006). It is important to realise that this kind of imagery might actually slow down the rehabilitation process (Ievleva and Orlick 1991), and should be avoided. Like with self-talk, the best way of avoiding unwanted images is to have a supply of vivid, positive and helpful images available. This "supply" will only exist if such images have been practised, and so knowing what images help is extra important for this reason. When rehearsed, the performer will gradually be able to identify debilitative images, and replace them with facilitative ones.

Some general points to consider

In order for an athlete or rehabilitator to implement successful PST programmes, there are some further points to note. These include self-determination and injury understanding, which are outlined in turn.

Self-determination

As self-determination theory is gaining more and more research support, it is becoming increasingly evident how important it is to encourage a sense of self-determination in all aspects of sports and performing, and the injury process is no exception (Podlog and Eklund 2007). As noted by Duda et al. (2005) and Hardy (2006), PST is likely to be more effective if performers feel that they have a say. A rehabilitator should therefore be knowledgeable enough about psychological skills to be able to advise performers, yet encouraging them to be working in an active, autonomous, self-determined way. This might involve an athlete actively generating their own self-statements, images, and goals that have relevance to them.

Injury understanding

As noted by numerous researchers and practitioners (Taylor and Taylor 1997; Green 1999; Williams et al. 2001; Morris et al. 2005; Evans et al. 2006), an accurate understanding of injury type, severity, prospects and anatomical structures should aid the rehabilitating athlete in numerous ways. For instance, it should enhance imagery accuracy, and therefore effectiveness (Heil 1993; Price and Andersen 2000). Similarly, a good understanding of one's injury will help the athlete set effective, self-determined goals related to injury rehabilitation (O'Connor et al. 2005). This may be achieved in a number of ways, and the rehabilitator should make every effort to interest the athlete in better understanding their relevant anatomy, physiology, and exercise requirements – skills that are likely to be useful to them even after the injury has healed, and helping to prevent re-injury. Pictures, anatomical models and drawings, websites, animations, interesting anecdotes, variations on exercises and a generally positive outlook are skills and techniques that may be beneficial. Francis, Andersen and Maley (2000) report that athletes often appreciate having a timeline, ranging from the present to full recovery; this will undoubtedly facilitate accurate goal setting.

Sample PST programme to rehabilitate from injury

1. *Set goals for recovery.* Use information from therapists and coaches to set up a realistic rehabilitation schedule. Then break down your long-term goal (e.g. being back to playing 100%) into shorter-term goals. You may need to do this in several steps; you know you have "done it" when your schedule tells you what to do tomorrow! Set goals that include work such as your rehabilitation exercises (e.g. leg circles each morning and evening), mental practice (e.g. 5 minutes of imagery 3x/day) and fun things (e.g. see teammates socially at least once/week). Get your coach and any rehabilitation personnel to give you regular feedback on how you are doing, and bear in mind that goals often need updating as you go along – so stay flexible. Again, a training diary can help you do all this.

2. *Use imagery daily.* Use imagery to rehearse sport skills and scenarios and to support the healing process. It is a good idea to imagine yourself as healed, strong, and back in action; however, it might be even better to create realistic, gradual images of being stronger and stronger, with less and less pain. Rehearse reactions to pain, and any functional limitations you may expect to have initially. Most importantly, imagine how you want to cope with such limitations. And don't forget to generate feelings of confidence throughout! Imagery is a great way to rehearse emotion as well as more "concrete" things like movements. For healing, try imagining ultrasound as a healing glow or your pain receptors shutting down as a result of ice treatment (Heil 1993), or your pain being washed away by water or your blood (Ievleva and Orlick 1991). You can also imagine "breathing into" an area to induce relaxation and a feeling of space in joints. An imagery script that also includes a relaxation procedure is available in Price and Andersen (2000).

3. *Use self-talk to your advantage.* Monitor your internal dialogue during rehabilitation, ensuring it stays constructive. Use it to improve performance in your rehabilitation exercises and perhaps while observing or imagining training sessions, and use it to boost your mood and to maintain your confidence throughout the day. For instance, telling yourself that "of course I am disappointed – but taking time out now to rehabilitation properly will teach me why I got this injury in the first place, and avoid re-injury in the future" acknowledges negative emotions such as disappointment, while staying constructive.

4. *Relax – you deserve it!* During rehabilitation, being able to relax is an important skill – especially for performers who are prone to stress. Not only can it help you identify (and release) any excess tension that you are carrying, but it can also help you tune into your body and thereby enhancing awareness – a valuable skill also after the injury is gone.

Table 15.1 Scrape (Hinderliter and Cardinal 2007)

Social support	Athletes with a high level of social support from team, family or friends demonstrate higher levels of adherence as well as higher motivation and self-esteem
Confidence	This is needed by the athlete towards both the practitioner and the rehabilitation process. Through successful accomplishments your athlete will feel a sense of achievement, autonomy and competence. It is helpful for an athlete to maintain an optimistic view especially as within the rehabilitative process there are often setbacks
Refer	This pertains to an ethical and moral sense of obligation by you, the sports rehabilitator, of the importance of referring to a sport psychologist or GP for more serious issues such as depression, anxiety or eating disorders. This is also another reason why studying psychology is important for rehabilitators: by understanding more about psychological issues, you will be in a better place to know when you can help an athlete, and when they need more specialist support
Accommodate	This refers to being flexible and adjusting to the patient's needs and wants. Beware of relying on standardised approaches, and make every effort to develop an individualised programme that suits the athlete
Psychological skills	These include strategies such as imagery, relaxation and goal setting. The use of a diary is beneficial in helping the athlete to keep a record of progress and feelings, and can also include instructions for physical and psychological exercises
Educate	The need to educate each client about their specific injury and rehabilitation process is paramount. Therefore, this part of the SCRAPE acronym refers to the giving of material (e.g. photocopied pictures of exercises) explaining the nature of the injury or what is required for the treatment or the length of the process

Bringing it all together: the SCRAPE model of psychological aspects of recovery

This chapter has contained a lot of information, and you may be wondering how to remember it all. Fortunately, help is at hand through *SCRAPE*: a model (Hinderliter and Cardinal 2007), which may help you as a sport rehabilitator to easily remember the psychological aspects of recovery. The acronym stands for six concepts representing sports rehabilitation recovery, including **s**ocial support, **c**onfidence and competence, **r**efer, **a**ccommodate, **p**sychological skills and **e**ducate. The model is based on Hinderliter and Cardinal's (2007) own research as well as their clinical experiences.

Finally, Williams (2001) summarises the importance of psychology within the sports rehabilitation field when he states that "the ultimate value of research dealing with the psychosocial risk factors is the potential for using the knowledge to reduce the tragedy and expenses caused by avoidable injuries". The integration of psychology within the field of sports rehabilitation is vital for you as a practitioner, for the teaching and training of sports rehabilitators and for the future development of this profession as a whole.

References

Arvinen-Barrow, B., Hemmings, D., Weigand, C. and Becker, C. (2007) Views of chartered physiotherapists on the psychological content of their practice: a follow-up 2007. *Journal of Sport Rehabilitation.* 16 (2), 111–121.

Beneka, A., Maliou, P., Teodorakis, Y. and Godolia, G. (2000) The effect of self talk and goal setting in muscular performance of knee injured athletes during the rehabilitation period. *Nauka, bezbednost, policija*, 5 (1), 109–122.

Brewer, B.W., Jeffers, K.E., Petitpas, A.J. and Van Raalte. J.L. (1994). Perceptions of psychological interventions in the context of sport injury rehabilitation. *The Sport Psychologist*, 8, 176–188.

Brewer, B.W. and Cornelius, A.E. (2003) Psychological factors in sports injury rehabilitation. In W.A. Frontera (Ed.) *Rehabilitation of Sport Injuries: A scientific basis* (pp. 160–183). Malden, MA: Blackwell Science.

Broucek, M.W., Bartholomew, J.B., Landers, D.M. and Linder, D.E. (1993) The effects of relaxation with a warning cue on pain tolerance. *Journal of Sport Behaviour*, 16 (4), 239–246.

Calmels, C., d'Arripe-Longueville, F., Fournier, J.F. and Soulard, A. (2003) Competitive strategies among elite female gymnasts: An exploration of the relative influence of psychological skills training and natural learning experiences. *International Journal of Sport & Exercise Psychology*, 1, 327–352.

Carson, F. and Polman, R. (2008) ACL injury rehabilitation: A psychological case study of a professional rugby union player. *Journal of Clinical Sport Psychology*, 2 (1), 71–90.

Carver, C.S. and Scheier, M.F. (2000) *Perspectives on Personality*, 4th edn. Boston: Allyn and Bacon, p.5.

Carver, C., Scheier, M. and Weintraub, J. (1989) Assessing coping strategies: A theoretically based approach. *Journal of Personality and Social Psychology* 56 (2), 267–283.

Chase, M., Magyar, M. and Drake, B. (2005) Fear of injury in gymnastics: Self-efficacy and psychological strategies to keep on tumbling. *Journal of Sports Sciences*, 23 (5), 465–475.

Christakou, A. and Zervas, Y. (2007) The effectiveness of imagery on pain, edema, and range of motion in athletes with a grade II ankle sprain. *Physical Therapy in Sport*, 8 (3), 130–140.

Cox, R.H. (2001) *Sport Psychology: concepts and applications*. London: McGraw-Hill.

Crossman, J. (1997) Psychological rehabilitation from sports injuries. *Sports Medicine*, 23, 3333–3339.

Crossman, J. (2001) *Coping with Sports Injuries: Psychological Strategies for Rehabilitation*. New York: Oxford University Press.

Cupal, D.D. and Brewer, B.W. (2001) Effects of relaxation and guided imagery on knee strength, reinjury anxiety, and pain following anterior cruciate ligament reconstruction. *Rehabilitation Psychology*, 46, 28–43.

Davis, J.O. (1991) Sport injuries and stress management. An opportunity for research. *The Sport Psychologist*, 5, 175–182.

Durso Cupal, D. (1998) Psychological interventions in sport injury prevention and rehabilitation. *Journal of Applied Sport Psychology*, 10, 103–123.

Evans, L., Hardy, L. and Fleming, S. (2000) Intervention strategies with injured athletes: an action research study. *Sport Psychologist*, 14, 188–206.

Fawkner, H.J., McMurray, N. and Summer, J.J. (1999) Athletic injury and minor life events: A prospective study. *Journal of Science and Medicine in Sport*, 2, 117–124.

Filby, W.C.D., Maynard, I.W. and Graydon, J.K. (1999) The effect of multiple-goal strategies on performance outcomes in training and competition. *Journal of Applied Sport Psychology*, 11 (2), 230–246.

Fletcher, D. and Fletcher, J. (2005) A meta-model of stress, emotions and performance: conceptual foundations, theoretical framework, and research directions. *Journal of Sports Science*, 23 (2), 157–158.

Folkman, S. and Lazarus, R.S. (1984) If it changes it must be a process: A study of emotion and coping during three stages of a college examination. *Journal of Personality and Social Psychology*, 48, 150–170.

Francis, S.R., Andersen, M.B. and Maley, P. (2000) Physiotherapists' and male professional athletes' views on psychological skills for rehabilitation. *Journal of Science and Medicine in Sport*, 3 (1), 17–29.

Gallagher, B. and Gardner, F. (2007) An examination of the relationship between early maladaptive schemas, coping, and emotional response to athletic injury. *Journal of Clinical Sport Psychology*, 1 (1), 47–67.

Gill, S., Kolt, G.S. and Keating, J. (2004) Examining the multi-process theory: An investigation of the effects of two relaxation techniques on state anxiety. *Journal of Bodywork and Movement Therapies*, 8, 288–296.

Gordon, S., Potter, M. and Ford, I.W. (1998) Toward a Psychoeducational Curriculum for Training Sport-Injury Rehabilitation Personnel. *Journal of Applied Sport Psychology*, 10, 140–156.

Green, L.B. (1992) The use of imagery in the rehabilitation of injured athletes. *The Sport Psychologist*, 6, 416–428.

Green, L.B. (1999) The use of imagery in the rehabilitation of injured athletes. In D. Pargman (Ed.), *Psychological Basis of Sport Injuries*, 2nd edn. Morgantown, WV: Fitness Information Technology, pp. 235–251.

Gucciardi, D.F., Gordon, S. and Dimmock J.A. (2008) Towards an understanding of mental toughness in Australian football. *Journal of Applied Sport Psychology*, 20 (3), 261–281.

Hamson-Utley, J.J., Martin, S. and Walters, J. (2008) Athletic trainers' and physical therapists' perceptions of the effectiveness of psychological skills within sport injury rehabilitation programs. *Journal of Athletic Training*, 43 (3), 258–264.

Hanrahan, C. and Vergeer, I. (2000) Multiple uses of mental imagery by professional modern dancers. *Imagination, Cognition and Personality*, 20 (3), 231–255.

Hanson, S., McCullagh, P. and Tonymon, P. (1992) The relationship of personality characteristics, life stress, and coping resources to athletic injury. *Journal of Sport and Exercise Psychology*, 14 (3), 262–272.

Hardy, J. (2006) Speaking clearly: A critical review of the self-talk literature. *Psychology of Sport and Exercise, 7* (1), 81–97.

Hardy, C., Richman, J. and Rosenfeld, L. (1991) The role of social support in the life stress/injury relationship. *The Sport Psychologist, 5* (2), 128–139.

Hardy, J., Hall, C.R. and Hardy, L. (2005) Quantifying athlete self-talk. *Journal of Sport Sciences, 23*, 905–917.

Heaney, C. (2006) Physiotherapists' perceptions of sport psychology intervention in professional soccer. *International Journal of Sport and Exercise Psychology, 4*, 67–80.

Heil, J. (1993) *Psychology of Sport Injury*. Champaign, IL: Human Kinetics Publishers.

Hemmings, B. and Povey, L. (2002) Views of chartered physiotherapists on the psychological content of their practice: A preliminary study in the United Kingdom. *British Journal of Sports Medicine, 36*, 61–64.

Hinderliter, C.J. and Cardinal, B.J. (2007) Psychological rehabilitation for recovery from injury: The SCRAPE approach. *Athletic Therapy Today*. Champaign, IL: Human Kinetics, *12* (6), 36–38.

Holmes, P.S. and Collins, D.J. (2001) The PETTLEP Approach to Motor Imagery: A Functional Equivalence Model for Sport Psychologists. *Journal of Applied Sport Psychology, 13* (1), 60–83.

Ievleva, L. and Orlick, T. (1991) Mental links to enhance healing. *The Sport Psychologist, 5* (1), 25–40.

Johnson, U. (2000) Short-term psychological intervention: a study of long-term-injured competitive athletes. *Journal of Sport Rehabilitation, 9*, 207–218.

Johnson, U., Ekengren, J. and Andersen, M.B. (2005) Injury prevention in Sweden: Helping soccer players at risk. *Journal of Sport and Exercise Psychology, 27*, 11.

Jones, L. and Stuth, G. (1997) The uses of mental imagery in athletics: An overview. *Applied and Preventive Psychology, 6* (2), 101–115.

Jones, G., Hanton, S., Connaughton, D. (2002) What is this thing called mental toughness? An investigation of elite sport performers. *Journal of Applied Sport Psychology, 14* (3), 205–218.

Kolt, G.S. and Roberts, P.D. (1998) Self esteem and injury in competitive field hockey players. *Perceptual Motor Skills, 87* (1), 353–354.

Kolt, G.S. (2000) Doing sport psychology with injured athletes. In M.B. Andersen (Ed.) *Doing Sport Psychology*. Champaign, IL: Human Kinetics, 223–236.

Korn, E.R. (1983) The use of altered states of consciousness and imagery in physical and pain rehabilitation. *Journal of Mental Imagery, 7*, 25–34.

Kowalski, K. and Crocker, P. (2001) Development and validation of the coping function questionnaire for adolescents in sport. *Journal of Sport and Exercise Psychology, 23* (2), 136–155.

Kübler-Ross, E. (1969) *On Death and Dying*. London: Macmillan.

Lazarus, R.S. and Folkman, S. (1984) *Stress, Appraisal and Coping*. New York: Springer.

Levy, A., Polman, R.C.J., Clough, P.J., Marchant, D.C. and Earle, K. (2006) Mental toughness as a determinant of beliefs, pain, and adherence in sport injury rehabilitation. *Journal of Sport Rehabilitation, 15* (3), 246–254.

Lind, E., Ekkekakis, P. and Vazou, S. (2008) The affective impact of exercise intensity that slightly exceeds the preferred level: 'pain' for o additional 'gain'. *Journal of Health Psychology, 13* (4), 464–468.

Lyngcoln, A., Taylor, N., Pizzari, T. and Baskus, C. (2005) The relationship between adherence and therapy and short-term outcome after distal radius fracture. *Journal of Hand Therapy, 18* (1), 2–8.

Maddison, R. and Prapvessis, H. (2005) A psychological approach to the prediction and prevention of athletic injury. *Journal of Sport and Exercise Psychology, 27* (3), 289–310.

Mann, B., Grana, W., Indelicato, P., Oneil, D. and George, S. (2007) A survey of sports medicine physicians regarding psychological issues in patients-athletes. *American Journal of Sports Medicine, 35* (12), 2140–2147.

Marchant, D. and Earle, K. (2006) Mental toughness as a determinant of sport injury beliefs, pain and rehabilitation adherence, *Journal of Sports Rehabilitation, 15* (3), 246–254.

May, J.R. and Brown, L. (1989) Delivery of psychological service to the U.S. Alpine ski team prior to and during the Olympics in Calgary. *The Sport Psychologist, 3*, 320–329.

Middleton, S., Marsh, H., Martin, A., Riches, J. and Perry, Jr, C. (2005) Making the leap from good to great: Comparisons between sub-elite and elite athletes on mental toughness. In Peter L. Jeffery (Ed.) *Proceedings of the Australian Association for Research in Education Conference*. Parramatta.

Morris, T., Spittle, M. and Watt, A. (2005) *Imagery in Sport*. Champaign, IL: Human Kinetics.

Murphy, S. (2005) *The Sport Psych Handbook*. Champaign, IL: Human Kinetics.

Nicholls, A., Polman, R., Levy, A., Taylor, J. and Cobley, S. (2007) Stressors, coping, and coping effectiveness: Gender, type of sport, and skill differences. *Journal of Sports Sciences, 25* (13), 1521–1530.

Nicol, M. (1993) Hypnosis in the treatment of repetitive strain injury. *Australian Journal of Clinical and Experimental Hypnosis, 21*, 121–126.

Ninedek, A. and Kolt, G.S. (2000) Sports Physiotherapists' perceptions of psychological strategies in sport injury rehabilitation. *Journal of Sport Rehabilitation*, *9*, 191–206.

Niven, A. (2007) Rehabilitation adherence in sport injury: sport physiotherapists' perceptions. *Journal of Sport Rehabilitation*, *16* (2). 93–110.

Noh, Y-E. and Morris, T. (2004) Designing research-based interventions for the prevention of injury in dance. *Medical Problems of Performing Artists*, *19*, 82–89.

Noh, Y.E., Morris, T. and Andersen, M.B. (2005) Psychosocial factors and ballet injuries. *International Journal of Sport and Exercise Psychology*, *1*, 79–90.

Noh, Y.E., Morris, T. and Andersen, M.B. (2007) Psychological intervention programs for reduction of injury in ballet dancers. *Research in Sports Medicine: an International Journal*, *15* (1), 13–32.

Nordin, S.M. and Cumming, J. (2005) Professional dancers describe their imagery: Where, when, what, why, and how. *The Sport Psychologist*, *19*, 395–416.

O'Connor, E., Heil, J., Harmer, P. and Zimmerman, I. (2005) Injury. In J. Taylor and Wilson, G. (Eds), *Applying Sport Psychology – Four perspectives*. Champaign, IL: Human Kinetics.

Pargman, D. (2007) *Psychological bases of sport injuries*. Morgantown, WV: Fitness Information Technology.

Perna, F.M., Antoni, M.H., Baum, A., Cordon, P. and Schneiderman, N. (2003) Cognitive behavioral stress management effects on injury and illness among competitive athletes: A randomized clinical trial. *Annals of Behavioral Medicine*, *25*, 66–73.

Petrie, T. (1992) Psychosocial antecedents of athletic injury: the effects of life stress and social support on female collegiate gymnasts. *Behavioural Medicine*, *18* (3), 127–138.

Pizzari, T., Taylor, N.F., McBurney, H. and Feller, J. (2005) Adherence to rehabilitation after anterior cruciate ligament reconstructive surgery: implications for outcome. *Journal of Sport Rehabilitation*, *14* (3), 201–214.

Podlog, L. and Eklund, R.C. (2007) The psychosocial aspects of a return to sport following serious injury: A review of the literature from a self-determination perspective. *Psychology of Sport and Exercise*, *8*, 535–566.

Price, P.L. and Andersen, M.B. (2000) Into the Maelstrom: A Five-Year Relationship From College Ball to the NFL. In M.B. Andersen (Ed.), *Doing Sport Psychology*. Champaign, IL: Human Kinetics, pp. 193–206.

Ross, M.J. and Berger, R.S. (1996) Effects of stress inoculation training on athletes' postsurgical pain and rehabilitation after orthopedic injury. *Journal of Consulting and Clinical Psychology*, *64*, 406–410.

Scherzer, C.B., Brewer, B.W., Cornelius, A.E., Van Raalte, J.L., Petitpas, A.J., Sklar, J.H., Pohlman, M.H., Krushell, R.J. and Ditmar, T.D. (2001) Psychological skills and adherence to rehabilitation after reconstruction of the anterior cruciate ligament. *Journal of Sport Rehabilitation*, *10*, 165–172.

Schomer, H.H. (1990) A cognitive strategy training program for marathon runners: Ten case studies. *South African Journal of Research in Sport, Physical Education and Recreation*, *13*, 47–78.

Sheppard, K. (2004) Foreword. In Kolt and Andersen (eds), *Psychology in the Physical and Manual Therapies*. London: Churchill Livingstone.

Short, S.E., Hall, C.R., Engel, S.R. and Nigg, C.R. (2004) Exercise imagery and the stages of change. *Journal of Mental Imagery*, *28*, 61–78.

Smith, R.E., Smoll, F.L. and Ptacek, J.T. (1990a) Conjunctive moderator variables in vulnerability and resiliency research: life stress, social support and coping skills, and adolescent sport injuries. *Journal of Personality Social Psychology*, *58* (2), 60–70.

Smith, R.E., Smoll, F. and Schutz, R. (1990b). Measurement and correlates of sport-specific cognitive and somatic trait anxiety: The Sport Anxiety Scale. *Anxiety Research*, *2* (4), 263–280.

Smith, R.E., Schutz, R.W., Smoll, F.L. and Ptacek, J.T. (1995) Development and validation of a multidimensional measure of sport-specific psychological skills: The Athletic Coping Skills Inventory-28. *Journal of Sport and Exercise Psychology*, *17* (4), 379–398.

Sordoni, C., Hall, C. and Forwell, L. (2000) The use of imagery by athletes during rehabilitation. *Journal of Applied Sport Psychology*, *3*, 329–338.

Sordoni, C., Hall, C. and Forwell, L. (2002) The use of imagery in athletic injury rehabilitation and its relationship to self-efficacy. *Physiotherapy Canada*, Summer, 177–185.

Steffen, K., Pensgaard, A.M. and Bahr, R. (2008) Self-reported psychological characteristics as risk factors for injuries in female youth football. *Scandinavian Journal of Medicine and Science in Sports*, *19* (3), 442–451.

Sthalekar, H.A. (1993) Hypnosis for relief of chronic phantom pain in a paralysed limb: a case study. *Australian Journal of Clinical Hypnotherapy and Hypnosis*, *14*, 75–80.

Suinn, R.M. (1975) *Behavior Modification for Athletic Injury*. Fort Collins, CO: Colorado State University.

Taylor, A.H. and May, S. (1996) Threat and coping appraisal as determinants of compliance with sports injury rehabilitation. *Journal of Sport Sciences*, *14* (6), 471–482.

Taylor, J. and Taylor, S. (1997) *Psychological Approaches to Sport Injury rehabilitation*. Gaitherburg, MD: Aspen.

Theodorakis, Y., Malliou, P., Papaioannou, A., Beneca, A. and Filaktakidou, A. (1996) The effect of personal goals, self-efficacy, and self satisfaction on injury rehabilitation. *Journal of Sport Rehabilitation*, 5, 214–223.

Theodorakis, Y., Beneka, A., Malliou, P. and Goudas, M. (1997) Examining psychological factors during injury rehabilitation, *Journal of Sport Rehabilitation*, 6, 355–363.

Tracey, J. (2003) The emotional response to the injury and rehabilitation process. *Journal of Applied Sport Psychology*, 15 (4), 279–293.

Udry, E., Gould, D., Bridges, D. and Beck, L. (1997) Down but not out: athlete responses to season ending injuries. *Journal of Sport and Exercise Psychology*, 19 (3), 229–248.

Walker, N., Thatcher, J. and Lavallee, D. (2007) Review: Psychological responses to injury in competitive sport: a critical review. *The Journal of the Royal Society for the Promotion of Health*. 127, 174–180.

Van Raalte, J.L., Brewer, B.W., Lewis, B.P., Linder, D.E., Wildman, G. and Kozimor, J. (1995). Cork! The effects of positive and negative self-talk on dart throwing performance. *Journal of Sport Behavior*, 18, 50–57.

Wethington, E. and Kessler, R.C. (1991) Situation and Processes of Coping. In J. Eckenrode (Ed.) *The Social Context of Coping*. New York: Plenum Press, pp. 13–30.

White, A. and Hardy, L. (1998) An in depth analysis of the uses of imagery by high-level slalom canoeists and artistic gymnasts. *The Sport Psychologist*, 4, 180–191.

Wiese. D.M., Weiss, M.R. and Yukelson, D.P. (1991) Sport psychology in the training room: A survey of athletic trainers. *The Sport Psychologist*, 5, 25–40.

Williams, J.M. (2001) Psychology of injury risk and prevention. In R.N. Singer, H.A. Hausenblas and C.M. Janelle (eds), *Handbook of Sport Psychology* (pp. 766–786). New York: Wiley.

Williams, J.M. and Andersen, M.B. (1998) Psychosocial antecedents of sport injury: review and critique of the stress and injury model. *Journal of Applied Sport Psychology*, 10 (1), 5–25.

Williams, M. and Andersen, J. 1988. A Model of Stress and Athletic Injury : prediction and prevention. *Journal of Sport and Exercise Psychology* 10 (3)

Williams, J.M., Tonymon, E. and Wadsworth, W.A. (1986) Relationship of stress to injury in intercollegiate volleyball. *Journal of Human Stress*, 12 (11), 38–43.

Williams, J.M., Hogan, T.D. and Andersen, M.B. (1993) Positive states of mind and athletic injury risk. *Psychosomatic Medicine*, 55 (5), 468–472.

Williams, J.M., Rotella, R.J. and Scherzer, C.B. (2001). Injury risk and rehabilitation: Psychological considerations. In J.M. Williams (Ed.), *Applied Sport Psychology: Personal growth to peak performance*, 2nd edn. Mountain View, CA: Mayfield Publishing, pp. 456–479.

16

Clinical reasoning

Earle Abrahamson
London Sport Institute, Middlesex University

Dr Lee Herrington
University of Salford, Greater Manchester

This chapter provides an overview, analysis and application of clinical reasoning and problem solving skills in the development of professional competencies within the healthcare profession generally and more specifically sports rehabilitation. It will help you develop your thinking skills as you progress your reading throughout the book. By the end of this chapter you will be able to locate and explain the role and efficacy of clinical reasoning skills within a professional practice domain. This will inform an appreciation for the complex nature of knowledge construction in relation to clinical explanation and judgement. By considering clinical reasoning as a functional skill set, you will further be in a position to explain different models of reasoning and ask structured questions in an attempt to better formulate and construct answers to clinical questions, issues and decisions. The chapter will further encourage the use of problem solving and clinical reasoning skills to justify substantially, through research evidence, professional practice actions and outcomes.

To adequately define, discuss and synthesise clinical reasoning and its application to healthcare practices, it is important to relate evidence-based learning and decision making to the skills and competencies of the sports rehabilitator.

Understanding clinical reasoning

It is what we think we know that keeps us from learning (Claude Bernard)

Members of the professions must build and maintain a formidable store of knowledge and skills; they must learn to absorb information through various senses and to assess its validity, reliability and relevance; and they must acquire the art and culture of their calling. And most importantly, they must learn to use these qualities to solve practical problems (Heath 1990, cited in Higgs and Jones 1995).

In considering the above it is apparent that clinical reasoning and the application and synthesis of clinical knowledge is at best an obscure and complex phenomenon. It involves a complex process of structuring meaning from confusing data and experiences occurring within a specific clinical setting and then making informed judgements based on understanding and evidence-based practices.

What is clinical reasoning and how best can it be explained and applied?

Clinical reasoning may be defined as "the process of applying knowledge and expertise to a clinical situation to develop a solution" (Carr 2004).

Several forms of reasoning exist and each has its own merits and uses. Reasoning involves the processes of cognition and metacognition. In sports rehabilitation, clinical reasoning skills are an expected component of expert and competent practice. Interprofessional health research, predominantly from nursing practices, have identified concepts, processes and thinking strategies that might underpin the clinical reasoning used by healthcare professionals. Much of the available research on reasoning is based on the use of the think aloud approach. Although this is a useful method, it is dependent on ability to describe and verbalise the reasoning process (Schon 1991; Hauer et al. 2007). Information-processing theory, developed by Newell and Simon (1972), is useful in explaining and describing how to organise information using knowledge, and experience, and the use of cognitive processes to resolve a problem. Rather than analysing how a problem ought to be solved, or a decision made, this theory describes decision making as an open system of interaction between a problem solver and a task (Ericsson and Simon 1984; Simmons et al. 2003). This theory is useful in helping to describe and categorise the organisation of problem solving and decision making for clinical reasoning development. According to Simmons et al. (2003), when applied in healthcare settings, the term information processing becomes interchangeable with clinical reasoning.

Is clinical reasoning simply the collection and collation of data in an attempt to provide answers to often complex questions? Is it the barrier between expertise knowledge and novice enterprise? Or is clinical reasoning and subsequent applications, a more involved process of knowing, appraising and deciding how best to answer?

In support of the later, the area of evidence-based learning captures the nature of clinical enquiry through a simple question: *What is the best possible answer that can be given based on current knowledge levels?* This question triggers a complex cerebral cascade that demands substantiated support, often through scientific and research driven processes, to arrive at an answer that fully and comprehensibly details a specific knowledge cluster. It is this process of thinking through reason that allows a clinician to consolidate, appreciate and apply knowledge in relation to clinical challenges.

The literature is flooded with examples and extensions of clinical and problem-solving skills and decisions, with an array of application and analysis. Despite this, there is no one accepted evidence-based research model to adequately explain; or account for all aspects of clinical reasoning practices. There are a number of models that attempt to explain the interdependent process of thinking through reasoning to arrive at an informed answer to substantiate action. Fleming and Mattingly (1994), argue that clinical reasoning in its most simple form is "judgement in action" leading to "action based upon judgement".

Models of clinical reasoning

Appreciating the challenge and process of clinical reasoning, often demands an understanding and analysis of models of clinical reasoning.

The generation of hypotheses based on clinical data and knowledge, coupled with the testing of these hypotheses through further inquiry, forms the basis of the *hypothetico-deductive reasoning* model (Elstein et al. 1978; Kassirer and Gorry 1978; Gale 1982). Hypothesis generation and testing involves both inductive (considering a specific observation and moving to a more general view), and deductive (moving from a generalised view to a more specific and pronounced outcome) reasoning (Ridderrikhoff, 1989). Induction is a reasoning process used in the formulation of hypotheses, whereas deduction is used to test the hypothesis. Inductive reasoning is probabilistic in nature, since a conclusion is reached and the presented evidence is evaluated in relation to existing knowledge (Albert et al. 1988). Deductive reasoning is widely used in sports rehabilitation practice to defend decisions and actions.

A second explanation of how reasoning is used to support actions and decisions is contained within a pattern recognition approach. Pattern recognition, or more precisely inductive reasoning, as an interpretation and expression of diagnostic reasoning has received much support in the literature (Hamilton 1966; Gorry 1970; Elstein et al. 1990).

This reasoning approach allows more experienced practitioners to arrange their thought processes into patterned formations. This development of pattern formation allows for heuristic analysis in clinical judgements. Pattern recognition further allows clinicians to draw on past treatments, to better evaluate their effectiveness, and categorise the success of the treatment into a management plan for the treatment of a similar or recurring condition. This approach

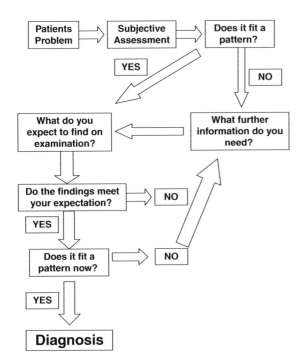

Figure 16.1 Conceptual flow diagram of clinical reasoning process.

includes a process of interpretation of patterns, such as the identification of set signs and symptoms in the assessment of pathology or injury, to better account and plan for the treatment, management and rehabilitation of the injury or condition.

A third approach to explain the reasoning process is best presented in the work of Boshuizen and Schmidt (1992), who developed a stage theory on the development of expertise that emphasises the parallel development of knowledge acquisition and clinical reasoning expertise. This model is based on the notion that the construction of knowledge and subsequent expertise is largely the result in changes to knowledge structure and development. The approach has been described as a knowledge-reasoning integration. Figure 16.1 outlines the reasoning approach and provides a reasoning map for clinical decision making.

Leaver-Dunn, Harrelson, Martin and Wyatt (2002), investigated the tendency of undergraduate athletic training students to think critically. Their research findings provide some useful considerations in terms of understanding critical thinking as a precursor to clinical reasoning. In their study, they employed an objective measure called the California critical thinking disposition inventory to quantitatively measure critical thinking. The inventory has a number of subscales, which depict the components of critical analysis. The weakest subscale was that of truth seeking, which highlights the ability of a critical thinker to reflect on knowledge and engage a quest to find the truth irrespective of current views. The reasons attributed to the weakness in the truth scale could be explained by emphasising the need to know the correct answer as opposed to understanding the reasons for the correct answer and the ability to reflect on the answers reached. Whilst this chapter does not explore critical thinking as a separate entity to clinical reasoning, knowledge of critical analysis and reflection is central to developing skills necessary for understanding clinical decisions and actions.

Development of clinical reasoning skills

Problem based learning

One possible approach to teaching and developing reasoning skills is problem based learning (PBL). PBL has its history firmly rooted within the teaching and education of medical practitioners although its application to other disciplines has been researched and documented (Wood 2003). PBL evolved through two assumptions: the first was that learning through problem solving is more effective than memory based learning for creating a usable body of knowledge; the second was that clinical skills, which are important for patient treatment are problem solving skills not necessarily memory skills. PBL is not a method as much as a total teaching approach and reflects the way learners learn in real life situations. It has been used in varied study contexts to develop critical analysis and thought, as well as problem solving skills (Duncan et al. 2007). Through the use of PBL within the sports rehabilitation curriculum, it is hypothesised that the cardinal skills of critical thinking, analysis and application will develop leading to greater levels of student intrinsic motivation. Using PBL with undergraduate students could enable the growth and evolution of critical analysis (Martin et al. 2008). One of the primary features of PBL is that it is student-centred. This refers to learning opportunities that are relevant to the students, the goals of which are at least partly determined by the students themselves. Creating assignments and activities that require student input also increases the likelihood of students being motivated to learn (Richardson 2005). A common criticism of student-centred learning is that students, as novices, cannot be expected to know what might be important for them to learn, especially in a subject to which they appear to have no prior exposure. The literature on novice-expert learning does not entirely dispute this assertion; rather, it does emphasise that our students come to us, not as the proverbial blank slates, but as individuals whose prior learning can greatly impact their current learning. Problem based learning encourages students to use and develop knowledge by examining problems or case study scenarios in a relevant, real-life and applied context (Martin et al. 2008). Like many of the teaching approaches, PBL needs to be used carefully, especially in the design of the problem. Savin-Baden (2003) noted that PBL is an approach to learning that is characterised by flexibility and diversity, in the sense that it can be implemented in a variety of ways, across subjects and disciplines in diverse contexts. As such it can therefore look different to different people at different times, depending on the staff and students using it. What is unique and yet similar is the locus of learning around problem scenarios rather than discrete subject areas (Duncan et al. 2007).

Savery (2006), notes that the widespread adoption of the PBL instructional approach by different disciplines, for different age levels, and in different content domains has produced some misapplications and misconceptions of PBL. Certain practices that are called PBL may fail to achieve the anticipated learning outcomes for a variety of reasons:

- confusing PBL as an approach to curriculum design with the teaching of problem-solving

- adoption of a PBL proposal without sufficient commitment of staff at all levels

- lack of research and development on the nature and type of problems to be used

- insufficient investment in the design, preparation and ongoing renewal of learning resources

- inappropriate assessment methods which do not match the learning outcomes sought in problem-based programmes

- evaluation strategies that do not focus on the key learning issues and which are implemented and acted on far too late.

Using PBL in developing clinical reasoning skills is useful. Unlike traditional information driven curricula, PBL begins with a problem, often based on real facts or simulations of real situations, and requires the student to work alone and in groups to find solutions. The advantage is that real problems do not have simple solutions and require comparison and analysis of resources. As such the student develops skills of retrieval, selection and discrimination and applies these to reason through answers and solutions to problems (Duncan et al. 2007). PBL is one way in which clinical reasoning skills can be developed.

To develop an appreciation of the dynamics of PBL within the evolution of clinical reasoning skills, it is important to briefly address two important concepts that directly impact this evolutionary process namely: Troublesome knowledge (TK) and threshold concepts (TCs). It is not the intention of this chapter to analyse these concepts in depth, nor to critique their application, but to rather introduce them as important consideration in the development of clinical reasoning knowledge.

Meyers and Land (2003) related a definitional construct of threshold concepts by outlining that in each discipline, such as sports rehabilitation, there are *conceptual gateways* or *portals* that must be negotiated to arrive at important new understandings. In crossing the portal or threshold transformation occurs in both knowledge and subjectivity. Meyers and Land (2008) and Land et al. 2008 expand this further by detailing the transformation as irreversible (no-going back) and integrative (involving the inclusion and fusion of different ideas and concepts in both detail and variation). Such transformation involves troublesome knowledge.

Perkins (1999) referred to troublesome knowledge as knowledge that is alien or counter-intuitive, ritualised, inert, tacit or academically challenging. Perkin further relates that threshold concepts could lead to troublesome knowledge within their own rights.

Sports rehabilitation and injury prevention as a subject area and professional practice is troublesome within itself. The scope of practice, content of learning, curriculum map and clinical competencies may,

DEVELOPMENT OF CLINICAL REASONING SKILLS 301

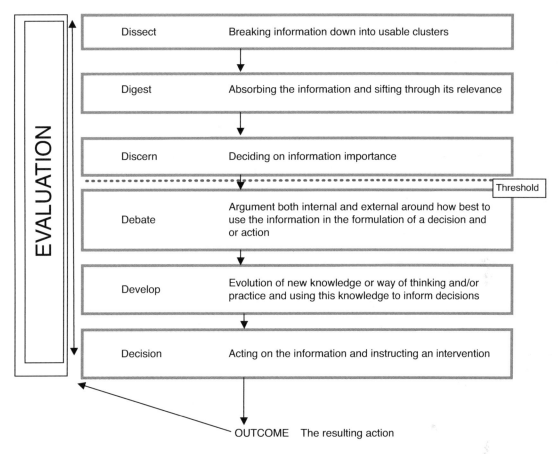

Figure 16.2 6D approach to knowledge development within a conceptual threshold framework (Abrahamson 2009).

at times, be conceptually and practically difficult to identify and embed into a distinct clinical practice model that stands unique from similar clinical and healthcare practices. The explanation of the field of sports rehabilitation involves defending a scope of practice that is construed by some professionals and clinicians as alien, or a subset of physiotherapy practice, yet surprisingly different. To fully consider action and decision making within sports rehabilitation and the ensuing development of clinical reasoning skills, it is fundamental that clinicians, academics and students, identify the threshold concepts and then use approaches such as PBL, to navigate TK and TCs to develop reflective ways of thinking and practice. The ability to think and reason like a professional clinician, is an important goal of the sports rehabilitation student.

The model in Figure 16.2 integrates TCs and TKs into a clinical reasoning development framework and considers a 6-D approach in defining conceptual difficulties and thresholds.

The model in Figure 16.2 develops a progressive strategy for thinking about action and clinical decision making. Säljö (1979), provides a useful analysis of learning as a developmental learning concept leading to a change in personal identity; that is the ability to master the strategies and competencies to think like a sports rehabilitator. To achieve this, learning must be aligned with understanding. The conceptual models align to illustrate the process of transformation and the traversing of TCs and TK.

The true characteristics of a proficient sports rehabilitator practitioner lie in the ability to cross thresholds, integrate new knowledge and construct bridges between the concepts to arrive at informed and evidence-based decisions and actions (Barrows and Pickell 1991).

In a recent study by Hauer et al. (2007) on the effect of causal knowledge on judgments of the likelihood of unknown features, the researchers reported that respondents perceived that technique problems in history taking and physical examination were readily correctable, but that poor performance resulting from inadequate knowledge or poor clinical reasoning ability was more difficult to ameliorate. Interpersonal skill deficiencies, which often manifested as detachment from the patient, and professionalism problems attributed to lack of insight, were mostly refractory to remediation. A possible explanation to this discrepancy could lie in the way in which problem solving/clinical reasoning skills are taught at an undergraduate level of training. Traditional approaches to teaching clinical skills are often based on the assumption that clinical reasoning is a skill, divorced from content knowledge. Although clinical reasoning skills and clinical knowledge could be developed and delivered separately, there is support for an integrated approach to improve the organisation and structure of relevant clinical knowledge and practices (Barrows and Pickell 1991).

Sports rehabilitation clinicians work within a framework of problematic situations. Many of these situations can be characterised by complexity, ambiguity, doubt and uniqueness. With this known entity, it may be better to conceptualise the skills required for professional practice and competency, in terms of smart action, as opposed to clinical reasoning. Differentiating the two at this level allows for a centred approach in dealing with judgment and decision making within a specific context and time. In other words smart action implies making the best decision under a given set of circumstances. It does not, however, mean always taking the right action. Smart action skills are often the catalyst for reflective analysis of performance. The evolution of this process provides the basis for professional development and critical awareness in decision making.

An important aspect of the development of higher cognitive skills and clinical reasoning ability is the ability to construct and use knowledge. The construction of knowledge requires an interpretation and processing of experience in order to appreciate reality. This often involves developing constructs to help understand reality and interpretation of experience. One cannot divorce thinking from the process, and that knowledge development requires thought, critical analysis and self-reflection on and of the knowledge construct. The next section will tease out the key elements that need consideration and development in mastering clinical reasoning skills.

How can we become better at clinical reasoning?

In order to become better, firstly we need to decide what needs improving. The majority of available research on clinical reasoning concludes that three processes interact in order to bring about good quality clinical reasoning (Higgs and Jones 2000; Higgs and Titchen 2001).

These three processes are knowledge, cognition and metacognition. These three interact throughout the process of receiving, interpreting, processing and utilising clinical information during decision making, clinical intervention and reflection on actions and outcomes.

Knowledge

Knowledge is essential for reasoning and decision making with knowledge and clinical reasoning being interdependent phenomena. There are two broad categories of knowledge. One is propositional knowledge ("knowing that"). This is achieved through research and scholarship (reading and being taught) and involves generalising information, looking for cause and effect relationships. The second type of knowledge is non-propositional ("knowing how"). Here knowledge is gained through practice experience. It would appear therefore that background knowledge is important but this must be task specific.

> It is not the way problems are tackled, nor the thoroughness of the investigations, nor the use of problem solving strategies, but the ability to activate the

pertinent knowledge as a consequence of situational demands, which distinguishes experienced from inexperienced physicians (Custers et al. 1992).

This ability to have and be able to recall task specific knowledge is one of the elements that marks out an experienced practitioner from a novice one. They are able to bring together many key elements of knowledge; anatomy, physiology, biomechanics, pathology, etc. and link them together to provide a rational explanation of the problem with which they are presented. This is often coupled with their practice experience (knowing how to do things), which then allows for a superior management of the patient. This use of knowledge is very much coupled to the process of reflection and this is covered in the next section.

Cognition and metacognition

Cognition is the act of thinking, with metacognition being the awareness and monitoring of cognition; that is, thinking about thinking. These two elements are integrated into clinical reasoning through the process of reflection. Reflection is an activity in which people recapture their experience, think about it, mull it over and evaluate it. Reflection itself involves two different aspects, one is reflection in action and the other is reflection about action. Reflection in action involves thinking about what you are doing, for example, whilst carrying out Lachman's test at the knee, the practitioner should be reflecting on, "is this what I expected to find? If it isn't, why isn't it?" Reflection about action, is essentially thinking about what has happened after it has happened, questions such as, "what was good about what I did? What was bad? What can I improve for next time? What do I need to take away from the experience and store to use again?"

Reflection is the key to information processing in clinical reasoning. From a strong knowledge base, patterns can be identified; by reflecting in action the strength of the patterns relationship to the presenting problem can be tested. Reflection in action allows the solving of problems when the presentation does not fit a pattern through hypothetico-deductive reasoning. Finally, to complete the process, reflection about action becomes a process whereby those experiences can be stored and used to generate future patterns.

The process of clinical reasoning involves the testing of hypothesis in order to both discover what a

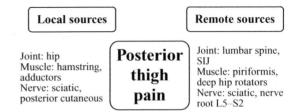

Figure 16.3 Component model of source of symptoms for posterior tight pain.

patient's problem is and develop an appropriate management strategy. There are a number of potential categories that this hypothesis can fit into, including: source of symptoms; mechanism of symptoms; contributing factors; precautions and contraindications; management and treatment; and finally prognosis.

The source of symptoms can be considered by using a component model (Figure 16.3). As can be seen from Figure 16.3 the source of symptoms can either be local to the site of pain or remote from the site of pain, but capable of referring pain to the area of the pain. The mechanism of symptoms is what has caused the pain, this could be from an extrinsic source such as a kick or fall, it also could be from an intrinsic source such as with overuse injuries, were the tissues become overloaded and break down. The contributing factors are those factors that predispose the patient to the mechanism of injury occurring, they could include poor proprioception, muscle imbalances, shortened soft tissue structures or even lack of fitness and skill. The precautions and contra-indications are those factors that might limit any chosen course of action because of potential harm to the patient. The management and treatment hypothesis category is generating a plan for the patient, which, where possible, takes into account all of the above. For instance, not just treating the local source of symptoms but also those remote symptoms, which could perpetuate the problem, whilst also addressing any contributing factors that may cause the problem to reoccur or not be able to be resolved. Finally, the prognosis, taking into account all of the above, of how long will it take for the patient to get better? How much better will they get? And how long will it take? As you can see some of this information will come from knowledge, some from reflecting on the findings of the examination (reflection in action) and the progression of any treatments (reflection in and about action) and finally from collating the experiences of treating other patients and applying that information (reflection about action).

In summary, open mindedness, the questioning of existing beliefs and reflective thinking are essential for good clinical reasoning to work.

Example

Using the example in Figure 16.3 it can be seen that the potential sources of the symptom of posterior thigh pain are multiple. If we take the simplest option in terms of differential diagnosis such as a strain of hamstring muscles as the injury, the first thing we need to do is discount the other sources of symptoms. For example, in order to discount the lumbar spine, sacroiliac and hip joints, these joints must have full pain-free range of movement, the spinal and peripheral nerves must on tensile loading show no mechano-sensitivity, and the other muscles must have full range of movement and no pain on contraction. The hamstring muscle group shows pain on contraction and elongation (stretch) along with pain on palpation. The typical mechanism of injury for hamstring injury is one of a sudden onset of pain during an eccentric contraction to decelerate knee extension during swing phase whilst sprinting, if any other mechanism occurred then this must be clarified to make sure it fits with one which would result in hamstring muscle injury. There are a number of factors that can contribute to the occurrence of an hamstring muscle injury, one is previous history of hamstring injury, another is age; older athletes are more likely to have an hamstring strain. Further contributory factors are strength imbalances between the quadriceps and hamstrings, altered mechano-sensitivity of the sciatic nerve and degenerative change in the lumbar spine.

Assuming there are no contra-indications and precautions to treatment, the management and treatment can be planned. The treatment needs to be directed not only at the local cause of symptoms; the hamstrings in this case (though remote causes may have to be dealt with in other cases) but also at the predisposing factors to prevent any reoccurrence. So the strength of the tissue (its tolerance to load) will need to be gradually increased along with its tolerance to elongation, in doing this its ability to tolerate eccentric load will need special attention and be brought to a level in balance with quadriceps concentric strength. Simultaneously, mobility of the sciatic nerve and lumbar spine must be maintained to reduce the influence of these factors on any future injury.

To conceptualise the clinical decisions considered within the above example, it may be useful to use the 6D approach depicted in Figure 16.2. The example considers posterior thigh pain from a multiple of perspectives and encourages a fusion of thought and argument to best decide on effective management. Using the 6D approach, the clinician firstly needs to dissect knowledge on posterior thigh pain aetiology and consider the numerous sources of the pain. This component will draw on anatomical landscapes of the posterior thigh region. The next phase will drive the clinician to consider which information sets are most useful in digesting the issues confronting the cause of the pain. Once considered, a process of discerning information, in this case, consideration of the hamstring mechanism of action and functionality may be most important clinically, to develop an appropriate treatment and management plan for the pain. These decisions, drawn from a sequential analysis of knowledge and anatomical architecture, allow the clinician to cross a threshold into a different and often new way of thinking about the presenting issues and problems. This threshold crossing may be transformative and irreversible and the clinician may now be forced to live with the decisions made and actions taken. The later phase of the 6D approach, teases out how the clinician chooses to defend the actions based on the knowledge organisation from the presenting issue. In summary, the example presented above can be divided into two important clinical reasoning processes. The first is understanding the issue or problem and using prior learning or experiential knowledge to develop a plan of action and treatment. The second involves the cognitive processes of higher order reasoning to defend and justify clinically the decisions made and actions taken. The entire process is fuelled by an evaluation of the outcome.

The model in Figure 16.4, used extensively in action research and developed by Susman (1983), is useful in providing an alternatively analysis and conceptual map for the example on posterior thigh pain. The model depicts a progress process-driven approach in helping the clinician organise thinking into a practice and management plan for action.

This chapter has provided an overview and application of clinical reasoning skills through the

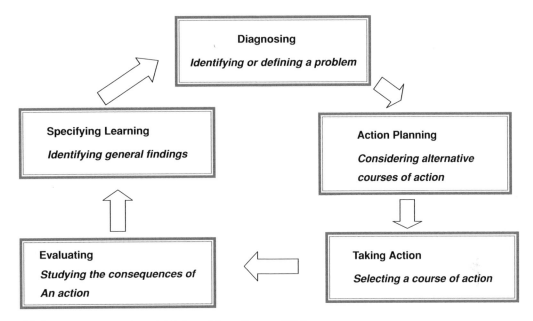

Figure 16.4 Action research model (adapted from Susman 1983).

use of applied examples to accentuate the process of clinical reasoning development. It is important that the tenets of clinical reasoning are practised and understood so that clinical competencies can further be developed and enhanced. Use this chapter to help you understand some of the decisions made in clinical practice that appear in the injury management chapters of this book.

References

Abrahamson, E.D. (2009) *6D Approach to knowledge development within a conceptual threshold framework*. Sports conference paper presentation, Middlesex University, London.

Albert, A.D., Munson, R. and Resnik, M.D. (eds) (1988) *Reasoning in Medicine: An introduction to clinical inference*. Baltimore, MD: John Hopkins University Press.

Barrows, H.S. and Pickell, G.C. (1991) *Developing Clinical Problem Solving Skills: A guide to more effective diagnosis and treatment*. New York: Norton and Comp.

Boshuizen, H.P.A. and Schmidt, H.G. (1992) On the role of biomedical knowledge in clinical reasoning by experts, intermediates and novices. *Cognitive Science*, *16*, 153–184.

Carr, S. (2004). A framework for understanding clinical reasoning in community nursing. *Journal of Clinical Nursing*, *13* (7), 850–857.

Custer, J.F.M., Boshuizen, H.P.A. and Schmidt, H.G. (1992) The relationship between medical expertise and the development of illness scripts. Paper presentation at the annual American educational research Association Conference, San Francisco, California, April 20–24.

Duncan, M., Lyons, M. and Al-Nakeeb, Y. (2007) 'You have to do it rather than being in a class and just listening.' The impact of problem-based learning on the student experience in sports and exercise biomechanics. *Journal of Hospitality, Leisure, Sport and Tourism Education*, *6* (1), 71–80.

Elstein, A.L., Shulman, L.S. and Sprafka, S.A. (1978) *Medical Problem Solving – An Analysis of Clinical Reasoning*. Cambridge: Harvard University Press.

Elstein, A.S., Shulman, L.S. and Sprafka, S.A. (1990) Medical problem solving: a ten year retrospective. *Evaluation and the Health Professions*, *13*, 5–36.

Ericsson, K.A. and Simon, H.A. (1984) *Protocol Analysis: Verbal reports as data*. Cambridge, MA: MIT Press.

Fleming, M.H. and Mattingly, C. (1994) *Action and Inquiry: Reasoned action and active reasoning. In clinical reasoning: forms of inquiry in a therapeutic practice*. Philadelphia, PA: F.A. Davis.

Gale, J. (1982) Some cognitive components of the diagnostic thinking process. *British Journal of Educational Psychology*, 52, 64–76.

Gorry, G.A. (1970) Modelling the diagnostic process. *Journal of Medical Education*, 45, 293–302.

Hamilton, M. (1966) *Clinicians and Decisions*. Leeds: Leeds University Press.

Hauer, K.E., Teherani, A., Kerr, K.M., O'Sullivan, P.S. and Irby, D.M. (2007) Student performance problems in medical school skills assessments. *Academic Medicine*, 82, 69–72.

Higgs, J. and Jones, M. (1995) *Clinical Reasoning in the Health Professions*. Oxford: Butterworth-Heinemann.

Higgs, J. and Jones, M. (2000) *Clinical Reasoning in the Health Professions*, 2nd edn. Oxford: Butterworth-Heinemann.

Higgs, J. and Titchen, A. (2001) *Practice Knowledge and Expertise*. Oxford: Butterworth Heinemann.

Kassirer, J.P. and Gorry, G.A. (1978) Clinical problem solving: a behavioural analysis. *Annals of Internal Medicine*, 89, 245–255.

Land, R., Meyers, J.H.F. and Smith, J. (2008) *Tresholds Concepts within the Disciplines*. Sense Publishers.

Leaver-Dunn, D., Harrelson, G.L., Martin, M. and Wyatt, T. (2002) Critical-thinking predisposition among undergraduate athletic training students. *Journal of Athletic Training*, 37 (4), 147–151.

Martin, L., West, J. and Bill, K. (2008) Incorporating problem based learning strategies to develop learner autonomy and employability skills in sports science undergraduates. *Journal of Hospitality, Leisure, Sport and Tourism Education*, 7 (1), 18–30.

Meyers, J.H.F. and Land, R. (2003) Threshold concepts and troublesome knowledge: linkages to ways of thinking and practising. In Rust, C. (Ed.), *Improving Student Learning – Theory and Practice Ten Years On*. Oxford: Oxford Centre for Staff and Learning Development (OCSLD), pp 412–424.

Meyers, J.H.F., Land, R. and Davies, P. (2008) Threshold concepts and troublesome knowledge (4): Issues of variation and variability. In Land, R., Meyer, J.H.F. and Smith, J. (eds), *Threshold Concepts within the Disciplines*. Sense Publishers, pp 59–74.

Newell, A. and Simon, H.A. (1972) *Human Problem Solving*. Englewood Cliffs, NJ: Prentice Hall.

Perkins, D. (1999) The many faces of constructivism. *Educational Leadership*, 57 (3), 6–11.

Richardson, J.T.E. (2005) Instruments for obtaining student feedback: a review of the literature. *Assessment and Evaluation in Higher Education*, 30 (4), 387–415.

Ridderrikhoff, J. (1989). *Methods in Medicine: A descriptive study of physicians behaviour*. Dordrecht: Kluwer Academic.

Säljö, R. (1979). *Learning in Learner's Perspective*. Gothenberg: University of Gothenberg.

Savery, J.R. (2006) Overview of problem based learning: Definitions and distinctions. *The Interdisciplinary Journal of Problem Based Learning*, 1 (1), 9–20.

Savin-Baden, M. (2003) *Facilitating Problem Based Learning*. Buckingham: SRHE/Open University Press.

Schon, D. (1991) *The Reflective Practitioner*, 2nd edn. New York: Basic Books.

Simmons, B., Lanuza, D., Fonteyn, M., Hicks, F. and Holm, K. (2003) Clinical reasoning in experienced nurses. *Western Journal of Nursing Research*, 25 (6), 701–719.

Susman, G.I. (1983) *Action Research a Sociotechnical Systems Perspective*. London: Sage, pp 95–113.

Wood, D.F. (2003) ABC of learning and teaching in medicine: problem based learning. *British Medical Journal*, 326, 328–330.

Part 5
Joint specific injuries and pathologies

17

Shoulder injuries in sport

Ian Horsley
English Institute of Sport

This chapter outlines the anatomy of the shoulder girdle and discusses commonly presenting pathology around this area. Common orthopaedic assessment tests are described, together with a presentation of the effectiveness of these tests in assessing for specific diagnoses of commonly presenting pathology, from currently available literature. The role of rehabilitation is covered with analysis of the function of commonly utilised exercise and the role of clinical reasoning in determining the diagnosis and formulating a safe and effective rehabilitation programme.

Incidence of shoulder injury

The glenohumeral joint is one of the most frequently injured areas of the upper extremity in competitive sports. Studies indicate that 8–20% of athletic injuries involve the glenohumeral joint (Hill 1983; Lo et al. 1990; Hutson 1996; Terry and Chopp 2000; Ranson and Gregory 2008).

Athletes whose sports require a large amount of time with their arms above the level of the shoulder, such as those playing racquet sports, sports involving throwing (baseball, cricket, American Football and water polo), swimmers and rugby players (due to their arm position within the tackle) commonly report a high incidence of shoulder pain with up to 43.8% reporting shoulder pain (Lo et al. 1990). Hutson (1996) reported that more than 40% of elite swimmers complained of shoulder pain at some point during their careers, and this was related to the fact that 90% of the propulsive force comes from the upper extremity (Counsilman 1977) with the main cause of pain being attributed to glenohumeral joint instability (Weldon and Richardson 2001), due to significantly increased humeral head translation (Tibone et al. 2002).

In American Football 15.2% of all injuries incurred by quarterbacks were shoulder injuries with direct trauma being responsible for 82.3% of the shoulder injuries (Kelly et al. 2004), and in professional cricket 23% of players in one study reported suffering a shoulder injury during one professional season (Ranson and Gregory 2008).

The epidemiology of Rugby Union and Rugby League injuries appears to suggest that injury to the shoulder accounts for approximately 12–16% of all injuries, with an incidence of 10–13 per 1000 game hours, with this statistic higher when compared to pre-professionalism incidence rates (Garraway and Macleod 1995; Bird et al. 1998; Gabbet 2000; Chalmers et al. 2001; Lee et al. 2001; Gissane et al. 2003; Junge et al. 2004; Handcock et al. 2005). With regards to Rugby Union, Bathgate et al. (2002) highlighted the upper limb as responsible for 15.4% of injuries, with 6.3% of overall injuries located at the shoulder.

Even within non-overhead sports, such as skiing, shoulder injuries have been reported as high as 11.4% of all injuries (Kocher 1996).

Table 17.1 Static stabilisers of the glenohumeral joint

Ligament	Description	Action
Superior glenohumeral ligament	Attaches from the supraglenoid tubercle of the glenoid labrum onto the proximal tip of the lesser tuberosity of the humerus	Resists inferior humeral translation with the arm adducted and in neutral rotation. Limits external rotation in conjunction with the coracohumeral ligament
Middle glenohumeral ligament	Attaches from the supraglenoid tubercle and anterior aspect of glenoid labrum onto the lesser tuberosity of the humerus, blending with the subscapularis tendon	Provides anterior humeral stability from humeral adduction to approximately 45 degrees abduction
Inferior glenohumeral ligament complex	Anterior band: from anterior labrum to the glenoid rim Middle band: is an axillary pouch Posterior band: form the posterior labrum to the glenoid rim. Not found in all patients	From 0 to 30 degrees humeral abduction the anterior band is the primary static stabiliser of the glenohumeral joint. It tightens with abduction and moves superiorly with combined external rotation to become the primary anterior humeral stabiliser in this position The primary static stabiliser with the arm in flexion and medial rotation, providing posterior stability. It tightens with abduction and moves superiorly with combined internal rotation
Coracohumeral ligament	Lateral aspect of the coracoid process of the scapula onto the upper facet of the greater tuberosity of the humerus, blending with the supraspinatus tendon	Resists posterior and inferior translation of the suspended shoulder, it is an inferior stabiliser and tightens with external rotation
Glenoid labrum	A fibrocartilaginous rim attached around the margin of the glenoid cavity attached to the circumference of the glenoid, while the free edge is thin and sharp. It is continuous with the tendon of the long head of biceps	It deepens the articular cavity, and protects the edges of the bone

Repetitive overhead stress within the overhead athlete challenges the functional, dynamic integrity of the glenohumeral joint within these athletes. As there is little bony contact between the head of the humerus and the glenoid fossa of the scapula, there is a great range of mobility at the joint with an inherent instability of the articulation (Armfield et al. 2003). Joint homeostasis is maintained by the harmonious static and dynamic interaction of the muscles, ligaments and joint capsule. The static stabilisers (Table 17.1) of the joint consist of the labrum, capsule and ligaments, and the dynamic stabilisers of the joint (Table 17.2) are the muscles of the rotator cuff, deltoid and scapular stabilisers (Terry and Chopp 2000; Woodward and Best 2000). Lack of ability to maintain the humeral head centred within the glenoid fossa during movement is defined as instability (Magarey and Jones 1992).

Hess (2000) adapted Panjabi's model proposed for spinal segmental stability (Panjabi 1992) for the glenohumeral joint, which states that joint stability is based on the interaction between the active, passive and neural control subsystems, with the rotator cuff muscles, activating at different positions, compressing the convex humeral head into the concave glenoid, thus resisting the shear force experienced by the humeral head (Lee et al. 2000). Receptors within the joint capsule contribute to a reflex arc, which will cause activation of the muscles which overlie the joint capsule (Guanche et al. 1995).

Table 17.2 Muscles of the shoulder girdle. Adapted from Horsley (2005) Assessment of shoulders with pain of a non-traumatic origin. Physical Therapy in Sport. 6:6–14 © Elsevier.

Muscle	Origin	Insertion	Action
Deltoid	Lateral one-third of clavicle, acromion and spine of scapula	Deltoid tuberosity of the humerus	Abducts the shoulder joint posterior fibres extend and laterally rotate humerus. Anterior fibres flex and medially rotate the humerus
Supraspinatus	Supraspinous fossa of the scapula	Upper facet of the greater tuberocity of the humerus	Abducts the humerus; stabilizes head of humerus in glenoid cavity. Medially rotates the humerus, draws it forward and down when arm is raised
Infraspinatus	Infraspinous fossa of the scapula	Middle facet of the greater tuberocity of the humerus	Laterally rotates, adducts, extends the humerus. Stabilises the head of humerus in glenoid cavity.
Teres minor	Superior half of the lateral border of the scapula	Lower facet of the greater tuberocity of humerus	Laterally rotates, adducts, extends the humerus, stabilises the head of humerus in the glenoid cavity
Subscapularis	Subscapular fossa of the scapula (anterior surface of scapula)	Lesser tubercle of the humerus	Medially rotates humerus, stabilises the head of the humerus in the glenoid cavity
Teres major	Inferior angle of the scapula	Medial lip of bicipital grove of the humerus. Inserts with Latissimus dosi	Adducts and medially rotates the humerus and draws it back
Serratus anterior	Outer surface of ribs 1–8	Anterio-medial border of the scapula	Abducts and upwardly rotates the scapula, holds the scapula against the thoracic wall
Pectoralis major	From the anterior surface of the sternal half of the clavicle; the anterior surface of the sternum; from the cartilages of the first seven ribs	The fibres converge to a flat tendon, about 5cm broad, which is inserted into the crest of the greater tubercle of the humerus	Clavicular head: flexes and adducts arm. Sternal head: adducts and medially rotates arm. Acts as an accessory muscle for inspiration
Pectoralis minor	From the upper margins and outer surfaces of the third, fourth, and fifth ribs, near their cartilage and from the aponeuroses covering the intercostalis	Converges to form a flat tendon, which is inserted into the medial border and upper surface of the coracoid process of the scapula	Depresses, abducts, downwardly rotates (inferior angle of scapula moves towards the spine), and anteriorly tilts the scapula. It also acts as an accessory muscle with inspiration
Trapezius	From the external occipital protuberance and the medial third of the superior nuchal line of the skull, from the ligamentum nuchæ, the spinous process of the seventh cervical, and the spinous processes of all the thoracic vertebræ and their supraspinal ligament	The superior fibres are inserted into the posterior border of the lateral third of the clavicle; the middle fibres into the medial margin of the acromion, and into the superior lip of the posterior border of the spine of the scapula; the inferior fibres are inserted into a tubercle at the medial end of the spine of the scapula	The whole Trapezius retracts the scapula and braces back the shoulder; if the head is fixed, the upper part of the muscle will elevate the point of the shoulder, when the lower fibres contract they assist in depressing the scapula. The middle and lower fibres of the muscle rotate the scapula, causing elevation of the acromion. If the shoulders are fixed, the Trapezii, acting together, will extend the cervical spine; or if only one side acts, the head is rotated to the same side

(Continued)

Table 17.2 (*Continued*)

Muscle	Origin	Insertion	Action
Latissimus dorsi	From the spinous processes of the lower six thoracic vertebræ and from the posterior layer of the lumbodorsal fascia and the posterior part of the crest of the ilium and from the three or four lower ribs	he tendon, passes in front of the tendon of the teres major, and is inserted into the bottom of the intertubercular groove of the humerus	Extends and medially rotates the humerus. If the humerus is fixed it can elevate the rib cage and assist in respiration, or can elevate the trunk as in a pull up
Rhomboideus major	From the spinous processes of the second, third, fourth and fifth thoracic vertebræ and the supraspinal ligament	The lower part of the root of the spine of the scapula; below to the inferior angle	The rhomboids move the inferior angle backward and upward producing downward rotation of the scapula and assist with retracting the scapula
Rhomboideus minor	The lower part of the ligamentum nuchæ on the skull and from the spinous processes of the seventh cervical and first thoracic vertebræ	The lower part of the root of the spine of the scapula; below to the inferior angle adjacent to rhomboideus major	The rhomboids move the inferior angle backward and upward producing downward rotation of the scapula and assist with retracting the scapula
Levator scapulae	From the transverse processes of the first and second cervical vertebrae and from the transverse processes of the third and fourth cervical vertebræ	The vertebral border of the scapula, at the medial angle and the root of the spine of the scapula	It raises the medial angle of the scapula if the head is fixed, if the shoulder is fixed, the muscle side flexes the neck to that side and rotates it in the same direction
Coracobracialis	Corocoid process of the scapula	Middle of the medial shaft of the humerus	Flexes and adducts the humerus
Biceps brachii	Short head - coracoid process of scapula Long head - supraglenoid tubercle of scapula and labrum	Tuberosity of the radius and aponeurosis of biceps brachii	Flexes elbow, supinates forearm, flexes shoulder joint
Triceps brachii	Long head - infraglenoid tubercle of the scapula Lateral head - posterior surface of proximal half of humerus Medial head - posterior surface of distal half of humerus	All heads - olecranon process of ulna	Long head - extends and adducts the shoulder All heads - extend the forearm (elbow)

Overhead athletes suffer repeated microtrauma resulting from repetitive use of the limb at extreme ranges of motions without increasing force. Instability can result from muscle imbalance, contracture, and ligamentous and capsular laxity (Cofield et al. 1993). Range of motion deficits will contribute to injury as this will produce a situation whereby some muscles become tight and some muscles become lax (Baltaci and Johnson 2001). Patients with chronic shoulder pain or instability are sometimes difficult to diagnose and treat. A thorough history and systematic clinical examination followed by a systematic approach to the use of investigating tools such as diagnostic ultrasound or MRI is essential for a successful outcome (Rolf 2008).

Assessment of injury risk

The assessment of posture within the domain of injury rehabilitation has traditionally been performed via visual observation of specific joints/bony landmarks, and the corresponding position they have to one another. Good posture has been described as a state of muscular and skeletal balance that protects the supporting structures of the body against injury or progressive deformity, irrespective of the attitudes in which the structures are resting or working (Kendall et al. 1993). Ideal alignment standards used in clinical practice have previously been highlighted (Kendall et al. 1993; Sahrmann 2002). The widely accepted description of normal standing posture is that proposed by Kendall and McCreary (1983) as a vertical line passing through the lobe of the ear, the seventh cervical vertebra, acromion process, greater trochanter and slightly anterior to the midlines of the knee and lateral malleolus. Deviations outside this theoretical plumb-line have been described as abnormal, and have been linked to numerous problems. Posture deviations frequently found in the cervical and thoracic spine have been suggested to affect the normal function of the glenohumeral joint (Ayub 1991; Kendall et al. 1993; Einhorn et al. 1997; Janda 2002; Sahrmann 2002; Lewis et al. 2005a).

Standing postures associated with a forward head are seen in association with combinations of increased lordosis in the cervical and lumbar regions, an increased kyphosis in the thoracic region, protracted shoulders (with elevation or depression) and abnormal scapula position (Ayub 1991; Greenfield et al. 1995; Grimsby and Gray 1997; McDonnell and Sahrmann 2002; Sahrmann 2002; McDonnell et al. 2005) (Figure 17.1), although not all studies have found this (Raine and Twomey 1997; Hanten et al. 2000). Several authors have suggested that muscle imbalances and shortening can occur in the sternocleidomastoid, upper trapezius and levator scapula with a forward head position. This will lead to elevated and abducted scapula, and increased thoracic kyphosis, increasing the risk of impingement (Ayub 1991; Grimsby and Gray 1997). Subjects with increased thoracic kyphosis have been shown to predispose altered scapular kinematics; when

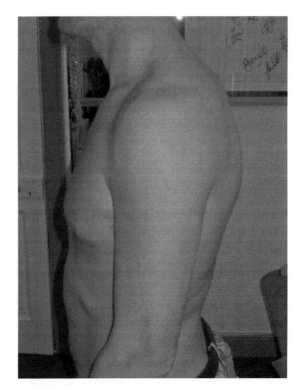

Figure 17.1 Posture.

asymptomatic subjects were positioned in a slouched posture when sitting and instructed to elevate their arm, there was a significant reduction in posterior tilt and upward rotation of the scapula, as well as an increase in the amount of scapular elevation and internal rotation (Kebaetse et al. 1999). When subjects who were experiencing sub acromial impingement improved their posture, it was not found to have a significant effect on the intensity of the pain, but increased the range of shoulder elevation before the pain was experienced (Lewis et al. 2005a). Thus thoracic posture needs to be optimised in patients with impingement-like symptoms, during all daily activities, and exercises directed at improving thoracic extension should be considered. Interventions to consider are, amongst others, thoracic spine joint mobilisation (Bang and Deyle 2000), corrective taping of the scapular and thoracic spine (Lewis et al. 2005b), facilitation scapulothoracic musculature (Konrad et al. 2006), and facilitate the activity of the rotator cuff (Magarey and Jones 2003).

Shoulder girdle, scapular and glenohumeral joint position

The role of the scapula is extremely important in providing a stable base from which the glenohumeral joint functions, as well as determining the overall position of the shoulder girdle (Kibler 1991; Paine and Voight 1993; Kibler 1998; Sahrmann 2002; Magarey and Jones 2003). The efficiency of muscular activity is dependent on the position of the scapula and the length-tension relationships of the scapular stabilisers and rotator cuff muscles, which originate on the scapula, cervical and/or thoracic spine (Einhorn et al. 1997; Mottram 1997; Magarey and Jones 2003). The scapula stabilisers, such as trapezius and serratus anterior, can be adversely affected by common abnormal postures, such as increased thoracic kyphosis and forward head positions (Greenfield et al. 1995; Ludewig and Cook 2000; Borstad and Ludewig 2005; Lewis et al. 2005a;). Certain muscle imbalances, particularly shortening, can occur in the sternocleidomastoid, upper trapezius and levator scapula with a forward head position, leading to increased thoracic kyphosis, and elevated or depressed, abducted scapula (Ayub 1991; Grimsby and Gray 1997). This increased thoracic kyphosis causes the scapular to become abducted due to lengthening of the rhomboid and lower trapezius muscles, whilst shortening the serratus anterior, latissimus dorsi, subscapularis, teres major and pectoralis major and minor muscles, and pulling the humerus into an anterior and/or internally rotated position, and further anteriorly tilting the scapula (Ayub 1991; Borstad and Ludewig 2005). This posture alters the scapulohumeral rhythm and perpetuates various forms of impingements, either in the subacromial space or inter-articular, during arm elevation, as the ability of the scapula to tilt posteriorly is inhibited by overactive pectoralis minor (Lewis et al. 2005a).

Functional examination

- Active movements: Active tests do not enable us to differentiate between inert and contractile structures. Active tests inform us about the patient's willingness to move.

- Passive movements: Test the integrity of the inert structures. Look for pain, range of movement and end-feel.

- Resisted tests (maximal isometric contractions from a neutral, generally mid range, position): Examine the contractile structures, assess pain and muscle strength.

Palpation

Abnormal findings:

- at rest: warmth, fluid, synovial thickening

- on movement: crepitus, end-feel.

End-feel

Normal/physiological:

- hard: e.g. elbow extension, knee extension

- capsular (elastic): e.g. rotations at shoulder, elbow, hip

- extra-articular (tissue approximation): flexion at elbow, hip.

Pathological:

- too hard: e.g. osteoarthrosis

- too soft: e.g. loose body in the elbow joint

- muscle spasm (involuntary muscle contraction): e.g. arthritis

- empty (voluntary muscle contraction, not always the same range): e.g. abscess

- springy block: e.g. meniscus subluxation.

There are many special tests for evaluation of the pathologies arising around the glenohumeral joint, and there have been numerous articles evaluating the sensitivity and specificity, as well the positive and negative likelihood ratios (Dinnes et al. 2003; Hegedus et al. 2008; Munro and Healy 2008). Sensitivity is the ability to identify *everyone* with a specific condition. Specificity is the proportion of patients without a specific condition who have a negative test. A positive likelihood ratio describes the impact that a positive test has on raising the suspicion that a condition actually exists. High values infer that the condition which is being tested for really exists. Conversely, a low negative likelihood ratio infers that the condition for which is being tested is likely not to exist.

Several authors (Razmjou et al. 2004; Boettcher et al. 2008; Hegedus et al. 2008; Munro and Healy 2009) have analysed the pooled results of studies and have come to the same conclusion; the commonly utilised diagnostic tests for shoulder pathology have a low diagnostic utility.

Below is a description of some of the more common tests for various pathologies arising around the shoulder. Since there are several tests described for the various pathologies, it is indicative that there is no superior test for any single pathology.

Anterior instability

Anterior load and shift test (Hawkins et al. 1996)

The humeral head is grasped with the one hand, while the other hand stabilises the scapula. The humeral head is loaded medially into the joint and then an anterior and posterior shearing force is applied. The direction and translation can be graded using Altchek and Dines classification (1993), a scale of 0 to 3.

Anterior drawer test (Gerber and Ganz 1984)

The patient is placed supine and the arm abducted over the edge of plinth. The examiner stabilises the scapula with one arm whilst the other grasps the humeral head and translates it in an anteromedial direction on the glenoid. Unilateral increases in humeral head translation of the symptomatic shoulder indicate anterior glenohumeral joint instability.

Apprehension test (Jobe et al. 1989)

This is performed with the humerus in 90 degrees of abduction, 90 degrees of elbow flexion and external rotation of the shoulder. The examiner exerts gentle pressure into progressive external rotation (Figure 17.2). A positive test is when the patient feels a sensation of impending dislocation.

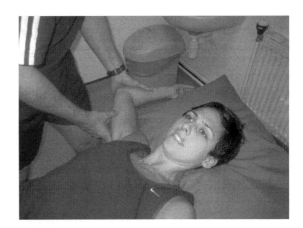

Figure 17.2 Apprehension test.

Relocation test

With the patient supine the arm is taken into abduction and external rotation. The test can be augmented by pushing the humeral head anteriorly from behind. The relocation test is performed by pushing posteriorly on the upper part of the humerus (Figure 17.3). The relocation test is positive if the apprehension or pain is relieved.

Posterior instability

Posterior load and shift – posterior drawer test (Gerber and Ganz 1984)

This test is similar to the anterior draw test, and the humeral head is translated in a posterolateral

Figure 17.3 Relocation test.

direction. A positive result is a unilateral increase in humeral head posterior translation on the glenoid.

Posterior apprehension test

This is a modification of the posterior draw test described by Gerber and Gantz (1984) where the is arm adducted and flexed to 90 degrees, whilst the examiner imparts an axial posterolaterally directed force to the humerus. A positive result is that of pain, apprehension and often the feeling of a click as the humerus rides over the posterior rim of the glenoid.

Inferior laxity

The sulcus sign (Neer and Foster 1980)

This is an examination to determine the extent and/or presence of inferior instability of the glenohumeral joint. This test can be administered with the patient either seated or standing with their arm relaxed at their side. The examiner palpates the shoulder by placing thumb and fingers on the anterior and posterior aspects of the humeral head. The examiner grasps the patient's elbow with their other hand and applies a downward distraction force. A positive test will result in a sulcus being formed between the acromion and the humeral head as the humeral head moves inferiorly while the force is being applied (Figure 17.4).

Figure 17.4 Sulcus test.

Figure 17.5 O'Brien's Test.

SLAP lesions

O'Brien test (O'Brien et al. 1998)

The patient's shoulder is held in 90 degrees of forward flexion, 30–45 degrees of horizontal adduction and maximal internal rotation. The examiner exerts a downward force distal to the patient's elbow which the patient tries to resist. The patient is asked to identify, if produced, the location of the pain. The test is repeated in the same position except that this time the humerus is externally rotated and the forearm supinated, so the palm faces up. Once again, a downward force is applied by the examiner, which the patient actively resists, and the patient is asked to identify the location of any pain provoked. The test is considered positive if pain produced during the first part of the test is abolished with the second part of the test (Figure 17.5). For indication of a SLAP tear the pain is located over the anterior aspect of the shoulder, and for AC joint pathology, the pain must be located over the AC joint.

Anterior slide (Kibler 1995b)

The patient stands with hands on hips. One of the examiner's hands is placed over the shoulder and the other hand behind the elbow. A force is then applied anteriorly and superiorly, and the patient is asked to push back against the force. The test is positive if pain is localised to the front of the shoulder or a click is experienced by the patient.

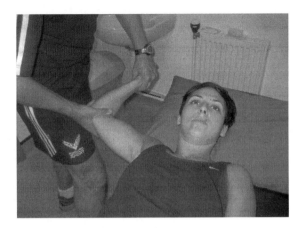

Figure 17.6 Biceps Load 1.

Posterior slide test
Biceps load test I (Kim et al. 1999)

The test is performed with the patient in the supine position. The examiner sits adjacent to the patient on the same side as the affected shoulder and gently grasps the patient's wrist and elbow. The arm to be examined is abducted at 90 degrees, with the forearm in the supinated position (Figure 17.6).

The patient is allowed to relax, and an anterior apprehension test is performed. When the patient becomes apprehensive during the external rotation of the shoulder, external rotation is stopped. The patient is then asked to flex the elbow while the examiner resists the flexion with one hand. If the apprehension is lessened, or if the patient feels more comfortable than before the test, the test is negative for a SLAP lesion. If the apprehension has not changed, or if the shoulder becomes more painful, the test is positive.

Biceps load test II (Kim et al. 2001)

The patient is tested in supine. The arm is abducted to 120 degrees, externally rotated maximally, elbow in 90 degrees flexion and forearm supinated. If this test position reproduces pain then perform active elbow flexion against resistance. If the active elbow flexion component of the test increases pain (or produces pain) the test is positive.

Crank test (Liu et al. 1996)

With the patient upright, or supine, and the arm elevated to 160 degrees in the plane of the scapula, joint load is applied along the axis of the humerus with one hand, whilst the other hand performs humeral rotation. A positive test is reproduction of the patient's overhead symptoms (with or without a click).

Pain provocation test (Mimori et al. 1999)

The patient is seated with the arm is in 90 degrees abduction and 90 degrees external rotation, and the elbow flexed to 90 degrees. The examiner places one hand over the scapula, whilst the other hand holds the patient's wrist. The patient is then asked to supinate and pronate the forearm. If the pain is worse on pronation, this is indicative of a SLAP tear.

The resisted supination external rotation test (Myers et al. 2005)

The patient is placed in the supine position on the examination bed with the scapula near the edge of the bed. The examiner stands at the patient's side, supporting the affected arm at the elbow and hand, with the shoulder abducted to 90 degrees, the elbow flexed 65–70 degrees, and the forearm in neutral or slight pronation. The patient then attempts to supinate the hand with maximal effort against the examiner's resistance. The patient forcefully supinates the hand against resistance as the shoulder is gently externally rotated to the end of range. They are then asked to describe the symptoms at maximum external rotation. The test is positive if the patient experiences anterior or deep shoulder pain, clicking or catching in the shoulder, or reproduction of symptoms that occurred during throwing. The test is negative if the patient described posterior shoulder pain, apprehension, or no pain.

Long head of the biceps
Yergason's test (Yergason 1931)

The patient is seated or standing with the elbow flexed to 90 degrees and forearm pronated. The examiner resistes active supination and elbow flexion whilst feeling for subluxation of the biceps tendon out of the bicipital groove (Figure 17.7). A positive

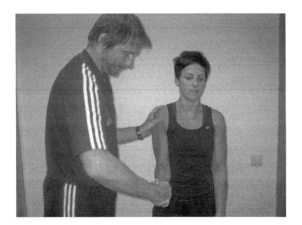

Figure 17.7 Yergason's Test.

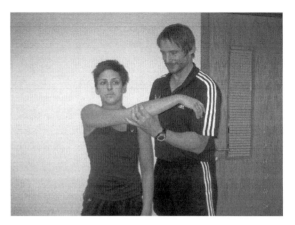

Figure 17.9 Scarf Test.

test is detection of movement of the tendon out of the groove.

Speed's Test (Bennett 1998)

The patient's supinated arm is held at 90 degrees elbow flexion and then flexed forwards against resistance (Figure 17.8). Pain felt in the bicipital groove indicates biceps tendon pathology.

AC joint

Anterior/posterior AC shear test (Davies et al. 1981)

With the patient sitting, the examiner cups the heels of both hands, one over the midpoint of the clavicle, anteriorly, and one over the spine of the scapula, posteriorly. With a compressive action both hands are squeezed towards each other. Several repetitions are applied with note being taken of the amount of movement compared with the opposite shoulder. Pain is also considered. A positive test is when the patient complains of superiorly located pain unilaterally.

Cross chest adduction (Scarf/Forced adduction test) (Silliman and Hawkins 1994)

The symptomatic shoulder is flexed to 90 degrees and then forcibly adducted across the chest (Figure 17.9).

Subacromial impingement

Neer impingement test (Neer and Welsh 1977)

In this test, there is forced elevation of the humerus in the scapula plane whilst the shoulder is internally rotated with the other hand on the top of the shoulder girdle to stabilise. A positive test gives rise to pain with passive abduction, which indicates impingement within the subacromial space (Figure 17.10).

Neer impingement injection test (Neer 1983)

The subacromial space is infiltrated with 8–10 mls of local anaesthetic, and the above test is repeated. If there is greater than a 50% reduction in the pain, then this indicates that the probable cause of the pain is the bursa or a rotator cuff tendon.

Figure 17.8 Speeds Test.

Figure 17.10 Neer Test.

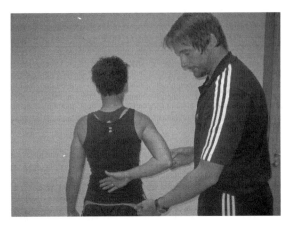

Figure 17.12 Positive IRLS.

Hawkin's–Kennedy test (Hawkins and Kennedy 1980)

The shoulder is placed in 90 degrees of forward flexion and then passive internal rotation of the humerus is applied by the examiner (Figure 17.11). A positive test is provocation of pain around the subacromial space. This test indicates internal impingement of the shoulder as the rotator cuff tendons are compressed by the coracoacromial arch.

Empty can test (Jobe and Moynes 1982)

Standing in front of the patient in order to monitor facial expression during the test, the patient elevates their arm in the scapular plane to 90 degrees with the arm in full internal rotation, so that the thumb is pointing downwards. The examiner then exerts a downward force and asks the patient to resist (Figure 17.12). A positive test produces pain, weakness, or both, and indicates involvement of the supraspinatus tendon.

Full can test

Carried out as the above test except that the thumbs are pointed upwards (Figure 17.13). The test has been shown to isolate the supraspinatus as well as the empty can test (Itoi et al. 1999).

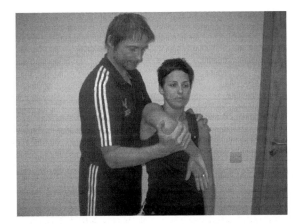

Figure 17.11 Hawkins Test.

Figure 17.13 Full Can.

Rotator cuff tear

Supraspinatus: *Drop arm test (Hoppenfield and Hutton 1976)*

The patient actively abducts the arm in the coronal plane with the thumb pointing forward. From the end of abduction, the patient is instructed to slowly, under control, lower the arm. If there is a lesion within the tendon of Supraspinatus, the patient will be unable to control the descent of the arm into adduction from approximately 90 degrees abduction. If the patient can hold the arm at 90 degrees abduction, then the examiner can lightly apply pressure in a downward direction to the hand, which – if a Supraspinatus lesion is present – will cause the arm to fall into adduction.

Figure 17.15 Empty Can.

Infraspinatus: *External rotation lag sign (Hertel et al. 1996)*

The examiner stands behind the patient with the elbow flexed to 90 degrees, and elevated to approximately 20 degrees in the plane of the scapula. The examiner passively externally rotates the shoulder, by holding around the wrist, to the onset of capsular tightening, whilst supporting the weight of the arm by placing a hand under the elbow, and asks the patient to actively maintain this position when the examiner lets go of the wrist, but maintaining support at the elbow. A positive test is recorded if the arm falls back into internal rotation, and the magnitude is recorded to the nearest 5 degrees (Figure 17.14).

Subscapularis: *Internal rotation lag sign test (Hertel et al. 1996)*

The patient is asked to position his hand behind his back so that the dorsum of the hand is on the lumbar region. The examiner passively lifts the hand away from the lumbar region, whilst maintaining glenohumeral internal rotation. The patient is then asked to voluntarily maintain this position with only elbow support from the examiner. A positive result is if the hand falls back towards the spine, indicating a lesion of the subscapularis (Figure 17.15). The magnitude of the fall back can be recorded to the nearest 5 degrees.

Gerber's lift off test (Gerber and Krushell 1991)

The dorsum of the patient's hand is positioned at the level of the midlumbar spine. The subject is then asked to lift the dorsum of the hand off the back as far as possible, by internally rotating the shoulder (Figure 17.16). The test is considered positive for subscapularis dysfunction if the subject cannot lift the hand off of the back or if the subject performed the lifting manoeuver with elbow or shoulder extension. The test can be repeated whereby the patient is asked to try and push the examiner's hand away from "hand behind back position". A positive test is inability with or without pain.

Figure 17.14 Positive ERLS.

Figure 17.16 Gerber's Lift off test.

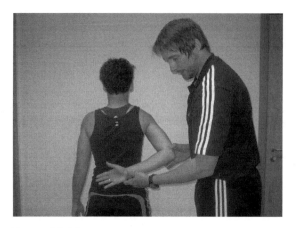

Figure 17.18 Internal rotation lag sign.

The external rotation lag sign

The patient is seated. The elbow is passively flexed to 90 degrees and the shoulder is held at 20 degrees elevation in the scapular plane in a position of near maximum external rotation (i.e. maximum external rotation minus five degrees to avoid elastic recoil). The examiner supports the elbow and holds the arm in external rotation at the wrist. The patient is asked to hold the position while the examiner supports the elbow but releases the hold at the wrist (Figure 17.17). The degree of movement is estimated and is referred to as the "lag" (i.e. the difference between active and passive ROM).

The internal rotation lag sign (Hertel et al. 1996)

The patient is seated. The patient is asked to bring the arm behind the back with the palm facing outward. The arm is held in near maximum internal rotation and with the hand away from the back by approximately 20 degrees of extension. The patient is asked to hold the position while the examiner supports the elbow but releases the wrist hold (Figure 17.18). If the patient is unable to hold the position, the lag sign is positive.

Table 17.3 gives the sensitivities, specificities and likelihood rations of special tests.

Hanchard et al. (2004) formulated Table 17.4 as a method of correlating the, often, confusing results gained from applying a battery of clinical orthopaedic tests in order to identify possible pathologies implicated.

Table 17.5 shows the intricacies of the body and the inter-relation between body parts local to the shoulder girdle. This table can be extrapolated to assess the role of the pelvic girdle position in posture, and how leg position can affect the pelvic girdle posture.

Although postural alterations have been shown to have some detrimental effects on shoulder girdle function; observed postural deviations should be taken in context with the "normal" posture of the patient. One way of assessing whether local postural alterations are responsible is to assess the patient in sitting, having placed them in an optimal posture, and see if positive results from tests are altered; for example, correction of forward head posture, reduction of

Figure 17.17 External rotation lag sign.

Table 17.3 Sensitivities, specificities and likelihood ratios of special tests

Test	Diagnosis	Sensitivity	Specificity	+ve LR	−ve LR
Neer's sign[1,2]	Sub acromial impingement	0.75–0.88	0.51	-	-
Hawkin's test[1,2]	External impingement	0.92	0.25–0.44	-	-
Drop arm test[12]	Rotator cuff tear	0.35	0.72	0.06	0.96
Cross arm test[12]	ACJ/Bursitis	0.77	0.79	0.20	0.98
Apprehension Test[3]	Anterior GH instability	0.68	1.00	-	-
Relocation Test[3,4]	Anterior GH instability	0.57	1.00	-	-
Sulcus sign[5]	Inferior GH instability	0.31	0.89	2.8	0.78
Yergason test[11]	Biceps tendon instability/tendinosis	0.12	0.86	-	-
Speed's test[11]	Biceps tendon instability/tendinosis	0.90	0.14	1.1	0.72
Clunk sign[5]	Labral tear	0.35	0.98	16	0.67
Anterior draw[5]	Anterior GH instability	0.54	0.78	2.5	0.59
Posterior draw[5]	Posterior GH instability	0.00	1.00	1.7	0.99
Compression-rotation test[5]	SLAP lesion	0.24	0.76	1.00	1.00
Anterior slide test[5]	Superior labral lesion	0.78	0.92	8.3	0.24
O'Brien's test[8]	Labral lesion	0.54	0.31	0.8	1.5
Crank test[9]	Labral lesion	0.91	0.93	13	0.10
Gerber's lift-off test[2]	Subscapularis lesion	0.62	100	>25	0.38
External rotation lag sign[10]	Supraspinatus/infraspinatus tendon tear	0.70	1.00	34.8	0.3
Internal rotation lag sign[10]	Subscapularis tendon tear	0.97	0.96	23.2	0.0

[1] Calis et al. 2000
[2] Macdonald et al. 2000
[3] Lo et al. 2004
[4] Speer et al. 1994
[5] Luime et al. 2004
[6] Calis et al. 2000
[7] Bennett 1998
[8] McFarland et al. 2002
[9] Liu et al. 1996
[10] Hertel et al 1996
[11] Holtby and Razmjou 2004
[12] Chronopoulous et al. 2004

thoracic kyphosis, optimal positioning of the scapulae on the chest wall, optimal positioning of the lumbar lordosis. The change from assessment in standing to assessment in sitting may effect a change in symptoms on testing. Certainly with sportsmen and women, assessment tends to involve breaking down the symptomatic sports-specific movement, and assessing the individual links within the chain. But this is beyond the scope of this chapter.

Rehabilitation

The rehabilitation strategies utilised will depend on the diagnosis made from a thorough clinical evaluation. The Table 17.5 above assesses the whole functional chain and its possible contribution to shoulder pathology. The days of diagnosing "rotator cuff tendinitis" are long gone, as this is an identification of the site of the pathology, but it does nothing to address the cause. Certainly if the cause of the pathology is not identified and rectified, then the outcome (injury) will return or not resolve completely.

Recent research has highlighted that common shoulder pathologies have a commonly presenting feature; loss of translational control (Lukasiewicz et al. 1999; Ludewig and Cook 2000; Magarey and Jones 2003; Ogston and Ludewig 2007). In addition to this there is an abundance of clinical research which has identified alterations in the dynamic and static positioning of the scapula within a cohort of

Table 17.4 Differential diagnosis: A summary of shoulder symptoms (adapted from Hanchard et al., 2004)

Sign		SIS	Instability	PSGI	SLAP	Capsulitis	ACJ Arthritis
Painful active	Arc	Possible	Possible				Possible
	Elevation	Possible	Possible	+	+	+	Possible
	Medial rotation	Possible	Possible			+	Possible
	Lateral rotation					+	Possible
	Horizontal adduction	Possible				N/a	+
Limitation of active	Elevation	Possible: with RCT may not achieve full ROM	Possible			+	
	Lateral rotation					+	
	Medial rotation	Possible	Possible			+	
Limitation of passive	Elevation	Possible	Possible			+	
	Lateral rotation					+	
	Medial rotation	Possible	Possible			+	
	Horizontal adduction	Possible	Possible			N/a	
Positive	Neer Test	+	Possible		Possible	N/a	N/a
	Hawkins-Kennedy test	+	Possible		Possible	N/a	N/a
	Load and shift test		Possible		Possible	N/a	
	Apprehension test		Possible		Possible	N/a	N/a
	Relocation test		+		Possible	N/a	N/a
	Sulcus sign		Possible		Possible	N/a	N/a
	Internal rotation resistance strength test		+			N/a	N/a

SIS = subacromial impingement syndrome
PSGI = posterior inferior glenoid impingement
SLAP = superior labrum anterior posterior
ACJ = acromio clavicular joint
RCT = rotator cuff tear
ROM = range of movement

individuals with shoulder pathology (Kibler 1998; Ludewig and Cook 2000; Moraes 2008).

These factors need to be identified and address along with the restoration of neuromuscular control. The rehabilitation will require that the individual's motor skills are trained back to pre-injury levels. Dynamic stability of the glenohumeral joint is aided by the sensorimotor system, due to the presence of mechanoreceptors within the joint which influence the patterns of muscle recruitment, reflex activity and joint stiffness. Without correct sensorimotor control there will be increased translation between the humeral head and glenoid, resulting in plastic deformation and laxity of the joint capsule, decreased rotator cuff facilitation and alterations in muscle sequencing and timing (Ogston and Ludewig 2007).

Ultimately, the management of the injured shoulder complex is a challenge that can be made easier if based on a thorough and exact clinical examination of the whole patient. Any approach to management of the shoulder will be optimally effective in the presence of good clinical reasoning, a sound knowledge of the clinical patterns associated with shoulder

Table 17.5 Postural deviations and possible musculoskeletal causes in relation to shoulder dysfunction (adapted from Horsley 2005). Adapted from Hanchard, N., Cummings, J., Jeffries, C. (2004) Evidended-based Clinical Guidelines for the Diagnosis, Assessment and Physiotherapy Management of Shoulder Impingement Syndrome. Chartered Society of Physiotherapy, London, UK. Page 33.

Static	Normal	Faults	Possible Cause
Clavicular resting position Scapular resting position	15 deg elevation distal end 3–5 deg lateral rotation inferior angle	Elevation	(i) over active levator scaplulae (ii) over active rhomboids (iii) over active upper trapezius (iv) neural sensitivity
		Winging	(i) tight pectoralis minor (ii) tight calvi-pectoral fascia (iii) weak/inhibited serratus anterior (iv) injury to lung thoracic nerve
		Depressed	(i) weak upper trapezius (ii) lengthened upper trapezius (iii) weak seratus anterior (iv) increased gleno-humeral joint laxity
		Protraction	(i) tight pectoralis minor (ii) tight clavi pectoral fascia (iii) tight serratus anterior (iv) tight latissismus dorsi (v) tight posterior cuff (vi) weak scaplular retractors (vii) increased thoracic kyphosis (viii) increased lumbar lordosis
		Abduction	(i) tight pectoralis major (ii) tight serratus anterior (iii) weak scaplular retractors (iv) increased thoracic kyphosis
	Normal medial border of scapula 7cm from spine	Adduction	(i) short serratus anterior (ii) short rhomboids (iii) long serratus anterior
	Inferior angle of scapula in contact with thorax	Anterior tilt	(i) shortness of short head biceps (ii) tight pectoralis minor
Humeral head position		Anterior	(i) tight posterior capsule (ii) lax/tight superior glenohumeral ligament (iii) lax/tight coracohumeral ligament
		Superior	(i) tight posterior capsule
		Posterior	(i) tight anterior capsule
		Medially rotated	(i) tight/over active pectoralis major (ii) tight/over active latissimus dorsi (iii) tight/over active Subscapularis (iv) weak/inhibited lateral rotators
Cervical spine posture	Plumb line passes	Forward head posture	(i) shortened cervical extensors (ii) over active cervical extensors (iii) elongated anterior cervical flexors (iv) weak deep cervical neck flexors

Table 17.5 (*Continued*)

Static	Normal	Faults	Possible Cause
			(v) tight ligamentum nuchae
			(vi) kypohosis – lordosis posture
			(vii) flat back posture
			(viii) sway back posture
			(ix) tight/over active hip flexors
			(x) weak external obliques
			(xi) weak thoracic extensors
			(xii) weak/lengthened hamstrings
			(xiii) weak internal obliques
			(xiv) poor core control
Thoracic spine posture	Plumb line should pass through shoulder joint and mid way through trunk	Increased kyphosis	(i) sway back posture
			(ii) kyphosis – lordosis posture
			(iii) shortened cervical extensors
			(iv) over active cervical extensors
			(v) weak/lengthened thoracic erector spinae
			(vi) elongated rectus abdominus
			(vii) lengthened hamstrings
			(viii) poor core control

dysfunction, coupled with critical reflective review and reassessment (Magarey and Jones 2004).

Integrated scapulothoracic rehabilitation

Table 17.6 is a very useful tool which can be utilised with any shoulder injury. In all cases, whether treatment involves surgical intervention or not, alterations in faulty posture can be addressed, and rehabilitation of other parts of the kinetic chain – trunk and pelvic girdle – can commence at a relatively high level. Once again, consideration of the kinetic chain links and myofascial slings will lead the therapist to areas distal to the shoulder girdle which will require soft tissue work in order to elongate shortened tissues. More local tissue work will need to be carried out under the advisement of the surgeon following surgical intervention, so that newly repaired tissues are not placed under excessive strain at too early a stage.

Ranges of movement for the exercises can be modified for the specifics of the patient, ensuring that the quality of the movement is correct from the outset, and that early substitution patterns are identified and correct, and that movement is fluent and pain free. Once again surgical intervention may require a little more lateral thinking in order to carry out specific exercises effectively and safely.

When rehabilitating a shoulder that has received surgical intervention, it is imperative that the therapist converses with the surgeon and understands what technique has been carried out, what type of fixation was used, what state the repaired tissue was in at the repair, and what tissues have been repaired. The surgeon and therapist can then formulate a patient-specific, injury-specific rehabilitation protocol, based on information such as at what ranges of movement during the surgery was the repaired tissue put on tension? This information can then be utilised as a guide for the protected range of movement during the early stages of rehabilitation.

At all times the therapist should bear in mind the histology and phases of healing – inflammatory stage, proliferation phase and remodeling phase – and adjust their rehabilitation programme accordingly. The table below gives some indication of the level of involvement of some of the muscles around the shoulder girdle in common rehabilitation exercises. This can be utilised to expedite recovery knowing that some exercises place more or less stress on certain muscles than others.

Table 17.6 Upper quadrant exercise progression (adapted from Kibler and McMullen 2004.) Adapted from Kibler and McMuller: Scapulothoracic Problems in Overhead Athletes, in The Shoulder and the Overhead Athlete: 2004. Krishnan, S. G., Hawkins, R. J., Warren, R. F. (Eds). Lippencott, Williams and Wilkins, Philadelphia.

	Weeks (Estimated)							
	1	2	3	4	5	6	7	8
Scapular motion								
Thoracic Posture Exercises	X	X	X					
Trunk Flexion Extension Rotation	X	X	X					
Lower Abdominal Hip extension Exercises	X	X	X	X	X			
Muscular flexibility								
Massage	X	X						
Electotherapy Modalities	X	X	X					
Stretching	X	X	X	X	X	X	X	X
Pectoralis minor stretch	X	X	X					
Sleeper stretch	X	X	X					
Closed chain co-contraction exercise								
Weight transfer	X	X						
Balance board	X	X						
Scapular clock exercise	X	X						
Rhythmic ball stabilisation		X						
Weight-bearing isometric extension	X	X						
Wall push-up		X						
Table push-up				X	X	X		
Modified to prone push-up						X	X	X	X
Axially loaded active ROM exercise								
Scaption		X	X	X	X			
Flexion slide		X	X	X	X			
Abduction glide			X	X	X			
Diagonal slides		X	X	X	X	X		
Integrated open kinetic chain exercises								
Scapular motion + arm elevation			X	X	X	X	X	X
Unilateral/bilateral resistance band pulls + trunk motion				X	X	X	X	X
Modified shoulder dump series			X	X	X	X	X	
Dumbbell punches + progressions						X	X	X
Lunges with dumbbell reaches					X	X	X	X
Plyometric sport/specific								
Weighted ball throw and catch						X	X	X
Resistance tubing plyometics						X	X	X

Table 17.7 Exercise progression for shoulder muscles (Adapted From Uhl and Kibler, 2009)

Exercise	Deltoid	Supraspinatus	Upper trapezius	Serratus anterior	Lower trapezius
Elastic tubing rows	n/a	39% MVC +/− 16%	34% MVC +/−23%	10% MVC +/− 6%	n/a
Unilateral rows	72% MVC +/− 20%	n/a	63% MVC +/− 17%	14% MVC +/− 6%	45% MVC +/− 17%
Standing press up	30% MVC +/− 11%	30% MVC +/− 17%	24% MVC +/− 8%	295 MVC +/− 13%	9% MVC +/− 5%
Forward punch	39% MVC +/− 23%	48% MVC +/− 83%	n/a	49% MVC +/− 14%	n/a
Prone flexion at 135 degrees abduction	n/a	n/a	79% MVC +/− 18%	43% MVC +/− 17%	97% MVC +/− 21%
Prone external rotation at 90 degrees	n/a	50% MVC	20% MVC +/− 18%	57% MVC +/− 22%	79% MVC +/− 21%
Unilateral supine protraction	n/a	n/a	7% MVC +/−3%	62% MVC +/− 19%	11% MVC +/− 5%
Scaption < 80 degrees	91% MVC +/− 26%	82% MVC +/− 27%	72% MVC +/− 19%	62% MVC +/− 18%	50% MVC +/− 21%
Military press	72% MVC +/− 24%	56% MVC /− 48%	64% MVC +/− 26%	82% MVC +/− 24%	n/a
Scaption >120 degrees	72% MVC +/− 13%	64% MVC +/− 28%	79% MVC +/− 19%	96% MVC +/− 24%	61% MVC +/− 19%
Diagonal flexion, horizontal adduction, external rotation	n/a	n/a	66% MVC +/− 10%	100% MVC +/− 24%	39% MVC +/− 15%
Push up with plus	n/a	n/a	50%	140%	30%
Diagonal exercise with flexion, horizontal flexion and external rotation	n/a	n/a	66% MVC +/− 10%	100% MVC+/− 24%	39% MVC +/− 15%

Case study

A 29-year-old, left-handed, professional tennis coach presented with a complaint of increased left shoulder pain following serving. This pain was located over the antero-superior aspect of his glenohumeral joint, and increased in intensity with continued overhead activity. He stated that he had recently increased the amount of overhead activity during his coaching sessions, as he was working to improve some of his pupils' service action. Apart from this he stated that he had not changed anything else concerned with his training. He stated that his health was good and that he was not taking any medication, and that he had not changed his racquet, or string tension recently.

Observation was taken from the front, back and side of the patient with the patient stripped down to the waist. Figure 17.1 illustrates a posture which deviates from the stated "ideal"; the left profile shows a forward head posture, and increased thoracic kyphosis, protracted shoulder girdle and anterior humeral head. He has an anterior tilted pelvis and sway back posture.

Active movements produced left shoulder pain on abduction at 100 degrees (Figure 17.19) and flexion at 120 degrees (Figure 17.20), which increased as elevation continued, and eased at the end of the available active range. Abduction demonstrated increased activity in the left upper Trapezius. Flexion demonstrated increased lumbar extension and anterior pelvic tilt.

Active medial rotation on the right was to T7 and left was to T8 (this range was further if scapular winging was allowed to take place). Active lateral rotation utilising Apply's Scratch test (subject was instructed to reach over shoulder to "scratch" between scapula and it was noted to which vertebrae the thumb reached) was to T2 on the right and T4 on the left.

Resisted tests elicited pain on the empty can and full can tests, and on resisted lateral rotation in neutral. Hawkins-Kennedy test was negative, as were all labral tests, but Neer's test was positive. Inner range serratus anterior strength and endurance was deficient when compared to the right, and middle and lower trapezius strength was deficient bilaterally.

Supine examination (Figures 17.21 and 17.22) showed that there was an increased distance between the posterior acromion on the left as compared to

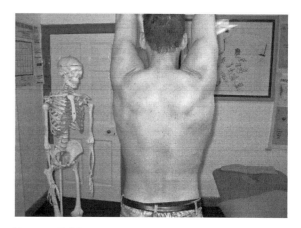

Figure 17.20 Note asymmetry of arms at the end range of flexion, left reduced.

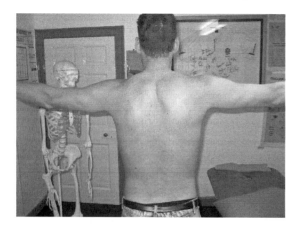

Figure 17.19 Increased left upper trapezius muscle activity at 90 degrees active abduction.

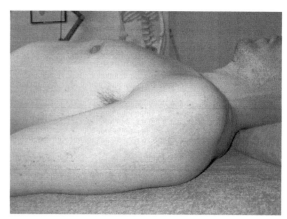

Figure 17.21 Identification of tight posterior glenohumeral joint structures on left.

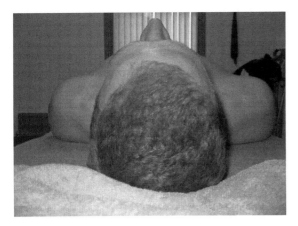

Figure 17.22 Bilateral comparison of humeral head position. Note increased distance of posterior acromion to bed on left.

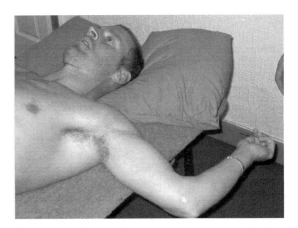

Figure 17.24 Active lateral rotation range at 90 degrees abduction.

the right which indicated possible posterior capsular tightness and/or tight pectoralis minor on the left.

Active medial and lateral rotation (Figures 17.23 and 17.24) at 90 degrees abduction appeared symmetrical, but when repeated with stabilisation of the shoulder girdle, it was shown that there was restriction of internal rotation of the left, which inferred a glenohumeral internal rotation deficit, that was greater than 10 degrees on the right, and the internal rotation deficit did not equal the external rotation gain.

Measurement of posterior capsular tightness indicated that the left was tighter than the right, and length testing of the pectoralis minor muscles indicated that the left was tighter, and responsible for anterior tilt of the scapula.

Single leg squat showed poor bilateral pelvic control with medial rotation and adduction of the femurs, and corkscrewing at the waist. Further lower limb examination showed that there was decreased inner range recruitment of the posterior fibres of gluteus medius on both sides, and an over dominance of hamstrings over gluteus maximus on active hip extension.

The Thomas test identified tightness of the tensor fascia lata, more so on the right than the left, and tightness in the iliopsoas muscle bilaterally. The Thomas Test position can be used to determine correct function of the iliopsoas muscle group, the rectus femoris, the tensor fascia latae and the sartorius muscle, and assess for their possible involvement in producing alterations in the sagittal pelvis orientation.

Rehabilitation focused on lengthening of the posterior capsule utilising the Sleeper stretch, and manually stretching pectoralis minor. Facilitation of the lower and middle fibres of trapezius was carries out in prone lying, and inner range facilitation of serratus anterior was carried out utilising manually resisted protraction in supine, then progressing to press-up with a plus. The initial focus was on endurance, with repetitions being in the 30–40 repetition range, followed by control through range.

Postural re-education was commenced, facilitating thoracic and lumbar flexion-extension in sitting, thoracic spine extension mobilisations were carried out to facilitate reduction of the thoracic kyphosis, and increase the recruitment of the middle and lower

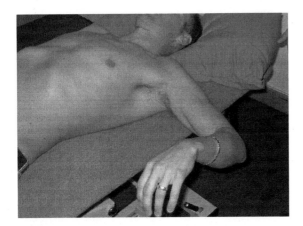

Figure 17.23 Active medial rotation range at 90 degrees abduction.

trapezius. Posterior pelvic tilting in crook lying was commenced to facilitate inner range holds of rectus abdominus and gluteus maximus, and lengthen the lumbar multifidus and interspinous ligaments.

Lower limb functional deficits were addressed with inner range gluteus medius holds, in side lying, and hip extension holds in prone lying. In addition to this functional movement patterns were commenced involving multiplanar movements executed with correct lower limb alignment.

The athlete ceased overhead activity, but was allowed to continue coaching ground strokes, until he had a full pain-free range of active flexion and abduction. Electromyographic feedback and video recording were used to reinforce the correct movement patterns. Closed kinetic chain exercises were utilised early on within the rehabilitation programme to facilitate rotator cuff co-activation, and postural taping was commenced at the outset to aid with proprioceptive awareness.

This athlete complied well with the rehabilitation programme, and was able to return to full tennis-related activities within one month, with the proviso that he further progress his rehabilitation, and include regular stretching exercises, and lower limb conditioning as a regular part of his training.

This case demonstrates the multi-factorial nature of shoulder dysfunction. The skill of the clinician is to identify *relevant* clinical findings that require addressing in order to establish a long-term recovery.

References

Altchek, D.W. and Dines, D.W. (1993) The surgical treatment of anterior instability: selective capsular repair. *Operative Techniques in Sports Medicine*, 1, 163–172.

Armfield, D.R., Stickle, R.L., Robertson, D.D., Towers, J.D. and Debski, R.E. (2003) Biomechanical basis of common shoulder problems. *Semin Musculoskelet Radiology*, 7 (1), 5–18.

Ayub, E. (1991) Posture and the upper quarter. In R.A. Donatelli (Ed.) *Physical Therapy of the Shoulder*, 2nd edn. New York: Churchill Livingstone, pp. 81–90.

Bagg, S.D. and Forrest, W.J. (1988) A biomechanical analysis of scapular rotation during arm abduction in the scapular plane. *American Journal of Physical Medicine and Rehabilitation*, 67, 238–245.

Baltaci, G. and Johnson, R. (2001) Shoulder range of motion characteristics in collegiate baseball players. *Journal of Sports Medicine and Physical Fitness*, 41, 236–242.

Bang, M.D. and Deyle, G.D. (2000) Comparison of supervised exercise with and without manual physical therapy for patients with shoulder impingement syndrome. *Journal of Orthopaedic and Sports Physical Therapy*, 30, 126–137.

Basmajian, J.V. and Bazant, F.J. (1959) Factors preventing downward dislocation of the adducted shoulder joint. *Journal of Bone and Joint Surgery*, 41-A, 1182–1186.

Bathgate, A., Best, J.P., Craig, G., Jamieson, M. and Wiley, J.P. (2002) A prospective study of injuries to elite Australian Rugby Union players. *British Journal of Sports Medicine*, 36, 265–269.

Bennett, W.E. (1998) Specificity of Speed's test: arthroscopic technique for evaluation of the biceps tendon at the level of the bicipital groove. *Arthroscopy*, 14, 789–796.

Bird, Y.N., Alsop, J.C., Chalmers, D.J., Gerrard, D.F., Marshall, S.W. and Waller, A.E. (1998) The New Zealand rugby injury and performance project: V. Epidemiology of a season of rugby injury. *British Journal of Sports Medicine*, 32, 319–325.

Blasier, R., Soslowsky, L. and Malicky, D. (1997) Posterior glenohumeral subluxation: Active and passive stabilisation in a biomechanical model. *Journal of Bone and Joint Surgery (Am)*, 79, 433–440.

Boettcher, C.E., Ginn, K.A. and Cathers, I. (2008) The 'empty can' and 'full can' tests do not selectively activate supraspinatus. *Journal of Science and Medicine in Sport*, 12 (4), 435–439.

Borstad, J.D. and Ludewig, P.M. (2005) The effect of long versus short pectoralis minor resting length on scapular kinematics in healthy individuals. *Journal of Orthopaedic and Sports Physical Therapy*, 35, 227–238.

Brossman, J., Preidler, K.K.W., Pedowitz, R.A., White, L.M., Trudell, D. and Resnick, D. (1996) Shoulder impingement syndrome: Influence of shoulder position on rotator cuff impingement – an anatomic study. *American Journal of Roentgenology*, 167 (6), 1511–1515.

Burkhart, S.S., Morgan, C.D. and Kibler, W.B. (2003a) The disabled throwing shoulder: spectrum of pathology part I: pathoanatomy and biomechanics. *Arthroscopy: The Journal of Arthroscopic and Related Surgery*, 19 (4), 404–420.

Burkhart, S.S., Morgan, C.D. and Kibler, W.B. (2003b) The disabled throwing shoulder: spectrum of pathology part III: the SICK scapula, scapular dyskinesis, the kinetic chain, and rehabilitation. *Arthroscopy: The Journal of Arthroscopic and Related Surgery*, 19 (6), 641–661.

Burkhead, W.Z. (1990) The biceps tendon. In C.A. Rockwood, and F.A. Matsen (Eds), *The Shoulder*. Philadelphia, PA: WB Saunders Co, pp. 791–833.

Cailliet, R. (1991) *Neck and Arm Pain* (3rd edn). Philadelphia: F.A. Davis Company.

Cain, P.R., Mutschler, T.A., Fu, F.H. and Kwon, L.S. (1987) Anterior stability of the glenohumeral joint: a dynamic model. *American Journal of Sports Medicine*, 15 (2), 144–148.

Calis, M., Akgun, K., Birtane, M., Karacan, I., Calis, H. and Tuzun, F. (2000) Diagnostic values of clinical diagnostic tests in subacromial impingement syndrome. *Annals of Rheumatic Disease*, 59, 44–77.

Carr, A.J. (1996) Biomechanics of shoulder stability. *Current Orthopaedics*, 10, 146–150.

Chalmers, D.J., Alsop, J.C., Bird, Y.N., Marshall, S.W., Quarrie, K.L. and Waller, A.E. (2001) The New Zealand rugby injury and performance project: VI. A prospective cohort study of risk factors for injury in rugby union football. *British Journal of Sports Medicine*, 35, 157–166.

Chronopoulous, E., Kim, T.K., Park, H.B., Ashenbrenner, D. and McFarland, E.G. (2004) Diagnostic value of physical tests for isolated chronic acromioclavicular lesions. *American Journal of Sports Medicine*, 32 (3), 655–661.

Clarke, J. and Harryman, D. (1992) Tendons, ligaments and capsule of the rotator cuff. *Journal of Bone and Joint Surgery* (Am), 74, 713–725.

Codman, E.A. (1934) *The Shoulder*. Boston: Thomas Todd Co.

Cohen, J. (1988) *Statistical Power Analysis for the Behavioural Sciences*. Hillsdale, NJ: Lawrence Erlbaum.

Cole, A., McClure, P. and Pratt, N. (1996) Scapular kinematics during arm elevation in healthy subjects and subjects with shoulder impingement syndrome. *Journal of Orthopaedic and Sports Physical Therapy*, 23, 68.

Counsilman, J.E. (1977) Swimming power. *Swimming-World and Junior Swimmer*, 18, 50–52.

Culham, E.G. and Peat, M. (1993) Functional anatomy of the shoulder complex. *Journal of Orthopaedic and Sports Physical Therapy*, 18, 342–350.

Davidson, P.A., El Attrache, N.S., Jobe, C.M. and Jobe, F.W. (1995) Rotator cuff and posterior-superior glenoid labrum injury associated with increased gleno humeral motion: a new site of impingement. *Journal of Shoulder and Elbow Surgery*, 4, 384–390.

Davies, G.J., Gould, J.A. and Larson, R.L. (1981) Functional examination of the shoulder girdle. *Physical Sports Medicine*, 9, 82–104.

Dillman, C.J., Fleisig, G.S. and Andrews, J.R. (1993) Biomechanics of pitching with emphasis upon shoulder kinematics. *Journal of Orthopaedic and Sports Physical Therapy*, 18 (2), 402–408.

Decker, M.J., Hintermeister, R.A., Faber, K.J. and Hawkins, R.J., (1999) Serratus anterior muscle activity during selected rehabilitation exercises. *American Journal of Sports Medicine*, 27 784–791.

Dinnes, J., Loveman, E., McIntyre, L. and Waugh, N. (2003) The effectiveness of diagnostic tests for the assessment of shoulder pain due to soft tissue disorders: a systematic review. *Health Technology Assessment*, 7, 29.

Doody, S.G., Freedman, L. and Waterland, J.C. (1970) Shoulder movements during abduction in the scapular plane. *Archives of Physical Medicine and rehabilitation*, 51, 595–604.

Duthie, G., Pyne, D. and Hooper, S. (2003) Applied physiology and game analysis of rugby union. *Sports Medicine*, 33 (13), 973–991.

Einhorn, A.R., Mandas, M., Sawyer, M. and Brownstair, B. (1997) Evaluation and treatment of the shoulder – in functional movement in orthopaedic and sports physical therapy. In B. Brownstair, and S. Bronner (eds), *Evaluation and Treatment Outcomes*. New York: Churchill Livingstone, pp. 89–140.

Ferrari, D.A. (1990) Capsular ligaments of the shoulder: Anatomical and functional study of the anterior and superior capsule. *American Journal of Sports Medicine*, 18, 20–24.

Field, A. (2000) *Discovering Statistics: Using SPSS for Windows*. London: Sage.

Finnoff, J.T., Doucette, S. and Hicken, G. (2004) Glenohumeral instability and dislocation. *Physical Medicine and Rehabilitation Clinics of North America*, 15, 575–605.

Flatlow, E.L., Saslowsky, L.J., Ticker, J.B., Pawlsk, R.J., Hepler, M. and Ark, J. (1994) Excursion of the rotator cuff under the acromion. Patterns of subacromial contact. *American Journal of Sports Medicine*, 22, 779–788.

Frame, M.K. (1991) Anatomy and biomechanics of the shoulder. In R.A. Donatelli (Ed.), *Physical Therapy of the Shoulder* 2nd edn.. New York: Churchill Livingstone, pp. 1–16.

Freedman, L. and Munro, R.R. (1966) Abduction of the arm in the scapular plane: Scapular and glenohumeral movements. *Journal of Bone and Joint Surgery (Am)*, 48, 1503–1510.

Fricton, J.R., Kroening, R., Haley, D. and Siegert, R. (1985) Myofascial pain syndrome of the head and neck: a review of clinical characteristics of 164 patients. *Oral Surgery, Oral Medicine and Oral Pathology*, 60 (6), 615–623.

Fuller, C.W., Brooks, J.H.M., Kemp, S.P.T. and Reddin, D.B. (2005) A prospective study of injuries and training amongst the England 2003 Rugby World Cup squad. *British Journal of Sports Medicine*, 39, 288–293.

Gabbett, T.J. (2000) Incidence, site and nature of injuries in amateur rugby league over three consecutive seasons. *British Journal of Sports Medicine*, 34, 98–103.

Garraway, W.M. and Macleod, D. (1995) Epidemiology of rugby football injuries. *Lancet*, 345, 1485–1487.

Gerber, G. and Ganz, R. (1984) Clinical assessment of instability of the shoulder with special reference to anterior and posterior draw tests. *Journal of Bone and Joint Surgery*, 66b (4), 551–556).

Gerber, C. and Krushell, R.J. (1991) Isolated rupture of the tendon of the subscapularis muscle. Clinical features in 16 cases. *Journal of Bone Joint Surgery*, 73, 389–394.

Gibbs, N. (1993) Injuries in professional Rugby League: a three year prospective study of the South Sydney professional Rugby League football club. *American Journal of Sports Medicine*, 21, 696–700.

Gibson, M.H., Goebel, G.V., Jordan, T.M., Kegerreis, S. and Worrell, T.W. (1995) A reliability study of measurement techniques to determine static scapular position. *Journal of Orthopaedic and Sports Physical Therapy*, 21 (2), 100–106.

Gissane, C., Jennings, D., Jennings, S., Kerr, K. and White, J. (2003) Health and safety implications of injury in professional rugby league football. *Occupational Medicine*, 53, 512–517.

Goldstein, B. (2004) Shoulder anatomy and biomechanics. *Physical Medicine and Rehabilitation Clinics in North America*, 15, 313–349.

Graichen, H., Hinterwimmer, S., von Eisenhart-Rothe, R., Vogl, T., Englmeier, K.H. and Eckstein, F. (2005) Effect of abducting and adducting muscle activity on glenohumeral translation, scapular kinematics and subacromial space width in vivo. *Journal of Biomechanics*, 38 (4), 755–760.

Greenfield, B., Catlin, P.A., Coats, P.W., Green, E., McDonald, J.J. and North, C. (1995) Posture in patients with overuse injuries and healthy individuals. *Journal of Orthopaedic and Sports Physical Therapy*, 21, 287–295.

Griegel-Morris, P., Larson, K., Mueller-Klaus, K. and Oatis, C.A. (1992) Incidence of common postural abnormalities in the cervical, shoulder and thoracic regions, and their association with pain in two age groups of healthy subjects. *Physical Therapy*, 72, 425–431.

Grimmer, K. (1997) An investigation of poor cervical resting posture. *Australian Journal of Physiotherapy*, 43 (1), 7–16.

Grimsby, O. and Gray, J.C. (1997) Interrelationship of the spine to the shoulder girdle. In R.A. Donatelli (Ed.), *Physical Therapy of the Shoulder*, 3rd edn. New York: Churchill Livingstone, pp. 95–129.

Guanche, C., Knatt, T., Solomonow, M. et al. (1995) The synergistic action of the capsule and the shoulder muscles. *American Journal of Sports Medicine*, 23, 301–306

Hanchard, N., Cummins, J. and Jeffreies, C. (2004) *Evidence-based Clinical Guidelines for the Diagnosis, Assessment and Physiotherapy Management of Shoulder Impingement Syndrome*. London: Chartered Society of Physiotherapy.

Handcock, P.J., Beardmore, A.L. and Rehrer, N.J. (2005) Return to play after injury: Practices in New Zealand rugby union. *Physical Therapy in Sport*, 6, 24–30.

Hanten, W.P., Olson, S.L., Russell, J.L., Lucio, R.M. and Campbell, A.H. (2000) Total head excursion and resting head posture: Normal and patient comparisons. *Archives of Physical and Medical Rehabilitation*, 81, 62–66.

Hardwick, D.H., Beebe, J.A., McDonnell, M.K. and Lang, C.E. (2006) A comparison of Serratus anterior muscle activation during a wall slide and other traditional exercises. *Journal of Orthopaedic and Sports Physical Therapy*, 36, 903–910.

Harryman, D.T., Sidles, J.A., Clark, J.M., McQuade, K.J., Gibb, T.D. and Matsen, F.A. (1990) Translation of the humeral head on the glenoid with passive glenohumeral motion. *Journal of Bone and Joint Surgery (Am)*, 72, 1334–1343.

Hawkins, R.J. and Kennedy, J.C. (1980) Impingement syndrome in athletes. *American Journal of Sports Medicine*, 8, 151–158.

Hawkins, R.J., Schult, J.P. and Janda, D.H. (1996) Translation of the glenohumeral joint with the patient under anesthesia. *Journal of Shoulder and Elbow Surgery*, 5, 286–292.

Hayes, K., Callanan, M., Walton, J., Tzannes, A., Paxinus, A. and Morrell, G.A.C. (2002) Shoulder instability: management and rehabilitation. *Journal of Orthopaedic and Sports Physical Therapy*, 32 (10), 497–509.

Hegedus, E.J., Goode, A., Campbell, S., Morin, A., Tamadoni, M.M., Moorman, C.T. and Cook, C. (2008) Physical examination tests of the shoulder: a symptomatic review with meta-analysis of individual tests. *British Journal of Sports Medicine*, 42, 80–92.

Hertel, R., Ballmer, F.T., Lambert, F.R.C.S. and Gerber, M.D. (1996) Lag signs in the diagnosis of rotator cuff rupture. *Journal of Shoulder and Elbow Surgery*, 5 (4), 307–313.

Hess, S. (2000) Functional stability of the glenohumeral joint. *Manual Therapy*, 5, 63–71.

Hill, J.A. (1983) Epidemiological perspective on shoulder injuries. *Clinical Sports Medicine*, 2 (2), 241–246.

Hoppenfield, S. and Hutton, R. (1976) Physical examination of the shoulder. In S. Hoppenfeld (Ed.), *Physical Examination of the Spine and Extremities*. Norwalk, CT: Appleton-Century-Crofts.

Holtby, R. and Razmjou, H. (2004) Accurcy of Speed's and Yergason's tests in detecting biceps pathology and SLAP lesions: comparisons with arthroscopic findings. *Arthroscopy*, 20 (3), 231–236.

Horsley, I. (2005) Assessment of shoulders with pain of a non-traumatic origin. *Physical Therapy in Sport*, 6, 6–14.

Howell, D.C. (1997) *Statistical Methods for Psychology* (4th edn). Belmont: Wadsworth Publishing Company.

Howell, S.M. and Galinat, B.J. (1989) The glenoid labral socket: A constrained articular surface. *Clinical Orthopaedics*, 243, 122–125.

Hutson, M.A. (1996) *Sports Injuries, Recognition and Management*, 2nd edn. Oxford: Oxford University Press.

Ihashi, K., Matsushita, N., Yagi, R. and Handa, Y. (1998) Rotational action of the supraspinatus muscle on the shoulder joint. *Journal of Electromyography and Kinesiology*, 8, 337–346.

Inman, V.T., Saunders, J.B. and Abbott, L.C. (1944) Observations on the function of the shoulder joint. *Journal of Bone and Joint Surgery*, 26, 1–30.

Itoi, E., Hsu, H. and An, K. (1996) Biomechanical investigation of the glenohumeral joint. *Journal of Shoulder and Elbow Surgery*, 5 (5), 407–424.

Itoi, E., Kido, T., Sano, A., Urayama, M. and Sato, K. (1999) Which is more useful the "full can test" or the "empty can test" in detecting the torn supraspinatus tendon? *American Journal of Sports Medicine*, 27 (1), 65–68.

Jakoet, I. and Noakes, T.D. (1998) A high rate of injury during the 1995 Rugby World Cup. *South African Medical Journal*, 87, 45–47.

Janda, V. (2002) Muscles and motor control in cervicogenic disorders. In R. Grant (Ed.): *Physical Therapy of the Cervical and Thoracic Spine*, 3rd edn. New York: Churchill Livingstone, pp. 182–199.

Jobe, F.W. and Moynes, D.R. (1982) Delineation and diagnostic criteria and rehabilitation program for rotator cuff injuries. *American Journal of Sports Medicine*, 10, 336–339.

Jobe, F.W., Kvitne, R.S. and Giangarra, C.E. (1989) Shoulder pain in the overhand or throwing athlete: the relationship of anterior instability and rotator cuff. *Orthopaedic Review*, 18, 963–975.

Jobe, C. (1990) Gross anatomy of the shoulder. In C.A. Rockwood, and F.A. Matsen (Eds), *The Shoulder*. Philadelphia, PA: WB Saunders, pp. 34–97.

Johnson, G., Bogduk, N., Nowitzke, A. and House, D. (1994) Anatomy and actions of trapezius muscle. *Clinical Biomechanics*, 9, 44–50.

Junge, A., Cheung, K., Edwards, T. and Dvorak, J. (2004) Injuries in youth amateur soccer and rugby players – comparison of incidence and characteristics. *British Journal of Sports Medicine*, 38, 168–172.

Kamkar, A., Irrgang, J. and Whitney, S. (1993) Non operative management of secondary shoulder impingement syndrome. *Journal of Orthopaedic and Sports Physical Therapy*, 17 (5), 212–224.

Kebaetse, M., McClure, P. and Pratt, N.E. (1999) Thoracic position effect on shoulder range of motion, strength and 3 dimensional scapular kinematics. *Archives of Physical and Medical Rehabilitation*, 80, 945–950.

Kelly, B.T., Barnes, R.P., Powell, J.W. and Warren, R.F. (2004) Shouler injuries to Quarterbacks in the National Football League. *American Journal of Sports Medicine*, 32 (2), 328–331.

Kendall, F.P. and McCreary, E.K. (1983) *Muscles: Testing and Function*, 3rd edn. Baltimore: Williams & Wilkins.

Kendall, F.P., McCreary, E.K. and Provance, P.G. (1993) *Muscles Testing and Function*, 4th edn. Baltimore: Lippincott, Williams and Wilkins.

Kibler, W.B. (1991) Role of the scapula in the overhead throwing motion. *Contemporary Orthopaedics*, 22, 525–533.

Kibler, W.B. (1995a) Biomechanical analysis of the shoulder during tennis activities. *Clinical Sports Medicine*, 14, 79–85.

Kibler, W.B. (1995b) Sensitivity and specificity of the anterior slide test in throwing athletes with superior glenoid labral tears. *Arthroscopy*, 11, 296–300.

Kibler, W.B. (1998) The role of the scapula in athletic shoulder function. *American Journal of Sports Medicine*, 26 (2), 325–337.

Kibler and McMullen Scapulothoracic Problems In Overhead Athletes, in *The Shoulder and the Overhead Athlete*; 2004. Krishnan, S.G., Hawkins, R.J., Warren, R.F (Eds). Lippencott Williams and Wilkins, Philadelphia.

Kim, S.H., Kwon, I.H. and Han, K.Y. (1999) Biceps load test: a clinical test for superior labrum anterior and posterior lesions in shoulders with recurrent anterior dislocation. *American Journal of Sports Medicine*, 27 (3), 300–303.

Kim, S.H., Ha, K.I. and Ahn, J.H. (2001) Biceps load test II: a clinical test for SLAP lesions of the shoulder. *Arthroscopy*, 17 (2), 160–164.

Kocher, M.S. (1996) Shoulder injuries in alpine skiing. *American Journal of Sports Medicine*, 24, 665–669.

Konrad, G.G., Jolly, J.T., Labriola, J.E., McMahon, P.J. and Debski, R.E. Thoracohumeral muscle activity alters glenohumeral joint biomechanics during active abduction. *Journal of Orthopaedic Research*, 24, 748–756.

Kronberg, M., Nemeth, G. and Brostrom, L. (1990) Muscle activity and control in the normal shoulder. *Clinical Orthopaedics*, 257, 76–85.

Kumar, V.P., Satku, K. and Balasubramaniam, P. (1989) The role of the long head of biceps brachii in the stabilisation of the head of the humerus. *Clinical Orthopaedics and Related Research*, 244, 172–175.

Kvitne, R.S. and Jobe, F.W. (1993) The diagnosis and treatment of anterior instability in the throwing athlete. *Clinical Orthopaedics an Related Research*, 291, 107–123.

Lee, H.W.M. (1995) Mechanics of neck and shoulder injuries in tennis players. *Journal of Orthopaedic and Sports Physical Therapy*, 21 (1), 28–37.

Lee, S.B., Kim, K.J., O'Driscoll, S.W., Morrey, B.F. and An, K.N. (2000) Dynamic glenohumeral stability provided by the rotator cuff muscles in the mid-range and end-range of motion. A study in cadaver. *Journal of Bone and Joint Surgery*, 82, 849–857.

Lee, A.J., Arneil, D.W. and Garraway, M. (2001) Influence of preseason training, fitness and existing injury on subsequent rugby injury. *British Journal of Sports Medicine*, 35, 412–417.

Lewis, J. (2004) Posture and subacromial impingement syndrome: does a relationship exist? *In Touch: The Journal of the Organisation of Chartered Physiotherapists in Private Practice*, 108, 8–17.

Lewis, J.S., Wright, C. and Green, A. (2005a) Subacromial impingement syndrome: the effect of changing posture on shoulder range of movement. *Journal of Orthopaedic and Sports Physical Therapy*, 35, 72–87.

Lewis, J.S., Wright, C. and Green, A. (2005b) Subacromial impingement syndrome: the role of posture and muscle imbalance. *Journal of Shoulder and Elbow Surgery*, 14 (4), 385–392.

Lippitt, S. and Matsen, F. (1993) Mechanisms of glenohumeral joint stability. *Clinical Orthopaedics and Related Research*, 291, 20–28.

Liu, S.H., Henry, M.H. and Nuccion, S.l. (1996) A prospective evaluation of a new physical examination in predicting glenoid labral tears. *American Journal of Sports Medicine*, 24 (6), 721–725.

Lo, Y.P., Hsu, Y.C. and Chan, K.M. (1990) Epidemiology of shoulder impingement in upper arm sports events. *British Journal of Sports Medicine*, 24, 173–177.

Ludewig, P.M. and Cook, T.M. (2000) Alterations in shoulder kinematics and associated muscle activity in people with symptoms of shoulder impingement. *Physical Therapy*, 80 (3), 267–291.

Lukasiewicz, A.C., McClure, P., Michener, L., Pratt, N. and Sennett, B. (1999) Comparison of 3-dimensional scapular position and orientation between subjects with and without shoulder impingement. *Journal of Orthopaedic and Sports Physical Therapy*, 29, 574–583.

Lo, I.K., Nonweiler, B., Woolfrey, M., Litchfield, R. and Kirkley, A. (2004) An evaluation of the apprehension, relocation and surprise tests for anterior shoulder instability. *American Journal of Sports Medicine*, 32, 301–307.

Luime, J.L., Verhagen, A.P., Miedema, H.S., Kuiper, J.L., Burdorf, A., Verhaar, J. and Koes, B.W. (2004) Does this patient have an instability of the shoulder or a labral lesion? *Journal of the American Medical Association*, 292, 1989–1999.

Macdonald, P.B., Clark, P. and Sutherland, K. (2000) An analysis of the diagnostic accuracy of the Hawkins and Neer subacromial impingement signs. *Journal of Shoulder and Elbow Surgery*, 9 (4), 299–301.

Magarey, M.E. and Jones, A. (1992) Clinical Diagnosis and management of minor shoulder instability. *Australian Journal of Physiotherapy*, 38, 269–279.

Magarey, M.E. and Jones, M.A. (2003) Dynamic evaluation and early management of altered motor control around the shoulder complex. *Manual Therapy*, 8 (4), 191–206.

Magarey, M.E. and Jones, M.A. (2003b) Specific evaluation of force couples relevant for stabilisation of the glenohumeral joint. *Manual Therapy*, 8 (4), 247–253.

Magarey, M.E. and Jones, M.A. (2004) Clinical evaluation, diagnosis and passive management of the shoulder complex. *New Zealand Journal of Physiotherapy*, 32 (2), 55–66.

Matsen, F.A. and Arntz, C.T. (1990) Subacromial impingement. In C.A. Rockwood, and F.A. Matsen (Eds.). *The Shoulder* (pp. 623–646). Philadelphia: WB Saunders Co.

McDonell, M.K. and Sahrmann, S. (2002) Movement-impairment syndromes of the thoracic and cervical spine. In R. Grant (Ed.), *Physical Therapy of the Cervical and Thoracic Spine*, 3rd edn. New York: Churchill Livingstone, pp. 335–354.

McDonell, M.K., Sahrmann, S. and Van Dillen, L. (2005) A specific exercise program and modification of postural alignment for treatment of cervicogenic headache: a case report. *Journal of Orthopaedic and Sports Physical Therapy*, 35, 3–15.

McFarland E.G., Tim, T.K. and Savino, R.M. (2002) Clinical assessment of three common tests for superior labrum anterior-posterior lesions. *American Journal of Sports Medicine*, 30 (6), 810–815.

Michell, L., Smith, A., Bachl, N., Rolf, C. and Chan, K. (2001) *International Federation of Sports Medicine: Team Physician Manual*. China: Lippincott, Williams and Wilkins.

Minagawa, H., Itoi, E., Konno, N., Kido, T., Sano, A., Urayama, M. and Sato, K. (1998) Humeral attachment of the supraspinatus and infraspinatus tendons: an anatomic study. *Arthroscopy: The Journal of Arthroscopic and Related Surgery*, 14 (3), 302–306.

Minori, K., Muneta, T., Nakagawa, T. and Shinomiya, K. (1999) A new pain provocation test for superior labral tears of the shoulder. *American Journal of Sports Medicine*, 27, 137.

Moore, K.L. (1980) *Clinically Oriented Anatomy*. Baltimore: Williams and Wilkins.

Moraes, G.F.S., Faria, C.D.C.M. and Teixeira-Salmela, L.F. (2008) Scapular muscle recruitment patterns and isokinetic strength ratios of the shoulder rotator muscles in individuals with and without impingement syndrome. *Journal of Shoulder and Elbow Surger*,; 17 (1), S48–53.

Moore, K.L. (1980) *Clinically Oriented Anatomy*. Baltimore: Williams and Wilkins.

Mottram, S.L. (1997) Dynamic stability of the scapula. *Manual Therapy*, 2 (3), 123–131.

Munro, W. and Healy, R. (2009) The validity and accuracy of clinical tests used to detect labral pathology of the shoulder-a systematic review. *Manual Therapy*, 14 (2), 119–130.

Myers, J.B. and Lephart, S.M. (2000) the role of the sensorimotor system in the athletic shoulder. *Journal of Athletic Training*; 35 (3): 351–363

Myers T.H., Zemanovic, J.R. and Andrews, J.R. (2005) The resisted supination external rotation test: a new test for the diagnosis of superior labrum anterior posterior lesions. *American Journal of Sports Medicine*, 33 (9), 1315–1320.

Neer, C.S. (1983) Impingement lesions. *Clinical Orthopaedics*, 173, 70–77.

Neer, C.S. and Welsh, R.P. (1977) The shoulder in sports. *Clinical Orthopeadics and Related Research*, 8, 583–590.

Neer, C.S. and Foster, C.R. (1980) Inferior capsular shift for involuntary and multidirectional instability of the shoulder. *Journal of Bone and Joint Surgery*, 62 (6), 897–908.

Neer, C.S. and Rockwood, C.A. (1984) Fractures and dislocations of the shoulder. In C.A. Rockwood and D.P. Green (Eds). *Fractures in Adults*. Philadelphia: J.B. Lippincott.

Neer, C.S. (1990) *Shoulder Reconstruction*. Philadelphia: WB Saunders.

Nicholl, J.P., Coleman, P. and Williams, B.T. (1995) The epidemiology of sports and exercise related injury in the United Kingdom. *British Journal of Sports Medicine*, 29, 232–238.

O'Brien, S.J., Neves, M.C., Arnoczky, S.P., Rozbruck, S.R., Dicarb, E.F., Warren, R.F., Schwartz, R. and Wickiewicz, T. (1990) The antomy and histology of the inferior glenohumeral ligament complex of the shoulder. *American Journal of Sports Medicine*, 18 (5), 449–456.

O'Brien, S.J., Pagnani, M.J., Fealy, S., McGlynn, S. and Wilson, J.B. (1998) The active compression test: a new and effective test for diagnosisng labral tears and acromial joint abnormality. *American Journal of Sports Medicine*, 26 (5), 610–613.

Ogston, J.B. and Ludewig, P.M. (2007) Differences in 3-dimensional shoulder kinematics between persons with multidirectional instability and asymptomatic controls. *American Journal of Sports Medicine*, 35 (8), 1361–1370.

Ovesen, J. and Nielsen, S. (1986) Anterior and posterior shoulder instability: a cadaver study. *Acta Orthopaedica Scandinavica*, 57, 324–327.

Paine, R.M. and Voight, M. (1993) The role of the scapula. *Journal of Orthopaedics and Sports Physical Therapy*, 18, 386–391.

Paine, R.M. (1994) The role of the scapula. In J.R. Andrews and K.E. Wilk (Eds). *The Athletes Shoulder* (pp. 495–512). New York: Churchill Livingstone.

Panjabi, M. (1992) The stabilizing system of the spine. Part I. Function, dysfunction, adaptation and enhancement. *Journal of Spinal Disorders*, 5, 383–389.

Panjabi, M.M., Oda, T., Crisco, J.J., Dvorak, J. and Grob, D. (1993) Posture affects motion coupling patterns of the upper cervical spine. *Journal of Orthopaedic Research*, 11, 525–536.

Perry, J. (1988) Biomechanics of the shoulder. In C. Rowe (Ed), *The Shoulder*. New York: Churchill Livingstone.

Petersson, C.J. and Redlund-Johnell, I. (1984) The subacromial space in normal shoulder radiographs. *Acta Orthopaedica Scandinavica*, 55, 57–58.

Poppen, N.K. and Walker, P.S. (1976) Normal and abnormal motion of the shoulder. *Journal of Bone and Joint Surgery (Am)*, 58 (2), 195–201.

Raine, S. and Twomey, L.T. (1997) Head and shoulder posture variations in 160 asymptomatic women and men. *Archives of Physical Medicine Rehabilitation*, 78, 1215–1223.

Ranson, C. and Gregory, P.L. (2008) Shoulder injury in professional cricketers. *Physical Therapy in Sport*, 9 (1), 34–39.

Razmjou, H., Holtby, R. and Myhr, T. (2004) Pain provocative shoulder tests: Reliability and validity of the impingement tests. *Physiotherapy Canada*, 56 (4), 229–236.

Roddey, T.S., Olson, S.L. and Grant, S.E. (2002) The effect of pectoralis muscle stretching on the resting position of the scapula in persons with varying degrees of forward head/rounded shoulder posture. *Journal of Manual and Manipulative Therapy*, 10 (3), 124–128.

Rolf, C. (2008) *The Sports Injuries Handbook; Diagnosis and Management.* London: A&C Black.

Rowe and Zarins (1981) Rowe CR, Zarins B. Recurrent transient subluxation of the shoulder. *J Bone Joint Surg Am* 1981; 63 (6): 863–72.

Rubin, B.D. and Kibler, W.B. (2002) Fundamental principles of shoulder rehabilitation: conservative to post-operative management. *Arthroscopy*, 18 (9 Suppl 2), 29–39.

Sahrmann, S.A. (1987) Posture and muscle imbalance. *Postgraduate Advances in Physical Therapy*, I-VIII, 2–21.

Sahrmann, S.A. (2002) *Diagnosis and Treatment of Movement Impairment Syndromes.* St Louis: Mosby.

Smith, L.K., Weiss, E.L. and Lehmkuhl, L.D. (1996) *Brunnstrom's Clinical Kinesiology* (5th edn). Philadelphia: F.A. Davis Company.

Silliman, J.F. and Hawkins, R.J. (1994) Clinical examination of the shoulder complex. In J.R. Andrews and K.E. Willk (eds), *The Athlete's Shoulder.* New York: Churchill Livingstone.

Solem-Bertoft, E., Thuomas, K.A. and Westerberg, C.E. (1993) The influence of scapular retraction and protraction on the width of the subacromial space. An MRI study. *Clinical Orthopaedics and Related research*, 296, 99–103.

Soslowsky, L., Carpenter, J., Bucchieri, J. and Flatlow, E. (1997) Biomechanics of the rotator cuff. *Orthopaedic Clinics in North America*, 28 (1), 17–29.

Speer, K.P., Hannafin, J.A., Altchek, D.W. and Warren, R.F. (1994) An evaluation of the shoulder relocation test. *American Journal of Sports Medicine*, 22 (2), 177–183.

Stephenson, S., Gissane, C. and Jennings, D. (1996) Injury in Rugby League: a four year prospective study. *British Journal of Sports Medicine*, 30, 341–345.

Struhl, S. (2002) Anterior internal impingement. *Arthroscopy: The Journal of Arthroscopic and Related Surgery*, 18 (1), 2–7.

Symeonides, P.P. (1972) The significance of the subscapularis muscle in the pathogenesis of recurrent anterior dislocation of the shoulder. *Journal of Bone and Joint Surgery (Br)*, 54 (3), 476–483.

Targett, S.G.R. (1998) Injuries in professional Rugby Union. *Clinical Journal of Sports Medicine*, 8, 280–285.

Terry, G.C. and Chopp, T.M. (2000) Functional anatomy of the shoulder. *Journal of Athletic Training*, 35 (3), 248–255.

Tibone, J.E., Lee, T.Q., Csintalan, R.P. et al. (2002) Quantitive assessment of glenohumeral joint translation. *Clinical Orthopaedics*, 400, 93–97.

Turkel, S.J., Panio, M.W., Marshall, J.L. and Girgis, F.G. (1981) Stabilising mechanisms preventing anterior dislocation of the glenohumeral joint. *Journal of Bone and Joint Surgery (Am)*, 63 (8), 1208–1217.

Twomey, L.T. and Taylor, J.R. (2000) Lumbar posture, movement and mechanics. In L.T. Twomey, and J.R. Taylor (Eds), *Physical Therapy of the Low Back* (3rd edn) (pp. 59–92). New York: Churchill Livingstone.

Vangsness, C.T., Ennis, M. and Taylor, J.G. (1995) Neural anatomy of the glenohumeral ligament, labrum and subacromial bursa. *Arthroscopy*, 11 (2), 180–184.

von Eisenhart-Rothe, R., Matsen, F A., Eckstein, F., Vogl, T. and Graichen, H. (2005) Pathomechanics in atraumatic shoulder instability: scapular positioning correlates with humeral head centring. *Clinical Orthopaedics and Related Research*, 433, 82–89.

Warwick, R. and Williams, P. (1989) *Gray's Anatomy.* London: Longman Group Ltd.

Watson, A.W.S. (1995) Sports injuries in footballers related to defects of posture and body mechanics. *Journal of Sports Medicine and Physical Fitness*, 35, 289–294.

Weiner, D.S. and MacNab, I. (1970) Superior migration of the humeral head: a radiological aid in the diagnosis of tears of the rotator cuff. *Journal of Bone and joint Surgery (Br)*, 52, 524–527.

Weldon, E.J. and Richardson, A.B. (2001) Upper extremity overuse injuries in swimming. A discussion of swimmer's shoulder. *Clinics in Sports Medicine*, 20 (3), 423–438.

Woodward, T.W. and Best, T.M. (2000) The painful shoulder: part I: clinical evaluation. *American Family Physician*, 61, 3079–3088.

Yergason, R.M. (1931) Supination sign. *Journal of Bone and Joint Surgery*, 13, 160.

18

The elbow

Angela Clough
Senior Lecturer, University of Hull

This chapter aims to identify common acute and overuse injuries of the elbow, and then discuss the application and principles of systematic assessment of musculoskeletal injuries of the elbow. The chapter will further detail acute management strategies of common elbow injuries and principles of rehabilitation through to return to sport. The use of appropriate exercises using single or multiple joints as opposed to the conceptually flawed concept of "open" and "closed" kinetic chains will be considered and debated.

An *"open* kinetic chain" exists when the foot or hand is *not* in contact with the ground or supporting surface. In a *"closed* kinetic chain", the foot or hand is weight-bearing and is therefore in contact with the ground or supporting surface

To further illustrate the management of these musculoskeletal injuries of the elbow, a case study will be used to highlight key assessment, treatment and rehabilitation strategies. This chapter draws together and analyses common approaches to treatment within an evidence-based framework.

Common elbow injuries/conditions

To fully appreciate the scope of injuries and pathologies common to the elbow joint, one needs to consider how the elbow functions in relation to upper limb kinematics. This chapter will focus on the injuries listed in Table 18.1 and will further provide guidelines on injury management techniques for a range of acute and overuse injuries to both the elbow and forearm. A systematic analysis will be detailed through assessment and treatment of these injuries, which then informs the nature of the rehabilitation.

Principles of assessment

Assessment relies on a good applied knowledge of anatomy; a systematic and applied approach to the assessment process. It is important, when assessing a client, to understand the functionality of the joint so that comparisons of dysfunction can be made. Good clinical assessment skills, such as the ability to listen to the client and record the appropriate assessment findings, will further enhance both the assessment and subsequent treatment of the client.

Assessment and treatment are often complex procedures that draw on a multitude of information processing techniques. Figure 18.1 provides an overview of the problem solving conceptual model, in relation to clinical management. It would be useful to refer to the chapter on clinical reasoning (Chapter 16) to better assist in understanding the process of clinical thinking and action.

Key principles of subjective history taking

The key aspects of assessing an elbow are: "active listening", ensuring we take a logical subjective

Table 18.1 Examples of acute and overuse injuries

Acute elbow injuries

Muscle lesions
Tendon ruptures
Acute rupture of the medial collateral ligament
" pulled elbow"
Fractures/ dislocations:
- Posterior dislocation
- Supracondylar fractures
- Radial head fractures
- Olecranon fracture
- Fracture of the radius and ulna
- Stress fracture

Overuse injuries to elbow and forearm
Tennis elbow/lateral epicondylitis/extensor tendinopathy
Entrapment of the posterior interosseous nerve (PIN)/radial tunnel syndrome
Olecranon bursitis
Radio-humeral bursitis
Osteochondritis dissecans of the capitullum
Panner's disease
Golfers elbow/medial epicondylitis/flexor/pronator tendinopathy
Medial collateral ligament sprain
Ulnar nerve compression
Muscle lesions (acute or overuse)
Osteoarthrosis (OA)

history of the onset of the problem and to guide the history taking but to avoid interrupting the client's flow of information. Prompts may be along the lines of:

- What brings you to see me today?
- What do you think I can do to help you?
- When did it happen?

Can the client recall how it happened? Did they "fall on an outstretched hand", commonly abbreviated to FOOSH. It is a constructive way of addressing taking a history if one includes a reflective practice approach and clearly identifies needs (Cole 2005). The goal of reflective practice is to help practitioners to continually improve their practice by identifying what they do well and what areas need improvement (Cross 2004; Hilliard 2006).

It is important to establish "informed consent" for the examination as well as treatment. Some questioning may be misinterpreted as being "personal" and all aspects of the assessment need to be clearly explained and the client given the opportunity to ask questions to clarify anything that they do not understand. Flory and Emanuel (2004) completed a systematic review on informed consent, comprehension or understanding and found that enhanced consent forms had limited success. They recommended that having a team member to spend time

KEY PRINCIPLES OF SUBJECTIVE HISTORY TAKING 339

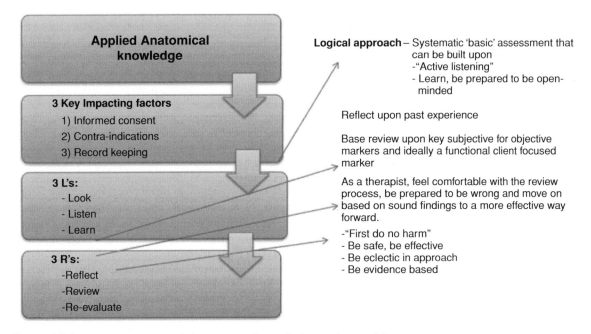

Figure 18.1 Clinical problem solving: systematic, applied reasoning model.

talking on a one-to-one basis seemed to be the best way of improving understanding. Lidz, Applebaum and Meisel (1988) discussed two different ways in which informed consent can be implemented. The "event" model treats informed consent as a procedure to be performed once in each treatment course, which must cover all legal elements at that time. The "process" model, in contrast tries to integrate informing the patient into the continuing dialogue between clinician and client that is a routine part of both diagnosis and treatment and has more benefits as a model to work on.

If they cannot recall an injury, was there a change in their training pattern? Had they undertaken any repetitive DIY type activities? Alternatively was there a prolonged pressure applied? How would they describe their symptoms? Did they occur straight away? Has the behaviour of the symptoms changed? Since the onset of symptoms are they "the same", "better" or worse"? This gives the clinician a guide as to the type of problem. Is it an acute trauma or an overuse/overload problem? Is it a pressure/impingement problem?

From the behaviour and pattern of the symptoms the clinician can get an idea of the stage of healing, from the descriptions the client uses the clinician can start to make some hypotheses by recognising, with reflection on clinical experience, "patterns" and also to localise the tissue most likely to be involved so that their objective testing part of the examination can be appropriate, logical and targeted at localising the target tissue for management. It is essential that the clinician "reflects" on what is being said and clarifies any potential misunderstanding.

Assessment is a dynamic process and it is important that we do not jump to hasty conclusions without first gathering sufficient evidence, reviewing it, in the light of previous experience and "pattern recognition" and clarifying with the client any areas of confusion. It is important that the clinician is clear about the demands of the client's occupation and sport and to work with the coach if appropriate.

It is absolutely essential to have a good knowledge of applied anatomy of the joint (Figures 18.2–18.5) and supporting soft tissues (Figures 18.6 and 18.7) as well as a working knowledge of "referred pain" from, for example, the cervical and thoracic spine, an applied knowledge of peripheral nerve pathways and muscles supplied by them and therefore affected by a

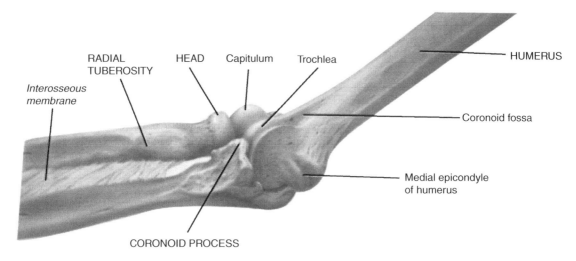

(a) Medial view in relation to humerus

Figure 18.2 Medial view of the elbow. Kuntzman, A.J., Tortora G.J. (2010) Anatomy and Physiology for the Manual Therapies. New Jersey, Wiley.

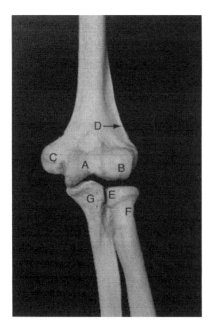

Figure 18.3 Bony landmarks – frontal view. A = Trochlea; B = Capitulum; C = Medial epicondyle; D = lateral supracondylar ridge; E = Radial head; F = Radial neck G = coronoid process. Harris, P.F., Ranson, C. (2008) Atlas of Living and surface Anatomy for Sports Medicine; London, Churchilll Livingston.

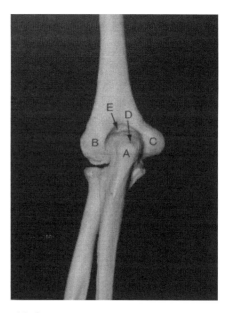

Figure 18.4 Bony landmarks – posterior view. A = Subcutaneous surface of olecranon; B = Lateral epicondyle of humerus; C = Medial epicondyle; D = Site of triceps tendon attachment; E = Olecranon fossa. Harris, P.F., Ranson, C. (2008) Atlas of Living and surface Anatomy for Sports Medicine; London, Churchilll Livingston.

KEY PRINCIPLES OF OBJECTIVE EXAMINATION   **341**

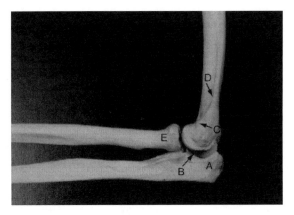

Figure 18.5 Bony landmarks – lateral view. A = Olecranon; B = Trochlea notch; C = Lateral epicondyle; D = Lateral Supracondylar ridge; E = Radial neck.

block to nerve supply. Also, it is important to have a knowledge of dermatomes (areas of skin supplied by peripheral nerves), an awareness of variations in dermatomes and also anomalies in dermatomes, which

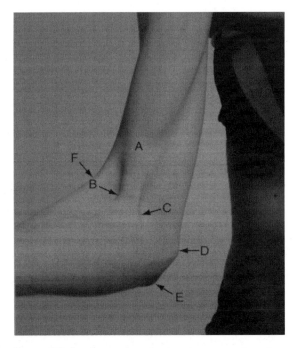

Figure 18.6 Anatomical landmarks – anterior view. A = Biceps brachii; B = Biceps tendon; C-= biceps aponeurosis (passing medially over common flexor tendon); D = Medial epicondyle; E = Olecranon tip; F = Brachioradialis.

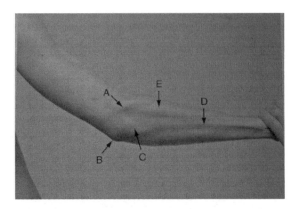

Figure 18.7 Anatomical landmarks – posterior view. A = Lateral epicondyle with common extensor tendon; B = medial epicondyle; C = Subcutaneous surface of olecranon; D = posterior subcutaneous border of Ulna; E = Extensor Carpi Ulnaris.

link to a wider and more consolidated knowledge of referred pain (Figure 18.8).

It is essential, to have an awareness of the variations of "normal" in terms of: range of movement, (ROM) is it within the normal limits or is it hyper-mobile/excessive motion? Is it stiff/limited in some way and if so, is that due to pain, apprehension, swelling, protective spasm. Application of these principals will facilitate a differential diagnosis.

Key principles of objective examination

Observation

Ideally a general observation is made of the client without the patient being aware, for example as they enter the reception area. The three key points to observe are: *face, posture and gait*. The face may indicate pain or lack of sleep. In terms of posture, there is an increased "carrying angle" in females (to clear the hips) than in males. The client may be protectively, "guarding" their elbow, they may be hypermobile (see Figure 18.9), or have a reduced arm swing.

Inspection

This should be completed with the client appropriately undressed so that the affected areas may be observed in a good light. The focus should be on:

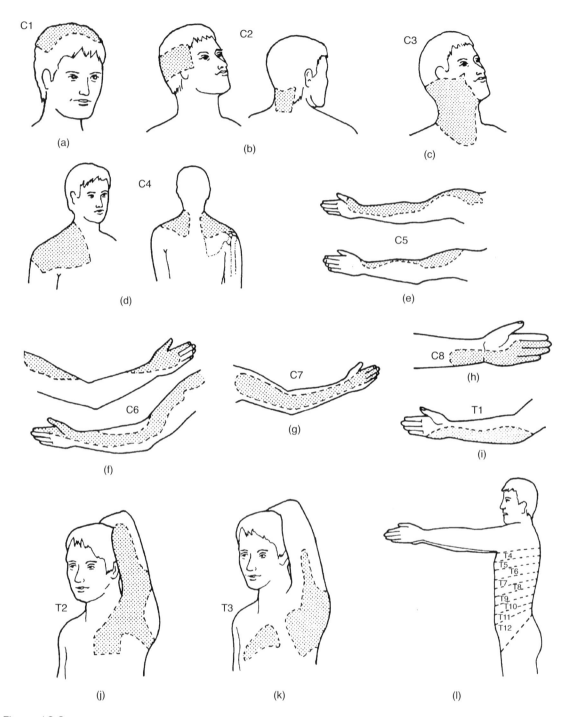

Figure 18.8 DERMATOMES. Kesson, M., Atkins, E. (2005).

KEY PRINCIPLES OF OBJECTIVE EXAMINATION

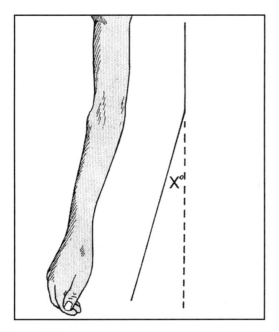

Figure 18.9 Hyper-extended elbow. McRae, R. (2003).

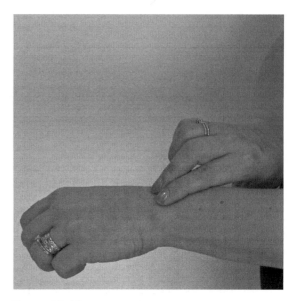

Figure 18.10 Palpatation of synovial thickening.

bony deformity, colour changes, muscle wasting or swelling.

The inspection would be completed after a thorough subjective history has been taken and reflected on so that a clinically reasoned approach may be taken as to what to test objectively and why it is being tested. Clinicians may add in additional tests but is essential to have a clear basic examination that is both logical and systematic. The approach taken by Society of Orthopaedic Medicine (SOM), which is based on the work of the late Dr James Cyriax is a good basic assessment approach. Additional tests can be added in as relevant to enhance clinical reasoning

If there has been a fall, there may well be an obvious visible distortion of the bones/joint following an injury, which may well indicate either a fracture or dislocation. Likewise there may be bruising evident or redness indicating inflammation. It is unusual to see muscle wasting in an acute injury as it often relates either to disuse or develops with a more chronic condition. Muscle wasting may also be an indication of neurological involvement. This may be due to reflex muscle inhibition associated with an effusion at the joint. The presence of swelling is indicative of inflammation from either overuse or trauma.

Palpation for: heat, swelling and synovial thickening

It is essential to establish "signs of activity" at this stage indicating presence of inflammation, using the back of the hand and comparing the symptom free side to the symptomatic side. Synovial thickening has a distinctive "boggy" feel and is relatively common in rheumatoid arthritis, particularly at the wrist (Figure 18.10), knee and ankle.

Establish state at rest

The symptoms at rest must be clarified *prior* to any objective testing requiring movement of joints and muscles. A baseline is established by asking an open question avoiding the use of the word "pain", to avoid leading the patient. An example may be "How are you feeling now?" Once this has been established it makes comparison of the state at rest with any potential change of symptoms on movement easier to clinically reason. It is helpful to use terms such as: *same, better or worse.* It may be also useful to utilise a 10-point Likart scale where the patient can draw a line, with "0" being symptom-free and "10" being worse symptoms they can imagine. A constructive suggestion may be to use a printed "smiley" face above the "0" on the scale and a "sad" face above the "10", on the scale, has a visual impact and helps

344 THE ELBOW

Table 18.2 Applying selective tension

Active	Passive	Resisted
Active movements assess the range of movement, the pain experienced by the client, strength in the client's muscle groups and it shows the willingness of the client to move and quality of that movement. The elbow is not an "emotional" joint, i.e. the reported signs and symptoms are normally specific and can be localised easily by the client. Unlike the cervical spine or shoulder that may have a more complex subjective history. Normal active movement of the elbow joint is: • elbow flexion: 0–150 degrees • elbow extension: 0–10 degrees of hyperextension	Passive movements test the inert structures, e.g. joint capsules and ligaments. Passive movements test pain, range and "end-feel". There are three normal "end-feels" to passive movement testing: • hard (bone to bone as in end of range elbow extension) • soft (approximation of soft tissue as in end of range elbow flexion) • elastic (it is the "elastic" resistance felt at end of range as in full elevation of the shoulder)	Resisted movement tests are used in order to test the contractile structures e.g. muscle, tendon. This is the minimum that would be appropriate depending on the experience and the clinical reasoning of the clinician.

(Loudon 2008)

The possible responses to resisted muscle testing are:
Normal response strong and painfree
Contractile lesion strong and painful
Neurological weakness weak but painfree
Partial rupture (or suspected more serious pathology, e.g. fracture or tumour) weak and painful
Claudication/provocation of an overuseinjury painful on repetition
Psychological component/serious pathology "juddering"/exaggerated response

the client to focus on giving accurate feedback to the rehabilitator.

Examination by application of selective tension

James Cyriax, developed a systematic approach to objective assessment, which is termed "applying selective tension" (Cyriax 1982; Cyriax and Cyriax 1983; Kesson and Atkins 2005). This means to apply: active, passive and resisted movements appropriately. Table 18.2 may clarify the application of selective tension.

With the latter response the clinician must heed the warning "beware the bizarre but consistent patient"!

Some clinicians will always start with active range of motion as it provides a guide to a client's "willingness" to move, the quality of movement and, more importantly, it is a movement within the control of the client at an early stage of the assessment procedure.

Table 18.3 Society of Orthopaedic Medicine's suggested order of selective tension tests for the elbow

Passive	Resisted
Elbow flexion (normally a "soft" end-feel)	Elbow flexion
Elbow extension (normally a "hard" end-feel)	Elbow extension
Pronation of the superior radioulnar joint (normally an "elastic" end-feel)	Pronation
Supination of the superior radioulnar joint (normally an "elastic" end-feel)	Supination

Elbow and radioulnar joints
Provocative tests for epicondylitis

These are:

- resisted wrist extension for tennis elbow

- resisted wrist flexion for golfers elbow

An additional test to be aware of if one suspects a peripheral nerve involvement is Tinel's test. A positive response reproduces the client's symptoms over the involved nerve sensory distribution. For the ulnar nerve, gently tap along the area where it is most superficial, where it travels along the groove between the olecranon and the medial epicondyle.

Novak et al. (1994) investigated provocative testing for cubital tunnel syndrome and found that this test had 0.70 sensitivity and 0.98 specificity. They had a sample of 32 patients with cubital tunnel syndrome (mean age 46, age range 24–81). Those with a previous history of nerve symptoms were excluded. In the test group 31 of the 32 had a positive Tinel's sign. The tester performed 4–6 taps over the ulnar nerve just proximal to cubital tunnel. Significant differences ($p < 0.0001$) between the group with cubital tunnel syndrome and the control group were found for all positive tests. In summary, this test accurately identifies the likelihood of cubital tunnel syndrome, given a positive test.

Within 30 seconds, the highest sensitivity, specificity and positive predictive value were found in the combined test. Within 60 seconds only the sensitivity for the pressure provocation and elbow flexion test increased to 0.98 in those subjects with cubital tunnel syndrome. The combined pressure and flexion test was performed by placing the subject's elbow in maximum flexion and whilst in this position pressure was placed on the ulnar nerve just proximal to the cubital tunnel. Subject symptom response was recorded at both 30 and 60 seconds. The clinical provocative evaluation techniques have been extrapolated to the cubital tunnel syndrome, although statistical verification of these tests is lacking (Buehler and Thayer 1988; Rayan 1992; Rayan, Jenson and Duke 1992).

The test has been adapted to gently tap over the mid-point of the flexor retinaculum at the wrist, which may reproduce tingling over the median nerve distribution consistent with carpal tunnel syndrome. If these tests are positive the client may be referred on

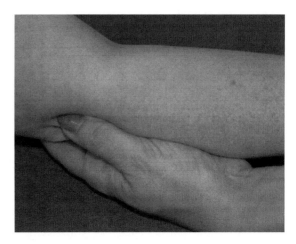

Figure 18.11 Palpatation of the elbow.

for nerve conduction study tests prior to consideration of surgical decompression of the affected nerve.

Palpation to confirm the lesion site

This is assuming there is nothing in the subjective history to suggest referred pain from the cervical spine. For example, altered sensation, "tingling", numbness and reduced or absent reflexes. The rehabilitator would then palpate for the exact site of the lesion (Figure 18.11).

Acute elbow injuries
Muscle lesions

Minor muscle tears commonly occur in muscles bellies around the elbow. Muscles likely to be involved extensor group presenting as tennis elbow on the lateral aspect. The pronator teres muscle may be tender 2–3cm distal to the medial epicondyle as a variation of the flexor group presenting as golfers elbow.

Assessment of involvement is by application of selective tissue tension. Pain is reported on resisted muscle contraction and involvement is confirmed by palpation. There is a good response to local transverse friction massage. If it is the muscle belly, the client is positioned with the muscle supported comfortably with the muscle in a shortened position. The transverse friction massage is performed at 90 degrees to the alignment of the muscle fibres. The application of gentle transverse friction massage applied in the initial inflammatory phase that may increase the rate of phagocytosis (Evans 1980). It is useful to apply the technique in the first days following injury provided the grade is appropriate for the stage of healing and the irritability of the tissue, and it avoids disruption to healing and increased bleeding (Kesson and Atkins 2005). This would normally decrease the pain and increase the range of movement. The increase in range should be followed up with exercise in the pain free range of movement.

Tendon ruptures

Acute avulsions of triceps or biceps are rare. Triceps tends to be affected more commonly with excessive deceleration force as in a fall. Biceps tendon is more associated with weight lifting activities. Acute ruptures of either require surgical repair.

Pulled elbow

This occurs quite frequently in the under 5s (often accompanied by a guilty and upset parent) as the most common mechanism is when a parent snatches the hand of a child misbehaving at the edge of a pavement, or when parents "swing" their child in play between them. Pitching in baseball, serving in tennis, spiking in volleyball, passing in American football and launching in javelin throwing can all produce elbow pathology by forceful valgus stress (usually during high velocity eccentric loading during the terminal deceleration of the limb), with medial stretching, lateral compression and posterior impingement. With the exception of baseball, there are few prospective cohort studies on the epidemiological trends of childhood elbow injuries in other sports. Delineating injury patterns to the elbow in children can be challenging, given the cartilaginous composition of the distal humerus and the multiple secondary ossification centres that appear and unite with the epiphysis at defined ages (Magra et al. 2007).

The joint at such a young age is lax. It is prone to recurrent injury if the annular ligament is subjected to repeated over-stretching (Illingworth 1975). The radial head easily slides from beneath the orbicular ligament, the child immediately complains of pain and there is a noticeable limit of supination. There is normally a spontaneous recovery if the arm is rested in a sling for 48 hours. It may be reduced by forced supination while pushing the radius in a proximal direction, by forced radial deviation of the hand (McRae 2003; Kesson and Atkins 2005).

In children and adolescents, the epiphyseal plate is weaker than the surrounding ligaments, predisposing them to epiphyseal plate injuries. On the other hand, post-pubescent or skeletally mature athletes are more prone to tendinous or ligamentous injury. Injuries may cause significant impact on the athlete, parents and healthcare system. (Magra 2007)

Fractures/dislocations

It is essential that fractures of the elbow region are diagnosed early and managed appropriately as the complication rate is higher than with fractures close to other joints. Unstable/displaced fractures should be promptly referred for surgical orthopaedic intervention. However, when the articular or cortical surface has less than 2mm of vertical or horizontal displacement, the fracture may be regarded as stable and as such treated conservatively (Shapiro and Wang 1995).

Over vigorous rehabilitation can be an issue with the elbow. Remember, safety of the client is paramount, "first do no harm!" A clear understanding of the applied anatomy and appropriate application of graded rehabilitation should result in there being no problem. Awareness and caution is essential in the musculoskeletal management of the elbow. It is therefore essential that the clinician has an awareness of myositis ossificans, which is a condition that may occur after supracondylar fractures and dislocations of the elbow.

Myositis ossificans/Hetertopic ossification

Myositis ossificans is a calcification which occurs within the haematoma that forms in the brachialis muscle covering the anterior aspect of the elbow joint. It is often attributed to inappropriate vigorous exercise after a supracondylar fracture or dislocation of the elbow. Gentle active, grade A exercise should always be within the painfree range of available movement. The ideal situation is to prevent it happening by avoiding over vigorous exercise. If it occurs it presents as a mechanical block to flexion with an abnormal "hard" end feel where the normal end feel to end of range flexion should be "soft". If it is discovered at an early stage and the joint is given complete rest this minimises the mass of calcified material formed in the muscle. In established cases it may be surgically excised after the lesion has appeared dormant for many months so that the range of flexion may be restored.

Often the terms are used interchangeably. However for accuracy, hetertophic ossification simply refers to the formation of trabecula bone outside the skeleton where as myositis ossificans is the specific pathology. So the first is a sign of the other. Kumar and colleagues (2009) describe the difference clearly in the authoritative text *Robbins and Cotran Pathological Basis of Disease*. Myositis ossificans is distinguished from other fibroblastic proliferations by the presence of metaplastic bone. It usually develops in athletic adolescents and young adults, and follows an episode of trauma in more than 50% of cases. The lesion typically arises in the musculature of the proximal extremities. The clinical findings are related to its stage of development; in the early phase, the involved area is swollen and painful, and within several weeks it becomes more circumscribed and firm. Eventually, it evolves into a painless, hard, well-demarcated mass.

Posterior dislocation

Posterior dislocation is often associated with a fracture of the coronoid process radial head and is probably the most serious acute injury to the elbow. The most common mechanism of injury is in contact sports.

Impairment of the vascular supply to the forearm is a major complication. The assessment of pulses distal to the dislocation is essential. Urgent reduction of the dislocation is required if pulses are absent. Reduction is normally relatively easy to do if it is done quickly before the onset of protective muscle spasm. The elbow is held in 45 degrees of flexion, the clinician applies longitudinal traction by applying a firm grip to the anterior aspect of the humerus. There is often an audible "clunk" when reduction occurs. However, surgical intervention is required as a matter of urgency if post reduction vascular impairment persists. A post reduction X-ray is advised and also assessing the stability of the collateral ligaments.

Undisplaced fractures of the radial head or small fractures of the coronoid process require conservative treatment in a supportive sling for 2–3 weeks.

A common complication is loss of extension following elbow dislocation. Ross et al. (1999) found that immediate active grade A mobilisation (in the pain-free range of movement) has been shown to result in less restriction of elbow extension with increase in instability of the elbow joint. Surrounding muscles should also be strengthened, statically initially then progress through range with a graduated progression in resistance. It is important that rehabilitation is carried out with care as myositis ossificans/heterotopic ossification may occur with over vigorous exercise

Supracondylar fractures

These are more common in children than adults. The mechanism is normally from a fall on an outstretched hand (FOOSH, as it is often written in notes). These fractures should be considered an orthopaedic emergency. They tend to have a high rate of neurovascular complications as they are rotationally unstable. The initial short-term management is closed reduction with pins. The arm is placed in a splint. The pins are removed after a 4–6 weeks. There is not normally any problems with recovery in children.

In an older population, Robinson et al. (2003) found the overall incidence of distal humeral fractures in adults was 5.7 cases per 100,000 per year, with an almost equal male to female ratio. There was a bimodal age distribution, with simple falls being the most common overall cause of fracture, and the majority of the fractures were extra-articular (AO/OTA type A) or complete articular fractures (AO/OTA type C). The risk of complications during treatment was generally low in most patients, and the majority healed their fractures uneventfully. Overall, 90.6% of fractures united within 12 weeks and just under half of the remaining 9.4% patients with union complications healed without requiring further operative intervention by 24 weeks. The risk of union complications was higher following high-energy injuries, open fractures, and non-operative treatment. Although the AO/OTA classification was not predictive of union complications, the low transcondylar (type A2.3 and A3) and simple intercondylar fracture (type C1.3) configuration had a greater risk of union complications than the high subtype. The rate of infection, myositis ossificans, and other implant-related complications were higher following operative treatment of type C fractures than type A and B fractures. They concluded that the epidemiology of a consecutive unselected series of adult distal humeral fractures is defined in this study. The majority of these fractures are best treated surgically by rigid open reduction and internal fixation, except for low Type A and C fractures, which have a higher risk of union complications. The role of total elbow arthroplasty to treat these more complex injuries requires further evaluation.

Radial head fracture

This is the most common fracture around the elbow. They are classified into types 1–4 and the mechanism is normally from a fall on an outstretched hand. Type 1 are "non" or minimally displaced. They respond to early aspiration, splinting (removable) active grade "A" (pain-free) mobilisation and normally are healed in 6–8 weeks. Type 2 are displaced fractures and respond to surgical intervention. Type 3 are managed by surgical excision. Type 4 are fractures in the presence of a dislocation and always require surgical treatment as they can be very unstable.

Olecranon fractures

The mechanism is normally from a fall on an outstretched hand. If it is stable the client should be able to extend the arm and lift it against gravity. Management would be 2–3 weeks in a splint then active grade "A" (pain-free ROM) mobilisation. If it is unstable open-reduction with internal fixation is required. In this instance active mobilisation can start within a week of surgery.

Fracture of the radius and ulna

The mechanism of injury to the forearm bones is a fall on an outstretched hand It is usually clinically obvious. X-rays should be taken for a post reduction comparison.

Two types of dislocation occur: the Monteggia injury, which is a fractured ulna with dislocated head of the radius at the elbow joint, and the Galeazzi injury, which is a fractured radius with dislocated head of the ulna at the wrist joint.

In the adult, perfect reduction by an orthopaedic surgeon for internal fixation of the radial and ulnar fracture is required in order to return to sport. Depending on the accuracy of reduction, either a cast or support is required post-operatively for 8–10 weeks. Monteggia and Galeazzi injuries are often displaced and should be referred to an orthopaedic surgeon for reduction. However, an isolated fracture of the ulna may be treated conservatively by the use of an above elbow cast in mid-pronation for 8 weeks.

In children, angulations of less than 10 degrees are considered acceptable. The favoured position for immobilisation is in pronation, although in proximal radial fractures and also in Smith's fractures at the wrist, the forearm should extend above the elbow and leave the metacarpophalangeal joints free.

Stress fracture

Stress fractures are a result of overload, overtraining or insufficient rest. Brukner et al. (1999) studied sportspeople (baseball, tennis and swimming) sustaining this type of fracture; they found that the key element of musculoskeletal management was rest primarily. It must also include correction of any predisposing factors and a multi-disciplinary approach to the correction of any faulty technique.

Overuse injuries to elbow and forearm

Tennis elbow/lateral epicondylitis/extensor tendinopathy

"Tennis elbow" is a traditional term. It is a common condition arising in the sporting context from overload in racquet sports. Overload in occupational and domestic settings that include repeated grasping movements particularly middle-aged DIY enthusiasts (Hutson 1990). It is a frequently reported condition, which is characterised by pain over the lateral epicondyle of the humerus and aggravation of the pain on resisted dorsiflexion of the wrist (Verhaar 1994). The incidence in general practice surgeries is about 4–7 per 1,000 patients per year (Kivi 1983). There is an annual incidence of 1–3% in the general population (Allander 1974; Chard and Hazelman 1989; Verhaar 1994). If left untreated, the complaint is estimated to last from 6 months to 2 years (Cyriax 1936; Bailey 1957; Binder et al. 1985; Hudak et al. 1996). (See Figure 18.12 for the possible sites of tennis elbow.)

Historically has been a difficult condition to manage with a wide range of procedures and management protocols advocated (Reid and Kushner 1993; Noteboom et al. 1994; Caldwell and Safran 1995; Putham and Cohen 1999; Jones and Rivett 2005).

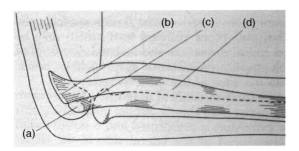

Figure 18.12 Sites of tennis elbow. Kesson, M., Atkins, E. (2005).

In general management is prolonged and long-term outcomes questionable.

When it is clinically indicated the mobilisation with movement (MWM) for tennis elbow as described by Mulligan (1999) is a simple but effective means of treating this condition. However, an indication for use is determined by trial application of the technique. The reasons for selection of a MWM for treatment were:

- immediate abolishment of pain during the trial/testing

- previous experience and knowledge of efficacy of the technique

- potential for integration into a home management programme.

(Mulligan 1999)

Struijs et al. (2001) conducted a systematic review of orthotic devices for tennis elbow. This study was done in Amsterdam as 21% of clients with lateral epicondylitis were prescribed an orthotic device as a treatment strategy (Verhaar 1992). They concluded that no definitive conclusions could be drawn concerning effectiveness of orthotic devices for lateral epicondylitis. They reported that no functional outcome measures such as the Pain Free Function Questionnaire were reported (Stratford et al. 1987). They recommended a set of valid and reliable outcome measures be included in management and also the cost effectiveness of a relatively inexpensive orthotic device as a treatment strategy, or as an addition to any other conventional treatment, since the use of orthotic devices may reduce costs of sick leave by reduction of the pain experienced during activities. Better designed and well conducted RCTs of sufficient power are warranted.

The pathological process is a degeneration of the ECRB tendon at the extensor origin of the lateral epicondyle. The mechanism of injury is multi-factorial. The key contributing factors being: "overuse"; a heavy racquet or too small grip; a recent change of racquet or too tight a grip between shots; muscle imbalance and loss of flexibility; poor blood supply of the 1–2 cm of the distal attachment of ECRB; and repeated excessive loading leading to degenerative changes in the tendon. If there is continued use there will be microscopic tearing and scarring within the tendon (Regan et al. 1992). This is a mechanical process primarily and as such it should be described as a tendonosis (Nirshel 1992) or tendinopathy rather than tendonitis. Histologically, there is an invasion of fibroblasts and vascular granulation tissue, which is known as angiofibroplastic hyperplasia (Brukner and Khan 2002).

With any overuse injury, "relative rest" is required; that is, maintain active full range of motion but limit the "aggravating" activity.

In view of the pathology, the treatment of lateral epicondylitis with corticosteroid injection is controversial (Labelle et al. 1992; Cameron 1995; Assendelft et al. 1996; Hay et al. 1999). It is accepted that in most cases this is degenerative rather than inflammatory. However, with correction of predisposing factors and appropriate staged rehabilitation injection is rarely required.

In a randomised controlled trial in Australia (Bisset et al. 2006) researchers found physiotherapy combining elbow manipulation and active exercise had a superior benefit in the first 6 weeks to corticosteroid injections after 6 weeks providing a reasonable alternative to injections in the mid to long term. The significant short-term benefits of corticosteroid injection are paradoxically reversed after 6 weeks, with high recurrence rates, implying that this treatment should be used with caution in tennis elbow. (See case study at the end of this chapter.)

Other common injuries

This section covers the following injuries: posterior impingement syndromes, thrower's elbow, stress fracture of olecranon, entrapment of the posterior interosseous nerve (PIN) and radial tunnel syndrome

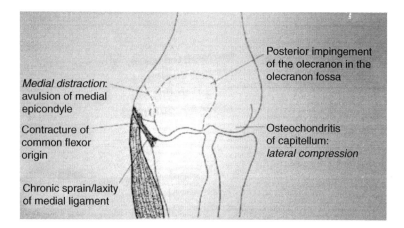

Figure 18.13 Sites of tennis elbow. Hutson, M.A. (1990).

Posterior impingement of the olecranon fossa, valgus stress to the medial structures of the elbow joint and compression injuries to the radiohumeral joint are all linked to repeated high velocity throwing, for example shot putting (Figure 18.13).

Often with throwing there is a forced hyper extension at the elbow which produces a shearing force between the olecranon and olecranon fossa, hence the blanket term "throwers elbow" which may give rise to elbow problems (Hutson, 1990). Initially there is soft tissue hypertrophy but if it is not addressed then osteophytes can form on the tip of the olecranon and within the fossa particularly in the supracondylar ridge area.

From a prevention perspective it is essential that the athlete's training technique is observed and their training schedule is discussed with both the athlete and coach in order to deploy the appropriate problem-solving approach to the management of this closely related range of conditions, which all have similar reported signs and symptoms. It may be that oseophytes appear due to over-use or there may be a stress fracture caused by the athlete having poor eccentric strength of the antagonistic muscle group. This would result in an inability to effectively decelerate once the object is released, therefore eccentric training of the antagonistic muscles is essential. The rehabilitator should keep an open mind, review and monitor intervention in order to clinically problem solve and utilise the most effective strategy for musculoskeletal management.

It is often difficult to differentiate between extensor tendinopathy and early stages of PIN entrapment. The latter is seen in patients who repetitively pronate and supinate the forearm as opposed to extensor tendinopathy, which is more frequently associated with repeated wrist extension. If this is taken account of in the subjective history taking it makes it easier to clinically reason the appropriate differential test to localise the target tissue.

As with any peripheral nerve involvement one would expect pain and parasthesia. Signs and symptoms that would indicate PIN entrapement may include:

- pain over the forearm extensor muscle group

- parasthesia of the hand and lateral aspect of the forearm

- aching of the wrist

- pain in the middle and/or upper third of the humerus

- maximum tenderness is reported over the supinator muscle (four fingers breadth distal to the lateral epicondyle). This differentiates it from extensor tendinopathy as the most common site is at the teno-osseous site

- resisted supination of the forearm with the elbow flexed to 90 degrees and the forearm fully pronated

- pain reproduced with resisted extension of the middle finger with the elbow in full extension. However, this may be a positive test in extensor tendinopathy as well.

Applying neural tension tests may also be included as they may reproduce the patients' symptoms. (The reader is directed to the works of Butler et al. 2000 and Shacklock 2005).

Management for extensor tendinopathy would include graded transverse friction massage (applied at 90 degrees to the target tissue). Appropriate neurodynamic mobilisation techniques may be added. This is normally successful; however, decompression may be required in stubborn, resistant cases and this has a good success rate (Lutz 1991). The key here is good differential diagnosis and early appropriate intervention as tendinopathy responds well to rehabilitation. However, if posterior impingement is not diagnosed early it is common for athletes to have a fixed flexion deformity and report pain at end-of-range extension. It is difficult to treat conservatively at that stage and requires arthroscopic surgery in order to remove the impinging bone. It often takes up to 3 months on average to return to throwing sports. Hay and Bell (1998) found that in a series of 100 elbow arthroscopies, 93% reported a satisfactory improvement in pain with an average of nine degrees improvement in extension.

Olecranon bursitis/"student's elbow"

Swelling of the olecranon bursa commonly occurs after repetitive trauma in clients who repeatedly traumatise the posterior aspect of the elbow, for example carpet fitters. It may occur after a single traumatic episode following a fall onto the elbow or indeed to those who rest their elbow on a hard surface, a common posture in students, hence the colloquial term "student's elbow". Any sport that involves either prolonged pressure on the elbow or a potential trauma directly to the elbow. The most common sport referencing olecranon bursitis is ice hockey. Also shooters/marksmen that lie prone to accurately aim their shot are prone to olecranon bursitis from a sustained pressure on the bursa. There is often a visible swelling, a reduced range of movement and pain on palpation of the bursa (Loudon et al. 2008).

It is a subcutaneous bursa and may become filled with blood and serous fluid (Figure 18.14). The condition is usually painless unless there is an associated bacterial infection. This is a serious complication and requires prompt drainage, strict rest and antibiotic therapy. This is in order to prevent the onset of oseomyelitis and septic arthritis. Surgi-

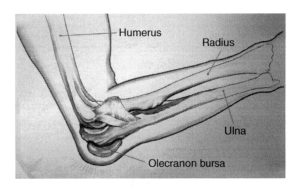

Figure 18.14 OLECRANON BURSITIS. Bahr, R., Maehulm, S. (2004) Clinical guide to Sports Injuries; Leeds, Human Kinetics.

cal excision is occasionally performed for cosmetic reasons.

Radio-humeral bursitis

Radio-humeral bursitis is seen very occasionally in athletes. It may be differentiated from extensor tendinopathy by the site of the tenderness which is anterior and distal to the lateral epicondyle. There is maximum tenderness over the anterolateral aspect of the head of the radius. It may be confirmed by using diagnostic ultrasound techniques. Clinically on presentation there is a "muddle" of signs on assessment. The most effective treatment is a corticosteroid treatment.

Osteochondritis dissecans of the capitullum

Osteochondritis dissicans is often seen in gymnasts and may respond well to rest in the early stages if it is diagnosed quickly. However, it is a localised area of avascular necrosis on the anterolateral aspect of the capitullum. The articular surface of the bone softens and may progress to subchondral collapse and even loose bodies forming in the elbow. It is a significant condition as it may cause an enlarged and deformed capitullum. In a paper on arthroscopic classification and treatment of osteochondritis dissecans of the capitellum, Baumgarten, Andres and Satterwhite (1996) investigated patients (n = 17) that underwent abrasion chondroplasty of the lesion and removal of any loose bodies and osteophytes. The average follow up was 48 months.

Three gave up sport but all the others returned to preoperative sport. They proposed an arthroscopic classification system (1998) including treatment recommendations:

- Symptomatic grade 1 should undergo drilling.

- Grade 2 lesions not responding to non-operative treatment should undergo removal of all affected cartilage back to a stable rim and then abrasion chondroplasty of the underlying bone.

- Grade 3 and 4 lesions should have removal of the osteochondral fragment.

- Grade 5 lesions should have abrasion. Chondroplasty of the exposed crater should include a diligent search of the remaining elbow joint for loose bodies, which should be removed. Any associated osteophyte or synovitis in other elbow compartments should be removed. An early active range of motion and strengthening programme should begin as soon as possible post-operatively.

Even though the authors advocate the arthroscopic treatment of this condition, they emphasise that it is a technically difficult procedure. It should be attempted only by surgeons with a high level of experience in elbow arthroplasty. If the lesion cannot be visualised or treated easily by arthroscopic means the authors advocate a formal, open arthrotomy to avoid damage to the surrounding normal joint space.

Panner's disease

Panner's disease is a self-limiting condition that presents in children under the age of 11. It is 90% male dominated (Loudon et al. 2008). Usually characterised by a fragmentation of the entire ossific centre of the capitullum bone. There are no loose bodies and surgery is not required (Brukner and Khan 2002). It is essential to ensure that growing children have adequate rests between their training sessions in the main. If there is adequate pacing of activity, combined with techniques being correctly coached and monitored by the most appropriate expert most overuse injuries in the young are preventable. It is often atraumatic and it presents with swelling, local tenderness, "clicking" and a decrease in motion of elbow extension. It responds well to relative rest and gentle grade "A" exercise in the pain-free range of motion.

Golfer's elbow/medial epicondylitis/ flexor/pronator tendinopathy

It is less common than lateral epiconylitis. It has many "labels", which all mean the same thing. Alternative names for this condition are: racquetball elbow, swimmer's elbow and Little League elbow.

It is an enthesis of the common flexor origin of the medial epicondyle. On assessment there is localised tenderness at or below the medial epicondyle. The patient reports pain on resisted wrist flexion and also on resisted forearm pronation. It is mainly the pronator teres that is affected. It results from repetitive micro trauma to flexor muscle group, mainly flexor carpi radialis during repeated pronation and flexion of the wrist (Prentice 2004).

Management should follow the same principles as tennis elbow with graded transverse friction massage to either the teno-osseous site or the musculotendinous junction. Alternatively a corticosteroid injection of tiamcinalone acetonide with local anaestheic (following principles outlined previously) to the teno-osseous site only, then 10 days relative rest. This may be supplemented with neural stretching. The philosophy of management of this overuse condition follows the same principles as tennis elbow.

Addressing any overuse issues in their sport is essential in the management. It is important that activity is appropriately "paced" in terms of frequency, duration and intensity. The initial step in management of his condition is altering any faulty performance mechanics to minimise the repetitive stress created by these activities. Plancher and Lucas (2001) noted that stressful components can be alleviated by altering the frequency, intensity or duration of play. One of the biggest errors in management is trying to progress too quickly in the exercise programme and hastily returning to sport rather than monitoring and managing the process, which should be individualised but following clear principles. Involved muscles must regain appropriate strength, flexibility and endurance with reduced inflammation and pain. Ideally in an overuse injury active movements should be pain-free before progressing to resisted exercises.

A good coach should be aware of preventable overuse injuries, like golfer's elbow, by ensuring that training is appropriately structured and that they work with the therapist and the client as a team. THe range of movement should be maintained in the pain-free range of motion with relative rest from any pain–provoking activities.

Medial collateral ligament (MCL) sprain

This may occur as either an overuse injury normally associated with excessive valgus stress due to repeated throwing activities, for example javelin and discus throwers. The common mechanism is throwers who "open up to early"; that is, they become front on at too early a stage in the throwing action. This results in inflammation of the ligament, which progresses to scarring, then calcification and very occasionally to ligament rupture. On assessment there is often a mild degree of instability on a valgus stress test. The client reports a localised tenderness over the ligament on palpation. There are often associated abnormalities such as a flexion contracture over the forearm muscles, synovitis and also loose body formation around the tip of the olecranon.

In America "Little Leaguer's elbow" (common in baseball pitchers) is the descriptive label for the adolescent complaining of a recurrent pain on the medial side of the elbow joint.

Repetitive trauma may cause a traction avulsion of the medial epicondyle (Slocum 1978). Plancher et al (2001) and Mehta and Bain (2004) noted that an MCL avulsion represents 10–30% of all injuries to the elbow.

Management in the early stages involves correction of any faulty technique and therefore reduction of biomechanical stress, grade A exercise (active, passive or resisted in the pain-free range of movement), graded transverse friction massage (TFM) applied at 90 degrees to the target tissues initially × 6 deep sweeps after analgesia progressing to 10 minutes of TFM after analgesia. Specific muscle strengthening should be started by focusing on the forearm flexors and pronators. Alternatively an injection of a small dose of corticosteroid, such as triamcinalone acetonide, mixed with local anaesthetic following principles mentioned earlier, followed by 10 days of relative rest then graded rehabilitation.

Advanced pathology may need arthroscopic removal of any loose bodies and bony spurs. Sometimes if timely, staged rehabilitation is not available significant instability may develop, which requires ligament reconstruction. Ideally this is a preventable situation and reconstruction is to be avoided as results have been disappointing.

Ulnar nerve compression

The ulnar nerve may be compressed at the elbow. Posterior to the medial epicondyle it is very superficial and is therefore at risk from direct trauma. It may also be compressed by the formation of osteophytes. This presents with numbness in the ulnar one and a half fingers and dorsal aspect of the ulnar border of the hand. It may respond to neurodynamic principles but often decompressive surgery gives best results, particularly if it has progressed to motor weakness of the thumb adductor, interossei and hypothenar muscles. The nerve may be compromised at the wrist in the canal of Guyon. This is common in cyclists and is often termed "handlebar neuropathy" at the wrist where there may be repeated pressure on the handlebars.

Muscle lesions

Minor muscle tears commonly occur in muscles bellies around the elbow. Acute lesions have been dealt with earlier. In "overtraining/overuse" situations muscles may be subjected to delayed onset muscle soreness, commonly referred to as DOMS.

There are two main theories relating to DOMS.

1. DOMS seems to be more likely with eccentric or isometric contractions. A primary component of DOMS is thought to be an inflammatory reaction in the tissues leading to minor connective tissue (Leiber and Friden 1999; Vickers 2001).

    Damage caused by eccentric exercise is thought to be mechanical damage to either the muscle or the connective tissue. Accumulation of oedema, and delay in the rate of glycogen replenishment are secondary reactions to mechanical damage (O'Reilly et al. 1987).

2. DOMS may be caused by damage to the elastic components of connective tissue at the musculotendinous junction. This damage results in the presence of hydroxyproline, a protein byproduct of collagen breakdown in blood and

urine (Clancy 1990). Muscle fibre damage results in blood serum levels including creatine kinase (Smith et al. 1994). This indicates that it is likely that there has been some damage to the muscle fibre often as a result of strenuous exercise (Evans 1987).

DOMS can best be prevented by grading activity. Athletes should be advised to begin at a moderate level of activity and gradually progress the intensity of exercise over time. So prevention is preferred. However, it responds well to local massage techniques. In general, and anecdotally, it is fair to say that massage/therapeutic touch will reduce the perception of DOMS. However, Smith, Keating, Holbert at al. (1994) hypothesied that athletic massage administered 2 hours after eccentric exercise would disrupt the initial, and crucial, accumulation of neutrophils during the acute inflammatory stage. This would result in a diminished inflammatory response and a concomitant reduction in delayed onset muscle soreness (DOMS) and serum creatine kinase (CK). In the study by Smith at al. (1994) untrained males were randomly assigned to a massage (N = 7) or control (N = 7) group. All performed five sets of isokinetic eccentric exercise of the elbow flexors and extensors. Two hours after exercise, massage subjects received a 30-minute athletic massage; control subjects rested. Delayed onset muscle soreness and CK were assessed before exercise and at 8, 24, 48, 72, 96, and 120 hours after exercise. Circulating neutrophils were assessed before and immediately after exercise, and at 30-minute intervals for 8 hours; cortisol was assessed before and immediately after exercise, and at 30-minute intervals for 8 hours. A trend analysis revealed a significant ($p < 0.05$) treatment by time interaction effect for (1) perception of DOMS; (2) CK, with the massage group displaying reduced levels; (3) neutrophils, with the massage group displaying a prolonged elevation; and (4) cortisol, with the massage group showing a diminished diurnal reduction. The results of this study suggest that sports massage will reduce DOMS and CK when administered 2 hours after the termination of eccentric exercise. This may be due to a reduced emigration of neutrophils and/or higher levels of serum cortisol.

Due to the ability of massage to encourage circulation and remove excess oedema, there are indications that massage is a helpful contribution to reducing the symptoms of DOMS (Rodenburg et al. 1994; Smith et al. 1994; Ernst 1998; Lowe 2003). However, these conclusions remain controversial, as other studies have questioned the conclusions that massage has made a significant contribution in the reduction of DOMS (Weber and Servedio et al. 1994; Tiidus and Shoemaker 1995; Field 1998; Lowe 2003).

Assessment of involvement is by application of selective tissue tension. Pain is reported on resisted muscle contraction and involvement is confirmed by palpation. There is a good response to local transverse friction massage and also to sports massage. To target the muscle belly, the client is positioned with the muscle supported comfortably with the muscle in a shortened position. By positioning muscles in a shortened position the tension is taken off the muscle. One of the key in indications for utilising a massage technique is for the reduction of muscle tightness (Lowe 2003). Muscles tend to tighten in response to excessive neuromuscular stimulation. Some research studies have shown that massage techniques have a significant impact on the reduction of that excessive neuromuscular stimulation (Morelli et al. 1991; Sullivan et al. 1991; Braverman and Schulman 1999). A range of conservative treatment modalities are often utilised in order to address excessive muscular tension and of those treatment selections. Liebenson (1989) found massage to be one of the most effective in achieving a reduction in muscle tension.

There have been a number of studies done with tendons rather than muscle bellies that have found strong evidence to support the use of massage to help in fibroblast proliferation in these tendons that have significant collagen degeneration (Davidson et al. 1997; Brosseau et al. 2002; Cook et al. 2000).

Massage with the muscle in a shortened position would normally decrease the pain and increase the range of movement by improving the circulation. The increase in range of movement that is achieved should be followed up with exercise in the pain-free range of movement. The analgesic effect of massage has been well documented (Field 2000; Yates 2004). Pain is a multi-factorial and complex phenomenon; the mechanisms for reducing pain using massage are also varied. Mechanisms include: muscle relaxation, improved circulation, deactivation of trigger points, the release of endorphins and serotonin, the pain gate theory and the promotion of restful sleep (Benjamin and Lamp 2005)

Osteoarthrosis (OA) of the elbow

Osteoarthrosis is often secondary to old fractures which have involved the articular surfaces of the elbow and may give rise to loose bodies forming which cause "pseudo locking" of the joint and restricts the movement. The client often develops the trick of unlocking the joint themselves. If it is noted on X-ray that loose bodies are evident they should be removed to reduce future re-occurrences of locking and also to reduce the risk of causing further damage to the articular surface of the elbow joint. Elbow replacements are relatively rare.

Acute treatment

Cryotherapy

Ice or cold treatment is often termed cryotherapy. Cold is the initial treatment of choice for virtually all acute conditions involving injuries to the musculoskeletal system (Cox 1993). This may take the form of cold packs, ice towels, cold baths or ice massage (Kitchen 2004). It can be used to control acute inflammation and thereby accelerate recovery from injury or trauma. Due to the low thermal conductivity of underlying subcutataneous fat tissues, applications of cryotherapy for short periods are ineffective in the cooling of deeper tissues. For this reason longer treatments of 20–30 minutes are recommended. Cold treatments are understood to be more effective and efficient in reaching deeper tissues than most forms of heat (Prentice 2004). Ice should be applied to the injured area until signs and symptoms of inflammation have gone. Ice should be used for at least 72 hours after an acute injury, (Calabrese and Rooney 1986; Cox 1993). Griffin (1997) suggests that re-warming begins about 20 seconds after the preceding application, and that statistically significant ($p < 0.05$) decreases in temperature can be produced with repeated applications of cryotherapy.

It is recommended that ice be applied following an injury and throughout the inflammatory phase (ACPSM 1999).

Cold applied to an acute injury will lower metabolism in the injured area and as a result lower the tissue demands for oxygen, thus reducing hypoxia. This benefit extends to uninjured tissue, preventing tissue death from spreading to adjacent normal cellular structures. Cold may be used in order to reduce the reflex muscle guarding and spastic conditions that accompany pain. Its pain relieving effects is probably one of its greatest clinical benefits (Prentice 2004).

With ice treatment applied for 20–30 minutes the athlete may report an uncomfortable feeling of cold, followed by a "burning" sensation, an aching and finally complete numbness (Calabres 1986; Cox 1993).

Cooling reduces swelling (Basur et al. 1976). It must be taken into consideration that often cooling is not done in isolation in an acute injury, it is often accompanied by compression, which means from a research and evidence base perspective it is difficult to state that the benefits are due to cooling alone. The cold treatment may lead to a reduction in bleeding which may be due to a resultant reduction in blood flow and is more likely to occur during the early phase of treatment. Sauls (1999) examined the evidence for the efficacy of cooling by undertaking a review of the effects of cold for pain relief; benefits were noted for certain orthopaedic procedures and for injections in adults.

Ultrasound

Ultrasound should be applied in a pulsed mode at a low intensity during the acute phase of inflammation to minimise the risk of aggravating the condition and to accelerate recovery.

The mast cells release histamine, which is one of the key chemicals to modify the wound after injury. Research has shown that a single treatment using therapeutic ultrasound in the early acute stage of the healing process can stimulate mast cells to degranulate and to release histamine into the surrounding tissues (Fyfe and Chahl 1982; Hamilton 1986). It is hypothesised that one of the effects of ultrasound is to stimulate the mast cell to degranulate by increasing the mast cell's permeability to calcium. Mast cells are one of the wound factors they have a key role in that they orchestrate the early repair sequences (Clark 1990). Histamine is released from mast cells, which causes vasodilation and an increase in the cell permeability due to the swelling of endothelial cells. Vasodilation and active hyperaemia are important in plasma formation and supplying leukocytes to the injured area. This helps in the early formation of a clot or plug. These plugs obstruct local lymphatic drainage and localise the injury response (Prentice 2004).

Young and Dyson (1990) found that using ultrasound at an intensity of 0.5W/cm2 and a frequency of 3.0 MHz appeared the most effective setting on the stimulation of fibroblast ("building-block") population growth.

For many years therapeutic ultrasound was mistakenly thought of an anti-inflammatory modality in the rehabilitator's "tool kit" (Reid 1981; Snow and Johnson 1988). It is easy to see why there was this error in clinical reasoning, as often a rapid resolution of oedema was noted after using ultrasound (El Hag et al. 1985). It is unfortunate that the evidence to enlighten our reasoning on this was available from as early as 1978 by Hustler, Zarod and Williams. At that time "evidence-based practice" was not part of the undergraduate programme in the way it is embedded into current practice. Goddard et al. (1983) identified in their research that ultrasound encourages oedema formation to occur more quickly and to accelerate the inflammatory phase, therefore accelerating the wound into the proliferative stage of healing more effectively and efficiently than "sham" irradiated ultrasound groups. A few years later, similar findings were reported by Fyfe and Chahl (1985).

Binder and colleagues (1985) reported significantly enhanced recovery in clients with lateral epicondylitis treated with ultrasound compared with those treated with sham ultrasound. Robertson and Baker (2001) in their review of the effectiveness of therapeutic ultrasound found it was difficult to compare studies as few of them had adequate methodology and covered a wide range of conditions. Also the dosages used in the studies varied considerably often for no discernable reason. This was further complicated when more than one electrotherapy modality was used (Herrera-Lasso and Fernandez Dominguez 1993; Nussbaum et al. 1994; van der Heijden et al. 1999). Surveys on the use of therapeutic physical agents in rehabilitation in USA, Canada, England, Australia and the Netherlands have all concluded that ultrasound therapy is by far the most widely used physical agent currently available to clinicians (Robinson and Snyder-Mackler 1988; ter Harr et al. 1988; Lindsey et al. 1990, 1995; Robertson and Spurritt 1998; Roebroeck et al. 1998). This was supported by a recent systematic review by Chinn and Clough (2007). Clinicians are utilising treatment modalities in a more enlightened way in clinical practice. *Note*: ultrasound is not a pain modulation therapy, but can be used clinically, depending on desired effect, to assist with the management of inflammation and initial healing.

Corticosteroids

An alternative intervention to therapeutic ultrasound would be the use of a corticosteroid injection. The suggested guide for a dose by the Society of Orthopeadic Medicine is a dosage of approx 10mg total volume, in a mixture of triamcinalone acetonide (in a small, 0.25 ml, concentrated dose of KENALOG or an equivalent mixed with 0.75 ml of local anaesthetic), which aims to reduce the inflammation. Good results have been shown in the short term by several different research groups (Price et al. 1991; Haker and Lundeberg 1993; Assendelft et al. 1996; Verhaar et al. 1996; Hay et al. 1999).

Different general practitioners and injecting therapists favour differing commercial drugs. Price and his team (1991) found a quicker relief from pain and a reduction in the requirement for repeat injections using 10mg of triamcinalone compared to using 25 mg of hydrocortisone or lidocaine alone in the short term. There is currently a move to increasing doses in some areas (particularly in the shoulder), however Price et al. (1991) found that an increased volume injection of 20 mg of triamcinalone produced similar clinical results to 10mg and the higher dosage was much more likely to produce skin atrophy.

Skin atrophy is a side effect of injection. However, the more significant but very rare complication of injection is anaphylaxis, which is described as a rapid and often unanticipated, life-threatening syndrome. It requires very prompt action in order to treat the resulting laryngeal oedema, bronchospasm, hypotension and associated tachycardia.

The CSP *Clinical Guideline for the Use of Injection Therapy by Physiotherapists* (CSP 1999), provides guidance on the management of anaphylactic shock:

- stop delivery of the drug

- call medical help

- administer adrenaline

- administer cardiopulmonary resuscitation.

Table 18.4 Aims and contra-indications of TFM (adapted from Keeson and Atkins 2005)

Key aims	Contra-indications
To produce therapeutic movement	Infection
To induce pain relief	Rheumatoid arthritis (RA)
To produce a traumatic hyperaemia in chronic lesions	
To improve functional movement	

(Kesson and Atkins 2005)

Hospitals and alternative health care settings have local policies and protocols in place and the reader is advised to check on local policies and procedures that are in place as injection therapy is increasing in practice but still has areas of controversy. Most physiotherapists in most health settings are not allowed to mix drugs (corticosteroid and local anaesthetic) as mixing drugs is outside of the scope of practice. They may inject but only having obtained an authorised prescription.

Graded transverse friction massage (TFM)

It is essential to have a high level of applied anatomy in order to achieve effective results with this technique. It requires accurate localisation to the exact side of the lesion to be graded (gentle or deep) transverse friction massage at a 90 degree/transverse sweep to the direction of the fibres in order to discourage the stationary attitude of fibres that promote anomalous cross-link formation.

There are four key aims of TFM and two contra-indications, as shown in Table 18.4.

Caution should taken and clinical judgement with regards to use of TFM with diabetic clients and also to clients who have are long term-steroid users due to potential effect on the connective tissue.

Grade A (pain-free) mobilisation

A grade A mobilisation is an active, passive or active/assisted movement performed within the client's *pain-free* range of movement. It is normally applied to painful or acutely inflamed lesions. Often it is made possible by preceding the grade A mobilisation with the application of graded transverse friction massage (Saunders 2000).

There are 5 key aims of transverse friction massage:

- To promote tissue agitation

- To prevent stationary attitude of fibres

- To apply longitudinal stress

- To promote normal function

- To reduce a loose body (it is recommended this is practised on a post graduate course)

(Kesson and Atkins 2005)

Kinetic chains and appropriate graded exercise

In the elbow to enhance muscular balance and the neuromuscular control, exercises of the surrounding agonists and antagonists should be included.

Exercise should be appropriate to the stage of healing and of the appropriate grade and type.

It is common practice to classify exercises as "closed" or "open" chains. An open chain exists when the hand is *not* in contact with the ground so the distal segment is mobile and is not fixed (Hillman 1994; Prentice 2004). In a closed chain the hand is weight-bearing, so the distal segment is fixed or stabilised.

The literature shows a clear trend to support the fact that "closed" chains should precede any progression to "open" chains but most of the research has been conducted using the lower limb (Cohen 2001). In the elbow, closed chain exercises should be used in order to improve the dynamic stability of the more proximal muscles surrounding the elbow in sports where the elbow is required to provide more proximal stability. Open chain exercises for strengthening flexion, extension, pronation and supination are essential to regain high-velocity dynamic movements of the elbow that are required in sports requiring throwing type activities (Prentice 2004).

There are advantages and disadvantages to both open and closed exercise. Fowler (2008, p18), sets out an argument in his paper that *"this classification is flawed and the rationale for it questioned"*. On reviewing the literature Fowler concludes on page 20 *"The critical differences between the exercises described in the majority of the literature is not that they are open or closed but that they involve either a single joint or a combination of joints."* This may be an accurate conclusion from a biomechanical standpoint but the concept of open and closed chains is a concept that is relatively easy to convey to an athlete and is widely used to good clinical effect in clinical practice. It is more appropriate that the clinician utilises their skills of applied anatomy and clinically reasons to decide on the most appropriate exercise and to review its effects than to become too focused on pedantic points of terminology to the detriment of applied clinical practice.

The Society of Orthopaedic Medicine uses grades: A, B and C.

Grade A, would be defined as a pain-free, active, passive or resisted exercise. Grade B is a mobilisation performed at the end of available range. It is a specific, cyclical, sustained stretching technique in the plastic range. The aim is to cause permanent elongation of connective tissue. Grade C (for example the mills manipulation) is a manipulative technique. It is a passive technique performed at the end of available range. It is a minimal amplitude high-velocity thrust (Kesson and Atkins 2005).

Factors relating to return to sport

Understanding the pathomechanics of injury

The chapter has outlined common acute and overuse injuries. It is essential that a thorough working knowledge of applied anatomy and the principles of biomechanics are appreciated in order to identify existing and potential adaptive or compensatory movements (Kirkendall et al. 2001).

Understanding of the relationship of the healing process to exercise intensity

Progression or adaptation of a rehabilitation programme should be based on stages of healing of the injury. Exercise that is too "high demand" in terms of time or intensity is likely to be detrimental to progress.

It is helpful to be aware of signs that the intensity of exercises being used in the programme are excessive for the stage of healing. Signs to look out for are:

- increase in the pain a client reports
- loss of range of movement
- plateau in progress
- increase in the laxity of a healing ligament.

(Tippett and Voight 1999)

Multi disciplinary team approach and establishing agreed short- or long-term goals

In order to achieve maximum potential in an effective manner it is key that the roles of the team are clear and understood to reduce conflict in management and avoid confusing the athlete. It is essential to review goals as progress is made. It is critical that the athlete is actively involved in the rehabilitation process (Piccininni and Drover 1999).

Restoration of muscular strength, endurance, power and neuromuscular control

All these elements are essential to restore the athlete to pre-injury status. A functional, strengthening programme should include exercises that include all three planes of movement and concentrate on a combination of concentric, eccentric and isometric exercise (Clark 2001). This must also ensure appropriate loading, velocity and muscle actions required by the individual's sport.

Therapeutic versus conditioning exercise

Therapeutic exercise focuses on the specific injury and exercises to facilitate and enhance recovery. It is also essential to maintain the "conditioning" of the client by ensuring there level of fitness is maintained. The term "therapeutic exercise" is more widely used to indicate exercise used in a rehabilitation programme (Kisner and Colby 1996).

Case study

Emily is a 38-year-old, university administrator who has developed, over the past month, a pain she rates as 7/10 on the lateral aspect of her elbow.

She reports an uncomfortable "ache" when she is typing, answering the phone, writing and also when holding her coffee mug. She is aware of some discomfort most of the time.

She is a member of the tennis league and finds playing tennis is particularly painful. She cannot recall an injury but on more focused history taking Emily recalls a forceful missed shot in the final game of a tournament about a month ago. She noted a local pain at the time. She had changed her racquet at about the same time and had been practising for longer periods to prepare for this important game. She is keen to play in a tournament in 2 months' time.

On presentation now Emily reports local pain over the lateral aspect of the elbow but this has now spread into the forearm, wrist and back of the hand.

The onsite University Sport Injury clinic has diagnosed lateral epicondylitis/"tennis elbow" of the teno-osseous site.

Principles of management of Emily

Rest

Ideally, Emily should rest from all the activities that aggravate her problem, in terms of work and sport. A compromise may have to be reached in reality. Rest is the single, most important factor in resolving the problem. An epicondylitis clasp or proprioceptive strapping to provide rest from end-of-range extension; or even a plaster back slab may be utilised to rest from aggravating movement. She should maintain active movement in the pain-free range.

Mills manipulation

This is performed following preparation by transverse friction massage (until analgesia is reached). The Mills manipulation is performed *only* if the source of the tennis elbow is at the teno-osseous site and if the patient has got full range extension of the elbow. The reason for this is that if full extension is available the manipulation force will fall on the adhesions and improve the range of movement and reduce pain. However, if full elbow extension is not available then the manipulation force would fall onto the elbow joint and potentially cause a traumatic arthritis. It is essential that the site of the problem is accurately located as it would be inappropriate for the Mills manipulation to be performed to treat symptoms at the other three sites. It is strongly advised that a Society of Orthopaedic Medicine Course is attended for training under supervision before adding this treatment option to your therapeutic "toolkit".

Graded transverse friction massage (TFM)

See above for principles of TFM.

Ice/cryotherapy or therapeutic ultrasound

Ice treatment is useful in the acute phase or indeed in acute exacerbation of a chronic elbow, which is common in tennis elbow. However, ice should be discontinued when acute inflammation has resolved to avoid slowing chemical reactions or impairing the circulation during the latter stages of healing when the effects of ice can impede recovery.

Research has shown that prophylactic use of ice AFTER exercise has been shown to reduce the severity of delayed-onset muscle soreness (DOMS). DOMS is thought to be the result of inflammation from muscle and connective tissue damaged caused directly by exercise (Jones et al. 1986). The prophylactic use of ice after exercise on an area with a pre-existing inflammation can be very effective in reducing post-activity soreness.

Low intensity, pulsed ultrasound is recommended in the acute phase. Continuous ultrasound at a high enough intensity to increase tissue temperature may be applied in conjunction with other manual approaches in order to assist the resolution of a chronic problem accompanied by soft tissue shortening due to adhesions and scarring.

Accessory movements

Accessory movements are movement of the joint that a person cannot perform actively by himself but when performed on the client by the rehabilitator. The principles are to fix close to the joint with one hand and to gently use graded passive movement in order to facilitate the restoration of normal joint range of motion.

Hydrocortisone injection

This is a localised anti-inflammatory injection (normally triamcinalone acetonide) that is more powerful than oral non-steroidal anti-inflammatory drugs (NSAIDs). In chronic injuries this is accepted more readily for its quick and effective pain relief, although it may only be temporary (Almekinders 1999). There is a risk of weakening the tendon particularly in weight-bearing tendons. The weakness could lead to rupture. The elbow is not at as much risk as the hip, knee or ankle joint, although in certain activities the upper limb could be considered weight-bearing but not to the same extent as the lower limb (Cameron 1995). In an acute injury it becomes a more controversial decision to use this type of injection as it can impede the healing process by eliminating the inflammatory response. As the inflammatory response part of the healing process most doctors consider the use of steroid injections in acute injuries to be inappropriate. The evidence on corticosteroids injections for the treatment of tennis elbows is not conclusive (Assendelft et al. 1996). In a randomised controlled trial in Australia (Bisset et al. 2006), researchers found that physiotherapy combining elbow manipulation and active exercise had a superior benefit in the first 6 weeks to corticosteroid injections after 6 weeks providing a reasonable alternative to injections in the mid to long term. The significant short term benefits of corticosteroid injection are paradoxically reversed after 6 weeks, with high recurrence rates, implying that this treatment should be used with caution in tennis elbow.

Re-education

This should include:

- A clear explanation of the impact of the aggravating factors on the present condition and potential for re-occurrence. To work through goals with Emily, focusing on a return to work and sport in an informed and constructive manner.

- To assess technique and approach to training

  o Looking at her tennis technique (working with the coach if possible).

  o Restore full pain-free range of movement, flexibility and endurance in the wrist extensor muscle group.

  o Pacing of her training and potential to overload. Discuss breaking up activities to avoid overuse of vulnerable structures. Caution her that overdoing the programme may aggravate the condition

  o To consider neurodynamic involvement due to the sensitising manoeuvres of long arm stretches and neck movements

  o To maintain Emily's general fitness, whilst her specific elbow injury is being rehabilitated.

  o Encourage general strengthening for stability, support and to correct any muscle imbalance, starting with static/isometric and progressing to dynamic/isometric strength. The dynamic contraction exercises can be sub-divided into lengthening (eccentric) and shortening (concentric) contractions of varying speed. Include early closed chain exercises such as press-ups.

- To support Emily in a sensitive and constructive manner through her rehabilitation and to explore alternative ways of doing her job

References

ACPSM (1999) *Guidelines for the management of soft tissue musculoskeletal injury with Protection, Rest, Ice, Compression and Elevation (PRICE) during the first 72 hours*. London: Chartered Society of Physiotherapy.

Allander, E. (1974) Prevalence, incidence and remission rates of some common rheumatic diseases or syndromes. *Scandinavian Journal of Rehabilitation Medicine*, 3, 145–153.

Almekinders, L.C. (1999) Anti- inflammatory treatment of muscular injuries in sports. An update on recent studies. *Sports Medicine*, 28, 383–388.

Assendelft, W.J., Hay, E.M., Adshead, R. et al. (1996) Corticosteroid injections for lateral epicondylitis: a systematic review. *British Journal of General Practice*, 46, 209–216.

Bahr, R., Maehulm, S. (2004) Clinical guide to Sports Injuries; Leeds, Human Kinetics.

Bailey, R.A. and Brock, B.H. (1957) Hydrocortisone in tennis elbow – a controlled series. *Journal of the Royal Society of Medicine*, 50, 389–390.

Basur, R., Shepherd, E. and Mouzos, G. (1976) A cooling method for the treatment of ankle sprains. *Practitioner*, 216, 708.

Baumgarten, T.E., Andrews, J.R. and Satterwhite, Y.E. (1996) The arthroscopic classification and treatment of osteochondritis dissecans of the capitullum Presented 22nd meeting of AOSSM, Lake Buena Vista, Florida, June.

Baumgarten, T.E., Andrews, J.R. and Satterwhite, Y.E. (1998) The arthroscopic classification and treatment of osteochondritis dissecans of the capitullum. *American Journal of Sports Medicine*, 26 (4), 520–523.

Benjamin, P.J. and Lamp, S.P. (2005) Understanding Sports Massage, 2nd edn. Champaign, IL: Human Kinetics.

Binder, A., Hodge, G., Greenwood, A.M., et al (1986) Is Therapeutic ultrasound effective in treating soft tissue lesions? *Br Med J 290*: 512–514

Bissett, L., Beller, B., Jull, G., Brooks, P., Darnell, R. and Vicenzino, B. (2006) Mobilisation with movement and exercise, corticosteroid injection, or wait and see for tennis elbow: randomised trial. *British Medical Journal*, 333, 939.

Braverman, D.L. and Schulman, R.A. (1999) Massage techniques in rehabilitation medicine. *Physical Medical Rehabilitation Clinics North America*, 10 (3), 631–649.

Brosseau, L., Casimiro. L., Milne, S., et al (2002) Deep transverse friction massage for treating tendonitis (Cochrane Review) Cochrane Database Syst Rev (1): CD003528

Brukner, P. and Khan, K. (2002) *Clinical Sports Medicine*, revised 2nd edition. New York: Mc Graw Hill.

Brukner, P.D., Bennell, K.L. and Matheson, G.O. (1999) Stress fractures. Melbourne: Blackwell Scientific, Asia

Buehler, M.J. and Thayer, D. (1988) The elbow flexion test: A clinical test for the cubital tunnel syndrome. *Clinical Orthopaedics*, 232, 213–216.

Butler, D.S., Matheson, J. and Boyaci, A. (2000) *The Sensitive Nervous System*. Adelaide: NeuroOrthopaedic Institute.

Calabrese, L. and Rooney, T. (1986) The use of non-steroidal anti-inflammatory drugs in sports. *Physician and Sports Medicine*, 14, 89–97.

Caldwell, G.L. and Safran, M.R. (1995) Elbow problems in the athlete. *Orthopaedic Clinics of North America*, 26, 465–485.

Cameron, G. (1995) Steroid arthropathy: myth or reality? A review of the evidence. *Journal of Orthopaedic Medicine*, 17, 51–55.

Chard, M.D. and Hazelman, B.L. (1989) Tennis elbow – a reappraisal. *British Journal of Rheumatology*, 28, 186–190.

Chinn, N. and Clough, A.E. (2007) A systematic review of the use of therapeutic ultrasound in musculoskeletal injuries. Presented at the Symposium of Society of Orthopaedic Medicine and British Institute of Musculoskeletal Medicine (SOM/BIMM), London.

Clancy, W. (1990) Tendon trauma and overuse injuries. In W. Leadbetter, J. Buckwalter and S.Gordon (eds), *Sports Induced Inflammation*. Park Ridge, IL: American Acadamy of Orthopaedic Surgeons.

Clark, R.A.F. (1990) Cutaneous wound repair. In L.E. Goldsmith (Ed.) *Biochemistry and Physiology of the Skin*. Oxford: Oxford University Press, pp 576–601.

Clark, M. (2001) *Integrated Training for the New Millennium*. Calabasas, CA: National Academy of Sports Medicine

Cohen, Z. (2001) Patellofemoral stresses during open and closed kinetic chain exercises. An analysis using computer simulation. *American Journal of Sports Medicine*, 29, 480–487.

Cole, M. (2005) Reflection in healthcare practice: Why is it useful and how might it be done? *Work Based Learning in Primary Care*, 3, 13–22.

Cook, J.L., Khan, K.M., Maffulli, N. and Purdam, C. (2000) Overuse tendinosis, not tendinitis part 2. Applying the new approach to patellar tendinopathy. *Physician Sports Medicine*, 28 (6), 31–36.

Cox, D. (1993) Growth factors in wound healing. *Journal of Wound Care*, 2 (6), 339–342.

Cross, V. (2004) Linking reflective practice to evidence of competence. *Reflective Practice*, 5 (1), 1–31.

Chartered Society of Physiotherapy (CSP) (1999) *Clinical Guideline for the Use of Injection Therapy by Physiotherapists*. London: CSP.

Cyriax, J.H. (1936) The pathology and treatment of tennis elbow. *Journal of Bone Joint Surgery*, 4A, 921–940.

Cyriax, J.H. (1982) *Textbook of Orthopaedic Medicine*, vol *1*, 8th edn. London: Balliere Tindall.

Cyriax, J.H. and Cyriax, P.J. (1983) *Illustrated Manual of Orthopaedic Medicine*. London: Butterworths.

Davidson, C.J., Ganion, L.R., Gehlsen, G.M., Verhoestra, B., Roepke, J.E. and Sevier, T.L. (1997) Rat tendon morphologic and functional changes resulting from soft tissue mobilisation. *Medicine and Science in Sport and Exercise*, 29 (3), 313–319.

El Hag, M., Coghlan, K., Christmas, P. et al. (1985) The anti-inflammatory effects of dexamethasone and therapeutic ultrasound in oral surgery. *British Journal of Oral Maxillofacial Surgery*, 23, 17–23.

Ernst, E. (1998) Does post exercise massage treatment reduce delayed onset muscle soreness- a systematic review. *British Journal of Sports Medicine*, 32 (3), 212–214.

Evans, P. (1980) The healing process at cellular level; a review. *Physiotherapy*, 66, 256–259

Evans, W.J. (1987) Exercise induced skeletal muscle damage. *Physician and Sports Medicine*, 15, 189–200.

Field, T.M. (1998) Massage therapy effects. *American Psychology*, 53 (12), 1270–1281.

Field, T. (2000) *Touch Therapy*. London: Churchill Livingston.

Flory, J. and Emanuel, E. (2004) Interventions to improve research participants' understanding in informed consent for research. *Journal of the American Medical Association*, 292, 1593–1601.

Fowler, N. (2008) When chains just don't add up. *SportEX Medicine*, 35 (Jan), 18–20.

Fyfe, M.C. and Chahl, L.A. (1982) Mast cell degranulation: A possible mechanism for of action for therapeutic ultrasound. *Ultrasound in Medicine and Biology*, 8 (suppl 1), 62.

Fyfe, M.C. and Chahl, L.A. (1985) The effect of single or repeated applications of "therapeutic" ultrasound on plasma extravasation during silver nitrate induced inflammation of the rat hindpaw ankle joint in vivo. *Ultrasound in Medicine and Biology*, 11, 273–283.

Goddard, D.H., Revell, P.A., Cason, J., Gallagher, S. and Currey, H.L.F. (1983) Ultrasound has no anti-inflammatory effect. *Annals of Rheumatic Diseases*, 42, 582–584.

Griffin, S. (1997) Study to examine the changes in skin temperature produced by. the application of ice spray on the ankle, BSc. Dissertation King's college London.

Haker, E. and Lundeberg, T. (1993) Elbow-band, splintage and steroids in lateral eicondylagia (tennis elbow). *The Pain Clinic*, 6 (2), 103–112.

Hamilton, P.G. (1986) The prevalence of humeral epicondylitis: a survey in general practice *Journal of the Royal College of General Practitioners*, 36, 464–465.

Harris, P.F., Ranson, C. (2008) Atlas of Living and surface Anatomy for Sports Medicine; London, Churchilll Livingston.

Hay, E.M., Paterson, S., Lewis, M. et al. (1999) Pragmatic randomised controlled trial of local corticosteroid injection and naproxen for treatment. *British Medical Journal*, 319, 964–968.

Hay, S. and Bell, S.N. (1998) Results of Therapeutic Elbow Arthroscopy (abstract) Sydney: International Congress on Surgery of the Shoulder, p.142.

Herrera-Lasso, I. and Fernandez Dominguez, L. (1993) Comparative effectiveness of packages of treatment including ultrasound or transcutaneous electrical nerve stimulation in painful shoulder syndrome. *Physiotherapy*, 79, 251–253.

Hilliard, C. (2006) Using structured reflection on a critical incident to develop a professional portfolio. *Nursing Standard*, 21 (2), 35–40.

Hillman, S. (1994) Principles and techniques of open chain rehabilitation. *Journal of Sport Rehabilitation*, 3 (4), 319–330.

Hudak, P.L., Cole, D.C. and Haines, A.T. (1996) Understanding prognosis to improve rehabilitation: the example of lateral elbow pain. *Archives of Physical Medicine and Rehabilitation*, 77, 586–593.

Hustler, J.E., Zarod, A.P. and Williams, A.R. (1978) Ultrasonic modification of experimental bruising in the guinea pig pinna. *Ultrasonics*, 16 (5), 223–228.

Hutson, M.A. (1990) *Sports Injuries, Recognition and Management*. Oxford: Oxford Medical Publications.

Illingworth, C.M. (1975) Pulled elbow: a study of 100 patients. *British Medical Journal*, 2, 672–674.

Jones, M. and Rivett, D. (2005) *Clinical Reasoning for Manual Therapists*. Amsterdam: Elsevier.

Jones, D., Newham, D., Round, J. et al. (1986) Experimental human muscle damage : morphological changes in relation to other indicies of damage. *Journal of Physiology*, 375, 435–448.

Kesson, M. and Atkins, E. (2005) *Orthopaedic Medicine: A practical approach*. Oxford: Butterworth-Heinemann.

Kirkendall, D.T., Prentice, W.E. and Garrett, W.E. (2001) Rehabilitation of muscle injuries. In G. Puddu, A. Giombini and A. Slevanetti (eds), *Rehabilitation of Sports Injuries: Current concepts*. Berlin: Springer.

Kisner, C. and Colby, A. (1996) *Therapeutic Exercise: Foundations and techniques*. Philadelphia, PA: F.A. Davies.

Kitchen, S. (2004) *Electrotherapy – Evidence based practice*, 11th edn. Amsterdam: Elsevier.

Kivi, P. (1983) The etiology and conservative treatment of humeral epicondylitis. *Scandinavian Journal of Rehabilitation Medicine*, 15, 37–41.

Kumar, V., Abbas, A.K. and Fausto, N. (2009) Robbins and Cotran Pathologic Basis of Disease, 8th edn. New York: Elsevier Saunders.

Kuntzman, A.J., Tortora, G.J. (2010) Anatomy and Physiology for the Manual Therapies. New Jersey, Wiley

Labelle, H., Gilbert, R., Joncas, J. et al (1992). Lack of scientific evidence for the treatment of lateral epicondylitis of the elbow – a meta-analysis. *Journal of Bone Joint Surgery*, 74b, 646–651.

Lieber, R.L. and Friden, J. (1999) Mechanisms of muscle injury after eccentric contraction. *Journal of Science and Medicine in Sport*, 2 (3), 253–262.

Liebenson, C. (1989) Active Muscular relaxation techniques. Part 1 Basic principles and methods. *Journal of Manipulative Physiology and Therapy*, 12 (6), 446–454.

Lidz, C.W., Applebaum, P.S. and Meisel, A. (1988) Two models of implementing informed consent. Archives of Internal Medicine, 148 (6), 1385–1389.

Lindsey, D., Dearnes, L. et al. (1990) A survey of electromodality usage in private practices. *Australian Journal of Physiotherapy*, 36, 348–356.

Lindsey, D., Dearnes, L. et al. (1995) Electrotherapy usage trends in private practice in Alberta. *Physiotherapy Canada*, *47*, 30–34.

Loudon, J., Swift, M. and Bell, S. (2008) *The Clinical Orthopaedic Assessment Guide*. Champaign, IL: Human Kinetics.

Lowe, W.W. (2003) *Orthopaedic Massage Theory and Technique*. New York: Mosby.

Lutz, F.R. (1991) Radial tunnel syndrome: an etiology of chronic lateral elbow pain. *Journal of Orthopaedics and Sports Physical Therapy*, *14* (i), 14–17

Magra, M., Caine, D. and Maffuli, N. (2007) A review of paediatric elbow injuries in sports. *Sports Medicine*, *38* (8), 717–735.

McRae, R. (2003) *Clinical Orthopaedic Examination*, 5th edn. London: Churchill Livingston.

Mehta, J.A. and Bain, G.I. (2004) *Surgical Approaches to the Elbow. Hand Clinics*. Amsterdam: Elsevier.

Morelli, M., Seabourne, D.E. and Sullivan, S.J. (1991) H-reflex modulation during manual muscle massage of human triceps surae. *Archive of Physical Medicine and Rehabilitation*, *72* (11), 915–919.

Nirshel, R.P. (1992) Elbow tendinosis/tennis elbow. Clinical Sports Medicine, *11*, 851–870.

Noteboom, T., Cruver, R., Keller, J. et al. (1994) Tennis elbow: a review. *Journal of Orthopaedic and Sports Physical Therapy*, *19*, 357–366.

Novak, C.B., Lee. G.W., MacKinnon, S.E. and Lay, L. (1994) Provacative testing for cubital tunnel syndrome. *Journal of Hand Surgery (American)*, *19* (5), 817–820.

Nussbaum, E.L., Biemann, I. and Mustard, B. (1994) Comparison of ultrasound/ultraviolet- C and laser for treatment of pressure ulcers in patients with spinal cord injury. *PhysicalTherapist*, *74*, 812–823.

O'Reilly, K., Warhol, M., Fielding, R.A. et al. (1987) Eccentric exercise induced muscle damage impairs muscle glycogen depletion. Journal of Applied Physiology, *63*, 252–256

Piccininni, J. and Drover, J. (1999) Athlete- patient education in rehabilitation: Developing a self-directed program. *Athletic Therapy Today*, *4* (6), 51.

Plancher, K.D., Halbrecht, D.J. and Lourie, G.M. (1996) Medial and lateral epicondylitis in the athlete. *Clinics in Sports Medicine*, *15* (2), 283–305.

Plancher, K.D. and Lucas, K.D. (2001) Fracture dislocation of the elbow in athletes. *Clinics in Sports Medicine*, *20* (1), 59–76.

Prentice, W.E. (2004) *Rehabilitation Techniques for Sports Medicine and Athletic Training*. New York: McGraw–Hill.

Price, R., Sinclair, H., Heinrich, I. and Gibson, T. (1991) Local injection treatment of tennis elbow-hydrocortisone, tiamcinalone and lidnocaine compared. *British Journal of Rheumatology*, *30*, 39–44.

Putnam, M.D. and Cohen, M. (1999) Painful conditions around the elbow. *Orthopaedic Clinics of North America*, *30*, 109–118.

Rayan, G.M. (1992) Proximal ulnar nerve compression: cubital tunnel syndrome. *Hand Clinics*, *8*, 325–336.

Rayan, G.M., Jenson, C. and Duke, J. (1992) Elbow flexion test in the normal population. *Journal of Hand Surgery*, *17A*, 86–89.

Regan, W.D., Wold, L.E., Coonrad, R. et al. (1992) Microscopic histopathology of chronic refractory lateral epicondylitis. *American Journal of Sports Medicine*, *20*, 746–749.

Reid, D.C. (1981) Possible contraindications and contraindications and precautions associated with ultrasound therapy. In A.J. Mortimer and N. Lee (eds) *Proceedings of the International Symposium on Therapeutic Ultrasound*. Winnipeg: Canadian Physiotherapy Association, p 274.

Reid, D.C. and Kushner, S. (1993) The elbow region. In R. Donatelli and M.J. Wooden (eds), *Orthopaedic Physical Therapy*. London: Churchill Livingston, pp. 203–232.

Robertson, V.J. and Spurritt, D. (1998) Electrophysical agents: Implications of EPA availability and use in private practices. *Physiotherapy*, *84*, 335–344.

Robertson, V.J. and Baker, K.G. (2001) A review of therapeutic ultrasound. *Effectiveness Studies*, *81* (7), 1339–1350.

Robinson, A.J. and Snyder-Mackler, L. (1988) Clinical application of electrotherapeutic modalities. *Physical Therapy*, *68*, 1235–1238,

Robinson, C.M., Hill, M.F., Jacobs, N., Dall, G., Court-Brown, C.M. (2003) Adult distal humeral metaphyseal fractures: Epidemiology and results of treatment. *Journal of Orthopaedics*, *17* (1), 38–47.

Rodenburg, J.B., Steenbeek, D., Schiereck, P. and Bar, P.R. (1994) Warm-up, stretching and massage diminish harmful effects of eccentric exercise. *International Journal of Sport Medicine*, *15* (7), 414–419.

Roebroeck, M.E., Dekker, J., et al. (1998) The use of therapeutic ultrasound by physical therapists in Dutch primary health care. *Physical Therapy*, *78*, 470–478.

Ross, G., McDevitt, E.R., Chronister, R. et al. (1999) Treatment of simple elbow dislocation using an immediate motion protocol. *American Journal of Sports Medicine*, *199*, (27), 308–311.

Sauls, J. (1999) *Worldviews on Evidence Based Nursing*, *6* (1), 103–111.

Saunders, S. (2000) *Orthopaedic Medicine Course Manual*. Saunders.

Shacklock, M. (2005) *Clinical Neurodynamics: A new system of neuromuscular treatment*. Amsterdam: Elsevier.

Shapiro, M.S., Wang, J.C., (1995) Elbow Fractures Treating to avoid complications. Phys. Sports. med 1995; *23*: 39–50

Slocum, D.B. (1978) Classification of the elbow injuries from baseball players. *American Journal of Sports Medicine*, *6*, 62–67.

Smith, L.L., Keating, M.N., Holbert, D., Spratt, D.J., McCammon, M.R. and Smith, S.S. (1994) The effects of athletic massage on delayed onset muscle soreness, creatikinase and neutrophil count: a preliminary report. *Journal of Orthopaedic and Sport Therapy*, *24*, 267–278.

Snow, C.J. and Johnson, K.J. (1988) Effect of therapeutic ultrasound on acute inflammation. *Physiotherapy Canada*, *40*, 162–167.

Stratford, P., Levy, D.R., Gauldie, S., Levy, K. and Miseferi, D. (1987) Extensor carpi radialis tendonitis:a validation of selected outcome measures. *Physiotherapy Canada*, *39*, 250–254.

Stuijs, P.A., Smidt, N., Arola, H., VanDijk, C.N., Bushbinder, R. and Assendelft, W.J. (2001) Orthotic devices for tennis elbow: a systematic review. *British Journal of General Practice*, *51* (472), 924–929.

Sullivan, S.J., Williams, L.R., Seabourne, D.E. and Morrelli, M. (1991) Effects of massage on alpha motoneuron excitability. *Physical Therapy*, *71* (8), 555–560.

Ter Harr, G., Dyson, M. and Oakley, E.M. (1988) Ultrasound in Physiotherapy in the United Kingdom: Results of a questionnaire. *Physiotherapy Theory and Practice*, *4*, 69–72.

Tiidus, P.M. and Shoemaker, J.K. (1995) Effleurage massage, muscle blood-flow and long-term postexercise strength recovery. *International Journal of Sport Medicine*, *16* (7), 555–560.

Tippett, S. and Voight, M. (1999) *Functional Progressions for Sport Rehabilitation*. Champaign, IL: Human Kinetics.

Van der Heijden, G.J.M.G., Leffers, P., Wolters, P.J. et al. (1999) No effect of bipolar interferential electrotherapy and pulsed ultrasound for soft tissue shoulder disorders: a randomised controlled trial. *Annals of Rheumatoid Diseases*, *58*, 530–540.

Verhaar, J.A. (1992) *Tennis Elbow*. Maastrit: University Press.

Verhaar, J.A. (1994) Tennis Elbow. Anatomical, epidemiological and therapeutic aspects. *International Orthopaedics*, *18*, 263–267.

Verhaar, J.A., Walenkamp, G.H.I.M., van Mameren, H. et al. (1996) Local corticosteroid injection versus Cyriax-type physiotherapy for tennis elbow. *Journal of Bone and Joint S*, *78-B*, 128–132.

Vickers, A.J. (2001) Time course of muscle soreness following different types of exercise. *BMC Musculoskeletal Disorders*, *2* (1), 5.

Weber, M.D., Servedio, F.J. and Woodhall, W.R. (1994) The effects of 3 modalities on delayed onset muscle soreness. *Journal of Orthopaedic Sports and Physical Therapy*, *20* (5), 236–242.

Yates, J. (2004) *A Physician's Guide to Therapeutic Massage* (3rd edn). Toronto: Curties- Overzet Publications.

Young, S.R. and Dyson. M. (1990) The effect of therapeutic ultrasound on the healing of full-thickness excised skin lesions. *Ultrasonics*, *28*, 261–269.

19

Wrist and hand injuries in sport

Luke Heath
Graduate Sports Rehabilitator

Wrist and hand injuries are common in a majority of sports. The wrist and hand combine to form an important functional unit of the upper limb. This chapter will review the functional anatomy of the hand and wrist; explain the causative factors of acute soft tissue injuries of the hand and wrist; and discuss the differences between acute and chronic injuries of the hand and wrist. The chapter will further provide an overview of common soft tissue injuries of the hand and wrist by detailing the assessment, management, and rehabilitation strategies and protocols.

Incidence of wrist and hand injuries

Wrist and hand injuries are common in all types of sports. These injuries ranging from acute traumatic fractures, which can be seen in contact sports, such as football, rugby, hockey and basketball (Aitken and Court-Brown 2008), to chronic stress and overuse injuries, such as those seen in golf, gymnastics and various types of racquet sports (Rettig 2004).

Over a 10-season period of American Football, more than 1385 injuries occurred to the hand, first ray[1] (metacarpals) and fingers (Mall et al. 2008). It was also reported in this study that offensive and defensive linemen were most likely to sustain hand injuries, due to the higher incidence of contact with opponents, accounting for over 80% of metacarpal fractures. In addition, wide receivers and secondary defensive linemen were most likely to sustain finger type injuries, due to tackling while handling the ball (Mall et al. 2008). In field hockey, the odds of sustaining a hand or finger type injury are significantly greater ($p < 0.01$) than those participating in gloved wearing sports (Bowers et al. 2008), an example being ice hockey or lacrosse. The same risks also apply to cricket, as players have to handle the ball with bare hands (Stretch 2003).

In basketball, the bulk of wrist and hands injuries can be clearly defined as either cumulative (overuse) or acute (traumatic) (Raissaki et al. 2007). Furthermore, the epidemiology of basketball injuries between 2005 and 2007 showed that basketball players were more likely to sustain injury during competition (3.27 per 1000 exposures), compared to that in training (1.94 per 1000 exposures) (Borowski et al. 2008). Tennis players are also prone to overuse injuries (Jacobson et al. 2005), with studies showing that the extensor carpi ulnaris (ECU) in particular, may become dysfunctional as a result of prolonged or faulty technique (Montalvan et al. 2006; Tagliafico et al. 2009). However, the differences seen in player handgrip position may offer some insight into such an injury, as the Eastern grip, Western or semi-Western grip, will determine either a radial or ulna pathology (Tagliafico et al. 2009). Golfing injuries of the wrist are indeed rare but when they do occur they can often be devastating

---

[1] Rays are the four radial grooves that separate the areas of the hand. The first ray indicates the metacarpals.

Sports Rehabilitation and Injury Prevention   Edited by Paul Comfort and Earle Abrahamson
© 2010 John Wiley & Sons, Ltd

towards the athlete. The majority of golfing injuries for the hand and wrist occur from overuse of the common flexor tendons (Batt 1992; Murray and Cooney 1996; Jacobson et al. 2005; McHardy and Pollard 2005). Furthermore, this type of injury accounts for nearly 20% of all wrist injuries in amateur golf and a further 20–27% in professional golf (McHardy and Pollard 2005).

Gymnastics poses another problem for athletes due to repetitive loading of the musculoskeletal system (Zlotolow and Bennett 2008), as the upper body is not specifically designed to cope with such high forces. As a result, the wrist is the second most injured site in gymnastics after the shoulder (Dobyns and Gabel 1990), which reflects the type of movement and apparatus used by athletes for this sport (Cainea and Nassarb 2005). By comparison, rock climbers face a specific type of injury to the hand and fingers. As a strain to the finger flexor system, termed a 'pulley disruption' (Zlotolow and Bennett 2008), is closely associated with the 'crimp-grip' hand position, which is commonly used to hold onto smaller ledges (Jebson et al. 1997; Klauser et al. 2002).

In the field of winter sports, it is not uncommon for snowboarders and skiers to sustain a different type of injury pattern (Sutherlandet al. 1996), which highlights the importance of individually assessing the biomechanics of each sport. Although, the injury pattern for both brain and spinal injuries is considered the same in each sport (McBeth et al. 2009). Despite this, the wrist (and forearm) and ankle are the most injured sites in snowboarding, at 23% (Biasca et al. 1995; McBeth et al. 2009), compared to that of the knee and thumb in alpine skiing, at 46%. Overall, the injury rate of snowboarders is comparatively higher, which has important implications for the instruction of proper technique, especially when beginners are involved (Sutherland et al. 1996).

Sporting injuries of the wrist and hand are altogether not that uncommon, accounting for around a fifth of all emergencies presented to medical units within the UK (Diasa and Garcia-Eliasb 2006). In addition, over £100 million is spent per year in treating those injuries (Diasa and Garcia-Eliasb 2006). Therefore, the sports rehabilitator should provide adequate assistance in successfully assessing and managing those injuries, if costs are to be minimalised and a greater understanding towards the process of injury prevention is developed, as each case history unfolds.

Anatomy of the wrist and hand

The anatomical structure of the wrist and hand is a complex network of superficial soft tissue and palpable bony landmarks, whose role primarily, is to perform specific motor tasks that pertain to the shoulder and that of the elbow. These types of motor tasks often include hand and wrist actions such as blocking, gripping, catching and throwing, which can be seen being performed by most of today's athletes in almost every sport.

The structure and function of the wrist and hand are unique, in that the muscles, joints, tendons and ligaments all work together to provide stability, whilst enabling the thumb and four fingers to perform intricate and often delicate movements. However, when optimally trained, the hand can also provide powerful clasping and gripping movements, such as those seen in gymnastics, martial arts and rock climbing.

The movements of the wrist and hand are not altogether exclusive, however, as the initial positioning and resultant joint actions are dependent on several muscles pertaining to the forearm. Therefore, careful consideration must be given to the whole kinetic chain when assessing and managing any type of injury to the upper extremity, as other joints may also be involved. Furthermore, a key understanding of the structure and function of the wrist and hand are vital for sports rehabilitator's who may be embarking on careers in professional sport, as the diagnosis and management of musculoskeletal injuries could mean the difference between retirement and recovery.

Bones

The wrist and hand are comprised of 27 bones and more than 20 joints, of which there are 8 carpal bones, 5 metacarpal bones and 14 phalanges (Figure 19.1). The motion of these bones is somewhat complex, as movement occurs in three dimensions, and remains incompletely defined (Moorea et al. 2007).

The scaphoid is the most commonly fractured bone of the carpals (Rizzo and Shin 2006), the mechanism of injury occurring from a compression of the wrist while in extension, as during a block or fall. Blood flow can also become disrupted, leading to complications in wound healing. Dislocations of the capitate with the lunate may also occur as a result of scaphoid fractures, termed a 'perilunate dislocation'

ANATOMY OF THE WRIST AND HAND

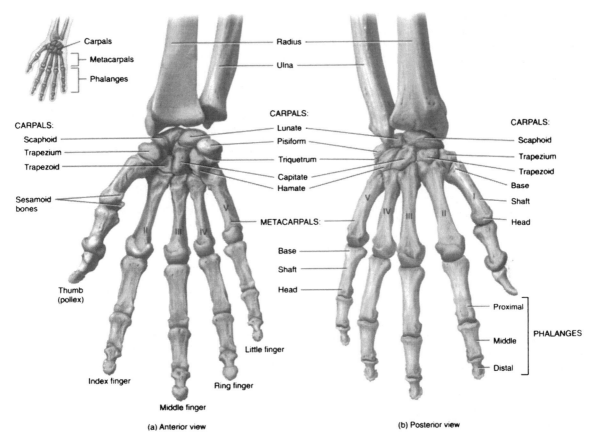

Figure 19.1 Bones of the wrist and hand (example). Reproduced, with permission, from Principles of Anatomy and Physiology, (11th ed). Tortora, G.J., & Derrickson, B., (2006)

(Bathala and Murray 2007). Here injury is as a result of forced extension, coupled with axial loading, and compression of the wrist.

Joints

The joints of the wrist and hand can be classified by their anatomical location and articulation with one another, to make up the radiocarpal joint (RC), midcarpal joint (MC), carpometacarpal joint (CMC), metacarpalphalangeal joint (MP), and interphalangeal joint (IP).

Radiocarpal joint

The radiocarpal joint of the wrist, which is formed by articulation of the radius with the scaphoid and lunate, and the triangular fibrocartilage complex (TFCC), with the lunate and triquetral, is often referred to as the common site for high intensive and chronic overuse injuries in sport (Bencardino and Rosenberg 2006; Tagliafico et al 2009). The triangular fibrocartilage is particularly susceptible to injury by way of forced extension and pronation, giving rise to issues of joint instability, disc degeneration and force distribution between the distal radial and ulnar joints (Melone and Nathan 1992). Several ligaments also aid in strengthening the radiocarpal joint (collateral, palmar and dorsal), allowing for movement in flexion, extension, abduction and adduction.

Midcarpal joint

The midcarpal joint is formed by two rows of carpals, the proximal row (scaphoid, lunate, triquetral and pisiform) and the distal row (trapezium, trapezoid, capitate and hamate). The carpal bones are closely packed together by intercarpal ligaments (dorsal, palmar and interosseous). As a result, flexion and abduction of the wrist occur more at the midcarpal joint, with extension and adduction greater at the radiocarpal joint.

Carpometacarpal joint

The carpometacarpal joints of the hand are formed by close articulation with the distal row of carpals and the five metacarpal bones, with the first carpometacarpal joint, being that of the thumb, classed as separate. The carpometacarpal joints are held together by dorsal and palmar ligaments, which allow for only a small degree of movement in joint glide. The thumb also has a lateral ligament, and due to its anatomical configuration, allows for a greater range of movement than the opposing fingers; allowing flexion, extension, abduction, adduction, opposition and reposition.

Metacarpalphalangeal joints

The metacarpalphalageal joints are where the fingers begin to become distinct from the hand. These are comprised of the five metacarpal bones articulating with the five proximal phalanges. The deep transverse metacarpal ligament holds all the MP joints together, allowing for flexion, extension, abduction and adduction. The second and third MP joints are quite rigid, making up the major stabiliser for the hand, whereas the fourth and fifth joints become increasingly mobile, in order to initiate the action that permits the closed grip of the hand.

The first metacarpalphalangeal joint of the thumb is separate from the other MP joints, to allow for greater abduction and adduction. In addition, the collateral ligament complex also provides stability, allowing for the thumb to adopt any position on the palmar aspect of the hand, and to provide a precision grip to the distal phalanges. An injury to the collateral ligament is particularly common in sports where the thumb is exposed, resulting in trauma from forced abduction and hyperextension, often from a direct contact with an opponent. In traumatic circumstances such as this, the stability of the joint can become compromised, leading to the need for medical intervention and wound repair.

Interphalangeal joints

The interphalangeal joints of the fingers are expressed as the proximal PIP and distal DIP interphalangeal joints, the DIP being furthest away from the hand. The interphalangeal joints are held closely together by collateral ligaments, which remain taught in both flexion and extension, unlike the collateral ligaments of the metacarpalphalageal joints, which are loose in extension. The collateral ligaments of the fingers are a common injury sustained in physical activity (Chomiak et al. 2000; Shewring and Matthewson 1993), occurring mainly in field and contact sports such as football and rugby.

Muscles

There are five muscles that act on the wrist joint. These are the flexor group; flexor carpi radialis, palmaris longus and flexor carpi ulnaris, and the extensor group; extensor carpi radialis longus and extensor carpi ulnaris (Figure 19.2). It is worth noting that a small percentage of individuals do not possess the palmaris longus, which aids to tighten the palmar fascia (Saied and Karamoozian 2009).

There are several intrinsic and extrinsic muscles of the hand, which co-exist to produce movement for the thumb and fingers. Three extrinsic muscles act on all four fingers. These include flexor digitorum superficialis, flexor digitorum profundus and extensor digitorum. Two smaller muscles also help to assist with extension of the index finger and fifth finger, respectively. These are extensor indicis and extensor digiti minimi. Disruption of the finger flexor tendon pulley is one of the most frequently occurring injuries in rock climbing, due to bowstringing of the tendon during the closed 'crimp-hand' position (Schöffl and Schöffl 2006). Considerable friction has also been shown to be apparent in a study by Moora, Nagya, Snedekera and Schweizer (2009), reporting that the fingers, particularly at 90-degree flexion, stress the PIP joint.

There are a total of 11 intrinsic muscles in the hand that act upon the proximal phalanges and middle and distal phalanges. These muscles include 4 lumbricales, 4 dorsal and 3 palmar interossei muscles that lie between the metacarpal bones (Figure 19.3). The muscles most commonly injured within the hand are the interossei, usually by overstretching the fingers. Pain and swelling are often localised, and restriction

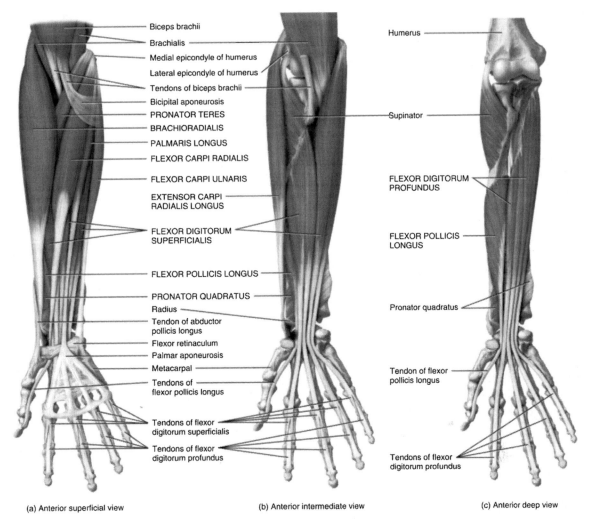

Figure 19.2 Extrinsic muscles of the wrist and hand (example). Reproduced, with permission, from Principles of Anatomy and Physiology, (11th ed). Tortora, G.J., & Derrickson, B., (2006)

in movement occurring in either abduction (dorsal interossei) or adduction (plantar interossei).

Eight muscles co-exist to produce the movements of the thumb. These include the extensor pollicis longus, extensor pollicis brevis, abductor pollicis longus, and flexor pollicis longus, which are extrinsic in origin, and the flexor pollicis brevis, opponens pollicis, abductor pollicis brevis and the adductor pollicis, which converge to form the thenar eminence. Three more intrinsic muscles of the hand also act on the little finger, including the abductor digiti minimi, flexor digiti minimi brevis and opponens digiti minimi, which converge on the medial aspect of the hand to form the hypothenar eminence.

Assessment and management of wrist and hand injuries

Clinical examination of the wrist and hand is necessary in order to delineate injured structures and to assess the need for referral. The protocol of assessment most commonly recognised in the UK for sports rehabilitation is SOAP (Brown et al. 2007), the acronym for subjective, objective, assessment

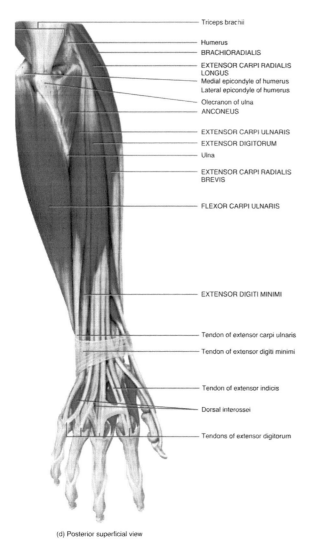

Figure 19.2 (Continued).

and prognosis. After completion of the subjective portion of the assessment, which includes detailing both personal details and medical history, the rehabilitator should now begin to conduct the objective portion, which involves the physical assessment.

Objective assessment

Palpation

Palpation should occur with the athlete seated, with the forearm, wrist and hand in a relaxed position.

Because of the superficial nature and complexity of the wrist and hand, a thorough knowledge of structural and functional anatomy should be gained beforehand in order to understand the extent of any musculoskeletal injury

Range of motion

Wrist and hand motions occur in a variety of different planes, which can often be thought of as corresponding to the joint actions of the elbow and the shoulder. Testing includes active and passive movements for the wrist in flexion, extension, radial and ulnar deviation, MCP joints in flexion, extension, abduction and adduction, and interphalangeal joints in flexion and extension. The thumb is also included in flexion and extension, and also abduction, adduction and circumduction. The purpose of range of motion assessment is to assess physiological and accessory motion, to differentiate injury between anatomical structures, and to assess end feel for abnormal movement.

Special testing

Special tests of the wrist and hand extend to physiological, neurological and also vascular examinations, to conclude or exclude pathology. These tests include examinations to both soft tissue and bony articulations, for the presence of acute and chronic inflammation, stress fractures and underlying medical conditions.

Functional assessment

Functional assessment is a component that is often overlooked during injury assessment, as it is important to gauge the physical capabilities of the athlete. Several exercises can be used here to assess functional capacity of the wrist and hand, which include sports-specific activities already performed by the athlete. In turn this will help assess motion, stability, strength, balance and coordination.

The rehabilitator may also perform an individualised approach towards injury assessment, through the use of a differential diagnosis. Here the assessment of other closely affected structures of the wrist and hand, namely the head, neck, shoulder and elbow, are conducted to measure the extent of those injuries. In addition, testing should also always be

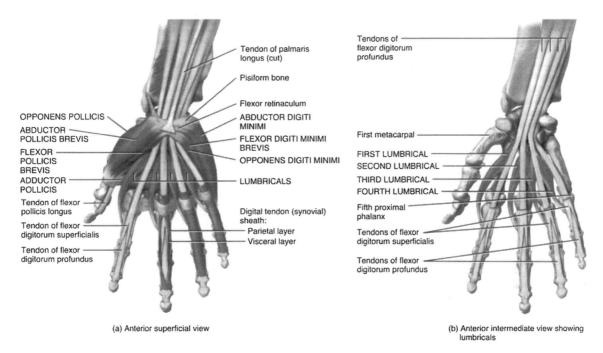

Figure 19.3 Intrinsic muscles of the hand (example). Reproduced, with permission, from Principles of Anatomy and Physiology, (11th ed). Tortora, G.J., & Derrickson, B., (2006)

conducted in comparison to the unaffected limb, to gain a comparable sign of what appears to be normal to the athlete.

Common UK sporting wrist and hand injuries

Research has shown that there are a number of common wrist and hand injuries that readily occur within UK sports (Rettig 2004). The bulk of these injuries sustained from the very nature of participating in these sports. For the purposes of review, the most common injuries will be discussed, together with assessment and conservative management, according to scope of practice for GRSs, who operate exclusively within the UK. The full list of injuries, however, can be seen illustrated as they are presented according to each sport (Table 19.1).

Acute soft tissue injuries

Athletes participating in high intensive and contact sports will no doubt be affected by acute soft tissue injuries. The role of the sports rehabilitator, therefore, is to work on minimising those risks, while focusing on rehabilitating those injuries. The classification of acute soft tissue injuries can be subdivided into sprains, strains and contusions.

Gamekeeper's thumb

A common injury sustained by players in football and hockey, the gamekeeper's thumb is a sprain of the ulnar collateral ligament, also termed the 'skier's thumb', for the mechanism in which the injury is sustained (Figure 19.4).

Signs and symptoms

An athlete with gamekeeper's thumb will present with pain and swelling on the palmar aspect of the thumb joint, between the web space of the thumb and the hand. Palpation will show a palpable defect in the instance of full rupture, with instability evident on bilateral comparison, with the thumb in abduction. Partial ruptures will elicit moderate joint laxity with a definite end feel. Further investigation by imaging can also confirm suspicion of tendon avulsions.

Table 19.1 UK sports and associated injuries (Rettig 2003, 2004)

Sport	Associated injuries	Occurrence
Rugby Union and Rugby League	Mallet finger DeQuervain's tenosynovitis Metacarpal fractures Scaphoid fractures Trigger finger/thumb Jersey finger	Direct trauma from player contact
American Football	Hyper extension wrist injury Flexor digitorum profundus ruptures Perilunate dislocation CMC and PIP joint injuries: collateral ligament tears, dislocations, fractures and volar plate injuries Intra-articular tears	Direct trauma and deviation forces
Boxing	Extensor tendon injury: Boxer's knuckle	Repetitive trauma
Tennis (other racquet sports such as squash and badminton)	Triangular fibrocartilage complex (TFCC) tears Extensor carpi ulnaris (ECU) dysfunction Carpal tunnel syndrome Ulnar wrist pain Hook of hamate fracture Kienbock's disease	Repetitive and overuse injury
Golf	Wrist flexor tendonitis	Overuse injury
Basketball	Sprained wrist Boutonniere deformity Finger fractures Gamekeeper's thumb	Direct trauma and falling
Gymnastics	Distal radius fractures Avascular necrosis of the capitate Ulnocarpal abutment syndrome Dorsal impingement	Repetitive loading and axial compression
Rock climbing	Pulley disruption	Falling
Hockey	Gamekeeper's thumb Hand/finger fractures Finger tendon and ligament sprains	Direct trauma and player contact
Cricket	Bennett's fracture Carpal tunnel Mallet finger Wrist arthritis	Repetitive trauma and overuse injury
Cycling	Guyon's canal syndrome	Repetitive loading and compression
Weightlifting	Subluxation of the ECU Intersection syndrome	Overuse syndrome

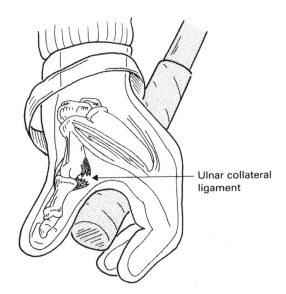

Figure 19.4 Skier's thumb (example). Reproduced, with permission, from Clinical Practice of Sports Injury Prevention and Care: Olympic Encyclopaedia of Sports Medicine, Volume V. Renstrom, P.A., F., H., (1994)

Management

In the immediate instance of injury to the ulnar collateral ligament, ice packs and compression are the best forms of treatment (Janoff 1999). Ultrasound can also be effective in the early stages to diagnose injury (Chuter et al. 2009; Malik et al. 2009), followed by massage and mobilisation, to aid in ligament repair and restore function (Alexy and De Carlo 1998). Thumb strength and dexterity can also be improved by using hand therapy balls and therapeutic putty (Alexy and De Carlo 1998; Shafer-Crane 2006; Wilson et al. 2008), whereas grip and thumb strengthening devices may be useful in restoring normal hand and thumb motion.

Complications in healing

Where there is a complete rupture of the ulnar collateral ligament, further medical intervention may be necessary. As in some cases the ruptured ligament may have become entangled in the soft tissue at the base of the thumb, known as a "Stener lesion", which will further complicate the procedure for conservative management and present a delay in wound healing (Ebrahim et al. 2006). Although, ruptured ulnar collateral ligaments without Stener lesions, have a good capacity to heal without complication. Following successful rehabilitation, return to sporting activities is possible. Returning to full sport is considered once the athlete has regained full functional ROM, with 90% strength of the unaffected part, and pain and swelling are minimal (Stracciolini et al. 2007).

Injury prevention

For football (soccer) goalkeepers and rugby players, preventative taping can be effective in aiding thumb joint stability and preventing further injury, and in sports where no catching is required, a thumb brace can also provide an alternative means to a permanent wrist support (Alexy and De Carlo 1998)

Mallet finger

Another common injury sustained in field and contact sports is mallet finger. Mallet finger is sustained from forceful flexion of an extended distal DIP joint, such as when a player has mistimed a shot or catching the ball.

Signs and symptoms

An athlete with mallet finger will present with pain at the dorsal DIP joint, with an inability to actively extend the joint, demonstrating a characteristic flexion deformity. If the tendon is only partially stretched, then movement may be restricted by 15–20 degrees extension. However, if a full rupture is present, then movement will be limited by 30–40 degrees extension, although full passive motion is typically preserved (Micheo 2003).

Management

The majority of mallet finger injuries are treated conservatively with rehabilitation. Ice packs can be useful to relieve pain in the early stages. The terminal interphalangeal joint (the joint in the finger closest to its tip) should be splinted in slight hyperextension (an overly straightened position), without immobilising any of the other joints of the injured finger (Janoff 1999). This position can then be maintained to allow time for wound healing.

Treatment

Treatment includes active DIP flexion exercises (making a full fist) that act to regain strength and mobility of the injured finger. This should be practised without the splint for 10 minutes every hour for the first two weeks (Teoh and Lee 2007), followed by several weeks of DIP flexion exercises with an extended PIP joint (Walshaw 2004).

Complications in healing

It is important to isolate the DIP joint during evaluation to ensure extension is from the extensor tendon and not the central slip, the absence of full passive extension possibly indicating a bony or soft tissue entrapment (Bach 1999; Lee and Montgomery 2002). Furthermore, bony avulsion fractures are present in one-third of patients with mallet finger, (Lairmore and Engber 1998; Palmer 1998).

Injury prevention

In sports such as basketball, cricket, and rugby it may be sensible to tape the fingers to provide further support to the joints. In individuals with a previous history, it is wise to wear a finger splint as protection.

Jersey finger

A disruption of the flexor digitorum profundus tendon, also known as jersey finger, commonly occurs when an athlete's finger catches on another player's clothing, usually while playing a team sport such as football or rugby. As the athlete pulls away, the finger is forcibly straightened while the profundus flexor tendon continues to contract. The ring finger is the weakest digit of the four fingers, accounting for 75% of all reported cases (Hankins and Peel 1990).

Signs and symptoms

An athlete with jersey finger will present with pain and swelling at the volar aspect of the DIP joint, and will be unable to bend the tip of the affected finger. Tenderness may also felt elsewhere along the finger or hand, if the profundus tendon has become retracted. The digitorum profundus tendon can be evaluated by holding the affected finger's MCP and PIP joints in extension while the rest remain in flexion, and performing a concentric contraction of the affected DIP joint (Hankins and Peel 1990). A positive sign for rupture to the digitorum profundus tendon is that the DIP joint should not move.

Management

Immediate management of jersey finger includes diagnostic imaging to confirm suspicion of an avulsion fracture, as complications can quickly arise in the case of tendon retractions (Mastey et al. 1997). Athletes with confirmed or suspected jersey finger should also be referred for medical consultation. Following medical intervention, rehabilitation should consist of passive range of motion exercises followed by a return to normal activity only after a period of several weeks, during which time movement is restricted in order to promote wound repair.

Boutonnière deformity

A common injury to the central slip extensor tendon (boutonnière deformity) occurs when the PIP joint is forcibly flexed while actively extended. It is a common injury among basketball players. Volar dislocation of the PIP joint can also cause central slip tendon ruptures (Perron et al. 2001).

Sign and symptoms

Signs and symptoms of boutonnière deformity include pain and localised swelling to the PIP joint. The PIP joint should be evaluated by holding the joint in a position of 15–30 degrees of flexion. If the PIP joint is injured, then the athlete will be unable to actively extend the joint, however, passive extension will be possible. Tenderness over the dorsal aspect of the middle phalanx will also be present.

Management

The PIP joint should be splinted in full extension for the first six weeks of healing, and in cases where no avulsion has taken place, or the avulsion involves less than one third of the joint. All available splints can be used to treat PIP injuries, except for the stack splint, which is used only for DIP injuries. As with mallet finger, extension of the PIP joint must be maintained continuously. If full passive extension is not possible, then the rehabilitator should refer the athlete.

Complications in healing

If an avulsion fracture is present on imaging then medical intervention may be necessary to prevent future complications, as a delay in the proper treatment may cause permanent deformity. A boutonnière deformity usually develops over a period of several weeks, as the intact lateral bands of the extensor tendon slip inferiorly. However, a boutonnière deformity will also occur more acutely.

Injury prevention

In sports such as basketball, cricket, and rugby it may be sensible to tape the fingers to provide further support to the joints. Athletes with PIP joint injuries may also continue to participate in athletic events during the splinting period, although some sports are difficult to play with a fully extended PIP joints.

Extensor pollicis longus rupture

A tendon rupture of the extensor pollicis longus (EPL) is a recognised complication of distal radial fractures and their fixation with dorsal radial plates and pins. A number of other conditions including internal fixation of wrist fractures and inflammatory arthropathies have also been reported as aetiological factors of EPL tendon rupture (Ansede et al. 2009).

Management

The EPL tendon is complex and if a rupture occurs then medical intervention may be necessary. The standard procedure for this type of injury is to use the extensor indicis proprius (EIP) tendon. The tendon is then transferred from its normal location to replace the function of the EPL. There however some disadvantages to using the EIP, as Bullon (2007) states that the EPL may not have sufficient tendons, therefore using the accessory abductor pollicis longus (AAPL) is preferred, as transference can then be conducted without affecting the function of the APL.

Complications in healing

Not all people have this tendon, as it is present in approximately 85% of the general population. However, Bullon (2007) states that the AAPL could be used successfully in restoring function of the ruptured EPL, as everyday newer surgical procedures are being conducted, which have many advantages over other tendon transfers for this type of injury. Wimsey, Kurian and Jeffery (2006) also state that there is a conservative approach to this condition, which would be to immobilise the wrist joint in a mallet splint for 12 weeks. However, this approach would only be advocated to the younger population, as complications can occur from immobilisation of the joints, such as muscle atrophy and weakness, which will hinder effective management.

Other acute soft tissue injuries
Triangular fibrocartilage complex tears

The triangular fibrocartilage complex (TFCC) sits between the distal end of the ulna and the triquetrum and part of the lunate. It consists of the triangular fibrocartilage, the ulnar meniscus homolog, the ulnar collateral ligament, carpal ligaments and the extensor carpi ulnaris tendon sheath. Testing for the TFCC is to place the wrist in extension and ulnar deviation and then rotate. This movement, as in the forehand volley, relates to overloading of the complex.

The integrity of the TFCC is closely bound up with the stability of the distal radio-ulnar joint (DRUJ). In a study by Adlercreutz, Aspenberg and Lindau (2000), it was shown that a complete tear of the TFCC is almost always associated with instability of the DRUJ. Instability of the DRUJ is also associated with generalised capsuloligamentous laxity. Often pain on the ulnar side of the wrist is multifactorial and stability-related. Another underlying cause of compression injuries to the TFCC is ulnar variance.

Collateral ligament sprains

Collateral ligament injuries of the MCP, PIP and IP joints can occur in many types of sport. Injuries of the collateral ligaments of the MCP joints are seldom but instability can present if the tear is complete. Injury to the PIP joint is quite common in athletes and team sports such as volleyball, basketball and rugby. The majority of collateral ligament injuries can be treated by splinting or taping the affected part to the adjacent fingers, known as 'buddy taping', in order to provide additional support (Sennet 2004). The IP joint of the thumb is similar to the PIP joint of the fingers, and as such, should be treated in much the same way.

Wrist sprains

Wrist sprains typically occur after a trip or fall, resulting in stretching or tearing of the ligaments of the wrist. Common causes of wrist sprain include falls during team sports, such as when a basketball player is tackled during a jump shot, or a rugby player barged from the side. Moreover, if the tissues are inflexible and weak, the risk of injury increases. However, the majority of these injuries can be treated conservatively, and seldom result in prolonged loss of sporting activity.

Contusions

Contusions of the wrist and hand are a common injury, because of the many superficial tendons and bony prominences that are exposed. Contusions are rarely serious, and can be treated conservatively over time. However, care must be taken not to rule out more serious injury, such as ligament sprains, tendon injuries and joint fractures.

Chronic and overuse injuries

Repetitive actions can take their toll in team and action sports, resulting in chronic inflammatory issues and degenerative pathologies. The repetitive cycle of injury that can occur from an acute injury that is poorly managed is also a source of chronic instability, resulting in faulty recruitment from scar tissue, chronic inflammation and irritation from repetitive forces. In sports such as tennis, gymnastics and rock climbing, the wrists are exposed to a greater frequency of joint irritation, from the dissipation of load through the upper extremity. Therefore, the role of the sports rehabilitator is to work on breaking the cycle of injury and promote effective wound healing, and in the instances where injuries are idiopathic in nature, to work on minimising those risks to aid in ensuring longevity for the athlete and their sport.

De Quervain's disease

De Quervain's disease (also known as Hoffmann's disease) is an inflammation and thickening of the synovial lining of the common sheath of the abductor pollicis longus and extensor pollicis brevis tendons (Figure 19.5).

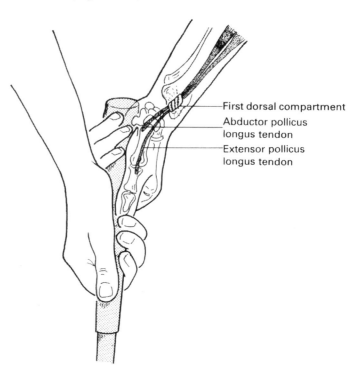

Figure 19.5 De Quervain's disease (example). Reproduced, with permission, from Clinical Practice of Sports Injury Prevention and Care: Olympic Encyclopaedia of Sports Medicine, Volume V. Renstrom, P.A., F., H., (1994)

Signs and symptoms

The common sheath provides support to the tendons to prevent bowstringing when the wrist is in extension (Rettig 2004). Therefore, space inside the sheath is limited. Thickening occurs particularly at the distal portion of the radial styloid. There is often pain in resisted thumb extension and abduction, and while passively moving the wrist in ulnar deviation, keeping the thumb fully flexed (Finkelstein's test) (Van Tulder et al. 2007). There is also local tenderness on palpation, and with the tendon on stretch, crepitus to repeated movements.

Management

De Quervain's disease responds well to frictional massage with the tendons in a lengthened position, and immobilisation of the thumb joint in a splint (McCarroll 2001; Coldham 2006). Corticosteroid injections may also be prescribed for persistent cases (Mason et al. 2008).

Complications in healing

De Quervain's disease can often be confused with rheumatoid arthritis (Daenen et al. 2004). However, imaging for tendon pathology, such as in the use of sonography, can accurately determine a number of inflammatory, metabolic and infectious wrist disorders (Daenen et al. 2004).

Injury prevention

Breaking the cycle of injury comes highly recommended for managing de Quervain's disease (Rettig 2004), as a wide variety of ergonomic factors are related to its onset. The cessation of repetitive activity is the first recommendation, followed by the reduction in movements that cause pain and the reporting of symptoms (Foye et al. 2002). In this instance, breaking the cycle is as much about educating the athlete on the long-term health benefits.

Other chronic and overuse injuries

Carpal tunnel syndrome

Carpal tunnel syndrome is a result of compression of the medial nerve as it passes beneath the flexor retinaculum and into the carpal tunnel. The condition may manifest as a result of swelling of the flexor tendon sheaths (tenosynovitis), as can be seen from the repetitive flexion actions caused by the wrist in sports such as gymnastics, cycling and weightlifting. The condition can also manifest itself as a result of arthritic degenerative changes from repetitive or previous impact traumas, such as wrist fractures.

Ulnar nerve compression

Compression injuries of the ulnar nerve can occur between the wrist space formed between the pisiform and the hamate, known as the tunnel of Guyon, termed "Guyon's canal syndrome". The symptoms are caused by compression or friction of soft tissue structures surrounding the ulnar nerve, resulting in pain, tingling and numbness. The condition commonly affects cyclists, as the wrist is compressed and extended against the handlebars (Figure 19.6).

Impingement syndrome

This condition can manifest as a result of forced compression of two carpal bones of the wrist. Such injuries include impaction between the scaphoid or, less commonly, the lunate and radius with forced extension, triquetrohamate impingement with forced extension and ulnar deviation, and radial styloid impaction with forced radial deviation. Impingement syndrome may also relate to chronic instabilities, whereby increased accessory joint motion refers to greater chances of impingement during physiological movement.

Trigger finger

Trigger finger, is a common name for finger tendon disruption that causes the joints to prevent from extending. As the finger bends a nodule on the tendon passes out of the synovial sheath coating the tendon and into the palm, but as the finger straightens the nodule may not pass back into the sheath, becoming lodged in its entrance. The athlete may attempt to forcibly straighten the finger but this should be avoided in all circumstances.

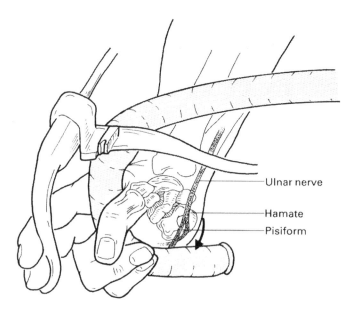

Figure 19.6 Ulnar nerve compression. Reproduced, with permission, from Clinical Practice of Sports Injury Prevention and Care: Olympic Encyclopaedia of Sports Medicine, Volume V. Renstrom, P.A., F., H., (1994)

Bone pathology

Scaphoid fracture

A scaphoid fracture is the most commonly injured bone of the carpals. Often occurring from sudden impact of a closed wrist or during a trip or fall, with the wrist in full extension. The injury is particularly common in contact sports, such as American football, rugby and martial arts, but it may also occur in other types of sports, such as gymnastics, skiing or snowboarding.

Signs and symptoms

An athlete with a scaphoid fracture will present with swelling at the base of the thumb, on the outside of the wrist (anatomical snuffbox). Palpation will elicit pain, particularly with passive movements during pronation and ulnar deviation, which will stress the scaphoid. It is also not uncommon for a scaphoid fracture to go undetected, and in the event, diagnostic imaging is highly recommended (Haisman et al. 2006).

Management

Acute management of scaphoid fractures is immobilisation of the wrist and thumb (Jones 2006), where a support would extend from the wrist to the IP joint of the thumb, but also sometimes from the elbow. Rehabilitation in the early stage begins with maintaining range of movement in the shoulder and the elbow, and, in less complicated fractures, healing can be resolved in as little as four weeks. In more complicated fractures, however, such as a fracture to the proximal bone, healing may take several months. Therefore, maintaining muscle function in the upper extremity is imperative, to prevent loss of movement.

Once healing has been approved, mobilisation and strengthening exercises using therapeutic putty and hand therapy balls can begin to restore full function (Vucekovich et al. 2004). During this period it may also be helpful to wear a wrist support when not performing exercise to aid further support.

Complications in healing

Early diagnosis of scaphoid injuries is important, as misdiagnosed or untreated fractures may lead to malunion or nonunion bone (Haisman et al. 2006). Athletes with these problems would most certainly present with later issues, as persistent wrist pain and abnormal kinematics can lead to wrist arthrosis. Medical intervention using screw fixation may be recommended in most cases, as evidence suggests that conservative management may lead to a higher rate of non-union complications during healing (McQueen et al. 2008).

Injury prevention

There is not much that can be done to prevent a fall on an outstretched hand in most sports. However, in snowboarding the incidence of scaphoid injury is less due to the measure of wearing wrist guards. As wrist guards have been shown to reduce the incidence of wrist injuries during falls (O'Neill 2003; Idzikowski et al. 2000).

Other bone pathologies

Metacarpal fractures

A punching type injury is the most common cause of metacarpal fractures. This may occur either when the hand strikes another object (hand off), such as in handball, or when a player falls with a closed fist hitting another solid object. Sports involving this type of injury include American football, boxing, basketball, soccer and in some cases cricket.

Bennett's fracture

A Bennett's fracture of the thumb (first metacarpal bone) is similar in mechanism to an MCP fracture. Except in this case impact is specifically targeted at the thumb. In this instance the pull of the abductor muscle causes displacement of the metacarpal bone, resulting in shearing of the bone at its base. Because this fracture occurs near the joint line, it can also present with a false joint with palpation and movement.

Finger fractures

Finger fractures are very common in basketball, and can be caused by a variety of mechanisms. While finger fractures are often considered minor trauma, without proper treatment they can also cause major problems. Conservative management involves buddy taping or splinting while the injured bone heals (Sennet 2004), and to prevent further injury. These types of fractures are also sometimes referred to as hairline fractures, small avulsion fractures, and non-displaced fractures.

PIP joint dislocations

Dislocation of the PIP joint is common in sports such as handball, basketball and cricket. The mechanism of injury can occur in axial loading and hyperextension of the joint, lateral loading and, more rarely, volar disruption of the PIP joint. Lack of treatment may cause permanent joint deformity, and in the case of more traumatic PIP dislocations, fractures may also be present.

Case study

A 31-year-old male rock climber comes to you complaining of pain in his hand and fingers whilst performing activity. He was indoor-climbing during the weekend and fell, whilst holding onto a rock wall with bolt-on features. Despite the pain he continued to climb for the rest of the afternoon and two days later after having applied athletic taping, as recommended by his training partner. Since then his symptoms have become much worse. The management of this athlete are discussed, following the headings of injury management, immediate treatment, restoration of motion, muscular strength endurance and power, and functional integration.

Injury management

Active rehabilitation methods are highly recommended, due to the increased instances of joint contractures and deformity for finger type injuries. Cessation of finger type activities is also desired at this point (Wyatt et al. 1996) to localise inflammation and reduce instances of pain. Athletic taping, however, is not recommended at this point, especially without prior medical consultation, as research is still not yet conclusive for the efficacy of taping for rock climbing injuries (Bollen and Gunson 1990). Diagnostic imaging can also be advised at this time, to determine the extent of trauma to the finger pulley (Klauser et al 2002).

Immediate treatment

In the instance of acute injury and soft-tissue strains, several therapeutic interventions are propagated, including ultrasound (US), interferential stimulation (IFS) and ice-massage (Janoff 1999). US and ice-massage may be used in combination with exercise to decrease inflammation, the duration and frequency depending upon the severity of injury. By using 10 minutes of ice application, treatment can then be followed by 10 minutes of therapeutic exercise, with the process repeated twice up to 3–6 times per day for the first week (Bleakley et al. 2004, 2007). Flexibility exercises may also be performed using either cryokinetic techniques or as inflammation decreases, heat, to facilitate motion.

Restoration of motion

Once pain-free motion and soft-tissue flexibility have been restored, close chain exercises can then be prescribed to improve dexterity. The exercises used here should be the same or similar to those used in functional testing, as this will mark any improvements and already be familiar to the athlete. The methods used can also include the precision-grip or power-grip, as used in rock climbing, with specific modifications selected based on detriments in performance.

Muscular strength endurance and power

Once range-of-motion and dexterity have been established, open-chain exercises specific to the physiological demands of rock climbing can then be prescribed, including a progression of movements using hand weights and body weight, building the intensity until the athlete can manage 25–30 repetitions (Read 2000). Athletic taping may also be used at this point but only a preventative measure, to provide further joint stability (Jebson et al. 1997). It is important to note that this phase of rehabilitation is now beyond the scope of most conventional allied health professionals and is specific to sports rehabilitators. Therefore, steps must be taken to ensure that the athlete has successfully completed the early stages of recovery to reduce the likelihood of re-injury.

Functional Integration

If, after a period of periodisation, the athlete is able to perform maximal strength and power tests bilaterally, the rehabilitator can then introduce movements that replicate the technical aspects of the sport. The biomechanics of indoor and sports climbing demand that most of the body is supported by the distal phalanx, and while the crimp-grip technique is mainly used, it is often the position of frequent ruptures (Vigourioux et al. 2008). Therefore, the rehabilitation goal here is to restore the ability for distal phalanx to cope with the transference and dissipation of load. Taking the athlete back to the onsite location

Figure 19.7 Observation of climbing technique. Courtesy of Fabio Jesus

of the initial injury will also be helpful to assess movement and correct technique, should the athlete wish to continue with a sports-specific training programme with the rehabilitator (Figure 19.7).

Furthermore, the rate of adaptation required will be unique to each individual and level of athletic ability. Therefore, the rehabilitator would do well to evaluate the needs of each athlete, and assessments and feedback should almost become second nature for each stage of the rehabilitation process.

References

Adlercreutz, C, Aspenberg, P. and Lindau T., (2000) Peripheral tears of the triangular fibrocartilage complex cause distal radioulnar joint instability after distal radial fractures. *Journal of Hand Surgery*, 25 (3), 464–468.

Aitken, S. and Court-Brown, C.M. (2008) The epidemiology of sports-related fractures of the hand. *Injury*, 39, 1377–1383.

Alexy, C. and De Carlo, M. (1998) Rehabilitation and use of protective devices in hand and wrist injuries. *Clinics In Sports Medicine* 17 (3), 635–655.

Ansede, G. Healy, J. and Lee, J. (2009) Extensor pollicis longus tendon rupture following avulsion fracture of the third metacarpal. *Skeletal Radiology 38* (1), 88–84.

Bach, A.W. (1999) Finger joint injuries in active patients: pointers for acute and late-phase management. *Physical Sports Medicine*,. 27, 89–104.

Bathala, E.A. and Murray, P.M. (2007) Long-term follow-up of an undiagnosed trans-scaphoid perilunate dislocation demonstrating articular remodeling and functional adaptation. *The Journal of Hand Surgery*. *32* (7), 1020–1023.

Batt, M.E. (1992) A survey of golf injuries in amateur golfers. *British Journal of Sports Medicine*, *26*, 63–65.

Bencardino, J.T. and Rosenberg, Z.S. (2006) Sports-related injuries of the wrist: An approach to MRI interpretation. *Clinics in Sports Medicine*, *25* (3), 409–432.

Bleakley, C., McDonough, S.and MacAuley, D. (2004) The se of ice in the treatment of acute soft-tissue injury. A systematic review of randomized controlled trials. *American Journal of Sports Medicine*, *32* (1), 251–261.

Bleakley, C.M., O'Connor, S., Tully, M.A., Rocke, L.G., MacAuley, C. and, McDonough, S.M. (2007) The PRICE study (Protection Rest Ice Compression Elevation): design of a randomised controlled trial comparing standard versus cryokinetic ice applications in the management of acute ankle sprain. *BMC Musculoskeletal Disorders*, *8*, 125.

Biasca, N., Battaglia, H., Simmen, H.P., Disler, P. and Trentz, O. (1995) An overview of snow-boarding injuries. *Unfallchirurg*, *98* (1), 33–9.

Bollen, S.R. and Gunson, C.K. (1990) Hand injuries in competition climbers. *British Journal of Sports Medicine*, *24* (1), 16–18.

Borowski, L.A., Yard, E.E., Fields, S.K. and Comstock, R.D. (2008) The epidemiology of US High School basketball injuries, 2005–2007. *The American Journal of Sports Medicine*, *36*, 2328–2335.

Bowers, A.L., Baldwin, K.D. and Sennett, B.J. (2008) Athletic hand injuries in intercollegiate field hockey players. *Medicine and Science in Sports and Exercise*, *40* (12), 2022–2026.

Brown, M.C., Kotlyar, M., Conway, J.M., Seifert, R. and St Peter, J.V. (2007) Integration of an internet-based medical chart into a pharmacotherapy lecture series. *American Journal of Pharmacology Education*, *71* (3), Article 3.

Bullón, A., (2007). Reconstruction after chronic extensor pollicis longus ruptures. *Clinical Orthopaedics and Related Research*, *462*, 93–98.

Cainea, D. and Nassarb, L. (2005) Gymnastics injuries. *Medicine and Sports Science*, *48*, 18–58.

Chomiak, J., Junge, A., Peterson, L. and Dvorak, J. (2000) Severe injuries in football players influencing factors. *The American Journal of Sports Medicine*, *28*, S58–S68.

Chuter, G.S.J., Muwangaa, C.L. and Irwin, L.R. (2009) Ulnar collateral ligament injuries of the thumb: 10 years of surgical experience. *Injury*, *40* (6), 652–656.

Coldham, F. (2006). The use of splinting in the non-surgical treatment of De Quervain's disease: A review of the literature. *Hand Therapy*, *11* (2), 48–55.

Daenen, B., Houben, G., Bauduin, E., Debry, R. and Magotteau, P. (2004) Sonography in wrist tendon pathology. *Journal of Clinical Ultrasound*, *32* (9), 462–469.

Diasa, J.J. and Garcia-Eliasb, M. (2006) Hand injury costs. *Injury*, *37* (11), 1071–1077.

Dobyns, J.H. and Gabel, G.T. (1990) Gymnast's wrist. *Hand Clinics*, *6* (3), 493–505.

Ebrahim, F.S., De Maeseneer, M., Jager, T., Marcelis, S., Jamadar, D.A. and Jacobson, J.A. (2006) US diagnosis of UCL tears of the thumb and stener lesions: Technique, pattern-based approach, and differential diagnosis. *RadioGraphics*, *26*, 1007–1020.

Foye, P.M., Cianca, J.C. and Prather, H. (2002) Industrial medicine and acute musculoskeletal rehabilitation. 3. Cumulative trauma disorders of the upper limb in computer users. *Archives of Physical Medicine and Rehabilitation*, *83*, (Suppl 1), S12–15.

Haisman, J.M., Rohde, R.S. and Weiland, A.J. (2006) Acute fractures of the scaphoid. *Journal of Bone and Joint Surgery (American)*, *88*, 2750–2758.

Hankins, F.M. and Peel, S.M. (1990) Sport-related fractures and dislocations in the hand. *Hand Clinics*, *6*, 429–453.

Idzikowski, J.R., Janes, P.C. and Abbott, P.J. (2000) Upper extremity snowboarding injuries ten-year results from the Colorado Snowboard Injury Survey. *American Journal of Sports Medicine*, *28*, (6), 825–832.

Jacobson, J.A., Miller, B.S. and Morag, Y. (2005) Golf and racquet sports injuries. *Seminars in Musculoskeletal Radiology*, *9* (4), 346–359.

Janoff, E. (1999) Rehabilitation and splinting of common upper extremity injuries. *Interscience Conference on Antimicrobial Agents and Chemotherapy*, *39*, 779 (abstract no.1984).

Jebson, P.J.L., Curtis, M., Steyers, C.M. (1997) Hand injuries in rock climbing: Reaching the right treatment. *The Physician and Sports Medicine*, *25*, 5.

Jones, G.L. (2006) Upper extremity stress fractures. *Clinics in Sports Medicine*, *25* (1), 159–174.

Klauser, A., Frauscher, F., Bodner, G., Halpem, E.J., Schocke, M.F., Springer, P., Gabl, M., Judmaler, W. and Zur Nedden, D. (2002) Finger pulley injuries in extreme rock climbers: Depiction with dynamic US. *Radiology*, *222*, 755–761.

Lairmore, J.R. and Engber, W.D. (1998) Serious, often subtle, finger injuries: Avoiding diagnosis and treatment pitfalls. *Physical Sports Medicine*, *26*, 57–69.

Lee, S.J. and Montgomery, K. (2002) Athletic hand injuries. *Orthopaedic Clinical North America*, 33, 547–554.

Malik, A.K., Morris, T., Chou, D., Sorene, E. and Taylor, E. (2009) Clinical testing of ulnar collateral ligament injuries of the thumb. *Journal of Hand Surgery (European)*, 34 (3), 363–366.

Mall, N.A., Carlisle, J.C., Matava, M.J., Powell, J.W. and Goldfarb, C.A., (2008) Upper extremity injuries in the National Football League part I: Hand and digital injuries. *American Journal of Sports Medicine*, 36, 1938–1944.

Mason, L.W. Oakley, J.E. and Wilson, I.E. (2008) Soft tissue atrophy following repeated corticosteroid injection for De Quervain's tenosynovitis. *Internet Journal of Orthopedic Surgery*, 10, 1.

Mastey, R.D., Weiss, A.P. and Akelman, E. (1997) Primary care of hand and wrist athletic injuries. *Clinical Sports Medicine*, 16, 705–724.

McBeth, P.B., Ball, C.G., Mulloy, R.H. and Kirkpatrick, A.W. (2009) Alpine ski and snowboarding traumatic injuries: incidence, injury patterns, and risk factors for 10 years. *American Journal of Surgery*, 197 (5), 560–563.

McCarroll, J.R. (2001) Overuse injuries of the upper extremity in golf. *Clinics in Sports Medicine*, 20 (3), 469–479.

McHardy, A.J. and Pollard, H.P. (2005) Golf and upper limb injuries: a summary and review of the literature. *Chiropractic and Osteopathy*, 13, 7.

McQueen, M.M., Gelbke, M.K., Wakefield, A., Will, E.M. and Gaebler, C. (2008) Percutaneous screw fixation versus conservative treatment for fractures of the waist of the scaphoid. *Journal of Bone and Joint Surgery (British)*, 90B (1), 66–71.

Melone, C.P. Jr. and Nathan, R. (1992) Traumatic disruption of the triangular fibrocartilage complex. Pathoanatomy. *Clinical Orthopaedics and Related Research*, 275, 65–73.

Micheo, W.F. (2003) Wrist, hand, and finger disorders. http://www.aapmr.org/ Accessed 21 Febuary 2009.

Montalvan, B., Parier, J., Brasseur, J.L., Viet, D.L. and Drape, J.L. (2006) Extensor carpi ulnaris injuries in tennis players; a study of 28 cases. *British Journal of Sports Medicine*, 40, 424–429.

Moorea, D.C., Crisco, J.J., Traftona, T.G. and Leventha, E.L. (2007) A digital database of wrist and bone anatomy and carpal kinematics. *Journal of Biomechanics*, 40 (11), 2537–2542.

Moora, B.K., Nagya, L., Snedekera, J.G. and Schweizer, A. (2009) Friction between finger flexor tendons and the pulley system in the crimp grip position. *Clinical Biomechanics*, 24 (1), 20–25.

Murray, P.M. and Cooney, W.P. (1996) Golf-induced injuries of the wrist. *Clinics in Sports Medicine*, 15 (1), 85–109.

O'Neill, D.F. (2003) Wrist injuries in guarded versus unguarded first time snowboarders. *Clin Orthop Relat Res*. 409, 91–5.

Palmer, R.E. (1988) Joint injuries of the hand in athletes. *Clinical Sports Medicine*, 17, 513–531.

Perron, A.D., Brady, W.J., Keats, T.E. and Hersh, R.E. (2001) Orthopedic pitfalls in the emergency department: closed tendon injuries of the hand. *American Journal Emergency Medicine*, 19, 76–80.

Raissaki, M., Apostolaki, E. and Karantanas, A.H. (2007) Imaging of sports injuries in children and adolescents. *European Journal of Radiology*, 62 (1), 86–96.

Read, M.T.F. (2000) *A Practical Guide To Sports Injuries*. London: Butterworth Heinemann.

Rettig, A.C. (2003) Athletic injuries of the wrist and hand: art 1: Traumatic injuries of the rist. *American Journal of Sports Medicine*, 31 (6), 1038–1048.

Rettig, A.C. (2004) Athletic injuries of the wrist and hand: part 2: Overuse injuries of the wrist and traumatic injuries to the hand. *American Journal of Sports Medicine*, 32 (1), 262–273.

Rizzo, M. and Shin, A.Y. (2005) Treatment of acute scaphoid fractures in the athlete. *Current Sports Medicine Reports*, 5 (5), 242–248.

Saied, A. and Karamoozian, S. (2009) The relationship of presence or absence of palmaris longus and fifth flexor digitorum superficialis with carpal tunnel syndrome. *European Journal of Neurology*, 16 (5), 619–623.

Schöffl, V.R. and Schöffl, I. (2006) Injuries to the finger flexor pulley system in rock climbers: Current concepts. *The Journal of Hand Surgery*, 31 (4), 647–654.

Sennet, B.J. (2004) Hand and wrist injuries related to hockey. *Annual Meeting of the American of Medical Society of Sports Medicine*, University of Pennsylvania.

Shafer-Crane, G.A. (2006) Repetitive stress and strain injuries: Preventive exercises for the musician. *Physical Medicine and Rehabilitation Clinics of North America*, 17 (4), 827–842.

Shewring, D.J. and Matthewson, M.H. (1993) Injuries to the hand in Rugby Union Football. *Journal of Hand Surgery*, 18 (1), 122–124.

Stracciolini, A., Meehan, W. P. and DHemecourt III, P.A. (2007) Sports rehabilitation of the injured athlete. *Clinical Pediatric Emergency Medicine*, *8* (1), 43–53.

Stretch, R.A. (2003) Cricket injuries: A longitudinal study of the nature of injuries to South African cricketers. *British Journal of Sports Medicine*, *37*, 250–253.

Sutherland, A.G., Holmes, J.D. and Myers, S. (1996) Differing injury patterns in snowboarding and alpine skiing. *Injury*, *27* (6), 423–425.

Tagliafico, A.S., Ameri, P., Michaud, J., Derchi, L.E., Sormani, M.P. and Martinoli, C. (2009) Wrist injuries in nonprofessional tennis players: Relationships with different grips *American Journal of Sports Medicine*, *37* (4), 760–767.

Teoh, L.C. and Lee, J.Y.L. (2007) Mallet fractures: A novel approach to internal fixation using a hook plate. *Journal of Hand Surgery*, *32E* (1), 24–30.

Van Tulder, M., Malmivaara, A. and Koes, B. (2007) Repetitive strain injury. *The Lancet*, *369* (9575), 1815–1822.

Vigourioux, L., Quaine, F., Paclet, F., Colloud, F. and Moutet, F. (2008) Middle and ring fingers are more exposed to pulley ruptuer than index and little during sport-climbing: A biomechanical explanation. *Clinical Biomechanics*, *23*, 562–570.

Vucekovich, K., Gallardo, G. and Fiala, K. (2004) Rehabilitation after flexor tendon repair, reconstruction, and tenolysis. *Hand Clinics*, *21* (2), 257–265.

Walshaw, L. (2004) Practical procedures for minor injuries: Mallet splint. *Accident and Emergency Nursing*, *12*, 182–184.

Wilson, L.M. Roden, P.W., Taylor, Y. and Marston, L. (2008) The effectiveness of Origami on overall hand function after injury: A pilot controlled trial. *Hand Therapy*, *13*, 2–20.

Wimsey, S., Kurian, J. and Jeffery, I.T.A. (2006) Spontaneous recovery of extensor pollicis longus tendon rupture following intra-articular distal radius fracture. *Injury Extra*, *37* (9), 331–333.

Wyatt, J.P., McNaughton, G.W. and Grant, P.T. (1996) A prospective study of rock climbing injuries. *British Journal of Sports Medicine*, *30*, 148–150.

Zlotolow, D.A. and Bennett, C. (2008) Athletic injuries of the hand and wrist. *Current Orthopedic Practice*, *19* (2), 206–210.

20

The groin in sport

John Allen and Stuart Butler
England Athletics

Introduction

As with many words used to describe parts of the human body, there is a lay usage and a more precise anatomical meaning for both the hip and the groin. The anatomy of the groin can be clinically difficult to define.

Both structures can be defined primarily by position concerning the groin, and by function regarding the hip, within the musculoskeletal structure of the body. The hip joint allows the lower limb to be moved through an exceptional range of motion, although the movement in the hip joint is more anterior and posterior due to anatomical structure and the functional requirement of gait.

The groin consists of the structures deep to the anterior and medial intersection of the leg and the lower abdomen, and includes the structures of the perineum. Movement of the hip underlies and effects the groin structures. Stretching and mobility exercises of the hip involve the groin muscles. The groin, therefore, particularly includes the lower rectus abdominis musculature, the inguinal region, the symphysis pubis, the upper portions of the adductor muscles of the thigh, and the genitalia, including the scrotum in males.

This chapter introduces common groin injuries and pathologies associated with sport and physical activity participation. The chapter provides an overview of the functionality of both the groin and hip and then details the injuries and pathologies through evidence based analyses of the literature. The chapter concludes with applied case studies, illustrating the management of groin injuries and recommending suitable rehabilitation techniques and exercise protocols. When reading this chapter it is important to understand injury, pain and pathology in the light of function, and more often dysfunction, of joint motion.

Groin pain in sport

Groin pain is one of the most poorly understood syndromes in clinical sports medicine despite its relatively common occurrence. There are many opinions and theories describing the causes of groin pain (Bradshaw et al. 2008), particularly in sport but many commonly held assumptions are unsupported by robust scientific evidence. There is much significant discussion in the sports medicine literature as to whether groin pain is caused predominantly by:

- a single pathology, presenting in various ways

- one of multiple distinct entities that need to be accurately diagnosed and treated differently

- several merging pathologies that coexist, resulting in a similar syndrome that requires almost the same treatment, no matter what the presenting symptomatology.

Sports Rehabilitation and Injury Prevention   Edited by Paul Comfort and Earle Abrahamson
© 2010 John Wiley & Sons, Ltd

Part of the explanation, for the resulting unsatisfactory causal and diagnostic opinions, lies in the complex anatomy of the groin. Patients very often become uncertain and despondent after receiving varied opinions from different highly respected clinicians only to receive conflicting and confusing information as to the diagnosis and the way the condition should be managed, treated and rehabilitated.

The groin is a relatively unstable area, acting as the link between structures that generate large forces through it. Groin pain can therefore be related to several different joints, particularly the lumbar spine, sacroiliac, hip and pubic symphysis; the adductor, hip flexor, gluteal, abdominal and lumbar extensor muscle groups; and even the obturator, ilioinguinal and genitofemoral neural structures (Noiseux and Tanzer 2008).

Functional anatomy (Figures 20.1 and 20.2)

The pelvis is a ring of bone with the symphysis pubis joint anteriorly and the two sacroiliac joints posteriorly. It links the hip joints at each side and the spine posteriorly. Instability in the pelvic ring may make it difficult to diagnose symptoms, and repetitive stress on the pelvic bone itself may cause a stubborn stress fracture.

The gluteal muscles provide stability and generate significant power during hip joint movement. They are important for the transference of power between the legs and the trunk in sport. The gluteus maximus extends the hip and therefore particularly contributes to propulsion and strong anti-gravity stability when balancing or pivoting on one leg. Gluteus medius and minimus move the leg away from the midline and are important stabilisers of the pelvis. Weakness of these muscles may well contribute to pain in the groin.

The hip flexor muscles bring the thigh forward during running and kicking. They are made up of the iliopsoas muscle combination. The Psoas muscle originates from the lumbar spine and travels through the posterior part of the abdomen and beneath the inguinal ligament to attach on the upper part of the thigh bone. Iliacus originates from inside the pelvic bone and joins psoas towards the thigh. Importantly the iliopsoas connects the lumbar spine directly to the thigh over the front of the hip joint. As a result of its position and activity groin injury or pain may well

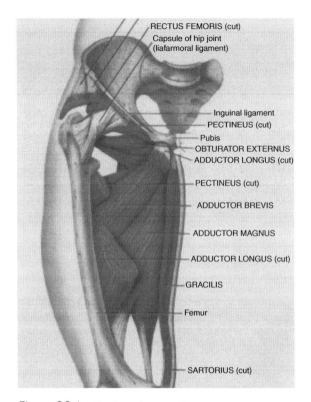

Figure 20.1 Groin and surounding anatomy.

have an effect on the function of the iliopsoas. Iliopsoas status and function may also influence groin and pelvic pain.

Hip adductor muscles connect the central pubic bone of the pelvis to the medial thigh. They move the active leg toward and past the midline of the body in activities like kicking diagonally across the body. They also stabilise the supporting limb during movements that require balancing on one limb, such as in the repetitive kicking in karate. The adductor muscles can be strained, torn or even ruptured, usually more proximal than distal (Tuite et al. 1998; Lohrer and Nauck 2007).

The rectus abdominis, internal and external oblique and transversus abdominis abdominal muscles all combine to perform pelvic movement and contribute towards stability in the trunk and pelvic region. Strength imbalance and dysfunction of these muscles (Maffey and Emery 2007) may contribute to groin related symptoms. Weakness of the lower abdominal muscles (Morales-Conde 2009) may be related in a condition known as 'sportsman's hernia'.

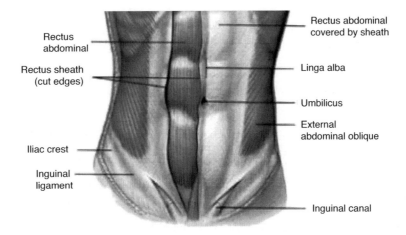

Figure 20.2 Groin and surounding anatomy.

The quadratus lumborum muscle runs down either side of the trunk and is attached to the posterior edge of the pelvis. Pain and dysfunction in other structures around the pelvis may cause secondary reactive tightness in this sheet of muscle (Kibler et al. 2006). The specific posterior attachment to the pelvis combined with tension in this muscle may result in an anterior tilt of the pelvis (Hibbs et al. 2008), causing impingement and torsion to the structures of particularly the anterior groin.

The biomechanics around the pelvis is such that dysfunction and imbalance of any one or more of these anatomical structures can therefore result in disconnection in the chain of movement, especially in the complex and stressful environment of sport (Meyers et al. 2008). Therefore once a sportsperson has had groin symptoms over a significant period, it is often the case that they will develop several related pathologies (Macintyre et al. 2006). This scenario requires patient examination and re-examination, sometimes over a period of time to identify the main areas of dysfunction and pathology to provide the sportsperson with pragmatic and appropriate rehabilitation objectives.

Overview of groin injuries

Groin pain can be difficult to examine and assess for a variety of reasons (Harmon 2007). The history and location of pain is often ambiguous, therefore the history and examination should be approached methodically (Wettstein et al. 2007) to achieve as accurate a diagnosis as possible (Fricker 1997) (refer to Chapter 2 on screening and assessment for further information of initial clinical management of injuries). Discussion of factors that aggravate or reduce pain, especially relating to the specific aspects of the athletes sport, is essential. For general guidance only, histories of duration longer than several months or one season are usually hip joint related, whereas less than three months can be various pathologies (Verrall 2008).

Histories must include:

- duration
- location
- onset details
- predisposing factors
- response to rest and treatment so far
- investigation history e.g. X-ray, ultrasound, MRI, CT, bone scan.

Incorporated within the varied clinical tests used, assessment must ideally include:

- inspection for asymmetry and anatomical irregularity, including poor habitual posture

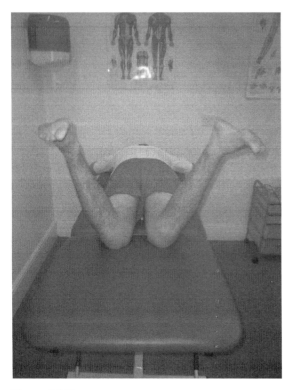

Figure 20.3 Medial hip rotation assessment looking for restriction or asymmetry.

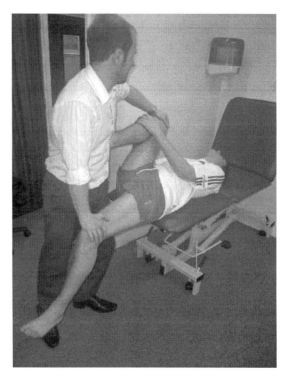

Figure 20.4 Modified Thomas test for hip extension, flexion and adduction.

- assessment of the range of motion of the joints near the area and particularly movement of the hip joint (Figures 20.3 and 20.4) and pubic symphysis stress tests

- observation for discrepancy of leg length

- evaluation of gait, including if possible the individual's specific sport activities that exacerbate the symptoms and training load

- muscle length and strength tests (Figures 20.5 and 20.6), including adductor weakness, abdominal/gluteal control and load transfer failure (Figures 20.7–20.10)

- palpation of the affected area.

Hip joint injuries are very often overuse, repetitive strain injuries, and sometimes caused by trauma (Paluska 2005). The hip is a very large joint that is stressed by weight bearing movement, as well as being dependant for stability on the soft tissue around it, so hip injuries are a common feature of many sports (Anderson et al. 2001). The range of motion

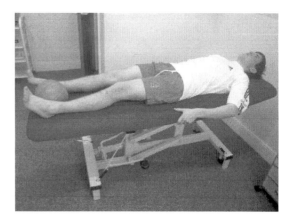

Figure 20.5 Squeeze test and exercise for adductors in neutral.

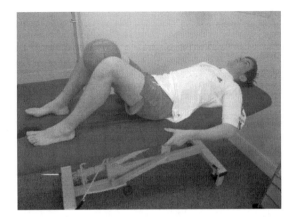

Figure 20.6 Squeeze test and exercise in flexion.

of the hip joint is second only to that of the shoulder joint; combined with the fact that the hip joint bears weight and is subjected to repetitive body weight stress which makes this joint extremely susceptible to injury. Persistent sports related groin pain is frequently caused by an intra-articular hip disorder (Bohnsack et al. 2006).

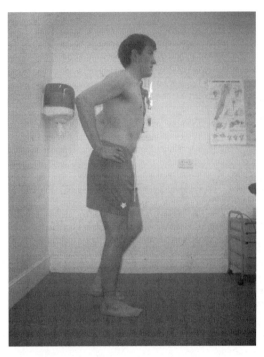

Figure 20.8 Poor single leg anterior pelvic tilt with compensatory trunk flexion.

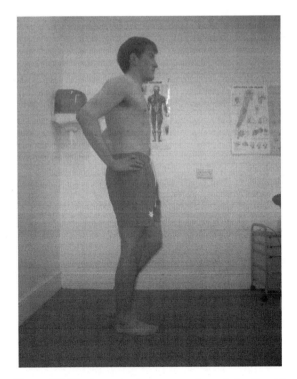

Figure 20.7 Poor single leg posterior pelvic tilt with compensatory extension.

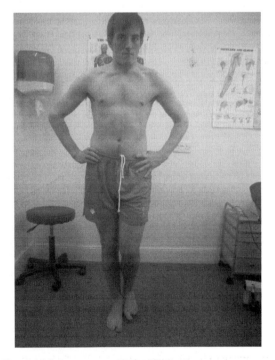

Figure 20.9 Inferior left lateral pelvic tilt.

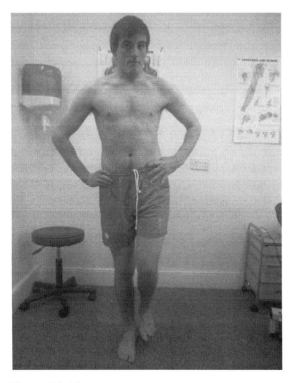

Figure 20.10 Superior left lateral pelvic tilt.

In an Australian study of elite football players, Verrall et al. (2007) suggested that the development of chronic groin injury may be preceded by loss of hip mobility. There are several guiding tests for range and symmetry of hip movement including medial rotation (Figure 20.3; Verrall 2005) and the modified Thomas test (Figure 20.4). The Thomas test in particular can be a differential indicator of hip flexor related pathology on the stretch side, or hip impingement on the compression side.

The most common injuries can be divided into acute and chronic accounting for approximately 5% of the cases attending sports injury clinics (Hackney 2008). The most common acute injuries are soft-tissue contusions and haematomas encountered in contact sports like football, ice hockey, basketball and rugby that result from direct impact but sometimes non-contact sports like gymnastics (Junge et al. 2006). Acute muscle strains are commonly encountered in activities in which loaded adduction of the hip occurs, like football (Ibrahim et al. 2007), rugby, ice hockey and breaststroke swimming, or in activities where forced abduction of the hip occurs as in any sporting activity in which the sportsperson may perform a split, either accidentally or deliberately. However, there is a need for consensus on acceptable study designs and methods of data analysis in sports epidemiology (Brooks and Fuller 2006).

The most common chronic conditions are strains of the muscle and tendon (Amaral 1997). Repetitive acute groin injuries may become the cause of chronic or overuse injuries (Holmich 2007). Chronic groin injuries tend to occur in those who participate in activities that promote overuse of the groin area like ice hockey, skating, football and running (Maffey and Emery 2007).

The iliotibial band, hip flexor, gluteal and adductor muscles run from the pelvis to the knee, aiding hip stability. Where an imbalance exists between these very strong muscle groups which are necessary to lift and propel the leg, and those of the abdomen, the sportsperson may experience imbalance detrimental to performance as well as being more injury susceptible (Machotka et al. 2009).

Sometimes, wear and tear on the hip leads to osteoarthritis (Spector et al. 1996), usually in the older athlete, but trauma in the younger athlete in as much as 50% can instigate increased wear early (Bradshaw et al. 2008). When treatment and management of this becomes progressively less effective, resurfacing of the hip, an alternative to replacement, has given some athletes further time in their sport.

Groin injuries are found in athletes of all ages at all levels of competition. They are particularly common in activities where strong and particularly repetitive adduction of the hip occurs, like football, skating and swimming. As many as 5% of soccer-related injuries are groin injuries (Gilmore 1998). They are often a result of quick, lateral movements (Ziltener and Leal 2007), usually in combination with acceleration made from a standing start (Johnson 1997). Football, rugby and tennis are sports where there is a significant incidence of this type of groin stress. The least debilitating of the groin injuries is often called a groin pull, or strain. Like other strains, this is usually caused by an overstretching under load, of the lower abdominal muscles or adductors, particularly where hips are restricted (Ibrahim et al. 2007), causing the muscle to stretch and sustain a minor or worse tear. A typical mild groin strain will usually resolve with rest and management over one or two weeks.

The more serious groin injury is a partial or complete tear of the tissue. A tear will cause essential modification of an athlete's training schedule for two or three months, or even longer. Gradual and progressive care must be taken with the rehabilitation. The particular type of groin tear called a sportsman's hernia, or Gilmore's Groin is a tear of sometimes several structures and surgery is required to repair the tear, with a very varied convalescence and recovery of up to six months, depending on the extent of the injury.

Special consideration must be given to females (Kark and Kurzer 2008), children and adolescents with groin pain, as they may have conditions that are more serious and need medical or surgical intervention. Pain in the adolescent athlete must take into consideration the susceptible growth plate in the hip. Apophysitis and apophyseal fractures are more common in adolescent athletes, where the growth plate may be a weak link in the area (Longacre 1994). Also, children, adolescents and adults may report knee pain that is actually referred from pathology in the hip, which must always be considered when no knee pathology is demonstrated on examination (Tomoko et al. 2006).

Differential diagnosis of groin injuries

Pain that is aggravated with activity and alleviated with rest suggests a structural problem. Constant pain suggests an infectious, inflammatory, or neoplastic (abnormal growth) process.

Differential diagnosis must take into consideration medical conditions that affect the groin in all individuals, not just athletes (Morelli and Weaver 2005):

- referred pain from lumbar spine pathologies

- hip joint disorders like osteoarthritis, Perthes disease, slipped epiphysis, osteochondritis desiccans

- intra-abdominal disorders like appendicitis

- genitourinary abnormalities like urinary tract infections; sexually transmitted diseases.

Many patients with groin pain have more than one injury, often leading to managing one injury whilst a second is missed and untreated (Putukian et al. 1997). Therefore the importance of a comprehensive differential diagnosis is clearly justified (Harmon 2007).

Impact injuries

Impact injuries, such as those that occur during football, hockey, or other contact sports, may result in contusions. However, such injuries may cause fractures of the pelvis; exacerbate previously asymptomatic inguinal hernias; and, in rare cases, produce bladder, testicular, or even urethral injuries. Any patient with lower abdominal or pelvic impact injury that causes severe groin pain, loss of function, or blood in the urine should be immediately medically investigated. If no bony injury is discovered, conservative measures may be implemented.

Hip pointer

The hip pointer can arise from both direct contusion causing impact and indirect strain injuries of the hip, primarily in contact sports. Forced extension of the hip, by for example a tackle from behind in rugby, may result in a sprain or avulsion of the sartorius muscle at its iliac crest attachment. These injuries are severely painful and make leg movement very painful, taking from one to several weeks to rehabilitate, depending on severity. Pain may be felt when walking, laughing, coughing, or even deep breathing. Direct contusions to the anterior superior iliac crest may also involve the attachment of the sartorius muscle.

Hip pointers are common in contact sports such as rugby and judo and proper protective equipment is essential. Additionally, developing appropriate skills and techniques may help avoid a hip pointer. Ultimately, however, there is not much an athlete can do to prepare for such an impact.

Treatment and rehabilitation

- Initial routine ice treatment and compression, and rule out bony injury or more significant soft tissue damage by X-ray or scan.

- Medication as required, including topical anti-inflammatory when the injury is superficial.

- Immobilisation, with crutches if necessary during the acute period, which may be days or weeks.

- Incremental progression to return to sport, including proprioceptive, mobility and strengthening exercises, including guidelines that can be applied to most groin injuries (Brukner and Khan 2006):

- exercising within painful thresholds
- isolating and reducing pelvic load sources
- developing lumbo-pelvic stability
- regional strengthening
- progressing activity according to repeated assessment
- static progressing to dynamic and finally sports specific.

It is important that the sportsperson does not return too quickly to activity. If they still have pain or tenderness, they are liable to compensate by altering the movement pattern and technique which may result in injury to another part of the body.

Classic groin strain, 'pull' or adductor tendinopathy

The most frequent acute strain of the groin may involve single or multiple muscles, including the iliopsoas, adductors and the gracilis, which attach to the femur or pubis and help to stabilise the legs and flex the thigh. The muscle most commonly strained and injured in the groin is the adductor longus (Harmon 2007). Other muscles that must be considered include the rectus femoris, the rectus abdominis and the sartorius.

Falling, running, changing direction, kicking or doing the splits may generate these injuries. Groin strains usually cause pain in the groin and radiate down the inside of the thigh, more proximal than distal, with pain on resisted adduction of the hip. There may be confusion on occasions as MRI and ultrasound studies will often report changes at the adductor attachment, when the patient is completely asymptomatic in that area.

Occasionally there may be haematoma formation which can prolong the healing time. The weakened area created by a strain may continue to be susceptible to repeated injury for a long time (Lynch and Renström 1999). In fact, in a review of ice hockey players, those with a previous groin injury had twice the risk of repeat injuries as that of athletes without a previous injury (Emery and Meeuwisse 2001). Older, experienced hockey players had an injury rate five times higher than that of novice players (Emery et al. 1999).

In another review in ice hockey, the statistics revealed that adductor strains occurred 20 times more frequently during training camps rather than during the playing season (Caudill et al. 2008), possibly related to the benefits of a strength-training programme and to the fact that out- of- season lack of maintenance conditioning may contribute to groin injuries.

Sometimes in explosive sports, the adductor tendon may fully rupture, and even pull away from its osseous attachment, taking a piece of the bone with it. These are called avulsion fractures and if they cause severe displacement, surgical repair may be indicated. Most groin strains are mild, however, and eventually respond to conservative treatment. Unfortunately, adductor tendon pathology often coexists with other dysfunction around the pelvic region.

Treatment and rehabilitation

Adductor tendon strains may be treated specifically if the symptoms are very localised and imaging is consistent with the clinical observations. Soft-tissue release techniques to reduce tightness can be combined with progressive strengthening of the adductor muscles. However, it is very important that treatment should be accompanied by attention to any other strength (Figures 20.5 and 20.6; Holmich et al. 2004; Verrall 2005) or flexibility deficiencies in the pelvic area. The adductors work with gluteus medius after propulsion in running; and with the hip abductors to maintain pelvic stability during the stance phase, therefore it is considered that pelvic stability prevents excessive eccentric loading of the adductors (Figures 20.7–20.10). Strain of the adductor longus should be managed depending on the location of the injury. Physical examination should reveal whether the injury lies within the muscle belly or within the teno-periosteal attachment (Schilders et al. 2007). Injuries to the muscle belly are best managed with strengthening, gentle stretching and progressive return to activity.

However, avulsion injury to the teno-periosteal attachment requires more conservative management with rest until the patient is pain free, then strengthening and very careful stretching over a significantly longer period of weeks; then running and sprinting; and, lastly, running and sprinting combined with rapid changes in direction if the sport demands it.

Injection of corticosteroid around the adductor origin may be beneficial in stubborn cases (Bentley 1981), as long as the athlete has, and continues to, conform to the guidance given. Proximal thigh strapping during activity will often be successful in dissipating stress away from the weakened area.

When all conservative measures (Verrall et al. 2007) have failed, surgical release of the tendon from the bone called an adductor tenotomy may be indicated. Although adductor tenotomy has been shown to leave a strength deficit in some studies, this does not appear to significantly be associated with adverse performance.

Iliopsoas syndrome

The iliopsoas muscle is comprised of two muscles that work together, the iliacus and psoas. The main function is flexion of the hip, a fundamental sport movement as in running and kicking. In between the proximal iliopsoas tendon and the hip joint lies the psoas bursa, which helps to reduce friction. Iliopsoas syndrome is inflammation of the iliopsoas bursa and/or iliopsoas tendonopathy, and is often found in conjunction with adductor abnormalities. It more frequently occurs in sports with repetitive hip flexion like gymnastics and athletics. This syndrome causes pain in the hip and upper thigh, as well as hip stiffness and sometimes a clicking or snapping hip. Diagnosis of this condition is usually evident through manual tests, but can be confirmed by an ultrasound (Deslandes et al. 2008) or MRI scan (Shabshin et al. 2005).

Treatment and rehabilitation

The condition will often settle down with ice treatment, rest, support and initial anti-inflammatory medication. When the acute pain has diminished, a programme of progressive flexibility (Figures 20.3 and 20.4) and strengthening exercises for the structures around the hip should be followed by an incremental return to full activity (Torry et al. 2006). Core strengthening exercises to help improve the stability of the trunk and pelvis (Figures 20.5–20.10) are an important component of the rehabilitation programme (Donatelli 2006) for this condition as it appears to be associated with hypomobility of the upper lumbar spine from which the iliopsoas originates.

Hernias

Hernias of the abdominal wall must always be considered in athletes with groin pain. Hernias are frequently overlooked in the athlete with only 8% of patients with hernias being detectable on initial physical examination, but 95% returning to sport after surgical repair and rehabilitation (Diesen and Pappas 2007).

The sportsman's hernia, or Gilmore's groin, was first described by O.J. Gilmore in 1980. It is a varied syndrome characterised by recurrent or chronic groin pain that is associated with a dilated superficial inguinal ring and weakness of the posterior wall of the inguinal canal although the exact cause of this injury is considered by many clinicians to be largely speculative and could be a combination of factors.

The true incidence of sportsman's hernia remains controversial; some authors (Swan and Wolcott 2007) believe it is only an obscure cause of groin pain in athletes, although others (Meyers et al. 2008) have now come to believe that it is the most common cause of chronic groin pain. The term 'sportsman's hernia' is considered by many to be a misnomer as there is typically no demonstrable hernia or defect in the groin or the abdominal wall. The definition of Gilmore's groin, therefore, could be any condition that causes persistent unilateral pain in the groin (Brannigan et al. 2000), without a conventional hernia. It is probably best described as a severe musculo-tendinous injury or disruption of the groin, which can be successfully treated by surgical repair towards the restoration of normal balanced and stable anatomy.

The symptoms are pain in the groin increased by running, sprinting, turning and kicking. After sport, the athlete is commonly temporarily stiff and sore. The day after a game, turning or getting out of bed or a car often causes pain, as may coughing, sneezing and sit-ups. There is a history of specific injury in only 30% of patients.

Treatment and rehabilitation

Effective treatment for ongoing hernia is ultimately surgical repair and reinforcement of the abdominal wall. In comparing the recovery times for patients after hernia repair, Stoker et al. (1994) and colleagues cited laparoscopic repair halved the time leading to resumption of sport participation in one week. For patients with a positive herniography (test for hernia) there are indications that surgery results in earlier return to sport with exercise therapy, and that laparoscopic intervention might result in an even earlier return compared with open surgery (Jansen et al. 2008).

Significantly 40% of patients diagnosed with Gilmore's Groin also have torn adductors; a fact that plays an important part in the rehabilitation, as minor and moderate tears usually respond to adductor exercises (Gilmore 1998); static (Figures 20.5 and 20.6), progressing to active balance and eventually multidirectional activities relating to the individual sport, combined with manual therapy. However, patients with severe adductor tears require adductor tenotomy or release.

Athletes hernia

Some believe that this condition cannot be detected by examination, while others believe that careful history and examination can result in this diagnosis (Caudill et al. 2008). Real time ultrasound of the groin, comparing both sides during increased intra-abdominal pressure has developed as an investigation for this condition. Robinson (2002) and Diesen and Pappas (2007) maintain the accuracy of ultrasound for detecting sports hernias is yet to be established.

In patients with the possibility of several different diagnoses, ultrasound can also evaluate surrounding muscles and joints. In athletes with groin pain ultrasound is a good initial imaging tool, but MRI may be necessary for specific conditions, although, according to Daigeler (2007) MRI findings do not always correlate with outcome in athletes with chronic groin pain.

Some clinicians contend that anterior groin pain is a result of biomechanical abnormalities, rather than an anatomical weakness (Meyers et al. 2007). They refer to studies like Farber and Wilckens (2007) demonstrating abnormal ultrasound findings in asymptomatic cases, and pathological findings in both those that respond well to surgical treatment and those who gain no benefit.

The pain is often described as variable and ongoing, near the pubic tubercle, maximal in the evening, after exercise or on the morning afterwards, and exacerbated by the type of activities that increase the intra-abdominal pressure requiring good core stability. Gilmore's groin is more common in male than female footballers (Hagglund et al. 2008).

On digital examination, the superficial inguinal ring is dilated. Evidence of herniation may or may not be palpable. The point of most tenderness is often the ipsilateral pubic tubercle. Pain can be elicited with a Valsalva manoeuvre, forcibly exhaling against a closed airway, or a resisted sit-up. Examination of the hip joint and evaluation of the athlete's gait often significantly reveal weak adductors.

Gilmore (1998) maintains the success of surgery depends on accurate diagnosis, meticulous repair of each element of the disruption and intensive rehabilitation according to a standard rehabilitation protocol. Surgery is indicated in athletes who are unable to play or fail to respond to conservative rehabilitation.

Operative repair

In order to rehabilitate an athlete after operative repair it is important to understand the procedure and its variables with ideally sight of the operative report.

The classic operative findings (Anderson et al. 2001) include laddering of the external oblique in conjunction with separation of the conjoint tendon from the ligament and laxity of the transversalis fascia. Some studies, such as Morelli and Weaver (2005), have suggested abnormalities with the rectus abdominis insertion, avulsions of the internal oblique muscle fibers at the pubic tubercle, or entrapment of the ilioinguinal or genitofemoral nerves.

Conservative therapies may only temporarily alleviate a patient's pain, causing definitive later surgical procedure and management to be recommended. Gilmore himself repairs injuries with a technique that uses a six-layered reinforcement of the weakened transversalis fascia. Gilmore and others (Meyers et al. 2008) maintain approximately 97% of their patients return to competitive sport by the tenth week after postoperative care, and placement of supportive meshes or patches, with or without neurectomy or ablation of the ilioinguinal nerve, has demonstrated success.

Others like Diesen and Pappas (2007) have claimed only a limited surgical success and maintain athletes who do not respond well with surgery for this condition are often subsequently provided with a different diagnosis for their groin pain and need to be highly motivated in their rehabilitation. However, this still demonstrates the success of operative intervention overall (Farber and Wilckens 2007).

Treatment and rehabilitation

The surgical treatment consists of restoring normal anatomy with a six-layered structural repair of the inguinal region. Adductor tenotomy may be performed in order to lengthen the tissue that has developed improperly, or become shortened and is resistant to stretch, and is indicated in patients with persistent and troublesome adductor tears which do not respond to conservative treatment. Patients are admitted on the day of operation and return home after initial physiotherapy guidance on early post-operative exercises, usually within 24 hours. The operation site discomfort takes approximately 8–12 weeks to settle completely.

A standard rehabilitation programme is then followed which can be adjusted in the later stages to account for specific sport requirements of the individual athlete. Rehabilitation exercises, especially those which activate the core stability muscles are combined with pelvic control (Figures 20.7 – 20.10) and adductor strengthening (Figures 20.5 and 20.6). Active repetitive drills, specific to the needs of the sport, to make sure there are no compensatory adaptations, are particularly beneficial.

The successful Gilmore's groin rehabilitation programme (www.108harleystreet.co.uk) emphasises the importance of activating the core stability muscles, particularly transversus abdominus with multifidus.

- *Week 1.* First day after operation: Essential to stand upright and walk 20 mins. Thereafter walk gently 4 times a day. Gentle stretching and core stability exercises given by a physiotherapist.

- *Week 2.* Jogging and gentle running in straight lines. Gentle sit-ups with knees bent. Adductor exercises. Step ups

- *Week 3.* Increase speed to sprinting. Increase sit-ups and adductor exercises. Cycling. Swimming (crawl)

- *Week 4.* Sprint. Twist and turn. Kick. Play

Nerve entrapment

It is pertinent to discuss entrapment after hernia surgery as both are involved in the discussion of surgical results. The nerves that have been involved in groin pain are the obturator, ilioinguinal and genitofemoral. The sportsperson will usually describe a difficult to localise intermittent sharp or burning pain in the medial thigh or the genital area. Due to hypersensitivity of the irritated nerve, pain may be aggravated by very light touch of the sensitised area. A local anaesthetic injection specifically in the area can be used as a confirmation of the diagnosis to some extent.

Treatment and rehabilitation

Conservative treatment by appropriate neural and hip stretching movements and exercises will often improve symptoms, but sometimes a surgical release procedure of the trapped nerve is required. Hydrotherapy and swimming can be effective to mobilise the hip through full range without weight bearing aggravation.

Hip dislocation or fracture

The most severe and debilitating groin injury is a hip dislocation or fracture. This usually results from a violent or high-speed impact, collision or fall, seen in sports like skiing or ice hockey. The pain is usually very acute and associated with an inability to even partially weight-bear with a visible shortening and rotation of one leg medially or laterally.

Treatment and rehabilitation

Immobilisation and immediate medical investigations with reduction, preferably within a few hours, are critical to maximise recovery. Hydrotherapy to mobilise the hip initially, then to regain stability by progressive non to partial weight bearing, then dynamic impact torsional and multi-directional weight bearing balance and strengthening exercises (Fig 20.11) are very beneficial.

Soft tissue release techniques combined with assisted mobilising will be necessary to reduce local imbalances of laxity and tightness.

Ligamentum teres tear

In relatively major trauma this ligament in the hip joint can cause hip instability with 'catching or clunking' symptoms. They can occur in sport such as gymnastics where there is extreme hip abduction.

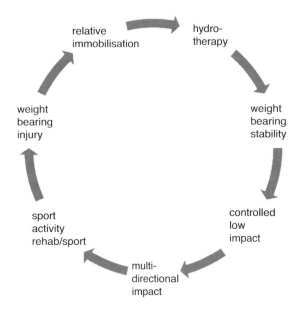

Figure 20.11 Fundamental cycle of rehabilitation in sport (copyright Allen JW 2009).

Arthroscopic surgery is usually necessary followed by steadily progressive rehabilitation.

Loose bodies in the hip

These are rare but may be as a result of cartilage pathology, causing catching and locking of the hip. The loose body usually needs to be removed arthroscopically. Recovery is steady but longer term degenerative implications must be a consideration in return to sport.

Stress fractures

Femoral neck and the pubic ramus are the most common stress fractures in the groin region (Maffulli and Bruns 2000). Just as in many other bones in the body, these stress fractures are caused by repetitive trauma to the bones or through muscular attachments. These stress fractures particularly occur secondary to running related sports, although additional risk factors can be intrinsic or quite diverse and include osteoporotic tendency in young female athletes secondary to nutritional or hormonal imbalances, changes in footwear and mode of training, or changes in intensity and/or duration of training.

Stress fractures in the groin or hip can be difficult to diagnose and treat. Femoral neck stress fractures are especially troublesome because if not curtailed they may lead to avascular necrosis of the femoral head which can cause long-term disability. These may appear as a cortical irregularity or haziness on plain X-rays, but a magnetic resonance image (MRI) or a bone scan is usually required for definitive diagnosis.

Treatment and rehabilitation

There must be modification of causative stress patterns, rest and usually reduction of weight bearing with crutches if necessary. Non-weight bearing progressing to recovery, although usually complete, may take several months. Hydrotherapy in the early stages, progressing to land-based minimal impact re-education and finally usual impact related specific activities.

Avulsion fractures

Groin pain from avulsion fractures of the hip must be considered particularly in young athletes who say they experienced acute, sudden-onset, specific pain. The relative weakness of the apophysis of the adolescent skeleton predisposes the young athlete to a variety of avulsion fractures. Two of which are forceful movement avulsion of the anterior inferior iliac spine by rectus femoris, and avulsion of the anterior superior iliac spine by sartorius.

Treatment and rehabilitation

Management for the majority of avulsion fractures (Koh and Dietz 2005) is usually by conservative rest and gradual incremental return to activity. Depending on the size and amount of displacement of the fracture fragment, the injury may sometimes need surgical repair.

Avascular necrosis of the femoral head

This is an insidious condition that usually affects individuals between the ages of 20–40. Disruption of the circulatory supply to the femoral head either acutely or chronically results in cell ischemia and necrosis, eventually damaging the hip joint. Apart from well-documented systemic causes, factors for its development include high loads and sudden or irregular impact, as required in many sports. X-rays may not show the condition for three months after the initial trauma. However, MRI will usually specifically indicate an affected area.

Treatment and rehabilitation

Caught early, rest and reduced weight bearing will help, but there will usually be a longer term consequence requiring careful guidance and management of everyday activity as well as training in order to continue in sport.

Osteitis pubis

The symphysis pubis is a fibrous joint between the two halves at the front of the pelvis. The adductor muscles attach either side and the abdominal muscles attach along the top of the pubic bones. Therefore the symphysis is subjected to significant shearing forces, especially during alternate single leg weight bearing with change of direction during activities like running and kicking (Lovell et al. 2006). The shearing forces can be increased by biomechanical limitations, such as restriction of internal hip rotation (Figure 20.3).

Sometimes called Gracilis syndrome, this is a chronic pathology affecting the bone or cartilage of the pubic symphysis and relates to repetitive stress and resulting pain and tenderness from activities like kicking, weightlifting, running and jumping. Torsional or rotational movement stressing the pubic symphysis seems to particularly aggravate, and postpartum women are more susceptible (Asian and Fynes 2007), possibly due to the laxity of the pelvic ligaments during and after pregnancy.

Osteitis pubis causes pain in the pubic region, radiating into the medial thigh of one or both sides. The pain may be specifically in the thigh, but is often present in the lower abdominal area.

There may also be associated instability at the symphysis, and the athlete may be aware of clicking in the region, often noticed when turning in bed. Commonly pain will diminish with warm up and be tolerable during activity.

X-ray will sometimes demonstrate a mottled appearance at the symphysis as a result of the inflammatory process taking place in the joint. Standing antero-posterior and single-leg-stance pelvic radiographs aid in the diagnosis of pelvic instability more effectively than standard radiographs of the pelvis in supine or a standing antero-posterior radiograph of the pelvis alone (Siegel et al. 2008). A bone scan usually shows increased uptake in the pubic bone and MRI (Kunduracioglu et al. 2007) will often show oedema in the bone.

Treatment and rehabilitation

Treatment of this condition is notoriously difficult (Choi et al. 2008). A conservative approach to treatment of osteitis pubis is considered the first choice of action, while warning the athlete that prolonged rest and prolonged absence from sport is likely. However, the condition may take several months to resolve, which can mean difficulty with the patience of athlete and coach.

Pain will often respond well to anti-inflammatory medication initially. However, symptoms often gradually or suddenly become more acute with pain that inhibits or stops activity, becoming unresponsive to conservative management.

Treatment and rehabilitation (Figure 20.12) should include strengthening exercises for the core and muscles around the pelvis and lumbar spine, but more importantly dealing with biomechanical influences like restricted movement of the hip joints and particularly rest from aggravating movement and activities (Wollin and Lovell 2006). All of these elements are interrelated and need to be considered in concert with each other.

Injection of corticosteroid or sclerosant and surgical curettage (Radic and Annear 2008) of the joint

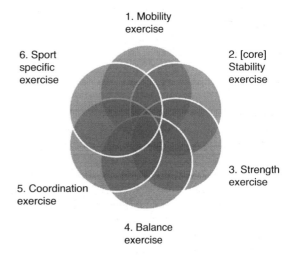

Figure 20.12 Fundamental elements of rehabilitation in sport (copyright Allen JW 2009).

have all been used with variable results (Mehin et al. 2006). There is very little sound evidence for any invasive treatment in osteitis pubis other than time, management and incrementally progressive rehabilitation.

Hip labral tear

The labrum is a lip of cartilage around the edge of the hip acetabulum, which serves to deepen the socket providing added stability to the joint. In jumping and kicking sports such as high jump and football, there are significant repetitive stresses placed on the labrum. This cumulative stress with anterior hip impingement (Noiseux and Tanzer 2008) can result in damage to the labrum, causing pain, restriction and sometimes clicking with hip movement, particularly anterior impingement on combined flexion and adduction, although some recent findings (Guevara et al. 2006) suggest abnormal hip morphology may be a risk factor for labral tears.

The athlete will usually complain of anterior hip pain which may radiate down the anterior thigh, commonly aggravated by sitting, squatting or a prolonged hip flexion position. It has been observed (Carreira et al. 2006) that 72% of athletes with a labral tear reported low back pain as one of their symptoms.

Examination may show decreased range of flexion, abduction, and internal rotation at 90° hip flexion, and positive impingement and grind (compression) test. Diagnosis can be confirmed on MRI and an injection of local anaesthetic into the hip area at the same time helps to confirm that MRI findings are related to the symptoms.

Treatment and rehabilitation

There has been some limited success with tractional mobilisation of especially the anterior hip structures to relieve compression, using 'seat-belt' mobilisation techniques that give the therapist more mechanical advantage over the very strong hip joint. Hydrotherapy for hip stability and mobility may diminish symptoms and is certainly prudent post operatively. Modification of technique or training that directly or indirectly impinges the anterior hip is essential.

Treatment in elite level athletes by arthroscopic surgery has been very successful. However, athletes who have undergone arthroscopy for labral tears will unfortunately frequently develop subsequent symptoms associated with an early degenerative hip joint (Murphy et al. 2006).

Synovitis of the hip

It is important to mention that synovitis may be a complication of hip pathologies. However, it may be a primary problem, especially when associated with a rheumatological condition. It usually responds well to rest and anti-inflammatory medication or injection.

Rehabilitation exercises for the groin

According to Jansen (2008) there is evidence that physical therapy aiming at strengthening (Figures 20.5 and 20.6) and coordinating the muscles stabilising hip and pelvis has superior results compared with passive physical therapy. Holmich et al. (1999) maintains that active rehabilitation gives the athlete over 10 times the outcome possibility of returning to active sport. The correct continually adapting balance between training and competition activity and recovery is critical to maximise the performance of athletes. A number of recovery sessions and modalities are now used as a fundamental element of the training programmes of elite athletes to help attain this critical balance and to reduce injury potential alongside optimising performance (Barnett 2006).

The pelvic mobility, control and proprioceptive stability of the athlete are important in preventing injury, but are also fundamental towards performance in sport. The performance benefits of strength and conditioning exercises can be a useful tool of persuasion for the therapist to convince athlete and coach of the benefits of certain seemingly pedantic exercises and drills.

Single leg tests and exercises can be a useful guide to the control available to the athlete in a relatively static situation as compared with expectations in the much harder and spontaneous active situation of sporting activity. These can take the form of reduced weight bearing tests and exercise in a pool (Vaile et al. 2008), particularly relevant to acute early stage rehabilitation; or land based with full weight bearing tests and exercises (Figures 20.7–20.10 and Mens et al. 2006).

Groin pain in children

Groin pain in children warrants a separate discussion as it includes situations that are rarely seen in adults; needing to take into consideration open epiphyses (Su et al. 2008) in the groin region, infection and developmental abnormalities.

Synovitis or inflammation of the hip joint is the most common cause of hip pain in children. The hip joint is far more prone to viral inflammation in children than in adults but fortunately is usually self-limiting with rest from sport and conservative management, and has no long-term effects. It is important however to differentiate this condition from septic arthritis where a bacterial infection can develop in the joint, which is far more serious and necessitates treatment in hospital with intravenous antibiotics.

Perthes disease

Perthes disease is a disturbance in the development of the hip, with the head of the femur becoming misshapen. For reasons that are not fully understood, the blood supply to the hip joint sometimes becomes compromised in children. Properly called Legg-Calve-Perthes disease it typically affects one hip, but sometimes it develops in both hips. Although it can affect children of nearly any age, it is most common among boys between the ages of four and eight years (Rosenfeld et al. 2007).

There are many degrees of severity, and of significant relevance to sport, some mild cases even remain undetected until later in life particularly exacerbated by the more intensive physical stress of sport. If diagnosed in childhood, sporting activities will have to be modified to minimise the chance of exacerbating the condition. In more severe cases, surgical intervention is indicated and sporting activity has to be significantly managed (Kocher and Tucker 2006).

Epiphysis related injury

This very significant and sometimes, especially initially, undiagnosed condition occurs particularly in sport; the epiphysis at the top of the femur moving out of position (Boardman et al. 2009). It affects boys more than girls, primarily between 8 and 16 years old and usually as a result of repeated cumulative stress, sometimes exacerbated by a fall or impact trauma. The pain may settle with rest but is normally accompanied by a limp. Less commonly a limp or odd walk may be the only abnormality present and a few children only complain of pain in the knee. This condition constitutes a medical emergency and can produce deformity in the child not diagnosed and treated promptly.

One area that is commonly injured in child athletes is the attachment of Rectus Femoris to the anterior inferior iliac spine. Children involved in explosive activities like sprinting, jumping and kicking are at higher risk due to excessive or repetitive loading on the origin of the quadriceps. The severity of the injury varies from a mild strain, where there is inflammation of the tendon attachment to avulsion, where the growth plate is pulled off the pelvis.

Most of these growth plate injuries can be treated conservatively with rest, gentle very careful stretching and progressive and controlled functional strengthening (e.g. slow controlled squatting). However, if there is significant displacement of the avulsed bony attachment, surgical reattachment is required.

It is important to emphasise that problems from the hip joint can present as knee pain, particularly in children, because of their particularly sensitive somatic radiation from the hip to the knee.

Treatment and rehabilitation

Hydrotherapy is particularly effective with children due to the novelty element of being in water doing what would otherwise be a strain on a short attention span. Exercises are used for hip mobility, strength and stability as well as to regain confidence to use the limb and progress into shallower water. Non-weight bearing exercises can be used on land, progressing to partial weight bearing with the use of crutches, and finally to full weight bearing and dynamic activity.

Gait re-education is particularly important in the developing child as asymmetries that can be addressed will be magnified with growth and become habitual and subsequently permanent. Proprioception and balance can also be inhibited by hip conditions in children and must also be improved as soon as possible to achieve optimum recovery. Hydrotherapy and exercise in game format can be structured by the therapist to maintain the child's interest and enthusiasm. Hip arthroscopy to investigate ongoing symptoms is becoming an established procedure for children and adolescents (Kocher and Lee, 2006).

Other causes of groin pain

Injuries to the groin that do not involve the bones or musculature are usually the result of a direct impact. Reasons to suspect these injuries are more significant include blood in the urine; severe abdominal pain and tenderness, usually in the event of a bladder injury; persistent nausea, vomiting, or abdominal distension; and swelling in the femoral triangle or inguinal area, usually in the event of a hernia. All of the soft-tissue structures of the groin, especially the penis, testis, urethra and bladder are susceptible to injury. All of these symptoms merit immediate medical evaluation, preferably in a hospital.

It should now be evident that musculoskeletal causes of groin pain can be difficult enough to differentiate from each other however, we must be aware that intra-pelvic and intra-abdominal conditions can mimic musculoskeletal groin injuries. Therefore, if there are any doubts referral for abdominal examination to an appropriate clinician should be part of the assessment of the athlete with groin pain.

Discussion

Although an athlete's symptoms may be in the area of the groin, the subjective history and objective examination in relation to the complexity of predisposing factors and specific sport often makes definitive diagnosis and prognosis uncertain. A recent study by Bradshaw et al (2008) showed a different breakdown of injuries in a sporting population presenting with groin pain than previously reported, which infers there may be diagnostic difficulties in the groin area. The high incidence of hip pathology and the poor prognosis that this confers are worthy of note.

It is important therefore to re-evaluate groin injury at regular intervals, as its evolution as a condition, and response to treatment and rehabilitation may often alter the initial diagnostic and prognostic opinion. Several rehabilitation principles are therefore paramount (Figure 20.11 and 20.12):

- monitoring of load modification
- re-establishing efficient weight transference
- retraining the global muscles supporting the 'sling' support system
- restoring joint mobility
- restoring muscle extensibility
- reducing adductor muscle strength deficits
- training sport-specific areas of strength, i.e the muscle groups predominantly used
- training relevant functional movement patterns, i.e. relating to the characteristic techniques of the sport.

Conservative management of athletic chronic groin injury results in an excellent outcome when assessed by the return to sport criterion. However, the results were only satisfactory if the criterion of ongoing symptoms after treatment was used. More research is needed to compare the efficacy of all treatments that are used in this troublesome condition (Verrall et al. 2007). Treatment should certainly be centered on the underlying cause.

Case study 1

A 19- year-old rugby player developed low grade right- sided groin pain over two months.

Pain was aggravated by training and playing and had built up to lasting 2–4 days after activity. He tried resting on two occasions, for two weeks and then three weeks, but pain returned on activity, especially sprinting and changing direction. Pain described as between 1 and 8 on a 0–10 scale, directly activity dependent. Stairs were uncomfortable up and down. Turning in bed was uncomfortable. There had been a feeling of tightness in the ipsilateral proximal 'hamstring'.

He had a consultation with GP, who referred for an X-ray, which was reported as negative.

He had good general health.

On examination

- In standing the ipsilateral leg was medially rotated in relation to the contralateral leg.

- Flexion and particularly flexion adduction of the ipsilateral hip was painful in the right groin.

- There was deep palpable tenderness medial to the centre of a line between the ASIS and the pubis.

- Mild pain and restriction of passive medial rotation compared to the contralateral hip.

- Mild pain with resisted adduction supine with straight legs; more pain with resisted adduction in 'turn-out' position.

- Double leg jump asymptomatic but single leg hop discomfort.

Differential diagnosis

- Adductor tendonopathy/periostitis/avulsion

- Nerve entrapment

- Labral tear

- Osteitis Pubis

Initial treatment plan

- Rest from sport training until relatively pain free.

- Static isometric specific adductor (Figures 20.5 and 20.6) and general hip movement strengthening (Figures 20.7–20.10).

- Specific and general stretching within discomfort, e.g bent knee turn out in crook lying.

- Hydrotherapy for trunk and hip stability, proprioception and controlled movement.

- Hip mobility, particularly medial rotation and extension (Figures 20.3 and 20.4).

- Lying 'clam' exercise progressing to standing flexed knee external rotation with cliniband resistance to strengthen hip lateral rotators.

- Non-impact progressing to impact and eventually controlled sport movement simulation, e.g. jogging and passing a rugby ball alternate sides.

Four weeks later

Tolerance to everything except change of direction which aggravated symptoms, therefore gradual progression jogging to running, weaving between cones, wide spaced at first, progressing to smaller spacing and tighter, faster turns.

Ultrasound scan showed a tear in the proximal adductor tendon.

X-rays of the hip/pelvis, including stork views to eliminate osteitis pubis/instability were normal.

MRI to eliminate stress fracture and to visualise source of persistent pain which confirmed ultrasound scan findings.

Further treatment plan

To continue with conservative treatment and rehabilitation, particularly specific progressive rehabilitation exercises.

Continue activity as tolerated, but should limit activities with discomfort awareness.

Using supportive reusable strapping to proximal thigh for activities.

Conclusion

There was full return to sport 5 months after the initial onset.

Discussion

- Were the number of investigations more than needed, or is it necessary to rule out certain conditions?

- The practitioner, athlete and coach must exhibit patience to allow recovery time. More time is required to recover than is often thought to be needed.

- Recognising the importance of incremental graded progressive rehabilitation is essential.

- A good knowledge of functional anatomy, particularly in relation to the sport, is essential in differential diagnosis.

- The source of pain can often be distant from the cause.

Case study 2

A 24-year-old male sprinter with left-sided groin discomfort since a plyometric session three months before this initial consultation had resulted in discomfort after every training session.

- Lower abdominal and medial anterior groin pain following activity that is becoming progressively longer to improve with rest.

- Becomes very low grade and almost unnoticeable with rest.

- There is irritable pain when coughing and sneezing.

- Feels 'sore' in the groin when sitting upright for a while.

- Pain in the deep inner groin when squeezing the legs together, particularly in bed.

Pain was described as exercise related and variable between 1 and 7 on the 10-point scale.

There were minimal impingement signs with hip flexion-adduction.

On inverting the scrotum and placing the little finger in both superficial inguinal rings, the left side appeared more tender and dilated than the right, with a cough impulse.

The left adductor was relatively weaker than the right and painful in resisted adduction lying with straight legs, but not with legs bent in flexion.

There was no discomfort on stretch.

Stork views of the pelvis, i.e. X-rays of standing on one leg and then the other, excluded pelvic instability, pubic symphysis and hip pathology.

The patient was referred to a surgeon for opinion. During surgery the following groin disruption was identified in the operative report:

- torn external oblique aponeurosis

- the conjoined tendon was torn from pubic tubercle

- dehiscence between conjoined tendon and inguinal ligament

Each element of this groin disruption was repaired surgically.

Treatment and rehabilitation

Normal protocol for the first day post operation included standing and walking with gentle stretching and stability exercises.

Five days post-operative ultrasound ascertained core stability to be poor and Transversus Abdominis activation (Cowan et al. 2004) was achieved with practice, using patient visualisation of the ultrasound real-time image for re-education.

Adductor exercises (Figures 20.5 and 20.6) were encouraged one week post op, several times per day.

Closed chain exercises for stability (Figures 20.7–20.10) combined with slow controlled squats progressing to single leg squats, were developed two weeks post op with hydrotherapy for flexibility and stability.

Swimming, cycling and cross-trainer elliptical exercise developed in the third week.

After four weeks he started straight line running build ups alternate days.

Conclusion

This athlete returned to relatively full training after two months and competed internationally six months after the surgery.

Discussion/critical review questions

- At what time should an athlete with groin discomfort be referred to a surgeon to consider operative intervention?

- Should a longer period of conservative treatment and rehabilitation take place before referral for surgery?

- Should the patient have been referred for other investigations, e.g. ultrasound scan or MRI?

- What other areas of the body may contribute towards this athlete's injury?

References

Amaral, J.F. (1997) Thoraco-abdominal injuries in the athlete. *Clinical Sports Medicine*, 16 (4), 739–753.

Anderson, K., Strickland, S.M. and Warren, R. (2001) Hip and groin injuries in athletes. *American Journal of Sports Medicine*, 29 (4), 521–533.

Asian, E. and Fynes, M. (2007) Symphysial pelvic dysfunction. *Current Opinion in Obstetrics and Gynecology*, *19* (2), 133–139.

Barnett, A. (2006) Using recovery modalities between training sessions in elite athletes: does it help? *Sports Medicine*, *36* (9), 781–796.

Bentley, S. (1981) The treatment of sports injuries by local injection. *British Journal of Sports Medicine*, *15* (1), 71–74.

Boardman, M.J., Herman, M.J., Buck, B. and Pizzutillo, P.D. (2009) Hip fractures in children. *Journal of the American Acadamy of Orthopedic Surgeons*, *17* (3), 162–173.

Bohnsack, M., Lekkos, K., Börner, C.E., Wirth, C.J. and Rühmann, O. (2006) Results of hip arthroscopy in sports related groin pain. *Sportverletz Sportschaden*, *20* (2), 86–90.

Bradshaw, C.J., Bundy, M. and Falvey, E. (2008) The diagnosis of longstanding groin pain: a prospective clinical cohort study. *British Journal of Sports Medicine*, *42*, 851–854.

Brannigan, A.E., Kerin, M.J. and McEntee, G.P. (2000) Gilmore's groin repair in athletes. *Journal of Orthopaedic Sports and Physical Therapy*, *30* (6), 329–332.

Brooks, J.H.M. and Fuller, C.W. (2006) The influence of methodological issues on the results and conclusions from epidemiological studies of sports injuries: Illustrative examples. *Current Opinion in Sports Medicine*, *36* (6), 459–472.

Brukner, P. and Khan, K. (2006) *Clinical Sports Medicine*, 3rd edn. New York: McGraw Hill.

Carreira, D.S., Mazzocca, A.D., Oryhon, J., Brown, F.M., Hayden, J.K. and Romeo, A.A. (2006) A prospective outcome evaluation of arthroscopic Bankart repairs: minimum 2-year follow-up. *American Journal of Sports Medicine*, *34* (5), 771–777.

Caudill, P., Nyland, J., Smith, C., Yerasimides, J. and Lach, J. (2008) Sports hernias: a systematic literature review. *British Journal of Sports Medicine*, *42*, 954–964.

Choi, H., McCartney, M. and Best, T. (2008) Treatment of osteitis pubis and osteomyelitis of the pubic symphysis in athletes: A systematic review. *British Journal of Sports Medicine*. Published online: 23 September 2008. doi:10.1136/bjsm.2008.050989.

Cowan, S.M., Schache, A.G., Brukner, P., Bennell, K.L., Hodges, P.W., Coburn, P., Crossley, K.M. (2004) Delayed onset of Transversus Abdominis in long-standing groin pain. *Medicine and Science in Sports and Exercise*, *36* (12), 2040–2045.

Daigeler, A., Belyaev, O., Pennekamp, W.H., Morrosch, S., Köster, O., Uhl, W., Weyhe, D. (2007) MRI findings do not correlate with outcome in athletes with chronic groin pain. *Journal of Sports Science and Medicine*, *6*, 71–76.

Deslandes, M., Guillin, R., Cardinal, E., Hobden, R. and Bureau, N.J. (2008) The snapping iliopsoas tendon: new mechanisms using dynamic sonography. *American Journal of Roentgenology*, *190* (3), 576–581.

Diesen, D.L. and Pappas, T.N. (2007) Sports hernias. *Advances in Surgery*, *41*, 177–187.

Donatelli, R.A. (2006) *The Anatomy and Pathophysiology of the Core. Sports Specific Rehabilitation*. Amsterdam: Elsevier Health Sciences.

Emery, C.A. and Meeuwisse, W.H. (2001) Risk factors for groin injuries in hockey. *Medicine and Science in Sports and Exercise*, *33* (9), 1423–1433.

Emery, C.A., Meeuwisse, W.H. and Powell, J.W. (1999) Groin and abdominal strain injuries in the National Hockey League. *Clinical Journal of Sports Medicine*, *9* (3), 151–156.

Farber A.J. and Wilckens, J.H. (2007) Sports hernia: diagnosis and therapeutic approach. *Journal of the American Academy of Orthopedic Surgeons*, *15* (8), 507–514.

Fricker, P.A. (1997) Management of groin pain in athletes. *British Journal of Sports Medicine*, *31* (2), 97–101.

Gilmore, J. (1998) Groin pain in the soccer athlete: fact, fiction, and treatment. *Clinical Sports Medicine*, *17* (4), 787–793.

Guevara, C.J., Pietrobon, R., Carothers, J.T., Olson, S.A. Vail, T.P. (2006) Comprehensive morphologic evaluation of the hip in patients with symptomatic labral tear. *Clinical Orthopaedics and Related Research*, *453*, 277–285.

Hackney, R. (2008) Groin pain in sport. In P. Abrams and M. Fall (eds), *Urogenital Pain in Clinical Practice*. London: CRC Press.

Hägglund, M., Waldén, M. and Ekstrand, J. (2008) Injuries among male and female elite football players. *Scandinavian Journal of Medical Science Sports*, *19* (6), 819–827.

Harmon, K.G. (2007) Evaluation of groin pain in athletes. *Current Sports Medicine Reports*, *6* (6), 354–361.

Hibbs, A.E., Thompson, K.G., French, D., Wrigley, A. and Spears, I. (2008) Optimizing performance by improving core stability and core strength. *Sports Medicine*, *38* (12), 995–1008.

Hölmich, P., Uhrskou, P., Ulnits, L., Kanstrup, I., Nielsen, M., Bjerg, A., Krogsgaard, K. (1999) Effectiveness of active physical training as treatment for long-standing adductor related groin pain in athletes. *Lancet 353*, 439–443.

Hölmich, P., Hölmich, L.R. and Bjerg, A.M. (2004) Clinical examination of athletes with groin pain: an intraobserver and interobserver reliability study. *British Journal of Sports Medicine*, 38 (4), 446–451.

Hölmich, P. (2007) Long-standing groin pain in sportspeople falls into three primary patterns, a "clinical entity" approach: a prospective study of 207 patients. *British Journal of Sports Medicine*, 41, 247–252.

Ibrahim, A., Murrell, G.A. and Knapman, P. (2007) Adductor strain and hip range of movement in male professional soccer players. *Journal of Orthopaedic Surgery (Hong Kong)*, 15 (1), 46–49.

Jansen, J.A.C.G., Mens, J.M.A., Backx, F.J.G., Kolfschoten, N., Stam, H.J. (2008) Treatment of long-standing groin pain in athletes: a systematic review. *Scandinavian Journal of Medicine and Science in Sports*, 18 (3), 263–274.

Johnson R. (1997) Ice hockey. In M.B. Mellion, W.M. Walsh and G.L. Shelton (eds), *The Team Physician's Handbook*, 2nd edn. Philadelphia, Pa: Hanley and Belfus.

Junge, A., Langevoort, G., Pipe, A., Peytavin, A., Wong, F., Mountjoy, M., Beltrami, G., Terrell, R., Holzgraefe, M., Charles, R. and Dvorak, J. (2006) Injuries in team sport tournaments during the 2004 Olympic Games. American Journal of Sports Medicine, 34 (4), 565–576.

Kark, A.E. and Kurzer, M. (2008) Groin hernias in women. *Hernia*, 12 (3), 267–270.

Kibler, W.B., Press, J. and Sciascia, A. (2006) The role of core stability in athletic function. Current pinion. *Sports Medicine*, 36 (3), 189–198.

Kocher, M.S. and Lee, B. (2006) Hip arthroscopy in children and adolescents. *Orthopedic Clinics of North America*, 37 (2), 233–240.

Kocher, M.S. and Tucker, R. (2006) Pediatric athlete hip disorders. Clinical Sports Medicine, 25 (2), 241–253.

Koh, J. and Dietz, J. (2005) Osteoarthritis in other joints (hip, elbow, foot, ankle, toes, wrist) after sports injuries. Clinical Sports Medicine, 24 (1), 57–70.

Kunduracioglu, B., Yilmaz, C., Yorubulut, M. and Kudas, S. (2007) Magnetic resonance findings of osteitis pubis. Journal of Magnetic Resonance Imaging, 25 (3), 535–539.

Lohrer, H. and Nauck, T. (2007) Proximal adductor longus tendon tear in high level athletes. A report of three cases. Sportverletz Sportschaden, 21 (4), 190–194.

Longacre, M.E. (1994) Hip, groin, and thigh problems. In M.B. Mellion (Ed.) Sports Medicine Secrets. Philadelphia, PA: Hanley and Belfus, pp 285–291.

Lovell, G., Galloway, H., Hopkins, W. and Harvey, A. (2006) Osteitis pubis andassessment of bone marrow edema at the pubic symphysis with MRI in anelite junior male soccer squad. *Clinical Journal of Sport Medicine*, 16 (2) 117–122.

Lynch SA, Renström PA. (1999) Groin injuries in sport: treatment strategies. Sports Medicine. Aug 1999;28 (2): 137–144.

Machotka, Z., Kumar, S. and Perraton, L.G. (2009) A systematic review of the literature on the effectiveness of exercise therapy for groin pain in athletes. *Sports Medicine Arthroscopy Rehabilitation Therapy and Technology*, 1, 5.

Macintyre, J., Johson, C. and Schroeder, E.L. (2006) Groin pain in athletes. Current Sports Medicine Reports, 5 (6), 293–299.

Maffey, L. and Emery, C. (2007) What are the risk factors for groin strain injury in sport? A systematic review of the literature. *Sports Medicine*, 37 (10), 881–894.

Maffulli, N. and Bruns, W. (2000) Injuries in young athletes. *European Journal of Pediatrics*, 159 (1–2), 59–63.

Mehin, R., Meek, R., O'Brien, P. and Blachut, P. (2006) Surgery for osteitis pubis. *Canadian Journal of Surgery*, 49 (3), 170–176.

Mens, J., Inklaar, H., Koes, B.W., Stam, H.J. et al. (2006) A new view on adduction –related groin pain. *Clinical Journal of Sport Medicine*, 16 (1), 15–19.

Meyers, W. Yoo, E.,, Devon, O., Jain, N., Horner, M., Lauencin, C. and Zoga, A. (2007) Understanding "sports hernia" (athletic pubalgia): The anatomic and pathophysiologic basis for abdominal and groin pain in athletes. *Operativetechniques in sports medicine*, 15 (4), 165–177.

Meyers, W.C., McKechnie, A., Philippon, M.J., Horner, M.A., Zoga, A.C. and Devon, O.N. (2008) Experience with "sports hernia" spanning two decades. *Annals of Surgery*, 248 (4), 656–665.

Morales-Conde, S. (2009) Sportsman's hernia: an entity to be defined, diagnosed and treated properly? *Videosurgery and Other Miniinvasive Techniques*, 4 (1), 32–41.

Morelli, V. and Weaver, V. (2005) Groin injuries and groin pain in athletes: part 1. *Primary Care*, 32 (1), 163–183.

Morelli, V. and Espinoza, L. (2005) Groin injuries and groin pain in athletes: part 2. *Primary Care*, 32 (1), 185–200.

Murphy, K.P., Ross, A.E., Javernick, M.A. and Lehman, R.A.Jr. (2006) Repair of the adult acetabular labrum. *Arthroscopy*, 22 (5), 567.e1-3

Noiseux, N. and Tanzer, M. (2008) Anterior hip impingement causes labral tears. *Journal of Bone and Joint Surgery (British)*, 90-B (Supp I), 85.

Paluska, S.A. (2005) An overview of hip injuries in running. *Sports Medicine*, 35 (11), 991–1014.

Putukian, M., Stansbury, N. and Sebastianelli, W. (1997) Specific sports: football. In M.B. Mellion, W.M. Walsh and G.L. Shelton (eds), *The Team Physician's Handbook*, 2nd edn. Philadelphia, PA: Hanley and Belfus, pp 654–656.

Radic, R. and Annear, P. (2008) Use of pubic symphysis curettage for treatment-resistant osteitis pubis in athletes. *The American Journal of Sports Medicine*, Jan; *36* (1), 122–8. Epub 2007 Aug 16.

Robinson, P. (2002) Ultrasound of groin injury. *Imaging*, *14* (3), 209–216.

Rosenfeld, S.B., Herring, J.A. and Chao, J.C. (2007) Legg-calve-perthes disease: a review of cases with onset before six years of age. *Journal of Bone and Joint Surgery*, *89* (12), 2712–2722.

Schilders, E., Bismil, Q., Robinson, P. O'Connor, P.J., Gibbon, W.W., Talbot, J.C. (2007) Adductor-related groin pain in competitive athletes. Role of adductor enthesis, magnetic resonance imaging, and entheseal pubic cleft injections. *The Journal of Bone and Joint Surgery*, *89* (10), 2173–2178.

Shabshin, N., Rosenberg, Z.S. and Cavalcanti, C.F. (2005) MR imaging of iliopsoas musculotendinous injuries. *Magnetic Resonance Imaging Clinics of North America*, *13* (4), 705–716.

Siegel, J., Templeman, D.C. and Tornetta, P. (2008) Single-leg-stance radiographs in the diagnosis of pelvic instability. *The Journal of Bone and Joint Surgery*, *90*, 2119–2125.

Spector, T.D., Harris, P.A., Hart, D.J., Cicuttini, F.M., Nandra, D., Etherington, J., Wolman, R.L. and Doyle, D.V. (1996) Risk of osteoarthritis associated with long-term weight-bearing sports: a radiologic survey of the hips and knees in female ex-athletes and population controls. *Arthritis and Rheumatism*, *39* (6), 988–995.

Stoker, D.L., Spiegelhalter, D.J., Singh, R. and Wellwood, J.M. (1994) Laparoscopic versus open inguinal hernia repair: randomised prospective trial. *The Lancet*, *343* (8908), 1243–1245.

Su, P., Li, R., Liu, S., Zhou, Y., Wang, X., Patil, N., Mow, C.S., Mason, J.C., Huang, D. and Wang, Y. (2008) Age at onset-dependent presentations of premature hip osteoarthritis, avascular necrosis of the femoral head, or Legg-Calvé-Perthes disease in a single family, consequent upon a p.Gly1170Ser mutation of COL2A1. *Arthritis and Rheumatism*, *58* (6), 1701–1706.

Swan, K.G. Jr and Wolcott, M. (2007) The athletic hernia: a systematic review. *Clinics in Orthopaedic Related Research*, *455*, 78–87.

Tomoko, K., Takahiro, U. Masahiko, I. Teruhiko, K. Norio, Y. Tatsunori, I.Toshikasu, T. and Makoto, K. (2006) Clinical study of hip joint referred pain. *Pain Research*, *21* (3), 127–132.

Torry, M.R., Schenker, M.L., Martin, H.D., Hogoboom, D. and Philippon, M.J. (2006) Neuromuscular hip biomechanics and pathology in the athlete. *Clinical Sports Medicine*, *25* (2), 179–197.

Tuite, D.J., Finegan, P.J., Saliaris, A.P., Renström, P.A.F.H., Donne, B. and O'Brien, M. (1998) Anatomy of the proximal musculotendinous junction of the adductor longus muscle. *Knee Surgery, Sports Traumatology, Arthroscopy*, *6* (2), 134–137.

Vaile, J., Halson, S. Gill, N. and Dawson, B. (2008) Effect of hydrotherapy on the recovery of exercise-induced fatigue and performance. Medicine and Science in Sports and Exercise: May 2008, 40, 5–S67.

Verral, G. (2008) Hip Groin Examination. ACPSM UK presentation November.

Verral, G.M. (2005) Hip joint range of motion reduction in sports-related chronic groin injury diagnosed as pubic bone stress injury. *Journal of Sports Science and Medicine*, *8* (1), 77–84.

Verrall, G.M., Slavotinek, J.P., Fon, G.T. and Barnes, P.G. (2007) Outcome of conservative management of athletic chronic groin injury diagnosed as pubic bone stress injury. *American Journal of Sports Medicine*, *35* (3), 467–474.

Wettstein, M., Mouhsine, E., Borens, O. and Theumann, N. (2007) [Differential diagnosis of groin pain] [French]. *Revue Medical Suisse*, *3* (138), 2882–2888.

Wollin, M. and Lovell, G. (2006) Osteitis pubis in four young football players: a case series demonstrating successful rehabilitation. *Physical Therapy in Sport*, *7* (4), 173–174.

Ziltener, J.L. and Leal, S. (2007) Groin pain in athletes. *Revue Medical Suisse*, *3* (120), 1784–1787.

21

The knee

Nicholas Clark
Integrated Physiotherapy and Conditioning Ltd

Dr Lee Herrington
University of Salford, Greater Manchester

Introduction

Exercise rehabilitation following knee joint injury can be controversial, with physical rehabilitation after anterior cruciate ligament (ACL) injury probably being the most debated. This perspective is clearly illustrated in Table 21.1, which summarises the basic differences in selected international clinical research groups' ACL-reconstruction (ACL-R) rehabilitation programmes. Following ACL-R, Table 21.1 demonstrates there is currently no consensus of opinion with regard to the use of a knee brace, the type of exercise performed, the timing of when an exercise is introduced in the rehabilitation process, the use of strength tests as objective rehabilitation progression criteria, or the use of hop tests as objective outcome measures. As such, this section will provide a rigorous evidence-based introduction to key concepts in exercise rehabilitation for ACL, posterior cruciate ligament (PCL), medial collateral ligament (MCL), lateral collateral ligament (LCL) and menisus injuries. For each injury, weight-bearing restrictions, knee bracing recommendations and rehabilitation strength training guidelines are presented in line with very specific rehabilitation concerns, concepts and exercise modifications unique to each injury type during the 'early' stages of rehabilitation (i.e. ≤ 12 weeks post-injury/post-surgery) when specific tissue healing time frames and the protection of surgically repaired or reconstructed injury sites may still be of significant concern. Since ideal knee and lower limb biomechanical function for high level sports performance is then going to be the same in the 'late' stages of rehabilitation (i.e. ≥ 12 weeks to 12 months) and at the point of full return-to-function regardless of the initial type of knee injury, generic exercise rehabilitation guidelines for all knee ligament and meniscus injuries are then presented with reference to proprioception, neuromuscular control, balance, deceleration, plyometric and agility training.

This means, for example, that towards the 'late' stages of rehabilitation it does not necessarily matter whether a rugby player has sustained an injured ACL, PCL or lateral meniscus – if the player wishes to return to competitive rugby the knee still has to perform in a multi-directional agility-biased sport with the same frequent rapid deceleration-acceleration cycles, as well as tolerating high impact forces with every foot-strike when running and jumping or as a result of collisions with opponents or the ground, regardless of the initial type of knee injury. As such, a major premise of this chapter's overall rehabilitation philosophy is that range-of-motion (ROM) restrictions and rehabilitation strength

Table 21.1 Summary of selected research groups' anterior cruciate ligament reconstruction rehabilitation programmes

Research group	Reference	Knee brace	Dynamic CKC quadriceps strength training	Dynamic OKC quadriceps strength training 90-30°	Dynamic OKC quadriceps strength training 30-0°	Earliest straight line running	Earliest Plyometric drills	Earliest sport specific agility drills	Earliest return to competitive sport	Strength test before starting straight line running	Strength test before starting plyometric drills	Formal functional performance test (Hop test) before discharge
Cincinnati Sports Medicine and Orthopaedic Centre Cincinnati, US	Barber et al. (1992) DeMaio et al. (1992) Mangine et al. (1992)	Y	Y at 2 wks	Y at 1 wk	N	20 wks	13 wks	25 wks	36 wks	Y	Y	Y
Hokkaido University School of Medicine Sapporo, Japan	Majima et al. (2002)	Y	Y at 2 wks	Y at 5 wks	N	6 wks	6 wks	10 wks	7 mths	Y	Y	N
Hospital for Special Surgery New York, US	Williams et al. (2004)	Y	Y at 4 wks	Y at 4 wks	Y at 4 wks	8-10 wks	N/S	N/S	4 mths	N	N	N
Isokinetic Rehabilitation Centre Bologna, Italy	Roi et al. (2005)	N	Y at 2 wks	Y at 2 wks	Y at 2 wks	5 wks	10 wks	8 wks	13 wks	N	N	N
Karolinska Hospital Stockholm, Sweden	Mikkelsen et al. (2000)	N	Y at 2 wks	Y at 6 wks	Y at 6 wks	12-16 wks	6-12 wks	4-6 mths	N/S	N	N	N
La Trobe University Melbourne, Australia	Feller and Webster (2003)	N	Y at 10 days	N	N	10 wks	N/S	3-4 mths	9 mths	Y	Y	N
Methodist Sports Medicine Centre Indianapolis, US	Shelbourne and Gray (1997) Shelbourne and Nitz (1990) Shelbourne and Rask (1998)	Y	Y at 10 days	Y at 4 wks	Y at 1 wk	6 wks	6 wks	10 wks	6 mths	Y	Y	Y
Naval Medical Centre California, US	Bynum et al. (1995)	Y	Y at 1 wk	Y at 6 wks	Y at 6 wks	8 wks	12 wks	6-7 mths	12 mths	N	N	N
Nicholas Institute of Sports Medicine New York, US	McHugh et al. (2002)	Y	Y at 2 wks	Y at 4 wks	Y at 4 wks	12 wks	4 wks	20 wks	24 wks	N	N	Y
Travis Air Force Base California, US	Howell and Taylor (1996)	N	Y at 4 wks	Y at 4 wks	Y at 4 wks	8-10 wks	N/S	N/S	4 mths	N	N	N
University of East London and King's College London London, UK	Morrissey et al. (2000) Perry et al. (2005a)	N	Y at wk 2-3	Y at wk 2-3	Y at wk 2-3	8 wks	8 wks	8 wks	N/A	N	N	N/A
University of Pittsburgh Pennsylvania, US	Fitzgerald et al. (2003)	Y	Y	Y	N	N/S	N/S	N/S	N/S	Y	Y	N/S
University of Queensland Queensland, Australia	Keays et al. (2003)	Y	Y at 1 wk	Y at 12 wks	Y at 12 wks	12 wks	5 mths	4 mths	9 mths	N	N	Y
University of Sydney Sydney, Australia	Pinczewski et al. (2002) Salmon et al. (2006)	N	Y at 1-2 wks	N	N	6 wks	N/S	12 wks	6-12 mths	N	N	N
University of Vermont Vermont, US	Beynnon et al. (2005)	Y	Y at 6 wks	Y at 2 wks	Y at 6 wks	8 wks	16 wks	16 wks	N/S	N	N	N

Table © Copyright Nicholas Clark. Reproduced with permission.
CKC = closed kinetic chain; OKC = open kinetic chain; 90-30° = 90-30° knee flexion-extension arc-of-motion; 30-0° = 30-0° knee flexion-extension arc-of-motion; Y = Yes; N = No; wk = week; wks = weeks; mths = months
US = United States; UK = United Kingdom; N/S = not stated; N/A = not applicable

training modifications specific to a unique knee injury type and its associated healing constraints are typically discontinued in ≤ 12 weeks post-injury or post-surgery, after which all tibiofemoral soft tissue injuries can often then be guided through rehabilitation according to the same progression criteria towards the same sport-specific functional drills and goals. Table 21.2 shows typical knee mechanisms of injury and common clinical diagnostic tests.

Early stage rehabilitation

The early rehabilitation phase following injury or surgical reconstruction of the knee ligaments or menisci is initially concerned with the resolution of inflammation, the reduction and removal of swelling, and regaining full activation of the inhibited muscles, principally the quadriceps. Use of the rest, ice, compression, elevation (RICE) principle and how it helps in the resolution of inflammation has been discussed elsewhere within this book. With all knee injuries it is critical that the injured structures are allowed time to recover from the trauma they have received as a result of the initial injury or as a result of any surgical intervention. It is important that in all but the most minor of injuries, the patient's level of weight-bearing during gait and functional activities in the first 72 hours post-injury is ideally pain-free or at least causes minimal pain. The use of crutches and partial weight bearing (PWB) is often critical to achieving this. In simplest terms, if the knee-injured patient presents with any antalgic gait or 'limp', crutches should be used for as long as is necessary to temporarily unload and protect sensitised healing tissues. Crutches should not be discontinued until a patient can demonstrate full active knee extension ROM (i.e. there is no 'quadriceps lag'). Table 21.3 illustrates more injury-specific weight-bearing guidelines. Minimising quadriceps inhibition and atrophy is of critical importance in the management of any knee injury, and this is covered below.

Knee rehabilitation concerns and concepts

There are several clinical concerns and concepts regarding the safe and expedient rehabilitation of any knee-injured patient, especially in the initial rehabilitation period (≤ 12 weeks post-injury/post-surgery). Specific questions that address these clinical concerns and concepts include:

- How to manage quadriceps muscle atrophy and overcome its inhibition?

- Should a surgical reconstruction be performed?

- Should a knee brace be used during rehabilitation?

- What are the rehabilitation strength training and joint loading considerations when treating the knee-injured patient?

- Does a surgical reconstruction prevent knee osteoarthrosis (OA)?

Quadriceps inhibition and muscle atrophy

Persistent quadriceps weakness is associated specifically with ACL injury (Palmieri-Smith et al. 2008) but also knee injuries in general, and presents a major rehabilitation challenge for patients and clinicians alike. Quadriceps strength deficits can still easily exceed 20% at 6 months after ACL-R, a time when many athletes are cleared to return to activity (Keays et al. 2007). The magnitude of quadriceps weakness can lessen with time, but most research suggests that quadriceps strength deficits following knee joint injury persist for many months or years following injury or surgery (Shelbourne and Nitz 1990; Snyder-Mackler et al. 1995; Natri et al. 1996; Holder-Powell and Rutherford 1999; Risberg et al. 1999b, 2007; Mikkelsen et al. 2000; Feller and Webster 2003), illustrating the long-term nature of this problem. As the quadriceps is critical to dynamic joint stability, weakness of this muscle group is related to poor functional outcomes and may contribute to the early onset of knee OA (Palmieri-Smith et al. 2008).

There is evidence to suggest that arthrogenic muscle inhibition (AMI) and perhaps muscle atrophy are primarily responsible for the decrements in quadriceps strength following knee injury (Palmieri-Smith et al. 2008). As such, AMI in the form of voluntary activation failure is hypothesised to result from reflex activity in which altered sensory information

Table 21.2 Mechanism of injury, incidence of injury and common clinical diagnostic tests

Injury	Mechanism of injury	Incidence	Orthopaedic tests
ACL	Greater than 70% ACL injuries are non contact injuries, involving a 'valgus collapse' of the knee during the first half of stance-phase when landing from a jump, an abrupt deceleration when running, cutting manoeuvres or pivoting manoeuvres. A valgus collapse is associated with combined movements of knee flexion + anterior tibial displacement + internal tibial rotation + knee valgus. Contact mechanism usually involves the pivoting (into valgus/rotation) whilst the tibia is fixed	Incidence can range between 0.04 and 3.07 injuries per 1,000 athlete-exposures. Females demonstrate a 3–6 times higher incidence	Lachmans test Anterior draw test Pivot shift test
PCL	The mechanism of injury consistently involves a high-velocity and 'high-impact' posteriorly-directed force applied to the anterior aspect of the proximal tibia. An example would be a front-on tackle to a flexed knee in rugby or American football. The combined movements of forced knee hyperflexion + knee varus-valgus are also common mechanisms of injury for the PCL	Incidence of injury can range between 0.01 and 0.90 injuries per 1,000 athlete exposures	Posterior Sag sign Reverse pivot shift Posterior draw test
MCL	Results from a collision with another person. Most frequently a lateral blow to the distal femur or proximal tibia whilst the foot is fixed on the ground, causing a valgus torque to the knee and overstretching the fibres of the MCL resulting in a partial or complete tear of the ligament substance	Most commonly injured knee ligament accounting for almost 30% of all knee ligament injuries, approximately 3.1 MCL injuries per 1,000 player hours	Abduction or valgus stressing of knee
LCL	It is quite difficult to induce an *isolated* acute sprain, since this either requires the knee to collapse into varus in a non-contact situation, or for the knee to be forced into varus in a contact/collision situation. However, isolated acute LCL sprains do occasionally occur in collision sports as a result of a low-velocity low-impact blow from the medial or anteromedial aspect of the knee. Far more common is an LCL injury *combined* with simultaneous PCL or PLC knee injury due to a high-impact collision in sports	Less than 2% of all knee ligament injuries even in collision sports with an incidence of less than 0.02 injuries per 1,000 player hours	Adduction or varus stressing of knee
Meniscal	Traumatic meniscus injuries are usually non-contact in nature, involving forced valgus in weight-bearing, in turn resulting in extreme rotary compression forces through the lateral tibiofemoral joint. It appears most lateral meniscus injuries are traumatic, whereas most medial meniscus injuries appear to be degenerative (i.e. non-traumatic). Radial, longitudinal, vertical and complex meniscal tears are more likely the result of trauma, whereas horizontal and oblique tears are typically due to degeneration	17% of all knee joint injuries sustained during competitive play with twice as many injuries to lateral meniscus. Non-traumatic tears of both the lateral and medial meniscus frequently discovered in asymptomatic athletes, so the clinician must carefully relate the patient's history of injury to the findings of the clinical examination before deciding whether the source of a patient's knee pain is likely to be the menisci	Appley's test McMurray's test Combined flexion adduction or extension adduction (medial meniscus), flexion abduction or extension abduction (lateral meniscus)

ACL = anterior cruciate ligament; PCL = posterior cruciate ligament; MCL = medial collateral ligament; LCL = lateral collateral ligament; PLC = posterolateral corner

QUADRICEPS INHIBITION AND MUSCLE ATROPHY

Table 21.3 Weight-bearing, range-of-motion and exercise rehabilitation guidelines

Injury	Weight bearing restrictions	Bracing and ROM	Type of training OKC/CKC	Frequency – intensity – sets
ACLD	Unrestricted[1]	Unrestricted	Both OKC and CKC, initial bias towards OKC to reverse isolated muscle weakness, then bias CKC to progression functional performance, single leg whenever possible. Progress CKC when isolated quads strength LSI > 80%	Initially daily 3–5 × 15–20RM (first eight weeks), progress every other day and 8–12RM (week 9–12) All without eliciting pain/swelling
ACLR	Unrestricted[1]	Unrestricted	As ACLD	As ACLD
PCLD	Unrestricted[1]	Partial tear braced 0–50° 2–4 weeks. Full tear locked in extension for four weeks, removed for ROM and quads exercises	OKC quads limited only to general ROM limitations. OKC hamstrings 0–20° first eight weeks, CKC exercises not until after week two when they are well below BW and progressed according to patient tolerance.	Initially daily 3–5 × 15–20RM (first eight weeks), progress every other day and 8–12RM (week 9–12) All without eliciting pain/swelling
PCLR	PWB 14 days then progressed as tolerated	Locked in extension for four weeks, removed for ROM (0–50° only) and quads exercises, unlocked brace (0–50°) next four weeks, then progressively increase range (15° per week) for CKC and OKC work	OKC quads limited only to general ROM limitations. OKC hamstrings 0–20° first eight weeks, CKC exercises not until after week two when they are well below BW and progressed according to patient tolerance	As PCLD
ACL/PCL	PWB for four weeks, If reconstruction of PLC or PMC also performed, PWB does not begin until week two at < 25% BW, very carefully progressed up to FWB by week 12	Locked in extension for four weeks	As PCLR except CKC quads training started four weeks after surgery, hams training is avoided for a minimum 12 weeks	As PCLD
MCL	Unrestricted[1]	Unrestricted	As ACLD/R	As ACLD/R

(Continued.)

Table 21.3 (*Continued*).

Injury	Weight bearing restrictions	Bracing and ROM	Type of training OKC/CKC	Frequency – intensity – sets
LCL	Grade 1–2 Unrestricted[1] Grade III isolated LCL injury and LCLR NWB from week one to two, then PWB at 25% BW from week three to four, 50% BW from week five to six, FWB achieved if tolerated by between week 8 and 12	Isolated LCL braced 0° one week, braced 0–90° for next 2–5 weeks. LCLR braced 0° six weeks, braced 0–110° for next two weeks, with progression increase in braced range to week 12	LCL as ACLD/R but with caution doing single leg CKC because of higher tensile loads on LCL LCLR running might not be commenced until 6–9 months post surgery	As ACLD/R
Meniscal	For resection (partial menisectomy) unrestricted[1] For repair PWB 25% BW from week 1–2, 50% BW from week 3–4, 75% BW from week 5–6, after which FWB can be allowed as tolerated. Longitudinal tears, FWB can be achieved by week four, whereas radial and complex tears must wait until week 6–7	Hinged knee brace is worn for six weeks, being immediately opened to permit gait as well as NWB PROM exercises at 0–90° knee flexion until the end of week two, 0–120° from week 3–4, and 0–135° from week 5–6	For repair; OKC quads exercises limited from 0–90° knee flexion, isometric hamstring exercises limited from 0–20° knee flexion from week 1–5 post-surgery, and dynamic OKC hamstring exercises from 0–90° knee flexion initiated after six weeks. CKC exercises beginning 0–60° knee flexion begin week five for longitudinal tears and week seven for radial and complex tears when FWB is permitted if tolerated. Sub-bodyweight CKC exercise can be initiated in line with the weight-bearing and ROM restrictions	As ACLD/R

[1]Between 5–14 days of non-weight bearing and even bed rest immediately following surgery/injury have been recommended. This is with a view to controlling the inflammatory reaction and resultant joint effusion than protecting the integrity of the graft complex itself, since the longer the inflammatory process and the duration of significant effusion the greater the incidence of joint stiffness, severe muscle weakness and poor overall outcomes. Aggressive pain and swelling control are the priorities of treatment during this timeframe (Shelbourne and Klotz 2006; Shelbourne and Rask 1998; Shelbourne et al. 1992).

ROM = range-of-motion; CKC = closed kinetic chain; OKC = open kinetic chain; RM = repetition maximum; LSI = limb symmetry index; BW = bodyweight; ACLD = anterior cruciate ligament deficient; ACLR = anterior cruciate ligament reconstruction; PCLD = posterior cruciate ligament deficient; PCLR = posterior cruciate ligament reconstruction; MCL = medial collateral ligament; LCL = lateral collateral ligament; NWB = non-weightbearing; PROM = passive range-of-motion from the injured knee joint leads to a diminished motor drive to its surrounding muscles. If clinicians can be effective in combating AMI early in the rehabilitation process, they will minimise the resulting strength deficits and diminish muscle atrophy, which should lead to a more complete more effective rehabilitation, and likely will result in quicker patient recovery and return to sport. The question is how can this muscle inhibition be addressed? In order to remove AMI, it will need an alteration in the inhibitory processes causing it. This can be achieved by one of two strategies: altering sensory feedback that signals something is wrong with the knee, or modifying the motor drive to the muscles.

Removing arthrogenic muscle inhibition: Altering sensory feedback

Arthrogenic muscle inhibition can be caused by increased abnormal sensory activity caused by joint effusion, or by a lack of normal sensory feedback due to the loss of mechanoreceptors subsequent to an ACL or other ligamentous rupture. So, one approach to removing AMI would be blocking/modifying the sensory signals responsible for initiating the inhibitory process. This might be achieved by removing abnormal sensory stimuli, minimising pain, or sending signals to the central nervous system (CNS) that may modify any pre-synaptic pathways contributing to AMI.

The most obvious method to reduce a joint effusion is aspiration of the joint. Though not without risks, it has been shown to be effective in significantly reducing AMI (Fahrer et al. 1988). Other modalities can also be used to reduce effusion such as compression, electrical muscle stimulation and elevation, but their effectiveness in reducing AMI is yet to be evaluated. Application of ice has been shown to decrease nerve conductance velocity and so slow the discharge of mechanoreceptors. Also following the application of ice, quadriceps activity has been found to be facilitated (Hopkins et al. 2002). Hopkins et al. (2002) also found that the application of transcutaneous electrical nerve stimulation (TENS) decreased AMI through the reduction of pre-synaptic inhibition. Stokes et al. (1985) found the effect of TENS was maximised when used whilst the patient exercised.

Removing arthrogenic muscle inhibition: Improving motor drive

Rather than blocking or modifying abnormal sensory feedback, an alternative approach to targeting AMI is to activate the inhibited motorneurons. As AMI prevents active recruitment of the quadriceps, electrical muscle stimulation (EMS) would allow for direct recruitment of motorneurons. Moreover, EMS has also been shown to preferentially activate type II muscle fibres compared with voluntary training of a similar intensity (Binder-Macleod et al. 1995). Research has also shown quadriceps EMS in combination with active quadriceps exercise has a superior effect than exercise or EMS alone (Snyder-Mackler et al. 1991).

Electromyography (EMG) and other forms of biofeedback may also be a useful adjunct to active exercise in overcoming AMI. Though the evidence is limited, Maitland et al. (1999) present a case study showing positive findings. Eccentric exercises have the potential to overload the muscle in a manner greater than that of concentric exercise. Gelber et al (2000) found positive results with the addition of specific eccentric exercise programmes, and that by using a graduated programme any issue with delayed onset muscle soreness (DOMS) was avoided.

Should a surgical reconstruction be performed

When an isolated ACL injury is present there is a great deal of evidence demonstrating that it is possible for an ACL-deficient (ACL-D) patient to return to high level physical activities following sufficient exercise rehabilitation (Herrington 2004; Neuman et al. 2008). However, there is no single subjective or objective predictor as to whether an ACL-D patient should or should not undergo reconstructive surgery (Herrington and Fowler 2006), and the amount of anterior knee laxity measured by anterior tibial displacement does not predict whether or not a patient will be able to perform high level physical activities (Clark 2001; Herrington 2004). Therefore, it seems at present the best way to decide on whether a patient should undergo ACL-R surgery is whether there are persistent repetitive episodes of giving way resulting in major functional disability. In other words, how a patient performs functionally having undergone an intensive rehabilitation period. However, if an ACL injury co-exists with a major tear of the MCL or LCL, then it is generally accepted that an ACL-R should be performed (Halinen et al. 2006; Shelbourne and Porter 1992).

There is also substantial evidence that a patient who is PCL-deficient (PCL-D) can perform high-level physical activities or return to sports participation with minimal functional limitation regardless of the amount of posterior laxity present (Fontbote et al. 2005; Shelbourne et al. 1999). Therefore, it seems the best way to decide whether a patient should undergo PCL-reconstruction (PCL-R) surgery is whether there are persistent episodes of giving way and knee pain or swelling. Of consideration as to whether a PCL-R is performed is the integrity of the remaining tibiofemoral capsuloligamentous structures. If there is significant post-injury laxity in any of the other main tibiofemoral ligaments (i.e. ACL, LCL, MCL) or the posterior-lateral complex (PLC) or posterior-medial complex (PMC), then a PCL-R is recommended (LaPrade and Wentorf 2002). A PCL-R is also recommended for the PCL-D knee if the injury has resulted in a bony avulsion of either of the PCL attachments (Wilk et al. 1999).

The MCL has a greater ability to heal than the other knee ligaments, with excellent tensile strength and overall functional outcome, even after a Grade III sprain, and its is generally now recommend that non-surgical treatment and management of acute isolated MCL injuries occurs (Halinen et al. 2006). Even in instances where a combined MCL-ACL injury exists, the majority opinion is that the MCL is treated conservatively whilst an ACL-R is performed (Halinen et al. 2006; Noyes and Barber-Westin 1995). Although the LCL heals more slowly than the MCL, Grade I and II LCL tears are treated conservatively where these injuries typically present as isolated LCL trauma (Meislin 1996). Since Grade III LCL tears rarely present without combined trauma to the PLC or PCL, surgical reconstruction of the LCL and any additionally damaged adjacent capsuloligamentous structures is recommended (Noyes and Barber-Westin 2007).

Surgical repair of the menisci is often recommended to relieve related symptoms, restore knee kinematics to as near normal as possible, and restore the patient's function, although outcomes vary considerably (Brindle et al. 2001). Multiple variables influence the decision to carry out an operative repair including the patient's age, health, lifestyle, willingness to undergo surgery, likely post-operative adherence with rehabilitation, and the location of a tear (Brindle et al. 2001; Lee et al. 2002). Healing of the menisci is influenced by the degree of vascular penetration, with only 25–30% of the menisci outer substance possessing a viable blood supply (Brindle et al. 2001). Consequently, the middle and inner thirds of the menisci substance possess a poor blood supply and are unlikely to heal easily, if at all (Brindle et al. 2001). The alternative to meniscal repair, partial meniscectomy, frequently causes altered knee kinematics and an earlier onset of tibiofemoral OA (Lohmander et al. 2007), and so meniscal repair is preferred wherever possible (Lee et al. 2002). However, partial meniscectomy is usually performed if the tear is in the inner one-third of the meniscus substance, if the tear has major tissue fragmentation or degeneration, or for tears where the edges cannot be approximated (McLaughlin and Noyes 1993).

Use of a knee brace during rehabilitation

The use of a knee brace following knee joint injury or surgery can depend on the the type of tissue injured or the surgical procedure that has been performed (Table 21.3). Research demonstrates there is no significant difference in clinical or functional outcome in braced versus non-braced ACL-R patients (McDevitt et al. 2004; Risberg et al. 1999a). As such, there appears to be no need to use a knee brace after isolated ACL-R. In contrast to ACL injury, routine use of a knee brace after PCL injury can be essential, although specific recommendations vary for the PCL-D and PCL-R knee.

For the PCL-D knee which is to be managed non-operatively, recommendations can differ according to whether there is a partial or complete tear of the PCL (Margheritini et al. 2002). Posterior shear forces begin to dramatically increase at approximately 60° knee flexion during closed kinetic chain (CKC) knee exercise such as the double-leg bodyweight (BW) squat (Figure 21.1), BW wall-squat (Figure 21.2), horizontal double-leg press (Figure 21.3), dumbbell single-leg squat (Figure 21.4), and horizontal single-leg press (Figure 21.5). So for the PCL-D knee with a partial tear of the ligament substance it would appear sensible to wear a long-leg hinged knee brace locked to allow 0–50° knee flexion for the first two weeks post-injury. This will protect the healing ligament from excessive tensile loads in the

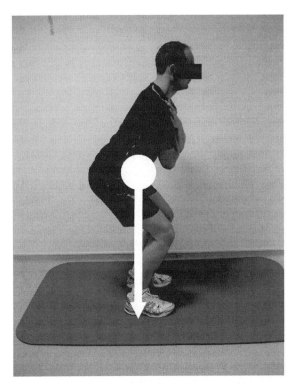

Figure 21.1 Double-leg bodyweight squat. The white arrow represents how the body's centre-of-mass can theoretically induce more relative hip, knee and ankle joint compression than joint shear. When the lower limb is extended, the body's centre-of-mass loads the lower limb along its longitudinal axis, inducing significant hip, knee and ankle joint compression. Photograph © Copyright Nicholas Clark. Reproduced with permission.

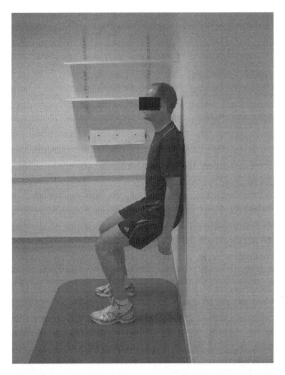

Figure 21.2 Bodyweight wall-squat. Photograph © Copyright Nicholas Clark. Reproduced with permission.

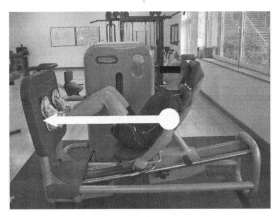

Figure 21.3 Horizontal double-leg press. White arrow represents how the leg press sled's approximate point-of-application and line-of-action can theoretically induce more relative hip, knee and ankle joint compression than joint shear. When the lower limb is extended, the lower limb is predominantly loaded along its longitudinal axis, inducing significant hip, knee and ankle joint compression. Photograph © Copyright Nicholas Clark. Reproduced with permission.

early stages after injury whilst permitting a functional ROM during gait, since the posterior tibial shear forces during level walking remain at relatively low loads (Shelburne et al. 2004). This ROM will also permit selected muscle strengthening exercises. The brace can then be discarded after four weeks (Margheritini et al. 2002).

For the PCL-D knee with a complete tear of the ligament substance, a long-leg hinged knee brace locked at 0° knee flexion during ambulation is considered necessary for four weeks (Harner and Hoher, 1998). This is because the injury mechanism causing a complete tear of the PCL involves greater forces than those for a partial tear, and so it is most likely that there will be injury to the knee's PLC or PMC also (Margheritini et al. 2002). During this time-

Figure 21.4 Single-leg dumb-bell squat. Photograph © Copyright Nicholas Clark. Reproduced with permission.

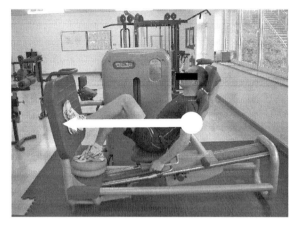

Figure 21.5 Horizontal single-leg press. White arrow represents how the leg press sled's approximate point-of-application and line-of-action can theoretically induce more relative hip, knee and ankle joint compression than joint shear. When the lower limb is extended, the lower limb is predominantly loaded along its longitudinal axis, inducing significant hip, knee and ankle joint compression. Notice inflatable cushions under the right foot in an effort to minimise 'cheating' with the right leg. Photograph © Copyright Nicholas Clark. Reproduced with permission

frame, the brace can be regularly unlocked to allow 50–60° knee flexion for controlled ROM and muscle strengthening exercises (Noyes and Barber-Westin 2007). For the PCL-R knee the patient can wear a hinged knee brace locked to allow 0–50° knee flexion for the first eight weeks post-surgery. The brace is limited to 50–60° knee flexion for *active* ROM activities for eight weeks because this is the timeframe thought necessary for minimal soft tissue graft integration with the bone tunnel interface to occur (Harner and Hoher 1998).

Following Isolated MCL injury, a long-leg knee brace with a robust medial hinge is recommended. For the first week post-injury the brace is locked at 0° at rest and for gait and ADLs (Yoshiya et al. 2005), being unlocked frequently for passive ROM (PROM) and active ROM (AROM) exercises every day (Indelicato 1995). From week two onwards the brace can be unlocked as symptoms allow up to 90° knee flexion (Reider et al. 1993), being discarded at three to six weeks post-injury or when 90° pain-free knee flexion is achieved (Reider et al. 1993; Lind et al. 2009). Some clinicians recommend that the patient also wear the brace at night for three weeks post-injury (Reider et al. 1993).

During single-leg stance in uninjured subjects, LCL tensile forces are normally up to 8.5 times higher than MCL tensile forces (Shelburne et al. 2005), ranging from 25% to 60% BW (Schipplein and Andriacchi 1991; Shelburne et al., 2005). As such, the LCL is clearly loaded far more than the MCL, and so a long-leg hinged knee brace is considered essential even for Grade I tears (Meislin 1996). For the first seven days after isolated LCL injury, the brace is locked at 0° at rest and for gait and ADLs, being regularly unlocked for PROM and AROM exercises up to six times per day. As for isolated MCL injuries, from week two onwards the brace can be unlocked as symptoms allow up to 90° knee flexion, being discarded at three to six weeks post-injury or when 90° pain-free knee flexion is achieved and there is no quadriceps lag. If an LCL-reconstruction (LCL-R) is performed with or without a PLC-reconstruction (PLC-R), the brace is worn for at least eight weeks (Noyes and Barber-Westin 2007).

Joint loading considerations in knee rehabilitation strength training

There has been a commonly held belief that open kinetic chain (OKC) exercises such as the single-leg resisted knee extension (Figure 21.6) generate 'excessive' anterior tibial shear forces and ACL graft tensile loads. The term 'excessive' refers to the concern that OKC quadriceps strength training (i.e. resisted knee extension) induces large anterior tibial shear forces which, in turn, impose large tensile forces and load on a new ACL-R that are of sufficient magnitude to damage the healing graft-fixation complex (Fitzgerald, 1997). Although this concern is historically grounded with the best intentions for the patient, recent research demonstrates it no longer has a solid foundation in clinical practice, and so normal lower limb OKC and CKC muscle strength levels are illustrated in Table 21.4 and Table 21.5, respectively. Between-limb differences in muscle strength should be no greater than 10% as determind by normative data obtained from uninjured subjects (Daniel et al. 1982; Greenberger and Paterno 1995; Ostenberg et al. 1998; Petschnig et al. 1998; Sapega 1990).

Anterior tibial shear forces are considered representative of ACL tensile forces (Escamilla et al. 1998) and vary considerably according to the type of exercise performed (Table 21.6). A resisted knee extension against a load of between 2kg and 39kg generates mean peak anterior tibial shear forces of between 50 Newtons (N) (Isaac et al. 2005) and 250N (Escamilla et al. 1998), which is equivalent to 200–250N generated by a BW forward step-up (Isaac et al. 2005). It is also significantly less than the 300–410N generated by just walking (Shelburne et al. 2004, 2005). So, based on these studies, it is difficult to reconcile why many ACL-R patients are encouraged to walk and climb stairs as soon as possible post-surgery but forbidden to perform resisted knee extensions, since walking actually generates higher anterior tibial shear forces (Table 21.6). Furthermore, as the vast majority of popular cruciate ligament reconstruction graft tissues fail at > 1900N (Table 21.7), and popular graft fixation methods for ACL-R fail at ≥ 500N post-surgery (Table 21.8), it is also clear that the 50–250N generated by sub-maximal resisted knee extensions are of insufficient magnitude to damage a properly applied graft and fixation method within correctly placed bone tunnels.

With regard to the clinical application of OKC and CKC strength training after ACL-R, five research groups have studied this (Bynum et al. 1995; Mikkelsen et al. 2000; Morrissey et al. 2000; Beynnon et al. 2005; Perry et al. 2005a). These five clinical research studies collectively clearly demonstrate no significant difference in anterior tibial displacement between patients performing CKC-biased quadriceps strength training or mixed CKC+OKC quadriceps strength training. Furthermore, ACL-R and ACL-D patients who performed the mixed CKC+OKC quadriceps strength training programmes consistently demonstrated significantly greater quadriceps strength versus the CKC-biased strength training patients when tested at follow-up (Mikkelsen et al. 2000; Tagesson et al. 2008). These studies all demonstrate that the clinical application of resisted OKC knee extensions with ACL-injured patients does not cause an increase in knee laxity as defined by anterior tibial displacement.

Figure 21.6 Single-leg dynamic resisted knee extension. Curved arrow represents how the mass of the lower leg and ankle-weight can theoretically induce more relative knee joint rotation than knee joint compression. This is because when the knee is extended, the ankle-weight externally loads the lower leg perpendicular to its long axis (straight arrow). Photograph © Copyright Nicholas Clark. Reproduced with permission.

Table 21.4 Open kinetic chain quadriceps muscle strength levels

Strength test	Reference	♂ Subjects	♀ Subjects	Mean BW (kg)	Mean 1RM (kg)	Mean kg/kgBW	Mean RSI (%)
Double-leg knee extension 1RM	Blackburn and Morrissey (1998)	—	University Students (n = 20)	65.3	20.6	0.32	32
	Stanforth et al. (1992)	Sedentary Adults (n = 40)	—	68.1	41.1	0.61	61
MEAN of Means		—	—	—	♂ = **41.1**	♂ = **0.61**	♂ = **61**
		—	—	—	♀ = **20.6**	♀ = **0.32**	♀ = **32**
Single-leg knee extension 1RM	Augustsson et al. (2004)	ACL-R (n = 19) *Uninjured Knee*	—	79.8	62.1	0.77	77
	Clark et al. (1999)	ACL-R (n = 12) *Uninjured Knee*	—	80.7	61.7	0.76	76
	Clark et al. (1999, 2001)	Recreational Athletes (n = 4)	—	75.5	53.5	0.71	71
	Clark et al. (1999, 2001)	—	Recreational Athletes (n = 6)	58.8	34.1	0.57	57
	Izquierdo et al. (1999)	Recreational Athletes (n = 26)	—	84.1	75.1	0.89	89
	Lemmer et al. (2007)	Recreational Athletes (n = 21)	—	82.5	48.7	0.59	59
	Lemmer et al. (2007)	—	Recreational Athletes (n = 18)	68.1	29.1	0.43	43
	Tagesson and Kvist (2007)*	Recreational Athletes (n = 16)	—	82.5	49.1	0.59	59
	Tagesson and Kvist (2007)*	—	Recreational Athletes (n = 11)	68.1	31.5	0.46	46
MEAN of Means		—	—	—	♂ = **58.4**	♂ = **0.72**	♂ = **72**
		—	—	—	♀ = **31.6**	♀ = **0.49**	♀ = **49**

Table © Copyright Nicholas Clark. Reproduced with permission.
BW = bodyweight; kg = kilograms; 1RM = one repetition maximum; kg/kgBW = kilograms per kilogram of bodyweight; RSI = relative strength index; ACL-R = ACL-reconstruction subjects
Relative Strength Index (%) = load lifted ÷ bodyweight × 100
* The authors of this study did not report subjects' bodyweight. Therefore, for this study, muscle strength calculations have been performed using the bodyweight data for similar subjects from the study by Lemmer et al. (2007)

Table 21.5 Closed kinetic chain lower limb muscle strength levels

Strength test	Reference	♂ Subjects	♀ Subjects	Mean BW (kg)	Mean 1RM (kg)	Mean kg/kgBW	Mean RSI (%)
Double-leg squat 1RM	Blackburn and Morrissey (1998)	—	University Students (n = 20)	65.3	75.1	1.2	120
	Fatouros et al. (2000)	Untrained Adults (n = 41)	—	82.3	129.7	1.6	160
	Jones et al. (2001)	Baseball Players (n = 25)	—	80.4	139.8	1.7	170
	Murphy and Wilson (1997)	Recreational Athletes (n = 30)	—	79.5	120.5	1.5	150
	Wisloff et al. (2004)	Soccer Players (n = 17)	—	76.5	171.7	2.2	220
MEAN of means	—	—	—	—	♂ = 140	♂ = 1.75	♂ = 175
	—	—	—	—	♀ = 75.1	♀ = 1.2	♀ = 120
Single-leg Machine squat 1RM	McCurdy et al. (2004)	Sedentary Adults (n = 8)	—	86.3	88.6	1.03	103
	McCurdy et al. (2004)	—	Sedentary Adults (n = 22)	62.7	45.8	0.7	70
	McCurdy et al. (2004)	Weight-Trained Adults (n = 10)	—	90.3	121.6	1.3	130
	McCurdy et al. (2004)	—	Weight-Trained Adults (n = 12)	68.4	55.3	0.8	80
	Tagesson and Kvist (2007)*	Recreational Athletes (n = 16)	—	82.5	82.5	1	100
	Tagesson and Kvist (2007)*	—	Recreational Athletes (n = 11)	68.1	47.5	0.7	70
MEAN of means	—	—	—	—	♂ = 97.6	♂ = 1.11	♂ = 111
	—	—	—	—	♀ = 49.5	♀ = 0.73	♀ = 73
Double-leg leg press 1RM	Fatouros et al. (2000)	Untrained Adults (n = 41)	—	82.3	179.3	2.2	220
	Hoffman et al. (1999)	Infantry Soldiers (n = 136)	—	67.1	98.5	1.5	150
	Kraemer et al. (1995)	—	Tennis Players (n = 38)	60	99	1.7	170
	Stockbrugger and Haennel (2003)	Volleyball Players (n = 20)	—	83	263	3.2	320
	Stockbrugger and Haennel (2003)	Wrestling (n = 20)	—	85	320	3.8	380
MEAN of means	—	—	—	—	♂ = 215.2	♂ = 2.7	♂ = 270
	—	—	—	—	♀ = 99	♀ = 1.7	♀ = 170

(*Continued*).

Table 21.5 (*Continued*).

Strength test	Reference	♂ Subjects	♀ Subjects	Mean BW (kg)	Mean 1RM (kg)	Mean kg/kgBW	Mean RSI (%)
Single-leg leg press 1RM	Clark et al. (1999)	ACL-R (n = 12) *Uninjured Limb*	–	80.7	179.4	2.22	222
	Clark et al. (1999, 2001)	Recreational Athletes (n = 4)	–	75.5	170.1	2.25	225
	Clark et al. (1999, 2001)	–	Recreational Athletes (n = 6)	58.8	97.7	1.66	166
	Clark and Rees (2008)	Infantry Soldiers (n = 10)	–	74.6	109.1	1.46	146
	Clark and Rees (2008)	Injured Infantry Soldiers (n = 24) *Uninjured Limb*	–	82.4	114.3	1.39	139
	Lemmer et al. (2007)	Recreational Athletes (n = 21)	–	82.5	306.7	3.7	370
	Lemmer et al. (2007)	–	Recreational Athletes (n = 18)	68.1	192.8	2.8	280
	Worrell et al. (1993)	Data presented as mixed group of ♂ & ♀ University Students (n = 38)*	–	67.8	140.9	2.1	210
MEAN of means	–	–	–	–	♂ = **170.1**	♂ = **2.2**	♂ = **220**
	–	–	–	–	♀ = **143.8**	♀ = **2.1**	♀ = **210**

Table © Copyright Nicholas Clark. Reproduced with permission.

BW = bodyweight; kg = kilograms; 1RM = one repetition maximum; kg/kgBW = kilograms per kilogram of bodyweight; RSI = relative strength index; ACL-R = ACL-reconstruction subjects

Relative Strength Index (%) = load lifted ÷ bodyweight × 100

*The author of this study did not report subjects' bodyweight. Therefore, for this study, muscle strength calculations have been performed using the bodyweight data for similar subjects from the study by Lemmer et al. (2007)

**Mixed male and female group data were used to calculate the mean of means for both males and females

Table 21.6 Rank comparison of mean peak tibiofemoral anterior shear forces during selected strength training exercises*

Exercise	Reference	Condition	Mean Peak Anterior Shear Force (N)	Mean Peak Anterior Shear Force (% BW)	Mean Peak Anterior Shear Force Knee Angle (°)
Power clean	Souza and Shimada (2002)	70% 1RM	850	94	–
Single-leg isokinetic knee extension	Baltzopoulos (1995)	30° · sec	706	90	–
	Nisell et al. (1989)	30° · sec	700	90	45
Walking	Harrington (1976)	BW	411	–	20
Single-leg isometric knee extension	Toutoungi et al. (2000)	MVE	396	55	30
Single-leg isokinetic knee extension	Toutoungi et al. (2000)	60° · sec	349	48	40
Single-leg isometric knee extension	Smidt (1973)	MVE	343	45	15
Walking	Shelburne et al. (2004)	BW	303	44	20
	Shelburne et al. (2005)	BW	303	44	20
Single-leg isometric knee extension	Lutz et al. (1993)	MVE	285	37	30
Single-leg isokinetic knee extension	Kaufman et al. (1991)	60° · sec	241	30	25
Double-Leg Anisometric** Knee Extension	Wilk et al. (1996)	78kg load	248	27	15
	Escamilla et al. (1998)	79kg load	158	18	15
Bodyweight Forward Step-Up	Isaac et al. (2005)	BW	200	–	30
Single-Leg Anisometric Knee Extension	Isaac et al. (2005)	16kg load	200	–	30
Single-Leg Bodyweight Squat	Toutoungi et al. (2000)	BW	142	18	40
Single-Leg Isometric Knee Extension	Yasuda and Sasaki (1987)	MVE	121	–	15
Barbell Back Squat	Hattin et al. (1989)	34kg load	112	14	25
Single-Leg Anisometric Knee Extension	Isaac et al. (2005)	2kg load	70	–	30
Single-Leg Dumb-Bell Squat	Escamilla et al. (2009)	10–15kg load	59	8.5	30
Double-Leg Bodyweight Squat	Toutoungi et al. (2000)	BW	28	3	40
	Ohkoshi et al. (1991)	BW	0	0	N/A
	Shelbourne and Pandy (1998)	BW	0	0	N/A
Single-leg bodyweight squat	Ohkoshi and Yasuda (1989)	BW	0	0	N/A
Barbell back squat	Wilk et al. (1996)	147kg load	0	0	N/A
	Escamilla et al. (1998)	146kg load	0	0	N/A
	Escamilla et al. (2001)	133kg load	0	0	N/A
	Stuart et al. (1996)	23kg load	0	0	N/A

(Continued).

Table 21.6 (Continued).

Exercise	Reference	Condition	Mean Peak Anterior Shear Force (N)	Mean Peak Anterior Shear Force (% BW)	Mean Peak Anterior Shear Force Knee Angle (°)
Barbell front squat	Stuart et al. (1996)	23kg load	0	0	N/A
Barbell lunge	Stuart et al. (1996)	23kg load	0	0	N/A
Dumb-bell wall squat	Escamilla et al. (2009)	57kg	0	0	N/A
Horizontal double-leg press	Escamilla et al. (1998)	146kg load	0	0	N/A
	Wilk et al. (1996)	146kg load	0	0	N/A
	Escamilla et al. (2001)	129kg load	0	0	N/A
Single-leg isokinetic knee flexion	Kaufman et al. (1991)	60° · sec	0	0	N/A
	Toutoungi et al. (2000)	60° · sec	0	0	N/A
Single-leg isometric knee flexion	Lutz et al. (1993)	MVE	0	0	N/A

Table © Copyright Nicholas Clark. Reproduced with permission.
* Tibiofemoral anterior shear forces are considered representative of Anterior Cruciate Ligament tensile loads (Escamilla et al. 1998, 2001, 2009; Wilk et al. 1996)
** Refers to what is historically termed an 'isotonic' muscle action
N = Newtons; BW = bodyweight; MVE = maximum voluntary effort; sec = second; N/A = not applicable

Tibiofemoral posterior shear forces during ADL, exercise and sport are frequently significantly higher than tibiofemoral anterior shear forces (Table 21.9). As such, exercise rehabilitation for the PCL-D or PCL-R knee requires greater care and slower progression than for the ACL-injured knee. This is because the normal posterior shear forces that are generated during physical activities are consistently of a greater magnitude than the lower limits of the mean ultimate load for common PCL graft fixation methods approximated at 500N (Table 21.8). Consequently, there is a high risk of damage to the PCL-R graft fixation site if exercise rehabilitation is progressed too aggressively or too quickly post-surgery. As outlined earlier, six to eight weeks is considered to be a sufficient period for initial 'graft protection' since this is the time-frame currently thought necessary for minimal fixation site healing and soft-tissue graft incorporation to occur at the bone tunnel interface (Harner and Hoher 1998).

With regard to OKC and CKC quadriceps strength training exercises, because posterior shear forces appear to rapidly increase beyond 500N at angles greater than 50–60° of knee flexion (Table 21.9), exercises such as, for example, resisted knee extensions, single-leg-press and BW wall-squats should be limited between 0–50° knee flexion. After eight weeks, knee flexion ROM can be progressively increased in a controlled manner (Wilk et al. 1999). The authors' preferred method of cautiously increasing ROM during controlled strength training exercises following PCL-R is by 15° knee flexion per week (Table 21.3).

With regard to OKC hamstring strength training exercise such as the prone hamstring curl (Figure 21.7a), some clinicians recommend complete abstinence from such exercises for eight weeks or more after PCL-R surgery (Wilk et al. 1999). This is due to the large posterior shear forces that can be generated during 'maximum-effort' resisted knee flexion exercises (Table 21.9). However, because tibiofemoral posterior shear forces do not approach 500N until well beyond 20–30° knee flexion during OKC resisted knee flexion exercises (Toutoungi et al. 2000), it is safe for the PCL-R patient to perform *sub-maximal* resisted isometric hamstring exercises within the limits of pain from 0–20° knee flexion (Figure 7b). This may deter progressive hamstring weakness secondary to the surgical trauma induced by harvesting the hamstring tendons.

Table 21.7 Mean ultimate load of cruciate ligament reconstruction

	Human graft tissues	
Graft type	Reference	Mean ultimate load (N)
Anterior cruciate ligament	Woo et al. (1991)	2160
Posterior cruciate Ligament	Race & Amis (1994)	4000
B-PT-B 7mm wide untwisted	Cooper et al. (1993)	2238
B-PT-B 10mm wide untwisted	Cooper et al. (1993)	3057
	Noyes et al. (1984)*	2900
	Staubli et al. (1999)	1965
	Wilson et al. (1999)	1784
B-PT-B 10mm wide twisted	Cooper et al. (1993)	2542
B-PT-B 15mm wide untwisted	Cooper et al. (1993)	4389
	Noyes et al. (1984)**	2734
B-PT-B 15mm wide twisted	Cooper et al. (1993)	3397
QT 10mm wide untwisted	Staubli et al. (1999)	2170
Single-strand semitendinosus	Hamner et al. (1999)	1060
	Noyes et al. (1984)	1216
Double-strand semitendinosus	Hamner et al. (1999)	2330
	Wilson et al. (1999)	2422
Single-strand gracilis	Hamner et al. (1999)	837
	Noyes et al. (1984)	838
Double-strand gracilis	Hamner et al. (1999)	1550
Iliotibial band	Noyes et al. (1984)	769
Quadruple-strand*** unbraided	Hamner et al. (1999)	4140
	Kim et al. (2003)	3000
	Millett et al. (2003)	3404
Quadruple-strand*** braided	Kim et al. (2003)	2215
	Millett et al. (2003)	2223

Table © Copyright Nicholas Clark. Reproduced with permission.
N = Newtons; * Central third of patellar tendon; ** Middle third of patellar tendon
B-PT-B = bone-patellar tendon-bone; QT = quadriceps tendon
*** Refers to four-strand composite graft formed by double-strand hamstring + double-strand gracilis graft

Mediolateral tibiofemoral joint shear forces during OKC and CKC rehabilitation strength training exercises are very low compared to anterior and posterior shear forces, being just 5% BW for OKC quadriceps exercise (Kaufman et al. 1991) and 12% BW for double-leg squats (Hattin et al. 1989). Since frontal plane knee alignment should be relatively neutral during these dynamic rehabilitation exercises with very little valgus rotation of the tibiofemoral joint, such exercises are clearly safe for all MCL injuries when performed within the limits of pain. Because it is normal for the knee to undergo some varus-valgus oscillation during dynamic CKC function (Kirtley 2006), the authors teach patients to maintain the tibial tubercle inbetween the first and third toes during the eccentric and concentric phases of CKC rehabilitation strength training exercises, thereby controlling excessive varus-valgus knee motion.

During standing *single*-leg CKC exercise LCL tensile forces dramatically increase from 60% BW in uninjured subjects up to 88% in knee injured patients (Schipplein and Andriacchi 1991; Shelburne et al. 2005). This is because it is normal for the lateral femoral condyle to 'lift-off' the lateral tibial plateau so that more compressive load is borne through the

Table 21.8 Mean ultimate load of common cruciate ligament reconstruction

Graft fixation methods		
Graft fixation method	Reference	Mean ultimate load (N)
Paired 4mm – 5mm diameter lag screws	Campbell et al (2007)*	762
	Gupta et al. (2009)*	638
7mm diameter interference screw	Adam et al. (2004)**	536
	Kohn and Rose (1994)*	461
	Pena et al. (1996)*	640
	Steiner et al. (1994)*	588
9mm diameter interference screw	Camillieri et al. (2004)*	497
	Gerich et al. (1997)*	678
	Honl et al. (2002)*	637
	Kitamura et al. (2003)**	835
	Kohn and Rose (1994)*	550
	Kurosaka et al. (1987)*	476
	Steiner et al. (1994)*	674
	Zantop et al. (2004)**	702
Endobutton	Ahmad et al. (2004)**	864
	Honl et al. (2002)*	572
	Kitamura et al. (2003)**	580
	Rowden et al. (1997)*	612
	Scheffler et al. (2002)*	505
Staple	Gerich et al. (1997)*	588
	Magen et al. (1999)**	705
Suture	Campbell et al (2007)*	582
	Honl et al. (2002)*	507
	Steiner et al. (1994)*	573
Washer	Magen et al. (1999)**	930
	Scheffler et al. (2002)*	554
Cross-pin	Ahmad et al. (2004)**	737
	Zantop et al. (2004)**	639
MEAN of means	–	**493.5**

Table © Copyright Nicholas Clark. Reproduced with permission.
* Human cadaver study; ** Animal study

medial tibiofemoral joint and greater tensile loads are imposed on the lateral tibiofemoral joint (Schipplein and Andriacchi 1991). Consequently, the clinician should take great care when choosing to implement standing single-leg CKC rehabilitation strength training exercises (e.g. Figure 21.4) without a long-leg hinged knee brace. Such single-leg brace-free CKC strength training exercises are not implemented after LCL-R until at least nine weeks post-surgery (Table 21.3).

Does a surgical reconstruction prevent knee OA?

Surgical reconstruction of a torn ACL has frequently been recommended with the intention of delaying or preventing the onset of tibiofemoral OA. To date, it is clear that an ACL-R does *not* protect the knee from developing OA (Myklebust and Bahr 2005; Roos 2005; Keays et al. 2007; Lohmander et al. 2007). As such, following a technically proficient ACL-R, appropriate and sufficient exercise rehabilitation

Table 21.9 Rank comparison of mean peak tibiofemoral posterior shear forces during selected strength training exercises*

Exercise	Reference	Condition	Mean peak posterior shear force (N)	Mean peak posterior shear force (% BW)	Approximate mean knee flexion angle at which posterior shear force rises above 500N**
Single-leg isometric knee flexion	Toutoungi et al. (2000)	MVE	3330	470	20
Double-leg bodyweight squat	Toutoungi et al. (2000)	BW	2704	350	50
Single-leg bodyweight squat	Toutoungi et al. (2000)	BW	2246	290	50
	Ohkoshi and Yasuda (1989)	BW	–	273	–
Barbell back squat	Escamilla et al. (2001)	133kg load	2212	240	30
	Escamilla et al. (1998)	146kg load	1868	203	30
Horizontal double-leg leg press	Escamilla et al. (1998)	146kg load	1866	203	50
Barbell back squat	Wilk et al. (1996)	147kg load	1783	194	40
Horizontal double-leg leg press	Escamilla et al. (2001)	133kg load	1726	186	50
	Wilk et al. (1996)	146kg load	1667	181	50
Single-leg isometric knee flexion	Smidt (1973)	MVE	1495	184	30
Double-leg bodyweight squat	Ohkoshi and Yasuda (1989)	BW	–	128	–
	Ohkoshi et al. (1991)	BW	–	128	–
Stair ascent	Morrison (1969)	BW	1229	–	–
Double-leg anisometric*** knee extension	Wilk et al. (1996)	78kg load	1178	128	60
	Escamilla et al. (1998)	78kg load	960	104	50
Dumb-bell wall-squat	Escamilla et al. (2009)	36–57kg load	786	114	60
Single-leg dumb-bell squat	Escamilla et al. (2009)	10–15kg load	414	60	N/A
Single-Leg Isometric Knee Extension	Yasuda and Sasaki (1987)	MVE	361	–	N/A
Walking	Morrison (1969)	BW	356	–	N/A
	Morrison (1970)	BW	333	–	N/A
Single-Leg Isokinetic Knee Extension	Toutoungi et al. (2000)	60° · sec	74	10	N/A

Table © Copyright Nicholas Clark. Reproduced with permission.
* Tibiofemoral posterior shear forces are considered representative of Posterior Cruciate Ligament tensile loads (Escamilla et al. 1998, 2001, 2009; Wilk et al. 1996)
** Approximate Mean Knee Flexion Angle read from graph in published article
*** Refers to what is historically termed an 'isotonic' muscle action
N = Newtons; BW = bodyweight; MVE = maximum voluntary effort; sec = second; N/A = not applicable

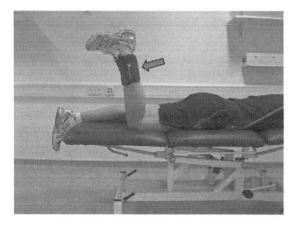

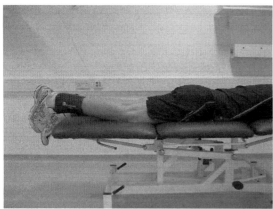

Figure 21.7 (a) Single-leg dynamic resisted hamstring curl. Arrow represents how, when the knee is extended, the ankle-weight externally loads the lower leg perpendicular to its long axis. (b) Outer-range resisted isometric hamstring curl: maximum 20° knee flexion. Photograph © Copyright Nicholas Clark. Reproduced with permission.

is critical since muscle functions to protect joint surfaces from excessive impact forces (Buckwalter, 2003). It is almost inevitable that any physically active young adult who sustains a significant tibiofemoral joint injury of any kind will develop premature knee OA during their lifespan (Gelber et al. 2000; Roos 2005; Lohmander et al. 2007). As is the case following ACL-R, PCL-R alone does not protect the tibiofemoral joint *or* PFJ from the development of OA (Sekiya et al. 2005; Jackson et al. 2008). Again, as for ACL-R, appropriate and sufficient exercise rehabilitation is critical following PCL-R to induce optimal neuromuscular function in order to protect joint surfaces from excessive impact forces (Buckwalter 2003; Fontbote et al. 2005).

Generic knee rehabilitation concepts

Basic concepts

Knee and lower limb biomechanics have been extensively studied with regard to running, jumping, hopping and leaping in sports (Hewett et al. 1996; Decker et al. 2002; Lewek et al. 2002; McLean et al. 2004; Fontbote et al. 2005; Noyes et al. 2005; Paterno et al. 2007; Ortiz et al. 2008). In other words, for all injured knees to perform safely and effectively in running, jumping, hopping and leaping activities without *re-injury*, all injured knees should eventually be able to perform the same necessary ROM and tolerate the same inherent joint compression and shear forces typical to such high-velocity high-impact movement patterns. As such, although different types of knee injury can have very different weight-bearing, ROM and rehabilitation strength training restrictions in the early stages of exercise rehabilitation (e.g. $\leq$ 12 weeks post-injury/post-surgery) (Table 21.3), all types of knee injury should eventually be progressed to tolerate, for example, the same approximate joint compression and shear forces natural to running, jumping, hopping and leaping in sports in the late stages of rehabilitation (e.g. $\geq$ 12 weeks to 12 months post-injury/surgery). So, there are 'generic' knee rehabilitation concepts from early to late stage exercise rehabilitation.

Effects of proximal muscles on knee function and injury

Basic concepts

It is well established that lumbo-pelvic-hip complex frontal plane alignment can have a powerful effect on whole lower limb alignment and, in particular, knee alignment in the frontal and transverse planes (Powers 2003). It has been identified how trunk lateral flexor and gluteal muscle weakness is related to excessive knee valgus and poor frontal plane alignment during single-leg CKC tasks (Willson et al. 2006; Jacobs et al. 2007), whilst others have statistically proven how impaired trunk and gluteal muscle function is consistently linked to non-contact knee injury (Nadler et al. 2000; Zazulak et al. 2007). Moreover, Bobbert and Van Zandwijk (1999) have demonstrated how increased gluteal muscle function directly enhances both quadriceps and hamstring muscle function during dynamic CKC tasks. Therefore, generic exercises for all tibiofemoral joint injuries should include strength training for the trunk lateral flexors, gluteus maximus, and gluteus medius.

Trunk lateral flexors

The trunk lateral flexors have a profound effect on maintaining pelvic alignment in the frontal plane (Neumann 2002), and weakness and dysfunction in these muscle groups predicts non-contact knee injury (Zazulak et al. 2007). An excellent exercise to specifically target these muscle groups is the isometric side-bridge in standing (Figure 21.8) which can

Figure 21.9 Side-laying isometric side-bridge. This exercise can be made more difficult by straightening the legs. Photograph © Copyright Nicholas Clark. Reproduced with permission.

later be progressed to side-laying (Figure 21.9). As an exercise this can be performed using 10 repetitions per set for up to eight seconds each to avoid the effects of cumulative ischemia over multiple repetitions (McGill 2007). In testing, normal mean total holding times are ≥ 80 seconds for each side (McGill et al. 1999).

Gluteus maximus

The gluteus maximus has the ability to limit excessive hip adduction and internal rotation of the femur (Neumann 2002), also playing a crucial role in preventing lower limb flexion collapse in single-leg stance (Liu et al. 2006). Unilateral weakness of the gluteus maximus is linked to eventual onset of lower limb injury and low back pain (LBP) (Nadler et al. 2000), and so exercises for this muscle group should also be included in all knee rehabilitation programmes. Effective exercises for the gluteus maximus include prone isometric setting (Figure 21.10) eventually progressing to straight-leg hip extension in four-point kneeling (Ekstrom et al. 2007). This exercise can be performed using 10 repetitions per set for up to eight seconds each to avoid the effects of cumulative ischemia over multiple repetitions (McGill 2007).

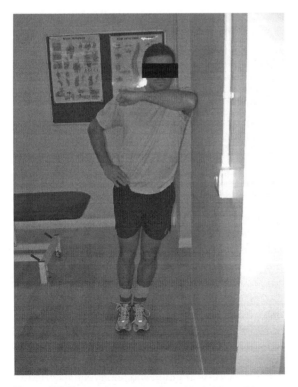

Figure 21.8 Standing isometric side-bridge. This exercise can be standardised between patients by placing the lateral border of the foot nearest the wall approximately one shoe-length from the wall. Photograph © Copyright Nicholas Clark. Reproduced with permission.

Gluteus medius

The gluteus medius is critical for both controlling frontal plane alignment of the pelvis on the femur

Figure 21.10 Prone isometric gluteus maximus setting. Pillows under the torso make the position more comfortable for the patient. Pillows under the ankles place the knee in flexion to relax the hamstrings. As the patient's ability to demonstrate good gluteus maximus isometric holding times improves, the ankle pillows can be removed and the patient can lift the leg and thigh $\leq$ 5cm from the floor, after which the patient can then be progressed to four-point kneeling (see text). Photograph © Copyright Nicholas Clark. Reproduced with permission.

Figure 21.11 Side-laying hip abduction and external rotation at 45° hip flexion. Note how the patient is against a wall to stabilize the pelvis. This exercise can be made more difficult by extending the hips to just 30° flexion. Photograph © Copyright Nicholas Clark. Reproduced with permission.

(Neumann 2002), and for controlling knee valgus via its ability to limit excessive femoral adduction and internal rotation (Neumann 2002; Powers 2003). Since valgus collapse of the knee is linked with traumatic non-contact injury to the ACL, MCL, and lateral meniscus (Boden et al. 2000; Hewett et al. 2005) as well as non-traumatic gradual onset knee joint pain (Powers 2003; Leetun et al. 2004), exercises for this muscle are extremely important in knee injury prevention and rehabilitation programmes. Effective exercises for this muscle group are side-lying hip abduction + external rotation in $\leq$ 45° hip flexion with the feet supported (Figure 21.11) and side-lying straight-leg hip abduction (Ekstrom et al. 2007; Distefano et al. 2009). Exercise training and testing for this muscle can be the same as described for the gluteus maximus.

Models and levels of sensorimotor control

Motor control is defined simply as "the control of both movement and posture" (Shumway-Cook and Woollacott 1995). There are at least nine models of motor control, most of which function on a 'sensory-motor' basis. This means that *before* effective motor output is executed, accurate sensory input must be received, and there is a critical need for normal proprioceptive *feedback* to the CNS since mechanoreceptor feedback modifies motor output at all three levels of the CNS, these being the spinal cord, brainstem and cerebral cortex (Ghez 1991a, 1991b).

Proprioception

Proprioception has been poorly defined and confused with other sensoriomotor functions such as balance (Riemann and Lephart 2002a, 2002b). Proprioception is correctly defined as the sense of position and movement of the joints and limbs, which correspond to static joint position sense (JPS) and kinaesthesia, respectively (Martin and Jessell 1991). Several authors have identified impaired JPS and kinaesthesia in ACL-injured (Borsa et al. 1997, 1998; Roberts et al. 1999), PCL-injured (Safran et al. 1999), and meniscal-injured (Jerosch et al. 1996) patients.

In view of these findings, it is clinically useful to identify exercise training methods that are capable of enhancing peripheral joint proprioception. If we consider that the muscle spindle is the most potent proprioceptor which is *always* stimulated with active movements as a consequence of alpha-gamma coactivation (Gordon and Ghez 1991), *any* active movement can then be considered 'proprioceptive training' since it generates a barrage of proprioceptive impulses from the most potent (muscle) proprioceptors *and* joint mechanoreceptors. Consequently, OKC training (Docherty et al. 1998; Friemert et al. 2006), CKC (axial loading) training (Rogol et al. 1998), balance training on a wobble board (Waddington et al. 1999, 2000) and plyometric training (Waddington et al. 2000) have all been shown to enhance peripheral joint proprioception defined by reduced errors in *active* static JPS repositioning tasks.

Neuromuscular control

Definition and classification

Neuromuscular control refers to any aspect of CNS control of muscle to function as a dynamic restraint and stress shield of inert tissues and with the specific aim of maintaining functional joint stability (FJS) (Riemann and Lephart 2002a, 200b). In other words, neuromuscular control is the *active* restraint of excessive joint motion and the coordinated dampening of joint loads in response to specific sensory feedback (Riemann and Lephart 2002a, 2002b), and can be classified into acute and learned neuromuscular control (Clark 2008).

Strength training alone does not consistently modify potentially dangerous whole lower limb and local knee joint kinematics or kinetics (Herman et al. 2008). As such, neuromuscular control is an important concept since different exercise training methods induce the different physiological and biomechanical adaptations that comprise its separate components. This is another major premise of this chapter – different training methods induce different adaptations, and so a comprehensive injury prevention or injury rehabilitation programme must be composed of several different types of exercise in order to thoroughly encompass all the different components of neuromuscular control. Neuromuscular control can be considered as the motor component of sensorimotor control.

Acute neuromuscular control

Acute neuromuscular control is an almost instantaneous motor response to 'at-that-moment-in-time' sensory information about sudden joint displacements, and is composed of the electromechanical delay (EMD), and the rate-of-force development (RFD). The EMD is the timeframe between the onset of reflex muscle electrical activity and the onset of measurable force development (Winter and Brooks 1991). The RFD is the timeframe between the onset of measurable force development and the achievement of a specific quantity of force (e.g. 25% 1RM) (Kaneko et al. 2002). Relatively shorter EMD and RFD timeframes equate to faster reactive muscle activity which is clearly desirable when resisting sudden potentially dangerous knee joint displacements. To date, there is no convincing evidence that the EMD can be shortened with training, whereas the knee muscle RFD can be significantly shortened with OKC and CKC strength training (Hakkinen and Komi 1983, 1986; Hakkinen et al. 1998; Aagaard et al. 2002; Bruhn et al. 2004), as well as balance training (Bruhn et al. 2004; Gruber and Gollhofer 2004; Ihara and Nakayama 1986), and agility training (Wojtys and Huston 1996). Therefore, a selection of different training methods can enhance knee acute neuromuscular control in the form of faster reactive muscle activity.

Learned neuromuscular control

Where acute neuromuscular control refers to reactive muscle activity at a single point-in-time, learned neuromuscular control refers to how the CNS uses repeated sensory feedback generated within multiple training sessions to induce long-term training adaptations that includes joint-protective motor programmes and more favourable knee and lower limb kinematics and kinetics. Learned neuromuscular control includes decreased vertical ground reaction force (VGRF), decreased knee joint reaction force (JRF), decreased knee joint shear forces, decreased knee abduction (valgus) forces, improved frontal plane hip-knee-ankle alignment, increased hip-knee-ankle flexion, improved hamstring:quadriceps (H:Q) ratios, improved feed-forward pre-impact muscle activation, and improved postural stability. All of these components contribute to improved whole lower limb alignment and decreased knee joint forces during highly dynamic tasks, and so are highly desirable for knee injury prevention and rehabilitation programmes.

Increased VGRF during the loading response phase of landing tasks results in increased tibiofemoral anterior shear forces (McNair and Marshall 1994; Yu et al. 2006; Sell et al. 2007) and, since the VGRF is a major component of the compression JRF (Enoka 2002) increased knee impact forces. This offers the potential for both ACL and tibiofemoral articular surface injury. Balance, deceleration (landing) and plyometric training have all been shown to reduce the VGRF during athletic tasks (Hewett et al. 1996; Cowling et al. 2003; Myer et al. 2006a), and so are essential training methods for reducing knee joint shear and compression forces.

Increased knee abduction forces during landing tasks co-exist with increased valgus hip-knee-ankle frontal plane alignment and reduced hip-knee-ankle saggital plane flexion (Hewett et al. 1996, 2005; Noyes et al. 2005), placing the ACL, MCL and lateral meniscus at high risk of injury during the rapid loading response phase of deceleration manoeuvres. Open kinetic chain and CKC strength training, balance training, deceleration training and plyometric training all show the ability to reduce knee abduction forces, improve whole lower limb frontal plane alignment, and increase whole lower limb saggital plane flexion excursion (Hewett et al. 1996; Myer et al. 2005, 2006b, 2007; Noyes et al. 2005), thereby reducing potentially dangerous forces imposed on the ACL, MCL, lateral meniscus, and tibiofemoral joint articular surfaces.

Poor postural stability (i.e. balance) has been implicated in non-contact knee injury because of its detrimental effects on whole lower limb alignment and local knee valgus (Hewett et al. 2005; Myer et al. 2006a). Closed kinetic chain strength training, balance board training, deceleration training, and plyometric training have been proven to enhance postural stability during balance tasks (Myer et al. 2006a), and so are also considered critical to a knee injury prevention and rehabilitation programme.

Summary

Based on the studies just cited, and as for acute neuromuscular control, many different training methods result in different training adaptations associated with favourable learned neuromuscular control of the knee. Therefore, comprehensive knee injury prevention and rehabilitation programmes should ideally include all of the training methods just discussed either in a single training session or certainly within a weekly periodised training plan (Myer et al.

Warm-up + dynamic stretches

Plyometrics

Deceleration drills

Balance drills

Closed kinetic chain strength training

Open kinetic chain strength training

No-impact cardiovascular training + aerobic cool-down

Isolated trunk muscle exercises

Passive stretches

Figure 21.12 End-stage rehabilitation single training session outline.

2005, 2006a, 2006b, 2007). An example of an 'end-stage' single-session knee rehabilitation programme sequence of exercises is presented in Figure 21.12, with exercise order moving from most complex to least complex in order to minimise the effects of fatigue (Fleck and Kraemer 1997).

Between-sex differences

There is no doubt that female athletes experience a greater relative number of knee injuries than males (Agel et al. 2005). This is also the case for low back pain (LBP) and foot, ankle and lower leg injuries (Strowbridge 2002, 2005). Compared to males, females demonstrate less knee muscle strength and less proprioceptive acuity (Rozzi et al. 1999; Lephart et al. 2002), lower reflex gluteal activation levels after foot-strike in landing tasks (Hart et al. 2004, 2007; Zazulak et al. 2005), and higher VGRF, higher knee anterior shear forces, higher knee abduction forces, and increased knee valgus alignment during deceleration manoeuvres (Hewett et al. 1996; McLean et al. 2005; Noyes et al. 2005), all of which are thought to increase females' predisposition to increased incidence of knee injury.

In an injury prevention context, most of the neuromuscular control variables just described can be significantly improved with different exercise methods that include strength training, balance training, deceleration training and plyometric training

(Lephart et al. 2005; Myer et al. 2005, 2007; Noyes et al. 2005), and such training methods have been shown to significantly reduce the incidence of non-contact male and female knee injury (Hewett et al., 1999; Mandelbaum et al. 2005; Olsen et al. 2005). In an injury rehabilitation context, even though females consistently demonstrate greater impairment in proprioception and neuromuscular control variables than males, the principles underlying female exercise rehabilitation progression can essentially be the same as for males. However, comparatively speaking, since females may well be starting their knee exercise rehabilitation at a relatively 'lower' proprioception and neuromuscular control 'baseline' than males, female knee rehabilitation and return to full function may then take longer than the time required for males.

Proposed knee exercise rehabilitation pathway

The existence of a knee exercise rehabilitation pathway (Figure 21.13) where rehabilitation is progressed in such a way that the complexity of exercise is progressed once a sufficient strength training base has been established would appear possible especially since many different training methods induce different favourable proprioception and neuromuscular control training adaptations (Clark, 2008). The rationale for focussing on isolated (OKC) and functional (CKC) muscle strength prior to the implementation of more dynamic exercise rehabilitation methods relates to ensuring injured tissues are able to first tolerate the higher tensile and compressive loads associated with more dynamic and impact exercises (Clark 2001, 2004) as well as correcting between-limb and within-limb compensations and facilitating the most desirable joint kinematics associated with enhanced force absorption during the loading response phase of walking, running, hopping and jumping gait (i.e. increased saggital plane knee flexion).

The authors present a 'multi-modal' knee exercise rehabilitation pathway (Clark, 2008) that has been employed with consistent success for more than seven years with recreational and professional athletes as well as frontline military operatives. The driving principle of this pathway is that patients do not progress from one exercise type to another until specific clinical criteria are achieved as goals of treatment.

These criteria will be outlined in more detail later. Such conditions ensure the patient is progressed through rehabilitation as safely and effectively as possible according to their own unique individual ability, and so goals of treatment also function as progression-criteria (Clark 2004, 2008). When implementing the pathway in Figure 21.13, it is important to remember that as the patient progresses and a new exercise type is added to a single training session, each new exercise type 'builds' on the type that preceded it, so that the patient is always performing all the previous exercise types for the remaining rehabilitation process.

Balance and perturbation training

Basic concepts

For the purposes of this chapter, balance training refers to single-leg CKC training performed on exercise devices that function as an unstable base-of-support (BOS). Such devices include, for example, rocker-boards (Figure 21.14), wobble-boards, foam rollers, inflatable equipment (Figure 21.15) and mini trampolines, although the clinician can obviously use any other method deemed useful for to creating an unstable BOS. The patient should be encouraged to maintain optimal pelvis-hip-knee-ankle alignment during balance training (Figure 21.15). The authors cue patients to this effect by instructing in keeping the anterior superior iliac spines (ASIS) as level as possible, and to keep the tibial tuberosity inbetween the first and third toes, at no time should the tibial tubercle deviate inside the first toe indicating valgus collapse of the knee and whole lower limb (Figure 21.15).

Perturbation training refers to orchestrated training situations where the clinician attempts to deliberately unbalance the patient by rhythmically tapping and knocking the balance device itself, pushing and pulling the patient at the shoulder or pelvic girdles (i.e. rhythmic stabilisations), or by distracting the patient with ball tosses (Figure 21.16). Since this type of training does not include impact forces, it seems sensible to become proficient in this method of training before introducing high-impact exercise drills (e.g. plyometrics).

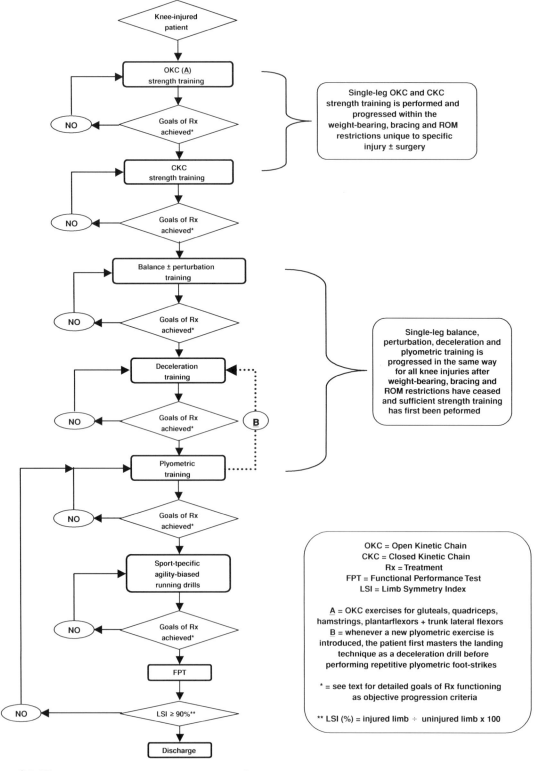

Figure 21.13 Knee exercise rehabilitation pathway. © Copyright Nicholas Clark (2008). Reproduced with permission.

Figure 21.14 Frontal plane-biased rocker-board balance training. Photograph © Copyright Nicholas Clark. Reproduced with permission.

Figure 21.15 Single-leg squat on inflatable balance training device. Note the anterior superior iliac spines are level (horizontal line) and the tibial tubercle does not deviate inside the first toe (downwards arrow). Photograph © Copyright Nicholas Clark. Reproduced with permission.

Entry Criteria

Since the intent of balance training is to perform single-leg CKC exercise on an unstable BOS, it seems appropriate to ensure patients can maintain optimal knee and whole lower limb alignment on a stable BOS. As such, patients should ideally demonstrate the ability to perform three sets of 15 repetitions of a single-leg squat to $\geq 45°$ knee flexion on the ground on *both* legs in order to demonstrate reasonable strength-endurance before undertaking balance training.

Methods

Patients can begin balance training using a rocker-board or foam roller to deliberately cause frontal plane varus-valgus oscillation of the knee (Figure 21.14), progressing to a round-base balance device for multi-planar training (Figure 21.15), and then to perturbation and distraction training (Figure 21.16). Patients should be able to perform three sets of 15 repetitions of a single-leg squat to $\geq 45°$ knee flexion on *both* legs before progressing from one balance training device to another.

Goals of treatment and objective progression criteria

The minimal goal of treatment for this type of training should be to perform three sets of 15 repetitions of a single-leg squat to $\geq 45°$ knee flexion on both legs on a round-base device (i.e. multi-planar balance training) before progressing to impact training (e.g. deceleration/plyometric drills).

Figure 21.16 Perturbation and distraction training using a ball toss. © Copyright Nicholas Clark. Reproduced with permission.

Deceleration training

Basic concepts

Deceleration training refers to exercise drills intended to teach correct landing technique for running, jumping, hopping and leaping activities in sports and high-level occupational tasks. Such training uses exercises designed to *decelerate* the momentum of the lower limb and trunk following foot-strike and the onset of *impact forces* emphasising eccentric-biased muscle control of saggital plane knee and lower limb joint flexion, followed by isometric-biased muscle activity to stabilise the knee, lower limb and trunk with optimal frontal and transverse plane pelvis-hip-knee-ankle alignment. As such, the sequence of muscle activity is eccentric to isometric – there is no consecutive concentric propulsion phase. Clinically, such eccentric-biased drills are applied to

RSI (%) = weight pushed (kg) ÷ bodyweight (kg) × 100

Figure 21.17 Calculation of the relative strength index (RSI)

dampen and shock-absorb high tensile and impact (compressive) forces away from joints by reducing the VGRF. In simplest terms, deceleration training is preparation for plyometric training, and always precedes the introduction of any new plyometric drill so that patients can first master correct landing technique.

Entry criteria

Since deceleration training involves high shear forces as well as high VGRF, ideal entry criteria are modified from Clark (2001, 2004, 2006), and include the normal muscle strength values outlined in Table 21.4 and Table 21.5 and Figure 21.17 and Figure 21.18, all of the trunk and gluteal muscle function recommendations described earlier in this chapter, as well as three sets of 15 repetitions of a single-leg squat to $\geq 45°$ knee flexion on a round-base balance device. At first glance this may seem like a high number of criteria, however many of these should already have been achieved much earlier in the rehabilitation process, and such criteria are necessary to ensure as safe and effective introduction as possible of the high joint shear and compression forces inherent in high-impact training.

Methods

Deceleration training sessions must be supervised by a clinician competent in such training techniques and able to use visual, verbal and tactile teaching methods. Training sessions never take place on concrete, yielding surfaces are most suitable (e.g. dance studio, sports hall, grass, etc.) (Chu 1998). Patients are initially taught the biomechanically correct and functionally stable final landing alignment where the head is up and eyes looking forwards, the trunk is inclined forwards approximately 30°, shoulders are over the knees, knees are over the mid- to rear-foot, and hands are in the 'athletic ready' position (Figure 21.19), whilst in the frontal plane the centre of the hip, knee and ankle joints are

LSI (%) = weight pushed by injured limb (kg) ÷ weight pushed by uninjured limb (kg) × 100

Figure 21.18 Calculation of the limb symmetry index (LSI)

approximately in a straight line, the tibial tubercle is in-between the first and third toes (Figure 21.20), and the tips of the toes and heels should be in a straight line (Figure 21.19) (Chu 1998; Hewett et al. 1996, 1999; Noyes et al. 2005).

On landing, foot-strike should ideally roll from the ball of the foot to the heel, followed by distal-to-proximal ankle, knee, and hip joint flexion to finish in ≥ 45° knee flexion with minimal if any forward-backward or side-to-side unsteadiness of the pelvis or overt muscle tremor (Chu 1998; Hewett et al. 1996, 1999; Kovacs et al. 1999; Noyes et al. 2005). Patients should be instructed to land as softly as possible (Devita and Skelly 1992; Zhang et al. 2000),

and the moment of landing should be as quiet as possible (Onate et al. 2001, 2005; Prapavessis et al. 2003). Following landing, it is entirely normal to demonstrate a degree of knee varus-valgus oscillation (i.e. 'wobbling'). The key concept is to control how much varus–valgus oscillation occurs (Clark 2006). This can be achieved by again verbally cueing the patient not to allow the tibial tubercle to deviate inside the first toe/metatarsal.

Alignment for single-leg deceleration drills should not differ greatly from that for double-leg drills, although the trunk and pelvis must move slightly laterally to the foot to maintain the body's centre-of-mass over its BOS, the foot. If the entry

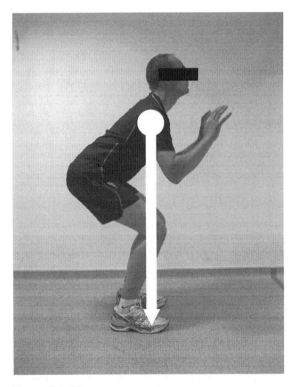

Figure 21.19 Correct double-leg deceleration training landing position – lateral view. Photograph © Copyright Nicholas Clark. Reproduced with permission.

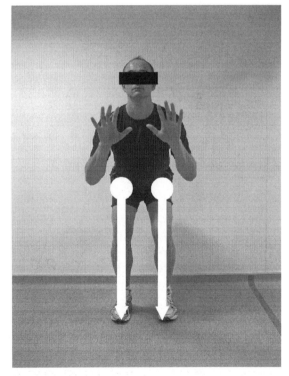

Figure 21.20 Correct double-leg deceleration training landing position – anterior view. Photograph © Copyright Nicholas Clark. Reproduced with permission.

criteria for this type of training have been fulfilled, it is possible for patients to immediately commence single-leg deceleration drills without having to undertake double-leg drills first. The clinician should frequently observe the patient from both the front and the side to ensure correct landing technique.

Once the eccentric phase has been completed, patients should maintain the ideal landing position (Figure 21.19 and 21.20) for three to five seconds. If the patient loses balance, the ideal landing position should be reassumed for three to five seconds to provide the CNS with a 'memory' of the correct versus incorrect final position. The intensity of a deceleration drill can be influenced by increasing distance or height covered, with a recommended 80–100 total foot-contacts per training session for an uninjured beginner (Chu 1998) and 40–50 foot-contacts for initial injury rehabilitation. A within-session work:rest ratio of 1:5–1:10 is advised, and there should be no more than three training sessions per week, separated by at least 48 hours between each session (Chu 1998). Signs of fatigue include loss of ideal frontal plane alignment, decreased saggital plane joint flexion, increased frequency of loss-of-balance, muscle tremor, and louder landings, at which point the drills should be stopped due to risk of new or re-injury. The clinician should be extremely vigilant for symptoms of pain and signs of swelling, redness or heat, which all signal that deceleration training should be terminated immediately pending reassessment of the patient's injury site.

Goals of treatment and objective progression-criteria

The minimal goal of treatment for this type of training should be to perform three sets of 10–15 repetitions of single-leg drills on both legs with correct technique before progressing to plyometric training. It is possible for a skilful patient to achieve this goal within one training session.

Plyometric training

Basic concepts

Plyometric training is a progression from deceleration training, and involves the ballistic three-phase stretch-shortening cycle of eccentric-isometric-concentric muscle actions (Chu 1998). The isometric muscle action is the transition phase between the eccentric and concentric phase, being termed the 'amortisation phase', and this phase should be as short as possible in order to avoid the elastic energy stored within the muscle's connective tissues being lost as heat (Chu 1998). Historically, plyometric training has been classified using an ascending system of intensity from least intense to most intense including jumps/hop-in-place, standing jumps/hops, multiple jumps/hops, bounding, box drills and finally depth jumps (Chu 1998). Although this classification system has been used for decades with un-injured athletes in a performance enhancement and injury prevention context (Chu 1998; Hewett et al. 1996, 1999), it is built from empirical observations rather than quantifiable forces (e.g. VGRF) (Clark 2006), and is difficult to transfer to an injury rehabilitation context since the ascension through the different classes of plyometric drill can be quite sudden with regard to the anticipated dramatic increase in VGRF and tissue loads (Clark 2006).

In view of this clinical concern, Clark (2006) proposed a different classification and progression of plyometric drills termed 'Clinical Plyometrics' (Figure 21.21) for more refined clinical application in an injury rehabilitation context. The ascending complexity of clinical plyometrics drills was based simply on a best rank ordering of joint shear forces (Table 21.10) and VGRF (Table 21.11) drawn from a wide variety of published research – the higher the forces the more intense the drill with regard to the magnitude of tissue loads (Clark 2006), as well as the notion that some types of drill naturally generate higher knee abduction forces (valgus moments) than others (Sell et al. 2006). Thus, the rank ordering of drills (Figure 21.21) was based on the anticipated tissue loading characteristics of a specific drill defined by the magnitude of shear and compression forces as well as knee valgus moments (Clark, 2006). This is because, for example, increased VGRF results in increased knee joint shear forces (McNair and Marshall 1994; Yu et al. 2006; Sell et al. 2007) and increased local JRF (Enoka 2002), whilst transverse plane knee motion induces greater ACL tensile loads than saggital plane knee motion (Markolf et al. 1995).

In-place drills predominantly load the tibiofemoral joint along its longitudinal axis and induce relatively low compression JRF (Clark 2006). Examples of in-place drills are up-and-down

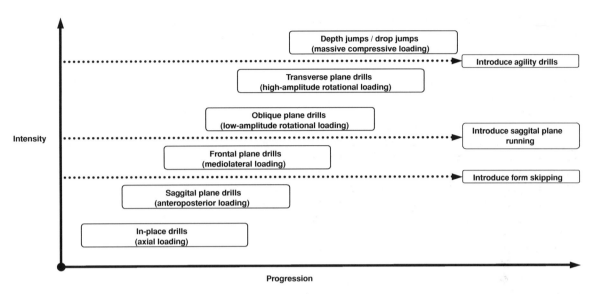

Figure 21.21 Intensity classification of clinical plyometrics.© Copyright Nicholas Clark (2006). Reproduced with permission.

Table 21.10 Rank comparison of mean peak tibiofemoral anterior shear forces during activities of daily living (ADL), running and plyometric drills*

Exercise	Reference	Condition	Mean peak anterior shear force (N)	Mean peak anterior shear force (% BW)	Mean peak anterior shear force knee angle (°)
Run and forward leap landing	Steele & Brown (1999)	BW	1566	217	–
Double-leg horizontal jump and lateral jump	Sell et al. (2006)	BW	753	111	21
50cm single-leg drop-landing	Kernozek et al. (2008)	BW	602	95	–
40cm single-leg drop-landing and vertical hop	Ortiz et al. (2008)	BW	562	90	–
Run and double-leg horizontal jump-landing	Yu et al. (2006)	BW	469	79	35
	Herman et al. (2008)	BW	321	51	–
Walking downhill	Morrison (1969)	BW	450	70	–
20cm box hop up-and-down	Ortiz et al. (2008)	BW	399	64	–
Sidestep cutting manoeuvre	McLean et al. (2004)	BW	472	63	–
Walking on level ground	Shelburne et al. (2004)	BW	303	44	20
	Shelburne et al. (2005)	BW	303	44	20
Double-leg horizontal jump landing	Sell et al. (2007)	BW	228	36	29
Double-leg anisometric knee extension	Wilk et al. (1996)	78kg load	248	27	15
Single-leg anisometric knee extension	Isaac et al. (2005)	16kg load	200	–	30
	Isaac et al. (2005)	2kg load	70	–	30

Table © Copyright Nicholas Clark. Reproduced with permission.
N = Newtons; BW = Bodyweight; cm = Centimetres

Table 21.11 Rank comparison of mean peak vertical ground reaction forces during different running and impact activities

Exercise	Reference	Condition	Mean Peak VGRF (% BW)
69cm single-leg drop landing	Irmischer et al. (2004)	BW	580
50cm depth jump and vertical jump	McKay et al. (2005)	BW	540
50cm depth jump landing	McKay et al. (2005)	BW	470
40cm single-leg drop-landing and vertical hop	Ortiz et al. (2008)	BW	462
30cm single-leg drop landing	McNair & Marshall (1994)	BW	460
50cm single-leg drop-landing	Kernozek et al. (2008)	BW	384
Side-to-side jumps	McKay et al. (2005)	BW	380
45° sidestep V-cut	Dayakidis & Boudolos (2006)	BW	358
Jumping jacks	McKay et al. (2005)	BW	350
Forward running	Johnson et al. (2005)	BW	350
20cm box hop up-and-down	Ortiz et al. (2008)	BW	333
Squat jump	Jensen & Ebben (2007)	BW	305
Double-leg horizontal jump and lateral jump	Sell et al. (2006)	BW	292
Forward hop	Jensen & Ebben (2007)	BW	289
Forward running	Threlkeld et al. (1989)	BW	270
Forward jump landing	Yu et al. (2006)	BW	267
Forward hop	Webster et al. (2004)	BW	260
	Gauffin & Tropp (1992)	BW	250
Backward running	Threlkeld et al. (1989)	BW	250
Form skipping	Johnson et al. (2005)	BW	250
In-place vertical hops	McKay et al. (2005)	BW	210
In-place jumps	Nisell & Mizrahi (1988)	BW	137

Table © Copyright Nicholas Clark. Reproduced with permission.
VGRF = Vertical ground reaction force; BW = Bodyweight

on-the-spot jumps and hops such as squat jumps, tuck jumps and vertical hops.

Saggital plane drills introduce greater anteroposterior shear forces (Clark 2006). Examples of saggital plane drills are forward jumps and forward hops. Form skipping involves typical low-impact drills usually employed for practicing running mechanics and technique (e.g. skipping high knee lift), and is a means of reintroducing running-like movements with low impact forces (Johnson et al. 2005).

Frontal plane drills introduce larger mediolateral loads, varus-valgus forces, VGRF, and anterior tibial shear forces than saggital plane drills (Sell et al. 2006), and so also potentially increase tensile forces on the cruciate and collateral ligaments (Clark 2006). Examples of frontal plane drills are side-to-side jumps and hops. Once frontal plane drills have safely been completed, saggital plane running (jogging) can be undertaken over relatively short distances. Running in itself, even at low speeds, is a plyometric activity that induces joint shear forces (Table 21.10) and VGRF (Table 21.11). Consequently, it is clinically judicious to only introduce running after some other form of deceleration and plyometric training has been performed first in order to prepare the patient's tissues and sensorimotor control systems, since running can be relatively 'uncontrolled' compared to structured deceleration and plyometric drills (Clark 2006).

Oblique plane drills introduce greater torsion forces through the knee, although they are relatively low amplitude rotational loads compared to other plyometric drills (Clark 2006). Even so, when knee combined movements such as flexion and internal rotation and anterior tibial shear are added together, greater ACL tensile forces are generated than in any single movement in isolation (Markolf et al. 1995). Examples of oblique plane drills are zig-zag jumps and hops.

Transverse plane drills introduce relatively high amplitude rotational loads and so are potentially very dangerous for the knee ligaments and menisci (Clark 2006). Examples of transverse plane drills are 180° and 360° on-the-spot spinning jumps and hops. These are some of the most dangerous single-leg hopping drills since there is a major risk of knee and lower limb high velocity varus–valgus collapse if the patient is not extremely cautious and proficient in their performance.

Depth jumps and hops induce some of the largest joint shear forces and VGRF (Table 21.10 and 21.11), and so these are also high risk drills which the patient must perform with meticulous technique in order to minimise the risk of re-injury (Clark 2006). Examples of depth-hops are single-leg drop-landing drills from a 40cm high box.

Entry criteria

Entry criteria for clinical plyometrics can be the same as for deceleration training, along with the successful completion of three sets of 10–15 repetitions of the chosen plyometric drill as a deceleration drill first – in other words, the patient emphasises mastering the drill's landing phase alone *before* converting it to a plyometric drill.

Methods

The methods described for deceleration training underpin the safe and effective execution of clinical plyometrics. Clinical plyometrics differ from deceleration drills in that after the eccentric-isometric muscle action sequence, a concentric propulsive phase is added, and as such the patient is encouraged to take-off as fast as possible after the landing (Clark 2006). Verbal cues used to teach the patient a rapid take-off include "tough 'n' go", "quick feet", and "fast feet" (Chu 1998). Clinical plyometrics intensity, within-session work:rest ratio, weekly frequency, and between-session rest duration are identical to those for deceleration training. Saggital plane, frontal plane and oblique plane drills can also be implemented using cones or barriers for patients to jump or hop over.

Goals of treatment and objective progression criteria

A specific goal of treatment is three sets of 10 to 15 repetitions of single-leg drills on both legs with correct technique. It is important to recognise, of course, that it is not necessary for all patients to progress through clinical plyometrics to depth-jumps/drop-jumps (Figure 21.21). For example, a marathon runner might only need to progress to frontal plane drills to be confident that saggital plane-biased long-duration running can be safely undertaken, whereas a rugby player should definitely progress to transverse plane drills, and a basket ball player should certainly progress to drop-jumps. For multi-directional agility-biased games players, three sets of 10 to 15 repetitions of single-leg *transverse plane drills* on both legs is recommended before implementing sport-specific agility-biased running drills.

Agility-biased running drills

Basic concepts

Agility-biased running drills' (e.g. Figure 8's, T run) main purpose is simply to integrate single-leg clinical plyometric training into functional running tasks demanding the cyclical alternation between legs (Clark 2006). Running tasks obviously involve the use of both legs, and so between-limb compensation and cheating is easily accomplished. Therefore, the clinician can only be confident that approximately equal use of both legs is being made during multi-directional running tasks if the patient has successfully completed single-leg clinical plyometric drills first.

Entry criteria

Entry criteria for sport-specific agility-biased running drills are the same as for deceleration training and clinical plyometrics, with the addition of the successful completion of three sets of 10–15 repetitions of single-leg transverse plane drills on both legs.

Goals of treatment and objective progression -criteria

Ideal goals of treatment for this type of training are simply the successful execution of timed agility tests specific to the athlete's sport within the drills' normal timed standard. The clinician should research sport-specific coaching texts for the appropriate normal values. At this stage of rehabilitation, the patient is almost ready to be discharged. Prior to this, best clinical practice recommends the administration of the hop functional performance test (FPT) to gather objective outcome measurement data (Clark 2001; Fitzgerald et al. 2001).

Functional performance tests
Basic concepts

Outcome measurement in sports rehabilitation is directed at identifying an athlete's ability to tolerate the physical demands inherent in sport-specific activity and prevent re-injury on return-to-competition (Clark 2001). The FPT recreates the knee shear, compression and torsion forces encountered during sport-specific activity under controlled clinical conditions, its use becoming increasingly popular since traditional clinical outcome measures such as knee joint laxity and isokinetic quadriceps muscle strength demonstrate weak to moderate and often insignificant relationships with functional tasks such as running and jumping (Clark 2001; Fitzgerald et al. 2001). Functional performance tests include hop, leap, and jump tests and all may be administered to an athlete following knee ligament injury (Clark 2001). In fact, the administration of an FPT to a knee-injured patient is considered essential before deciding if discharge is appropriate, and so many clinical and research groups employ FPTs as objective outcome measures (Table 21.1).

The hop FPT simultaneously measures joint laxity/mobility, muscle extensibility, muscle strength and power, proprioception, neuromuscular control, dynamic balance, agility, pain, and athlete-confidence, and so represents a 'cumulative effect' since it is unable to identify impairments in single physical variables (Clark 2001). However, the hop FPT is still a useful measurement tool for the clinician because it is a quantitative measure utilised to *define* function or outcome, it simulates the forces encountered during sport-specific activity under controlled clinical conditions, indirectly assesses the extent to which pain inhibits the execution of functional tasks and the ability of previously injured tissues to safely absorb force, quantifies between-limb differences that may predispose re-injury and assesses progress within rehabilitation, provides psychological reassurance to the patient, and correlates with subjective assessment of knee function (Clark 2001; Fitzgerald et al. 2001). Therefore, hop FPTs generate valuable information for the clinician regarding a knee-injured athlete's physical and psychological status.

Entry criteria

When employing a hop FPT for clinical decision making with regard to potential patient discharge, a great deal of exercise rehabilitation will already have been completed. Since hop FPTs are essentially plyometric in nature (Clark 2001), in order to decide if it is first safe to administer an FPT to a knee-injured patient, FPT entry criteria are the same as for clinical plyometrics, and include the successful completion of single-leg clinical plyometric drills.

Methods

With regard to choice of FPT, there is good evidence that multi-directional hop FPTs such as the adapted crossover hop for distance (Figure 21.22) as presented by Clark et al. (2002) are more sensitive to detecting between-limb differences in knee FJS than uni-directional FPTs such as the single hop-for-distance (Clark 2001; Noyes et al. 1991). This is because multi-directional hop FPTs are considered to be most challenging for patients since they impose frontal and transverse plane forces on the knee in addition to the saggital plane-biased forces generated by most horizontal hop FPTs (Clark 2001, 2002).

Discharge criteria

A mean LSI $\geq$ 90% (Figure 21.18) is well established as the normal value for hop testing (Daniel et al. 1982; Ostenberg et al. 1998). Moreover, a mean LSI $\geq$ 90% is consistently demonstrated following multi-directional hop tests in asymptomatic ligament-injured knees where patients are regularly participating in high-level sports or occupational activities without functional limitation (Eastlack et al. 1999; Hopper et al. 2002). As such, assuming the uninjured limb is achieving normal values, an LSI $\geq$ 90% can be considered an ideal discharge criteria providing that all hop FPT entry criteria as described previously are also fulfilled.

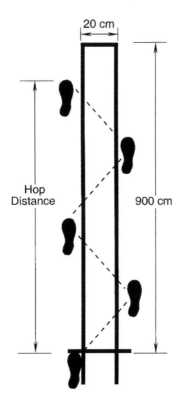

Figure 21.22 Adapted crossover hop for distance (Clark et al. 2002).

Anterior knee pain, differential diagnosis and treatment

Anterior knee pain

Anterior knee pain (AKP) is a common clinical entity in patients of all ages and activity levels. The category of conditions placed within the grouping AKP could be defined as involving pain, inflammation, muscle imbalance and/or instability of any component of the extensor mechanism of the knee. This disturbance of the extensor mechanism of the knee has been regarded as one of the commonest disorders of the knee affecting between 5–15% of all patients reporting for treatment (Devereaux and Lachmann 1984; Kannus et al. 1987; Milgrom et al. 1991). Once present it frequently becomes a chronic problem forcing the patient to stop sport and other activities.

Differential diagnosis

The classification of symptoms into AKP is confusing with AKP being present in many clinical conditions. The commonest clinical conditions which have symptoms of AKP are; patellofemoral pain syndrome (PFPS), patella tendonopathy, fat pad syndrome, traction apophysitis (Osgood Schlatters/Sinding Larsen Johansson disease), plica syndrome, iliotibial band friction syndrome (ITBS) and nerve entrapment. Taunton et al. (2002) in a retrospective review of a sports medicine clinic patient's found AKP to be 29.2% of all running injuries, a figure which is very similar to the 28% Clement et al. (1981) found two decades earlier. Of the AKP patients found in the study of Taunton et al. (2002), 56.5% had PFPS, 28.8% ITBS and 16.4% patella tendonopathy.

For treatment of AKP to be successful appropriate rehabilitation programmes need to be established. These can only be developed if accurate diagnosis of the underlying cause of the AKP is recognised. It is the purpose of this section to describe the common clinical conditions which present with AKP, how to ascertain their differential diagnosis and their own particular management. The list below shows the commonest clinical problems which can present as AKP. Table 21.12 shows the distinguishing differences on examination between the most common causes of AKP; PFPS, patella tendonopathy, ITBS and fat pad syndrome. As can be seen in Table 21.12 with careful examination it becomes relatively easy to distinguish between these different causes of AKP, this becomes important as we shall see in the treatment section where each of these conditions require very specific interventions.

Potential causes of anterior knee pain are:

- patellofemoral joint
- patella tendon (patella tendonopathy)
- iliotibial band (Iliotibial band friction syndrome
- fat pad (fat pad syndrome)
- plica (plica syndrome)
- traction apophysitis (Osgood Schlatters disease, Sinding Larsen Johanson disease)
- referred pain from lumbar spine, sacroiliac joint or hip joint

Table 21.12 Distinguishing features on examination between potential causes of anterior knee pain

Sign or symptom	PFPS	Patella tendonopathy	Fat pad syndrome	Iliotibial band friction syndrome
Aggravating Factor	Running, stairs, eccentric quads prolonged load	Jumping, landing, eccentric quads	Standing, prolonged load	Repetitive flexion/extension
Pain	Retro-patella, local or non-specific	Infra-patella, localised	Infra-patella, diffuse	Lateral patella tibial plateau
Tender	Peripatella	Inferior pole patella	Fat pad	Gerdy's tubercle or lateral femoral condyle
Giving way	Pseudo (quads inhibition)	None	None	None
Effusion	Occasional small,	Tendon thickened	Fat pad	Rarely
Clicks, clunk and crepitus	Older patients occasional	None	None	Occasional catch last 30°
ROM	Decreased flexion particularly squat	Decreased flexion particularly squat	Decreased extension	Decreased extension
Patella mobility	Decreased medial caudally	Normal	Decreased caudal and cephalic	Decreased medial
Quads	Decreased/inhibited	Decreased/inhibited	Normal	Normal

- local Nerve entrapment (lateral cutaneous nerve, infra-patella branch of saphenous nerve).

The list above highlights that alongside the major causes of AKP a number of others occur with some regularity and so are worth discussing briefly here. A common cause of AKP in adolescences is traction apophysitis (Osgood Schlatters/Sinding Larsen Johansson disease). These two conditions occur when excessive loading has been placed through the growth plates at the tibial tubercle in the case of Osgood Schlatters disease and the inferior pole of the patella (Sinding Larsen Johansson disease); this results in inflammation and pain. In both cases the patients have usually recently undergone or in the middle of a growth spurt. The pain is activity related, often showing a linear relationship; the greater the activity the greater the pain, pain is localised to the respective growth plates and they a very painful on palpation. Treatment here is simply to reduce the level of loading on these tissues to one which the patient can tolerate and then gradually re-introduce loading to the tissues over a period which may extend into a number of months. Secondary factors such as soft tissue length and muscle strength imbalances may also need addressing (see the treatment section).

The entrapment of two peripheral nerves has been reported as potential causes of AKP within the literature, these are the lateral cutaneous nerve and the infra-patella branch of the saphenous nerve. Problems with the lateral cutaneous nerve are often mistaken for ITBS as the pain is on the lateral side of the knee often following the course of the Iliotibial band. But this pain is most often superior to the lateral femoral condyle, even though it is irritated by similar actions as ITBS; the pain the patients describe is often shooting and burning in nature which allows further differentiation. The infra-patella branch of the saphenous nerve most often gives patients pain that shoots from medial (around the vastus medialis area) across the infra-patella area of the knee to the lateral side. Injury to this nerve is most often associated with trauma, or through surgery (ACL or arthroscopy) damaging or irritating the nerve. It has also been seen in ballet dancers, who in failing to achieve full turn out excessively externally rotate and abduct the tibia stressing the nerve.

AKP may also be referred from the lumbar spine, sacroiliac or hip joints. Whenever assessing a patient with AKP it is vital that these other joints are screened to see if they are involved. This is especially the case if the patient reports having pain in any of those areas or if the patient reports the pain "going up from the knee or coming down to the knee". Along with if the patient reports numbness or altered sensation in and around the knee anterior, medial or lateral thigh.

Causes of altered loading

In the above discussion we have seen that numerous structures could become injured and cause AKP. Regardless of the particular structure which becomes injured there is one feature common to all these injuries, the injury itself is caused by a overloading of the tissue concerned, which is either acute and usually traumatic in nature or chronic (long-term) low loads that eventually cause the tissue to break down – a "dripping tap effect". The categories of potential causes of tissue stress are:

- abnormal biomechanics
- shortened soft tissue
- muscle imbalances and strength deficits
- training/environmental.

Abnormal biomechanics

Understanding of the Q angle (Figure 21.23) and its effect on patellofemoral joint (PFJ) loading is crucial to the understanding of how abnormal biomechanics affects the PFJ. The Q or quadriceps angle represents the force vector (direction of pull) of the quadriceps during their contraction, if during contraction the quadriceps causes the patella to be drawn medially or laterally from its normal course this will potentially increase the stress and loading of the PFJ and the structures associated with it. Decreasing the Q angle by 10° significantly reduces load on the lateral structures of the PFJ (Elias et al. 2004). The Q angle can be affected by both soft tissue tightness and muscle weakness which will be discussed later; it can also be effect by mal-alignment within the lower limb such as anteriorly rotated pelvis or pronation of the foot. If the foot over pronates (the longitudinal arch of the foot flattens), it will cause the leg to in-

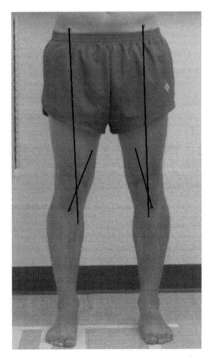

Figure 21.23 Q angle.

ternally rotate excessively, causing the knee to point inwards, thus changing the Q angle. Anterior pelvic rotation causes one leg to appear longer and the body must compensate for this. Typically one way it compensates is to overly flatten (pronates the foot of the longer leg) in an attempt to shorten it, thus changing Q angle.

Shortened soft tissues

A variety of shortened soft tissues can influence Q angle. At the hip shortened hip flexors (rectus femoris, iliopsoas and ITB) can cause the pelvis to be held in an anteriorly rotated position. If the adductor (groin) muscles are short; principally adductor longus, this will cause the femur to be held in an internal rotated and adducted position, increasing the Q angle. A short ITB through its attachment onto the tibia can cause the tibia to be held in an externally rotated position moving the tibial tubercle laterally so changing Q angle. If the gastrocnemius or soleus (triceps surae complex) are short this limits the ability to dorsi-flex at the ankle, in order to still allow full movement the foot will compensate for this lack of movement by pronating excessively.

Muscle imbalances and strength deficits

In the research into AKP considerable attention has been paid to achieving increased activity and strength in the vastus medialis oblique (VMO) muscle with the aim of drawing the patella medially against the pull of the laterally attached vastus lateralis. The problem is the majority of the literature has failed to find either problems with VMO in patients with AKP or a means of specifically training VMO in isolation (Herrington et al. 2006). What is consistent in the literature is that patients with AKP have weak quadriceps as a whole (Witvrouw et al. 2002) with a number of studies showing successful resolution of symptoms upon strengthening of the quadriceps (Witvrouw et al. 2002; Herrington and Al-Shehri 2007). A second group of muscles whose weakness is consistently reported within the literature to be associated with AKP are the gluteal muscles (gluteus maximus, medius and minimus) (Mascal et al. 2003). Weakness of these muscles causes the thigh to drop into a more adducted and internal rotated position during weight bearing, this increases the Q angle and so loading on the PFJ (Figure 21.24).

Training or environmental triggers

All of the above problems can be found in many members of the public and yet they do not have AKP, what these predisposing factors need is a trigger that effects the tissue in a negative way reducing its tolerance to loading. There are many potential triggers to this change in tissue load tolerance. One example would be direct trauma from a blow or surgery. Another example would the change in loading brought about by new training shoes or boots, or a change of training surface. A further example is a too rapid increase in loading following a period of de-training (decreased loading of the tissues, so loss of tolerance) caused by illness or even holiday.

Treatment

The aims of treatment are:

- Relief of pain
- Control of lower limb rotation

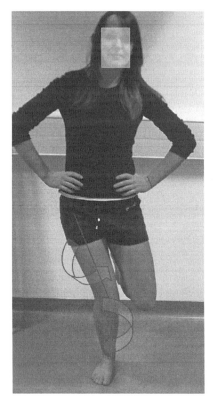

Figure 21.24 Poor rotation control during squatting and landing.

- Strengthening of quadriceps
- Stretching of shortened soft tissues

Relief of pain

Taping of the patella (Figures 21.25a and 21.25b) has consistently been shown to relieve pain (see Aminaka and Gribble (2005) for review). The mechanism by which taping brings about the relief of pain has often been questioned, but it would appear to bring about enough of a change in local tissue loading to alter tissue homeostasis (Dye 2005). Figure 21.25a shows taping of the patella medially, to change loading on the medial and lateral sides of the PFJ, Figure 21.25b shows an alternate method taping the patella to correct the lateral tilt. Figure 21.25c shows the taping used to relieve loading on the infra-patella fat pad. Relief of pain could also be achieved through the use of joint mobilisation. Figure 21.26 shows a

ANTERIOR KNEE PAIN, DIFFERENTIAL DIAGNOSIS AND TREATMENT

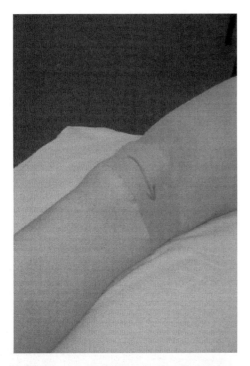

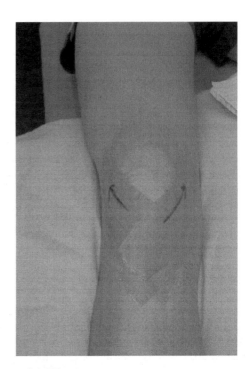

Figure 21.25 (*Continued*).

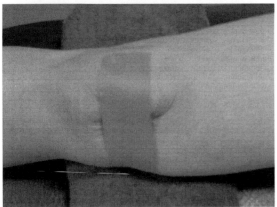

Figure 21.25 (a) Patella taping medially. (b) Patella taping to correct lateral tilt. (c) Patella taping to relieve loading on the fat pad.

number of examples of techniques that may prove useful in the treatment of AKP. These techniques are initially done without going into resistance; the techniques are then progressed into resistance.

Strengthening of quadriceps

Strengthening of the quadriceps has been shown in a number of papers to bring about improvements in AKP (Witvrouw et al. 2002; Herrington and Al-Shehri 2007); these authors have also demonstrated that the mode of strengthening does not appear to be important. Previously, both open kinetic chain exercises such as seated knee extension and closed kinetic chain exercises, such as leg press or squatting, have been advocated as being individually better for

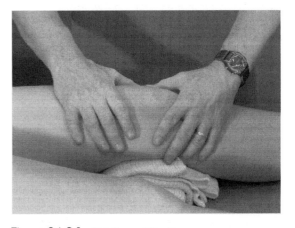

Figure 21.26 Patella mobilisation.

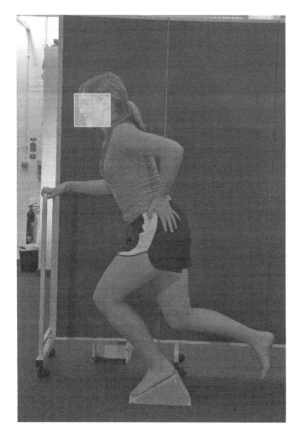

Figure 21.27 Incline squat.

Figure 21.28 Extension control squat.

rehabilitating AKP patients, this would appear not to be the case, with progressive loading during either form of exercise being more important than the mode of exercise itself (Herrington and Al-Shehri 2007). Both the studies of Witvrouw et al. (2002) and Herrington and Al-Shehri (2007) brought about significant improvement in function and relief of pain within 6–8 weeks of starting the strengthening programme, the significant feature of both these programmes was the constant reassessment and adjustment of load on the quadriceps, not necessarily the specifics of the exercise programme.

A number of studies (Jonasson and Alfredson 2005 being the most recent) have noted a specific and highly useful variation on the squat exercise for patients with patella tendonopathy, this is the incline squat (Figure 21.27). This exercise would initially be done bilaterally and then the patient would be progressed to unilateral squatting as pain permitted.

In a similar vein, because of the lack of control in the final degrees of extension in patients with fat pad syndrome, the squat exercise can be modified to improve control of extension as shown in Figure 21.28.

Stretching of shortened soft tissues

As mentioned in the section above a number of tissues may become shortened and alter the loading on the PFJ via changing the Q angle. Figures 21.29–21.32 show the techniques for stretching rectus femoris, ITB, adductor longus, gastrocnemius and soleus respectively. The key to these stretching exercises is that one end of the muscle (either the origin or the insertion) is fixed whilst the other end moves, this allows the stretch to be isolated to the respective muscle and prevents compensatory movements occurring at other joints.

ANTERIOR KNEE PAIN, DIFFERENTIAL DIAGNOSIS AND TREATMENT

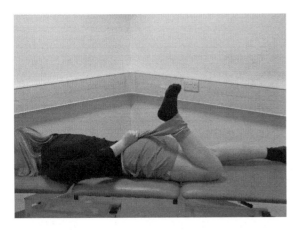

Figure 21.29 Technique for stretching rectus femoris.

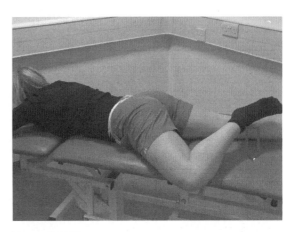

Figure 21.31 Technique for stretching adductor longus.

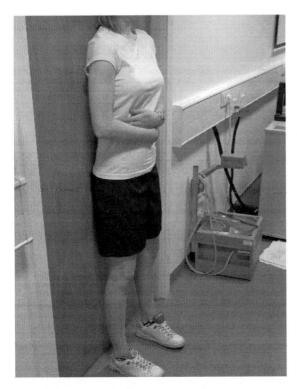

Figure 21.30 Technique for stretching ITB.

Figure 21.32 (a) Technique for stretching gastrocnemius. (b) Technique for stretching soleus.

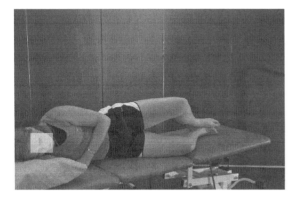

Figure 21.33 Clam exercise.

Control of lower limb rotation

Failure to control lower limb rotation is very much associated with increasing the Q angle (Figure 21.23), as mentioned above strengthening of the gluteal muscles could improve this pattern of movement by decreasing the adduction and internal rotation of the femur during loading. Figure 21.33 shows an exercise for strengthening the gluteal muscles and improving control of femur adduction and internal rotation. The exercise for facilitation of gluteal contraction and strengthening during the more functional movement of squatting is shown in Figure 21.34.

Figure 21.34 Gluteal contraction incorporated into a squat.

Summary

Knee joint exercise rehabilitation can be a complex and controversial task with regard to weight-bearing and ROM restrictions, tissue biomechanics, between-limb and within-limb compensations, evidence-based progression-criteria, the restoration of normal gait, and therapists' and surgeons' opinions regarding a specific injury type. This chapter has presented a comprehensive, meticulously clinically reasoned and evidence-based approach to the rehabilitation of different types of knee ligament and meniscus injury or surgery, including normative data for different testing procedures to be used as goals of treatment and objective progression-criteria. The use of such data frees the clinician from the uncertainty associated with tissue healing timeframes, and outdated personal opinions frequently expressed by colleagues, by accounting for the individual patient's ability and unique physiological status. It is the safest and most effective way to administer exercise rehabilitation for tibiofemoral and patellofemoral joint injury.

References

Aagarrd, P., Simonsen, E., Andersen, J., Magnusson, P. and Dyhre–Poulsen, P. (2002) Increased Rate of Force Development and Neural Drive of Human Skeletal Muscle Following Resistance Training. *Journal of Applied Physiology*, 93, 1318–1326.

Adam, F., Pape, D., Schiel, K., Steimer, O., Kohn, D. and Rupp, S. (2004) Biomechanical Properties of Patellar and Hamstring Graft Tibial Fixation Techniques in Anterior Cruciate Ligament Reconstruction. *American Journal of Sports Medicine*, 32, 71–78.

Agel, J., Arendt, L. and Breshadsky, B. (2005) Anterior Cruciate Ligament Injury in National Collegiate Athletic Association Basketball and Soccer. A 13–Year Review. *American Journal of Sports Medicine*, 33, 524–530.

Ahmad, C., Gardner, T., Groh, M., Arnouck, J. and Levine, W. (2004) Mechanical Properties of Soft Tissue Femoral Fixation Devices for Anterior Cruciate Ligament Reconstruction. *American Journal of Sports Medicine*, 32, 635–640.

Anderson, F. and Pandy, M. (2003) Individual Muscle Contributions to Support in Normal Walking. *Gait and Posture*, 17, 159–169.

Andriacchi, T. and Birac, D. (1993) Functional Testing in the Anterior Cruciate Ligament–Deficient Knee. *Clinical Orthopaedics and Related Research*, 288, 40–47.

Arendt, E. and Dick, R. (1995) Knee Injury Patterns Among Men and Women in Collegiate Basketball and Soccer. *American Journal of Sports Medicine*, 23, 694–701.

Arnason, A., Sigurdsson, S., Gudmundsson, A., Holme, I., Engebretsen, L. and Bahr, R. (2004) Risk factors for Injury in Football. *American Journal of Sports Medicine*, 32, Supplement, 5S–16S.

Augustsson, J., Thomee, R. and Karlsson, J. (2004) Ability of a New Hop Test to Determine Functional Deficits After Anterior Cruciate Ligament Reconstruction. *Knee Surgery Sports Trauamatology Arthroscopy*, 12, 350–356.

Baker, K., Xu, L., Zhang, Y., Nevitt, M., Niu, J., Aliabadi, P., Yu, W. and Felson, D. (2004) Quadriceps Weakness and its Relationship to Tibiofemoral and Patellofemoral Osteoarthritis in Chinese. *Arthritis and Rheumatism*, 50, 1815–1821.

Baltzopoulos, V. (1995) Muscular and Tibiofemoral Joint Forces During Isokinetic Concentric Knee Extension. *Clinical Biomechanics*, 10, 208–214.

Bandy, W. (1992) Functional Rehabilitation of the Athlete. *Orthopaedic Physical Therapy Clinics of North America*, 1, 269–281.

Barber, S., Noyes, F., Mangine, R. and DeMaio, M. (1992) Rehabilitation After ACL Reconstruction: Function Testing. *Orthopedics*, 15, 969–974.

Barber–Westin, S., Noyes, F., Heckman, T. and Shaffer, B. (1999) The Effect of Exercise and Rehabilitation on Anterior–Posterior Knee Displacements After Anterior Cruciate Ligament Autograft Reconstruction. *American Journal of Sports Medicine*, 27, 84–93.

Berchuk, M., Andriacchi, T., Bach, B. and Reider, B. (1990) Gait Adaptations by Patients Who Have a Deficient Anterior Cruciate Ligament. *Journal of Bone and Joint Surgery*, 72A, 871–877.

Beynnon, B., Uh, B., Johnson, R., Abate, J., Nichols, C., Fleming, B., Poole, A. and Roos, H. (2005) Rehabilitation After Anterior Cruciate Ligament Reconstruction. A Prospective, Randomized, Double–Blind Comparison of Programs Administered Over 2 Different Time Intervals. *American Journal of Sports Medicine*, 33, 347–359.

Blackburn, J. and Morrissey, M. (1998) The Relationship Between Open and Closed Kinetic Chain Strength of the Lower Limb and Jumping Performance. *Journal of Orthopaedic and Sports Physical Therapy*, 27, 430–435.

Bobbert, M. and Van Zandwijk, J. (1999) Dynamics of Force and Muscle Stimulation in Human Vertical Jumping. *Medicine and Science in Sports and Exercise*, 31, 303–310.

Boden, B., Dean, G., Feagin, J. and Garrett, W. (2000) Mechanism of Anterior Cruciate Ligament Injury. *Orthopedics*, 23, 573–578.

Borsa, P., Lephart, S. and Irrgang, J. (1998) Comparison of Performance–Based and Patient–Reported Measures of Function in Anterior Cruciate Ligament Deficient Individuals. *Journal of Orthopaedic and Sports Physical Therapy*, 28, 392–399.

Borsa, P., Lephart, S., Irrgang, J., Safran, M. and Fu, F. (1997) The Effects of Joint Position and Direction of Joint Motion on Proprioceptive Sensibility in Anterior Cruciate Ligament Deficient Athletes. *American Journal of Sports Medicine*, 25, 336–340.

Brand, J., Pienkowski, D., Steenlage, E., Hamilton, D., Johnson, D. and Caborn, D. (2000) Interference Screw Fixation Strength of a Quadrupled Hamstring Tendon Graft is Directly Related to Bone Mineral Density and Insertion Torque. *American Journal of Sports Medicine*, 28, 705–710.

Brindle, T., Nyland, J. and Johnson, D. (2001) The Meniscus: Review of the Basic Principles With Application to Surgery and Rehabilitation. *Journal of Athletic Training*, 36, 160–169.

Bruhn, S., Kullmann, N. and Gollhofer, A. (2004) The Effects of Sensorimotor Training and Strength Training on Postural Stabilisation, Maximum Isometric Contraction and Jump Performance. *International Journal of Sports Medicine*, 25, 56–60.

Buseck, M. and Noyes, F. (1991) Arthroscopic Evaluation of Meniscal Repair After Anterior Cruciate Ligament Reconstruction and Immediate Motion. *American Journal of Sports Medicine*, 19, 489–494.

Bush–Joseph, C., Hurwitz, D., Patel, R., Bahrani, Y., Garretson, R., Bach, B. and Andriacchi, T. (2001) Dynamic Function After Anterior Cruciate Ligament Reconstruction With Autologous Patellar Tendon. *American Journal of Sports Medicine*, 29, 36–41.

Bynum, E., Barrack, R. and Alexander, A. (1995) Open Versus Closed Chain Kinetic Exercises After Anterior Cruciate Ligament Reconstruction. A Prospective Randomized Study. *American Journal of Sports Medicine*, 23, 401–406.

Buckwalter, J. (2003) Sports, Joint Injury, and Posttraumatic Osteoarthritis. *Journal of Orthopaedic and Sports Physical Therapy*, 33, 578–588.

Camillieri, G., McFarland, E., Jasper, L., Belkoff, S., Kim, T., Rauh, P. and Mariani, P. (2004) A Biomechanical Evaluation of Transcondylar Femoral Fixation of Anterior Cruciate Ligament Grafts. *American Journal of Sports Medicine*, 32, 950–955.

Campbell, R., Torrie, A., Hecker, A. and Sekiya, J. (2007) Comparison of Tibial Graft Fixation Between

Simulated Arthroscopic and Open Inlay Techniques for Posterior Cruciate Ligament Reconstruction. *American Journal of Sports Medicine*, 35, 1731–1738.

Chappell, J., Yu, B., Kirkendall, T. and Garrett, W. (2002) A Comparison of Knee Kinetics Between Male and Female Recreational Athletes in Stop–Jump Tasks. *American Journal of Sports Medicine*, 30, 261–267.

Chen, C., Chen, W., Shih, C. and Chou, S. (2004) Arthroscopic Posterior Cruciate Ligament Reconstruction With Quadriceps tendon Autograft: Minimal 3 Years Follow–Up. *American Journal of Sports Medicine*, 32, 361–368.

Chimera, N., Swanik, K., Swanik, C. and Straub, S. (2004) Effects of Plyometric Training on Muscle–Activation Strategies and Performance in Female Athletes. *Journal of Athletic Training*, 39, 24–31.

Chmielewski, T., Wilk, K. and Snyder–Mackler, L. (2002) Changes in Weight–Bearing Following Injury or Surgical Reconstruction of the ACL: Relationship to Quadriceps Strength and Function. *Gait and Posture 16*, 87–95.

Chu, D. (1998) *Jumping Into Plyometrics*, 2nd edn. Champaign, IL: Human Kinetics.

Clark, N. (2001) Functional Performance Testing Following Knee Ligament Injury. *Physical Therapy in Sport*, 2, 91–105.

Clark, N. (2002) Functional Rehabilitation of the Lower Limb. Concepts and Clinical Application. *SportEx Medicine*, 18, 16–21.

Clark, N. (2004) Principles of Injury Rehabilitation. *SportEx Medicine*, 19, 6–10.

Clark, N. (2006) *Clinical Plyometrics in Knee Injury Rehabilitation: Basic Science and Practical Applications*. Keynote Presentation. British Association of Sports Rehabilitators and Trainers Annual Clinical Symposium. Birmingham.

Clark, N. (2008) *Proprioception and Neuromuscular Control in Exercise Rehabilitation for the Lower Limb*. 3rd edn. London: Integrated Physiotherapy and Conditioning Ltd.

Clark, N., Gumbrell, C., Rana, S. and Traole, C. (1999) *The Correlation of Short-Term Clinical Measures After Anterior Cruciate Ligament Reconstruction to Long-Term Outcome* [dissertation]. London, UK: University of East London.

Clark, N., Gumbrell, C., Rana, S., Traole, C. and Morrissey, M. (2002) Intratester Reliability and Measurement Error of the Adapted Crossover Hop for Distance. *Physical Therapy in Sport*, 3, 143–151

Clark, N., Gumbrell, C., Rana, S., Traole, C. and Morrissey, M. (2001) The Relationship Between Vertical Hop Performance and Isotonic Open and Closed Kinetic Chain Muscle Strength of the Lower Limb. *Journal of Sports Sciences*, 19, 18–19.

Clark, N. and Rees, M. (2008) Unpublished Clinical Data. British Army.

Cooper, D., Deng, X., Burstein, A. and Warren, R. (1993) The Strength of the Central Third Patellar Tendon Graft. A Biomechanical Study. *American Journal of Sports Medicine*, 21, 818–824.

Cooper, C., Snow, S., McAlindon, T., Kellingray, S., Stuart, B., Coggon, D. and Dieppe, P. (2000) Risk Factors for the Incidence and Progression of Radiographic Knee Osteoarthritis. *Arthritis and Rheumatism*, 43, 995–1000.

Cooper, D. and Stewart, D. (2004) Posterior Cruciate Ligament Reconstruction Using Single–Bundle Patella Tendon Graft With Tibial Inlay Fixation. *American Journal of Sports Medicine*, 32, 346–360.

Covey, D. and Sapega, A. (1993) Injuries of the Posterior Cruciate Ligament. *Journal of Bone and Joint Surgery*, 75A, 1376–1386.

Cowling, E., Steele, J. and McNair, P. (2003) Effect of Verbal Instructions on Muscle Activity and Risk of Injury to the Anterior Cruciate Ligament During Landing. *British Journal of Sports Medicine*, 37, 126–130.

Cross, M. and Powell, J. (1984) Long–Term Follow–Up of a Posterior Cruciate Ligament Rupture: A Study of 116 Cases. *American Journal of Sports Medicine*, 12, 292–299.

Dallalana, R., Brooks, J., Kemp, S. and Williams, A. (2007) The Epidemiology of Knee Injuries in English Professional Rugby Union. *American Journal of Sports Medicine*, 35, 818–830.

Daniel, D., Malcom, L., Stone, M., Perth, H., Morgan, J. and Riehl, B. (1982) Quantification of Knee Stability and Function. *Contemporary Orthopaedics*, 5, 83–90.

Daniel, D., Stone M., Dobson, B., Fithian, D. Rossman, D. and Kaufman, K. (1994) Fate of the ACL–Injured Patient: A Prospective Outcome Study. *American Journal of Sports Medicine*, 22, 632–644.

Dayakidis, M. and Boudolos, K. (2006) Ground Reaction Force Data in Functional Ankle Instability During Two Cutting Movements. *Clinical Biomechanics*, 21, 405–411.

De Carlo, M. and Sell, K. (1997) Normative Data for Range of Motion and Single–Leg Hop in High School Athletes. *Journal of Sport Rehabilitation*, 6, 246–255.

Decker, M., Torry, M., Noonan, T., Riviere, A. and Sterett, W. (2002) Landing Adaptations After ACL Reconstruction. *Medicine and Science in Sports and Exercise*, 34, 1408–1413.

de Loes, M., Dahlstedt, L. and Thomee, R. (2000) A 7–Year Study on Risks and Costs of Knee Injuries in

Male and Female Youth Participants in 12 Sports. *Scandinavian Journal of Medicine and Science in Sports*, 10, 90–97.

Delp, S., Hess, W., Hungerford, D. and Jones, L. (1999) Variation of Rotation Moment Arms With Hip Flexion. *Journal of Biomechanics*, 32, 493–501.

DeMaio, M., Mangine, R., Noyes, F. and Barber, S. (1992) Advanced Muscle Training After ACL Reconstruction: Weeks 6 to 52. *Orthopedics*, 15, 757–767.

Devita, P. and Skelly, W. (1992) Effect of Landing Stiffness on Joint Kinetics and Energetics in the Lower Extremity. *Medicine and Science in Sports and Exercise*, 24, 108–115.

Diamantopolous, A., Lorbach, O. and Paessler, H. (2008) Anterior Cruciate Ligament Revision Reconstruction: Results in 107 Patients. *American Journal of Sports Medicine*, 36, 1896–1902.

Distefano, L., Blackburn, T., Marshall, S. and Padua, D. (2009) Gluteal Muscle Activation During Common Therapeutic Exercises. *Journal of Orthopaedic and Sports Physical Therapy*, 39, 532–540.

Docherty, C., Moore, J. and Arnold, B. (1998) Effects of Strength Training on Strength Development and Joint Position Sense in Functionally Unstable Ankles. *Journal of Athletic Training*, 33, 310–314.

Drouin, J., Houglum, P., Perrin, D., Gansneder, M. (2003) Weight–Bearing and Non–Weight–Bearing Knee–Joint Reposition Sense and Functional Performance. *Journal of Sport Rehabilitation*, 12, 54–66.

Eastlack, M., Axe, M. and Snyder–Mackler, L. (1999) Laxity, Instability, and Functional Outcome After ACL Injury: Copers Versus Noncopers. *Medicine and Science in Sports and Exercise*, 31, 210–215.

Ekstrom, R., Donatelli, R. and Carp, K. (2007) Electromyographic Analysis of Core Trunk, Hip, and Thigh Muscles During 9 Rehabilitation Exercises. *Journal of Orthopaedic and Sports Physical Therapy*, 37, 754–762.

Enoka, R. (2002) *Neuromechanics of Human Movement*, 3rd edn. Champaign, IL: Human Kinetics.

Escamilla, R., Fleisig, G., Zheng, N., Barrentine, S., Wilk, K. and Andrews, J. (1998) Biomechanics of the Knee During Closed Kinetic Chain and Open Kinetic Chain Exercises. *Medicine and Science in Sports and Exercise*, 30, 556–569.

Escamilla, R., Fleisig, G., Zheng, N., Lander, J., Barrentine, S., Andrews, J., Bergemann, B. and Moorman, C. (2001) Effects of Technique Variations on Knee Biomechanics During the Squat and Leg Press. *Medicine and Science in Sports and Exercise*, 33, 1552–1566.

Escamilla, R., Zheng, N., Imamura, R., Macleod, T., Edwards, W., Hreljac, A., Fleisig, G., Wilk, K., Moorman, C. and Andrews, J. (2009) Cruciate Ligament Force During the Wall Squat and the One–Leg Squat. *Medicine and Science in Sports and Exercise*, 41, 408–417.

Fanelli, G. (2000) Treatment of Combined Anterior Cruciate Ligament–Posterior Cruciate Ligament–Lateral Side Injuries of the Knee. *Clinics in Sports Medicine*, 19, 493–502.

Fatouros, I., Jamurtas, A., Leontsini, D., Taxildaris, K., Aggelousis, N., Kostopoulos, N. and Buckenmeyer, P. (2000) Evaluation of Plyometric Exercise Training, Weight Training, and Their Combination on Vertical Jumping Performance and Leg Strength. *Journal of Strength and Conditioning Research*, 14, 470–476.

Faude, O., Junge, A., Kinderman, W. and Dvorak, J. (2005) Injuries in Female Soccer Players. A Prospective Study in the German National League. *American Journal of Sports Medicine*, 33, 1694–1700.

Feller, J. and Webster, K. (2003) A Randomized Comparison of Patellar Tendon and Hamstring Tendon Anterior Cruciate Ligament Reconstruction. *American Journal of Sports Medicine*, 31, 564–573.

Fitzgerald, G. (1997) Open Versus Closed Kinetic Chain Exercise: Issues in Rehabilitation After Anterior Cruciate Ligament Reconstructive Surgery. *Physical Therapy*, 77, 1747–1754.

Fitzgerald, G., Lephart, S., Hwang, J. and Wainner, R. (2001) Hop Tests as Predictors of Dynamic Knee Stability. *Journal of Orthopaedic and Sports Physical Therapy*, 31, 588–597.

Fitzgerald, G., Piva, S. and Irrgang, J. (2003) A Modified Neuromuscular Electrical Stimulation Protocol for Quadriceps Strength Training Following Anterior Cruciate Ligament Reconstruction. *Journal of Orthopaedic and Sports Physical Therapy*, 33, 492–501.

Fleck, S. and Kraemer, W. (1997) *Designing Resistance training Programs*, 2nd edn. Champaign, IL: Human Kinetics.

Fontbote, C., Sell, T., Laudner, K., Haemmerle, M., Allen, C., Margheritini, F., Lephart, S. and Harner, C. (2005) Neuromuscular and Biomechanical Adaptations of Patients With Isolated Deficiency of the Posterior Cruciate Ligament. *American Journal of Sports Medicine*, 33, 982–989.

Fowler, P. and Messieh, S. (1987) Isolated Posterior Cruciate Ligament Injuries in Athletes. *American Journal of Sports Medicine*, 15, 553–557.

Fredericson, M., Cookingham, C., Chaudhari, A., Dowdell, B., Oestreicher, N. and Sahrmann, S. (2000) Hip Abductor Weakness in Distance Runners With

Iliotibial Band Syndrome. *Clinical Journal of Sport Medicine*, 10, 169–175.

Friemert, B., Bach, C., Schwarz, W., Gerngross, H. and Schmidt, R. (2006) Benefits of Active Motion for Joint Position Sense. *Knee Surgery, Sports Traumatology, Arthroscopy*, 14, 564–570.

Fukuda, Y., Woo, S., Loh, J., Tsuda, E., Tang, P., McMahon, P. and Debski, R. (2003) A Quantitative Analysis of Valgus Torque on the ACL: A Human Cadaveric Study. *Journal of Orthopaedic Research*, 21, 1107–1112.

Gastel, J., Bergfeld, J., Calabrese, G. and Gray, R. (1999) Surgical Management for the Athlete With a PCL–Deficient Knee. *Journal of Sport Rehabilitation*, 8, 289–303.

Gauffin, H. and Tropp, H. (1992) Altered Movement and Muscular Activation Patterns During the One–Legged Jump in Patients With an Old Anterior Cruciate Ligament Rupture. *American Journal of Sports Medicine*, 20, 182–192.

George, M., Dunn, W. and Spindler, K. (2006) Current Concepts Review: Revision Anterior Cruciate Ligament Reconstruction. *American Journal of Sports Medicine*, 34, 2026–2037.

Gelber, A., Hochberg, M., Mead, L., Wang, N., Wigley, F. and Klag, M. (2000) Joint Injury in Young Adults and Risk for Subsequent Knee and Hip Arthritis. *Annals of Internal Medicine*, 133, 321–328.

Gerich, T., Cassim, A. and Latterman, C. (1997) Pullout Strength of Tibial Graft Fixation in Anterior Cruciate Ligament Replacement With a Patellar Tendon Graft: Interference Screw Versus Staple Fixation in Human Knees. *Knee Surgery Sports Traumatology Arthroscopy*, 5, 84–89.

Ghez, C. (1991a) The Control of Movement. In E. Kandel, J. Schwartz and T. Jessell (eds), *Principles of Neural Science*, 3rd edn. London: Prentice–Hall International, pp 533–547.

Ghez, C. (1991b) The Cerebellum. In E. Kandel, J. Schwartz and T. Jessell (eds), *Principles of Neural Science*, 3rd edn. London: Prentice–Hall International, pp 626–646.

Gordon, J. and Ghez, C. (1991) Muscle Receptors and Spinal Reflexes: The Stretch Reflex. In E. Kandel, J. Schwartz and T. Jessell (eds), *Principles of Neural Science*, 3rd edn. London: Prentice–Hall International, pp 564–580.

Grassmayr, M., Parker, D., Coolican, M. and Vanwanseele, B. (2008) Posterior Cruciate Ligament Deficiency: Biomechanical and Biological Consequences and the Outcomes of Treatment. A Systematic Review. *Journal of Science and Medicine in Sport*, 11, 433–443.

Greenberger, H. and Paterno, M. (1995) Relationship of Knee Extensor Strength and Hopping Test Performance in the Assessment of Lower Extremity Function. *Journal of Orthopaedic and Sports Physical Therapy*, 202–206.

Gruber, M. and Gollhofer, A. (2004) Impact of Sensorimotor Training on the Rate of Force Development and Neural Activation. *European Journal of Applied Physiology*, 92, 98–105.

Gupta, A., Letterman, C., Busam, M., Riff, A., Bach, B. and Wang, V. (2009) Biomechanical Evaluation of Bioabsorbable Versus Metallic Screws for Posterior Cruciate Ligament Inlay Graft Fixation. *American Journal of Sports Medicine*, 37, 748–753.

Hakkinen, K., Kallinen, M., Izquierdo, M., Jokelainen, K., Lassila, H., Malkia, E., Kraemer, W., Newton, R. and Alen, M. (1998) Changes in Agonist–Antagonist EMG, Muscle CSA, and Force During Strength Training in Middle–Aged and Older People. *Journal of Applied Physiology*, 84, 1341–1349.

Hakkinen, K. and Komi, P. (1983) Changes in Neuromuscular Performance in Voluntary and Reflex Contraction During Strength Training in Man. *International Journal of Sports Medicine*, 4, 282–288.

Hakkinen, K. and Komi, P. (1986) Training Induced Changes in Neuromuscular Performance Under Voluntary and Reflex Conditions. *European Journal of Applied Physiology*, 55, 147–155.

Halinen, J., Lindahl, J., Hirvensalo, E. and Santavirta, S. (2006) Operative and Nonoperative Treatments of Medial Collateral Ligament Rupture With Early Anterior Cruciate Ligament Reconstruction: A Prospective Randomized Study. *American Journal of Sports Medicine*, 34, 1134–1140.

Hamner, D., Brown, C., Steiner, M., Hecker, A. and Hayes, W. (1999) Hamstring Tendon Grafts for Reconstruction of the Anterior Cruciate Ligament: Biomechanical Evaluation of the Use of Multiple Strands and Tensioning Techniques. *Journal of Bone and Joint Surgery*, 81A, 549–557.

Harner, C. and Hoher, J. (1998) Evaluation and Treatment of Posterior Cruciate Ligament Injuries. *American Journal of Sports Medicine*, 26, 471–482.

Harner, C., Fu, F., Irrgang, J. and Vogrin, T. (2001) Anterior and Posterior Cruciate Ligament Reconstruction in the New Millennium: A Global Perspective. *Knee Surgery, Sports Traumatology, Arthroscopy*, 9, 330–336.

Harrington, I. (1976) A Bioengineering Analysis of Force Actions at the Knee in Normal and Pathological Gait. *Biomedical Engineering*, 11, 167–172.

Hart, J., Garrison, J., Kerrigan, D., Boxer, J. and Ingersoll, C. (2004) Gender Differences in Gluteus Medius Muscle Activity Exist in Soccer Players Performing a Forward Jump. *Journal of Athletic Training*, *39*, Supplement, S35.

Hart, J., Garrison, J., Kerrigan, D., Palmieri–Smith, R. and Ingersoll, C. (2007) Gender Differences in Gluteus Medius Muscle Activity Exist in Soccer Players Performing a Forward Jump. *Research in Sports Medicine*, *15*, 147–155.

Hattin, H., Pierrynowski, M. and Ball, K. (1989) Effect of Load, Cadence, and Fatigue on Tibiofemoral Joint Force During the Half–Squat. *Medicine and Science in Sports and Exercise*, *21*, 613–618.

Herman, D., Weinhold, P., Guskiewicz, K., Garrett, W., Yu, B. and Padua, D. (2008) The Effects of Strength Training on Lower Extremity Biomechanics of Female Recreational Athletes During a Stop–Jump Task. *American Journal of Sports Medicine*, *36*, 733–740.

Herrington, L. and Al–Sherhi, A. (2007) A Controlled Trial of Weight–Bearing Versus Non–Weight–Bearing Exercises for Patellofemoral Pain. *Journal of Orthopaedic and Sports Physical Therapy*, *37*, 155–160.

Herrington, L. and Fowler, E. (2006) A Systematic Literature Review to Investigate If We Can Identify Those Patients Who Can Cope With Anterior Cruciate Ligament Deficiency. *The Knee*, *13*, 260–265.

Hewett, T., Lindenfeld, T., Riccobene, J. and Noyes, F. (1999) The Effect of Neuromuscular Training on the Incidence of Knee Injury in Female Athletes. *American Journal of Sports Medicine*, *27*, 699–706.

Hewett, T., Myer, G., Ford, K., Heidt, R., Colosimo, A., McLean, S., van den Bogert, A., Paterno, M. and Succop, P. (2005) Biomechanical Measures of Neuromuscular Control and Valgus Loading of the Knee Predict Anterior Cruciate Ligament Risk in Female Athletes: A Prospective Study. *American Journal of Sports Medicine*, *33*, 492–501.

Hewett, T., Stroupe, A., Nance, T. and Noyes, F. (1996) Plyometric Training in Female Athletes. Decreased Impact Forces and Increased Hamstring Torques. *American Journal of Sports Medicine*, *24*, 765–773.

Hiemstra, L., Webber, S., MacDonald, P. and Kriellaars, D. (2000) Knee Strength Deficits After Hamstring Tendon and Patellar Tendon Anterior Cruciate Ligament Reconstruction. *Medicine and Science in Sports and Exercise*, *32*, 1472–1479.

Hoffman, J., Chapnik, L. and Shamis, A. (1999) The Effect of Leg Strength on the Incidence of Lower Extremity Overuse Injuries During Military Training. *Military Medicine*, *164*, 153–156.

Holder–Powell, H. and Rutherford, O. (1999) Unilateral Lower Limb Injury: Its Long–Term Effects on Quadriceps, Hamstring, and Plantarflexor Muscle Strength. *Archives of Physical Medicine and Rehabilitation*, *80*, 717–720.

Holder–Powell, H., Di Matteo, G. and Rutherford, O. (2001a) Do Knee Injuries Have Long–Term Consequences for Isometric and Dynamic Muscle Strength? *European Journal of Applied Physiology*, *85*, 310–316.

Holder–Powell, H., Rutherford, O. and Bartlett, G. (2001b) The Long–Term Effects of Unilateral Lower Limb Musculoskeletal Injury on Bone Mineral Density and Isometric Quadriceps Strength. *Physiotherapy*, *87*, 451–457.

Honl, M., Carrero, V., Hille, E., Schneider, E. and Morlock, M. (2002) Bone–Patellar Tendon–Bone Grafts for Anterior Cruciate Ligament Reconstruction. *American Journal of Sports Medicine*, *30*, 549–557.

Hooper, D., Morrissey, M., Drechsler, W., Clark, N., Coutts, F. and McAuliffe, T. (2002) Gait Analysis 6 and 12 Months After Anterior Cruciate Ligament Reconstruction Surgery. *Clinical Orthopaedics and Related Research*, *403*, 168–178.

Hopper, D., Goh, S., Wentworth, L., Chan, D., Chau, J., Wootton, G., Strauss, G. and Boyle, J. (2002) Test–Retest Reliability of Knee Rating Scales and Functional Hop Tests One Year Following Anterior Cruciate Ligament Reconstruction. *Physical Therapy in Sport*, *3*, 10–18.

Howell, S. and Taylor, M. (1996) Brace–Free Rehabilitation, With Early Return to Activity, for Knees Reconstructed With a Double–Looped Semitendinosus and Gracilis Graft. *Journal of Bone and Joint Surgery*, *78–A*, 814–825.

Hurd, W., Chmielewski, T. and Snyder–Mackler, L. (2006) Perturbation–Enhanced Neuromuscular Training Alters Muscle Activity in Female Athletes. *Knee Surgery Sports Traumatology Arthroscopy*, *14*, 60–69.

Ihara, H. and Nakayama, A. (1986) Dynamic Joint Control Training for Knee Ligament Injuries. *American Journal of Sports Medicine*, *14*, 309–315.

Indelicato, P. (1995) Isolated Medial Collateral Ligament Injuries in the Knee. *Journal of the American Academy of Orthopaedic Surgeons*, *3*, 9–14.

Ingram, J., Fields, S., Yard, E. and Comstock, R. (2008) Epidemiology of Knee Injuries Among Boys and Girls in US High School Athletics. *American Journal of Sports Medicine*, *36*, 1116–1122.

Irmischer, B., Harris, C., Pfeiffer, R., DeBeliso, M., Adams, K. and Shea, K. (2004) Effects of a Knee Ligament Injury Prevention Exercise Program on Impact

Forces in Women. *Journal of Strength and Conditioning Research*, 18, 703–707.

Isaac, D., Beard, D., Price A., Rees, J., Murray, D. and Dodd, C. (2005) In–Vivo Saggital Plane Knee Kinematics: ACL Intact, Deficient, and Reconstructed Knees. *The Knee*, 12, 25–31.

Izquierdo, M., Ibanez, J., Gorostiaga, E., Garrues, M., Zuniga, A., Anton, A., Larrion, J. and Hakkinen, K. (1999) Maximal Strength and Power Characteristics in Isometric and Dynamic Actions of the Upper and Lower Extremities in Middle–Aged and Older Men. *Acta Physiologica Scandinavica*, 167, 57–68.

Jackson, W., van der Tempei, Salmon, L., Williams, H. and Pinczewski, L. (2008) Endoscopically–Assisted Single–Bundle Posterior Cruciate Ligament Reconstruction. Results at Minimum Ten–Year Follow–Up. *Journal of Bone and Joint Surgery*, 90B, 1328–1333.

Jacobs, C. and Mattacola, C. (2005) Sex Differences in Eccentric Hip Abductor Strength and Knee Joint Kinematics When Landing from a Jump. *Journal of Sport Rehabilitation*, 14, 346–355.

Jacobs, C., Uhl, T., Mattacola, C., Shapiro, R. and Rayens, W. (2007) Hip Abductor Function and Lower Extremity Landing Kinematics: Sex Differences. *Journal of Athletic Training*, 42, 76–83.

Jensen, R. and Ebben, W. (2007) Quantifying Plyometric Intensity Via Rate Of Force Development, Knee Joint, and Ground Reaction Forces. *Journal of Strength and Conditioning Research*, 21, 763–767.

Jerosch, J., Prymka, M. and Castro, W. (1996) Proprioception of Knee Joints With a Lesion of the Medial Meniscus. *Acta Orthopaedica Belgica*, 62, 41–45.

Johnson, S., Golden, G., Mercer, J., Mangus, B. and Hoffman, M. (2005) Ground Reaction Forces During Form Skipping and Running. *Journal of Sport Rehabilitation*, 14, 338–345.

Jones, K., Bishop, P., Hunter, G. and Fleisig, G. (2001) The Effects of Varying Resistance Training Loads on Intermediate- and High-Velocity-Specific Adaptations. *Journal of Strength and Conditioning Research*, 15, 349–356.

Jonkers, I., Stewart, C. and Spaepen, A. (2003) The Study of Muscle Action During Single Support and Swing Phase of Gait: Clinical Relevance of Forward Simulation Techniques. *Gait and Posture*, 17, 97–105.

Juris, P., Phillips, E., Dalpe, C., Edwards, C., Gotlin, R. and Kane., D. (1997) A Dynamic Test of Lower Extremity Function Following Anterior Cruciate Ligament Reconstruction and Rehabilitation. *Journal of Orthopaedic and Sports Physical Therapy*, 26, 184–191.

Kaneko, F., Onari, K., Kawaguchi, K., Tsukisaka, K. and Roy, S. (2002) Electromechanical Delay After ACL Reconstruction: An Innovative Method for Investigating Central and Peripheral Contributions. *Journal of Orthopaedic and Sports Physical Therapy*, 32, 158–165.

Kaufman, K., An, K., Litchy, W., Morrey, B. and Chao, E. (1991) Dynamic Joint Forces During Knee Isokinetic Exercise. *American Journal of Sports Medicine*, 19, 305–316.

Keays, S., Bullock–Saxton, J., Keays, A., Newcombe, P. and Bullock, M. (2007) A 6–Year Follow–Up of the Effect of Graft Site on Strength, Stability, Range of Motion, Function, and Joint Degeneration After Anterior Cruciate Ligament Reconstruction. *American Journal of Sports Medicine*, 35, 729–739.

Keays, S., Bullock–Saxton, J., Newcombe P. and Keays A. (2003) The Relationship Between Knee Strength and Functional Stability Before and After Anterior Cruciate Ligament Reconstruction. *Journal of Orthopaedic Research*, 21, 231–237.

Kernozek, T., Torry, M. and Iwasaki, M. (2008) Gender Differences in Lower Extremity Landing Mechanics Caused by Neuromuscular Fatigue. *American Journal of Sports Medicine*, 36, 554–565.

Kibler, W. (2006) The Role of Core Stability in Athletic Function. *Sports Medicine*, 36, 189–198.

Kim, D., Wilson, D., Hecker, A., Jung, T. and Brown, C. (2003) Twisting and Braiding Reduces the Tensile Strength and Stiffness of Human Hamstring Tendon Grafts Used for Anterior Cruciate Ligament Reconstruction. *American Journal of Sports Medicine*, 31, 861–867.

Kirtley, C. (2006) *Clinical Gait Analysis. Theory and Practice*. Edinburgh: Elsevier Churchill Livingstone.

Kitamura, N., Yasuda, K., Yamanaka, M. and Tohyama, H. (2003) Biomechanical Comparisons of Three Posterior Cruciate Ligament Reconstruction Procedures With Load–Controlled and Displacement–Controlled Cyclic Tests. *American Journal of Sports Medicine*, 31, 907–914.

Kohn, D. and Rose, C. (1994) Primary Stability of Interference Screw Fixation: Influence of Screw Diameter and Insertion Torque. *American Journal of Sports Medicine*, 22, 334–338.

Kovacs, I., Tihanyi, J., Devita, P., Racz, L., Barrier, J. and Hortobagyi, T. (1999) Foot Placement Modifies Kinematics and Kinetics During Drop Jumping. *Medicine and Science in Sports and Exercise*, 31, 708–716.

Kowalk, D., Duncan, J., McCue, F. and Vaughan, C. (1997) Anterior Cruciate Ligament Reconstruction and Joint Dynamics During Stairclimbing. *Medicine and Science in Sports and Exercise*, 29, 1406–1413.

Kraemer, W., Triplett, N., Fry, A., Koziris, L., Bauer, J., Lynch, J., McConnell, T., Newton, R., Gordon, S., Nelson, R. and Knuttgen, H. (1995) An In–Depth Sports Medicine Profile of Women College Tennis Players. *Journal Sport Rehabilitation*, 4, 79–98.

Kumagai, M., Mizuno, Y., Mattessich, S., Elias, J., Cosgarea, A. and Chao, E. (2002) Posterior Cruciate Ligament Rupture Alters In Vitro Knee Kinematics. *Clinical Orthopaedics and Related Research*, 395, 241–248.

Kurosaka, M., Yoshiya, S. and Andrish, J. (1987) A Biomechanical Comparison of Different Surgical Techniques of Graft Fixation Anterior Cruciate Ligament Reconstruction. *American Journal of Sports Medicine*, 15, 225–229.

LaPrade, R. and Wentorf, F. (2002) Diagnosis and Treatment of Posterolateral Knee Injuries. *Clinical Orthopaedics and Related Research*, 402, 110–121.

LaStayo, P., Woolf, J., Lewek, M., Snyder–Mackler, L., Reich, T. and Lindstedt, S. (2003) Eccentric Muscle Contractions: Their Contribution to Injury, Prevention, Rehabilitation, and Sport. *Journal of Orthopaedic and Sports Physical Therapy*, 33, 557–571.

Lee, J., Allen, C. and Fu, F. (2002) Natural History of the Post–Meniscectomy Knee. *Sports Medicine and Arthroscopy Review*, 10, 236–243.

Leetun, D., Ireland, M., Willson, J., Ballantyne, B. and Davis, I. (2004) Core Stability Measures as Risk factors for Lower Extremity Injury in Athletes. *Medicine and Science in Sports and Exercise*, 36, 926–934.

Lehnhard, R., Lehnhard, H., Young, R. and Butterfield, S. (1996) Monitoring Injuries on a College Soccer Team: The Effect of Strength Training. *Journal of Strength and Conditioning Research*, 10, 115–119.

Lemmer, J., Martel, G., Hurlbut, D. and Hurley, B. (2007) Age and Sex Differentially Affect Regional Changes in One Repetition Maximum Strength. *Journal of Strength and Conditioning Research*, 21, 731–737.

Lephart, S., Abt, J., Ferris, C., Sell, T., Nagai, T., Myers, J. and Irrgang, J. (2005) Neuromuscular and Biomechanical Characteristic Changes in High School Athletes: A Plyometric Versus Basic Resistance Program. *British Journal of Sports Medicine*, 39, 932–938.

Lephart, S., Ferris, C., Riemann, B., Myers, J. and Fu, F. (2002) Gender Differences in Strength and Lower Extremity Kinematics During Landing. *Clinical Orthopaedics and Related Research*, 401, 162–169.

Lewek, M., Rudolph, K., Axe, M. and Snyder–Mackler, L. (2002) The Effect of Insufficient Quadriceps Strength on Gait After Anterior Cruciate Ligament Reconstruction. *Clinical Biomechanics*, 17, 56–63.

Li, G., Gill, T., DeFrate, L., Zayontz, S., Glatt, V. and Zarins, B. (2002) Biomechanical Consequences of PCL Deficiency in the Knee Under Simulated Muscle Loads – An In Vitro Experimental Study. *Journal of Orthopaedic Research*, 20, 887–892.

Lind, Jakobsen, B. and Lund, B. (2009) Anatomical Reconstruction of the Medial Collateral Ligament and Posteromedial Corner of the Knee in Patients With Chronic Medial Collateral Ligament Instability. *American Journal of Sports Medicine*, 37, 1116–1122.

Lindenfeld, T., Schmitt, D., Hendy, M., Mangine, R. and Noyes, F. (1994) Incidence of Injury in Indoor Soccer. *American Journal of Sports Medicine*, 22, 364–371.

Liow, R., McNicholas, M., Keating, J. and Nutton, R. (2003) Ligament Repair and Reconstruction in Traumatic Dislocation of the Knee. *Journal of Bone and Joint Surgery*, 85B, 845–851.

Liu, M., Anderson, F., Pandy, M. and Delp, S. (2006) Muscles That Support the Body Also Modulate Forward Progression During Walking. *Journal of Biomechanics*, 39, 2623–2630.

Liu, M., Chou, P., Liaw, L. and Su, F. (2007) Lower Extremity Adaptations During the Squat After Posterior Cruciate Ligament Rupture. *Journal of Biomechanics*, 40, S530.

Lohmander, L., Englund, P., Dahl, L. and Roos, E. (2007) The Long–Term Consequences of Anterior Cruciate Ligament and Meniscus Injuries. Osteoarthritis. *American Journal of Sports Medicine*, 35, 1756–1769.

Lohmander, L., Ostenberg, A., Englund, M. and Roos. (2004) High Prevalence of Knee Osteoarthritis, Pain and Functional Limitations in Female Soccer Players Twelve Years After Anterior Cruciate Ligament Injury. *Arthritis and Rheumatism*, 50, 3145–3152.

Louw, Q., Grimmer, K. and Vaughan, C. (2006) Biomechanical Outcomes of a Knee Neuromuscular Exercise Programme Among Adolescent Basketball Players: A Pilot Study. *Physical Therapy in Sport*, 7, 65–73.

Lundberg, M. and Messner, K. (1996) Long–Term Prognosis of isolated Partial Medial Collateral Ligament Ruptures. A Ten–Year Clinical and radiographic Evaluation of a Prospectively Observed Group of Patients. *American Journal of Sports Medicine*, 24, 160–163.

Lundberg, M. and Messner, K. (1997) Ten–Year Prognosis of Isolated and Combined Medial Collateral Ligament Ruptures. A Matched Comparison in 40 Patients Using Clinical and Radiographic Evaluations. *American Journal of Sports Medicine*, 25, 2–6.

Lutz, G., Palmitier, R., An, K. and Chao, E. (1993) Comparison of Tibiofemoral Joint Forces During Open–Kinetic–Chain and Closed–Kinetic–Chain

Exercises. *Journal of Bone and Joint Surgery*, *75A*, 732–739.

MacKinnon, C. and Winter, D. (1993) Control of Whole Body Balance in the Frontal Plane During Human Walking. *Journal of Biomechanics*, *26*, 633–644.

Magee, D. (1997) *Orthopedic Physical Assessment*, 3rd edn. Philadelphia, PA: WB Saunders.

Magen, H., Howell, S. and Hull, M. (1999) Structural Properties of Six Tibial Fixation Methods, for Anterior Cruciate Ligament Soft Tissue Grafts. *American Journal of Sports Medicine*, *27*, 35–43.

Majima, T., Yasuda, K., Tago, H., Tanabe, Y. and Minami, A. (2002) Rehabilitation After Hamstring Anterior Cruciate Ligament Reconstruction. *Clinical Orthopaedics and Related Research*, *397*, 370–380.

Mandelbaum, B., Silvers, H., Watanabe, D., Knarr, J., Thomas, S., Griffen, L., Kirkendall, T. and Garrett, W. (2005) Effectiveness of a Neuromuscular Training Program in Preventing Anterior Cruciate Ligament Injuries in Female Athletes: 2 Year Follow Up. *American Journal of Sports Medicine*, 1003–1010.

Mangine, R and Kremchek, T. (1997) Evaluation–Based Protocol of the Anterior Cruciate Ligament. *Journal of Sport Rehabilitation*, *6*, 157–181.

Mangine R., Noyes, F. and DeMaio, M. (1992) Minimal Protection Program: Advanced Weight Bearing and Range of Motion After ACL reconstruction – Weeks 1 to 5. *Orthopedics*, *15*, 505–515.

Margheritini, F., Mauro, C., Rihn, J., Stabile, K., Woo, S. and Harner, C. (2004) Biomechanical Comparison of Tibial Inlay Versus Transtibial techniques for Posterior Cruciate Ligament Reconstruction. *American Journal of Sports Medicine 32*, 587–593.

Margheritini, F., Rihn, J., Musahl, V., Mariani, P. and Harner, C. (2002) Posterior Cruciate Ligament Injuries in the Athlete. *Sports Medicine*, *32*, 393–408.

Markolf, K., Burchfield, D., Shapiro, M., Shepard, M., Finerman, G. and Slauterback, J. (1995) Combined Knee Loading States That Generate High Anterior Cruciate Ligament Forces. *Journal of Orthopaedic Research*, *13*, 930–935.

Martin, J. and Jessell, T. (1991) Modality Coding in the Somatic Sensory System. In E. Kandel, J. Schwartz, T. Jessell (eds), *Principles of Neural Science*, 3rd edn. London: Prentice–Hall International, pp 341–352.

McCrory, J., Lephart, S., Ferris, C., Sell, T., Abt, J., Irrgang, J. and Fu. F. (2003) Adaptations to an Intervention Program for the Prevention of Female ACL Injuries. *Journal of Orthopaedic and Sports Physical Therapy*, *33*, 8, A–29.

McCurdy, K., Langford, G., Cline, A., Doscher, M. and Hoff, R. (2004) The Reliability of 1– and 3–RM Tests of Unilateral Strength in Trained and Untrained Men and Women. *Journal of Sports Science and Medicine*, *3*, 190–196.

McDevitt, E., Taylor, D., miller, M., Gelber, J., Ziemke, G., Hinkin, D., Uhorchak, J., Arciero, R. and St Pierre, P. (2004) Functional Bracing After Anterior Cruciate Ligament Reconstruction. A Prospective, Randomized, Multicenter Study. *American Journal of Sports Medicine*, *32*, 1887–1892.

McFarland, E. (1993) The Biology of Anterior Cruciate Ligament Reconstructions. *Orthopedics*, *16*, 403–410.

McGill, S. (1998) Low Back Exercises: Evidence for Improving Exercise Regimes. *Physical Therapy*, *78*, 754–765.

McGill, S. (2007) *Low Back Disorders*, 2nd edn. Champaign, IL: Human Kinetics.

McGill, S., Childs, A. and Liebenson, C. (1999) Endurance Times for Low Back Stabilization Exercises: Clinical Targets for Testing and Training From a Normal Database. *Archives of Physical medicine and Rehabilitation*, *80*, 941–944.

McGinn, P., Mattacola, C., Malone, T., Johnson, D. and Shapiro, R. (2006) Strength Training for 6 Weeks Does Not Significantly Alter Landing Mechanics of Female Collegiate Basketball Players. *Journal of Orthopaedic and Sports Physical Therapy*, *37*, 2, A24.

McHugh, M., Tyler, T., Browne, M., Gleim, G. and Nicholas, S. (2002) Electromyographic Predictors of Residual Quadriceps Muscle Weakness After Anterior Cruciate Ligament Reconstruction. *American Journal of Sports Medicine*, *30*, 334–339.

McKay, H., Tsang, G., Heinonen, A., MacKelvie, K., Sanderson, D. and Khan, K. (2005) Ground Reaction Forces Associated With an Effective Elementary School Based Jumping Intervention. *British Journal of Sports Medicine*, *39*, 10–14.

McLaughlin, J. and Noyes, F. (1993) Arthroscopic Meniscus Repair: Recommended Surgical Techniques for Complex Meniscal Tears. *Techniques in Orthopaedics*, *8*, 129–136.

McLean, S., Huang X., Su, A. and van den Bogert, A. (2004) Saggital Plane Biomechanics Cannot Injure the ACL During Sidestep Cutting. *Clinical Biomechanics*, *19*, 828–838.

McLean, S., Huang X., Su, A. and van den Bogert, A. (2005) Association Between Lower Extremity Posture at Contact and Peak Knee Valgus Moment During Sidestepping: Implications for ACL Injury. *Clinical Biomechanics*, *20*, 863–870.

McNair, P. and Marshall, R. (1994) Landing Characteristics in Subjects With Normal and Anterior Cruciate

Ligament Deficient Knee Joints. *Archives of Physical Medicine and Rehabilitation, 75*, 584–589.

Meislin, R. (1996) Managing Collateral Ligament Tears of the Knee. *Physician and Sports Medicine, 24*, 67–80.

Messina D., Farney, W. and DeLee, J. (1999) The Incidence of Injury in Texas High School Basketball. A Prospective Study Among Male and Female Athletes. *American Journal of Sports Medicine, 27*, 294–299.

Mikkelesen, C., Werner, S. and Eriksson, E. (2000) Closed Kinetic Chain Alone Compared to Combined Open and Closed Kinetic Chain Exercises for Quadriceps Strengthening After Anterior Cruciate Ligament Reconstruction With Respect to Return to Sports: A Prospective Matched Follow–Up Study. *Knee Surgery Sports Traumatology Arthroscopy, 8*, 337–342.

Millet, P., Miller, B., Close, M., Sterett, W., Walsh, W. and Hawkins, R. (2003) Effects of Braiding on Tensile Properties of Four–Strand Human Hamstring Tendon Grafts. *American Journal of Sports Medicine, 31*, 714–717.

Miyasaka, K., Daniel, D. and Stone, M. (1991) The Incidence of Knee Injuries in the Population. *American Journal of Knee Surgery, 4*, 3–8.

Morrissey, M., Drechsler, W., Morrissey, D., Knight, P., Armstrong, P. and McAuliffe, T. (2002) Effects of Distally Fixated Versus Nondistally Fixated Leg Extensor Resistance Training on Knee Pain in the Early Period After Anterior Cruciate Ligament Reconstruction. *Physical Therapy, 82*, 35–43.

Morrissey, M., Hooper, D., Drechsler, W. and Hill, H. (2004) Relationship of Leg Muscle Strength and Knee Function in the Early Period After Anterior Cruciate Ligament Reconstruction. *Scandinavian Journal of Medicine and Science in Sports, 14*, 360–366.

Morrissey, M., Hudson, Z., Drechsler, W., Coutts, F., Knight, P. and King, J. (2000) Effects of Open Versus Closed Kinetic Chain Training on Knee Laxity in the Early Period After Anterior Cruciate Ligament Reconstruction. *Knee Surgery Sports Traumatology Arthroscopy, 8*, 343–348.

Morrison, J. (1969) Function of the Knee Joint in Various Activities. *Biomedical Engineering, 4*, 573–580.

Morrison, J. (1970) The Mechanics of the Knee Joint in Relation to Normal Walking. *Journal of Biomechanics, 3*, 51–61.

Mountcastle, S., Posner, M., Kragh, J. and Taylor, D. (2007) Gender Differences in Anterior Cruciate Ligament Injury Vary With Activity. *American Journal of Sports Medicine, 35*, 1635–1642.

Mueller, M. and Maluf, K. (2002) Tissue Adaptation to Physical Stress: A Proposed "Physical Stress Theory" to Guide Physical Therapist Practice, Education, and Research. *Physical Therapy, 82*, 383–403.

Murphy, A. and Wilson, G. (1997) The Ability of Tests of Muscular Function to Reflect Training–Induced Changes in Performance. *Journal of Sports Sciences, 15*, 191–200.

Myer, G., Ford, K., Palumbo, J. and Hewett, T. (2005) Neuromuscular Training Improves Performance and Lower Extremity Biomechanics in Female Athletes. *Journal of Strength and Conditioning Research, 19*, 51–60.

Myer, G., Ford, K., Brent, J., Hewett, T. (2006a) The Effects of Plyometric vs. Dynamic Stabilization and Balance Training on Power, Balance, and Landing Force in Female Athletes. *Journal of Strength and Conditioning Research, 20*, 345–353.

Myer, G., Ford, K., McLean, S. and Hewett, T. (2006b) The Effects of Plyometric Versus Dynamic Stabilization and Balance Training on Lower Extremity Biomechanics. *American Journal of Sports Medicine, 34*, 445–455.

Myer, G. Ford, K., Brent, J. and Hewett, T. (2007) Differential Neuromuscular Training Effects on ACL Injury Risk Factors in "High–Risk" Versus "Low–Risk" Athletes. *BMC Musculoskeletal Disorders, 8*, 39.

Myklebust, G. and Bahr, R. (2005) Return to Play Guidelines After Anterior Cruciate Ligament Surgery. *British Journal of Sports Medicine, 39*, 127–131.

Nadler, S., Malanga, G., DePrince, M., Stitik, T. and Feinberg, J. (2000) The Relationship Between Lower Extremity Injury, Low Back Pain, and Hip Muscle Strength in Male and Female Collegiate Athletes. *Clinical Journal of Sport Medicine, 10*, 89–97.

Natri, A., Jarvinen, M., Latvala, K. and Kannus, P. (1996) Isokinetic Muscle Performance After Anterior Cruciate Ligament Surgery. *International Journal of Sports Medicine, 17*, 223–228.

Neitzel, J., Kernozek, T. and Davies, G. (2002) Loading Response Following Anterior Cruciate Ligament Reconstruction During the Parallel Squat Exercise. *Clinical Biomechanics, 17*, 551–554.

Neuman, P., Englund, M., Kostogiannis, I., Friden, T., Roos, H. and Dahlberg, L. (2008) Prevalence of Tibiofemoral Osteoarthritis 15 Years After Nonoperative Treatment of Anterior Cruciate Ligament Injury. *American Journal of Sports Medicine, 36*, 1717–1725.

Neumann, D. (2002) *Kinesiology of the Musculoskeletal System*. Missouri: Mosby.

Nicholas, J., Strizak, A. and Veras, G. (1976) A Study of Thigh Muscle Weakness in Different Pathological States of the Lower Extremity. *American Journal of Sports Medicine, 4*, 241–248.

Nisell, R., Ericson, M., Nemeth, G. and Ekholm, J. (1989) Tibiofemoral Joint Forces During Isokinetic Knee Extension. *American Journal of Sports Medicine*, 17, 49–54.

Nisell, R. and Mizrahi, J. (1988) Knee and Ankle Joint Forces During Steps and Jumps Down From Two Different Heights. *Clinical Biomechanics*, 3, 92–100.

Noyes, F. and Barber–Westin, S. (1995) The Treatment of Acute Combined Ruptures of the Anterior Cruciate and Medial Ligaments of the Knee. *American Journal of Sports Medicine*, 23, 380–391.

Noyes, F. and Barber–Westin, S. (1997) Reconstruction of the Anterior and Posterior Cruciate Ligaments After Knee Dislocation. Use of Early Protected Postoperative Motion to Decrease Arthrofibrosis. *American Journal of Sports Medicine*, 25, 769–778.

Noyes, F. and Barber–Westin, S. (2000) Arthroscopic Repair of Meniscus Tears Extending Into the Avascular Zone With or Without Anterior Cruciate Ligament Reconstruction in Patients 40 Years of Age and Older. *Arthroscopy*, 16, 822–829.

Noyes, F. and Barber–Westin, S. (2002) Arthroscopic Repair of Meniscal Tears Extending Into the Avascular Zone in Patients Younger than 20 Years of Age. *American Journal of Sports Medicine*, 30, 589–600.

Noyes, F. and Barber–Westin, S. (2005a) Posterior Cruciate Ligament Revision Reconstruction, Part 1. Causes of Surgical Failure in 52 Consecutive Operations. *American Journal of Sports Medicine*.

Noyes, F. and Barber–Westin, S. (2005b) Posterior Cruciate Ligament Revision Reconstruction, Part 2. Results of Revision Using a 2–Strand Quadriceps Tendon–Patellar Bone Autograft. *American Journal of Sports Medicine*, 33, 655–665.

Noyes, F. and Barber–Westin, S. (2007) Posterolateral Knee Reconstruction With an Anatomical Bone–Patellar–Tendon–Bone Reconstruction of the Fibular Collateral Ligament. *American Journal of Sports Medicine*, 35, 259–273.

Noyes, F., Barber–Westin, S. and Albright, J. (2006) An Analysis of the Causes of Failure in 57 Consecutive Posterolateral Operative Procedures. *American Journal of Sports Medicine*, 34, 1419–1430.

Noyes, F., Barber–Westin, S., Fleckenstein, C., Walsh, C. and West, J. (2005) The Drop–Jump Screening Test. Difference in Lower Limb Control by Gender and Effect of Neuromuscular Training in Female Athletes. *American Journal of Sports Medicine*, 33, 197–207.

Noyes, F., Barber–Westin, S. and Hewett, T. (2000). High Tibial Osteotomy and Ligament Reconstruction for Varus Angulated Anterior Cruciate Ligament Deficient Knees. *American Journal of Sports Medicine*, 28 (3), 282–296.

Noyes, F., Barber, S. and Mangine, R. (1991) Abnormal Lower Limb Symmetry Determined by Function Hop Tests After Anterior Cruciate Ligament Rupture. *American Journal of Sports Medicine*, 19, 513–518.

Noyes, F., Butler, D., Grood, E., Zernicke, R. and Hefzy, M. (1984) Biomechanical Analysis of Human Ligament Grafts Used in Knee Ligament Repairs and Reconstructions. *Journal of Bone and Joint Surgery*, 66A, 344–352.

Noyes, F. and Grood, E. (1976) The Strength of the Anterior Cruciate Ligament in Humans and Rhesus Monkeys. *Journal of Bone and Joint Surgery*, 58A, 1074–1082.

Noyes, F., Schipplein, O., Andriacchi, T., Saddemi, S. and Weise, M. (1992). The Anterior Cruciate Ligament Deficient Knee with Varus Alignment. An Analysis of Gait Adaptations and Dynamic Joint Loadings. *American Journal of Sports Medicine*, 20 (6), 707–716.

Noyes, F., Torvik, P., Hyde, W. and Delucas, J. (1974) Biomechanics of Ligament Failure: II. An Analysis of Immobilization, Exercise, and Reconditioning Effects in Primates. *Journal of Bone and Joint Surgery*, 56A, 1406–1418.

Ohkoshi, Y. and Yasuda, K. (1989) Biomechanical Analysis of Shear Force Exerted on Anterior Cruciate Ligament During Half Squat Exercise. *Transactions of the Annual Meeting of the Orthopaedic Research Society*, 35, 193–194.

Ohkoshi, Y., Yasuda, K., Kaneda, K., Wada, T. and Yamanaka, M. (1991) Biomechanical Analysis of Rehabilitation in the Standing Position. *American Journal of Sports Medicine*, 19, 605–611.

Olsen, O., Myklebust, G., Engebretsen, L., Holm, I. and Bahr, R. (2005) Exercises to Prevent Lower Limb Injuries in Youth Sports: Cluster Randomised Controlled Trial. *British medical Journal*, 330, 449–455.

Onate, J., Guskiewicz, K., Marshall, S., Giuliani, C., Yu, B. and Garrett, W. (2005) Instruction of Jump–Landing Technique Using Videotape Feedback. Altering Lower Extremity Motion Patterns. *American Journal of Sports Medicine*, 33, 831–842.

Onate, J., Guskiewicz, K. and Sullivan, R. (2001) Augmented Feedback Reduces Jump Landing Forces. *Journal of Orthopaedic and Sports Physical Therapy*, 31, 511–517.

Orchard, J., Seward, H., McGivern, J. and Hood, S. (2001) Intrinsic and Extrinsic Risk Factors for Anterior Cruciate Ligament Injury in Australian Footballers. *American Journal of Sports Medicine*, 29, 196–200.

Ortiz, A., Olson, S., Libby, C., Trudelle–Jackson, E., Kwon, Y., Etnyre, B. and Bartlett, W. (2008) Landing Mechanics Between Noninjured Women and Women With Anterior Cruciate Ligament Reconstruction During 2 Jump Tasks. *American Journal of Sports Medicine*, 36, 149–157.

O'Shea, J. and Shelbourne, K. (2003) Repair of Locked Bucket–Handle Meniscal Tears in Knees With Chronic Anterior Cruciate Ligament Deficiency. *American Journal of Sports Medicine*, 31, 216–220.

Ostenberg, A., Roos, E., Ekdahl, C. and Roos, H. (1998) Isokinetic Knee Extensor Strength and Functional performance in Healthy Female Soccer Players. *Scandinavian Journal of Medicine and Science in Sports*, 8, 257–264.

Parolie, J. and Bergfeld, J. (1986) Long–Term Results of Nonoperative Treatment of Isolated Posterior Cruciate Ligament Injuries in the Athlete. *American Journal of Sports Medicine*, 14, 35–38.

Patel, R., Hurwitz, D., Bush–Joseph, C., Bach, B. and Andriacchi, T. (2003) Comparison of Clinical and Dynamic Knee Function in Patients with Anterior Cruciate Ligament Deficiency. *American Journal of Sports Medicine*, 31, 68–74.

Paterno, M., Ford, K., Myer, G., Heyl, R. and Hewett, T. (2007) Limb Asymmetries in Landing and Jumping 2 Years Following Anterior Cruciate Ligament Reconstruction. *Clinical Journal of Sport Medicine*, 17, 258–262.

Pena, F., Grontvedt, T. and Brown, G. (1996) Comparison of Failure Strength Between Metallic and Absorbable Interference Screws. Influence of Insertion Torque, Tunnel–Bone Block Gap, Bone Mineral Density, and Interference. *American Journal of Sports Medicine*, 24, 329–334.

Perry, M., Morrissey, M., King, J., Morrissey, D. and Earnshaw, P. (2005a) Effects of Closed Versus Open Kinetic Chain Knee Extensor Resistance Training on Knee Laxity and Leg Function in Patients During the 8– to 14–Week Post–Operative Period After Anterior Cruciate Ligament Reconstruction. *Knee Surgery Sports Traumatology Arthroscopy*, 13, 357–369.

Perry, M., Morrissey, M., Morrissey, D., Knight, P., McAuliffe, T. and King, B. (2005b) Knee Extensors Kinetic Chain Training in Anterior Cruciate Ligament Deficiency. *Knee Surgery Sports Traumatology Arthroscopy*, 13, 638–648.

Petschnig, R., Baron, R. and Albrecht, M. (1998) The Relationship Between Isokinetic Quadriceps Strength Test and Hop Tests for Distance and One–Legged Vertical Jump Test Following Anterior Cruciate Ligament Reconstruction. *Journal of Orthopaedic and Sports Physical Therapy*, 28, 23–31.

Pinczewski, L., Clingeleffer, A., Otto, D., Bonar, S. and Corry, I. (1997) Integration of the Hamstring Tendon Graft With Bone in Reconstruction of the Anterior Cruciate Ligament. *Arthroscopy*, 13, 641–643.

Pinczewski, L., Deehan, D., Salmon, L., Russell, V. and Clingeleffer, A. (2002) A Five–Year Comparison of Patellar Tendon Versus Four–Strand Hamstring Tendon Autograft for Arthroscopic Reconstruction of the Anterior Cruciate Ligament. *American Journal of Sports Medicine*, 30, 523–536.

Prapavessis, H., McNair, P., Anderson, K. and Hohepa, M. (2003) Decreasing Landing Forces in Children: The Effect of Instructions. *Journal of Orthopaedic and Sports Physical Therapy*, 33, 204–207.

Race, A. and Amis, A. (1994) The Mechanical Properties of the Two Bundles of the Human Posterior Cruciate Ligament. *Journal of Biomechanics*, 27, 13–24.

Rankin, M., Noyes, F., Barber–Westin, S., Hushek, S. and Seow, A. (2006) Human Meniscus Allografts' In Vivo Size and Motion Characteristics: Magnetic Resonance Imaging Assessment Under Weightbearing Conditions. *American Journal of Sports Medicine*, 34, 98–107.

Reid, A., Birmingham, T., Stratford, P., Alcock, G. and Giffin, J. (2007) Hop Testing Provides a Reliable and Valid Outcome Measure During Rehabilitation After Anterior Cruciate Ligament Reconstruction. *Physical Therapy*, 87, 337–349.

Reider, B. (1996) Medial Collateral Ligament Injuries in Athletes. *Sports Medicine*, 21, 147–156.

Reider, B., Sathy, M., Talkington, J., Blyznak, N. and Kolias, S. (1993) Treatment of Isolated Medial Collateral Ligament Injuries in Athletes With Early Functional Rehabilitation. A Five–Year Follow–Up Study. *American Journal of Sports Medicine,* 22, 470–477.

Richter, M., Kiefer, H., Hehl, G. and Kinzi, L. (1996) Primary Repair for Posterior Cruciate Ligament Injuries: An Eight Year Followup of Fifty–Three Patients. *American Journal of Sports Medicine*, 24, 298–305.

Riemann, B. and Lephart, S. (2002a) The Sensorimotor System. Part I: The Physiologic Basis of Functional Joint Stability. *Journal of Athletic Training*, 37, 71–79.

Riemann, B. and Lephart, S. (2002b) The Sensorimotor System. Part II: The Role of Proprioception in Motor Control and Functional Joint Stability. *Journal of Athletic Training*, 37, 80–84.

Rihn, J., Cha, P., Groff, Y. and Harner, C. (2004) The Acutely Dislocated Knee: Evaluation and Management. *Journal of the American Academy of Orthopaedic Surgeons*, 12, 334–346.

Risberg, M., Holm, I., Myklebust, G. and Engebretsen, L. (2007) Neuromuscular Training Versus Strength Training During the First 6 Months After Anterior Cruciate Ligament Reconstruction: A Randomized Clinical Trial. *Physical Therapy*, 87, 737–750.

Risberg, M., Holm, I., Steen, H., Eriksson, J. and Ekeland, A. (1999a) The Effect of Knee Bracing After Anterior Cruciate Ligament Reconstruction. A Prospective, Randomized Study With Two Years' Follow–Up. *American Journal of Sports Medicine*, 27, 76–83.

Risberg, M., Holm, I., Tjomsland, O., Ljunggren, E. and Ekeland, A. (1999b) Prospective Study of Changes in Impairments and Disabilities After Anterior Cruciate Ligament Reconstruction. *Journal of Orthopaedic and Sports Physical Therapy*, 29, 400–412.

Roberts, D., Friden, T., Zatterstrom, R., Lindstrand, A. and Moritz, U. (1999) Proprioception in People With Anterior Cruciate Ligament Deficient Knees: Comparison of Symptomatic and Asymptomatic Patients. *Journal of Orthopaedic and Sports Physical Therapy*, 10, 587–594.

Roi, G., Creta, D., Nanni, G., Marcacci, M., Zaffagnini. S. and Snyder–Mackler, L. (2005) Return to Official Italian First Division Soccer Games Within 90 Days After Anterior Cruciate Ligament Reconstruction: A Case Report. *Journal of Orthopaedic and Sports Physical Therapy*, 35, 52–66.

Roos, E. (2005) Joint Injury Causes Knee Osteoarthritis in Young Adults. *Current Opinion in Rheumatology*, 17, 195–200.

Rowden, N., Sher, D. and Rogers, G. (1997) Anterior Cruciate Ligament Graft Fixation: Initial Comparison of Patellar Tendon and Semitendinosus Autografts in Young Fresh Cadavers. *American Journal of Sports Medicine*, 25, 472–478.

Rozzi, S., Lephart, S., Gear, W. and Fu, F. (1999) Knee Joint Laxity and Neuromuscular Characteristics of Male and Female Soccer and Basketball Players. *American Journal of Sports Medicine*, 27, 312–319.

Rubman, M., Noyes, F. and Barber–Westin, S. (1998) Arthroscopic Repair of Meniscal Tears That Extend Into the Avascular Zone. A Review of 198 Single and Complex Tears. *American Journal of Sports Medicine*, 26, 87–95.

Safran, M., Allen, A., Lephart, S., Borsa, P., Fu, F. and Harner, C. (1999) Proprioception in the Posterior Cruciate Ligament Deficient Knee. *Knee Surgery, Sports Traumatology, and Arthroscopy*, 7, 310–317.

Salmon, L., Pinczewski, L., Russell, V. and Refshauge, K. (2006) Revision Anterior Cruciate Ligament Reconstruction With Hamstring Tendon Autograft. *American Journal of Sports Medicine*, 34, 1604–1614.

Sapega, A. (1990) Muscle Performance Evaluation in Orthopaedic Practice. *Journal of Bone and Joint Surgery*, 72A, 1562–1574.

Scheffler, S., Norbert, S., Gockenjan, A., Hoffman, R. and Weiler, A. (2002) Biomechanical Comparison of Hamstring and Patellar Tendon Graft Anterior Cruciate Ligament Reconstruction Techniques: The Impact of Fixation Level and Fixation Method Under Cyclic Loading. *Arthroscopy*, 18, 304–315.

Schipplein, O. and Andriacchi, T. (1991). Interaction Between Active and Passive Knee Stabilizers During Level Walking. *Journal of Orthopaedic Research*, 9, 113–119.

Schmidt, R. and Lee, T. (1999) *Motor Control and Learning*, 3rd edn. Champaign,IL: Human Kinetics.

Schneider, G., Bigelow, C. and Amoroso, P. (2000) Evaluating Risk of Re–Injury Among 1214 Army Airborne Soldiers Using a Stratified Survival Model. *American Journal of Preventive Medicine*, 18 (3, Suppl 1), 156–163.

Sekiya, J., West, R., Ong, B., Irrgang, J., Fu, F. and Harner, C. (2005) Clinical Outcomes After Isolated Arthroscopic Single–Bundle Posterior Cruciate Ligament Reconstruction. *Arthroscopy*, 21, 1042–1050.

Sell, T., Ferris, C., Abt, J., Tsai, Y., Myers, J., Fu, F. and Lephart, S. (2006) The Effects of Direction and Reaction on the Neuromuscular and Biomechanical Characteristics of the Knee During Tasks That Simulate the Noncontact Anterior Cruciate Ligament Injury Mechanism. *American Journal of Sports Medicine*, 34, 43–54.

Sell, T., Ferris, C., Abt, J., Tsai, Y., Myers, J., Fu, F. and Lephart, S. (2007) Predictors of Proximal Tibia Anterior Shear Force During a Vertical Stop–Jump. *Journal of Orthopaedic Research*, 25, 1589–1597.

Shapiro, M., Markolf, K., Finerman, G. and Mitchell, P. (1991) The Effect of Section of the Medial Collateral Ligament on Force Generated in the Anterior Cruciate Ligament. *Journal of Bone and Joint Surgery*, 73A, 248–256.

Shelbourne, K. and Carr, D. (2003) Meniscal Repair Compared With Meniscectomy for Bucket–Handle Medial Meniscal Tears in Anterior Cruciate Ligament Reconstructed Knees. *American Journal of Sports Medicine*, 31, 718–723.

Shelbourne, K., Davis, T. and Patel, D. (1999) The Natural History of Acute, Isolated, Nonoperatively Treated Posterior Cruciate Ligament Injuries. *American Journal of Sports Medicine*, 27, 276–283.

Shelbourne, K. and Gray, T. (1997) Anterior Cruciate Ligament Reconstruction With Autogenous Patellar Tendon Graft Followed by Accelerated Rehabilitation. A

Two- to Nine-Year Follow-Up. *American Journal of Sports Medicine*, 25, 786–795.

Shelbourne, K., Klootwyk, T. and DeCarlo, M. (1992) Update on Accelerated Rehabilitation After Anterior Cruciate Ligament Reconstruction. *Journal of Orthopaedic and Sports Physical Therapy*, 15, 303–308.

Shelbourne, K., Klootwyk, T., Wilckens, J. and DeCarlo, M. (1995) Ligament Stability Two to Six Years After Anterior Cruciate Ligament Reconstruction With Autogenous Patellar Tendon Graft and Participation in Accelerated Rehabilitation Program. *American Journal of Sports Medicine*, 23, 575–579.

Shelbourne, K. and Klotz, C. (2006) What I Have Learned About the ACL: Utilizing a Progressive Rehabilitation Scheme to Achieve Total Knee Symmetry After Anterior Cruciate Ligament Reconstruction. *Journal of Orthopaedic Science*, 11, 318–325.

Shelbourne, K. and Nitz, A. (1990) Accelerated Rehabilitation After Anterior Cruciate Ligament Reconstruction. *American Journal of Sports Medicine*, 18, 292–299.

Shelbourne, K. and Nitz. (1991) The O'Donoghue Triad Revisited: Combined Knee Injuries Involving Anterior Cruciate and Medial Collateral Ligament Tears. *American Journal of Sports Medicine*, 19, 474–477.

Shelbourne, K. and Patel, D. (1995) Management of Combined Injuries of the Anterior Cruciate and Medial Collateral Ligaments. *Journal of Bone and Joint Surgery*, 77A, 800–806.

Shelbourne, K. and Porter, D. (1992) Anterior Cruciate Ligament–Medial Collateral Ligament Injury: Nonoperative Management of Medial Collateral Ligament Tears With Anterior Cruciate Ligament Reconstruction: A Prospective Randomized Study. *American Journal of Sports Medicine*, 20, 283–286.

Shelbourne, K. and Rask, B. (1998) Controversies With Anterior Cruciate Ligament Surgery and Rehabilitation. *American Journal of Knee Surgery*, 11, 136–143.

Shelburne, K., Pandy, M., Anderson, F. and Torry, M. (2004) Pattern of Anterior Cruciate Ligament Force in Normal Walking. *Journal of Biomechanics*, 37, 797–805.

Shelburne, K. Torry, M. and Pandy, M. (2005) Muscle, Ligament, and Joint–Contact Forces at the Knee During Walking. *Medicine and Science in Sports and Exercise*, 37, 1948–1956.

Shumway-Cook, A. and Woollacott, M. (1995) *Motor Control. Theory and Practical Applications*. Baltimore: Williams & Wilkins.

Slemenda, C., Heilman, D., Brandt, K., Katz, B., Mazzuca, S., Braunstein, E. and Byrd, D. (1998) Reduced Quadriceps Strength Relative to Bodyweight. A Risk Factor for Knee Osteoarthritis in Women? *Arthritis and Rheumatism*, 41, 1951–1959.

Smidt, G. (1973) Biomechanical Analysis of Knee Flexion and Extension. *Journal of Biomechanics*, 6, 79–92.

Snyder-Mackler, L., Delitto, A., Bailey, S. and Stralka, S. (1995) Strength of the Quadriceps Femoris Muscle and Functional Recovery After Reconstruction of the Anterior Cruciate Ligament. *Journal of Bone and Joint Surgery*, 77A, 1166–1173.

Souza, A. and Shimada, S. (2002) Biomechanical Analysis of the Knee During the Power Clean. *Journal of Strength and Conditioning Research*, 16, 290–297.

Stanforth, P., Painter, T. and Wilmore, J. (1992) Alterations in Concentric Strength Consequent to Powercise and Universal Gym Circuit Training. *Journal of Applied Sport Science Research*, 6, 249–255.

Staubli, H., Schatzmann, L., Brunner, P., Rincon, L. and Nolte, L. (1999) Mechanical Tensile Properties of the Quadriceps Tendon and Patellar Ligament in Young Adults. *American Journal of Sports Medicine*, 27, 27–34.

Steele, J. and Brown, J. (1999) Effects of Chronic Anterior Cruciate Ligament Deficiency on Muscle Activation Patterns During an Abrupt Deceleration Task. *Clinical Biomechanics*, 14, 247–257.

Steiner, M., Hecker, A. and Brown, C. (1994) Anterior Cruciate Ligament Graft Fixation: Comparison of Hamstring and Patellar Tendon Grafts. *American Journal of Sports Medicine*, 22, 240–247.

Stockbrugger, B. and Haennel, R. (2003) Contributing Factors to Performance of a Medicine Ball Explosive Power Test: A Comparison Between Jump and Non–Jump Athletes. *Journal of Strength and Conditioning Research*, 17, 768–774.

Strowbridge, N. (2002) Musculoskeletal Injuries in Female Soldiers: Analysis of Cause and Type of Injury. *Journal of the Royal Army Medical Corps*, 148, 256–258.

Strowbridge, N. (2005) Gender Differences in the Cause of Low Back Pain in British Soldiers. *Journal of the Royal Army Medical Corps*, 151, 69–72.

Stuart, M., Meglan, D., Lutz, G., Growney, E. and An, K. (1996) Comparison of Intersegmental Tibiofemoral Joint Forces and Muscle Activity During Various Closed Kinetic Chain Exercises. *American Journal of Sports Medicine*, 24, 792–799.

Tagesson, S. and Kvist, J. (2007) Intra– and Inter–Rater Reliability of the Establishment of One Repetition Maximum on Squat and Seated Knee Extension. *Journal of Strength and Conditioning Research*, 21, 801–807.

Tagesson, S., Oberg, B., Good, L. and Kvist, J. (2008) A Comprehensive Rehabilitation Program With Quadriceps Strengthening in Closed Versus Open Kinetic Chain Exercise in Patients With Anterior Cruciate Ligament Deficiency. A Randomized Clinical Trial Evaluating Dynamic Tibial Translation and Muscle Function. *American Journal of Sports Medicine*, 36, 298–307.

Threlkeld, J., Horn, T., Wojtowicz, G., Rooney, J. and Shapiro, R. (1989) Kinematics, Ground Reaction Force, and Muscle balance Produced by Backward Running. *Journal of Orthopaedic and Sports Physical Therapy*, 11, 56–63.

Toutoungi, D., Lu, T., Leardini, A., Catani, F. and O'Connor, J. (2000) Cruciate Ligament Forces in the Human Knee During Rehabilitation Exercises. *Clinical Biomechanics*, 15, 176–187.

Uhorchak, J., Scoville, C., Williams, G., Arciero, R., Pierre, P. and Taylor, D. (2003) Risk Factors Associated With Noncontact Injury of the Anterior Cruciate Ligament. A Prospective Four–Year Evaluation of 859 West Point Cadets. *American Journal of Sports Medicine*, 31, 831–842.

Waddington, G., Adams, R. and Jones, A. (1999) Wobble Board (Ankle Disc) training Effects on the Discrimination of Inversion Movements. *Australian Journal of Physiotherapy*, 45, 95–101.

Waddington, G., Seward, H., Wrigley, T., Lacey, N. and Adams, R. (2000) Comparing Wobble Board and Jump–Landing Training Effects on Knee and Ankle Movement Discrimination. *Journal of Science and Medicine in Sport*, 3, 449–459.

Webster, K., Gonzalez–Adrio, R. and Feller, J. (2004) Dynamic Joint Loading Following Hamstring and Patellar Tendon Anterior Cruciate Ligament Reconstruction. *Knee Surgery, Sports Traumatology, Arthroscopy*, 12, 15–21.

Wilk, K. and Andrews, J. (1992) Current Concepts in the Treatment of Anterior Cruciate Ligament Disruption. *Journal of Orthopaedic and Sports Physical Therapy*, 15, 279–293.

Wilk, K., Andrews, J., Clancy, W., Crockett, H. and O'Mara, J. (1999) Rehabilitation Programs for the PCL–Injured and Reconstructed Knee. *Journal of Sport Rehabilitation*, 8, 333–361.

Wilk, K., Escamilla, R., Fleisig, G., Barrentine, S., Andrews, J. and Boyd, M. (1996) A Comparison of Tibiofemoral Joint Forces and Electromyographic Activity During Open and Closed Kinetic Chain Exercises. *American Journal of Sports Medicine*, 24, 518–527.

Wilk, K., Romaniello, W., Soscia, S., Arrigo, C. and Andrews, J. (1994) The Relationship between Subjective Knee Scores, Isokinetic Testing, and Functional Testing in the ACL–Reconstructed Knee. *Journal of Orthopaedic and Sports Physical Therapy*, 20, 60–73.

Wilk, K., Zheng, N., Fleisig, G., Andrews, J. and Clancy, W. (1997) Kinetic Chain Exercise: Implications for the Anterior Cruciate Ligament Patient. *Journal of Sport Rehabilitation*, 6, 125–143.

Wilkerson, G., Colston, M., Short, N., Neal, K., Hoewischer, P. and Pixley, J. (2004) Neuromuscular Changes in Female Collegiate Athletes Resulting From a Plyometric Jump–Training Program. *Journal of Athletic Training*, 39, 17–23.

Williams, R., Hyman, J., Petrigliano, F., Rozental, T. and Wickiewicz, T. (2004) Anterior Cruciate Ligament Reconstruction With a Four–Strand Hamstring Tendon Autograft. *Journal of Bone and Joint Surgery*, 86–A, 225–232.

Willson, J., Ireland, M. and Davis, I. (2006) Core Strength and Lower Extremity Alignment During Single–Leg Squats. *Medicine and Science in Sports and Exercise*, 38, 945–952.

Wilson, T., Zafuta, M. and Zobitz, M. (1999) A Biomechanical Analysis of Matched Bone–Patellar Tendon–Bone and Double–Looped Semitendinosus and Gracilis Tendon Grafts. *American Journal of Sports Medicine*, 27, 202–207.

Winter, E. and Brooks, F. (1991) Electromechanical Response Times and Muscle Elasticity in Men and Women. *European Journal of Applied Physiology*, 63, 124–128.

Wisloff, U., Castagna, C., Helgerud, J., Jones, R. and Hoff, J. (2004) Strong Correlation of Maximal Squat Strength With Sprint Performance and Maximal vertical Jump Height in Elite Soccer Players. *British Journal of Sports Medicine*, 38, 285–288.

Wojtys, E., Huston, L., Taylor, P. and Bastian, S. (1996) Neuromuscular Adaptations in Isokinetic, Isotonic, and Agility Training Programs. *American Journal of Sports Medicine*, 24, 187–192.

Woo, S., Hollis, M., Adams, D., Lyon, R. and Takai, S. (1991) Tensile Properties of the Human Femur–Anterior Cruciate Ligament–Tibia Complex. *American Journal of Sports Medicine,* 19, 217–225.

Worrell, T., Borchert B., Erner, K., Fritz, J. and Leerar, P. (1993) Effect of a lateral step-up exercise protocol on quadriceps and lower extremity performance. *Journal of Orthopaedic and Sports Physical Therapy*, 18, 646–653.

Yard, E., Schroeder, M., Fields, S., Collins, C. and Comstock, R. (2008) The Epidemiology of United States High School Soccer Injuries, 2005–2007. *American Journal of Sports Medicine*, 36, 1930–1937.

Yasuda, K. and Sasaki, T. (1987) Exercise After Anterior Cruciate Ligament Reconstruction. *Clinical Orthopaedics and Related Research*, 220, 275–283.

Yoshiya, S., Kuroda, R., Mizuno, K., Yamamoto, T. and Kurosaka, M. (2005) Medial Collateral Ligament Reconstruction Using Autogenous Hamstring Tendons. *American Journal of Sports Medicine*, 33, 1380–1385.

Yu, B., Lin, C. and Garrett, W. (2006) Lower Extremity Biomechanics During the Landing of a Stop–Jump Task. *Clinical Biomechanics*, 21, 297–305.

Zanetti, M., Pfirrmann, C., Schmid, M., Romero, J., Seifert, B. and Hodler, J. (2003) Patients With Suspected Meniscal Tears: Prevalence of Abnormalities Seen on MRI of 100 Symptomatic and 100 Contralateral Asymptomatic Knees. *American Journal of Roentgenology*, 181, 635–641.

Zantop, T., Weimann, A., Rummler, M., Hassenpflug, J. and Petersen, W. (2004) Initial Fixation Strength of Two Bioabsorbable Pins for the Fixation of Hamstring Grafts Compared to Interference Screw Fixation. *American Journal of Sports Medicine,* 32, 641–649.

Zazulak, B., Hewett, T., Reeves, N., Goldberg, B. and Cholewicki, J. (2007) Deficits in Neuromuscular Control of the Trunk Predict Knee Injury Risk. *American Journal of Sports Medicine*, 35, 1123–1130.

Zazulak, B., Ponce, P., Straub, S., Medvecky, M., Avedisian, L. and Hewett, T. (2005) Gender Comparison of Hip Muscle Activity During Single–Leg Landing. *Journal of Orthopaedic and Sports Physical Therapy*, 35, 292–299.

Zebis M., Bencke J., Andersen, L., Dossing, S., Alkjaer, T., Magnusson, S., Kjaer, M. and Aagaard, P. (2008) The Effects of Neuromuscular Training on Knee Joint Motor Control During Sidecutting in Female Elite Soccer and Handball Players. *Clinical Journal of Sports Medicine*, 18, 329–337.

Zeller, B., McCrory, J., Kibler, W. and Uhl, T. (2003) Differences in Kinematics and Electromyographic Activity between men and Women During the Single–Legged Squat. *American Journal of Sports Medicine*, 31, 449–456.

Zhang, S., Bates, B. and Dufek, J. (2000) Contributions of Lower Extremity Joints to Energy Dissipation During Landings. *Medicine and Science in Sports and Exercise*, 32, 812–819.

22

Ankle complex injuries in sport

David Joyce
Blackburn Rovers and The University of Bath

Epidemiology

Ankle injuries are one of the most common injuries seen by the sports medicine clinician, accounting for up to 30% of all sports injuries (Fong et al. 2007). Volleyball reports ankle injuries as 44% (Agel et al. 2007b), basketball 25% (Agel, Olsen et al. 2007a) and soccer 23% (Nelson et al. 2007) of all injuries. The lateral ligaments are primarily implicated in these injuries with the anterior tibiofibular ligament (ATFL) being particularly susceptible to damage. Woods et al. (2003) reported that 73% of ankle injuries in English football involved ATFL trauma. Moreover, sprains to the lateral ligaments of the ankle are responsible for more time lost from sports participation than any other injury (Hertel 2006). Whilst the vast majority of injuries to the ankle involve the lateral ligament complex, an accurate diagnosis taking into account all structures must be made in order to ensure that the appropriate management and rehabilitation is instituted.

This chapter will begin by giving a brief outline of the anatomy and biomechanics of the sporting ankle before discussing in greater detail clinical assessment and rehabilitation principles of ankle injuries. We will look at acute ankle sprains and chronic ankle instability as well as some of the other pathologies that are found in the region.

Functional anatomy and biomechanics of the sporting ankle

Functionally, the ankle complex is composed of three joints (Figures 22.1 and 22.2):

1. the talo-crual joint (TCJ – between the talus and the inferior aspect of the tibia)

2. the inferior tibiofibular joint (ITFJ – between the distal ends of the tibia and the fibula)

3. the subtalar joint (STJ – between the talus and the calcaneus).

Whilst it is certainly possible to damage one of these joints in isolation and it is true to say that the TCJ is implicated in the majority of lateral ankle sprains (Hertel 2002), concomitant STJ sprains may be present in as many as 80% of people presenting to sports medicine clinics with a lateral ankle sprain (Meyer et al. 1986). This behoves us to become adept at assessing all the structures in the ankle.

Stability of the ankle complex

The bony architecture of the ankle is partially responsible for joint stability but only in certain situations. The medial and lateral malleoli prove a certain degree of side-to-side stability and in full weight

---

Sports Rehabilitation and Injury Prevention   Edited by Paul Comfort and Earle Abrahamson
© 2010 John Wiley & Sons, Ltd

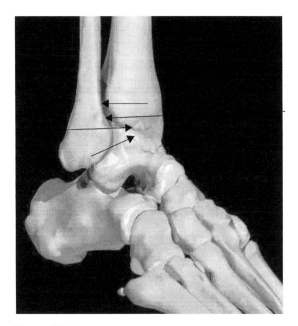

Figure 22.1 Ankle (lateral view). Courtesy and copyright Primal Pictures Ltd.

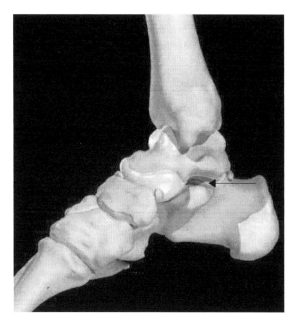

Figure 22.2 Ankle (medial view). Courtesy and copyright Primal Pictures Ltd.

bearing, all ligaments are in a relaxed state and so it is the congruency of the talo-crual joint that is all-important. The ankle is least stable in plantarflexion (the loose packed position) and so it comes as no surprise that this is the position most widely implicated in acute ankle injuries.

There are several ligaments that provide support to the ankle complex. The major ones are listed in Table 22.1 and illustrated in Figures 22.3 and 22.4.

Dynamic stability of the ankle complex is provided by the peroneal muscle group, in particular the peroneus brevis and longus. The action of these muscles has been shown to be anticipatory, preparing the ankle to accept body weight at foot strike (Konradsen and Hojsgaard 1993). In fact, it is the loss of the pre-emptive function of the peroneals that is thought to be one of the sequelae of CAI, leading to the foot accepting load in a more plantarflexed and inverted position. The load-to-failure point of the peroneal tendons is thought to be greater than that of the ATFL and equal to the CFL (Attarian et al. 1985).

History of acute injury

Mechanisms of ankle injury

As with all injury assessment, determining the mechanism of injury and the position of the ankle at the time of injury is particularly important when trying to deduce the injured structures(s). The most commonly reported method of spraining the lateral ligaments is when the ankle is "rolled" into plantarflexion and inversion (Wolfe et al. 2001). This is a common occurrence in sport such as when landing from a jump onto an opponent's shoe in basketball or when reaching for a wide backhand volley in tennis. In these situations, forefoot contact often precedes calcaneal contact, leading to a plantarflexion moment on the talus, thrusting its narrow body between the malleoli, and exposing the lateral ligaments to a stretching force that may exceed its tensile strength (Ritchie 2001). Whilst it is possible to sustain an isolated CFL injury, it is infrequent. It may occur in an incident that involves inversion in dorsiflexion or external rotation when fully weight bearing. This mechanism may also injure the syndesmosis or the talar dome. A combined ATFL and CFL injury is said to occur in 20% of ankle sprain cases (Brostrom 1964). PTFL injuries are usually found in the severely injured ankle where both the ATFL and the CFL have been injured as well. An isolated PTFL lesion is exceptional. The sequence of ligament rupture under inversion stresses are:

ATFL → anterolateral capsule (resulting in the ecchymosis commonly seen with an ankle sprain)

Table 22.1 Ankle ligaments

Name	Origin	Insertion	Action	Interesting facts	Implications for rehabilitation
Anterior talofibular ligament (ATFL)	Distal end of the fibula	Runs anteriorly to the neck of the talus	Limits excessive plantarflexion and inversion as well as anterior movement of the foot in relation to the leg	Intracapsular but the weakest of the lateral ankle ligaments and most frequently injured. Blends with the joint capsule	Following an ankle sprain, the ATFL needs to be protected from both inversion and inversion
Calcaneofibular ligament (CFL)	Distal end of the fibula	Runs inferiorly and slightly posteriorly to the upper lateral aspect of the calcaneus	Helps to limit inversion particularly during weight bearing	The CFL is cord-like and is the largest lateral ligament	Needs to be rehabilitated in plantar grade ankle position and in weight bearing
Posterior talofibular ligament (PTFL)	Distal end of the fibula	Runs horizontally to the lateral tubercle of the posterior fossa of the talus	Taught only in end of range dorsiflexion	The strongest and deepest of the lateral ligaments. Isolated rupture of the PTFL does not lead to ankle instability.	If this is injured, check for damage to other structures because it rarely is implicated in isolation
Inferior tibiofibular ligaments	Distal tibia	Distal fibula shaft and malleolus	Talar stabilisation	Consists of anterior and posterior parts (posterior is significantly stronger). The ITFL together with the interosseous ligament forms the syndesmosis	Can be assessed with the talar tilt test and may require immobilisation in a walking boot to allow to repair
Cervical ligament	Superior aspect of the calcaneus	Runs supero-medially to the inferior surface of the talus	Helps restrain ankle inversion	Acts with the interosseous talocalcaneal ligament to stabilise the STJ.	Frequently injured in ATFL strains but missed in assessment and a cause of rotary instability if injured

Table 22.1 Ankle ligaments (*Continued*)

Name	Origin	Insertion	Action	Interesting facts	Implications for rehabilitation
Interosseous talocalcaneal ligament	Talar neck	Superior aspect of the calcaneus	Helps restrain excessive talo-calcaneal rotary motion	Situated next to the cervical ligament within the sinus tarsi	Along with the cervical ligament, needs to be protected from rotary forces initially and then progressively stressed into inversion and eversion.
Deltoid ligament	Medial malleolus	(a) Runs anteriorly and inferiorly to the navicular and spring ligament (b) Runs vertically downwards to attach to the sustentaculum tali (c) Runs down and backwards to attach to the medial aspect of the talus	Provides medial ankle stability against eversion forces	Composed of three parts (anterior, intermediate and posterior) although in reality they all blend to form one unit. The deltoid ligament is incredibly strong and difficult to injure	Will often take much longer to rehabilitate due to the fact that it the force required to injure it is so high and it is often found with damage to other structures

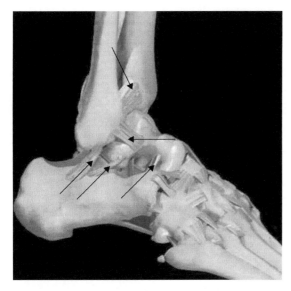

Figure 22.3 Ligaments of the anterolateral ankle. Courtesy and copyright Primal Pictures Ltd.

→ inferior tibiofibular ligament → CFL
→ PTFL.

Should the inversion stress continue, the ankle complex may dislocate or fracture (Safran et al. 1999).

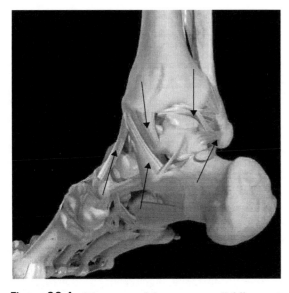

Figure 22.4 Ligaments of the posteromedial ligament. Courtesy and copyright Primal Pictures Ltd.

It is also possible to injure the deltoid ligament on the medial side of the ankle when the weight of the body is forced over an everted foot. Given the high forces that are required to do this however, the injury is infrequent and a lateral malleolus fracture or syndesmosis injury should be excluded in this instance.

Onset of pain

The most important thing to exclude in an acute ankle injury is the presence of a fracture. It is for this reason that the Ottawa Ankle Rules were formed (see below). It is possible for an athlete to sustain an ankle ligament sprain and not notice it at the time (particularly in a combat or collision sport where there may be a strong stress-induced analgesia response) although it is more common for any injury to be accompanied by immediate pain. An ankle fracture will almost invariably not allow the athlete to put any weight on it at all whereas a sprain may allow weight bearing but may get worse as it is continued. The athlete may recall a 'pop' or a 'crack' at the time of injury but not too much importance should be placed on these sensations, as they do not necessarily relate to the extent of injury.

Area of pain

Determining the area of pain can be particularly helpful when determining the injured structure because many of the ligaments are quite superficial and can therefore be located with ease. It is often useful to ask the athlete to "point with one finger" where it is most painful. The most frequent area for pain is around the antero-lateral ankle. The medial aspect of the ankle may be painful if there was an eversion force that injured the medial ligament complex although it is also not uncommon for it to be painful following an inversion injury if there was some medial joint compression causing some bone bruising. The pain from syndesmosis and talar dome injuries tends to be more centrally located, deeper and more diffuse. Ankle fractures tend to be exquisitely painful on palpation. The most common sites of ankle fractures are the distal fibula and the 5th metatarsal. These sites should be carefully examined (see the Ottawa Ankle Rules, below). Whilst the site, extent and intensity of pain, swelling and bruising can be a useful early indicator of the location and extent of damage, injury severity is more reliably judged by the degree of disability following the incident (how much the ankle can be actively moved, if weight bearing is possible).

Radiographic investigations and the Ottawa Ankle Rules

In the majority of cases, imaging of the acutely injured ankle is not necessary, as it does not alter the management of the athlete. There are some key exceptions, however, and these mainly revolve around the need to exclude an ankle fracture. In 1992, the Ottawa Ankle Rules (OAR) were published, which provide clinicians with guidance regarding referring an injured ankle for radiographic investigation (Stiell et al. 1992). According to these rules, it is only necessary to seek an X-ray of an injured ankle in those patients with:

- pain along the posterior edge of the distal 6cm of the fibula (including the tip of the malleolus); or

- an inability to walk more than four steps either immediately or at a subsequent closer assessment within 10 days; or

- pain at the base of the 5th metatarsal.

Should the patient present with any of these signs, they should be referred for an X-ray with anteroposterior (AP), lateral and mortise views. In the mortise view, the talus should be equidistant between the malleoli before a syndesmosis disruption can be excluded (Lynch 2002). Implementation of the OAR have resulted in a dramatic reduction in the number of x-rays taken with no reduction in the detection of ankle fractures (Leddy et al. 1998).

Other investigations such as magnetic resonance imaging (MRI), computerised tomography (CT) and ultrasongraphy (US) are typically unnecessary in the acutely injured ankle because usually they do not change clinical management, which is primarily based on symptoms, although they may be helpful in excluding an osteochondral lesion/talar dome fracture. Down the line, should the athlete's progress not be as swift as expected, these imaging techniques may be employed. CT is best for imaging osteochondral injuries, whereas US is the preferred technique for the imaging of tendon pathology (Shalabi et al. 2007). US imaging is growing in popularity due to the fact that it can provide dynamic imaging of the foot and ankle, and in particular the soft tissues. For example, it is possible to view the flexor hallucis longus tendon slide round the medial malleolus and through the fibro-osseous tunnel as the ankle moves through its range of motion in the sagittal plane. This has the advantage of the athlete being able to interact with the ultrasonographer or radiologist and they can demonstrate the movement and indeed the stage of the movement that provokes their pain. This is particularly useful when the athlete has a *functional*, rather than just an anatomical or structural pathology. At this stage, MRI and CT scans are still limited to static imaging.

Physical examination

Like every joint, the ankle should be examined in sequence. Table 22.2 details the specific areas to be examined.

Stress tests

Manual stress tests are often used in the clinical setting to determine the type and extent of ankle ligament injury. The two most common stress tests used for the ankle are the talar tilt test and the anterior draw test.

Anterior Draw Test

This test assesses the anterior translation of the talus with respect to the tibia. It is best performed with the patient sitting up with their legs over the edge of the bed in order to relax the gastrocnemius. The tibia is stabilised with one hand and the other hand grasps the talus and draws it forward in the sagittal plane. A positive test is said to be one where there is greater anterior displacement of the foot with respect to the lower leg when compared to the uninjured side. It has been stated that sagittal plane displacement of greater than 3mm is indicative of an ATFL lesion (Anderson et al. 1952) although like many manual stress tests, the results remain largely subjective and should therefore be used as a guide only. It can also frequently provide false negative results if there is guarding of the ankle due to pain or a thickened fat pad at the posterior calcaneal tuberosity (Johnson and Markolf 1983).

Table 22.2 Areas of examination

Test	Purpose of test/looking for	Notes
Observation	Swelling, bruising, deformations, alterations in weight bearing	Performed in supine, standing and walking if able
Active and resisted movements	Range of motion, quantity and location of pain	Examine PF, DF, inversion, eversion, heel raise
Passive movements	Range of motion, quantity and location of pain, end feel	Examine PF, DF, inversion and eversion + overpressure
Stress tests	Examines integrity of ligaments. Looking for amount of ROM and pain	Compare to the opposite side
Functional tests	Examines the ability of the ankle to cope with integrated tasks such as walking, running and changing directions, jumping and hopping.	Compare to the opposite side if appropriate and look for antalgic movement patterns
Palpation	Specific tenderness and swelling	Palpate each structure (see Table 22.3) to see if it hurts, rather than touch where it hurts and then try to determine the structure

Talar tilt test

The talar tilt (TT) test is used to assess anterolateral rotation of the ankle joint, a movement that is normally restrained by a combination of the ATFL and the CFL. The TT test is performed with the ankle in slight plantarflexion, the tibia and fibula held stable. The calcaneus and talus is grasped with the other hand and "tilted" into inversion. This is then compared to the uninjured side with a talar tilt of greater than 15° when compared to the uninjured side has been correlated with a dual ATFL and CFL strain (Gaebler et al. 1997) although there is a wide range of values for both injured and uninjured ankles reported in the literature which makes interpretation difficult (Frost and Amendola 1999).

The current consensus of opinion is steered towards functional, non-operative treatment, irrespective of the amount of ligamentous laxity demonstrated on stress testing. As a consequence, the treatment of the injured ankle is contingent on the functional status of the ankle and so the exact degree of ligamentous laxity may indeed be moot.

Palpation of structures

Palpation of the ankle after an injury can be especially useful when trying to localise an injury to a

---

On field assessment of the injured ankle

The key thing to note is the position of the ankle at the point of injury. If the ankle was in a fully weight bearing position when it was rolled, the chances are there is more extensive damage and the player is unlikely to be able to continue playing. It is important to check the un-injured ankle to determine what is "normal" and thus comparison of end feel can be made more authoritatively. If the player has an immediate effusion or bruising and/or is unable to weight bear, they should be removed from the field. If the decision is made to allow the player to continue playing, they should be observed to ensure that he is able to run freely, change directions and decelerate rapidly. If they cannot perform these tasks, there is a likelihood of a more serious injury and should be taken from the field for a more thorough assessment and immediate management.

Table 22.3 Order of palpation

Structure	Location
Distal fibula	Along the posterior edge of the fibula from 6cm above the tip of the fibula. If there is marked tenderness, the athlete should be referred for an X-ray to exclude a fracture.
Tip of the lateral malleolus	At the furthermost tip of the fibula to help determine the presence of an avulsion fracture
ATFL	From the lateral malleolus down to the talar neck (best done in plantarflexion and inversion)
CFL	From the lateral malleolus vertically down to the calcaneus (best done in inversion)
PTFL	From the back of the lateral malleolus across to the posterior midline
Peroneal tendons	Posterior to the lateral malleolus to the base of the 5th metatarsal (peroneus brevis) and underneath the foot to the medial cuneiform and great toe (peroneus longus)
Base of 5th metatarsal	Lateral aspect of the 5th metatarsal base to assess for avulsion of the peroneus brevis
Deltoid ligament	In a fan shape extending down from the medial malleolus to the navicular, sustentaculum tali and talus
Sustentaculum tali	A bony ridge inferior and slightly anterior to the distal end of the medial malleolus
Sinus tarsi	A 'cavity' or tunnel located anterior and inferior to the lateral malleolus leading to the subtalar joint
Anterior tibiotalar joint line	Best assessed in with the ankle at rest in plantarflexion
Talar dome	Slightly inferior to the tibial plafond when the foot is plantar flexed to $45°$
Anterior inferior tibiofibular ligament	Anterior border of the distal fibula shaft obliquely up to the tibia to assess for the 'high ankle sprain'
Flexor hallucis longus tendon	Posterior to the medial malleolus and it courses down through the talus' fibro-osseous tunnel
Posterior ankle recess	Posterior ankle between the malleoli and deep to the achilles tendon. Palpate from the calcaneus up to the talus.

structure. A good habit to get into is to systematically palpate structures in order and see if this elicits pain. This is better and more reliable than going straight for where it is sore and then working backwards to deduce exactly what it is that is sore underneath your finger. This will help minimise injured structures not being detected. The order of palpation should be as shown in Table 22.3.

An on-pitch assessment may need to be limited to range of motion (ROM), a couple of stress tests and ability to weight bear. In a clinical setting however, a full assessment should be conducted in order to be more diagnostically precise.

Grading of ankle ligament injury severity

Grading an injury to the lateral ankle ligaments remains a controversial topic because, by-and-large, grading scales are subjective and hard to validate. The three main scales used to grade severity are listed, below:

1. Grading based on number of ligaments injured (ie, a 1,2 or 3 ligament injury). The downfall of this method is that it does not assess to what extent each ligament is injured.

2. Traditional grading based on the extent of structural damage. A grade I injury exhibits microscopic unfurling of the crimped pattern of the ligament without any macroscopic damage. A grade II injury has macroscopic stretching without ligament rupture. A grade III injury is a complete rupture. The downfall of this method is that

it is almost impossible to validate in the clinical setting.

3. Clinical judgement grading based on functional ability. In this method, a grade I injury is assessed as involving no or minimal functional loss, no extra joint ROM, little or no pain, swelling or bruising. A grade II injury is where the patient's ankle ROM is reduced although accessory movement testing reveals extra joint movement but a firm end-feel, moderate pain over the injured ligament, swelling and bruising. A grade III injury implies complete rupture where there is no longer an end-feel on accessory movement testing, overall joint motion is markedly reduced, there is marked swelling and bruising and function has been lost.

Syndesmosis strains

The ankle syndesmosis is composed of the anterior inferior tibiofibular ligament (AITFL), the posterior inferior tibiofibular ligament (PITFL) and the interosseous membrane. The mechanism of injury is forced external rotation of the weight bearing foot or forced plantarflexion such as rolling over the front of the ankle. Injuries to the syndesmosis have historically been underdiagnosed and are a significant source of chronic ankle pain in the athlete. Like other ankle ligament sprains, injuries to the syndesmosis are often graded in terms of amount of damage sustained with a Grade I indicating AITFL stretch, Grade II indicating AITFL partial tear and Grade III indicating complete AITFL rupture (O'Donoghue 1984). The AITFL is tender on palpation, particularly when it is stressed with the external rotation stress test. For this test, the athlete is seated with their legs overhanging the couch, hips and knees flexed to 90 degrees. The lower limb is stabilised with one hand whilst the other hand dorsiflexes and externally rotates the foot. A grade I-II injury should be immobilised in a walking boot until pain free (typically up to 2 weeks) and needs to be followed by a progressive active rehabilitation programme. A weightbearing X-ray that demonstrates widening of the mortise may require a surgical opinion with a view to internal fixation. Following surgical fixation of a grade III syndesmosis strain, an athlete may be back to athletic competition in around six weeks following a progressive rehabilitation programme (Taylor et al. 2007).

Other causes of anterior ankle pain

Anterior impingement

Anterior impingement occurs when either anterioraly located hypertrophied soft tissue, boney exostosis on the talus or tibia, or loose body limits tibio-talar dorsiflexion. The athlete complains of persistent pain at the front of the ankle, particularly when lunging. It is also commonly seen in football in the non-dominant kicking leg. It may be caused by repeated loading at the extremes of dorsiflexion as seen in football or ballet. It is often seen after a lateral ankle sprain where there may be arthrokinematic changes that limit the posterior glide of the talus in the mortise (Hubbard and Hertel 2006). There may be tenderness on palpation of the anterior joint line and in advanced cases, it may be possible to feel a boney spur. Manual therapy to restore normal accessory talus motion and anti-inflammatory medications/modalities are often helpful and a period of rest from loaded dorsiflexion activities is recommended. Tape can be applied to limit the extent of talo-crual dorsiflexion.

Sinus tarsi syndrome

The sinus tarsi is a small tunnel that is located near the talar neck and the calcaneus at the antero-lateral aspect of the ankle, slightly antero-inferior to the lateral malleolus. The sinus tarsi is densely packed with synovial tissue that is easily inflamed. In the majority of cases, the athlete will have a history of at least one ankle sprain and present complaining of diffuse pain slightly below and anterior to the ATFL (Oloff et al. 2001). They may describe pain when walking on uneven ground, jumping or hopping to the side of the injury, landing from a jump or running off-line. Manual therapy of the subtalar joint can be very effective and whilst frequently the athlete will receive great benefit from a corticosteroid injection into the sinus, a proprioceptive, strength and biomechanical rehabilitation programme is imperative to correct the underlying causes of the pathology.

Osteochondral lesions of the talar dome

Osteochondral defects (OCD) of the talar dome are commonly thought to be due to impact of the talar dome as it shears across the tibial plafond as either a single trauma or as a result of repetitive plantarflexion + inversion ankle sprains. A large fracture of the chondral surface can be diagnosed at the time of injury but in many cases they are a diagnosis of exclusion as the athlete complains of recalcitrant, deep and diffuse anterior ankle pain that, despite appropriate conservative rehabilitation, lasts much longer than would be expected for a "normal" ankle sprain. Talar dome injuries can be tested clinically by palpating the talar dome (best located just distal to the tibial plafond when the ankle is plantar flexed). They are best diagnosed with an MRI scan (Mosher 2006) but CT scanning may be required as an adjunct to determine the extent of the chondral surface damage and, dependent on the extent of the lesion, may need surgical stabilisation, after which a comprehensive and extensive rehabilitation programme is necessary.

Lateral ankle pain

Peroneal tendinopathy

Peroneal tendinopathy is usually an overuse injury and is one of the most common causes of non-traumatic lateral ankle pain. The athlete will usually present with an insidious onset of lateral ankle pain, often located either posterior to the lateral malleolus, at the base of the 5th metatarsal.

A retrospective review of 40 patients who underwent peroneal tendon repair for peroneal tendon tears demonstrated that the peroneus brevis was by far the most commonly implicated tendon (88% of cases) with the rest being made up of peroneus longus tears. In 37% of cases there was a combined brevis and longus tear. The average age of the patient was 42 years and the most commonly reported mechanism of injury was a lateral ankle sprain (Dombek et al. 2003). These findings are consistent with those reported by (Saxena and Cassidy 2003).

Treatment of peroneal tendinopathy will depend on whether the pathology is acute and proliferative or chronic and degenerative. Both will require a period of rest from aggravating activities and biomechanical correction with footwear analysis and potentially orthotic insertion. Soft tissue therapy is useful to reduce muscle tone in the peroneals and gastro-soleus complex but there is some evidence to suggest that aggressive stretching of a pathological tendon is not advisable. Whilst strengthening of the peroneals is vital in both the acute and degenerative condition, the more chronic the situation, the more likely it is to respond to eccentric loading (Alfredson and Cook 2007).

Iselin's disease

Iselin's disease is a traction apophysitis of the peroneus brevis at its attachment onto the base of the 5th metatarsal. It is found in adolescents, particularly those with a prominent 5th metatarsal that participates in sports that require sudden changes in direction such as football, lacrosse and cross country running. Treatment for Iselin's disease is always conservative and may involve rest from provocative activities, anti-inflammatory interventions and medications followed by a progressive rehabilitation programme.

Posterior ankle pain

Posterior ankle impingement (PAI) syndrome

The posterior aspect of the ankle is a common site of pain in the athlete who participates in activities that requires extremes of talo-crual plantarflexion such as taekwondo, ballet, running and football. PAI is labelled as a "syndrome" because it is an umbrella term that is used to describe pain anywhere in the posterior region of the ankle joint. There are, however, a number of factors that can contribute to this condition. The athlete will commonly report pain upon plantarflexion, particularly if it is forced. PAI can arise from overuse or traumatically.

The structures in this area of the ankle that can be compressed and thus cause pain are:

- os trigonum (an unfused part of the back of the talus, present in 10–15% of the population).

- thickened posterior joint capsule or PTFL

- enlarged posterior talar process or distal tibial osteophyte

- posterior joint synovitis

- flexor hallucis longus tendon (Lee et al. 2008a)

In the overuse scenario, repeated forced plantarflexion (such as kicking in martial arts, downhill running or standing "en pointe" in ballet) gradually increases the normal range of motion of talo-crual joint plantarflexion, thereby reducing the "clearance distance" between the talus and the tibia at the end of range. If this space is further compromised by any or a combination of the above factors a compressive pathology may result.

Traumatic hyperplantarflexion can result in a fracture of the posterior talar process or traumatic posterior joint synovitis. Post-traumatic calcification of the joint capsule can occur as a late consequence of a plantarflexion plus inversion ankle sprain.

The athlete will usually describe a pain at the back of the ankle that they confuse with the Achilles tendon. Palpation of the Achilles tendon, however, is usually unremarkable although palpation of the posterior aspect of the talus will usually reproduce their pain. The pain is usually aggravated by forced plantar flexion. In the painful position, a medial glide on the calcaneus will usually compress the medial structures and a lateral glide will compress the lateral structures, helping to differentially diagnose between posteromedial and posterolateral impingement.

Rehabilitation should address any contributory talo-crual, mid foot and sub-talar joint hypomobility, hypertonic or tight FHL, triceps surae and peroneals and improving calf muscle and intrinsic muscle endurance. This is in addition to correcting any biomechanical/ alignment issues. A posterior recess steroid injection will often be required followed by 7–10 days of avoidance of plantarflexed positions. Occasionally, surgical excision of any structural abnormalities/loose bodies may be required (Lee et al. 2008b).

Medial ankle pain

Deltoid ligament sprain

The most commonly reported mechanism of injury to the deltoid ligament is when the athlete has their foot planted on the ground in a pronated position and then falls outward, placing a large abduction force on the ligament (Lynch 2002). Given the amount of force required to injure the deltoid ligament, there is a high probability of injury to associated structures. For this reason, it is reasonable to expect rehabilitation to be a much lengthier process than for a lateral ankle sprain, although the principles of injury management are very similar.

Flexor hallucis longus pathology

FHL tenosynovitis is an important cause of posteromedial ankle pain. It is particularly common in ballet dancers and kicking athletes due to the extraordinary range of motion that the tendon must travel from a fully dorsiflexed position to a fully plantar flexed one. The athlete will often complain of posterior ankle pain on landing from a jump or when striking a kick/ball impact. It can be differentially diagnosed by medially gliding the calcaneus whilst in end range talo-crual plantarflexion. Also, passive extension of the great toe whilst the patient is in a weight-bearing lunge position will draw the tendon or a thickened and low muscle belly through the fibro-osseous tunnel of the talus, reproducing their pain. This is often referred to as stenosing tenosynovitis (Hamilton et al. 1996). A steroid injection may be useful in the short term to reduce any inflammatory component but the athlete will also need assessment and rehabilitation of any biomechanical abnormalities when they are in a plantar flexed position. Frequently, if the calcaneus is abducted (known as 'sickling out') when a heel raise is performed, compression of the FHL will occur and re-education of this movement is required. Should conservative measures fail, surgical intervention may be required.

Tibialis posterior pathology

Tibialis posterior tendon injury is classically an entity of overuse and is characterised by medial ankle pain may track down to its insertion on the navicular tubercle (although it does also have distal attachments onto the cuboid, all three cuneiforms, the plantar calcaneonavicular (spring) ligament and the 2nd and 4th metatarsals). The function of the tibialis posterior is to invert the subtalar and midfoot joints and to help stabilise the medial longitudinal arch. It can become subjected to an overuse injury in the athlete with excessive subtalar joint pronation but is also susceptible to direct trauma. Resisted inversion will often be weak and painful and the tendon may be painful on palpation. MRI or USS is helpful in confirming the diagnosis and rehabilitation usually involves rest, anti-inflammatory modalities (particularly if a traumatic incident), eccentric strengthening and biomechanical correction (often involving tape of podiatric involvement).

Tarsal tunnel syndrome

The tarsal tunnel is a fibro-osseous tunnel formed by the flexor retinaculum as it arises from the medial malleolus and attached to the medial aspect of the calcaneus and onto part of the abductor hallucis longus fascia. Tarsal tunnel syndrome is a painful condition involving entrapment of the posterior tibial nerve in the tarsal tunnel and is characterised by a burning pain in the toes, plantar aspect of the foot or around the tarsal tunnel itself which is worsened by load-bearing (Kinoshita et al. 2006). Occasionally there may be a positive Tinel's sign but may also require nerve conduction studies and/or MRI. Conservative treatment involves rest, orthotic or tape intervention to reduce excessive STJ pronation and sometimes, steroid injection. Should it fail to resolve, surgical decompression may be indicated.

Stress fractures in the ankle complex

Stress fractures involving the ankle complex are not common but it is important that they are not missed in an assessment. Table 22.4 outlines the more common sites of stress fractures.

Treatment of a stress fracture in the ankle will most often require non-weight bearing or sometimes surgical internal fixation. Following this, a period of rehabilitation will be required as will correction of any of the factors that contributed to the problem in the first place, such as faulty lower limb biomechanics, technique faults, excessive training load, metabolic or nutritional factors.

Acute treatment and rehabilitation of the lateral ankle sprain

Following an ankle sprain, it is necessary to embark upon a comprehensive and progressive rehabilitation programme. A recent study by Aiken and colleagues (2008) demonstrated that although there was a natural recovery of strength and dorsiflexion ROM within 1 month following a grade I or II ankle sprain, more sensitive measures of ankle performance demonstrated ongoing deficits in those patients that did not receive rehabilitation intervention. The authors suggested that the ongoing disability following an ankle sprain in people who received only standard emergency care provided evidence of the need for more comprehensive rehabilitation plan. This result has been augmented by a study which demonstrated the superiority of a structured 4-week rehabilitation programme compared to a non-intervention control group on postural control and lower limb function on individuals with chronic ankle instability (Hale et al. 2007). A comprehensive rehabilitation aims to control pain and swelling, restore full joint ROM and kinematics, proprioceptive function, muscle strength and functional performance.

Acute management

The principles of acute ankle injury management are similar to those of other joints, namely to protect the joint, and to reduce pain and effusion. Protecting the joint primarily involves preventing further damage. This may involve removal from the field of play, or reducing weight bearing with crutches. Severe ankle sprains may benefit from short-term immobilisation

Table 22.4 Common sites of stress fractures

Site	Predominantly seen in	Characteristics
Medial malleolus	Distance runners	Persistent medial ankle pain that is aggravated by cyclical load-bearing.
Navicular	Distance runners and jumping athletes	Poorly localised antero-medial pain. Requires strict adherence to a NWB protocol.
Posterolateral talus	Track and field athletes	Lateral ankle pain aggravated by running and jumping. Secondary to excessive plantarflexion and subtalar joint instability.
Calcaneus	Distance runners and jumping athletes	Gradual onset of heel pain made worse by weightbearing. Pain may be reproduced by squeezing the posterior aspect of the calcaneus.

in an Aircast boot in order to facilitate an optimal repairing environment (Lamb et al. 2009).

The use of cryotherapy is widely practised in the acute management of acute soft tissue injuries. The purported benefits of this are the early control of oedema and haemorrhage from damaged vessels, as well as pain relief. Whilst a theoretical argument can be made to support these claims, as yet there is very little substantive proof that the use of cryotherapy improves clinical outcomes following ankle or other soft tissue injury (Collins 2008). The most common method of cryotherapy is applying crushed ice directly to the injured area although plunging the injured ankle into a bucket of iced water is potentially more effective as it combines both the reduced temperature and the compression provided by the water required to restrict vessel diameter and thereby reduce plasma leakage from the vessels. The cooling effect of the ice also helps to reduce pain and may potentially lower cell metabolism, thereby reducing the hypoxia that can lead to cell death. In the first 48 hours, immersing the ankle into an ice bath should be for a maximum of 20 minutes (intermittently) but should be encouraged every 2 hours. In between times, a horseshoe fabricated from felt or foam can be placed around the malleolus and secured with a compressive bandage. Anti-inflammatory medications are discouraged initially because the inflammatory process is necessary for the healing process to commence but may be commenced after 48 hours to ensure that the inflammatory process is not excessive, in particular, to control the collateral damage that may be caused by excessive neutrophil-mediated activity. Recently, there has been some encouraging signs that anti-inflammatory medications, either taken orally or applied topically can be safely consumed/applied and that there is a corresponding decrease in pain and improvements in short-term ankle function (Banning 2008; Bleakley et al. 2008).

Restoration of joint ROM and kinematics

Full weight bearing should be encouraged as soon as possible so long as the gait pattern is not antalgic. This may necessitate an initial period of partial weight bearing with crutches or walking in a pool (to reduce loading). Manual therapy to restore normal talo-crual kinematics and motion at the subtalar and midtarsal joints can be beneficial. It has been suggested that there may be a positional fault at the inferior tibio-fibualr joint following an acute ankle sprain (Kavanagh 1999) and this may need to be assessed and corrected with mobilisations. Active mobilisations should be commenced as soon as is comfortable and this may involve the use of a stationary bike or cross-trainer, hourly range of motion exercises and lunging (particularly in a bucket of iced water).

Proprioceptive retraining

Ankle injuries can disrupt the body's neuromuscular feedforward and feedback mechanisms, which may be displayed as reduced proprioception. Proprioception is the internally generated afferent information arising from peripheral areas of the body that contributes to postural control and joint stability. It is made up of joint position sense, kinaesthesia and resistance/force sense (Riemann and Lephart 2002). There is strong evidence that people with chronically unstable ankles demonstrate proprioceptive deficits whether it be in errors in detecting ankle positions prior to ground contact (Konradsen 2002), failure to accurately replicate passively positioned joint angles (Willems et al. 2002), or an inability to accurately set appropriate muscle force levels to provide joint stability prior to landing from a jump (Docherty and Miller 2002; Docherty and Arnold 2008). It seems that these proprioceptive deficits impair the athlete's ability to *prepare* the ankle to accept and transfer load in challenging athletic tasks such as changing directions and landing from a jump.

There is no overwhelming argument for a definitive cause of such proprioceptive deficits although alterations in gamma-motoneuron activity affecting muscle spindle sensitivity and reducing the "dynamic defence system" of the ankle are thought to predispose the athlete to repeated bouts of instability (Hertel 2002). Indeed, a recent study showed that proprioceptive training was more effective than orthotics or strength training in reducing the rates of ankle sprains in male soccer players (Mohammadi 2007).

Given these proprioceptive deficits are noted in the chronically unstable ankle, it is important to commence proprioceptive training very early on in the rehabilitation programme. Initially this may involve single leg standing and can be made more challenging by performing this with eyes closed and on less stable surfaces such as a trampoline. Wobble boards

Table 22.5 Examples of proprioceptive drills and their progressions

Variable	Example starting exercise	Example progression
Surface	• Hopping onto floor	• Hopping onto inflatable balance mat
	• Running in gym	• Running on grass
Speed	• Jogging through flags	• Sprinting through flags
Sight	• Dolly steps on balance beam with eyes open	• Dolly steps on balance beam with eyes closed
	• Carioca running looking down at feet	• Carioca running looking at roof
Direction	• Straight line running	• Slalom running with increasingly tight turns
	• Forwards ladder tasks	• Backwards ladder tasks
Attention	• Jumping ¼ turns	• Jumping ¼ turns whilst catching and throwing ball
	• Uncontested basketball tip-offs	• Contested basketball tip-offs
Decision making	• Sprint to a cone and cut to left and then right	• Sprint to a cone and cut to left and then right depending on instruction
	• Turn and run on a predicable metronome	• Turn and run on a random metronome
Energy levels	• Sandpit hopping when fresh	• Sandpit hopping when fatigued

are useful as is performing the star excursion balance test (SEBT). The SEBT is a dynamic lower limb reaching test that has been shown to be sensitive in detecting subjects with chronic ankle instability (Olmsted et al. 2002). It involves standing on the ground with the affected foot and reaching as far as possible to touch but not weight-bear on eight diagonal lines positioned 45 degrees from each other in a star shape. Essentially, any activity that challenges the body's proprioceptive system is useful and the more input the system receives the better. There are several variables that can be adapted to make the tasks more challenging (see Table 22.5 for examples).

The proprioceptive programme should be progressive, fun, challenging, functional and goal-orientated towards the sport-specific demands. Additionally, as healing progresses, the stress that the ankle is put under during these proprioceptive drills needs to be increased to include jumping, hopping, turning and cutting tasks. There are no specific rules about progressing the drills but each of the variables (above) should be taken into consideration. It may be that an athlete has no problems with any of them but their ankle feels unstable when they are fatigued. In this case, this is the variable that needs to be concentrated on. The athlete is allowed back to competition once they can demonstrate that they can complete the requisite parts of their sports-specific proprioceptive tasks during rehabilitation.

Muscle strengthening

All directions of motion should be considered when prescribing exercises for muscle strengthening and conditioning around the ankle. This may involve the use of elastic tubing or weighted cables. The ankle should be stressed both concentrically and eccentrically. As pain permits the ankle's range of motion to improve, it is important to strengthen into the new range. This will involve working into progressively more plantarflexion and inversion. Interestingly, the current prevailing thought is that whilst strength is an important consideration during ankle rehabilitation, deficits in ankle strength are not highly correlated with chronic ankle instability (Kaminski 2006) although recently it has been found that there may be a deficit in eccentric plantar flexion strength in subjects with functional ankle instability (Fox et al. 2008). Therefore, towards the later stages of a rehabilitation programme, it is advisable that eccentric strengthening is included. Additionally, it has recently been reported there may also be a deficit in knee flexion and knee extension average torque values in subjects with a history of ankle sprains, indicating that proximal neuromuscular adaptations should be assessed and appropriately managed (Gribble and Robinson 2009). The clinical significance of these findings is that muscle strength should be assessed but not be used as the only return to sport benchmark. It is also important to note that if the athlete is having to take an extended period of time out from their sport due to this injury, it is vital that their cardiovascular system and total body strength work is not neglected (these should be maintained, if not improved during the rehabilitative process).

Functional retraining

Once the athlete can demonstrate that they have a return of full ankle ROM and have appropriate levels of muscle strength, fatigue resistance and proprioception (as determined by the SEBT), they can progress to more functional, sports-specific tasks. This will involve a careful analysis of the requirements of the sport and the impacts it will have on the ankle. For example, for a basketballer, their rehabilitation programme will need to progress from jumping on the spot, to jumping for distance and height, landing in a contest for the ball as well as sudden deceleration and changing of direction when running. For a squash player, there may be more of an emphasis on lunging and rapid pivoting on the foot. Frequently, the athlete will require the application of tape to support the ankle when returning to more vigorous activities. Dependent on the structures that require supporting/protecting, this will often involve positioning the ankle in dorsiflexion (or at least neutral) and eversion, and the application of a combination of anchors, stirrups, calcaneal slings, heel locks, figure 6s and sometimes syndesmosis strips. Taping has been shown to restrict the amount of ankle inversion (Arnold and Docherty 2004) but the benefit is likely to be largely proprioceptive or even placebo (Sawkins et al. 2007). Whilst there are no high-quality studies on how long taping should be used for, clinical experience dictates that it should be used for as long as the athlete finds it beneficial. Nonetheless, the athlete should be counselled that it is probably better not to develop a reliance on taping in case for some reason or other taping is not possible down the track. Ankle braces are commonly used as a prophylactic aid to prevent ankle inversion injuries but they may in fact disrupt balance in non-injured athletes (Hardy et al. 2008).

The appendix to this chapter is a booklet that can be given to athletes on the day of their acute ankle sprain. It is, by no means, meant as a recipe, nor is it exhaustive. What it does provide, however, is a logical, progressive pathway upon which an athlete can be led so that they complete all the tasks required in a rehabilitation programme. There is room for adaptation to include sport-specific drills as required. An outcome measure (such as the Foot and Ankle Score, the Standardised Orthopaedic Assessment Tool or the Chronic Ankle Instability Scale) can be given at the start and at the conclusion of the programme so that the athlete can gauge their progress.

Risk factors for ankle sprains

Several factors have been proposed as risks to ankle sprains over the years. One recent published study suggested that slower running speed, poor cardiorespiratory endurance, impaired joint position sense and increased postural sway, decreased tibialis anterior strength, decreased dorsiflexion ROM and increased activity of the gastro-soleus complex were associated with an increased risk of ankle inversion sprains in males (Willems et al. 2005). A slightly more recent study has stated that a decrease in ankle dorsiflexion ROM is strongly associated with an increased risk for ankle sprains (de Noronha et al. 2006). Given the high rates of chronic ankle instability (see below), a previous ankle sprain is perhaps the most potent predictor of an ankle sprain. All these variables should be taken into account if designing a screening programme for the identification of ankle sprains or pre-disposers.

Chronic ankle instability

The rate of recurrence of an ankle sprain has been reported to be as high as 80% (Yeung et al. 1994) and yet the reasons why this rate is so high are still somewhat mysterious. Chronic ankle instability (CAI) is the term used to describe the occurrence of repeated bouts of instability leading to numerous ankle sprains (Hertel 2002). It has been estimated that CAI occurs in 30–40% of individuals who suffer first time ankle sprains but the occurrence of residual symptoms and decreased function are reported to be much higher (Hertel 2006). CAI gives rise to repeated complaints of pain, swelling, giving way, and degenerative joint disease. Interestingly, it appears that CAI is independent of the severity of the original injury and of the treatment received (Kaminski and Hartsell 2002). It is unclear why some individuals go on to suffer CAI. The mechanism of recurrent ankle sprains is not thought to be dramatically different to that of an initial sprain. Clearly, there is no such thing as a simple ankle sprain.

Ankle instability can be due to mechanical causes or functional causes. Mechanical Ankle Instability (MAI) refers to repeated episodes of 'giving way' due to structural abnormalities within the ankle complex. This can be due to:

- pathological ligamentous laxity

- athrokinematic abnormalities (such as restricted posterior talar glide in the mortise that limits the ankle's ability to reach a fully dorsiflexed (i.e. close-packed) position)

- synovial inflammation and degenerative joint changes (such as degenerative osteochondral lesions).

Functional Ankle Instability (FAI) is said to occur when there are repeated episodes of 'giving-way' without a specific mechanical cause (Kaminski and Hartsell 2002). FAI is a complex matrix of contributing factors that may include:

- impaired proprioception (joint position sense, kinaesthesia and joint resistance sense) (Hertel 2002)

- muscle weakness and arthrogenic muscle inhibition

- altered postural equilibrium sense (Riemann 2002)

- reduction in preparatory muscle activity to stabilise the ankle prior to ground contact (Hertel 2006)

It appears that CAI may be the product of an interaction between MAI and FAI although despite burgeoning research into this area, the precise nature and interaction of all the contributing factors is still unclear (Kaminski and Hartsell 2002). An interesting area of current research is concentrating on the possibilities that delayed trunk reflexes may predispose an individual to CAI. It seems that central nervous system adaptations in the manner of delayed stabilisation times are found in individuals with FAI, which is a finding somewhat akin to those found in individuals with low back pain (Marshall et al. 2009). A deeper appreciation of the features of CAI will help provide the sports clinician with a framework to structure a comprehensive rehabilitation plan aiming at not just the management of the acute ankle sprain but also the avoidance of the potential chronic sequelae. Certainly, a four-week rehabilitation plan designed to address the strength, range of motion, neuromuscular control and functional deficits seen in patients with CAI has been shown to be beneficial in reversing these trends (Hale et al. 2007) although as yet there have not been enough in the way of longitudinal studies undertaken to determine the long-term effect of such interventional strategies on recurrence rates.

Ankle injury prevention

There are many interventions that are purported to be beneficial in the prevention of ankle injuries. It follows that only the modifiable risk factors for ankle injuries could be prevented. Clearly, a past history of ankle sprains, the biggest risk factor, cannot be modified. A high body mass index is another variable that has been identified as a risk factor for non-contact ankle sprains (Gomez et al. 1998). It has not been conclusively proven why this is the case but it has been theorised that it may be due to an inability to control the body during dynamic movements. It follows that an all-encompassing ankle injury prevention programme should address this variable.

A reasonably well-conducted study by McHugh and colleagues in 2007 found that foam balance mat training for five minutes per day during the pre-season and two times per week during the season was effective in eliminating the risk of non-contact ankle sprains in high body mass high school American Football athletes (McHugh et al. 2007). This study was based on previous findings that have demonstrated the effectiveness of single leg ankle disc training in reducing the prevalence of ankle injuries (Bahr et al. 1997).

Whole body vibration is often used in gymnasiums and in some professional sports settings and is marketed as being very good for increasing lower limb proprioceptive awareness. It is unclear if this claim can be proven, and as yet it is unsubstantiated in a high quality study. Indeed, a recent study has shown that it is not protective against ankle inversion sprains (Melnyk et al. 2009).

It is unclear if prophylactic ankle taping or bracing is effective in preventing ankle injuries. Both Sitler and Horodyski (1995) and Surve et al., (1994) reported excellent reductions in ankle sprains using tape and braces, but McHugh et al. (2007) found no difference in American footballers.

There are, therefore, no clear-cut strategies that are guaranteed to prevent ankle sprains. It probably makes sense to ensure that some ankle stabiliser "activation" strategies are employed prior to athletic activity. This may involve practising landing strategies off a height and off-line (activities such as hopping into diagonally opposed hoops placed on the ground), pre-activation of the peroneal muscles with resistance tubing and single leg loading on balance disc/unstable surfaces. The integration of landing strategy training may also assist with the prevention/reduction of knee injuries (see Chapter 21).

Case study

Assessment

History

Allan, a 21year-old semi-professional footballer, presents with right ankle sprain, sustained 24 hours ago during training. The injury mechanism was a plantarflexion-inversion trauma when he landed awkwardly following a jumping drill. He felt pain immediately around the lateral ankle region but was able to limp unaided to the sidelines where he applied ice immediately.

He has 'turned' his right ankle on several occasions in the past three years, once every three months on average. Mostly these incidents are trivial in nature and not accompanied by any significant dysfunction although he has had to miss about five games in total over the past three years due to ankle sprains.

Allan's first ankle sprain occurred almost four years ago when he slipped on a rock when fishing. He was unable to play sport properly for about two months following this original incident. No specific diagnosis was made at the time although he remembers it being markedly swollen and he could not run for about a month. He reports his ankles as feeling "weak" since then but has not seen a clinician about it before. He generally does not tape his ankle although does occasionally borrow his brother's ankle brace.

He has no other injury or medical history, does not smoke and is generally fit and well.

Observations

Allan was able to limp into the clinic unaided but his gait was clearly antalgic. Walking with more weight through his right leg increased his distress. There was a marked effusion around the lateral ankle that extended into the postero-lateral gutter. Some bruising was evident distal to the lateral malleolus.

Area of pain

Allan was asked to point with one finger to his area of pain. He indicated a moderately large area around the antero-lateral ankle. The therapist then gently passively plantarflexed and inverted his ankle and he was able to more precisely locate the pain around the area of the ATFL.

Active movements

The most painful movement directions were PF+Inv, a combined movement that is known to stress the ATFL. He was unable to actively lunge due to pain on weight bearing. Resisted eversion was moderately painful indicating that the peroneal tendons may have been implicated by a traction-type force. Knee and hip ROM were full and pain free.

Passive movements

Corresponding with the active movements, PF+inv was painful. Eversion and DF were pain free. Knee and hip ROM were full and pain free.

Stress tests

Anterior draw was painful and showed and increased excursion although the amount of swelling precluded a definitive grading. Talar tilt test showed no increase in movement.

Functional tests

Due to the fact that this was an acute injury and Allan was struggling to perform the most basic of tasks, walking, functional tests were not considered to be appropriate although a note was made to assess these as his status improved. A knee-to-wall measurement was taken however as a objective benchmark. This was shown to be 0.5cm, compared with 9cm on the uninjured left side.

Palpation

Palpation of the ATFL reproduced Allan's pain. His distal fibula was non-painful, as was the base of the 5th metatarsal. These findings along with the fact that he could walk into the clinic meant that the need for immediate radiological investigations was negated.

Analysis

It was concluded that Allan had sustained an acute ATFL injury. Whilst the clinician was unable to assess the degree of laxity in the ATFL, it was decided that he was functioning as a Grade III injury. Given Allan's past history of ankle sprains, it was concluded that this was an acute episode of CAI.

Management (days 1-3)

Initially, the priority for management was to protect the injured area and limit an excessive inflammatory reaction. Consequently the interventions were:

- Educating Allan regarding the diagnosis, the pain mechanisms involved, the plan for treatment and the likely prognosis. It was also important that he be educated regarding CAI and the need to rehabilitate fully in order to reduce the risk of further injuries.

- RICE regimen. This comprised iced water immersion for 20 minutes and after every 5 minutes he was encouraged to perform 10 forward leans with the ankle immersed in the water. Following this an elasticised compressive bandage with a padded felt "horse shoe" was applied around the lateral malleolus to ensure uniform compression around the injured area.

- Basic balance exercises were prescribed.

- He was provided with a set of elbow crutches and instructed to only put as much weight through his injured side as was pain free. The motto of "it is preferable to walk well with an aid than poorly without". He was instructed in the safe usage of crutches on the flat and on stairs.

Assessment (day 3)

Allan had improved greatly over the course of the two days following the injury. He was now able to walk well without the aid of crutches. The ankle was swollen although much less than 48 hours previously. He still had pain when the ankle was plantarflexed had not attempted to jog. His functional status was reclassified to grade II.

Management (days 4-7)

The aims for management now moved to being about functional restoration. Allan had access to a local gymnasium that contained a pool and so he was allowed to walk in the pool as a means of gait re-education. It was also the ideal environment to commence lunging and provided a challenging environment for balance re-training.

Resistive hip strengthening as well as pelvic control and functional balance work was commenced during this period. Manual therapy was concentrating on normalising muscle tone around the ankle complex and ensuring normal talo-crual joint accessory mobility. He was given a home exercise programme that was to be completed twice daily.

Assessment (day 8)

At the beginning of week 2, Allan's pain was absent during all ADLs and he was able to jog in a straight line. Additionally, whilst there was still some swelling evident around the lateral aspect of his ankle, the postero-lateral gutter had cleared and bruising was no longer evident. His knee-to-wall measurement was 6cm. He was considered to have a functional grade I+ strain by this stage.

Management (days 8–14)

Week two's programme showed a progressive increase in lower limb load, both statically and dynamically. There was a gradual increase in gait speed from walking to jogging, initially in straight lines and then in gentle curves. Some agility work was introduced using ladders and other proprioceptive drills. He was asked to monitor his symptoms (pain and swelling) after each session and if there was an increase in symptoms, he was not allowed to progress to the next day's programme. Towards the end of the week, he was able to complete slalom agility courses at up to 80% speed. He was permitted to return to non-competitive, sub-threshold football training. He was a winger and so a running speed programme was given to him by the coach and he was permitted to re-commence striking a non-moving ball over progressively longer distances.

Manual therapy continued throughout this week and by day 14, talo-crual joint mechanics had been restored. By day 14, he was considered to have a grade I strain. This was because he was still getting a small amount of swelling following the second loading session of the day.

Assessment (day 15)

Allan had regained full physiological and accessory movement of his ankle complex by this stage. There was approximately 2–3mm of extra anterior draw. It was unable to be ascertained whether this was due to the most recent injury or was pre-existing.

Management (days 15–21)

The aims of the 3rd week of the programme were to integrate specific functional tasks into his rehabilitation. He was joining in with training drills and specific emphasis was placed on tackling, jumping/hopping, and deceleration drills. Additionally, specific gluteal strength-endurance tasks were included and high-level proprioceptive drills were performed. He completed a full training session on day 17 without any latent pain or refractory swelling. He played a competitive cup match at day 22.

Ongoing management

Given that Allan had a long history of ankle sprains, resisted ankle strengthening work as well as proprioceptive drills focusing on landing mechanics (for example, hopping onto a wobble board/landing from a step into a sandpit) were continued three times per week for the next two months.

The programme that Allan followed is included in the Appendix.

Appendix – 21-day ankle sprain rehabilitation booklet

21 day balance protocol following acute ankle sprains

booklet contents

1) **acute management information**
   - ✓ assessment
   - ✓ ice
   - ✓ compression

2) **rehabilitation program day 1–21**

   Week 1: recovery
   Acute care strategies aimed at the reduction of swelling, commencement of balance re-training, stimulation of muscle activity and maintenance of fitness.

   Week 2: return to modified training,
   Aim to progress balance work, increase muscle activity and return to modified sporting activity.

Week 3: skill and prevention of further injury
Sport specific drills, functional balance retraining, high-level muscle activity and return to full training.

ACUTE MANAGEMENT

This refers to the first 48 hours after your ankle injury and is vital in your overall recovery. Following an injury and subsequent tissue damage, your body responds through a process of inflammation. Inflammation is good for us and is a necessary part of the healing process. The problem is, the response can be a bit exaggerated and so we look to reduce the amount of swelling. That is why you are required to ice and compress the ankle in the early stages.

Assessment

A physiotherapy and perhaps a medical review is necessary to assess the extent of your injury and proceed with any further investigations (such as an x-ray or MRI) if deemed necessary.

The physiotherapist will also inform you when you can start this rehabilitation program. It is expected that you will complete every level as described, if you have any questions about or problems with the exercises you must discuss these with your physiotherapist at the time.

Ice

Ice applied to the injury site helps limit bleeding and controls inflammation in the ankle and can also helps in pain relief. The best manner of cooling the injury is the use of an ice bath because the water provides compression as well. It also reduces the risk of an ice burn that may occur by applying ice directly onto the skin.

> Because excessive cooling may be counterproductive, icing needs to be limited to 20-minute periods.

How to use an ice bath:

Half fill a suitable container with cold water and add ice. Place your foot into the container; making sure the site of injury is well covered. Every 5 minutes, take your foot out of the water and perform 10 ankle lunges so long as they are not too painful. Repeat this 4 times. After this process, dry your foot and apply the compression bandage. Repeat this process every hour if possible.

Compression

Compression can be very effective in controlling the swelling that often accompanies an ankle sprain. You will also be provided with a foam horseshoe to compress the areas around the outside of your ankle bone. Make sure you don't apply the bandage too tight or it may restrict blood flow.

If necessary, you will be provided with an aircast boot to reduce the risk of re-injury in the acute phase. It is important that you follow your physio's advice regarding wearing the boot, even if you feel that it is too bulky.

WEEK 1

Date programme started:_____
day 1 (day of injury)

The following exercises should be completed TWICE on day 1.

Balance

1) Single leg balance with eyes closed.
   Standing on floor
   8 × 30 seconds

2) STAR Excursion
   Standing on injured leg.
   Touch down (but don't weight bear) on all 8 points.
   Repeat 5×

Strength

3) Double leg calf raises (knees straight and bent)
   From the floor and within pain limits
   3 × 12

4) Calf stretches (pain free)
   3 × 30 seconds

Measurement

5) Knee-to-wall
   Injured foot in front, keeping heel on floor
   cm

Exercises completed
√ am
√ pm

day 2

Continue with physiotherapy, ankle ice baths, compression and elevation as often as possible.

The following exercises should be completed TWICE on day 2.

Balance

1) Single leg balance on trampette with eyes open.
   8 × 15 seconds

2) Single leg balance with eyes closed, standing on floor.
   8 × 15 seconds

3) Karate Kids/sports specific balance
   Blue theraband, standing on injured leg
   2 × 10 kicks each way

4) STAR Excursion
   Standing on injured leg.
   Touch down (but don't weight bear) on all 8 points.
   Repeat 5×

Strength

5) Double leg calf raise (knees straight and bent) and calf stretches
   From a low step within pain limits.
   As per day 1

6) Hip abduction with green tubing
   Standing on injured leg
   3 × 8

7) Hip extension with green tubing
   Standing on injured leg.
   3 × 8

Exercises completed
√ am
√ pm

day 3

Continue with acute ankle management. This includes physiotherapy, ankle ice baths and compression. Continue with CV work.

The following exercises should be completed TWICE on day 3.

Balance

1) Single leg balance. with eyes closed and head up
   Standing on floor
   8 × 30 seconds

2) BOSU standing (blue side up), eyes open/then closed.
   2 × 20 seconds

3) STAR Excursion
   Standing on injured leg and touch down (but don't weight bear) on all 8 points.
   Repeat 5×

Strength

4) Single leg calf raise
   4 × 10 reps

5) Resisted eversion
   Green theraband, control exercise.
   5 × 6 reps

6) Hip abduction AND extension with green tubing
   Standing on injured leg.
   3 × 8

7) Calf stretches (pain free)
   3 × 30 seconds

Gait

8) In pool (15 minutes 1 × per day if pain free)

Exercises completed
√ am
√ pm

day 4

Physiotherapy management as directed
   The following exercises should be completed TWICE on day 4.

Balance

1) Single leg balance with eyes closed
   Standing on floor
   3 × 10 seconds

2) BOSU walking (blue side up)
   Eyes open, and single leg stops.
   50 steps/10 stops

3) Double leg wobble board with eyes closed
   6 × 10 seconds

4) STAR Excursion
   Standing on injured leg
   Touch down (but don't weight bear) on all 8 points.
   Repeat 5×

Strength

5) Single leg calf raise (knees straight and bent) over step
   4 × 10 reps

6) Resisted eversion
   Green tubing-control exercise.
   5 × 6 reps

7) Single leg bridge
   3 × 8 each leg

Gait

8) In pool (15 minutes 1× per day if pain free)

Exercises completed
√ am
√ pm

day 5

Physiotherapy management continued if directed.
   The following exercises should be completed TWICE on day 5.

Balance

1) Single leg balance
   With eyes closed, standing on floor.
   3 × 12 reps

2) Karate kids on ground
   2 × 15 each leg

3) Wobble board
   Single leg balance with knee bent
   6 × 30 seconds each leg

Strength

4) Single leg calf raise over a low step (knees straight and bent)
   3 × 10 reps

5) Single leg step downs
   Control pelvic position – keep belt line level
   3 × 8 reps

6) Resisted eversion.
   Blue tubing with control
   5 × 6 reps

7) Hip abduction and extension with green tubing
   3 × 8 both legs

Running

8) In pool (10 minutes 1× per day if pain free)

Exercises completed
√ am
√ pm

day 6

Physiotherapy management continued if directed.
   The following exercises should be completed TWICE on day 6.

Balance

1) Mini-tramp
   Single leg balance with eyes closed
   6 × 20 seconds

2) Wobble board
   Full circle, edge rolls
   6× each way

3) Tape walking
   Walk forwards and sideways.
   10× each

4) STAR Excursion
   Both sides with green tubing on pointing leg
   Repeat 5×

Strength

5) Double leg forward jump
   3 × 10 reps

6) Eversion over a step
   3 × 8 each foot

7) Single leg bridge
   3 × 8 each leg

Running

8) In pool (15 minutes 1× per day if pain free)

Exercises completed
√ am
√ pm

day 7

The following exercises should be completed TWICE on day 7 without taping.

Balance

1) Karate kids on balance mat
   8 × 10 seconds

2) BOSU jogging
   5 × 30 sec

3) Lunge walking on line
   5 × 10m forward and back

4) Single leg balance on BOSU (black side up)
   8 × 10 seconds each leg – eyes open

Strength

5) Double leg heel raises (knee straight) from floor with dropouts
   3 × 10 reps

6) Eversion over a step
   3 × 12 reps

7) Single leg step downs
   Control pelvic position – keep belt line level
   3 × 8 reps

Running

8) Straight line 15 mins 1× per dayt if pain free

Measurement

9) Knee-to-wall
   Injured foot in front, keeping heel on floor
   cm

Exercises completed
√ am
√ pm

WEEK 2

day 8

Aim to progress balance re-training, increase strength and return to modified training. The programme should be completed TWICE on day 8 and ice should be applied afterwards. Monitor pain and swelling.

Balance

1) Double leg wobble board balance with medicine ball catches above head.
   4 × 10 reps

2) Line walking with medicine ball lifts and twists
   20 cycles

3) Floor jumps
   Single leg, forwards/backwards and side to side.
   3 × 10 reps each way

4) STAR
   Both sides with green tubing on pointing leg
   Repeat 5×

Strength

5) Walk on toes, return on heels
   5 × 10 metres

6) Resisted eversion with concentric speed and eccentric control
   3 × 10 with blue tubing

7) Monster walks with black tubing
   3 × 10 metres forward and back

Running

8) Straight line 15 mins 2× per day if pain free

Exercises completed
√ am
√ pm

day 9

The following exercises should be completed TWICE on day 9 without tape.

Balance

1) Single leg stance on BOSU (blue side up)
   44 × 10 catches each leg

2) Line walking
   Sideways, add hand claps front and back.
   20×

3) Karate kids on BOSU (blue side up)
   3 × 15

Strength

4) Single leg calf raise from the floor
   2 sets to fatigue each leg (perform 2× per day)

5) Walking on the outside of feet.
   5 × 10 metres

6) Single leg press
   2 × 15 @70% (1× per day)

7) Eversion over a step
   3 × 8 each foot

Agility

8) Ladders
   Single feet/double feet/sideways/ickeys
   10× at 70% speed

Running

9) Off-line running 3 × 10 mins 1 × per day if pain free

Exercises completed
√ am
√ pm

day 10

The following exercises should be completed TWICE on day 10.

Balance

1) Mini-tramp hopping with ¼ turns.
   25× each way

2) Line jogging
   20 × 10m

3) Walking on the outside of feet
   5 × 10 metres

4) STAR
   Both sides with green theraband on pointing leg
   Repeat 5×

Strength

5) Double leg heel raises over step with controlled dropouts
   3 × 8 reps

6) Resisted eversion
Blue tubing with concentric speed and eccentric control
3 × 10 reps

7) Monster walk with black tubing
3 × 10 metres forward, back and sideways

Agility

8) Ladders
Single feet/double feet/sideways/ickeys
10× at 70%

Running

9) Off-line running 3 × 10 mins 2 × per day if pain free

Exercises completed
√ am
√ pm

day 11

The following exercises should be completed TWICE on day 11 without taping.

Balance

1) Mini-tramp hopping with eyes closed
50×

2) BOSU (black side up)
Double leg with med ball lifts
6 × 30 seconds

Strength

3) Walking
On toes, return on heels.
5 × 10 metres

4) Theraband eversion
Black tubing- control
4 × 10 reps

5) Single leg press
2 × 15 @70% (1× per day)

Running

6) Slalom course at 70% speed 10 × 20 metres 2 × per day if pain free

Exercises completed
√ am
√ pm

day 12

The following exercises should be completed TWICE on day 12 without taping.

Balance

1) Tape walking with med ball throws
6×

2) Floor jumps.
Alternate $1/4$ and $1/2$ turns.
30 each direction

3) BOSU small knee bends (black side up)
3 × 8 reps

4) STAR
Both sides with blue theraband on pointing leg
Touch down (but don't weight bear) on all 8 points
Repeat 5×

Strength

5) Single leg calf raises (knee straight and bent) over a step
2 sets to fatigue

6) Walking.
On inside of feet.
5 × 10 metres

7) Eversion over a step
3 × 8 each foot

Running

8) Slalom course at 80% speed 10 × 20 metres 2 × per day if pain free

Exercises completed
√ am
√ pm

day 13

The following exercises should be completed TWICE on day 13 without taping.

Balance

1) Jump onto BOSU (blue side up) and $1/4$ turns
   2×10 each

2) Karate kids on BOSU (black side up)
   6×10 reps each leg

Strength

3) Resisted eversion with blue tubing
   Eversion and dorsiflexion controlled.
   4×10 reps

4) Single leg heel raises with drop outs
   3×15 reps

Running

5) Slalom course at 90% speed 10×20 metres $2 \times$ per day if pain free

Exercises completed
√ am
√ pm

day 14

The following exercises should be completed TWICE on day 14 without taping.

Balance

1) Tape walking
   Forwards and sideways, with eyes closed.
   $6\times$ each

2) Jump onto BOSU (blue side up) and $1/4$ turns
   Eyes closed
   10×4 reps

Strength

3) Heel raise and drop walking
   5×10 metres

4) Single leg press
   2×15 @70% $\times$ 1RM ($1\times$ per day)

5) Eversion over a step
   3×8 each foot

Running

6) Slalom course at 90% speed with tighter turns
   10×20 metres $2 \times$ per day if pain free

Exercises completed
√ am
√ pm

WEEK 3

day 15

You are now 2 weeks post injury. You need to continue treatment as directed by your physio. The aim in this week is to return you to full training. Running drills will be directed by your training demands. All exercises should be completed ONCE per day without taping (except for sport specific drills). You should monitor your pain and any increases in swelling and report any increases immediately.

Balance

1) Jumping onto BOSU (black side up)
   3×20

2) Small knee bends on BOSU (black side up) with med ball
   2×20 each leg

3) Ickeys around mat
   3 clockwise + 3 anti-clockwise

Strength

4) Kangaroo jumps
   Double leg, standing long jump.
   3×15 metres

Agility

5) Ladders
   Single feet/double feet/sideways/ickeys
   10× at 85%

Measurement

6) Knee-to-wall
   Injured foot in front, keeping heel on floor
   cm

Exercises completed

day 16

The following exercises should be completed ONCE on day 16.

Balance

1) 30 second lateral hop test on judo mats
   2× each leg

2) BOSU squats with external perturbations
   4 × 15

Strength

3) Single leg press at speed
   2 × 15 @70% × 1RM

4) Resisted eversion and dorsiflexion with blue tubing
   4 × 15 reps

Exercises completed

day 17

The following exercises should be completed ONCE on day 17.

Balance

1) Lateral ladder hopping on gymnastics mats
   5 × 2

2) Small knee bends on BOSU (black side up) and catching ball
   3 × 20 catches each leg

3) Ladders
   Single feet/double feet/sideways/ickeys
   10× at 100%

Strength

4) Eversion over a step
   3 × 8 each foot

5) Walking on toes with bounding on sounding of whistle
   5 × 25m

Exercises completed

day 18

The following exercises should be completed ONCE on day 18.

Balance

1) Hopping in sand pit
   5 × 20sec each leg

2) Small knee bends on BOSU (black side up) and catching ball
   3 × 20 catches each leg

3) Ladders
   Single feet/double feet/sideways/ickeys
   10× at 100%

Agility

4) Sprint + cutting on sound of whistle
   10 reps

Exercises completed

day 19

The following exercises should be completed ONCE on day 19

Balance

1) Hops from varying heights onto gymnastics mats
   5 × 5

2) Hopping onto wobble board
   3 × 20 catches each leg

Strength

3) Heel raise and drop walking
   5 × 10 metres

4) Resisted eversion with blue tubing
   Eversion and dorsiflexion controlled.
   4 × 10 reps

5) Walking on toes with bounding on sounding of whistle
   5 × 25m

Exercises completed

day 20

The following exercises should be completed ONCE on day 20 without taping.

Balance

1) Floor jumps and sprints
   6 jumps (5 metre sprint, 10 reps
   Direction of jump and sprint according to instruction

2) Balance beam
   Single leg balance, with ball passing.
   50 catches

3) Mirroring
   Quick feet
   10 × 15 sec

Strength

4) Eversion reaction on decline board
   No tape, no shoes, eyes closed.
   30×

5) Bounding
   Inside step bounds
   6 × 10m

Exercises completed

day 21

The following exercises should be completed ONCE on day 21

Balance

1) Mini-tramp.
   Single leg jumps, with one hand passing.
   50×

2) Jump, turn and sprint complex
   Double leg line jumps and sideway sprints.
   6 jumps/2 metre sprint/10 reps

3) Floor running.
   Sprints and change direction.
   5 metre sprint/cut left or right/25 reps

Strength

4) Eversion over a step
   3 × 8 each foot

5) Bounding.
   Alternate leg bounds
   6 × 20 metres

Measurement

6) Knee-to-wall
   Injured foot in front, keeping heel on floor
   cm

Exercises completed

References

Agel, J., Olsen, D.E. et al. (2007a) Descriptive epidemiology of collegiate female's basketball injuries: National Collegiate Athletic Association Injury Surveillance System, 1988–1989 through 2003–2004. *Journal of Athletic Training*, 42, 202–210.

Agel, J., Palmieri-Smith, R.M. et al. (2007b) Descriptive epidemiology of collegiate female's volleyball injuries: National Collegiate Athletic Association Injury Surveillance System, 1988–1989 through 2003–2004. *Journal of Athletic Training*, 42. 295–302.

Aiken, A.B., Pelland, L. et al. (2008) Short-term natural recovery of ankle sprains following discharge from emergency departments. *Journal of Orthopaedic and Sports Physical Therapy, 38*, 566–571.

Alfredson, H. and Cook, J. (2007) A treatment algorithm for managing Achilles tendinopathy: new treatment options. *British Journal of Sports Medicine, 41*, 211–216.

Anderson, K.J., Lecocq, J.F. et al. (1952) Recurrent anterior subluxation of the ankle joint: A report of two cases and an experimental study. *Journal of Bone and Joint Surgery, 34A*, 853–860.

Arnold, B.L. and Docherty, C.L. (2004) Bracing and rehabilitation – what's new?" *Clinics in Sports Medicine, 23*, 83–95.

Attarian, D. E., McCracken, H.J. et al. (1985) A biomechanical study of human lateral ankle ligaments and autogenous reconstructive grafts. *The American Journal of Sports Medicine, 13*, 377–381.

Bahr, R., Lian, O. and Bahr, I.A. (1997) A twofold reduction in the incidence of acute ankle sprains in volleyball after the introduction of an injury prevention program: a prospective cohort study. *Scandinavian Journal of Medicine and Science in Sports, 7*, 172–177.

Banning, M (2008) Topical diclofenac: clinical effectiveness and current uses in osteoarthritis of the knee and soft tissue injuries. *Expert Opinion on Pharmacotherapy, 9*, 2921–2929.

Bleakley, C.M., McDonough, S.M. and MacAuley, D.C. (2008) Some conservative strategies are effective when added to controlled mobilisation with external support after acute ankle sprain: a systematic review. *Australian Journal of Physiotherapy, 54*, 7–20.

Brostrom, L. (1964) Sprained ankles, I: anatomic lesions in recent sprains. *Acta Chirurgica Scandinavica, 128*, 483–495.

Collins, N.C. (2008) Is ice right? Does cryotherapy improve outcome for acute soft tissue injury? *Emergency Medicine Journal, 25*, 65–68.

de Noronha, M., Refshauge, K M., Herbert, R D., Kilbreath, S L. and Hertel, J. (2006) Do voluntary strength, proprioception, range of motion, or postural sway predict occurrence of lateral ankle sprain? *British Journal of Sports Medicine, 40*, 824–828.

Docherty, C.L. and Arnold, B. (2008) Force sense deficits in functionally unstable athletes. *Journal of Orthopaedic Research, 26*, 1489–1493.

Docherty, C.L. and Miller III, S.J. (2002) Can chronic ankle instability be prevented? Rethinking management of lateral ankle sprains. *Journal of Athletic Training, 37*, 430–435.

Dombek, M.F., Lamm, B.M. et al. (2003) Peroneal tendon tears: a retrospective review. *The Journal of Foot & Ankle Surgery, 42*, 250–258.

Fong, D., Hong, Y. et al. (2007) A systematic review on ankle injury and ankle sprain in sports. *Sports Medicine, 37*, 73–94.

Fox, J., Docherty, C. L., Schrader, J. and Applegate, T. (2008) Eccentric plantar-flexor torque deficits in participants with functional ankle instability. *Journal of Athletic Training, 43*, 51–54.

Frost, S.C.L. and Amendola, A. (1999) Is stress radiography necessary in the diagnosis of acute or chronic ankle instability? *Clinical Journal of Sport Medicine, 9*, 40–45.

Gaebler, C., Kukla, C. et al. (1997) Diagnosis of lateral ankle ligament injuries: comparison between talar tilt, MRI and operative findings in 112 athletes. *Acta Orthopaedica Scandinavica, 68*, 286–290.

Gomez, J.E, Ross, S.K., Calmbach, W.L., Kimmel, R.B., Schmidt, D.R. et al. (1998) Body fatness and increased injury rates in high school football linemen. *Clinical Journal of Sports Medicine, 8*, 115–120.

Gribble, P.A. and Robinson, R. H. (2009) An examination of ankle, knee, and hip torque production in individuals with chronic ankle instability. *Journal of Strength and Conditioning Research, 23*, 395–400.

Hale, S., Hertel, J. et al. (2007) The effect of a 4-week comprehensive rehabilitation program on postural control and lower extremity function in individuals with chronic ankle instability. *Journal of Orthopaedic and Sports Physical Therapy, 37*, 303–311.

Hamilton, W. G., Geppert, M.J. et al. (1996) Pain in the posterior aspect of the ankle in dancers. *Journal of Bone and Joint Surgery, 78A*, 1491–1500.

Hardy, L., Huxel, K. et al. (2008) Prophylactic ankle braces and star excursion balance measures in healthy volunteers. *Journal of Athletic Training, 43*, 347–351.

Hertel, J. (2002) Functional anatomy, pathomechanics, and pathophysiology of lateral ankle instability. *Journal of Athletic Training, 37*, 364–375.

Hertel, J. (2006) Overview of the etiology of chronic ankle instability. 3rd International Ankle Symposium, Dublin.

Hubbard, T. and Hertel, J. (2006) Mechanical contributions to chronic lateral ankle instability. *Sports Medicine, 36*, 263–277.

Johnson, E.E. and Markolf, K.L. (1983) The contribution of the anterior talofibular ligament to ankle laxity. *Journal of Bone and Joint Surgery, 65*, 81–88.

Kaminski, T.W. and Hartsell, H.D. (2002) Factors contributing to chronic ankle instability: a strength perspective. *Journal of Athletic Training, 37*, 394–405.

Kaminski, T.W. (2006) The relationship between muscle strength deficits and chronic ankle instability. Paper presented at the 3rd International Ankle Symposium, Dublin.

Kavanagh, J. (1999) Is there a positional fault at the inferior tibiofibular joint in patients with acute or chronic ankle sprains compared to normals? *Manual Therapy*, 4, 19–24.

Kinoshita, M., Okuda, R. et al. (2006) Tarsal tunnel syndrome in athletes. *American Journal of Sports Medicine*, 34, 1307–1312.

Konradsen, L. (2002) Factors contributing to chronic ankle insufficiency: kinaesthesia and joint position sense. *Journal of Athletic Training*, 37, 381–385.

Konradsen, L. and Hojsgaard, C. (1993) Pre-heel-strike peroneal muscle activity during walking and running with and without an external ankle support. *Scandinavian Journal of Medicine and Science in Sports*, 3, 99–103.

Lamb, S.E., Marsh, J.L., Hutton, J.L., Nakash, R. and Cooke, M.W. (2009) Mechanical supports for acute, severe ankle sprain: a pragmatic, multicentre, randomised controlled trial. *Lancet*, 373, 575–581.

Leddy, J., Smolinski, R.J. et al. (1998) Prospective evaluation of the Ottawa Ankle Rules in a University Sports Medicine Centre. With a modification to increase specificity for identifying malleolar fractures. *The American Journal of Sports Medicine*, 26, 158–165.

Lee, J.C., Calder, J.D.F. et al. (2008a) Posterior impingement syndromes of the ankle. *Seminars in Musculoskeletal Radiology*, 12, 154–169.

Lee, K.B., Kim, K.H. et al. (2008b) Posterior arthroscopic excision of bilateral posterior bony impingement syndrome of the ankle: a case report. *Knee Surgery, Sports Traumatology, Arthroscopy*, 16, 396–399.

Lynch, S.A. (2002) Assessment of the injured ankle in the athlete. *Journal of Athletic Training*, 37, 406–412.

Marshall, P.W, McKee, A.D. and Murphy, B.A. (2009) Impaired Trunk and Ankle Stability in Subjects With Functional Ankle Instability. Medicine and Science in Sports and Exercise, 41 (8), 1549–1557.

McHugh, M., Tyler, T. F., Mirabella, M. R., Mullaney, M.J. et al. (2007) The Effectiveness of a Balance Training Intervention in Reducing the Incidence of Noncontact Ankle Sprains in High School Football Players. *American Journal of Sports Medicine*, 35, 1289–1294.

Melnyk, M., Schloz, C. et al. (2009) Neuromuscular ankle joint stabilisation after 4-weeks WBV training. *International Journal of Sports Medicine*, 30, 461–467.

Meyer, J.M., Garcia, J. et al. (1986) The subtalar sprain: a roentgeenographic study. *Clinical Orthopaedics and Related Research*, 2226, 169–173.

Mohammadi, F. (2007) Comparison of 3 preventive methods to reduce the recurrence of ankle inversion sprains in male soccer players. *American Journal of Sports Medicine*, 5, 922–926.

Mosher, T.J. (2006) MRI of osteochondral injuries of the knee and ankle in the athlete. *Clinics in Sports Medicine*, 25, 843–866.

Nelson, A.J., Collins, C.L. et al. (2007) Ankle injuries among United States high school sports athletes, 2005–2006. *Journal of Athletic Training*, 42, 381–387.

O'Donoghue, D.H. (1984) *Treatment of Injuries to Athletes*. Philadelphia, PA: WB Saunders.

Olmsted, L.C., Carcia, C.R. et al. (2002) Efficacy of the star excursion balance tests in detecting reach deficits in subjects with chronic ankle instability. *Journal of Athletic Training*, 37, 501–506.

Oloff, L.M., Schulhofer, S.D. et al. (2001) Subtalar joint arthroscopy for sinus tarsi syndrome: a review of 29 cases. *Journal of Foot and Ankle Surgery*, 40, 152–157.

Riemann, B.L. (2002). Is there a link between chronic lateral ankle instability and postural instablity? *Journal of Athletic Training*, 37, 386–393.

Riemann, B.L. and Lephart, S.M. (2002) The sensorimotor system, part I: The physiologic basis of functional joint stability. *Journal of Athletic Training*, 37, 71–79.

Ritchie, D.H. (2001) Functional instability of the ankle and the role of neuromuscular control. *The Journal of Foot and Ankle Surgery*, 40, 240–251.

Safran, M.R., Benedetti, R.S. et al. (1999) Lateral ankle sprains: a comprehensivce review Part 1: etiology, pathoanatomy, histopathogenesis and diagnosis. *Foot and Ankle*, 37, S429–S437.

Sawkins, K., Refshauge, K. et al. (2007). The placebo effect of ankle taping in ankle instability. *Medicine and Science in Sports and Exercise*, 39, 781–787.

Saxena, A. and Cassidy, A. (2003) Peroneal tendon injuries: An evaluation of 49 tears in 41 patients. *The Journal of Foot and Ankle Surgery*, 42, 215–220.

Shalabi, A., Kristoffersen-Wilberg, M. et al. (2007) Eccentric training of the gastrocnemius-soleus complex in chronic Achilles tendinopathy results in decreased tendon volume and intratendinous signal as evaluated by MRI. *The American Journal of Sports Medicine*, 32, 1286–1296.

Sitler, M.R. and Horodyski, M. (1995) Effectiveness of prophylactic ankle stabilisers for prevention of ankle injuries. *Sports Medicine*, 20, 53–57.

Stiell, I.G., Greenberg, G.H. et al. (1992) A study to develop clinical decision rules for the use of radiography in acute ankle injuries. *Annals of Emergency Medicine*, 21, 384–390.

Surve, I., Schwellnus, M. P., Noakes, T. and Lombard, C. (1994) A five-fold reduction in the incidence of recurrent ankle sprains in soccer players using the Sport-Stirrup orthosis. *American Journal of Sports Medicine*, 22, 601–606.

Taylor, D.C., Tenuta, J.J. et al. (2007) Aggressive surgical treatment and early return to sports in athletes with grade III syndesmosis sprains. *The American Journal of Sports Medicine*, 35, 1833–1838.

Willems, T., Witvrouw, E. et al. (2002) Proprioception and muscle strength in subjects with a history of ankle sprains and chronic instability. *Journal of Athletic Training*, 37, 487–493.

Willems, T.M., Witvrouw, E., Delbaere, K., Mahieu, N., De Bourdeaudhuij, I. and De Clercq, D. (2005) Intrinsic risk factors for inversion ankle sprains in male subjects: a prospective study. *American Journal of Sports Medicine*, 3, 415–423.

Wolfe, M. W., Uhl, T.W. et al. (2001) Management of Ankle Sprains. *American Family Physician*, 63, 93–104.

Woods, C., Hawkins, R.D. et al. (2003) The Football Association Medical Research Programme: an audit of injuries in professional football. An analysis of ankle sprains. *British Journal of Sports Medicine*, 37, 233–238.

Yeung, M. S., Chan, K.M. et al. (1994) An epidemiological survey on ankle sprains. *British Journal of Sports Medicine*, 28, 112–116.

23
The foot in sport

John Allen
England Athletics

Introduction

Foot injuries can be challenging to properly diagnose, treat and rehabilitate. Taking the forces of the whole body, plus additional impact sometimes reaching several times body mass, the foot is under tremendous stress. In sport the foot absorbs loading and shearing forces, and to manage the disorders associated there must be a good understanding of the anatomy and biomechanics of the foot. To understand treat and rehabilitate foot problems, a working knowledge of the individual sports and the injuries that are commonly associated with techniques and training methods is essential.

A study by Vereecke and Aerts 2008 on gibbons, the most bipedal of non-human primates and the most distant relatives of humans within the ape lineage, indicated the human foot evolved from one that was primarily of an arboreal (tree dweller) nature, to one that was a kind of hybrid, and finally to the very flexible, yet simultaneously force absorbent locomotive mechanism adapted to bipedalism. Vereecke et al. (2005) maintained an arboreal compliant foot may not be the most effective mechanism for push-off during bipedal locomotion, particularly when compared to the more rigid arched foot of humans, but it does contribute to propulsion in as much as it uses stored elastic energy from passive stretch and recoil of the tendons and ligaments on the plantar side of the foot.

We do still possess extremes of foot posture which tend to predispose to increased potential for injury in sport (Burns et al. 2005), coupled with the tendency for some foot types being more suited to certain sports and susceptible to types of injury (Barnes et al. 2008). In general terms, higher arched semi-rigid feet are more prevalent in short-contact, endurance based sports like distance running, and flatter flexible feet to more powerful longer contact sport like sprinting and rugby. Burns et al. (2005) demonstrate that athletes with a conflicting foot type to their discipline appear to be more prone to injury, as with any biomechanical preferences of certain sports. As with many other parts of the human anatomy, that are functioning and adapting constantly, the foot is only noticed when it fails to function in its subliminal, reliable way.

High arched semi-rigid feet (Figure 23.1)

In extreme form these feet have been described as Pes Cavus or even Claw Foot. This primarily genetic defect in the foot has a high arch which is relatively inflexible. This will often be associated with tight calf muscles combined with reduced dorsiflexion, and predispose to toe running even at slow speeds with little or no heel contact. There is also a tendency to 'roll' the foot from lateral to medial, which may

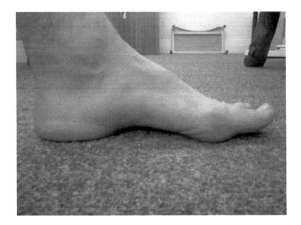

Figure 23.1 High arched semi-rigid foot.

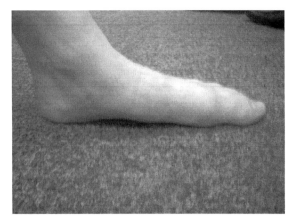

Figure 23.2 Flat flexible foot.

be a necessary compensation to dissipate impact, often associated with wearing out the outside of the shoe.

Athletes find they need to stretch the calf muscles before and particularly after activity for around 30 seconds, repeated three times, once a day (Bandy et al. 1997) and maintain dorsiflexion, although stretching before and after exercise has not been shown to necessarily reduce risk of injury (Herbert and Gabriel 2002). Shoe type selection is difficult as there needs to be support to counter the impact and flexibility to allow some compensatory movement (Orendurff et al. 2008).

Flat flexible feet (Figure 23.2)

Again primarily inherent, this foot type requires shoe selection to provide motion control to avoid stress on structures in some activities, but allow enough movement to tolerate longer contact impact in others (Chuckpalwong et al. 2008).

Injuries caused by either overuse or as a result of a structural non-conformity to the sport are similar in that only through significant activity do pre-existing or underlying structural problems usually reveal themselves. The development of an extremely wide choice of shoes, insoles and orthoses means there is the potential to influence the right compensatory combination to reduce injury potential, but also increases the need for more experience and knowledge of sport specifics, shoe construction and supplemental biomechanical device prescription.

Foot injuries can be divided into three anatomical and structural areas:

1. rearfoot

2. midfoot

3. forefoot.

Injuries to the rearfoot

Plantar fasciitis

Buchbinder (2004) maintains plantar fasciitis is not only the most common cause of pain in the inferior heel, but is responsible for approximately 10% of all running related injuries. Plantar fasciitis is inflammation of the plantar fascia, which may radiate into the arch of the foot aggravated by weight bearing and propulsive activity. It is an overuse injury usually causing plantar proximal and medial pain which may radiate distally to the mid-arch. A strain or sometimes rupture with either gradual or sudden overload is quite common, especially in the active sport population.

Plantar fasciitis can be related to a 'heel spur', which is a bony exostosis at the attachment of the plantar fascia to the calcaneus. A heel spur is believed to be caused by repetitive stress from the plantar fascia attachment and can be present on the asymptomatic foot; or a painful heel may have no heel spur (Ryan 2007).

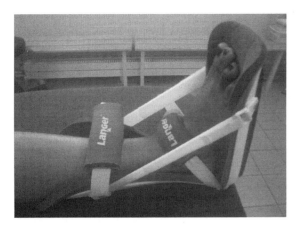

Figure 23.3 Night splint to maintain dorsiflexion.

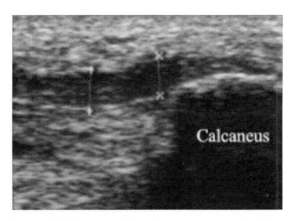

Figure 23.4 Ultrasound scan showing plantar fasciitis inflammation.

Predisposing factors also include: increased body weight, pronation and/or ankle eversion, prolonged periods of standing, reduced ankle dorsiflexion, and high arches. In sport it is often associated with harder training surfaces or increased intensity or duration of training (Huerta et al. 2008).

The lateral plantar nerve passes through the tarsal tunnel between the abductor hallucis muscle and the quadratus planus muscle. It may become compressed causing similar pain under the heel, and sometimes up towards the medial ankle.

Treatment and rehabilitation

Rest including unloading the foot with supportive taping, and even crutches in the early stages, is essential (Cole et al. 2005). A night splint to hold the foot stretched into dorsiflexion can reduce the overnight tendency to tighten (Figure 23.3). Manual tissue release techniques and needling can be effective to reduce tension in the usually tight tissue. Ultrasound guided (Figure 23.4) needle fasciotomy is a relatively new minimally invasive procedure where a needle is inserted into the plantar fascia to possibly disrupt the fibrous tissue that forms as a result of the chronic inflammation.

A gradual return to loading with incremental strengthening and rehabilitation can be frustrating for the athlete as it may take several weeks to strengthen. Many coaches have found movement drills in sandpits to be beneficial and preventative. After biomechanical evaluation, insoles or orthotics may help reduce stress, but may be too compressive during the acute phase. Injection may be helpful short term (Selman 1994) but must be combined with stretching and biomechanical influence to prevent recurrence. Extracorporeal shock therapy, inducing significant vibratory impact directly to the area, has conflicting research evidence (Buchbinder 2002; Sems et al. 2006).

Heel contusion

Repetitive impact on the heel can cause the fat pad that protects the calcaneus to be pushed up the side of the heel leaving less of a protective layer. Sometimes called 'policeman's heel' it is common in sport requiring repetitive impact on the heel like the hopping foot in triple jump. If managed early it should recover within a few days. If not, and it becomes chronic, it can be a long recovery as continued weight bearing tends to sustain the injury. This is considered to be related to a bursa between the calcaneum and the fat pad (Irving et al. 2007).

Treatment and rehabilitation

Insoles are more effective than heel pads as they dissipate rather than concentrate pressure; plus pads raising the heel without supporting the arch, destabilise the foot, increasing stress on the arch (Gosky et al. 2006). This is particularly important for high arched feet where an unsupported arch leaves the heel and forefoot exposed to more impact stress. Taping the heel can compress the soft tissue under the heel during activity, providing increased protection (Hyland et al. 2006).

Calcaneal stress fracture

A stress fracture can occur in the calcaneum as a result of cumulative impact. It is common in sports like long distance running (Fredricson et al. 2006), particularly with athletes who predominantly heel strike rather than mid to forefoot strike, which tends to be the heavier jogger or slower running athlete. The majority of plantar heel pain is probably diagnosed as plantar fasciitis or as heel spurs. When the history or examination findings are atypical or when routine treatment is ineffective, consideration should be given to atypical causes of heel pain (Webber et al. 2005). Stress fractures of the calcaneus are possibly a frequently unrecognised source of heel pain. In some cases they can continue to go unrecognised because the symptoms of calcaneal stress fracture sometimes improve with treatments aimed at soft tissue injuries. Calcaneal stress fractures can occur in any population of adults and are common among active people, such as running athletes. It is very likely that the incidence of diagnosed calcaneal stress fractures will rise with an increased consideration of their possibility. Commonly (Webber et al. 2005) there is an insidious, gradual onset pain, made worse by weight bearing. Pain is reproduced by squeezing the back of the heel from both sides. X-ray will frequently not show any sign of fracture at all, or until after approximately two weeks.

Treatment and rehabilitation

A stress fracture should be suspected if there is inactive or night pain, and requires rest from impact for 6–8 weeks, often including crutches in the early stages (McRae and Esser 2008). Return to sporting impact must be gradual. Ideally starting with minimal heel weight bearing activities, like hydrotherapy, cycling and rowing; progressing to elliptical trainer and treadmill intervals before running. Conversely it may be more appropriate to run quicker intervals on toes rather than slower runs with more heel impact initially. Appropriate shoes with insoles or orthotics to dissipate impact are advisable.

Nerve entrapment

This can be described as 'tarsal tunnel syndrome' or 'medial calcaneal nerve entrapment'. The two conditions are similar with variable symptoms of pain that can be in the heel, toes or the arch of the foot, and numbness and/or pins and needles may be felt in the sole of the foot. There is usually pain when standing and more when running. 'Tinel's test', tapping the nerve just behind the medial malleolus, is uncomfortably sensitive. Ankle eversion and foot pronation appear to stress and irritate the medial and lateral plantar nerves, causing the 'tarsal tunnel syndrome' variation (Kinoshita et al. 2006).

Treatment and rehabilitation

Slow controlled non weight-bearing exercise progressing to weight bearing in positions that do not aggravate; passive mobilisation and manipulation, cryotherapy (ice treatment) or transcutaneous electrical nerve stimulation (TENS) which may reduce sensitivity. Insoles or orthotics to alter the mechanical loading pattern are sometimes beneficial (DiDomenico and Masternick 2006). However, when conservative measures are not successful, surgical decompression (Mullick and Dellon 2008) or excision may result in relief of pain, fortunately with few potential complications

Sever's disease

This condition which is sometimes called calcaneal apophysitis occurs primarily in active children around the age of 9–14 years (Malanga and Ramirez-Del Toro 2008). Although opinion and poorly conducted retrospective case series make up the majority of evidence on this condition, (Scharfbillig et al. 2008) it is usually associated with a rapid growth spurt combined with repetitive weight bearing, propulsive and braking activity and tight calf muscles. Athletes typically complain of heel pain or soreness that improves with rest and increases with running. There is usually tenderness at the back of the heels, especially if you press in or give it a squeeze from the sides.

Treatment and rehabilitation

Rest from impact and propulsion usually resolves the condition, but this may be slow when the growth of the calcaneum is active. Use of a brace or cast does help, probably by inhibiting activity as well as weight bearing. A biomechanical assessment may indicate potential reasons for increased stress on the heel, which if reduced by insoles or orthoses, may assist recovery (Hendrix 2005). The most consistent difference between normal and Sever's patients is related to the more fragmented nature of the symptomatic epiphysis, which may suggest a mechanical etiology for that condition (Volpon and de Carvalho Filho 2002). Modified types of training with less impact are therefore usually necessary for prevention or resolution.

Retrocalcaneal bursitis and Haglund's deformity (Figure 23.5)

Pain at the back of the heel caused by an inflamed bursa which is situated between the Achilles tendon and the calcaneum (Haglund 1972). This is a common condition in athletes, particularly in running related sports, and is often associated with Achilles tendinopathy. A MRI study on Achilles tendonopathy enthesitis found retrocalcaneal bursitis in three-quarters of cases (D'Agostini and Olivieri 2006). The key to successful management is a proper understanding of the anatomy and pathological processes (Solan and Davies 2007). Sensitivity and swelling are especially aggravated when running uphill or on uneven surfaces. Wearing particular shoes that put pressure on the area can aggravate the condition, as can pivoting on the heel when driving.

A bony growth or exostosis can develop at the posterior calcaneum called 'Haglund's deformity'. If conservative management fails then surgery may be necessary to repair the Achilles tendon and reduce the exostosis (Stephens 1994).

Treatment and rehabilitation

Insoles or orthotics to alter the calcaneal alignment can reduce irritation from the side of the Achilles that may be causing avulsion stress, as well as diminish pressure and movement that may be causing frictional irritation (Heneghan and Pavlov 1984). Anti-inflammatory medication orally and even topically, as part of the inflammation is superficial, with protective dressing, usually assists recovery, however, injection or even surgery are used in persistent cases.

Achilles tendinopathy (Figure 23.6)

The Achilles tendon may be damaged and inflamed primarily through overloading during the propulsive phase of the gait cycle. It can be injured cumulatively with overtraining without sufficient recovery, or suddenly, for example by increased quality of training on a harder surface (Paavola et al. 2002). It is estimated that it may account for about 10% of all running related pathologies.

The Achilles tendon has a poor blood supply, which is probably why it is slow to recover. Chronic Achilles tendinopathy is a difficult condition to treat

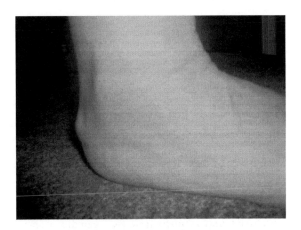

Figure 23.5 Haglund's deformity of the calcaneum.

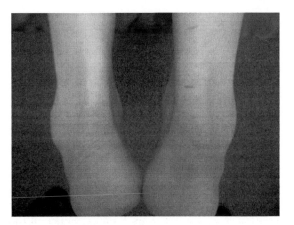

Figure 23.6 Right Achilles tendinopathy with thickening and predisposing calcaneal inversion.

because it is difficult to rest, and older athletes tend to be more susceptible.

A primary reason for the pain becoming chronic is that discomfort experienced often reduces significantly with the start of activity. This is probably due to reactive loss of sensitivity, giving the athlete a false impression with pain returning after the activity. A thickening in the tendon, particularly 2–3 cm proximal to the calcaneum may then develop. An extension of the blood supply may develop (Knobloch et al. 2006), which is now considered to be detrimental to recovery. Ohberg and Alfredson (2002) analysed blood flow in painful Achilles tendons and found some evidence that new blood vessels in the tendon were the cause of tendon pain. MRI or ultrasound scans only have moderate correlation with clinical assessment (Khan et al. 2003), indicating accurate assessment is more significant in determining the degree of Achilles pathology and therefore outcome expectation.

Treatment and rehabilitation

Wearing an insole/orthotic (Pope et al. 2009), rather than a heel raise that destabilises, will raise the heel and reduce dorsiflexion and propulsive stress. This should only be temporary during recovery or loss of length and restricted dorsiflexion may occur. Anti-inflammatory medication may assist in the early stages, but it must not be a mechanism just to allow more activity, except in the case of carefully managed rehabilitation exercises like Alfredson's regime.

Biomechanical causes may be positively influenced by orthotics alongside modifying technique and possibly the training programme. Mobilisation of sub-talar or talo-calcaneal joint eversion may help in improving alignment (Harper 1991) and creating more tolerant stress variability. Taping to support the tendon may assist recovery, and in acute or stubborn cases bracing or a cast can be used initially. Tissue release massage techniques can assist in reducing tightness in the calf complex, especially when the athlete is going through the protective phase (Davidson et al. 1997). Poor clinical evidence supported the use of deep friction massage in a Cochrane review in 2002 showing no obvious benefit, although the reviewers were not evaluating Achilles tendons in particular (Brosseau et al. 2002).

McCrory et al. (1999) concluded that strengthening exercises (see below for specific details) and orthotics to control the degree of rear foot pronation should be utilised for athletes suffering from Achilles tendinopathy. Loss of calcaneal movement, especially eversion, is probably related and needs more research. Corticosteroid injection may be used in chronic situations, however an injection directly into the tendon is not recommended as it may increase the risk of a total rupture. Rehabilitation is essential otherwise the condition is likely to return with increased loading. Wallmann's (2000) and Alfredson et al. (1998) regimes are now used widely but there is some debate as to whether the exercises are strengthening or reflect a specific stretching programme directed at the structures (Allison and Purdam 2009).

Wallmann's regime

Starting with a daily 5–10 minute warm-up of gentle cardiovascular exercise, preferably non-weight bearing, such as cycling. This is followed by a basic calf raise exercise using body weight on the floor, progressed by increasing the load and speed of the eccentric lowering.

The exercise should feel hard, but not painful. If the workout feels the same or easier the next day, then increase to level two difficulty on the next day. Progress until the athlete can achieve level five, which may take weeks to months.

Level 1: Straight leg heel raise with the uninjured leg. Then put the toes of the injured leg down and lower both legs slowly over 5 seconds ×10 repetitions ×3 sets with 30 seconds recovery (this remains the same for all the levels).

Increase the lowering speed to 2 seconds, and to 1 second as discomfort allows (this remains the same for all levels).

Progress to 20 degrees bent knee heel raise, to load the Soleus more.

Level 2: Progress to both legs for lowering and raising.

Level 3: Progress to the uninjured leg alone during the raising phase and the injured leg alone lowering.

Level 4: Progress to both legs during the raising phase and the injured leg alone lowering.

Level 5: Finally, lowering and raising only with the injured leg.

Alfredson's regime (Figure 23.7)

Twice a day for 12 weeks.

Calf raise with the forefoot placed on a step. The athlete shifts body weight pushing up on the uninjured leg and lowers down on the injured leg, slowly and in control, e.g. 5 seconds up and 10 down. Three sets ×15 repetitions of straight-leg calf raise and then the same bent-leg. Progress the eccentric loading by adding weights to a back-pack or a calf-raise resistance machine over the 12 weeks.

Alfredson progresses by increased loading rather than speed and emphasises eccentric lowering through a full range of movement. This means that Wallmann's regime may be more appropriate for faster, lighter shorter contact, and Alfredson's more powerful and longer contact progression. There is also current ongoing research between indicating that Wallmann's exercises in inner range are better for distal enthisopathy, rather than Alfredson's which have the outer range stretch element. Alfredson's research compared two groups. One group had surgery, while the other group used the twelve week programme. Both groups were able to return to normal activity afterwards with the same level of pain relief. This strongly indicated that a progressive emphasis on eccentric loading programme is a very effective alternative to surgery.

In summary, if you look after this injury early enough and rehabilitate the tendon properly you should make a good recovery. Investigators have evaluated the close correlation between good clinical results with eccentric training (Niesen-Vertommen et al. 1992) and a marked reduction in neovascularisation of the tendon (Ohberg et al. 2004; Knobloch 2007). Magnussen et al. (2009) found eccentric exercises still have the most evidence of effectiveness in treatment of midportion achilles tendinopathy Evidence for the role of heel pads, heparin injection, and peritendinous steroid injections is weak. As research continues (Almekinders and Temple 1998), newer approaches are being developed including the use of glyceryl trinitrate (GTN) patches (Paoloni et al. 2004) and the direct or indirect transfer of genes for platelet-derived growth factor and other growth factors that modulate responses to healing and promote type I collagen growth in the tendon (Maffulli et al. 2002).

Achilles tendon rupture (Figure 23.8)

Most individuals with Achilles tendon rupture have had prior tendinopathy (Cetti et al. 2003). The Achilles tendon can partially tear or totally rupture. A total rupture is more common in recreational older athletes.

Sometimes occurring following a history of inflammation, the classic description is of a sudden sharp pain with a distinct bang as if something has hit the back of the leg. Surprisingly, there may not be as much subsequent pain as expected. Movement is not as painful due to there not being a connection,

Figure 23.7 Alfredson's regime with backpack to add more weight as a progression.

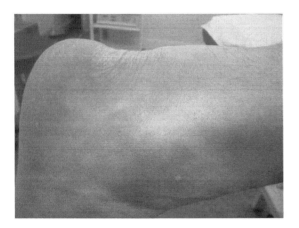

Figure 23.8 Thickening in healed tendon after an Achilles rupture.

and although limping, it is still possible to walk with other lesser plantarflexing muscles. There may be a palpable gap in the tendon usually with significant swelling and a positive result with Thomson's calf squeeze test resulting in loss of foot plantarflexion (Figure 23.9).

Treatment and rehabilitation

It is recommended that the sooner operated on to suture the ends together; the more chance you have of making a full recovery. The reported surgical success rates in surgery vary between 75–100% (Paavola et al. 2002). Progressive plastering without surgery has been used to good effect but tends not to be advocated as there is evidence of a higher re-rupture rate. However, there is a school of thought that believes the differences in rehabilitation used with the two methods may be significant and more research is needed.

Athletes are usually out of competition for 6–12 months, depending on rehabilitation accuracy, motivation and the type of sport (Maffulli and Ajis 2008). Progressive rehabilitation is essential for maximum recovery. Starting with non- and partial weight-bearing exercise using hydrotherapy, cycling, etc., progressing to weight bearing, propulsive and finally explosive activity (Figure 23.1).

Peroneal tendon injuries

Athletes with peroneal tendon dislocation typically present with acute pain and swelling behind the lateral malleolus (Brandes and Smith 2000). These symptoms are caused by a dorsiflexion-inversion stress injury of the ankle that pulls the peroneal retinaculum off the lateral malleolus (Slater 2007). Athletes usually complain of snapping and sudden sharp pain when changing direction or pushing forward with subsequent loss of eversion strength. Problems arising with the peroneal tendons may be tenosynovitis or tendonopathy. The os perineum, a sesamoid bone that may be present within the peroneus longus tendon at about the level of the calcaneo-cuboid joint, may be involved with the degenerative process or as a singular disorder being fractured or fragmented.

On physical examination, there is usually tenderness to palpation along the course of the peroneal tendons, often with swelling. A provocative test for peroneal pathology is to have the athletes foot hanging in a relaxed position with the knee flexed at 90 degrees. Slight pressure is applied to the peroneal tendons posterior to the fibula. The patient is then asked to dorsiflex and evert the foot against

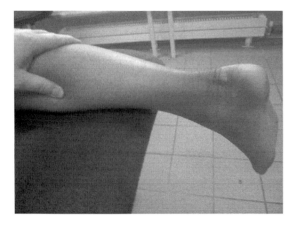

Figure 23.9 Thomson's squeeze test is positive when foot fails to plantarflex.

Table 23.1 Achilles rupture rehabilitation guide (copyright Allen JW 2009)

Months	0	1–2	2–4	3–6	4–8	5–10	6–12
NWB							
PWB							
FWB							
Propulsive							
Explosive							

resistance. Pain may be elicited, or the tendons may be felt to sublux.

Treatment and rehabilitation

Treatment depends on the type of peroneal tendon injury. A cast or splint may be used to immobilise the foot and ankle and allow the injury to heal (van Zoest et al. 2007). Oral and topical anti-inflammatory medication may be used to reduce swelling and pain. As symptoms improve, exercises can be added to strengthen the muscles and improve range of motion and particularly balance. A brace may initially be supportive for a short while, or during activities requiring repetitive ankle movement, or may be an option to allow more activity when a patient is not a candidate for surgery.

In some cases, surgery may be needed to repair the tendon or tendons and perhaps the supporting structures of the foot (Heckman et al. 2008). The foot and ankle surgeon will determine the most appropriate procedure for the patient's condition and lifestyle. After surgery, progressive rehabilitation is still essential.

Injuries to the middle of the foot

Midfoot stress fractures

Stress fractures occur as a result of prolonged repeated loads on the foot in the navicular, talus and metatarsals, usually the second, third and fourth (Snyder et al. 2006). Long distance runners are particularly susceptible to this type of injury and experience pain in the affected bone during and when established, after exercise when resting and even pain at night. The nature of the running process, especially with an increased forefoot loading may help to explain the incidence of stress fractures of the metatarsals under fatiguing loading conditions (Weist et al. 2004). There may be palpable tenderness and swelling at the specific point of the injury, but not always.

The navicular bone is subject to one of the most common stress fractures seen in athletes, especially those involved in sprinting and higher foot impact sports like high jump. It presents with a poorly localised midfoot ache particularly during and after activity. Pain resolves with rest but soon returns with resumed activity. It is usually sensitive to direct pressure on the navicular, and can be picked up with a bone scan, CT or MRI, but may not be evident on X-ray.

The talus is more susceptible to developing a stress fracture in activities where the ankle is repeatedly everted and plantarflexed as in footballers changing direction and jumping athletes. Surgery to remove the lateral process of the talus bone is sometimes done, which can speed up the healing and rehabilitation process. There tends to be a gradual onset pain on the outside of the ankle with palpable tenderness and sometimes swelling over the sinus tarsi.

The sinus tarsi is a small osseous canal which runs into the ankle under the talus bone and can become inflamed, leading to discomfort anterior to the lateral malleolus (Frey et al. 1999), Sinus tarsi syndrome can be caused from overuse in conjunction with poor foot biomechanics usually associated with a past ankle inversion sprain (Breitenseher et al. 1997). Running a curve or bend on an athletics track makes the left foot painful. Passive inversion of the subtalar joint elicits pain. An anaesthetic injection into the sinus tarsi will confirm the diagnosis by relieving symptoms.

Treatment and rehabilitation

Treatment for stress fractures is initially very light to non-weight bearing with crutches and possibly immobilisation in a boot or cast for several weeks until the palpable tenderness disappears (Dugan and Weber 2007). Insoles or orthotics, which support and change the line of stress though the foot, are prudent (Finestone et al. 1999). Gradual progressive loading over several weeks, depending on the intensity of the sport back towards full activity is essential. It is important to mobilise the stiff ankle, subtalar and midtarsal joints. Increase in dorsiflexion by controlled lunging exercises and hydrotherapy helps unload pressure on the midfoot. Oral analgesics, as required, are considered preferable to anti-inflammatories, as they may inhibit repair (Harder and An 2003).

Midfoot tendinopathy

The extensor tendons on the dorsum of the foot can become inflamed causing pain and swelling, especially when the tendons are passively stretched. Inflammation of tibialis anterior is the most common and extensor digitorum is the least. Shoes that are laced too tight, putting increased pressure on the top of the foot, can cause pain and swelling on the dorsum of the foot. It is usually exacerbated by running related sports, particularly by repeated braking and hills, where the extensors lift the foot, or downhill where they work eccentrically to slow the foot (Scott et al. 2005). It can also be triggered by a change in training regime, terrain or slippery surfaces. There is always the possibility of an underlying metatarsal stress fracture indicated by resting and night pain, or when the forefoot is passively deviated.

Flexor tendinopathy of flexor digitorum longus and the flexor hallucis longus can also occur in propulsive and repetitive jumping sports like basketball. With peroneal tendinopathy the peronius brevis and longus assists plantarflexion and eversion. Its tendon attaches to the fifth metatarsal on the lateral aspect of the foot and may become very tight in distance runners.

There may be swelling on the outside of the ankle or heel and pain is directly activity related, especially with sports requiring stability and sideways movement like raquet sports or movement on uneven ground and slopes. Apart from palpable pain, as the tendons are quite superficial, pain is exacerbated by passive invertion and dorsiflexion, and resisted eversion and plantarflexion.

Recent studies have indicated that the causes of tendinopathy may not be as simple as was once thought (Almekinders et al. 2003). Causal factors may include overuse, inflexibility, and equipment problems, but tendon degeneration and biomechanical considerations should be included. More research is needed to determine the significance of stress-shielding and compression in tendinopathy. This finding may significantly alter the approach in both prevention and treatment through exercise therapy. Current biomechanical studies indicate that certain joint positions are more likely to place tensile stress on the area of the tendon commonly affected by tendinopathy (Almekinders et al. 2003).

Treatment and rehabilitation

Rest, reducing aggravating movement is effective as it is primarily an overuse injury. A rehabilitation schedule is essential to strengthen the extensor muscles and stretch the antagonistic calf muscles. Anti-inflammatory medication tends to be very effective, even topical application, as the tendons are very superficial (Forslund et al. 2003). If it becomes established a steroid injection into the area is usually effective.

Tibialis posterior syndrome

The tibialis posterior muscle comes from the posterior tibia and its tendon passes behind the medial malleolus, passing under the foot to help support the medial arch. With cumulative impact, inflammation may occur around the posterior medial malleolus and under the foot where the tendon attaches, with pain as the tendon slides in the sheath (tenosynovitis) during exercise. The syndrome is usually associated with an increase in weight bearing ankle eversion and foot pronation. Athletes with this condition may present with plano-valgus (flattened arch) deformity and often play sports with sudden stop-start or push-off activity, such as soccer, football, and basketball. Patients typically complain of pain inferior to the medial malleolus and decreased range of movement.

Treatment and rehabilitation

Rest for a few of weeks may resolve acute episodes but support by insole or orthotic may be necessary longer term (Kulig et al. 2009). Steroid injection into the tendon sheath or immobilising in a brace or cast can be beneficial. Operation is sometimes necessary to repair a ruptured tibialis posterior tendon as it is essential in supporting the arch of the foot.

Tarsal coalition

This is a congenital fusion of either the calcaneo-navicular or talo-calcaneal joints. It usually affects adolescents as the joints between the bones consolidate. This may cause decreased range of motion in the rearfoot resulting in compensatory pain in the midfoot after weight bearing training and often becomes more obvious after an ankle sprain when the pain does not improve. An X-ray can show an osseous coalition but an MRI may be needed to demonstrate a fibrous one.

Treatment and rehabilitation

The goal of non-surgical treatment of tarsal coalition is to relieve the symptoms and reduce the motion at the affected joint. Treatment and rehabilitation depends on the severity of the condition and the response to treatment (Saxena and Erickson 2003). Sometimes the foot is immobilised to give the affected area a rest. Anaesthetic injection into the leg may be used to relax spasm and is often performed before immobilisation. The foot may benefit from being placed in a boot to immobilise, and crutches to reduce weight on the foot. Management primarily to mobilise the joints around the painful complex, alongside orthotic devices to distribute the weight away from the joint, limiting motion at the joint, may relieve pain.

Non-steroidal anti-inflammatory drugs may be helpful in reducing the pain and inflammation. If symptoms are not adequately relieved with non-surgical treatment, surgery could involve removal of the abnormal connection, or arthrodesis (permanent fusion) of the joint. The surgeon will determine the best surgical approach based on the patient's age, condition, arthritic changes, and activity level (Saxena and Erickson 2003).

Midtarsal joint sprain

The midtarsal joint consists of the calcaneo-cuboid and talo-navicular joints. A sprain of these joints is more likely to be seen in sport with landing impact like gymnastics, or uneven surfaces and change of direction like football. Calcaneo-cuboid injury presents with pain and swelling on the lateral dorsum of the foot often occuring in association with an ankle sprain. Passive inversion of the foot is uncomfortable. X-ray should ideally rule out a fracture.

Treatment and rehabilitation

Taping the foot, and longer term insoles or orthotics, help support the joint recovery (Willems et al. 2005). Exercises to improve and maintain dorsiflexion are important but may have to be patiently employed because of impingement discomfort. If symptoms persist, a steroid injection may aid recovery.

Sometimes after a severe ankle injury there may also be a fracture of the anterior process of the calcaneum. Combined active, and especially passive plantarflexion and eversion is painful. Treatment is similar to that of the calcaneo-cuboid sprain but with a longer period of reduced weight bearing and immobilisation, and surgery may be considered.

Cuboid syndrome

Particularly following an ankle inversion sprain, the peroneus longus may have excessively pulled on the cuboid, causing it to sublux (partially dislocate) resulting in pain when weight bearing on the outside of the foot. The mechanism is often associated with peroneal tendinopathy which should also be treated. It is involved with locking the foot for strength during various stages of the gait cycle. Any instability or dysfunction around the cuboid inhibits functional stability in the foot during propulsion. Athletes with cuboid syndrome will tend to evert and pronate more. They will avoid forcefully pushing off with the foot. Lateral, side-to-side sports, such as tennis or squash, place the greatest strain on this area.

Treatment and rehabilitation

There is some debate (Patterson 2006) whether the cuboid can or needs to be passively manipulated back into position. Non weight bearing hydrotherapy and swimming can be very beneficial, especially in the early stages. Taping to alter the mechanical stress and support the mid-foot may be beneficial for pain relief and recovery (Patterson 2006). Trimming a 0.5 cm thick felt pad approximately 2cm square and taping it under the cuboid may help to support. This is the area under the lateral border of the foot, just behind the base of the fifth metatarsal. Insoles and orthotics may not be tolerated unless relatively soft and flexible.

Lisfranc's fracture/dislocation

This injury to the tarso-metatarsal joints in the foot may be associated with midfoot sprain. If left untreated it may have significant consequences. It presents with midfoot pain and difficulty weight bearing, swelling on the dorsum. Passive plantarflexion causes pain, especially if rotated at the same time. Lisfranc is often missed, even with an X-ray, which is more definitive in a weight bearing position. MRI or bone scan is necessary to confirm the diagnosis.

Treatment and rehabilitation

A plaster cast with a toe plate extending under the toes is applied below the knee to immobilise the joint. Sometimes the bones require fixing with pins or wires. Surgical reduction of the bones is required if there is displacement, with immobilisation for several weeks followed by hydrotherapy and rehabilitation exercises to restore movement and stability (Latterman et al. 2007). Orthoses may be used to correct the intrinsic alignment of the foot and support the second metatarsal base.

Injuries to the forefoot

Metatarsal injuries

Metatarsalgia is an inflammatory condition which occurs in the joints between the metatarsals and phalanges (MTP joints), usually in the second, third or fourth MTP joints. It usually presents with gradual onset forefoot pain, exacerbated by weight bearing (Cohen 2007). Inferior palpation pressure, or flexion, of the MTP joint elicits pain. Sometimes the metatarsal head (MTP) descends or 'drops' causing focal weight bearing pressure. A high arched semi-rigid foot with tight extensor tendons of the toes may increase the potential for metatarsalgia. A short first metatarsal causing increased compensatory pressure through the second metatarsal is called Morton's metatarsalgia. X-ray may show a degree of joint degeneration, which will reduce improvement outcomes.

Treatment and rehabilitation

Anti-inflammatory medication may be prescribed to reduce pain and inflammation, but should only be a short term temporary measure. Material, insoles or orthotics to protect and dissipate pressure can be applied (Kang et al. 2006). Anti-inflammatory or 'lubricating' (Pons et al. 2007) injection into the joint has been successful in chronic cases (Courtney and Doherty 2005).

Neural signs

The metatarsal bones sometimes appear to compress the nerve between the metatarsal bones causing it to be painful or sensory loss particularly between the 3rd and 4th metatarsals. This may be caused by a benign mass (neuroma) on the plantar digital nerve between the toes. Narrow, close fitting shoes, like sprinting spikes, sometimes appear to compress the nerve between the metatarsals. These symptoms are particularly prevalent in sports where there is repeated twisting on the ball of the foot such as golf and squash (Nunan and Giesy 1997). Pain is induced by squeezing the forefoot or pressing between the metatarsals on the dorsal aspect.

Treatment and rehabilitation

A 'metatarsal pad' which elevates and spreads the metatarsals may reduce pressure on the nerve (Kang et al. 2006). These can be incorporated into an orthotics device. Circumferential taping of the foot, although compressive may relieve pain and pressure on the nerve by altering active posture and reducing mechanical compression. There is little evidence to suggest specific foot exercises influence symptoms (Headlee et al. 2008). Decompression operation has been successful in stubborn cases.

Metatarsal fractures (Figure 23.10)

The metatarsals can be fractured through direct impact from, for example solid ground or football studs (Reeder et al. 1996). If the bones are not

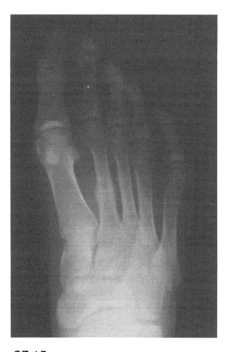

Figure 23.10 Strengthening callus forming around 2nd metatarsal stress fracture.

displaced then a short cast or boot should be fitted for the first three weeks. After six weeks the foot should be X-rayed again to ensure it has healed before a gradual return to sport can start.

More often there is a gradual onset causing a stress fracture (Weinfeld et al. 1997). According to Prentice and Arnheim (2005), this type of injury most commonly occurs due to structural deformities of the foot, changes in training surfaces, training errors, or wearing inappropriate shoes. Additional causes may include a sudden change in training patterns, hill running, running on hard surfaces, or increasing the amount of mileage.

The second, third and fourth metatarsals are the most susceptible (Brukner et al. 1999). When the second toe is longer than the first there is increased risk of damage or fracture of the second metatarsal sometimes classified as 'Frieberg's disease', which on X-ray appears as a flattening of the metatarsal head and epiphysis. It tends to include damage to the articular surfaces (osteochondroses) and most commonly occurs in females between ages 11 and 17.

The increased loading of high heeled shoes may be another predisposing factor, plus it is very common that athletes wear shoes that are too small, increasing compression on the toe (Farber 2007). When it is first noticed there is usually gradual onset pain and restricted movement at the joint, made worse with movement. There may also be swelling and tenderness in the area.

Treatment and rehabilitation

A decrease in activity and occasionally brace or plaster cast may be required to allow recovery. Insoles or orthotics can alter the weight bearing load and assist recovery. Surgical intervention is variable in procedure and success and is evolving.

Fifth metatarsal fractures

Fractures of the base of the fifth metatarsal may occur as a result of pivoting, acceleration and deceleration but particularly twisting the foot and ankle. The ankle inverts and because there is a powerful ligament attached to the base of the metatarsal a small bone fragment avulses. This injury can be treated without surgery, but walking should be in a supportive boot initially.

However, if the junction between the base of the metatarsal and the shaft is fractured, this particular area of the bone has a poor blood supply, and the rate of recovery is longer and may even require surgical intervention. These fractures occur as a result of twisting of the foot when landing, for example in basketball and soccer players.

Treatment and rehabilitation

It is important to differentiate between the acute traumatic fracture sometimes called a 'Jones' fracture in the slower healing area of the metatarsal and the chronic repetitive fatigue stress fracture as management is significantly different (Fetzer and Wright 2006). Although surgery can now be more readily considered, as after a small puncture in the skin, and possibly bone graft and added blood cells to promote healing, a pin is put into the metatarsal across the fracture and athletes are allowed to mobilise and partially weight-bear in a boot immediately following surgery, which reduces stiffness and rehabilitation time (Casillas and Strom 2006).

Rehabilitation is an essential part of recovery, maintaining mobility of the foot (Brukner and Bennell 1997). Exercise, starting in a single plane with bike, elliptical trainer and hydrotherapy, monitored according to swelling and discomfort, progressing to full activities with twisting and pivoting by around 8–10 weeks.

There may be some predisposition in the alignment of the leg and the foot, particularly in genu varum or slightly bow legged individuals incurring increased stress on the outside of the foot including the fifth metatarsal. Also, if the sub-talar and particularly talo-calcaneal joint is stiff, the foot cannot adapt to uneven ground or leg to surface angles. This, combined with an unstable or loose ankle, stress to the peroneal tendons, and stress fracture of the fifth metatarsal are the result of repetitive inversion sprains. There is evidence (Yu et al. 2007) that using off-the-shelf foot orthoses with medial arch support may cause increased plantar forces and pressures on the fifth metatarsal, thereby increasing the risk for proximal fracture of the fifth metatarsal Being aware of soreness along the lateral foot is important, particularly following activity which usually dissipates by the following day. Then reduction of stress by management can contain the situation and avoid further injury.

Hallux injuries

The big toe will sometimes deviate towards the other toes with the first MTP joint becoming medially prominent. This 'bunion' can become quite painful, with bone exostosis and enlargement medially and superiorly. Predisposing, often genetically inherited, biomechanical factors like ankle eversion and pronation are significant (Figure 23.11). Shoes that compress and rub against the first MTP joint may add to the pain experienced. Padding to separate the first and second toes or dissipate the pressure around the medial joint will help temporarily. More permanent relief can be obtained by orthotics or surgery in advanced cases.

Sprain of the first MTP joint can occur with forceful extension of the big toe. Artificial surfaces may cause this injury, where the shoe grips more and extends the toe up further, particularly in soft flexible shoes. It is also common in barefoot sports like judo and dance (Kadel 2006). In repetitive situations over a period of time the articular surface of the MTP joint, or even the bone can be damaged. There may be swelling and pain at the joint of the big toe and metatarsal bone in the foot. Passively extending the toe upwards elicits sharp pain.

Treatment and rehabilitation

Taping or bracing the joint coupled with a firm soled shoe to restrict movement will help in the early stages. Rest and sometimes crutches are needed to reduce weight bearing. If pain does not improve an X-ray is prudent to rule out a fracture. Recovery from this injury can take several weeks. It is essential to mobilise the toe passively with lightly activity in the later stages of recovery to reduce the onset of MTP joint degenerative stiffness called hallux limitus, or potentially the even more restricted hallux rigidus. In some stubborn cases, injection has been considered beneficial in reducing secondary changes.

Hallux sesamoid injuries

Hallux sesamoiditis is manifested by pain beneath the first metatarsal head with weight bearing on the ball of the foot or with motion at the first MTP joint. Common complaints include pain with jumping and with pushing off to run. The two sesamoid bones are embedded in the Flexor Hallucis Brevis tendon, which helps exert propulsive pressure from the big toe against the ground (Dedmond et al. 2006).

The sesamoid bones (Figure 23.12) absorb impact forces in the forefoot during propulsion through a series of attachments to other structures in the forefoot. Although they are separated by a bony ridge on the plantar aspect of the first metatarsal head, they are connected to one another by an inter-sesamoid ligament. They are also attached to other tendons and ligaments in the forefoot.

This complex of attachments enables the sesamoids to disperse some of the impact of the foot as it strikes the ground. The connecting ligaments,

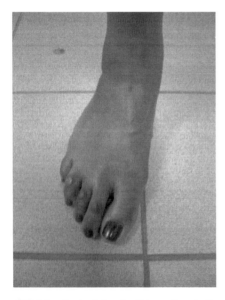

Figure 23.11 Typical flat flexible sprinter's foot with early signs of hallux valgus.

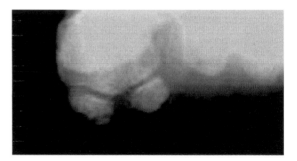

Figure 23.12 Sesamoid bones under the 1st metatarsal head.

the first MTP joint capsule, and the sesamoid bones act as a fulcrum, providing the flexor tendons with a mechanical advantage as they pull the big toe down against the ground during propulsion (Glascoe and Coughlin 2006).

Sesamoiditis typically develops when the structures of the first MTP joint are subjected to chronic pressure and tension and the surrounding tissue is irritated and inflamed. It is a common problem among gymnasts, putting constant force on the ball of the bare or relatively unprotected foot. Repeated stress to the sesamoid bones may also result in fine stress fractures. The pain usually begins as a mild discomfort and increases gradually if the aggravating activity is continued, building to an intense pain with minimal observable signs, although palpably sensitive. The pain may limit the ability of the first MTP joint to dorsiflex, causing restricted big toe extension and propulsive inhibition.

Sesamoid fracture, usually results from traumatic injury to the ball of the foot like when the athlete lands heavily on the feet, fracturing one or both sesamoids. This produces swelling throughout the forefoot and usually a bruise in the area of the big toe's MTP joint. X-ray may not be definitive as some people have bipartite (divided) sesamoids.

Treatment and rehabilitation

If mild, a strict period of reduced weight bearing rest will suffice. Insoles or pad with a cut-out to reduce pressure on the affected area helps recovery. The big toe may be taped to immobilise the joint as much as possible and allow healing to occur. More severe episodes may require oral anti-inflammatories, a below-knee boot for several days to weeks, or an injection. Sesamoid fracture involves keeping the injured foot immobilised and non-weight bearing for six or more weeks. The first MTP joint must also be fully immobilised. In persistent cases a sesamoidectomy (removal) may be necessary, but this can severely compromise normal foot function, especially in sport, and should be avoided.

Summary

To avoid a significant foot and lower leg injury in sport requires the development a comprehensive incrementally progressive training plan. Planning a training programme requires preparation, which stems from a comprehensive assessment of physical needs. The identification of structural problems and their management by appropriate footwear selection, possible orthotic use, and the maintenance of mobility, stability and strength as a preventative measure are significant parts of the athlete's planning and injury prevention process (Whiting and Zernicke 2008). Our lower limbs have evolved in such a way as to allow us to perform complex activities, absorb impacts and generate significant amounts of force. The articulation of our lower limbs assist in force production and absorption through use of angular momentum. The coordinated combination of muscles, levers and joints can allow us to kick a ball, sprint or jump high. Altered biomechanics which may occur through foot injury influencing gait play a role in the genesis of exercise related lower leg pain (Willems et al. 2005) and should always be considered in prevention and rehabilitation. Screening systems have developed that monitor fundamental movement (Kiesel et al. 2007). The information gathered enables the development of potentially corrective exercise programs to positively influence individual movement patterns. They are equally effective in general fitness and specific sports conditioning because they target the weak link in movement and may not only reduce injury susceptibility, but improve athletic performance.

Case study 1

A 26-year-old female elite middle distance runner presented with left foot pain around the distal, dorsal foot, which she had had for one month. She had no causative trauma but had increased her mileage in training from two months prior to the onset of symptoms. She described her pain as a constant dull ache, even when resting and at night, which had been getting progressively worse.

She complained of exquisite tenderness with flexion and extension of her left second toe. She had tenderness to palpation over the dorsum of the distal second metatarsal and MTP joint. She was able to walk but wanted to invert her foot to relieve pressure by putting weight through the lateral foot and avoiding extending the medial toes at push off. There was dorsal swelling present proximal to and across the second to third metatarsals.

Examination of trainers, walking shoes and spikes revealed that they were too small and the second toe was compressed. Damage to the second toe-nail and callus over the toe tip were also observed.

An X-ray revealed slight flattening of the second metatarsal head and joint space narrowing, indicating Freiberg's disease of the left second metatarsal head.

Initial treatment plan

- Firm orthotic insole with a metatarsal bar and cut-out around the second metatarsal head, to be worn as much as possible.

- Non-steroidal anti-inflammatory medication to help relieve discomfort from peri-articular tissue swelling.

- Firmer soled tennis shoes rather than running trainers. One size larger than previous shoes.

- Hydrotherapy, as near daily as possible with deep water light contact running to maintain toe extension without significant loading.

- No running and modified training to avoid propulsive movement but maintain aerobic fitness with increased body weight circuits.

- Existing weight training programme to continue.

Later plan and recommendations

- Larger shoes to allow normal movement without compression of the longer second toe.

- Training to develop as symptoms allowed with progression in running in mileage and intensity.

- Avoiding excessive braking, uneven ground and hills to reduce biomechanical loading.

- Passive mobilising of toe, daily on waking to maintain movement before walking.

The athlete was able to train fully and compete internationally three months from the first assessment. Although symptom free the athlete was warned there will probably be increased degenerative change in the joint in later life.

Case study 2

A 38-year-old male triathlete with right posterior heel pain for three weeks after a 'half-ironman' event, presented with pain on palpation of the distal lateral Achilles tendon insertion. Soreness was first noticed, for approximately one year, as low-grade irritability in the mornings, particularly after longer cycles or runs, and competitions.

Since the half-ironman he had been more aware of a 'lump' on the back of his heel which was sore to prod and was uncomfortable when driving to the extent he had to change the shoes he used to drive. He had noticed he was wearing his shoes down on the outside of the sole heel more than before.

On examination there was palpable discomfort at the peak of a proximo-lateral exostosis/bursa, continuing superior into the Achilles. The prominent osseous bursal projection in association with the Achilles enthisopathy was consistent with a diagnosis of Haglund's syndrome. Talo-calcaneal joint movement was decreased in inversion, and particularly eversion.

MRI highlighted a bony prominence on the postero-superior aspect of the calcaneum. It also showed a high signal consistent with a distal Achilles tendinopathy.

Initial treatment plan

- Initial non-steroidal anti-inflammatory medication followed by self applied topical anti-inflammatory and dressings over the calcaneal prominence to protect until acuteness subsides.

- The footwear he was wearing was changed for trainers with more motion control.

- Off-the-peg insoles with more support were fitted to the trainers to replace the removable less supportive ones that came with the shoe.

- Three-quarter length insoles with heel elevation to alter the relationship between the calcaneal bursal projection, Achilles tendon and shoes were supplied for everyday lace-up shoes. When the heel is elevated, the plantar calcaneal pitch angle decreases and shifts the osseous calcaneal bursal projection away from the shoe, decreasing friction and irritation. The reduction of pronation and assisted re-supination about the sub-talar joint axis also reduces frictional irritation.

- Mobilisation of talocalcaneal eversion to reduce proximo-lateral pressure and friction in shoe and decrease avulsing tension of lateral Achilles. Tissue release techniques to maintain pliability of calf-Achilles.

- Hydrotherapy as near daily as possible for foot and ankle stability, plus early partial weight bearing eccentric emphasis exercises three times daily.

- Full swimming training and slightly modified cycle programmes within discomfort levels.

- No running but to maintain aerobic fitness with increased body weight circuits. Existing weight training programme to continue.

Later plan and recommendations

- Prescription orthotics for increased control.

- Alfredson's eccentric regime to progressively strengthen the calf and Achilles.

- Progress cycling programme to more hills with out of saddle conditioning.

- Start progressive easy running, alternate days initially, straight line build-up interval runs, progressing to longer intervals and then steady runs; avoiding uneven ground.

The triathlete competed in an olympic distance triathlon five months after the ironman competition that caused the main exacerbation of symptoms, still needing to protect the calcaneal prominence during the run with a skin-like dressing. He was advised to have someone passively mobilise the talo-calcaneal joint on a weekly basis.

References

Alfredson, H., Pietilä, T., Jonsson, P. and Lorentzon, R. (1998) Heavy-load eccentric calf muscle training for the treatment of chronic Achilles tendinosis. *American Journal Sports Medicine*, 26 (3), 360–366.

Allison, G.T. and Purdam, C. (2009) Eccentric loading for Achilles tendinopathy – strengthening or stretching? *Br. J. Sports Med*, 43, 276–279.

Almekinders, L.C., Weinhold, P.S. and Maffulli, N. (2003) Compression etiology in tendinopathy. *Clinical Sports Medicine*, 22 (4), 703–710.

Almekinders, L.C. and Temple, J.D. (1998) Etiology, diagnosis, and treatment of tendonitis: an analysis of the literature. *Medicine and Science in Sports and Exercise*, 30 (8), 1183–1190.

Bandy, W.D., Irion, J.M. and Briggler, M. (1997) The effect of time and frequency of static stretching on flexibility of the hamstring muscles. Physical Therapy, 77 (10), 1090–1096.

Barnes, A., Wheat, J. and Milner, C. (2008) Association between foot type and tibial stress injuries: a systematic review. *British Journal of Sports Medicine*, 42, 93–98

Brandes, C.B. and Smith, R.W. (2000) Characterization of patients with primary peroneus longus tendinopathy: a review of twenty-two cases. *Foot Ankle International*, 21 (6), 462–468.

Breitenseher, M.J., Haller J., Kukla C., Gaebler C., Kaider A., Fleischmann D., Helbich T. and Trattnig S. (1997) MRI of the sinus tarsi in acute ankle sprain injuries. *Journal of Computer Assisted Tomography*, 21 (2), 274–279.

Brosseau, L., Casimiro, L., Milne, S., Welch, V., Shea, B., Tugwell, P. and Wells, G.A. (2002) Deep transverse friction massage for treating tendinitis. *Cochrane Database Systematic Reviews 2002, Issue 4*.

Brukner, P. and Bennell, K. (1997) Stress fractures in female athletes. Diagnosis, management and rehabilitation. *Sports Medicine*, 24 (6), 419–429.

Brukner, P.D., Bennell, K.L. and Matheson, G.O. (1999) *Stress Fractures*. Melbourne: Blackwell Scientific.

Buchbinder, R. (2004) Plantar fasciitis. *The New England Journal of Medicine*, 350 (21), 2159–2167.

Buchbinder R et al. (2002) Ultrasound guided extracorporeal shock wave therapy for plantar fasciitis. The Journal of the American Medical Association, 2002, 288 (11), 1364–72.

Burns, J., Keenan, A.M. and Redmond, A. (2005) Foot type and overuse injury in triathletes. *Journal of the American Podiatric Medicine Association*, 95 (3), 235–241, 220.

Casillas, M. and Strom, A. (2006) Internal fixation of fifth metatarsal metaphyseal–diaphyseal stress fractures with an intramedullary screw. *Operative Techniques in Sports Medicine*, 14 (4), 232–238.

Cetti, R., Junge, J. and Vyberg, M. (2003) Spontaneous rupture of the Achilles tendon is preceded by widespread and bilateral tendon damage and ipsilateral inflammation: a clinical and histopathologic study of 60 patients. *Acta Orthopaedica Scandinavica, 74* (1), 78–84.

Chuckpalwong, B., Nunley, J.A., Mall, N.A. and Queen, R,M, (2008) The effect of foot type on in-shoe plantar pressure during walking and running. *Archives of Podiatric Medicine and Rehabilitation, 28* (3), 405–411.

Cohen, J. (2007). Metatarsalgia – Causes, Symptoms, and Treatment Methods. 2 Sep. 2007. EzineArticles.com.

Cole, C., Seto, C. and Gazewood, J. (2005) Plantar fasciitis: Evidence-based review of diagnosis and therapy. *American Family Physician, 72* (11), 2237–2243.

Courtney, P. and Doherty, M. (2005) Joint aspiration and injection. *Best Practice and Research Clinical Rheumatology, 19* (3), 345–369.

D'Agostino, M.A., Olivieri, I. (2006) *Enthesitis. Best practice and research. Clinical Rheumatology, 20* (3), 473–486.

Davidson, C.J., Ganion, L.R., Gehlsen, G.M. et al. (1997) Rat tendon morphologic and functional changes resulting from soft tissue mobilization. *Medicine and Science in Sports and Exercise, 29* (3), 313–319.

Dedmond, B.T., Cory, J.W. and McBryde, A. (2006) The hallucal sesamoid complex. *Journal of American Academic Orthopedic Surgery, 14* (13), 745–753.

DiDomenico, L.A. and Masternick, E.B. (2006) Anterior tarsal tunnel syndrome. *Clinical Podiatric Medicine and Surgery, 23* (3), 611–620.

Dugan, S.A. and Weber, K.M. (2007) Stress fractures and rehabilitation. *Physical Medicine and Rehabilitation Clinics of North America, 18* (3), 401–416.

Farber, D.C. (2007) Foot injuries in the sports population. *Current Opinion in Orthopedics, 18* (2), 97–101.

Fetzer, G.B. and Wright, R.W. (2006) Metatarsal shaft fractures and fractures of the proximal fifth metatarsal. *Clinical Sports Medicine, 25* (1), 139–150.

Finestone, A., Giladi, M. and Elad, H. (1999) Prevention of stress fractures using custom biomechanical shoe orthoses. *Clinical Orthopaedics, 360*, 182–190.

Forslund, C., Bylander, B. and Aspenberg, P. (2003) Indomethacin and celecoxib improve tendon healing in rats. *Acta Orthopaedica Scandinavica, 74* (4), 465–469.

Fredericson, M., Jennings, F., Beaulieu, C. and Matheson, G. (2006) Stress fractures in athletes. *Topics in Magnetic Resonance Imaging, 17*, (5), 309–325.

Frey, C., Feder, K.S. and DiGiovanni, C. (1999) Arthroscopic evaluation of the sub-talar joint: does sinus tarsi syndrome exist? *Foot Ankle International, 20* (3), 185–191

Glascoe, W.M. and Coughlin, M.J. (2006) A critical analysis of Dudley Morton's concept of disordered foot function. *The Journal of Foot and Ankle Surgery, 45* (3), 147–155.

Goske S, A Erdemir, M Petre, S Budhabhatti, P Cavanagh, (2006) Reduction of plantar heel pressures: Insole design using finite element analysis. Journal of Biomechanics, Volume 39, Issue,13, 2363–2370

Haglund, P. (1927) Beitrag zur Klinik der Achillessehne. *Z Orthop Chir., 49*, 49–58.

Harder, A.T. and An, Y.H. (2003) The mechanism of the inhibitory effects of nonsteroidal anti-inflammatory drugs on bone healing: a concise review. *Journal of Clinical Pharmacology, 43*, 807–815.

Headlee, D.L., Leonard, J.L., Hart, J.M., Ingersoll, C.D. and Hertel, J. (2008) Fatigue of the plantar intrinsic foot muscles increases navicular drop. *Journal Electromyography and Kinesiology, 18* (3), 420–425.

Heckman DS, S Reddy, D Pedowitz, KL. Wapner, SG. Parekh. Operative Treatment for Peroneal Tendon Disorders. *The Journal of Bone and Joint Surgery*. 2008; 90: 404–418.

Harper, M.C. (1991) The lateral ligamentous support of the subtalar joint. *Foot Ankle, 11*, 354–358.

Hendrix, C.L. (2005) Calcaneal apophysitis (Sever disease). *Clinical Podiatric Medical Surgery, 22* (1), 55–62, vi.

Heneghan, MA, Pavlov H. (1984) The Haglund painful heel syndrome: experimental investigation of cause and therapeutic implications. Clinical Orthopaedics.

Herbert, R.D. and Gabriel, M. (2002) Effects of stretching before and after exercising on muscle soreness and risk of injury: systematic review. *British Medical Journal, 325*, 468.

Huerta, J.P., García, J.M.A., Matamoros, E.C., Matamoros, J.C. and Martínez, T.D. (2008) Relationship of body mass index, ankle dorsiflexion, and foot pronation on plantar fascia thickness in healthy, asymptomatic subjects. *Journal of the American Podiatric Medicine Association, 98* (5), 379–385.

Hyland, M.R., Webber-Gaffney, A., Cohen, L. and Lichtman, P.T. (2006) Randomized controlled trial of calcaneal taping, sham taping, and plantar fascia stretching for the short-term management of plantar heel pain. *Journal of Orthopedic Sports Physical Therapy 36* (6), 364–371.

Irving, D.B., Cook, J.L., Young, M.A. and Menz, H.B. (2007) Obesity and pronated foot type may increase the risk of chronic plantar heel pain: a matched case-control study. *BMC Musculoskeletal Disorders, 8*, 41.

Kadel, N.J. (2006) Foot and ankle injuries in dance. *Physical Medicine and Rehabilitation Clinics of North America*, 17 (4), 813–826, vii.

Kang Jiunn-Horng, Min-Der Chen, Chen Shih-Ching, Wei-Li His (2006) Correlations between subjective treatment responses and plantar pressure parameters of metatarsal pad treatment in metatarsalgia patients: a prospective study. *BMC Musculoskeletal Disorders*, 7, 95.

Kiesel, K., Plisky, P. and Voight, M. (2007) Can serious injury in professional football be predicted by a preseason Functional Movement Screen? *North American Journal of Sports Physical Therapy*, 2 (3), 147–158.

Knobloch, K. (2007) Eccentric training in Achilles tendinopathy: is it harmful to tendon microcirculation? *British Journal of Sports Medicine*, 41 (6), e2. Epub 2006 Nov 24.

Knobloch, K., Kraemer, R., Lichtenberg, A., Jagodzinski, M., Gossling, T., Richter, M., Zeichen, J., Hufner, T. and Krettek, C. (2006) Achilles Tendon and Paratendon Microcirculation in Midportion and Insertional Tendinopathy in Athletes 2006 *American Orthopaedic Society for Sports Medicine*, 2006 Jan; 34 (1), 92–7.

Khan, K.M., Forster, B.B., Robinson, J., Cheong, Y., Louis, L., Maclean, L., Taunton, J.E. (2003) Are ultrasound and magnetic resonance imaging of value in assessment of Achilles tendon disorders? A two year prospective study. *British Journal of Sports Medicine*, 2003 Apr; 37 (2), 149–53.

Kinoshita, M., Okuda, R., Yasuda, T. and Abe, M. (2006) Tarsal tunnel syndrome in athletes. *American Journal of Sports Medicine*, 34, (8), 1307–1312.

Kulig, K., Reischl, S.F., Pomrantz, A.B., Burnfield, J.M., Mais-Requejo, S., Thordarson, D.B. and Smith, R.W. (2009) Nonsurgical management of posterior tibial tendon dysfunction with orthoses and resistive exercise: A randomized controlled trial. *Physical Therapy*, 89 (1), 26–37.

Lattermann, C., Goldstein, J.L., Wukich, D.K., Lee, S, and Bach, B.R. (2007) Practical management of lisfranc injuries in athletes. *Clinical Journal of Sport Medicine*, 17 (4), 311–315.

Magnussen, R.A., Dunn, W.R., Thomson, A.B. (2009) Nonoperative treatment of midportion Achilles tendinopathy: A systematic review. *Clinical Journal of Sport Medicine*, 19 (1), 54–64.

Maffulli, N. and Ajis, A. (2008) Management of chronic ruptures of the Achilles tendon. *The Journal of Bone and Joint Surgery (American)*, 90, 1348–1360.

Maffulli, N., Moller, H.D. and Evans, C.H. (2002) Tendon healing: can it be optimised? *British Journal of Sports Medicine*, 36 (5), 315–316.

Malanga, G.A. and Ramirez-Del Toro, J.A. (2008) Common injuries of the foot and ankle in the child and adolescent athlete. *Physical Medicine and Rehabilitation Clinics of North America*, 19 (2), 347–371.

McCrory, J.L., Martin, D.F., Lowery, R.P., Cannon, D.W., Curl, W.W., Read, H.M. (1999) The effect of eccentric versus concentric exercise in the management of Achilles tendonitis. *Medicine Science in Sport and Exercise*, 1999; 31, 1374–1381.

McRae, R. and Esser, M. (2008) *Practical Fracture Treatment*, 5th edn. Amsterdam: Elsevier Health Sciences.

Mullick, T. and Dellon, A.L. (2008) Results of decompression of four medial ankle tunnels in the treatment of tarsal tunnels syndrome. *Journal of Reconstructive Microsurgery*, 24 (2), 119–126.

Niesen-Vertommen, S.L., Taunton, J.E. and Clement, D.B. (1992) The effect of eccentric versus concentric exercise in the management of Achilles tendonitis. *Clinical Journal of Sports Medicine*, 2 (2), 109–113.

Nunan, P.J. and Giesy, B.D. (1997) Management of Morton's neuroma in athletes. *Clinical Podiatric Medical Surgery*, 14 (3), 489–501.

Ohberg, L. and Alfredson, H. (2002) Ultrasound guided sclerosis of neovessels in painful chronic Achilles tendinosis: pilot study of a new treatment. *British Journal of Sports Medicine*, 36 (3), 173–175.

Ohberg, L., Lorentzon, R. and Alfredson, H. (2004) Eccentric training in patients with chronic Achilles tendinosis: normalised tendon structure and decreased thickness at follow up. *British Journal of Sports Medicine*, 38 (1), 8–11.

Orendurff, M.S., Rohr, E.S., Segal, A.D., Medley, J.W., Green, J.R. and Kadel, N.J. (2008) Regional foot pressure during running, cutting, jumping, and landing. *American Journal of Sports Medicine*, March 1, 2008 36, 566–571.

Paavola, M., Kannus, P., Järvinen, T.A.H., Khan, K., Józsa, L. and Järvinen, M. (2002) Achilles tendinopathy. *The Journal of Bone and Joint Surgery*, 2002 Nov; 84-A (11), 2062–76.

Paavola M, Kannus P, Paakkala T, Pasanen M, Järvinen M. (2000) Long-term prognosis of patients with achilles tendinopathy. An observational 8-year follow-up study. *American Journal of Sports Medicine*. Sep-Oct; 28 (5): 634–42.

Paoloni, J.A., Appleyard, R.C., Nelson, J., Murrell, G.A.C. (2004) Topical glyceryl trinitrate treatment of chronic non-insertional Achilles tendinopathy. *Journal of Bone and Joint Surgeons of America*, 86, 916–922.

Patterson, S.M. (2006) Cuboid syndrome: A review of the literature. *Journal of Sports Science and Medicine*, 5, 597–606.

Pope, J., Schache, A., Crossley, K., Payne, C., Child, S. and Creaby, M. (2009) Effects of medially posted orthotics on ankle joint complex kinematics and torques in runners with symptomatic Achilles tendinopathy. *Journal of Science and Medicine in Sport*, 12, S19–S19.

Pons, M., Alvarez, F., Solana, J., Viladot, R., Varela, L. (2007) Sodium hyaluronate in the treatment of hallux rigidus. A single-blind, randomized study. *Foot Ankle International*, 28 (1), 38–42.

Prentice, W. and Arnheim, D. (2005) *Essentials of Athletic Injury Management*, 6th edn. New York, NY: McGraw-Hill.

Reeder, M.T., Dick, B.H. and Atkins, J.K. (1996) Stress fractures. Current concepts of diagnosis and treatment. *Sports Medicine*, 22 (3), 198–212.

Ryan, G. (2007) *Evidence-Based Sports Medicine*. Oxford: Blackwell.

Saxena, A. and Erickson, S. (2003) Tarsal coalitions; activity levels with and without surgery. *Journal of the American Podiatric Medical Association*, 93 (4), 259–263.

Selman, J.R. (1994) Plantar fascia rupture associated with corticosteroid injection. *Foot Ankle International*, 15 (7), 376–381.

Sems, A., Dimeff, R. and Iannotti, J.P. (2006) Extracorporeal shock wave therapy in the treatment of chronic tendinopathies. *Journal of the American Acadamy of Orthopedic Surgery*, 14 (4), 195–204.

Scharfbillig, R.W., Jones, S., Scutter, S.D. (2008) Sever's disease: what does the literature really tell us? *Journal of the American Podiatric Medical Association*, 98 (3), 212–223.

Scott, A., Khan, K.M., Heer, J., Cook, J.L., Lian, O. and Duronio, V. (2005) High strain mechanical loading rapidly induces tendon apoptosis: an ex vivo rat tibialis anterior model. *British Journal of Sports Medicine*, May; 39 (5), e25

Slater, H.K. (2007) Acute peroneal tendon tears. *Foot Ankle Clinician*, 12 (4), 659–674, vii.

Snyder, R.A., Koester, M.C. and Dunn, W.R. (2006) Epidemiology of stress fractures. *Clinical Sports Medicine*, 25 (1), 37–52, viii.

Solan, M. and Davies, M. (2007) Management of insertional tendinopathy of the Achilles tendon. *Foot Ankle Clinician*, 12 (4), 597–615, vi.

Stephens, M.M. (1994) Haglund's deformity and retrocalcaneal bursitis. *Orthopedic Clinician of North America*, 25 (1), 41–46.

Vereecke, E & Aerts, P, (2008), 'The mechanics of the gibbon foot and its potential for elastic energy storage during bipedalism', Journal of Experimental Biology, vol. 211, no. 23, pp. 3661–3670.

Volpon,, J.B. and de Carvalho Filho G. (2002) Calcaneal apophysitis: a quantitative radiographic evaluation of the secondary ossification center. *Archives of Orthopaedic and Trauma Surgery*, 122 (6), 338–341.

Wallmann, H. (2000) Achilles tendinitis: eccentric exercise prescription *Health and Fitness Journal*, 4 (1), 7–16.

Weber, J.M., Vidt, L.G., Gehl, R.S. and Montgomery, T. (2005) Calcaneal stress fractures. *Clinical Podiatric Medical Surgery*, 22 (1), 45–54.

Weinfeld, S.B., Haddad, S.L. and Myerson, M.S. (1997) Metatarsal stress fractures. *Clinical Sports Medicine*, 16 (2), 319–338.

Weist, R., Eils, E., Rosenbaum, D. (2004) The influence of muscle fatigue on electromyogram and plantar pressure patterns as an explanation for the incidence of metatarsal stress fractures. *The American Journal of Sports Medicine*. Dec; 32, 1893–8.

Whiting, W.C. and Zernicke, R.F. (2008) *Biomechanics of Musculoskeletal Injury*, 2nd edn. Champaign, IL: Human Kinetics.

Willems, T., Witvrouw, E., Delbaere, K., De Cock, A. and De Clercq, D. (2005) Relationship between gait biomechanics and inversion sprains: a prospective study of risk factors. *Gait Posture*, 21 (4), 379.

Yu, B., Preston, J.J., Queen, R.M., Byram, I.R., Hardaker, W.M., Gross, M.T., Davis, J.M., Taft, T.N. and Garrett, W.E. (2007) Effects of wearing foot orthosis with medial arch support on the fifth metatarsal loading and ankle inversion angle in selected basketball tasks. *Journal of Orthopedic Sports Physical Therapy*, 37 (4), 186–191.

van Zoest, W.J.F., Janssen, R.P.A. and Tseng, C.M.E.S. (2007) An uncommon ankle sprain. *British Journal of Sports Medicine*, 41, 849–850.

Vereecke, E., D'Août K., Van Elsacker, L., De Clercq, D., Aerts, P. (2005) Functional analysis of the gibbon foot during terrestrial bipedal walking: plantar pressure distributions and three-dimensional ground reaction forces. *Am J Phys Anthropol*, 2005 Nov; 128 (3), 659–69.

Index

1/3/5RM *see* one/three/five repetition maximum
6D approach 301–2, 304
21-day ankle sprain rehabilitation booklet 483–92

abdominal muscles 386
abnormal biomechanics 443
acceptance stage 279–80
accessory abductor pollicis longus (AAPL) 375
accessory movements 359
accommodating patients' needs 292
Achilles tendon 475, 501–4
action research model 304–5
active knee extension (AKE) test 30
active listening 337–8
active movement
    ankle injuries 481–2
    assessment 188–9, 193–4
    elbow injuries 344, 358
    knee injuries 416
    shoulder injuries 314, 328–9
active rest 149–50
acute neuromuscular control 429
acute sport injuries 163–84
    ankle injuries 466–73
    case studies 167–71, 177–9
    concussions 166, 174–9
    decision-making 171, 179
    elbow injuries 337–8, 345–8, 355
    first aid and initial therapeutics 171–2
    follow-up/return to sport 179
    groin injuries 390–1
    history-taking 166, 168, 178
    incidence and prevalence 163–5, 166, 175
    initial assessment 165–71, 176–7
    inspection/observation 166–9, 178
    pain management 172–4
    palpation 167, 169, 177–8
    pharmacological agents 174
    special testing 167, 169–71, 178
    sport-specific injuries 165
    tendons 85–7, 90
    wrist and hand injuries 371–6
adaptation 228–33
adductor muscles 386, 392–3
adenosine tri-phosphate (ATP) 40, 69
adherence issues 275, 280–2, 288
AED *see* automated external defibrillation
aerobic performance 40, 45, 53
age effects 191
agility testing 45, 46, 211, 439–40, 488–91
AKE *see* active knee extension
AKP *see* anterior knee pain
alcohol 247, 251, 278
Alfredson's regime 503, 513
altered loading 443–4
amenorrhea 113
AMI *see* arthrogenic muscle inhibition
amino acids 247–8, 258
anabolic steroids 86
anaerobic performance 40, 44–5, 50–1
anaesthesia 192
anger stage 279–80

ankle injuries 465–95
    21-day rehabilitation booklet 483–92
    acute sport injuries 167–72, 466–73
    acute treatment 476–80, 484
    anatomy 465–6
    anterior ankle pain 472–4
    assessment and management 470–3, 481–92
    case study 481–3
    grading of injury severity 472–3
    incidence and prevalence 465
    injury prevention 26, 29, 480–1
    lateral ankle pain 474
    medial ankle pain 475–80
    Ottawa Ankle Rules 469–70
    palpation 471–2, 482
    posterior ankle pain 474–5
    rehabilitation 465–95
    risk factors 479–80
    sprains 100, 166–72, 465, 475–80, 483–4
    stress tests 470–1
annual training plans 148
anterior ankle pain 472–4
anterior cruciate ligament (ACL)
    incidence and prevalence of injury 166
    initial assessment 167
    mechanism of injury 410
    muscular strength and conditioning 236–7, 238
    needs analysis 41, 52, 54
    normal changes through life 99
    progressive rehabilitation 201, 207, 209

Sports Rehabilitation and Injury Prevention Edited by Paul Comfort and Earle Abrahamson
© 2010 John Wiley & Sons, Ltd

anterior cruciate ligament (ACL) (cont.)
  psychology 278, 280–1, 283, 288
  rehabilitation 151–3, 158–9, 407–9, 411, 428–30
  screening 20–1, 22
  surgical reconstruction 407–9, 413–14, 417, 423–6
  treatment of injuries 99, 100–1
anterior draw test 169–70, 470
anterior instability 310, 315
anterior knee pain (AKP) 22, 441–8
anterior primary rami (APR) 134
anterior superior iliac spines (ASIS) 431
anterior talofibular ligament 100, 169–70
anterior tibiofibular ligament (ATFL) 169–70, 465–7, 470–3, 481–2
anterolateral rotary instability 167
antibiotics 86
anticoagulants 192
antioxidants 263
anxiety 289
appendicular skeleton 105–6
applied reasoning model 337, 339
apprehension tests 315–16
appropriate care 10
arthrogenic muscle inhibition (AMI) 409, 413
assessment
  active, passive and resisted movements 188–9, 193–4
  acute sport injuries 165–71, 176–7
  ankle injuries 470–3, 481–92
  clinical orthopaedic examination 190
  clinical procedures 185–6
  elbow injuries 337
  emergency pitchside assessment 194
  fundamental principles 186–7
  groin injuries 387–91, 400–2
  musculoskeletal injuries 185–97
  objective examination 192–4, 195–7, 341–5, 370
  peripheral nerve injuries 121–7
  primary decision-making 187–8
  progressive rehabilitation 199–202
  referred pain 189–90, 191
  regulatory frameworks/documentation 194–7
  shoulder injuries 313–22
  subjective 190–2, 195–7, 337–41
  wrist and hand injuries 368–71, 373–4, 376–7
  see also fitness testing; needs analysis
asymmetry 387–8, 390, 435–6, 440
ATC see Certified Athletic Trainers
athlete's hernia 394–5
athletic trainers 5–6
autogenic training 285
automated external defibrillation (AED) 6
autonomic nervous system 119
avascular necrosis of the femoral head 396–7
avoidable injuries 237–8
avulsion stress 501, 513
axial skeleton 105–6
axontmesis 121, 128

balance
  ankle injuries 484–92
  fitness testing 43, 49–50, 478
  knee injuries 431–6
  progressive rehabilitation 208–9
balanced and coordinated training 100
balanced diets 251–2
bargaining stage 279–80
basic life support (BLS) 6
behaviour of injury 191–2
behavioural responses 275, 280–3, 288
Bennett's fracture 379
beta-endorphin 173
biceps brachii 312, 317–18
Bleep Test see Multi Stage Fitness Test
blood supply 96
bone mineral density (BMD) 112–14
bones see skeletal
Boutonnière deformity 374–5
brachial plexus neuropathy 127–30
bracing 171–2
bridging hold 29–30
British Association of Sport Rehabilitators and Trainers (BASRaT) 4–8
bursitis 351, 501

caffeine 258
CAI see chronic ankle instability
calcaneal stress fractures 500
calcaneofibular ligament (CFL) 466–7, 469, 471–2
calcium 108–9, 249–51, 264
calf heel raises 26, 29
calluses 114, 115
calorific value 246
capsular patterns 186–7
carbohydrates 247, 253–8, 261–2, 264
cardiovascular fitness 45, 58
carpal tunnel syndrome (CTS) 120, 133, 134, 345, 377
carpometacarpal joints 368
cartilage 109–10, 289
cascade membranes 87
Certified Athletic Trainers (ATC) 5–6
cervical ligament 467
cervical spine posture 324–5
chiropody felt/pads 172
chronic ankle instability (CAI) 480, 482
chronic injuries
  groin injuries 390–1
  tendons 83–4, 87–90
  wrist and hand injuries 376–8
circumferential compression 172
clavicular resting position 324
clean lifts 150, 154
clinical orthopaedic examination 190
clinical reasoning 297–306
  cognition/metacognition 303–4
  definition and key concepts 297–8
  example 304–5
  knowledge 300–3
  models 298–9, 303–5
  problem based learning 299–302
closed kinetic chain (CKC) exercises 41, 414–27, 429–33
Code of Ethics 7–8
cognition 303–4
cognitive appraisal 278–9
collagen 79–83, 86, 89, 95, 105
collateral ligament sprains 375–6
combined elevation test 29
common peroneal nerve (CPN) 137
competition periods 148–9
complete proteins 247
complex carbohydrates 247
compliance 281, 288

complications in healing 373–8
component model 303–4
compression *see* rest, ice, compression and elevation
computerised tomography (CT) 470, 474
concussions 166, 174–9
conditioning 223–44
   adaptation potential 232–3
   detraining 234
   elbow injuries 358
   explosive strength 223, 224, 239
   injury prevention 35, 237–8
   intensity/overload 228–31
   isometric training 226, 228
   long response training 226, 227–8
   mechanical demands 235–7
   periodisation 147–8, 150–1
   rate of adaptation 232–3
   rate of force development 224–6
   rehabilitation 238–40
   short response training 226, 227
   specificity of training 232–3, 234–7
   sport-specific training 236–7, 239
   timing 230–2
   training parameters 232–3
confidence 277, 279, 281, 289, 292
confidentiality 7, 10, 197
continuing professional development (CPD) 6, 9
contra-indications 303–4
contralateral extremities 167, 169
contusions 69, 376, 499–500
coping resources 277–8, 286
coracobracialis 312
coracohumeral ligament 310
corticosteroids 86, 192, 356–7, 359–60, 393, 397–8, 502
cortisol 354
crank tests 317
creatine kinase (CK) 354
creep, ligaments 98
cross chest adduction 318
cross-training 402
crutches 409
cryotherapy
   acute sport injuries 171–2, 173
   ankle injuries 477, 482, 484
   elbow injuries 355, 359
   foot injuries 500
   musculoskeletal injuries 71, 74
   progressive rehabilitation 202–4
CT *see* computerised tomography

cubital tunnel syndrome 120, 131–2, 345
cuboid syndrome 507
cyclo-oxygenase 2 (COX-2) inhibitors 174
cytokines 89

de Quervain's disease 376–7
deactivation 25–7
deceleration training *see* plyometrics
decorin 95
deductive reasoning 298
dehydration *see* fluid intake
delayed onset muscular soreness (DOMS) 213–14, 353–4, 358–60, 413
deltoid ligament 311, 468, 469, 475
denial stage 279–80
depression stage 279–80
dermatomes 122–3, 342
detraining 234
dexterity exercises 380
dietary fibre 247
dislocations
   elbow injuries 346–8
   foot injuries 507–8
   groin injuries 395–6
   peripheral nerve injuries 130
   wrist and hand injuries 379
distal radio-ulnar joint (DRUJ) 375
distraction training 433–4
documentation 10–11, 194–7
DOMS *see* delayed onset muscular soreness
drawer tests 315–16
drop tests 20–1
drugs testing 266–7
duration of injury 191, 286–7, 387
duration of training 232–3, 258
dynamic movement 72, 235–6
dynamic proprioceptive training 208–9
dynamic restraint system 207
dynamic stability 49–50, 357, 360

EAST *see* elevated arm stress test
eating disorders 113
eccentric exercise
   knee injuries 436
   musculoskeletal injuries 74
   periodisation 154
   progressive rehabilitation 214
   tendons 89
ECRB tendon 349

educating patients 292
elastic cartilage 109–10
elbow injuries 337–64
   acute sport injuries 337–8, 345–8, 355
   anatomy 339–41
   assessment principles 337
   case study 358–60
   history-taking 337–41
   objective examination 341–5
   overuse injuries 337–8, 348–55
   rehabilitation 346
   return to sport 358–9
   treatment 355–8
electrical nerve stimulation (ENS) 413
electromechanical delay (EMD) 429
electromyography (EMG) 413
electrotherapy 72–4, 173–4, 413, 500
elevated arm stress test (EAST) 129
elevation *see* rest, ice, compression and elevation
emergency pitchside assessment 194
emotional responses 275, 276–80, 283
empty can test 319
end-feels 189, 314–15
endomysium 67–8
endoneurium 119, 121
endurance *see* muscular endurance
energy requirements/balance 245–6, 252–3, 260
enhanced healing 288–9
enthesis 95–6
enthesopathy 85
entrapment neuropathies 120
epicondylitis 345
epimysium 67–8
epineurium 119, 121
epiphysis related injury 399
ergogenic aids 7
essential amino acids 247
ethical considerations 3, 7–8, 197
eversion stress tests 169–70
evidence-based learning 298
explosive strength *see* power output
explosive-ballistic training 225–6
Extended Nordic Musculoskeletal Questionnaire (NMQ-E) 16–17, 35–6
extension control squats 446
extensors 348–51, 365, 375
external rotation lag sign tests 320–2

extracellular matrix (ECM) 79–80, 83, 89–90

facility safety 10
FAI *see* functional ankle instability
fall on an outstretched hand (FOOSH) 338, 347–8
fascicles 95, 119
fat 248, 257–8
fat pad syndrome 441–2
fatigue
  fractures 111–12
  ligaments 97
  musculoskeletal injuries 73, 75
  nutrition 261, 264–5
  progressive rehabilitation 201
  skeletal injuries 112
  tendons 83
fear 289
female triad 113
fibrin 70, 73
fibroblasts 79, 100, 355
fibrocartilage 109–10
field testing 31–3, 44, 53, 471
finger fractures 379
first aid 171–2
fitness testing
  laboratory versus field testing 44, 53
  needs analysis 39, 42–53
  selection and purpose 45–53
  test order 44–5
  validity, reliability and objectivity 42–3, 46–52
  see also individual tests
five repetition maximum (5RM) test 52
flat flexible feet 498
flexibility 199–200, 202, 205–7, 215, 379
flexors 352–3, 386, 427, 475, 506
floating clip systems 22
fluctuating overload 228–30
fluid intake 251, 256–7, 260–2, 264–5
food guide pyramid 252
FOOSH *see* fall on an outstretched hand
foot injuries 497–516
  case studies 511–13
  forefoot injuries 508–11
  midfoot injuries 505–8
  posture 497–8
  rearfoot injuries 498–505

  treatment and rehabilitation 499–513
force acceptance 236
force plates 22, 23
force-time curves 225–6
force-velocity curves 214, 228
forearm injuries 348–55
forefoot injuries 508–11
fractures 110–12
  elbow injuries 346–8
  foot injuries 507–9
  groin injuries 395–6
  wrist and hand injuries 377–8
  see also stress fractures
frequency of injury 286
frequency of training 232–3
frontal plane drills 438
fructose 255
full can test 319
functional ankle instability (FAI) 480
functional integration 380
functional rehabilitation 239–40
  see also progressive rehabilitation
functional tests
  ankle injuries 479, 482
  injury prevention 26, 29–30, 32–5
  musculoskeletal injuries 74–5
  performance 440–1
  periodisation 149–50, 153
  shoulder injuries 314
  wrist and hand injuries 370–1

Gait Arms Legs Spine (GALS) tests 23–6
gait exercises 485–6
gamekeeper's thumb 371–3
gastrocnemius 205
general adaptation syndrome (GAS) 213
general conditioning periods 147–8, 150–1
general practitioners (GPs) 16, 292
Gerber's lift off test 320–1
Gilmore's groin 391, 393–4
glenohumeral joint 130, 309–12, 314–15, 329
glenoid labrum 310
gluteal muscles 26, 386, 427–8
glycaemic index (GI) 247, 255–7
glyceryl trinitrate (GTN) patches 503
glycogen 247, 253–8, 265
glycolytic system 40

goal setting 282–4, 286–7, 289, 291, 358
golfer's elbow 352–3
golgi tendon organs (GTOs) 205–6
graded exercise test (GXT) 53, 357–8
Graduate Sport Rehabilitators (GSR) 4–5
grief-loss model 275, 279–80
groin injuries 385–405
  acute sport injuries 390–1
  anatomy 386–7
  assessment 387–91, 400–2
  avascular necrosis of the femoral head 396–7
  avulsion fractures 396
  case studies 400–2
  children 399–400
  chronic injuries 390–1
  differential diagnosis 391, 401
  hernias 391, 393–5
  hip dislocation/fracture 395–6
  hip labral tears 398
  hip pointers 391–2
  iliopsoas syndrome 393
  impact injuries 391
  incidence and prevalence 67
  nerve entrapment 395
  osteitis pubis 397–8
  pathology of pain 385–6
  rehabilitation 391–402
  strain injuries 390, 392–3
  stress fractures 396
  treatment 391–402

haematomas 70–3, 114
Haglund's deformity 501
hallux/hallux sesamoid injuries 510–11
hamstrings
  clinical reasoning 303–4
  incidence and prevalence of injury 67
  injury prevention 26, 75
  knee injuries 424–6
  muscular strength and conditioning 236–7
  needs analysis 41, 52, 54–6, 58
  pathophysiology 73–5
  peripheral nerve injuries 136, 139
  progressive rehabilitation 201–2, 206–7, 209
  rehabilitation 154
  treatment 74–5

INDEX

hand injuries *see* wrist and hand injuries
Hawkins–Kennedy test 319, 328
head injuries *see* concussions
Health Care Index 165
health screening questionnaires 16–20
heart rate (HR) 53
heel contusions 499–500
heel spurs 498–9
hernias 391, 393–5
hetertrophic ossification 346–7
high arched semi-rigid feet 497–8
high-density lipoprotein (HDL) cholesterol 248
high-energy diets 246
high-voltage, low-frequency electrical stimulation 173–4
hip injuries 388–92, 395–6, 398
histamine 355
history of stressors 277–8
history-taking
    acute sport injuries 166, 168, 178
    ankle injuries 481
    elbow injuries 337–41
    groin injuries 387, 400
    musculoskeletal injuries 190–1, 197
hop tests 43, 48–9
humeral head position 324
hyaline cartilage 109–10
hydration *see* fluid intake
hydrotherapy 398, 500, 504–5, 512–13
hydroxyapatite 105
hyper-extension 341, 343, 349
hypercalorific diets 246
hyperextensions 21
hypertrophy 149, 151, 155–9
hypocalorific diets 246
hypothesis testing 298, 303
hypothetico-deductive reasoning model 298, 303
hysteresis 98

ice treatment *see* cryotherapy; rest, ice, compression and elevation
iliopsoas 26, 393
iliotibial band 22, 441–2
Illinois agility test 46–7
imagery 284–91
immediate care 4–5
immediate tendon injuries 85–7
impact injuries 391

impingement 313, 318–19, 377, 473–5
in-place drills 436, 438
incline squats 446
incomplete proteins 247
induction 298
inductive reasoning 298–9
inferior laxity 316
inferior tibiofibular joint (ITFJ) 465
inferior tibiofibular ligaments (ITFL) 467, 469, 472–3
inflammation
    ankle injuries 477
    assessment 193
    elbow injuries 354, 355–6, 359–60
    foot injuries 499, 501, 505–7, 512
    groin injuries 393
    ligaments 99
    musculoskeletal injuries 70
    progressive rehabilitation 202–4
    skeletal injuries 114–15
    tendons 88, 89
information-processing theory 297
informed consent 338–9
infraspinatus 311, 320
injury prevention
    ankle injuries 480–1
    clinical reasoning 300–1
    muscular strength and conditioning 223, 237–8
    musculoskeletal injuries 75
    needs analysis 54–5, 58
    nutrition 264–7
    recommendations 36
    risk assessment 32–5
    screening 15–37
    sport rehabilitators 4–5
    wrist and hand injuries 373–4, 377, 379
Injury Surveillance System (ISS) 164–5
injury understanding 290–1, 292
inspections 166–9, 178, 193, 341, 343
integrated scapulothoracic rehabilitation 325–6
integrins 80–1
intensity of training 228–30, 232–3, 258, 358
interferential current 174
internal rotation tests 30, 320–1
interosseous talocalcaneal ligament 468

interphalangeal joints 368, 373–6, 379
inter-practitioner reliability scores 30
inversion stress tests 169–70
iron 250–1, 263–4, 265
Iselin's disease 474
isocalorific diets 246
isokinetic testing 20–4
    fitness testing 43, 52
    knee injuries 421–2
    muscular strength and conditioning 238–9
isometric testing 52–3
    knee injuries 421–2
    muscular strength and conditioning 226, 228, 238–9
    musculoskeletal injuries 71–2, 74
    peripheral nerve injuries 130
    progressive rehabilitation 212–14
    tendons 83
isotonic fluids 261
isotonic testing 51–2, 72, 74, 212–14

jersey finger 374
joint reaction force (JRF) 429–30, 436
joints 367–8, 379, 417–24
jump tests 33, 43, 47–8, 55, 57, 211, 439

kinematic testing 22, 23, 25, 48, 477
kinetic chains 357–8
knee braces 414–17
knee injuries 407–63
    agility-biased running drills 439–40
    altered loading 443–4
    anterior knee pain 22, 441–8
    arthrogenic muscle inhibition 409, 413
    balance and perturbation training 431–4
    between-sex differences 430–1
    deceleration training 434–6
    differential diagnosis 441–3
    functional performance tests 440–1
    joint loading 417–24
    knee exercise rehabilitation pathway 431–2
    muscle atrophy 409, 413
    muscular strength and conditioning 238

knee injuries (*cont.*)
  needs analysis 58
  neuromuscular control 429–31
  normal changes through life 99
  osteoarthritis 424–6
  plyometric training 436–9
  progressive rehabilitation 209
  proprioception 428–9, 431
  proximal muscles 426–8
  psychology 289
  quadriceps inhibition 409, 413
  rehabilitation 407–9, 413–48
  sensimotor control 428
  surgical reconstruction 407–9, 413–14, 417, 423–6
  see also anterior cruciate ligament
knowledge 6–7, 300–3
Kübler–Ross stage model 279–80

laboratory testing 44
lacerations 69
lateral ankle injuries 100, 166–72, 474
lateral collateral ligament (LCL) 407, 410–14, 416–17
lateral epicondylitis 133, 348–9
latissimus dorsi 312
laws of strength training 149–50
LBP *see* lower back pain
learned neuromuscular control 429–30
leg raise tests 30–1, 125–6, 138
legal frameworks *see* regulatory frameworks
Leisure Accident Surveillance System (LASS) 163–4
length–tension relationships 84
levator scapulae 312
ligaments 95–103
  acute sport injuries 166–70
  anatomy 95–6
  ankle injuries 465–75
  blood supply 96
  elbow injuries 352–3
  healing processes 95, 99–100
  knee injuries 99, 407
  nerve supply 96–7
  normal changes through life 98–9
  pathology 99
  physiology 97–8
  shoulder injuries 310, 312
  sprains 100–1, 166–70
  treatment of injuries 99–100
  wrist and hand injuries 375–6
  see also anterior cruciate ligament
ligamentum teres tears 395–6
limb symmetry index (LSI) 435–6, 440
linoleic/linolenic fatty acids 248
Lisfranc's fracture/dislocation 507–8
LLI *see* lower limb injuries
load and shift tests 315–16
load tests 317
location of injury 191
long response training 226, 227–8
long thoracic nerves 130–1
loose bodies in the hip 396
low threshold training 34
low-density lipoprotein (LDL) cholesterol 248
lower back pain (LBP) 19–20, 22, 427
lower limb injuries (LLI) 30
lower limb neurodynamic tests 125–6, 139
lower limb rotation 448
LSI *see* limb symmetry index
lumbosacral plexus 135
lumbosacral radiculopathy 138

macro-cycles 146–7, 150, 158
macrophages 70
magnetic resonance imaging (MRI)
  ankle injuries 470, 474–5
  foot injuries 502, 506–7, 512
  groin injuries 392, 394, 396–8, 401–2
  peripheral nerve injuries 136–7
  shoulder injuries 313
  tendons 85–7, 89–90
maintenance stress injury 120
mallet finger 373–4
massage 345, 354, 357, 502
mast cells 355
maximal oxygen consumption (VO_2max) 53–5, 254–5, 263
maximal voluntary contraction (MVC) 206
mechanical demands 40–1, 235–7
mechanoreceptors 96–7, 208
medial ankle pain 475–80
medial collateral ligament (MCL) 99, 353, 407, 410–17, 428
medial epicondylitis 352–3
medial hip rotation 388
median nerves 123–4, 133–4
  see also carpal tunnel syndrome
medical history 192
meniscal injuries 167, 407, 410, 412, 414
mental imagery 284–91
mental skills training (MST) 275, 283–91
mental toughness 275, 282
meso-cycles 146–7, 149–50, 155–9
MET *see* muscle energy technique
metabolic demands 40
metacarpal fractures 379
metacarpalphalangeal joints 368
metacognition 303–4
metalloproteinases 86
metatarsal head position 22
metatarsal injuries 508–11
micro-cycles 146–7
micronutrients 248–51, 263–5
mid-exercise nutrition 255–7, 261–2
midcarpal joints 367–8
midfoot injuries 505–8
midtarsal joint sprain 507
Mills manipulation 359
minerals 250–1, 263–4, 265
mitogen activated protein kinase (MAPK) 80
mobilisation 71, 349, 357, 444–5
mobilisation with movement (MWM) 349
mobiliser muscles 72
Modified Thomas Test (MTT) 26, 29, 388, 390
monosaccharides 247
motor drive 413
movement pattern specificity 40–1, 235
MRI *see* magnetic resonance imaging
MST *see* mental skills training
Multi Stage Fitness Test (Bleep Test) 44, 53
multi-planar training 433
multidisciplinary teams 358
muscle energy technique (MET) 206
muscle fibres 67–71
muscle imbalances 444
muscle stimulation 73
muscular atrophy 120, 409, 413
muscular endurance
  elbow injuries 358
  fitness testing 45, 51–3
  nutrition 258–9, 260
  periodisation 149, 151, 155–9

progressive rehabilitation 199, 212–14
wrist and hand injuries 380
muscular hypertrophy 130
muscular lesions 345–6, 353–4
muscular strength 223–44
  adaptation potential 232–3
  ankle injuries 478–9, 484–92
  detraining 234
  elbow injuries 358
  explosive strength 223, 224, 239
  fitness testing 45, 51–3
  injury prevention 237–8
  intensity/overload 228–31
  isometric training 226, 228
  knee injuries 421–2, 424–5, 429, 444
  long response training 226, 227–8
  mechanical demands 235–7
  nutrition 259, 260
  periodisation 149–51, 155–9
  progressive rehabilitation 199, 212–14
  rate of adaptation 232–3
  rate of force development 224–6
  rehabilitation 238–40
  short response training 226, 227
  specificity of training 232–3, 234–7
  sport-specific training 236–7, 239
  timing 230–2
  training parameters 232–3
  wrist and hand injuries 380
musculoskeletal injuries 67–78
  anatomy 67–8
  assessment 185–97
  destruction/injury phase 70
  electrotherapy 72–4
  groin injuries 386–7
  incidence and prevalence 67
  injury prevention 75
  muscle action/fatigue 112
  needs analysis 41
  pain management 172–4
  pathophysiology 69–70, 73–5
  physiology 68–9
  repair and regeneration 70
  screening 15–37
  stretching 72, 74
  treatment 71–2, 74–5
  wrist and hand injuries 368–71
MVC *see* maximal voluntary contraction

MWM *see* mobilisation with movement
mylenated axons 119
myofibrils 68
myosin 68–9
myositis ossificans 346–7
myotendinous junctions 67
myotomes 122

National Athletic Trainers' Association 5–6
National Collegiate Athletic Association 164–5
National Strength and Conditioning Association (NSCA) 45
neck stiffness 129
needs analysis 39–63
  aerobic performance 40, 45, 53
  anaerobic performance 40, 44–5, 50–1
  demands of sport 39–42
  direction and velocity of force 41–2
  fitness testing 39, 42–53
  football 54–6
  mechanical demands 40–1
  metabolic demands 40
  muscle action 41
  rugby league 56–8
  sport-specific tests 53–8
Neer's tests 318, 328
negative adaptation 231
negligence 9–10
negotiation stage 279–80
neovascularisation, tendons 89, 90
nerve entrapment 395, 500
nerve gliding 132, 134
nerve injuries *see* peripheral nerve injuries
nerve supply 96–7
neural signs 508
neurodynamic testing 122–7, 135, 137–8
neuromuscular control
  elbow injuries 358
  knee injuries 429–31
  muscular strength and conditioning 239–40
  progressive rehabilitation 207–12
neuropraxia 121, 128
neutrophils 89, 354
NF-kappaB *see* nuclear factor
non-capsular patterns 186–7
non-coping tendons 87

non-essential amino acids 247
non-mylenated axons 119
non-steroidal anti-inflammatory drugs (NSAIDs)
  acute sport injuries 174
  elbow injuries 359
  foot injuries 507, 512
  peripheral nerve injuries 127, 128, 129–34, 137–8
  skeletal injuries 115
  tendons 90
Nordic eccentric exercise 74, 154
Nordic hamstring lowers 54–6, 237
nuclear factor (NF-kappaB) 80
nutrients 246–51
nutrition 245–73
  alcohol 247, 251
  balanced diets 251–2
  calorific value and energy 246, 252–3, 260
  carbohydrates 247, 253–8, 261–2, 264
  energy balance 245–6
  fat 248, 257–8
  fluid intake 251, 256–7, 260–2, 264–5
  fundamentals 245–52
  injury prevention 264–7
  mid-exercise 255–7, 261–2
  minerals 250–1, 263–4, 265
  nutrients 246–51
  performance 252–64
  post-exercise 256–7, 259, 262
  pre-exercise 254–5, 257, 259, 261–2
  proteins 247–8, 256–7, 258–60
  rehabilitation 265–7
  skeletal injuries 115
  sport-specific requirements 252–3, 258–60
  supplementation 263, 265–7
  vitamins 248–50, 263–4

objective examination 192–4, 195–7, 341–5, 370
oblique plane drills 438
O'Brien's test 316
observation
  acute sport injuries 166–9, 178
  ankle injuries 481
  elbow injuries 341
  musculoskeletal injuries 190
  progressive rehabilitation 199–202

occupational effects 191
OCD *see* osteochondral defects
olecranon 348, 351
Olympic lifts 149, 150–3, 226
omega 3/6 fatty acids 248
one repetition maximum (1RM) test 42, 51–2, 55, 57
  knee injuries 418–20
  periodisation 149, 151, 158–9
onset of injury 191
open kinetic chain (OKC) exercises 41, 417–18, 422–4, 429, 431–2
oral analgesics 174
Orebro Musculoskeletal Pain Screening Questionnaire (OMPSQ) 16, 19
orthopaedic examination 190
ossification 107–8
osteitis pubis 397–8
osteoarthritis (OA) 110, 355, 390, 424–6
osteoblasts 107–8
osteochodritis dissecans of the capitullum 351–2
osteochondral defects (OCD) 473–4
osteoclasts 107–8
osteocytes 107–8
osteopenia 112
osteoporosis 112–14, 249, 264
Ottawa Ankle Rules (OAR) 170, 469–70
overcompensation 35
overhead stress injuries 309–10, 312–13, 328–30
overload 145, 228–31
overreaching 231–2
overtraining syndrome 232
overuse injuries 337–8, 348–55, 376–8

pain management
  acute sport injuries 172–4
  ankle injuries 469–70, 476–7
  assessment 192
  elbow injuries 350
  groin injuries 397, 401–2
  knee injuries 444–5
  progressive rehabilitation 202–4
  psychology 289
pain perception 189–90
pain provocation tests 317
palpation
  acute sport injuries 167, 169, 177–8

ankle injuries 471–2, 482
assessment 188–9, 194
elbow injuries 343, 345
shoulder injuries 314
wrist and hand injuries 369–70
palsy 130–1
panner's disease 352
para-medical roles 4
Paracetomol 174
paraesthesia 192
paratendonitis 85
parathyroid hormone (PTH) 108–9, 115
partial tendon rupture 85, 503–4
partial weight bearing (PWB) 409, 411–12
passive movement
  ankle injuries 482
  assessment 188–9, 193–4
  elbow injuries 344, 358
  knee injuries 416
  shoulder injuries 314
passive recovery 256
passive stretching 72, 205–6
patella injuries 87, 167, 441–2
patellofemoral pain syndrome (PFPS) 22, 441–2
pattern recognition 298–9
PBL *see* problem based learning
pectoralis major/minor 311
pelvis 386, 389–90
performance
  functional performance tests 440–1
  goals 284
  groin injuries 398
  matrix 32–5
  needs analysis 40, 44–5, 50–1, 53
  nutrition 252–64
  periodisation 150–1
  psychology 276, 284, 287
perimysium 67–8
perineurium 119
periodisation 145–61
  competition periods 148–9
  conditioning periods 147–8, 150–1
  muscular strength and conditioning 223, 225
  performance 150–1
  progressive overload 145
  rehabilitation 151–9
  sport-specific training 150
  training cycles 145, 146–50

transition/recuperation periods 148, 149–50, 155–7
peripheral nerve injuries 119–41
  anatomy 119–20
  assessment 121–7
  axillary nerves 130
  brachial plexus neuropathy 121, 127–30
  case study 139
  classification 120–1
  long thoracic nerves 130–1
  lower limb nerve injuries 134–8
  median nerves 123–4, 133–4
  neurodynamic testing 122–7, 135, 137–8
  peroneal nerves 137
  posterior thigh injury 136–7
  radial nerves 124, 133
  sciatic nerves 135
  sliding/tensioning techniques 126–7, 137–8
  suprascapular nerves 131
  tibial nerves 138
  treatment 127–39
  ulnar nerves 124–5, 131–2
peritendinitis 85
peroneal injuries 137, 474, 504–5
perosteum 107
persisting injuries 7
personality 276–8, 281
Perthes disease 399
perturbation training 431–4
PETTLEP model 290
PFPS *see* patellofemoral pain syndrome
Phalens test 134
pharmaceutical therapies 7
phosphagen system 40
phosphocreatine (PCr) 40
phosphorus 249–50
Physical Stress Theory 120
physiotherapy
  ankle injuries 485–92
  elbow injuries 349
  psychology 276, 281, 285
piriformis syndrome 135
pitchside assessment 194
plantar fasciitis 138, 498–9
plyometrics
  knee injuries 434–9
  needs analysis 41, 54
  periodisation 149, 151, 154, 158–9

progressive rehabilitation 209,
214–16
shoulder injuries 326
PNF *see* proprioceptive
neuromuscular facilitation
Polidocanol 89
polysaccharides 247
post-exercise nutrition 256–7, 259,
262
posterior ankle pain 474–5
posterior cruciate ligament (PCL)
407, 410–11, 414–17, 426, 428
posterior dislocations 347
posterior instability 310, 315–16
posterior interosseous nerve (PIN)
349–51
posterior thigh injury 136–7, 303
posterior tibiofibular ligament
(PTFL) 466–7, 469, 472–3
posture 313, 321–5, 328, 497–8
power output
elbow injuries 358
fitness testing 47–8
periodisation 149
strength and conditioning 223,
224, 239
pre-competition periods 148, 155–7
pre-exercise nutrition 254–5, 257,
259, 261–2
pre-habilitation 15
pre-participation physical
examinations 9
prevention *see* injury prevention
PRICE *see* protection, rest, ice,
compression, elevation
problem based learning (PBL)
299–302
problem solving model 337, 339
professionalism 3, 4–5
progressive overload 145, 228–31
progressive rehabilitation 199–221
assessment and observation
199–202
balance tests 208–9
inflammation 202–4
muscular endurance/strength 199,
212–14
pain management 202–4
plyometric training 209, 214–16
proprioceptive/neuromuscular
control 199, 207–12
shoulder injuries 325–7
sport-specific injuries 201–2,
209

study outcomes 203–4
systematic overview 199, 200
progressive relaxation 285
pronation of the foot 20–1
pronators 133–4, 352–3
prone four-point hold test 29
proprioceptive control
ankle injuries 477–80
knee injuries 428–9, 431
ligaments 100
muscular strength and
conditioning 239–40
progressive rehabilitation 199,
207–12
proprioceptive neuromuscular
facilitation (PNF) 205, 206–7
protection, rest, ice, compression,
elevation (PRICE) 171, 179
protective equipment 10, 175, 179
proteins 247–8, 256–7, 258–60
proteoglycans 95
provocative tests 345
psychology 275–96
adherence issues 275, 280–2, 288
behavioural responses 275,
280–3, 288
emotional responses 275, 276–80,
283
grief-loss model 275, 279–80
injury understanding 290–1, 292
interventions 278–9
mental toughness 275, 282
pre-injury period 283, 286–7
psychological skills training 275,
283–90
rehabilitation 275–96
SCRAPE model 275, 292
sport rehabilitators 275–6, 278–9,
283, 292
stress–injury relationship model
275, 276–9
pulled elbows 346
pulsed shortwave diathermy 72–3
PWB *see* partial weight bearing

Q angle 21
quadratus lumborum muscle 387
quadriceps 409, 413, 445–6

radial nerves 124, 133
radial tunnel syndrome 133, 349–50
radial/radial head fractures 348
radio-humeral bursitis 351
radiocarpal joints 367

radioulnar joints 345
range-of-movement (ROM)
ankle injuries 472, 476–7, 479
elbow injuries 341, 348
groin injuries 388
injury prevention 30
knee injuries 407–9, 411–12, 416,
422, 426
ligaments 100
muscular strength and
conditioning 239
needs analysis 54
periodisation 149, 153, 159
progressive rehabilitation 200,
202, 205–7, 209, 215
shoulder injuries 312–13, 325
tendons 85
wrist and hand injuries 370, 380
rate of adaptation 232–3
rate of force development (RFD)
224–6, 228, 429
rearfoot injuries 498–505
reasoning *see* clinical reasoning
reciprocal inhibition 210
recovery
injury prevention 36
muscular strength and
conditioning 230–3
nutrition 256
psychology 288
rectus femoris 26
recuperation periods 148, 149–50,
155–7
re-education 360, 399
reference nutrient intakes (RNI)
249, 251–2, 258
referrals to other professionals 292
referred pain 189–90, 191, 339–41,
342
reflection 303–4
regulatory frameworks 3, 9–11,
194–7
rehabilitation
ankle injuries 465–95
clinical reasoning 300–1
deactivation 25–7
elbow injuries 346
fitness testing 50
foot injuries 499–513
groin injuries 391–402
knee injuries 407–9, 413–48
muscular strength and
conditioning 238–40
musculoskeletal injuries 67, 75

rehabilitation (cont.)
  nutrition 265–7
  periodisation 151–9
  psychology 275–96
  shoulder injuries 322–7, 330
  skeletal injuries 114–16
  sport-specific injuries 239–40
  tendons 90
  see also progressive rehabilitation; sport rehabilitators
re-injury
  musculoskeletal injuries 67, 73, 75
  progressive rehabilitation 202
  psychology 288
relative rest 349
relaxation techniques 285–9, 291
religion 278
relocation tests 315
remodelling processes 114–16
reparative phase 115
resistance training 224–6, 260
resisted movement assessment 188–9, 193–4
  elbow injuries 344–5, 358
  shoulder injuries 314, 317, 328
rest, ice, compression and elevation (RICE)
  acute sport injuries 171–2, 179
  ankle injuries 482, 484
  elbow injuries 353, 355, 359, 377
  knee injuries 409
  ligaments 100
  muscular strength and conditioning 232–3
  musculoskeletal injuries 71, 74
  progressive rehabilitation 202–4
retrocalcaneal bursitis 501
Revels model 20
RFD *see* rate of force development
rhomboideus major/minor 312
RICE *see* rest, ice, compression and elevation
RNI *see* reference nutrient intakes
rocker boards 431–4
ROM *see* range of motion
Romanian deadlift 236–7
rotator cuff tears 320–2

saddle position 22
saggital plane drills 438
salt 261
SALTAPS 194
saturated fatty acids 248
Saturday night palsy 130–1
scaphoid fractures 378–9
scapula 131, 314–15, 324, 326
Scarf test 318
Schwann cells 119, 121
sciatic nerves 135
SCRAPE model 275, 292
screening 15–37
  functional tests 26, 29–30
  health screening questionnaires 16–20
  isokinetic testing 20–4
  kinematic analysis 22, 23, 25
  methods 16–30
  recommendations 36
  reliability and validity 16–18, 30–3
  risk assessment 32–5
selective tension 344
self-determination 290
self-esteem 277
self myofascial release (SMR) 207
self-talk 285–7, 289–91
SEM *see* standard error measurement
sensimotor control 428
sensory feedback 413
serratus anterior 311
sesamoid injuries 109, 510–11
severity, irritability, nature, stage (SINS) notation 199–201, 215
Sever's disease 500–1
shear tests 318, 421–2, 424–5
shock training 147, 231–2
short response training 226, 227
shortened soft tissues 443, 446–7
shoulder injuries 309–36
  case study 328–30
  differential diagnosis 321, 323
  incidence and prevalence 309–13
  peripheral nerve injuries 130
  posture 313, 321–5, 328
  rehabilitation 322–7, 330
  risk assessment 313–22
  test properties 322
side spring tests 33
signalling pathways 80–2
simple carbohydrates 247
SINS *see* severity, irritability, nature, stage
sinus tarsi 473, 505
sit and reach test 30–1
site of injury 191

situational anxieties 276–7
skeletal injuries 105–17
  acute sport injuries 169
  anatomy 105–6
  bone on bone forces 26, 28
  bone covering 107
  bone formation and growth 107–8
  bone structure/types 106–7
  cartilage 109–10
  classification of bones 107, 109
  female triad 113
  fractures 110–12
  functional aspects 105
  healing and remodelling processes 114–16
  nutrition 249, 264, 265
  osteoporosis 112–14
  wrist and hand injuries 366–7, 378–9
  see also musculoskeletal injuries
skier's thumb 373
SLAP lesions 316
Sliding Filament Theory 68–9
sliding techniques 126–7, 135, 137–8, 316–17
slump tests 125, 139
SMART *see* specific, measurable, accurate, realistic and timed
SMR *see* self myofascial release
snatch balances 150, 153
SOAP *see* subjective, objective, assessment, plan
social support 278, 292
Society of Orthopaedic Medicine 343, 344, 358
special testing 167, 169–71, 178, 370
specific conditioning periods 148, 150–1
specific, measurable, accurate, realistic and timed (SMART) 284, 287
specificity of training 232–3, 234–7
speed testing 45, 46–7, 486–90
Speed's test 318
spinal cord injury 120
splinting 134, 171
split jerks 150, 154
spontaneous tendon rupture 85
sport educators 4
sport rehabilitators
  acute sport injuries 165, 167, 171, 175–7
  assessment 185–6, 197

clinical reasoning 302
Code of Ethics 7–8
continuing professional development 6
ethical considerations 7–8
evaluation 4–5, 7, 9
knowledge, ability and wisdom 6–7
progressive rehabilitation 212
psychology 275–6, 278–9, 283, 292
regulatory frameworks 9–11
roles 3–7
team members 3–4
sport scientists 4
sports drinks/gels 256–7, 261, 267
sprains
　ankle injuries 100, 166–72, 465, 475–80, 483–4
　elbow injuries 352–3
　foot injuries 507
　wrist and hand injuries 375–6
spread of injury 191
sprint tests 47, 55, 57, 211
squats 150, 153–4
squeeze tests 388–9
SSC *see* stretch-shortening cycle
stabilisation (RICES) 171
stabiliser muscles 72, 310–12
stability 49–50, 357, 360
stable surface training 211
standard error measurement (SEM) 30–1
state anxieties 276–7
state at rest 343–4
static proprioceptive training 208
static stretching 205–6
Stener lesions 373
Stinger Syndrome 121, 127–9
strain injuries
　groin injuries 390, 392–3
　musculoskeletal injuries 69–70, 73–5
　tendons 83
strength *see* muscular strength
stress fractures 110–12, 348, 396, 476, 500, 505
stress relaxation 98
stress tests 470–1, 482
stress–injury relationship model 275, 276–9
stress–strain curves 82, 97
stretch-shortening cycle (SSC) 84, 214–15

stretching exercises
　knee injuries 446–7
　musculoskeletal injuries 72, 74
　peripheral nerve injuries 134, 135
　progressive rehabilitation 205–6
student's elbow 351
subacromial impingement 318–19
subjective assessment 190–2, 195–7, 337–41
subjective, objective, assessment, plan (SOAP) notation 199–201, 215, 369–70
subscapularis 311, 320
subtalar joints (STJ) 20–1, 465
sulcus test 316
super-compensation 228–30
supination external rotation tests 317
supplementation 263, 265–7
supracondylar fractures 347–8
suprascapular nerves 132
supraspinatus 311, 320
surgical interventions 4
　foot injuries 506, 508–9
　groin injuries 393–5
　knee injuries 407–9, 413–14, 417, 423–6
　ligaments 100–1
syndesmosis strains 473
synovial thickening 193, 343
synovitis of the hip 398

T-tests 43, 46, 211
T-tubules 68–9
talar dome injuries 474
talar tilt (TT) test 471
talo-crual joint (TCJ) 465
talus injuries 505
targeted training 35
tarsal condition 506–7
tarsal tunnel syndrome (TTS) 138, 476
tenascin C 87
tendinitis 85
tendinopathy 85, 87–9
　ankle injuries 474
　elbow injuries 348–51, 352–3
　foot injuries 501–3, 505–6
　groin injuries 392–3
tendinosis 85
tendons 79–93
　acute sport injuries 85–7, 90, 169
　anatomy 79–81
　ankle injuries 474–5
　chronic injuries 83–4, 87–90

elbow injuries 346, 348–51, 352–3
foot injuries 501–6, 509
functional aspects 81–4
groin injuries 392–3
healing processes 89
management of injuries 84–5
physiology 79–81
treatment 89
wrist and hand injuries 375
tennis elbow 348–9
TENS *see* transcutaneous electrical nerve stimulation
tensioning techniques 126–7, 135, 137–8
teres major/minor 311
test-retest reliability scores 31–2, 48
TFM *see* transverse friction massage
therapeutic exercise 358
therapeutic ultrasound 73, 355–6, 359, 373, 379, 499
Thomas test 329
thoracic outlet syndrome (TOS) 129–30
thoracic spine posture 325
three repetition maximum (3RM) test 52
threshold concepts 300–2
thrombin 70
thrower's elbow 350
tibial nerves 138
tibialis posterior 475, 506
tibiofemoral shear forces 417, 421–2, 424–5, 437–9
timing 230–2, 254–7, 259–62
Tinel's test 132–4, 138, 345, 500
tissue release massage 502
topical analgesics 174
TOS *see* thoracic outlet syndrome
total tendon rupture 503–4
training cycles 145, 146–50
training years 146–8
trait anxieties 276–7
trans-fatty acids 248
transcutaneous electrical nerve stimulation (TENS) 173–4, 187, 413, 500
transition periods 148, 149–50, 155–7
transverse friction massage (TFM) 345–6, 353, 357
transverse plane drills 439
trapezius 311
trapped nerves 500

triangular fibrocartilage complex (TFCC) 367, 375
triceps brachii 312
trigger finger 377
triglycerides 254, 258
TT *see* talar tilt
TTS *see* tarsal tunnel syndrome

UK Sport 266–7
ulnar injuries 124–5, 131–2, 348, 353, 377–8
ultrasound
  ankle injuries 470, 475
  foot injuries 502
  groin injuries 392, 394
  shoulder injuries 313
  tendons 85–6, 88, 90
  therapeutic 73, 355–6, 359, 373, 379, 499
unavoidable injuries 237–8
unsaturated fatty acids 248
unstable surface training (UST) 210–11
upper limb neurodynamic tests 123–5
upper quadrant exercise progression 325–6

varus–valgus oscillation 435, 438
vastus 26
vastus medialis oblique (VMO) 444
vertebral artery dissection (VAD) 20
vertebral compression 111
vertical ground reaction force (VGRF) 429–30, 434, 436–9
Victorian Institute of Sport Assessment Questionnaire (VISA) 88
video analysis 23, 31–3
viscoelasticity 98
Visual Analog Scale (VAS) 19–20, 36
vitamins 108–9, 115, 248–50, 251, 263–5
VO_2max *see* maximal oxygen consumption

Wallmann's regime 502–3
water intake *see* fluid intake
weight bearing injuries 112
well-being 289–90
whole body vibration 481

Wingate anaerobic test (WAnT) 50–1
wobble boards 100, 431–3
Wolff's law 105, 115
World Anti-Doping Code (WADC) 266–7
wrist and hand injuries 365–83
  acute sport injuries 371–6
  anatomy 366–8
  assessment 368–71, 373–4, 376–7
  case study 379–80
  chronic and overuse injuries 376–8
  incidence and prevalence 365–6
  sport-specific injuries 365–6, 371
  treatment and management 373–80

X-ray examination
  acute sport injuries 170–1
  assessment 186–7
  foot injuries 506–10
  groin injuries 396–7, 401–2

Yergason's test 317–18
yo-yo tests 43, 53

Printed and bound by CPI Group (UK) Ltd, Croydon, CR0 4YY